Human Perspectives in Health Sciences and Technology

Volume 2

Series Editor
Marta Bertolaso, Campus Bio-Medico University of Rome, Rome, Italy

The Human Perspectives in Health Sciences and Technology series publishes volumes that delve into the coevolution between technology, life sciences, and medicine. The distinctive mark of the series is a focus on the human, as a subject and object of research. The series provides an editorial forum to present both scientists' cutting-edge proposals in biomedical sciences that are able to deeply impact our human biological, emotional and social lives, and thought-provoking theoretical reflections by philosophers and scientists alike on how those scientific achievements affect not only our lives, but also the way we understand and conceptualize how we produce knowledge and advance science, so contributing to refine the image of ourselves as human knowing subjects. The series addresses ethical issues in a unique way, i.e. an ethics seen not as an external limitation on science, but as internal to scientific practice itself; as well as an ethics characterized by a positive attitude towards science, trusting the history of science and the resources that, in science, may be promoted in order to orient science itself towards the common good for the future. This is a unique series suitable for an interdisciplinary audience, ranging from philosophers to ethicists, from bio-technologists to epidemiologists as well to public health policy makers.

More information about this series at http://www.springer.com/series/16128

Mariano Bizzarri
Editor

Approaching Complex Diseases

Network-Based Pharmacology and Systems Approach in Bio-Medicine

Springer

Editor
Mariano Bizzarri
Department of Experimental Medicine
Sapienza University of Rome
Rome, Italy

ISSN 2661-8915 ISSN 2661-8923 (electronic)
Human Perspectives in Health Sciences and Technology
ISBN 978-3-030-32856-6 ISBN 978-3-030-32857-3 (eBook)
https://doi.org/10.1007/978-3-030-32857-3

This Springer imprint is published by the registered company Springer Nature Switzerland AG.
The registered company address is: Gewerbestrasse 11, 6330 Cham, Switzerland

Foreword: Future Perspective in Drug Discovery

A relevant body of evidence suggests that since the 2000s, the pharmaceutical industry is experiencing a true crisis in developing innovative new drugs. Indeed, although the yearly number of new medicines remained substantially unchanged, research and development (R&D) investment per drug is escalating at a marked rate, thus substantiating the so-called innovation gap (Fig. 1) [1].

However, aside from the striking results in the treatment of cardiac, infectious, endocrinological, and cerebrovascular diseases attained since the 1950s, no comparative achievements have been made in the management and treatment of cancer, metabolic, autoimmune, and rare diseases, just to mention a few [3]. Undeniably, pharmacological research did not meet with the expectations raised by a widespread

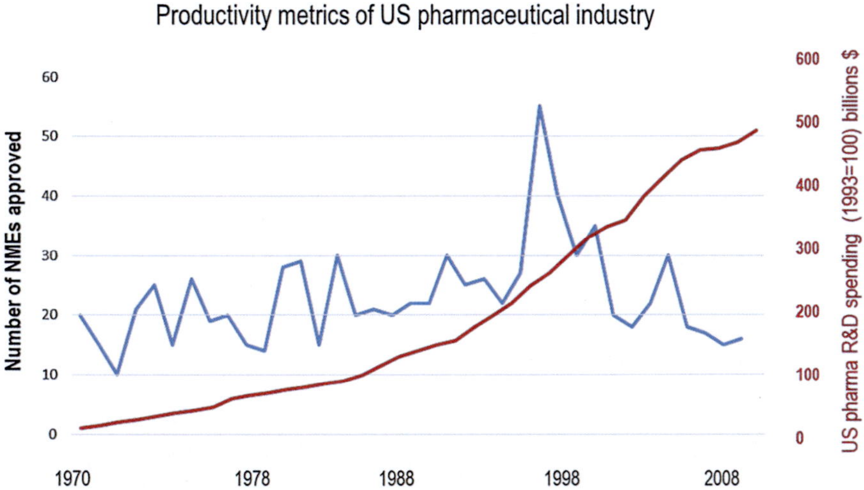

Fig. 1 *Productivity metrics of US pharmaceutical industry*. NME: new medicinal entities. Elaborated on data from The Changing Landscape of Research and Development Innovation, Drivers of Change, and Evolution of Clinical Trial Productivity [2]

propaganda who was claiming that the "solution" of devastating diseases was, finally, "at hand." In the meantime, modern medicine, through over prescription, paradoxically has become a major threat to public health [4]. Behind this "failure," we can recognize a number of different industrial, theoretical, and methodological concerns, some of which are extensively investigated in the present book.

In the first place, the ultimate problem may not be technological or even scientific but rather "philosophical." There are pivotal issues with the core assumptions that frame our approach to drug discovery that are probably wrong. It is not by accident that the increase in the rate of drugs failing in late-stage clinical development has been concurrent with the dominance of the assumption that the goal of drug discovery is to design exquisitely selective ligands that act on a single target. This philosophy – the "one gene, one drug, one disease" paradigm– arose from the congruence between genetic reductionism and new molecular biology technologies that enabled the isolation and characterization of individual "disease-causing" genes. However, contrary to popular beliefs, neither "magic bullets" have been so far discovered nor expectations have been kept.

As a consequence, criticism on current, prevailing paradigms in drug discovery is gaining momentum given that even network-based models – their heuristic and epistemological value notwithstanding – only partially explain the unfathomable complexity behind the outbreak of a disease. This bias is mostly because our models assign the predominant "causative" ground in the cell – specifically in its genome – where mathematical modelling of genetic and biochemical circuits can be "easily" fashioned. As such, those models discard the contribution of nongenetic elements, microenvironmental constraints, and other factors belonging to levels *higher* than the cellular one. Since the 1980s, the agenda of pharmacology discovery was thus dictated by aiming at discovering "relevant" molecules, abstracting from the true, physiological behavior of cells, tissues, and organism. In this perspective, genes assume the fundamental causal role, while cells simply act as causal proxies, dispensable because they represent an irrelevant intermediate level between the molecular input and the organismal output. Such framework is no longer tenable, from both a theoretical and a methodological point of view, though. Thereby, several attempts have been pursued to capture the complexity behind the disease development. These studies allowed to recognize a different spatially (multilevel) and timely distributed array of different targets, as well as appreciating the nonlinear dynamics that drives pharmacological interactions. This is a critical issue, given that drug-target interactions go far from the classical and reductionist lock-and-key model. Moreover, given that a disease is properly a *dynamic process* and not a steady state, treatments should also be diversified according to the timing of the disease evolution. This approach can help in detecting pre-disease state or critical transition points from which the illness might access different attractors, leading ultimately to very different outcomes. We should thus design *systems-oriented drugs* that tackle both the multifactorial pathogenic determinants of the disease and the intrinsic robustness of the living organism. This issue would probably require a polyvalent-based approach, i.e., the use of multiple drugs or drugs affecting several targets localized at different levels. Additionally, beyond classical pharmacodynamics, unconventional mechanisms of action – usually underestimated by current pharmacological studies – urge to be investigated and integrated within the network modelling. Overall, this means that treatment options ought to shift from *targets* to *processes* in "redrawing" the disease-related landscape – borrowed from Waddington – favoring the system displacement from preclinical state or true disease states toward healing "pathways."

Reconsidering the very basic premises on which our methodological and theoretical tools rely is the preliminary step in establishing a network-based pharmacology. A pivotal tenet of this system approach lies indeed on the awareness that drug-target interactions are shaped by a complex, nonintuitive, nonlinear dynamics. As pointed out in the chapter by Panos Macheras ("Dynamical Aspects of Pharmacokinetic/Pharmacodynamics and Quantitative Systems Pharmacology Models"), physiological systems are inherently complex dynamical and nonlinear systems. The erratic nature and the irregular behavior of these systems constitute to object of several studies in the field of systems biomedicine. The observation of chaotic phenomena (e.g., heart rate variability) in human physiology is often associated with the maintenance of homeostasis, whereas loss of complexity is perceived as indication of pathology. However, complexity has not been considered worth of investigation in pharmacology until recent times. Yet, complex nonlinear response, bistability, chaotic/erratic processes, and many other features of nonlinear behavior are increasingly discovered and appreciated downstream of drug administration. Currently, the response of physiological system to external perturbations (i.e., drug) is by now studied with pharmacokinetic/pharmacodynamic (PK/PD) and quantitative systems pharmacology (QSP) models in order to acquire additional – often "hidden" – information. However, analysis of the underlying dynamics of such models is frequently ignored or underestimated. In the Macheras's contribution, the presence of nonlinear dynamics in physiological and pharmacological systems is instead highlighted through a summary of specific applications in various therapeutic areas. In parallel, the additive benefit of implementing tools borrowed from dynamical systems' theory to gain insight into the behavior of PK/PD and QSP models is also underlined. Such models have been demonstrated to improve our knowledge about the physiological role of a number of endocrine signals (like melatonin and cortisol), as well as the relevance of some often unrecognized elements (including sensitivity on the initial conditions and participation of hidden factors) in modulating the pharmacological response, even toward opposite, unexpected outcomes. Thereby, a thorough understanding of the behavior of all the individual components is not only crucial but also imperative to increase confidence on the value, usefulness, and performance of such models.

As aptly discussed here by Alan Talevi ("The Efficiency of Multitarget drugs: A Network Approach"), the adhesion to such an approach challenges some deep-rooted concepts in the field of pharmaceutical research, e.g., the target-driven discovery and the notion that "potent," single target-based, drug candidates should be pursued. Whereas these might still be valid concepts in some scenarios, they should be (and are being) revised, as strict adherence to existing paradigms could result in loosing valuable opportunities for improved therapeutics. Polytarget pharmacology is therefore gaining momentum. This perspective is not novel, indeed, as most herbal drugs and many old drugs discovered through phenotypic screening are multitarget agents. However, we are now in a position to approach it in a very rational manner and with increased "awareness" of the complex relationships that such treatment could involve. As a result, multitarget agents have attracted great attention in the last 15 years, as they are expected to provide more efficacious and safer therapeutic solutions. This approach implies we have to establish a proper design for tailoring multitarget drug design, including target combination choice, optimization of potency ratio to the different targets, and peculiarities of computer-guided drug discovery. Moreover, as we have already been taught with the use of antibiotics, by hitting different targets, we can overcome the increasing burden of drug resistance phenomena in a wide range of diseases, including cancer, degenerative, and rare disease.

Indeed, there is an urgent need for the development of treatments or cures for rare diseases, as discussed in the chapter authored by Anthony Hickey ("Mining Complex Biomedical Literature for Actionable Knowledge on Rare Diseases"). In the field of rare disease, the complex biological systems and nature of drug discovery make iterative mechanistic strategies costly and inefficient. Current developments in database development, text mining, and machine learning tools allow efficient and inexpensive navigation through inferences to the identification of novel or repurposed drug candidates. The same principles apply when we address the complexity of drug delivery systems and biopharmaceutical principles that result in optimal drug disposition to achieve the desired therapeutic effect. In this manner, "the development of novel pharmaceutical treatment options can focus on the generation of data suited to regulatory scrutiny and positive clinical outcomes without investment in the tangential iterative data generation that has historically been required to support statistical validation of the action, process, or clinical observations that surround the optimal approach." Conclusively, the opportunity now exists to extract knowledge from reading sources using modern text mining to rapidly and affordably identify and develop new or repurposed drug candidates, especially for the treatment of rare diseases.

Yet, this is not a matter of "big data" but requires new tools and unexplored strategies. The current crisis in pharmacology discovery, in fact, stems also from the naïve idea that an (acritical) application of statistical methods would provide amazing and successful information on biological mechanisms and processes. As aptly discussed by Giuliani et al. in their chapter ("Big Data, Personalized Medicine and Network Pharmacology: Beyond the Current Paradigms"), analysis of huge amount of raw data, notwithstanding how sophisticated this process could be, cannot provide useful insights if and when the statistical tool is not supported and integrated within a robust theoretical framework. Indeed, "each data analysis choice is strictly dependent from theoretical assumptions and each theoretical assumption is in turn influenced and modulated along the process by the emerging results." Therefore, "the entanglement of 'content' and 'methodological' knowledge is the basic methodological novelty made necessary by the actual information crisis (and consequent lack of practical efficacy) of biomedical sciences. The classical separation of scientific enterprises into a linear sequence made of: 'hypothesis setting'- 'experimental methods' – 'data analysis'– 'hypothesis verification/falsification' is untenable in the high-throughput era."

The advocated new strategy requires encompassing new concepts and a shift from the cellular/molecular to the cell/microenvironment (tissue) level. Especially in the study of anticancer drugs, the research should focus on *epigenetics control*. In the last decades, this objective was achieved by using small molecules. As extensively discussed by Tomohiro Kozako and his team, abnormalities in epigenetic control are usually observed in human cancers, and correction/modulation of such "defects" has been proven successful in controlling several malignant features. A very relevant issue is constituted by the fact that a number of epigenetic modifiers are small, "manageable" molecules – frequently obtained from natural sources – that promise to open new, unforeseen perspectives.

Some attempts have been made to deal with these challenging hurdles, even if a rational strategy is still lacking. At the industry level, it is likely that new, small molecular entities will still dominate in drug innovation for the next decade. This strategy is primarily thought to reduce the burden of financial investments. However, still confusing is the class of compounds on which we have to focus. Currently, this approach mostly relies on the perspective of "industrial synergy," aimed at multichannel integration of small-/medium-size enterprises, specifically involved in studying properties and polyvalent functions of small molecular components [5]. It is worth of note that small molecules are considered to act primarily through epigenetic mechanisms, by hitting a number of different targets. This approach introduces a discontinuity facing the dominant paradigm in drug discovery based on designing maximally selective ligands (thought) to act on an individual drug targets. However, many effective drugs act via modulation of *multiple molecular factors* rather than by modulating a single target. Moreover, advances in systems biology are revealing a phenotypic robustness and a network structure that strongly suggest that highly selective compounds, compared with multitarget drugs, may exhibit lower than desired clinical efficacy. This new appreciation of the role of polypharmacology (meaning both single agents acting on different targets or different pharmacological factors involving several targets at the same time) has significant implications for tackling the two major sources of attrition in drug development: efficacy and toxicity. Integrating network biology and polypharmacology holds the promise of expanding the current opportunity space for druggable targets [6]. Moreover, it should be stressed that such targets are distributed along different levels. This is because complex diseases can be properly managed only when the complex interactions distributed across different, hierarchical levels (from organelles to the tissue/organ level) are considered into a unified, dynamical network. Mark Chaplain specifically addresses this issue ("Multiscale Modelling of Cancer: Micro-, Meso- and Macro-scales of Growth and Spread") by proposing a multiscale approach in understanding cancer dynamics, which holds out the best possibility for the development of optimal, individualized patient-based therapy. However, personalized and precision oncology medicines – two key concepts recently introduced into the scientific arena – are usually interpreted according to wrong, gene-centered premises; as such, they need to be "reframed" by considering that the emergence of a "specific" disease-associated perturbed network can be identified only if the specific microenvironmental field is concurrently contemplated.

In cancer management, this approach finally leads to contemplate very different objectives from the traditional one, i.e., the killing of transformed cells. Precision oncology was initially designed to antagonize the effects of those mutations supposed to orchestrate tumorigenesis; however, these attempts almost universally fail to eradicate the tumor that usually recurs in the short term and in a more malignant fashion. Consequently, current beliefs recognize different and "alternative" objectives, including the inhibition of tumor invasiveness and metastatic potential, induction of tumor dormancy, and even triggering of the so-called tumor reversion.

Jab Brabek ("Migrastatics: Anti-metastatic Drugs Targeting Cancer Cell Invasion") reports how the treatment of solid tumors can be successfully supplemented with drugs that suppress the ability of tumor cells to invade through the surrounding extracellular matrix and form secondary tumors. These anti-metastatic and anti-migrating ("migrastatics") compounds can suitably complement cytotoxic therapy and increase its benefit. Inhibition of the invasive capability of cancer can be fruitfully coupled with the strategy of "tumor dormancy," aimed at prolonging the survival of patients together with preserving a good quality of life. As aptly discussed by Aranda-Anzaldo ("Precision Oncology Vs Phenotypic Approaches in the Management of Cancer: A Case for the Postmitotic State"), it is well known that terminally differentiated, postmitotic cells cannot become cancerous. Postmitotic state, however, does not depend on genetic factors; instead, that condition is a consequence of a natural process driven by thermodynamic constraints, ultimately resulting in a high structural stability within the cell nucleus. Such stability becomes an insurmountable energy barrier for karyokinesis and mitosis. It is worth of interest that the induction of a postmitotic state in cancerous cells has already been obtained with small molecules and such an approach deserves to be explored in depth. Similarly, tumor reversion induced by morphogenetic factors extracted from zebra fish or amphibian/mammalian eggs represents a new, reliable alternative to standard chemotherapy-based treatments, as reported in the chapter authored by S. Proietti, A. Pensotti, and A. Cucina ("Tumor Reversion Induced by Embryo and Oocyte Extracts").

The possibility that a tumor can be induced to "revert" to a "normal" phenotype was first suggested by the preliminary studies carried out by B. Pierce in the 1970s [7]. Yet, only in the last decades an increasing number of reports have ascertained the occurrence of cancer reversion, both in vitro and in vivo, providing useful insights into some of the basic mechanisms involved. This process encompasses mandatorily a change in the cell-stroma interactions, leading to profound modification in tissue architecture. As cancer can be successfully "reprogrammed" through the modification of the dynamical cross talk with its microenvironment, the cell-stroma interactive network must be recognized as a target for pharmacological intervention [8]. Yet, understanding cancer reversion remains challenging, and refinement in modelling such processes in vitro as well as in vivo is urgently warranted. This new approach bears huge implications, from both a theoretical and clinical perspective, as it may facilitate the design of a novel anticancer strategy focused on mimicking or activating the tumor reversion pathway. Overall, these new approaches in cancer treatment devote a special attention to the cell-microenvironment relationships and, namely, on some critical transitions – like the epithelial mesenchymal transition – occurring at such level.

These issues are widely discussed in the chapters from Evgeny Denisov ("Critical Steps in Epithelial-Mesenchymal Transition as Target for Cancer Treatment") and Julia Kzhyshkowska ("Targeting the Tumor-Associated Macrophages for "Normalizing" Cancer"), in which the tumor microenvironment – including matrix components, fibroblasts, and immune cells – is believed to play an essential role in modulating malignancy. By focusing on the cancer-immune system relationships at the level of the microenvironment, it became evident in recent years that the interplay between tumor and immunity is far more complex than previously thought. Indeed, inappropriate activation of myeloid cells infiltrating tumor tissues promotes – rather than stop – cancer progression. Namely, tumor-associated macrophages are abundantly present in the microenvironment surrounding cancer, and here, they trigger and perpetrate a state of chronic inflammation which ultimately supports cancer proliferation and distant spreading as well as contributes to an immune-suppressive milieu. It is worth noting that trabectedin, a natural compound extracted from the tunicate *Ecteinascidia turbinata* – a marine organism – presents the unique feature of being able to simultaneously kill cancer cells and to affect several features of the inflammatory microenvironment, most notably inducing the rapid and selective apoptosis of monocytes and macrophages. As reported in the chapter from Paola Allavena ("Trabectedin: A Drug Acting on Both Cancer Cells and the Tumor Microenvironment"), trabectedin is the perfect specimen of those drugs that can hit several targets ("polytarget approach") at different levels (involving both cancerous cells and their microenvironment). Involvement of several targets at different levels in no way can be understood through classical, reductionist approaches. Here is where systems biology can provide fruitful insights.

It is worth of interest that several natural compounds – obtained from both plants and animals share both these critical features: polytarget and multilevel activity. These effects have been documented in the treatment of different diseases – including cancer, metabolic ailments, diabetes, immunological disorders, and neurological illness – and are currently supported by a compelling body of experimental and clinical evidence. Natural products and their derivatives have historically been invaluable as a source of therapeutic agents. Recent updates and technological advances, coupled with unrealized expectations from current lead-generation strategies, fostered a renewed interest in natural products in drug discovery, as witnessed by the increasing number of scientific published papers (Fig. 2) [9].

This is the case for melatonin – as sharply reviewed by a pioneer on that field like Russ Reiter in the chapter "Advances in Characterizing Recently-identified Molecular Actions of Melatonin: Clinical Implications" – as well as for inositol, a basic polyol with protean activities, wisely discussed in Ivana Vucenik's review, "Multitarget Activities of Inositol and Inositol Hexakisphosphate." Melatonin and inositol share the remarkable feature to interact with several targets, at both the

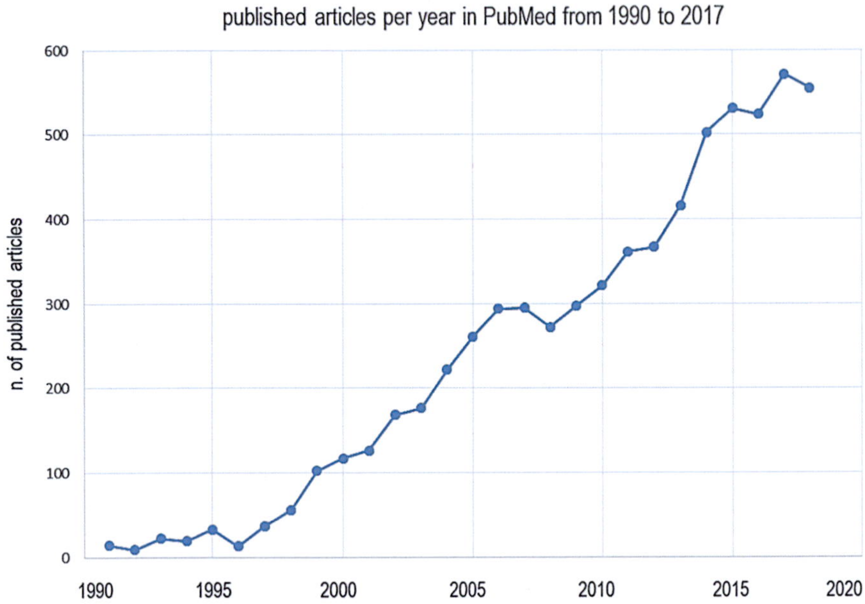

Fig. 2 Increase in the number of published articles on nutraceuticals and cancer, as recorded from PubMed (1990–2017)

cellular and the microenvironment levels, displaying a number of appreciable therapeutic effects in the treatment of cancerous as well as nonneoplastic diseases.

Undeniably, the study of pharmaceuticals extracted from natural sources is entering into maturity, given that they are currently investigated through sophisticated methods, as those provided by omics technologies. Moreover, as for synthetic drugs, "natural" molecules interact in a complex manner, showing wide (and unexpected) differences among individuals. Yet, a proper appraisal of their dynamical behavior allows us to set the cooperativity effect they can trigger when associated to conventional treatments, as aptly discussed in the chapter authored by Thomas Efferth (*Integration of Phytochemicals and Phytotherapy into Cancer Precision Medicine*). A specific case in point, specifically centered on gynecology and obstetrics, is instead provided by the review from Iñaki Lete ("Medicinal Herbs: Its Therapeutic Use in Obstetrics and Gynecology"). This statement holds especially for Chinese herbal medicine (CHM), which usually relies on mixtures of different herbs. These "traditional" combinations ("remedies") are in themselves "complex systems," and their interactions with living organisms are complex too, as they are mostly ruled by nonlinear dynamics, as pointedly discussed in the chapter from Xian Zhou (*Synergistic Effects of Chinese Herbal Medicine and Biological Networks*).

Emergence of synergy – a remarkable feature of complex herbal remedies – lies specifically on such kind of dynamical interplay, nonetheless the level of evidence remains low, given that randomized, multicentric, clinical trials are still warranted, the few seminal studies performed notwithstanding. However, relevance of CHM is likely to tacking off, as its application is increasing worldwide – namely, after the World Health Organization (for the first time) has recognized traditional Chinese medicine in its influential global medical compendium [10]. It should be noticed that controversies are rising too: no doubts on that [11].

Finally, the present book does not leave aside to address the so-called antibiotic question, a critical issue discussed in the chapter ("Overcoming Antibiotic Resistance: New Perspectives") authored by Matteo Bassetti. While various antibiotics are currently available to target resistant Gram-positive pathogens, fewer new molecules that are active against multidrug resistant Gram-negative bacteria, especially carbapenemase producers, have been approved for clinical use. Novel glycopeptides, oxazolidinones, and new-generation cephalosporins as well as novel β-lactam/β-lactamase inhibitors have been recently approved for clinical use and display appreciable activity against Gram-negative bacteria. Nevertheless, the emergence of untreatable, resistant microbial strain and infections still represents a deadly threat, while the flow of new classes of antibiotic has substantially declined at a time when resistance rates and new problems have increased significantly [12]

To sum up, the decline in productivity and in innovation path witnessed by modern pharmacology requires a profound rethinking of the basic conceptual premises on which pharmacological research is based. Yet, overcoming such hurdles is a complex and difficult task, and it implies that both academic research and industrial innovation could find a different way to cooperate altogether. The development of systems biology can reliably support such endeavor. Pharmaceutical companies may well have to go back to academia or, at least, to academics studying new and unexplored routes. Namely, systems biology, which today is still largely an enterprise of "academic" (i.e., noncommercial) interest, may find itself increasingly incorporated into the research programs of industrial enterprises.

Adopting industrial and theoretical strategies that are "divergent" in respect to those we have experienced in the last 50 years is urgently needed if we would efficiently cope with the increasing burden of degenerative diseases and the reappearance of old, resistant, infectious ailments. As suggested by this book, time is ripe to plan a systematic research program for addressing such issues.

References

1. Burrill & Company. Biotech. 2010. *Life sciences: Adapting for success*. San Francisco: Burrill & Company.
2. The Changing Landscape of Research and Development Innovation, Drivers of Change, and Evolution of Clinical Trial Productivity. April 23, 2019. https://www.iqvia.com/institute/reports/the-changing-landscape-of-research-and-development
3. Rishton, G.M. 2005. Failure and success in modern drug discovery: Guiding principles in the establishment of high probability of success drug discovery organizations. *Medicinal Chemistry* 1 (5): 519–527.

4. Malhotra, A. Why modern medicine is a major threat to public health, *The Guardian*, August 30, 2018. https://www.theguardian.com/society/2018/aug/30/modern-medicine- major-threat-public-health
5. Haskell, R., P.P. Constantinides, D. Sun. 2012. *Perspectives in pharmaceutical nanotechnology*. AAPS News Mag.
6. Csermely, P., T. Korcsmáros, H.J. Kiss, G. London, and R. Nussinov. 2013. Structure and dynamics of molecular networks: A novel paradigm of drug discovery. A comprehensive review. *Pharmacology & Therapeutics* 138 (3): 333–408.
7. Pierce, G.B. 1968. Teratocarcinoma: model for a developmental concept of cancer. *Current Topics in Developmental Biology* 2: 223–246.
8. Bizzarri, M., A. Cucina, and S. Proietti. 2014. The tumor microenvironment as a target for anticancer treatment. *Oncobiology and Targets* 1: 3–11.
9. Atanasov, A.G., B. Waltenberger, and E.M. Pferschy-Wenzig. 2015. Discovery and resupply of pharmacologically active plant-derived natural products: A review. *Biotechnology Advances* 33: 1582–1614.
10. Ciranosky, D. 2018. The big push for Chinese medicine. *Nature* 561: 448–450.
11. Break with tradition. 2019. *Nature* 570: 5.
12. Finch, R. 2007. Innovation—Drugs and diagnostics. *The Journal of Antimicrobial Chemotherapy* 60 (Suppl 1): i79–i82.

Contents

Revisiting the Concept of Human Disease

Rethinking the Causality Concept in Pathogenesis for Establishing a Different Pharmacological Strategy

Mariano Bizzarri, Mirko Minini, and Noemi Monti

Shortcomings and Challenges in Drug Discovery

The pharmaceutical industry is currently facing unparalleled challenges to develop innovative new drugs. Although the yearly number of new drugs remained substantially unchanged, research and development (R&D) investment per drug is escalating at a marked rate. The estimated cost of developing a new drug is approximately $1 billion [1]. This phenomenon, the increase in R&D investment without the corresponding increase in the number of new drug approval, is known as the "innovation gap" [2]. After the Thalidomide [3] and Vioxx [4] incidents, regulatory bodies throughout the world are demanding more safety data, which in turn increases the development costs. However, this is only the tip of the iceberg of an even worst situation.

Response to Health Problem

While an impressive result – chiefly in terms of reduced mortality – has been witnessed in the last six decades as for the death rate for cardiovascular, cerebrovascular and infective diseases – no proportional benefits have been recorded in cancer cure rates (Fig. 1) [5–7]. Moreover, the burden of drug-insensitive infectious

M. Bizzarri (✉) · M. Minini
Department of Experimental Medicine, Sapienza University, Rome, Italy

Systems Biology Group Lab (http://www.sbglab.org), Rome, Italy
e-mail: mariano.bizzarri@uniroma1.it

N. Monti
Systems Biology Group Lab (http://www.sbglab.org), Rome, Italy

Department of Experimental Medicine, Sapienza University, Rome, Italy

M. Bizzarri (ed.), *Approaching Complex Diseases*, Human Perspectives in Health Sciences and Technology 2, https://doi.org/10.1007/978-3-030-32857-3_1

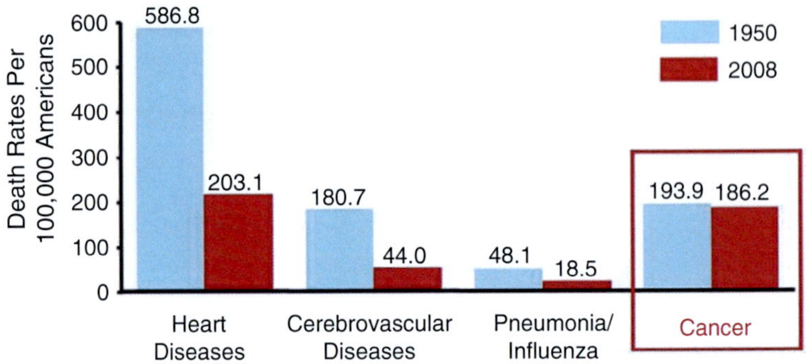

Fig. 1 Mortality rates (years 1950–2008) for infectious, heart, cerebrovascular diseases (From American Cancer Society (ACS) 2010 Cancer Facts & Figures; Atlanta, USA, 2014, modified)

diseases, is increasing [8], whereas diabetes, metabolic and degenerative diseases, autoimmune and allergic pathologies are reaching an epidemic profile [9, 10]. Overall, those data strongly evidence that many health problems, besides the astonishing progress earned in several fields, are still waiting a satisfactory answer [11]. Focusing on cancer, for instance, it is nowadays widely believed that "the $105 billion spent on cancer research since President Richard Nixon declared the war on cancer in 1971 has been poorly spent. At the same time, considering that age standardized death rates from adult cancers in America today are only very modestly lower now than they were in the early 1970s, this is no great success story" [12]. A thoughtful and threatening example has been provided in the last years by antibiotics [13]. Indeed, the flow of new classes of antibiotic has substantially declined at a time when resistance rates [14] and new problems have increased significantly [13].

The aforementioned considerations should be put in correlation with the increasing negative attitude of a growing number of people that perceives modern medicine as a substantial failure. Indeed, since the 90s, the use of natural/alternative drugs steadily increased in western countries, while American patients paid more visits to alternative health practitioners (425 million) than to primary-care physicians (388 million visits) [15]. In US conventional medicine had lost over half of the market share for primary health services to so-called snake-oil vendors. The report revealed that patients were visiting alternative doctors for ten top health problems: back pain, anxiety, headache, sprains or strains, insomnia, depression, arthritis, digestive problems, high blood pressure and allergies [16]. Regardless of any other consideration, we should ask ourselves why in western countries so many people rely on (frequently unproven) "alternative" medical supports [17]. Moreover, it is a matter of concern that patients referring to non-conventional medicaments have a higher level of education than those who do not use them [15]. Thereby, we are legitimate in asking, "if the scientific message that alternative therapies don't work is so "loud and clear," why do so many people, physicians included, use them?" [18].

Industry Productivity

Contrary to the expectations, pharmaceutical innovation has led to a decline in industry productivity, a phenomenon noticed since the early eighties [19]. Despite the increased investment in R&D by the industry, the number of new molecular entities (NME) achieving marketing authorization is not increasing. Over the past 20 years, the number of Investigational new drugs approved by regulatory agencies did not augmented as predicted, nor quality control level, safety assessment, and identification of new molecular targets was improved. Therefore, high investment, development of new technologies and conceptual approaches – likes "-omics" methods, including transcriptomics, proteomics and genomics – have neither reduced the R&D risk, nor have enhanced efficiency [20].

By now it is widely recognized that Big Pharma challenges include: (1) R&D spending is growing faster than sales growth, (2) drug discovery is lagging relative to industry growth needs, (3) increased presence of large molecules in big pharma's pipeline, (4) increasing need for in-licensing products and technologies, and (5) blockbuster drugs are going off patent (approximately 40% of patents owned by top 20 pharmaceutical companies are set to expire during 2009–2013). Efficacy and safety issues are actually the main causes of failure at the stage of phase III, given that two out of three among new proposed drugs are currently discarded for their side-effects [17]. Moreover, sixty-six of the 100 greatest companies studied by Herper [21], will launch only one drug this decade. The costs absorbed by these companies can be taken as a rough estimate of what it takes to develop a single drug. The median cost per drug for these singletons was $350 million, but for companies with more drugs approved, the cost per drug went up – until it hit $5.5 billion for companies that have brought to market between eight to 13 drugs over a decade [21].

As a result, as the main driver for its growth is innovation, biomedicine R&D is becoming increasingly challenged due to lower productivity and thus pharmaceutical companies have opened their R&D organizations to external innovation [22]. On this respect, it is noticeably that dynamics of NMEs release into the market markedly diverges among large and medium companies, reaching high efficiency almost only in medium-little companies. The decline in the output of large companies has been mostly driven by the diminishing number of large pharmaceutical companies, which has decreased by 50% over the past 20 years [20].

The Reason Behind the Failure

Three drug-discovery fads have driven the industry's R&D programs in the past 20 years computer aided drug design, combinational chemistry linked to high throughput screening and genomics. As a result, current trends in biomedical research have led to a decreased emphasis on the physiology-driven approach, supporting "a trend toward qualitative rather than quantitative science, with the implicit assumption that *all* targets represent a viable starting point for drug discovery efforts" [23]. No doubts that this approach will likely focus on non-essential targets, thus producing more failures through lack of efficacy. Furthermore, analysis of structure–activity relationship pattern evolution, drug–target network topology and literature mining studies, all indicate that more than 80% of the new drugs rely on previously discovered targets and strive in affecting – perhaps in a different/subtle way – the same network [24]. Sadly, "there are no evidence that any of these is or will be capable of replacing the old techniques" [25]. Consequently, the most fruitful basis for the discovery of a new drug – still today – is to start with an old drug [26], besides a renewed interest for natural compounds and complex herbal mixtures has been noticed in the last 20 years [27, 28].

The fundamental problem may not be technological, environmental or even scientific but rather "philosophical". There are pivotal issues with the core assumptions that frame our approach to drug discovery. In fact, the increase in the rate of drugs failing in late-stage clinical development has been concurrent with the dominance of the assumption that the goal of drug discovery is to design exquisitely selective ligands that act on a single target. This philosophy – the 'one gene, one drug, one disease' paradigm – arose from the congruence between genetic reductionism and new molecular biology technologies that enabled the isolation and characterization of individual 'disease-causing' genes. "Genetic determinism and reductionism emerge as significant research traps and a chasm-like separation might arise between molecular medicine and the sick patient. Furthermore, the newly added 'translational research' and 'functional genomics' cannot remedy this dichotomy" [29]. In fact, we should look at genes in a very different perspective, i.e. by embracing the global dynamics of networks that will reveal higher-order, collective behavior of the interacting genes [30].

These basic premises shaped the way upon which pharmacological strategies have been designed and developed in very recent times.

However, contrary to outlooks, pharmacological treatments established in the last 40 years did not achieve the expected success [31]. This state of affairs stands in stark contrast to prior decades of achievement in which a classical physiological approach led to identify effective treatments for several 'simple' diseases – like infections – for which at least a well-recognizable, dominant causative factor has been previously identified. Complex diseases – cancer, degenerative illness – have proven much less tractable and results provided by the combining interplay between epidemiological association and gene-based investigations led to contradictory outcomes. Findings proven to be hardly replicated, when not contradicted by new studies, undermining risk estimates that are highly variable or inconsistent or upon meta-analyses converge on little or no effect, and so forth [32].

Basic Paradigms in Pharmacodynamics

It is widely accepted that, fundamentally, drugs act by (a) mimicking or inhibiting normal biochemical processes, or inhibiting pathological processes in animals; or, (b) through inhibition of vital processes of microbial organisms. These effects are usually believed to be mediated by specific chemical interactions (i.e., covalent interactions) involving the pharmacological molecule and a "receptive" biological structure, to which we generally refer to as "receptor" or, broadly speaking, a "target" molecule [33]. This theoretical framework can be traced back to the seminal work of Paul Ehrlich who demonstrated that the selective and active uptake of pharmacological compounds must be dependent on the chemical binding of drugs to intracellular target molecules [34]. Currently, we recognize almost four types of target molecule for drug binding: receptors, as those classically involved in the transduction of endocrine effects (including both membrane and intracellular receptors), enzymes, ion channels (upon the membranes of both cells and organelles) and transport proteins. The interaction between the drug (the "ligand") and the target triggers an intertwined series of events, which eventually leads to the (desired) biological effect. Depending on the characteristics of the transduction machinery, a different time lag (spanning from milliseconds to hours) is required for activating a complex cascade of molecular events supporting the biological response (enzyme inhibition/activation, opening of ion channels, release of intracellular messengers, modification of gene expression, among others). Ligands fall into two main classes: agonists and antagonists. The former bind to the target molecule and promotes its activity through conformational changes, while the antagonist occupies the affinity side of the receptor without producing any conformational modification, thus preventing the activation of the target.

Hence, pharmacodynamics is usually considered a matter of ligand/receptor interactions following the equation, $L + R \rightleftharpoons LR$, basically dependent on the law of mass action at equilibrium.

This picture is, however, a very simplistic one and fails in explaining a number of consolidated evidences. Indeed, despite the fact that the ligand/receptor model of pharmacodynamics introduced by Fisher [35] has been experimentally vindicated and extensively studied by mathematical modelling [36], several issues needs to be addressed in order to afford some controversial results [37].

First, when the drug–receptor interaction involves feedbacks, the system becomes more complex, displaying emergent properties, non-linearity, and even chaotic behavior. These features are dependent on many factors including drug bioavailability and environmental constraints, thus inextricably linking pharmacodynamics with pharmacokinetics [38].

Second, drug dissolution, transport, and uptake are *heterogeneous* processes since they take place at interfaces of different phases, i.e., liquid–solid and liquid–membrane boundaries, where diffusion is regulated by physical and topological *constraints* [39]. Overall, those factors belong to the biological (morphogenetic) field and contribute in providing a fractal-like structure of the medium in which drug activity occurs. This aspect is usually overlooked, notwithstanding it could significantly influence drug efficacy by modulating the kinetic of the reaction. Indeed, kinetic orders in some cases reflect the fractal dimension of the physical surface on which the reaction occurs and every factor that modify the fractal topology of the medium can eventually inhibit or alternatively foster the pharmacological result [40].

Third, the active site of protein receptors is a rather flexible structure and it is continuously "reshaped" when it interacts with substrates or drugs. Since the sixties, it was then proposed that the reaction between a receptor and its ligand could involve an "induced fit" mechanism [41], where the active site undergoes "conformational changes". Conformational selection postulates that all the potential conformations of a given protein *preexist* and that once the ligand selects the most favored conformation, induced fit occurs and conformational change takes place [42, 43]. Overall, these results support the hypothesis that the molecules of water filling the active site of a protein, and surrounding the ligand, are as important as the contact interactions between the protein and the ligand for biomolecular recognition, and in determining the thermodynamics of binding. Conversely, any solute able to modify the solvation around the receptor could modify – in principle – the kinetics of the drug/receptor complex in a very unpredictable way. It is noteworthy that solutes and low-molecular weight compounds that can modify the configuration of water around enzymes and their substrates can also significantly influence the mechanical lock-and-key picture [44]. Indeed, in some binding processes occurring in aqueous solutions, the involvement of hydrophilic effects, as well as the biophysical constraints provided by the specific state of the solution [37], might be so profound that the lock-and-key model becomes irrelevant, then modifying thoroughly the way one approaches the problem of drug design [45]. Moreover, the displacement of free-energetically unfavorable water [46] or the presence of solute molecules, may substantially affect binding processes given that the major part of the Gibbs energy of binding could be due to interactions mediated through the solvent molecules [47].

Fourth, several compounds that display medical/biological effects do not act trough conventional pharmacodynamics, i.e. through establishing covalent bonds with their putative targets. This means that these substances exert non-canonical chemical interactions (belonging to the so-called supramolecular chemistry) [48], and a number of physically-mediated effects, including enhanced solubilization/absorption of other active factors [49, 50], physical disruption/distortion of cell membranes [51, 52], osmolarity properties [53], physical modulation of microtubule aggregation [54], physical distortion of biological fibers [55], protein surface binding modifications [56], physical sequestering/chelating effects on calcium/hydroxyapatite ions [57], changes in ionic strength, pH and surface tension [58].

Furthermore, it can be surmised that some active compound can modify the biological field [59], by acting through subtle modulation of its physical strength, i.e. via quantum effects on enzyme dynamics and on protein structure. Quantum tunneling effects on enzyme activity [60] and quantum-dependent coherent remodeling of cytoskeleton proteins [61, 62] have been noticed and there is some evidence suggesting that this happens *in vitro* following natural compounds treatments [63]. Finally, a number of drug mechanisms are still unknown (or only barely known), notwithstanding the drug still "functions". It is noteworthy that this class of compounds includes relevant drugs like Methocarbamol, Paracetamol, Phenytoin, PRL-8-53, Metformin, Thalidomide, Acamprosate, Armodafinil, Cyclobenzaprine, Demeclocycline, Fabomotizole, Lithium, and Meprobamate [64]. Overall, the relevance of those, still unexplored, mechanisms is underestimated, and grossly misunderstood, especially when mixtures of active compounds (like herbal formulas) are considered.

Limits of the Current Reductionist Paradigm

Besides the relevant achievements performed in the last 50 years, the above-sketched model of pharmacological activity not only has left aside some not negligible mechanisms of action, but ultimately it has shaped the philosophical underpinnings on which the current pharmacological research is rooted. By this way, the entire scientific inquiry, including its methodological models, became mainly focused on some well-known mechanisms, chiefly confined within the molecular level of interaction, considered as the privileged level of causality, i.e. the place where the pharmacological activity starts.

Focusing to the drug-receptor interaction has consequently driven the search for parameter efficacy by mostly considering factors that can influence the dynamics of this association: affinity (i.e., the reciprocal of the dissociation constant of the drug/receptor complex), efficacy (the ability to induce a *molecular* response) and potency (the lowest concentration at which the ligand elicits a response). This approach has been proven useful in explaining a relevant body of data – especially when dealing with hormone-receptors-mediated effects – but it[1] underestimates non-classical pharmacodynamics effects and restricts the recognition of response parameters by only considering those directly related to the ligand/receptor complex. Pleiotropic drugs effects – i.e., those exerted on different molecular targets or at levels higher that the molecular one – are generally left unnoticed. In other words, drug's efficacy became a matter of "molecular effect", thus disregarding the physiological effect(s) that for centuries has been the cornerstone for estimating the effectiveness of a cure. This approach has led to embarrassing outcomes, given that molecular effects are sometimes "translated" in an unpredictable way to the higher levels (cells/organs). Furthermore, by assuming a reductionist stance in appreciating drug's effectiveness has finally contribute to forget the true aim of every treatment: the well-being of the

[1] We are referring here to the Morphogenetic field, an epistemological/scientific concept, conceiving the field as a pattern of forces – such as mechanical or bioelectromagnetic – that constrains molecular "signals" and contributes in driving the overall behavior of the system. An excellent review addressing the multifaceted aspects associated to the debatable identification of such a field has been recently provided by SEB Tyler [ref. 59].

patients and the improvement in life expectancy we are expecting from that therapy. This conundrum has been specifically outlined in clinical management of tumors, where the classical approach was almost focused to establish the "objective tumor response rate" (ORR), i.e. the observable changes in tumor size/number.

This happened despite the fact that, since the eighties, it already became apparent that a proper estimation of drug efficacy should have taken into consideration the parameters belonging to the overall system, i.e. the patients. Indeed, Food and Drug Administration (FDA) has recommended that cancer drug estimation should be based on more direct evidence of clinical benefit, such as improvement in survival, improvement in a patient's quality of life, improved physical functioning, or improved tumor-related symptoms. These benefits may not always be predicted by, or correlate with, ORR [65]. This example raises several key questions that deserve to be discussed in detail: what is the disease? How to identify the target (s) for planning a proper treatment? And, finally, how to set the endpoints of a treatment?

Disease as a Controversial Concept

Constructivism Versus Naturalism

Shortcomings of current therapies, especially in some well-defined areas of medical inquiry (degenerative and neoplastic diseases), call into question the concept of human disease on which curative attempts rely for establishing a treatment strategy. During the last two centuries, the debate dealing with this subject has mostly polarized between the two alternative approaches represented by Constructivism and Naturalism [66]. The former, albeit hardly definable, essentially denies the naturalist thesis that disease necessarily encompasses bodily malfunction. Indeed, the principal Constructivism claim consists in that "disease judgments appeal to biological processes that are to be understood in terms of human practices rather than membership in some biologically definable class of abnormalities or malfunctions" [67]. Ultimately, "malfunction is not a necessary condition for disease", as even bodily malfunctions cannot be identified independently of human values and thus fall into the "normative" class of judgements [68]. Constructivists advocate that medical concepts (not only the definition of disease) should be "reformed" in order to reframe the meaning and the perception of ordinary events – including "disease" – by considering that our life is prominently shaped by our thoughts. This approach would finally end in delivering people from *avoidable* suffering, mostly to be ascribed to *prejudices* or cultural/societal misconceptions [69].[2] This framework

[2]Constructivism was instrumental in overturning the psychiatric view, dominant until the 1970s, that homosexuality is a mental illness. Activists argued that homosexuality was diagnosed for offensive moral reasons and not for medical ones and the classification of homosexuality as a

was eagerly adopted in a few domains, as in psychiatry [70]. However, it is quite clear that it can hardly accommodate with the daily experience of "true" organic diseases, like cancer. Instead, the Naturalistic system [71] conceives the human body as constituted by several sub-systems (organs and apparatus) that have natural functions from which they can depart in many ways, some of which are harmless and then deserve to be recognized as "diseases". This definition implies that recognizing a disease requires identifying both an altered functioning of some organism's apparatus and the resulting harmful effect on health. This definition represents a step forward the definition introduced in the XVII century by Sydenham [72], who identified the disease as a collection of symptoms (representing the "disease phenotype"), which should be ascribed to a unique "pathogenic" cause. In modern times, the "pathogenic factor" has been chiefly ascribed to external agents (microbes, toxins, iatrogenic factors) or to gene networks [73], as their deregulation can consequently lead to the perturbation of many biochemical/metabolic pathways downstream.

Overall, controversies among these two different approaches, probably reflect also the lack of a compelling theory of living organism [74]. As suggested by Lemoine [75], philosophic inquiry should focus to establishing a naturalistic-based concept of disease, by recognizing the basic theoretical premises of the medical science, and then looking for perspicuous accounts of different disease types within such framework. This approach will allows considering diseases as putative natural kinds, while leaving open the possibility that some diagnoses represent contingent historical outcomes.

From the naturalistic perspective – by far the only to which the daily medical practice actually relies – a disease necessarily involves biological malfunction, which either can or cannot be perceived as such, i.e. by complaining symptoms (some conditions, like hypertension, are equated to "disease" even if they neither produce symptoms nor are associated to malfunctions).

However, in the last 30 years, models of human diseases have been mostly "reduced" to the "malfunctioning" of a few, critical pathways. Consequently, drug discovery has been dominated by reductionism aiming to identify drugs that activate or inhibit specific molecular targets.

A target is usually a single gene or molecular mechanism that has been identified on the basis of genetic analysis or biological observations. Genetic targets represent genes or gene products that are deregulated or carry mutations.

---

disease was changed because of lobbying on moral grounds rather than based on any new discovery. By analogy, constructivists believe that a wide range of conditions – including pain, obesity, alcoholism, just to mention a few – have been counted as diseases mostly because they fall in the category of deprecated social behaviors. This quite an odd situation in which (questionable) political/philosophical considerations prevail on the proper scientific evaluation of a disease. Indeed, constructivism advocates are mainly interested in tracing the social/cultural processes by which categories are formulated and changed over time. Yet, it is beyond doubt that societies have at times estimated that some human behaviors (namely in the psychiatric field) were pathological because of values deprived of scientific evidence.

Mechanistic targets constitute enzymes or biochemical pathways for which a specific or broad malfunctioning has been put in causative correlation with the observed medical illness.

However, biology is complex, and it is increasingly clear that even discrete biological functions – or malfunction in case of a disease – can rarely be attributed to an individual molecular factor like a protein or a gene. As a result, approaches based on such a simplistic reductionist paradigm often showed either unforeseen toxicity or lack of efficacy when tested in clinical trials [76]. This breakdown stems, basically, from three (unsubstantiated) theoretical premises: (1) all diseases have a single (dominant) underlying cause; (2) disease features – signs and symptoms, i.e. the disease phenotype – are correlated with the causative factor in a linear fashion; (3) removal/correction of the underlying, putative "cause" will restore the healthy condition.

Unfortunately, evidence exists that all three assumptions are wrong [77]. Furthermore, the assumption that each illness is sustained by a specific "disease", i.e. mechanistically induced by quantitative changes in the molecular/physiological phenotype of the living system, still should be demonstrated beyond any doubt in several conditions (especially in psychiatric disorders). This unproven assumption has led to the "medicalization" of a wide range of conditions perceived as anomalous, besides any demonstrable disease process could be ascertained. Yet, perception of the relevance of symptoms, recognized as such, i.e. as belonging to a "disease state", is the result of cognitive process, highly influenced by cultural conditioning and societal/scientific model of illness [78]. Indeed, notions of health are highly context dependent, as human diseases only exist in relation to people, and people live in varied cultural contexts. Moreover, new clinical "entities", barely identifiable according to current medical rules, are often welcomed primarily as opportunities for market growth, the lack of compelling evidence notwithstanding [79]. This is especially true when we are dealing with "preventive medicine". Are presumptive markers of a "future" disease condition reliable enough to ask for a "preventive cure"? Namely, is someone with a genetic predisposition to an illness already sick? While no effective guarantee exists that a genetic predisposition or a biochemical anomaly will unavoidably lead to an overt disease, it is instead questionless that being aware of the probability of the (future) occurrence of such a threatening disease may be so traumatic to trigger a major psychological distress. As a result, quite sad to notice that, a number of new diseases have been 'created' simply to fit the ability to diagnose them and for opening new avenues in the drugs market [80].

The above sketched examples highlight that the model of disease, i.e. the conceptual meaning of a so widespread used category, is only rarely explicitly debated or defined in the scientific literature [81], besides still being the subject to extensive philosophical debate [82]. The model that dominated until the first half of the past century mostly originates from Virchow's conclusion that all diseases result from cellular abnormalities [83]. Since the discovery of the double helix in the 50s, however, this model was relentlessly superseded by an even more reductionist approach, as that provided by the New Genetics. According to this theoretical approach, every disease can be traced back to the malfunctioning of a discrete number of genes and, at least in principle, every change in gene activity can lead to a pathological state [84]. In the same time, the clinical entity (the "disease phenotype") was recognized by the association between the set of signs/symptoms and circulating/tissue markers, conceived to be "mechanistically" linked to the hypothesized pathogenic mechanism. Identification of disease according to these rules allowed clinicians in establishing a disease taxonomy that was instrumental in performing medical practice up to the current time. However, this approach has progressively shown shortcomings that reflect a lack of specificity (i.e., inability in defining disease unequivocally), as well as a lack of sensitivity (i.e., incapacity in recognizing preclinical, true causative state of disease). Ultimately, this model proven to be confounding, as it often posits wrong correlations between the disease-associated biological parameters (usually identified only when illness reach a "stable-state") and the alleged causative processes, thereby prejudicing efficient treatment strategies.

These limitations can principally be ascribed to the reductionist approach on which medical research has grounded its agenda. Increased awareness of such inadequacies, prompted for revisiting the concept of human disease during the last decades, striving to conjugate advantages offered by the molecular/reductionist stance with the opportunities of a physiological/systems-based framework [85]. Namely, such considerations apply when we refer to the target-based drug discovery framework that has largely replaced the traditional physiology-based approach, since the completion of the Human Genome Project [86]. Conclusively, the main outcome of that program, was *philosophical*, as it prompted to consider that every disease can be singled out by identifying a set of few genetic/biochemical targets. Development of the so called 'omic' technologies led to a more sophisticated reconnaissance of those targets, shifting the focus on their complex (eventually non-linear) interactions [39], however without questioning the basic assumption on which the disease model has been built.

Simple Diseases Are Not Simple, Indeed

Reassessing meaning and boundaries of human disease should be considered a main task that became even more complicated by the development of 'omics technologies during the post-genomic era, rather than solved. In fact, the attempt to bring back the definition of a disease to its genetic determinants, along with the triggered mechanistic pathways downstream gene modulation, has emptied of meaning the classical genetic approach, highlighting how even simple monogenic disorders are supported by a net of causative factors that, ultimately, are responsible of the disease phenotype. As an example, sickle cell anemia, a classic monogenic disorder due to a single point mutation, turned out to be a very complex condition characterized by several different clinical features [87]. Indeed, the pathogenesis of this classic Mendelian disease shows a bewildering intricacy, which ultimately ends up into different (almost six) disease phenotypes, for each of which a diverse treatment strategy is needed. The mandatory conclusion is that, even in monogenic disease, "the genotype simply cannot invariably predict the phenotype of patients with the disease", as extensively discussed by Loscalzo et al. [88]. As a second example, let us consider Epstein Barr virus (EBV) related diseases. EBV is one of the five recognized human herpesviruses, the others being herpes simplex virus types 1 and 2, cytomegalovirus, and varicella-zoster virus [89]. Undoubtedly, B cells are the primary targets of EBV infection [90], and infection of B cells leads to the expression of a limited set of viral gene products, which drive the cells into proliferation. Usually, in healthy inhabitant of Western countries, proliferation of infected B cells is limited by CD8+ and CD4+ T cells, thus leading to the onset of infectious mononucleosis. On the contrary, in children living in malaria endemic regions of the world (i.e., equatorial Africa, Brazil, and Papua New Guinea), EBV infection is more likely to induce Burkitt's lymphoma, a tumor of the lymphoreticular system. For a while, explaining the ways in which a single agent can evoke such different responses in different hosts has represented a challenging task [91]. It is now widely recognized that a critical factor that can switch the outcome from a mild form of influenza-like syndrome to an aggressive lymphoma is the host response to EBV infection. Indeed, a wide array of immunocompromised conditions (endemic malaria infection or immune-deficiency syndrome like HIV infection) are recognized triggering the shift from a transient lymphoproliferative reaction to a true, malignant transformation [92]. It is startling that EBV infection is currently known as the main "causative" factor of a number of disparate diseases, including pharyngeal carcinomas [93], gastric cancer [94], leiomyosarcoma, undifferentiated type I nasopharyngeal cancer [95], as well as non-malignant illness, such as the childhood disorders of Alice in Wonderland Syndrome [96], systemic lupus erythematosus [97], and acute cerebellar ataxia [98]. Those examples highlight how misleading can be the hypothesized link between a putative causal factor (the viral infection) and the associated disease, as the same "causal" factor can sustain very different pathogenic phenotypes. Instead, lesson learned from the EBV-related illness shows that the disease is the unpredictable outcome emerging from the complex interaction between "stressing" (rather than "causal") conditions and the host reactivity [99].

Conversely, as advocated by the case of familial pulmonary arterial hypertension [100], or hypertrophic cardiomyopathy [101], a common disease phenotype may be sustained by many different genotypes yielding it. Namely, hypertrophic cardiomyopathy is associated with mutations in several different genes that code for different sarcomeric (myosin heavy chain, myosin light chain, tropomyosin, and troponin C) and non-sarcomeric proteins [100]. In this case, a common pathophenotype is "produced" by very different "genetic" disease. The aforementioned examples highpoint that the molecular determinants – and especially the genetic "defects" – thought to be the cause of a disease, simply cannot predict the phenotype of patients with the disease, which ultimately emerges as a result of a complex interplay among different factors. These factors entail different levels – from the cell to the organism in its wholeness – and they cannot be "reduced" only to the molecular tier. Therefore, characterizing disease by establishing a nosology almost substantially based on putative molecular determinants has shown significant shortcomings.

Mechanistic/Genetic Versus the Physiological Approach

The above-mentioned issues are clearly epitomized by the so-called target-based medicine that constitutes the pre-requisite on which the personalized medicine relies. However, despite astonishing claims, it has already been noticed that the decline in drug discovery just coincided – to a large extent with the introduction of target-based drug discovery [102].

In oncology, the possibility of finding so-called synthetic-lethal drug targets, which are only essential in cancer cells that carry mutations in specific tumor suppressor genes, is attractive in theory [103]. However, the search for such genes – if they exist, as normal tissues has been proven to carry the same mutated genes as their malignant counterpart [104] – might be frustrated by tumor heterogeneity [105], given that such heterogeneity, arising by a hierarchical pattern of stem-cell divisions, yields a mosaic of different cells and, ultimately, can hamper cancer treatment [106]. This is why seemingly identical cells respond differently to treatments, given that phenotypic and genotypic differences provide differentiated response by activating even opposite outcomes in cell behavior and ultimately escaping the drug-induced inhibition on specific targets [107].

In addition, for three main reasons a genetic approach is unlikely to be a solution to common diseases in the near future. The first is the great importance of environmental circumstances in determining health, the second reason is the great complexity of gene/gene, gene/environment interactions, and the third reason is the high individual variability [108]. Conclusively, despite having identified "hundreds of common variants whose allele frequencies are statistically correlated with various illnesses and traits, the vast majority [of those studies] have no established biological relevance to disease or clinical utility for prognosis or treatment" [109].

Contrary to previous expectations, nearly all the genes associated with diseases are non-essential genes [110]. This means that they do not encode hub proteins and are localized in the functional periphery of the entire network. Mutations involving central, essential genes are likely to induce severe impairment of pivotal physiologic functions, hence leading to increased lethality during the early developmental stage or in the extra uterine life. As such, from an evolutionary point of view, such genes are under higher selective pressure that would ultimately end in deleting them from the population. It is a matter of probability that only "less-threatening" mutations – as those affecting "peripheral genes" – can be preserved within a population from such selection. Therefore, even the physiological relevance of such mutations is concurrently abridged.

In the early 1980s, the development and broad implementation of molecular methods, as well as the later development of genomics, significantly increased our understanding of the individual actors participating in cellular processes, ultimately providing a prodigious list of molecular factors. That exhausting collection of data facilitates the reductionist strategy in drug discovery by converging on (presumed) targets and designing compounds to interfere with them. Inevitably, this approach removed the targets from their physiological context to study them at quasi-atomic level and focused on the optimization of the target-compound cross talk, placing an almost exclusive emphasis upon the drug-target interaction parameters (i.e., binding affinity and target selectivity) [111]. Therefore, "the criteria to evaluate the potential of a novel molecule shifted from a strict physiological observation of the results obtained with the assayed compounds to a molecular one, where the best lead chemicals were those displaying a strong binding with the target protein and a good specificity profile (i.e. binding to only one target)" [112].

A target is usually defined as a single gene, gene product or molecular mechanism that has been identified as a putative causative factor in disease pathogenesis. A target-based drug would be a compound that selectively modulates the activity of the disease associated gene or mechanism, without directly involving other pathways.

These conditions are fulfilled only in a very few cases, in which the ailment must be attributed in a predominant manner to a genetic mutation or to a specific biochemical mechanism. Second, the causative factor must contribute to the disease process at the time of treatment. The first condition applies only for few diseases, as the more common illness have a multifactorial origin.

The second condition deserve special attention, as the mechanism/gene responsible for the onset of the illness might have exerted its action during early pathogenic steps and could no longer be active during the steady state of the disease, when diagnosis is usually reached. Some developmental-based diseases like schizophrenia [113], mental illness, or depression (for which no direct relationship between target and therapeutic effect has been so far evidenced) [114], falls within this category, as well as some cancers that lose their mutated, "driver" oncogenes before to progress [115, 116]. Indeed, as far as cancer treatment is concerned, it is widely recognized that 'inactivation' or inhibited expression of oncogenes (like BCR-ABL1, c-Myc, c-ras) is not mandatory for achieving tumor inhibition [117]. Moreover, a major obstacle for establishing an effective "precision" based therapeutic approach in oncology is represented by the genomic heterogeneity of tumor – even within the same tumor of a single patient [118] – a condition that get worse after chemotherapy as the treatment can likely select more aggressive and resistant clones [119]. This picture would suggest that it is virtually impossible portraying the tumor "genomic fingerprint", aiming at identifying key-druggable targets, as the targets are disparate, and change accordingly concurrently after the "evolution" of the gene-expression pattern. Current treatments are unable to cope with such an overwhelming complexity, and their acknowledged failure in curing cancer cannot be viewed as an unannounced surprise [12, 120, 121]. Ultimately, improvement in cancer survival observed in the last 40 years cannot be ascribed to anticancer drugs [122], and even drugs approved on the basis of better progression-free survival have been subsequently found not to produce better overall survival than the comparator drug [123]. Overall, "we overdiagnose, overtreat, and overpromise, with high costs and without clear benefits" [124].

This picture even gets into more complexity when mechanisms of action are considered. Mechanism-targeting drugs should be highly specific (i.e., acting only by hitting a single enzyme/pathway) and must affect only a fraction of the overall activity of the target, as a complete blockade of an enzymatic pathway would lead to undesirable, potentially lethal effects. However, the story of the imatinib mesylate – the "magic bullet" *par excellence* – taught us that even high specific target-based drugs are not as successful as initially thought. Indeed, despite Imatinib was designed to act on a single aberrant protein (BCR-ABL) expressed in cancerous cells, it was later shown to inhibit other targets (c-KIT and platelet-derived growth factor receptor), thus leading to unwarranted side effects [125]. In simple terms, target-based drugs rarely bind specifically to an only single target, therefore challenging the concept of magic bullet. Moreover, the use of knockout animals – where the target has been deleted for vindicating the causative role of that target – demonstrated later to be a debatable approach, leading in many circumstances to equivocal results. Indeed, during development, due to the redundancy characterizing biochemical pathways, the organism might activate compensatory contrivances and adaptive strategies. Thus, the outcome of a gene deletion can validate the hypothesis and the putative target only in few, more than ideal situations. Undeniably, it is far from being infrequent that knockout animals display unexpected and very surprising effects [126], which can be confidently attributed to the intertwined network among genes and pathways. In addition, when an adult animal – in which the target has already contributed to the development of the animal – is given a suppressive drug – the partial inhibition of the same target will have completely different consequences when compared to the situation in which the target has been "deleted" since the early developmental steps.

Additional complexity stems also from the fact that the observed changes in the putative causative target associated to the disease rarely disclose genuine causative relations. These targets are usually identified through statistical associations that provide no reliable information about the causative links [127].[3] Indeed, modification in target levels and/or their activity may likely arise as a part of the pathogenic process or, alternatively, they can represent adaptive and even antagonistic measures deployed by the system.

This is why we are currently dealing with too many targets and not enough target validation. "Target validation, crucial to rational drug design, is a concept often discussed but rarely defined [. . .] to develop innovative drugs, we need smarter and faster target validation, not increasing numbers of new targets" [128]. Yet, demonstrating a causal role of the putative target is an unavoidable task that cannot be overcome by adopting more or less refined biometric strategies, as those suggested by the advocates of the so-called Big Data approach.

---

[3] We are witnessing a true "epidemic" of biometric-based studies striving to support the strength of mere statistical associations between disease and the observed parameter/target by only adopting more or less sophisticated mathematical modelling. Indeed, causal relations are not hallmarks that can be directly read off from the data but have to be inferred. Causal relations can be identified in an experimental setting, and the common mantra that "correlation does not imply causation" is still valid.

The Big Data Illusion

The current interest in big data has generated the widespread illusion that complex algorithms and number manipulation could solve problems without pursuing experimental investigations. Data handling does not produce any new information by itself. Correlation is not enough, though. Moreover, mere computational brute force cannot compensate for the lack of theory into which information from experiments need to fit. Definitely, computationally intensive tools for the exploitation of huge data sets are not designed to model the structural characteristics of the underlying system but only as very efficient 'exploratory statistical tools' that (at their best) can only act as aid for generating hypothesis. It is thus "vital to use theory as a guide to experimental design for maximal efficiency of data collection and to produce reliable predictive models and conceptual knowledge" [129].

Proper target validation would require a very different model (closest as possible to the physiological one), a suitable modulation of the target and – even more important – an adequate theory for understanding what we are doing and the meaning of data we are gathering. We need models and theoretical insights to help guide the collection and interpretation of data. The relatively meagre initial returns from the human genome project demonstrate that data do not translate readily into understanding, let alone treatments [130]. Commendable as these efforts are, they are fundamentally flawed as the relevance of differences in gene-expression pattern – viewed as the driver causal factor of the disease – is grossly overestimated, namely in oncology [131, 132].

Even using a rigorous predictive statistical framework, characterizing average behavior from big genomics data will not deliver 'personalized medicine' [133]. This is because correlations observed in different sets of data are not necessarily evidence of dependency. The problem of spurious correlations is familiar when it comes to the use of quantitative structure–activity relationship models and machine learning to predict the biological activity of molecules. Correlations may not tell us precisely why something is happening, but they alert us that it is happening. However, no matter their 'depth' and sophistication, machine-learning algorithms merely fit model forms to "selected"[4] data. They may be capable of effective interpolation, but not of extrapolation beyond their training domain. They offer no structural explanations of the correlations they reveal, many of which are likely to be false-positives. Furthermore, most correlations are spurious, i.e. very large databases have to contain arbitrary correlations

[4]The collection of data is not a merely empirical activity. Science does not collect data randomly. Experiments are designed and carried out by choosing "fact" that are deemed to be worth of the definition implied in the concept of "data". This is made usually based on pre-existing pre-conceptions, as already remarked by Kant. In other words, as highlighted by Kuhn in his seminal book, we must cast on doubt the possibility of accessing the real world in a neutral way. Conclusively, as aptly stated by Mazzocchi, "We look at the world through the lens of a particular vantage point, and the possibility to speak of—or even perceive—certain facts, data and objects depends on this vantage point" [ref. 137].

[134]. Finally, diverse applications of Big Data Theory have met with limited success in scientific domains, up to now [135].

What is the illusion behind this all? Correlation can supersede causation, and science can advance even without coherent models, unified theories, or any mechanistic explanation at all. This claim assumes the primacy of correlations over causal explanation or, even more radically, the replacement of the latter with the former. Therefore, we are legitimate in asking if do we really come to "the end of the theory"? [136].[5] Indeed, "Big Data science renews the primacy of inductive reasoning in the form of technology-based empiricism and has inspired a view of the future in which automated data mining will lead directly to new discoveries [but] more data do not necessarily generate more knowledge. Data by themselves are meaningless. The idea that with enough data, the numbers speak for themselves hardly makes sense." [137].

We are reminded of the story of the blind men and the elephant: local data are difficult to interpret without a (previous) mental model of a pachyderm. In a similar way, various big data initiatives are blindly groping about that great beast that we know as biology. We need theory to help envisage it in all its meaning.

Overall, these issues contribute in explaining why so many clinical trials ultimately provide scarcely reproducible results or – even worst – false findings [138, 139], thus supporting what is currently known as the "reproducibility crisis" in biology and medicine [140, 141]. Similar considerations apply for the so-called "precision medicine", a disguised avatar of the target-based medicine, which has been brutally portrayed as an "overall failure", almost in oncology [142].

What Should We Treat?

Disease as a Process Entailing Non-linear Networks and Environmental Influences, Spanning from Different Levels

As previously stressed, inadequacies of theoretical models of human diseases play a major role in explaining the difficulties encountered in pharmacological research. Indeed, the increase in the rate of drugs failing in late-stage clinical development over the past decade "has been concurrent with the dominance of the assumption that the goal of drug discovery is to design exquisitely selective ligands that act on a

---

[5]At long last, the drift we are witnessing is almost entirely contained and forecasted in Bacon's heritage (F. Bacon, *Novum Organum*, London, 1620). Bacon uttered that knowledge should not be based on preconceived notions ("premises"), which would constrain the reasoning (confirming or invalidating the basic presumptions), according to Aristotle's tradition, basically framed by theory-driven experiments. Instead, the King's Chancellor proposed an "inductive" method, based on generalized inferences from data merely based on an empirical approach. The Baconian inductive method has been widely criticized, namely by P. B. Medawar (see: *Induction and Intuition in Scientific thought*, London: Methuen, 1969).

single disease target" [143]. This approach does not only overlook the relevance of multifactorial etiology of the disease, but also underestimates the robustness and resilience of the (pathologic) phenotype when stressed by a (pharmacological) perturbation. For instance, single gene knockout or complete silencing, have shown little, contradictory or even null effect on phenotype [144].

Robustness of the pathophenotype can be understood in terms of redundancy and alternative (compensatory) pathways, highly "structured" into scale-free networks, which are usually "triggered" in response to perturbations [145]. It has been observed that large transcriptional regulation networks act upon targets via different and alternative regulatory molecules. Indeed, multiple alternative pathways between regulator and target pairs are the rule rather than the exception. The "selective" activation of a pathway among many others should be ascribed to changing requirements of the context in which the system belongs [146]. Therefore, robustness in complex dynamical systems can be appropriately understood by considering the existence of multiple attracting domains (multistability), to which the system can suddenly switch (performing a transition remnant of the 1st phase transition observed in chemical physics). This behavior allows the system to access previously unexplored attractors, in response to environmental stresses, physical/chemical stimulation or small random perturbations. This property convincingly explain how cancer or bacterial infections can easily develop drug resistance by accessing new attractors (new stable states), thus making "precise" and targeted therapies, a futile attempt. Indeed, "by looking at the rich history of failures in targeting individual pathways, it is undoubtedly that targeting individual pathways may never be entirely successful" [147].

Intrinsic robustness has relevant implications for drug discovery, given that it put a special emphasis on the perturbations that can lead to several changes in the network activity/configuration associated to each disease [148]. However, it should be outlined that when tested in non-ideal conditions (i.e., in presence of "unfavorable" environmental milieu or when a small molecule/drug is added to the culture), the system displays an unexpected sensitivity. Indeed, nearly all genes (97%) are needed to ensure proper functioning in at least one condition when the cell-microenvironment cross talk is perturbed, namely when a genetic perturbation is combined with a chemical insult to a biochemical pathway [149]. These finding evidences that the relevance of genes and connected hubs within the network can be properly assessed only when the system is challenged by perturbing the microenvironment. To put the question in another way, the emergence of the disease-associated perturbed network can be identified only if the specific microenvironmental field is *concurrently* contemplated. Therefore, the dominant assumption that a critical, single target may suffice for obtaining a valuable therapeutic effect, once again, is cast on doubt [150]. Previously examples we mentioned – EBV-related diseases, sickle cell anemia and many others – clearly demonstrated that several factors participated in the genesis of the ailment, involving different levels of organism's organization (from organelles to organs and systems, like the immune system). Those factors and levels are tightly intertwined and therefore a successful therapeutic strategy should embrace all of them if the aim is properly to cure the patients, and not only "to fix" a "singled out" pathway.

Furthermore, it is widely recognized that the current narrative of the natural history of a disease overlooks the participation of associated factors/systems in shaping different pathophenotypes (and their underlying mechanisms), which may likely explain subtle, but potentially important differences among clinical manifestations. A comprehensive approach to this problem would enable in providing a complex network structure, constituted by modular sub-systems, whose (non-linear) interaction will drive the organism response toward emergent properties, i.e. disease or health. Accordingly, as advocated by many scholars, human disease can be conceptualized as a (pathological) phenotype, i.e. an emergent property of the human body as a complex system [151, 152]. Therefore, those considerations prompted for a rethinking of the taxonomy of human diseases [153], whereby data obtained by investigating diverse levels (from the molecular one to the systemic response) would be embraced and incorporated into a unified operational model, in which response parameters should be provided by the overall system estimate, rather than on singled-out molecular target (Fig. 2a, b) [154]. Nonetheless, the identification of such gene-based regulatory networks may be insufficient to understand the emergence of cellular functions as well as the three-dimensional organization of living structure [155].

Additionally, diseases as well as their "causative" targets are usually recognized and defined by their late-appearing manifestations [94]. Unreliability stems here from the erroneous and artefactual reproduction of the chronological steps through which the disease develops, while diagnostic parameters and putative causative factors are frequently (only) those associated with the steady state of the disease. This approach entails the obvious risk to consider a late emerging symptom/target as the driver-causative element of the pathogenic process.

Such shortcomings are deeply rooted on the disease model we usually adopted, where illness is merely considered as an almost "stable state", characterized by a predictable, linear dynamics, with established well recognizable steps, from the onset up to the end (death or healing).

Yet, being a complex system, a disease would be better depicted as a non-linear dynamic *process*. As such, it displays classical features of complex systems, including resilience, sensitivity to initial conditions and multi-attractor accessibility. Namely, the latter point deserves to be explored as the evolution of the disease is conditioned by travelling across a landscape in which the system can enters into different attractors (i.e., different clinical outcomes), downstream a critical transition point. Around such points, the systems display an astonishing sensitivity to environmental cues, showing an increased fluctuation of a set of critical parameters [156]. The interaction among these components allows the system to overcome the (energetic) boundary of the attractor, moving towards a diverse stable state. The transition can be smooth or, quite more often, abrupt, remnant of the first phase transitions occurring in chemical physics. There is compelling evidence demonstrating that, as reported for many other fields in the natural world, such transition states

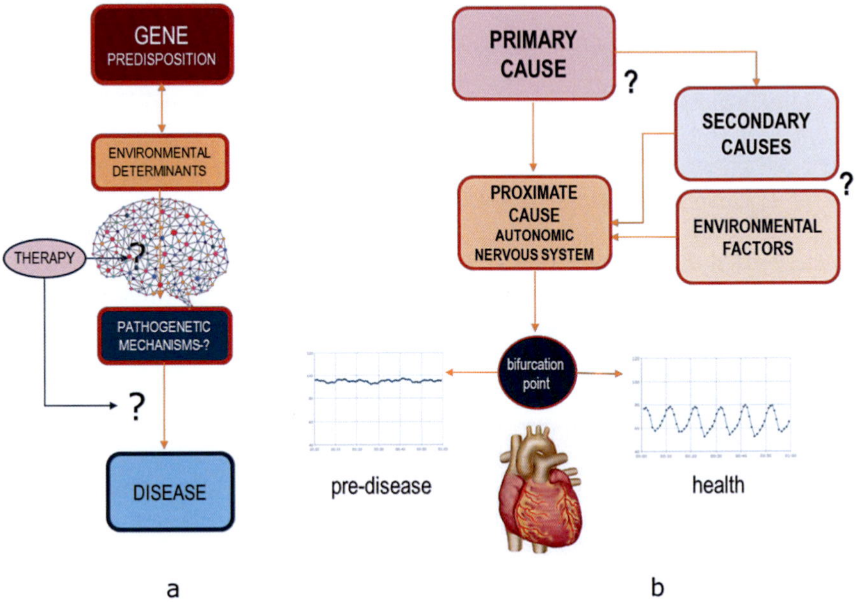

Fig. 2 *Causative factors in complex diseases.* (**a**) Schizophrenia has a strong genetic component and the risk factor is 50% for monozygotic twins. It is likely that the disease can be ascribed by a developmental deficit during the prenatal period or around birth. Causes include diverse factors such as viral infections in utero or hypoxia during birth, implying that some pathogenic factors may only have been present only during the early developmental steps (and then are gone forever), while the genetic environment only confers an increased susceptibility to these environmental cues. Therefore, a treatment cannot influence those factors acting at the beginning of the disease process, nor can modify the "genetic predisposition". Ultimately, the disease process cannot be reversed in the adult because the brain cannot be rewired back to the connectivity it should have had in absence of the pathogenic insults [ref. 112]. (**b**) Heart disease generally are supported by pathogenic cues largely unknown (the typical case is represented by idiopathic hypertension). On the contrary, a number of secondary causes and environmental factors converge in shaping the so-called proximal causes, mostly acting through modulation of the autonomic nervous system (ANS). In this case, there is no room for a treatment aiming at hitting the "primary causes", whilst efforts are usually aimed at modulating the activity of ANS. Furthermore, the parameter of efficacy basically relies on the overall system estimate (electrocardiographic tracing, heart rate variability), rather than on singled-out molecular target [ref. 153]

exist in clinical medicine. The evolution of such a process can be modeled likewise a time-dependent non-linear dynamical system, in which abrupt deterioration is viewed as a phase transition at a bifurcation point [157, 158]. Depending on the progression level of the disease across the metaphorical landscape, the process can schematically describe different stages: i.e., a normal state, a pre-disease state, a disease state (characterized by steady-state conditions), and a number of critical states at which the disease can alternatively move towards progression or healing. Critical states, altogether with the preclinical state, display bifurcation points, i.e. a set of both internal and external conditions that can drive the process into a very different fate. Recognizing such critical points by constructing a dynamical network will likely help in understanding the logic of the process [159]. Moreover, identification of biomarkers ("early-warning signals") indicating an imminent bifurcation or sudden deterioration before the critical transition occurs, can help in planning an appropriate management of the disease [160].

A Poly-Target Approach to Modify the "Pathogenic Field"

To cope with this complex challenge, in the last decade a new pharmacological strategy has been proposed on the basis of reconstructed, scale-free network of intracellular reactions. These networks are relatively insensitive to random damage [161], despite being rather vulnerable to attacks targeted to their most-connected elements (hubs). In assessing such an effect, experimental studies require the complete suppression of an element from the network to assess network stability [162]. Nevertheless, this strategy is highly controversial as lethal effects usually follow it [163]. On the contrary, evidence shows that the partial inactivation of more than one single target ("poly-target approach") is more efficient than the complete inactivation of a single target [164]. As demonstrated by a number of studies [165, 166], this approach is gaining momentum, and it represents a promising area for further drug development.

This is especially true when the mechanisms of action of herbal mixtures, belonging to the so-called complementary and alternative medicine (CAM), come into play. CAM-related activities involve both classical and non-conventional mechanisms of action, nonlinear multiple interactions [167], poly-targets hitting [168, 169], supra-cellular effects [170], and detoxifying properties [171]. Overall, CAM mechanisms of activity embrace interconnectedness among different levels of organization of living organisms (from the molecular to the organ plane) in a contextual view of human beings that are inseparable from and responsive to their environments [172]. CAM treatment lies in a different theoretical approach, entailing clinical criteria, treatment targets and patterning of response. These beliefs and principles run counter to the assumptions of reductionism and conventional biomedical research methods that dismantle and test only single aspects of the CAM system. Instead, complex herbal remedies are in themselves true complex systems that interact with another complex system (the living organism), displaying a nested network of relationships [173]. Changes elicited by herbal mixtures can modulate self-organization and emergence in living organism by encompassing a wide array of processes – already acknowledged by the theory of complex systems – like synergetic (e.g., multicomponent global coherences) and critical phase transition across rugged landscapes, which allow the system accessing multiple "basin" and interactive multistability among a variety of attractors [174]. This dynamic pattern establishes bidirectional feedback across scales, explaining how small stimuli often result in large effects and how seemingly catastrophic events can, at times, result in merely a ripple effect across the system [175].

All of these entwined factors may help explaining the astonishing synergistic properties displayed by complex mixtures in respect to the effects triggered by individual components in isolation. It noteworthy that synergy cannot be full explained by the multi-target-based mechanism of action displayed by mixtures of natural compounds, given that pharmacological polyvalence of many plant constituents might explain an amplification effect by a factor 2 or 3 but not by a factor 10 or more, as reported by several studies [176, 177]. Such findings claim for a very different mechanism on which synergy relies.

Synergy

Here the Synergy is an "emergent" property – susceptible to precise mathematical definition [178] – and as such, belonging to the whole system, not predictable by properties of the parts [179]. Undeniably, attempts to identified mechanisms responsible for synergistic effects by studying components in isolation, provided elusive responses. Nonetheless, by using standardized mono- and multi-extract combinations against well- known standard drugs, synergy is clearly established for a number of herbal mixtures, tested *in vitro* as well as *in vivo* conditions [180–182]. In some instances, these synergistic effects can be explained by either previously unrecognized factors or "indirect" mechanisms of action, as such exerted by herbal component on patient microbiota. For instance, several Traditional Chinese Medicines Remedies (TCMR) comprise both soluble, active small molecules (including saponins, iridoid glycosides and flavone glycosides), as well as complex polysaccharides, which are usually considered as "irrelevant" given that is they are generally indigestible by oral administration and hardly absorbable in the gastrointestinal tract [183]. Accordingly, in modern industrialized TCMR preparation, polysaccharides are habitually removed to meet the requirements on purity and dosage amounts of the final products. Likewise, scientific research on TCM decoctions also excluded polysaccharides from biologically key chemicals [184]. However, such products not only lack efficacy when clinically tested, but also are deprived of scientific evidences. These findings suggest that such "irrelevant" polysaccharides must have an effect, in spite of everything. Indeed, convincing evidence have been provided demonstrating that such components (usually represented in the diet) may trigger complex therapeutic effects by targeting the host microbiota, selectively stimulating the growth of a subset of beneficial gut bacteria, and consequently to sustain the homeostasis of gut microbial community as well as the host health [185, 186]. These findings are worth of note, giving that the relationships between microbiota homeostasis and human diseases is currently an area of extensive research, and developing strategies to modulate microbiota function and composition could likely represent a reliable option in the management of several illness [189]. Undoubtedly, studies are warranted to explore in depth those mechanisms in whose synergy relies. On that field, we are just moving the first, tentative steps.

Conclusion

The network-based models, as those extensively studied by Csermely [165] and his team, despite their heuristic and epistemological value, only partially explain the unfathomable complexity behind the outbreak of a disease. Current network models of human diseases suffer from an excess of "specificity", as they are habitually centered on an integrated representation of the cell biochemical and genetic pathways, and, as such, they discard the contribution of non-genetic factors and micro-environmental constraints. In fact, notwithstanding the sophisticated, even non-linear models adopted to represent complex genomic-proteomic networks, those approaches lie fundamentally on a "preformationist" view, being centered on a "genetic program", which operate deterministically by itself. Accordingly, since the eighties, the agenda of pharmacology discovery was then dictated by aiming at discovering "relevant" molecules (along with their classical rules of interaction), abstracting from the true, physiological response of cells, tissues and organism. In this perspective, genes assume the fundamental causal role while cells simply act as causal proxies, dispensable because they represent an irrelevant intermediate level between the molecular input and the organismal output. Such framework, both theoretically and methodologically, is no longer tenable [187].

First, we need a comprehensive model able in capturing the complexity on which the disease relies (Fig. 3).

This approach should point at ascertaining different targets (whose concurrence is mandatory for shaping the specific patho-phenotype we are dealing with). These putative targets are spatially *distributed in diverse cells and tissues*, thus involving different tiers (from sub-organelles to organs). Delocalization of the potential targets within different tissues and organs may likely affect a differential accessibility to drugs, whose bioavailability is tissue-dependent. Moreover, given that a disease is properly a *dynamic process and not a steady state*, treatments should be diversified according to a target selection dictated by the timing of the disease process. This approach can help in detecting pre-disease state or, alternatively, critical transition points from which the illness might access different attractors, leading ultimately to different outcomes (spontaneous healing, chronicity, disease exacerbation, death).

Second, the question now is how to design systems-oriented drugs that tackle both the multifactorial pathogenic determinants of the disease as well as the intrinsic robustness of the living organism [188]. Just to start with, this attempt requires an in-depth understanding of cellular- and organism-level dynamics, combined with advanced high-throughput screening and computational analysis tools. Instead of single target, pharmacology research should consider a polyvalent-based approach, i.e. the use of multiple drugs or drugs affecting several targets localized at different levels. Additionally, above and beyond classical pharmacodynamics, unconventional mechanisms of action urge to be investigated. Thereby, those non-canonical mechanisms of action as well as different, hierarchically structured level of "causation" should be integrated within network models currently in use.

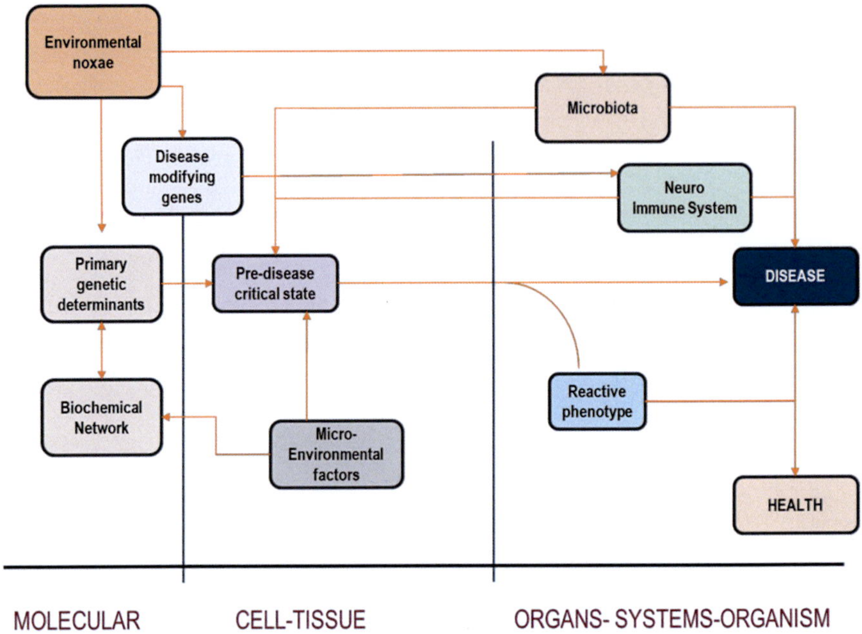

Fig. 3 Hypothetical diagram sketching the non-linear dynamics interaction among different causative factors distributed along different hierarchical levels, including both internal as well as environmental determinants

Systems Biology has already recognized that, the main challenge, in both a biological and a mathematical sense, is to find out how to control complex network systems. Highly distributed nonlinear network systems are still quite "intractable", in respect to simple feedback systems usually understood by means of classical and modern control theory. Yet, the most relevant hurdle stems from the still controversial nature of such system. The vast majority of studies deal with molecular or cellular-based model, and consequently do not consider the influence of the micro-environment or of the higher control-levels (immune, neuro-endocrine system) neither. Such a limitation is specifically exemplified by the kind of parameter we usually choose to ascertain the responsiveness to treatment. While disease response is habitually simplified by describing changes in a single (or a few) parameter, we instead have to move from target-related parameters to system-parameters, which could capture those modifications that could likely impact on the whole systems dynamics. Such an approach is partly ensured by metabolomics studies [189, 190], given that metabolites fluctuations usually amplify subtle modulation of the genome/proteome network, thus representing a more sensitive criteria for grasping changes in complex systems dynamics [191, 192]. By this way, "metabolomics represents more closely the phenotype of an organism" [193].

Third, we should shift from targets to processes, therefore pointing to influence several targets (poly-target treatments), conceivably by entailing different mechanisms of action. Modulation of processes implies we should be able to "redraw" the disease-related landscape, favoring the system displacement from pre-clinical state or true disease-states towards healing pathways. Disease should be "reverted", eventually involving also a "reprogramming" of the gene-regulatory network. A remarkable case in point is that of tumor reversion, a promising field of investigation [194]. An increasing number of reports has ascertained the occurrence of cancer reversion, both *in vitro* and *in vivo* [195–197]. This process encompasses mandatorily a change in the cell-stroma interactions, leading to profound modification in tissue architecture. As cancer can be successfully 'reprogrammed' through the modification of the dynamical cross talk with its microenvironment, cell-stroma interactive network must be recognized as a target for pharmacological intervention [198]. It is worth noting that several natural compounds as well as morphogenetic factors obtained from eggs [199] or animal embryo [200–202], demonstrated to be able in inducing a significant reversion of the tumor phenotype in a wide array of cancer types.

Fourth, we have to look at compounds displaying "pleiotropic" effects, i.e. able to tackle several targets. A dominant paradigm in drug discovery is the concept of designing maximally selective ligands to act on individual drug targets. However, many effective drugs act via modulation of multiple proteins rather than single targets, and we previously recalled that exquisitely selective compounds, compared with multitarget drugs, might exhibit lower than desired clinical efficacy. It is worth noting that a number of molecules from natural herbs and foods share this property [203]. Natural products and their derivatives have historically been invaluable as a source of therapeutic agents. Despite the disbelief that such class of potential drugs encompassed in the last decades, recent updates and technological advances, coupled with unrealized expectations from current lead-generation strategies, fostered a renewed interest in natural products in drug discovery [28, 204]. Indeed, many natural molecules, prone to be eventually engineered to amplify their efficacy, have already recognized to be effective in the treatment of several diseases [205]. High throughput techniques and new extraction strategies can be helpful in identifying a class of beneficial compounds. A remarkable case in point is that of the anti-malaria properties of Artemisia extract by relying on the 'traditional' efficiency recorded for this plant, while the 'rest of the World' was searching for a hypothetic 'synthetic magic bullet'. In fact, the pharmacological principle was extracted according a truly ancient protocol dating from 300 BC (according to the *Handbook of Prescriptions for Emergencies*) [206] because the 'modern' purification methods were ineffective. The rationale for extracting this specific principle was suggested by oral medical tradition dating back to the medieval age. The extracted drug introduced into treatment in the '70s was not 'recognized' by the Western world until a few years ago and is still not patentable (only the extraction procedure has been marketed). This is an outstanding example of technological failure not only in achieving the required 'goal', but also demonstrates that *'technology' (in excess!) may delay the discovery of new solutions*. *Artemisa annua* is currently deemed a pivotal asset in malaria management and it worth of notice that such a result was achieved by working far from current scientific mainstreams, in scientific structures with loose links to the so-called 'Big Science', and originally published only in Chinese journals by Dr. You-You. However, such an 'unconventional' approach led to her being – finally – awarded the Nobel Prize [206].

Thereby, what should we do? Clearly, many ventures in the biotechnology industry, from the earliest recombinant DNA based schemes for particular products to certain of the agricultural genetically modified organisms, have been successful (though not always uncontroversial). Yet, the looming difficulties will be primarily on the premises on which therapies are planned. And, for these, the companies may well have to go back to academia or, at least, to academics studying new and unexplored paths. Namely, systems biology, which today is still largely an enterprise of "academic" (i.e. non-commercial) interest may find itself increasingly incorporated into the research programs of industrial enterprises. How to maximize creativity in biological science is a topic rarely discussed and yet critical to success in improving health. We believe that the needed approaches are not simply to flog individuals to try harder but to build systems and infrastructures that enhance creative effort. Lateral thinking can and should be taught. Probably, as happened in the past, new avenues new theoretical approaches and different putative drugs – should be explored to counteract the decline in drug discovery we are facing nowadays. Indeed, time is gone to address such challenging issues and to restore both confidence and efficiency to the pharmaceutical industry. Time is ripe to move on this direction.

References

1. Morgan, S., P. Grootendorst, J. Lexchin, C. Cunningham, and D. Greyson. 2011. The cost of drug development: A systematic review. *Health Policy* 100: 4–17.
2. Burrill & Company. 2010. *Biotech 2010 life sciences: Adapting for success*. San Francisco: Burrill & Company.
3. Kim, J.H., and A.R. Scialli. 2011. Thalidomide: The tragedy of birth defects and the effective treatment of disease. *Toxicological Sciences* 122 (1): 1–6.
4. Krumholz, H.M., J.S. Ross, A.H. Presler, and D.S. Egilman. 2007. What have we learnt from Vioxx? *BMJ* 334 (7585): 120–123.
5. https://www.cancer.org/content/dam/cancer-org/research/cancer-facts-and-statistics/annual-cancer-facts-and-figures/2014/cancer-facts-and-figures-2014.pdf
6. Bailar, J.C., III, and H.L. Gornik. 1997. Cancer undefeated. *The New England Journal of Medicine* 336: 1569–1574.
7. https://www.cancer.org/research/cancer-facts-statistics/all-cancer-facts-figures/cancer-facts-figures-2014.html
8. Hay, S.I., et al. 2018. Measuring and mapping the global burden of antimicrobial resistance. *BMC Medicine* 16 (1): 78.
9. Zimmet, P.Z. 2017. Diabetes and its drivers: The largest epidemic in human history? *Clinical Diabetes and Endocrinology* 3: 1.
10. Hersoug, L.G., and A. Linneberg. 2007. The link between the epidemics of obesity and allergic diseases: Does obesity induce decreased immune tolerance? *Allergy* 62 (10): 1205–1213.

11. Le Fanu, J. 1999. *Rise and fall of modern medicine*. London: Abacus.
12. Ness, R.B. 2010. Fear of failure: Why American science is not winning the war on cancer. *Annals of Epidemiology* 20 (2): 89–91.
13. Finch, R. 2007. Innovation – Drugs and diagnostics. *The Journal of Antimicrobial Chemotherapy* 60 (Suppl 1): i79–i82.
14. Laxminarayan, R., et al. 2013. Antibiotic resistance – The need for global solutions. *The Lancet Infectious Diseases* 13: 1057–1098.
15. Astin, J.A. 1998. Why patients use alternative medicine: Results of a national study. *JAMA* 279 (19): 1548–1553.
16. Eisenberg, D.M., et al. 1993. Unconventional medicine in the United States. *The New England Journal of Medicine* 328: 246–252.
17. Fisher, P., and A. Ward. 1994. Complementary medicine in Europe. *BMJ* 309: 107–111.
18. Hoey, J. 1998. The arrogance of science and the pitfalls of hope. *CMAJ* 159: 803–804.
19. No Authors. 1980. A dearth of new drugs. *Nature* 283 (5748): 609
20. Liu, C.X. 2013. Biomedical development strategy inventory: Strategy and tactics looking back and forward [1]. *China Science Daily, Biology Week*
21. Herper M (2013) The cost of creating a new drug now $5 billion, pushing big pharma to change. *Forbes pharma healthcare*. Available from: http://www.forbes.com/sites/matthewherper/2013/08/11/how-the-staggering-cost-of-inventing-new-drugs-is-shaping-thefuture-of-medicine/
22. Liu, C., P.P. Constantinides, and Y. Li. 2014. Research and development in drug innovation: Reflections from the 2013 bioeconomy conference in China, lessons learned and future perspectives. *Acta Pharmaceutica Sinica B* 4 (2): 112–119.
23. Williams, M. 2011. Productivity shortfalls in drug discovery: Contributions from the preclinical sciences? *The Journal of Pharmacology and Experimental Therapeutics* 336: 3–8.
24. Iyer, P., Y. Hu, and J. Bajorath. 2011. SAR monitoring of evolving compound data sets using activity landscapes. *Journal of Chemical Information and Modeling* 51 (3): 532–540.
25. Horrobin, D.F. 2000. Innovation in the pharmaceutical industry. *Journal of the Royal Society of Medicine* 93 (7): 341–345.
26. Chong, C.R., and D.J. Sullivan Jr. 2007. New uses for old drugs. *Nature* 448: 645–646.
27. Kiuru, P., et al. 2014. Exploring marine resources for bioactive compounds. *Planta Medica* 80 (14): 1234–1246.
28. Koehn, F.E., and G.T. Carter. 2005. The evolving role of natural products in drug discovery. *Nature Reviews. Drug Discovery* 4 (3): 206–220.
29. Persson, C.G., J.S. Erjefält, L. Uller, M. Andersson, and L. Greiff. 2001. Unbalanced research. *Trends in Pharmacological Sciences* 22 (10): 538–541.
30. Huang, S. 2004. Back to the biology in systems biology: What can we learn from biomolecular networks? *Briefings in Functional Genomics & Proteomics* 2 (4): 279–297.
31. Sams-Dodd, F. 2013. Is poor research the cause of the declining productivity of the pharmaceutical industry? An industry in need of a paradigm shift. *Drug Discovery Today* 18: 211–217.
32. Davey Smith, G., and S. Ebrahim. 2001. Epidemiology – Is it time to call it a day? *International Journal of Epidemiology* 30: 1–11.
33. Rosenbaum, S. 2011. *Basic pharmacokinetics and pharmacodynamics: An integrated textbook and computer simulations*. Hoboken: Wiley.
34. Ehrlich, P. 1909. Über den jetzigen Stand der Chemotherapie. *Berichte der Deutschen Chemischen Gesellschaft* 42: 17–47.
35. Fischer, E. 1894. Einfluss der configuration auf die Wirkung der enzyme influence of configuration on the action of enzymes. *Berichte der Deutschen Chemischen Gesellschaft* 27: 2985–2993.
36. Lees, P., F.M. Cunningham, and J. Elliott. 2004. Principles of pharmacodynamics and their applications in veterinary pharmacology. *Journal of Veterinary Pharmacology and Therapeutics* 27 (6): 397–414.

37. Schneider, H.J. 2015. Limitations and extensions of the lock-and-key principle: Differences between gas state, solution and solid state structures. *International Journal of Molecular Sciences* 16 (4): 6694–6717.
38. Macheras, P., and A. Iliadis. 2016. *Modeling in biopharmaceutics, pharmacokinetics and pharmacodynamics: Homogeneous and heterogeneous approaches*. New York: Springer.
39. Macheras, P., and A. Dokoumetzidis. 2000. On the heterogeneity of drug dissolution and release. *Pharmaceutical Research* 17 (2): 108–112.
40. Siepmann, J., and N. Peppas. 2001. Modeling of drug release from delivery systems based on hydroxypropyl methylcellulose (HPMC). *Advanced Drug Delivery Reviews* 48 (2–3): 139–157.
41. Koshland, D.E. 1958. Application of a theory of enzyme specificity to protein synthesis. *Proceedings of the National Academy of Sciences of the United States of America* 44 (2): 98–104.
42. Boehr, D.D., R. Nussinov, and P.E. Wright. 2009. The role of dynamic conformational ensembles in biomolecular recognition. *Nature Chemical Biology* 5 (11): 789–796.
43. Pan, R., et al. 2010. Substrate-induced changes in protease active site conformation impact on subsequent reactions with substrates. *The Journal of Biological Chemistry* 285 (30): 22950–22956.
44. Ben-Naim, A. 2006. *Molecular theory of solutions*. Oxford: Oxford University Press.
45. Wang, H., and A. Ben-Naim. 1997. Solvation and solubility of globular proteins. *The Journal of Physical Chemistry. B* 101: 1077–1086.
46. Breiten, B., et al. 2013. Water networks contribute to enthalpy/entropy compensation in protein-ligand binding. *Journal of the American Chemical Society* 135 (41): 15579–15584.
47. Ben-Naim, A. 2002. Molecular recognition – Viewed through the eyes of the solvent. *Biophysical Chemistry* 101–102: 309–319.
48. Safont-Sempere, M.M., G. Fernández, and F. Würthner. 2011. Self-sorting phenomena in complex supramolecular systems. *Chemical Reviews* 111 (9): 5784–5814.
49. Fang, J.Y., C.F. Hung, H.C. Chiu, J.J. Wang, and T.F. Chan. 2003. Efficacy and irritancy of enhancers on the in-vitro and in-vivo percutaneous absorption of curcumin. *The Journal of Pharmacy and Pharmacology* 55 (8): 1175.
50. Gao, S., and J. Singh. 1997. Mechanism of transdermal transport of 5-fluorouracil by terpenes: carvone, 1,8-cineole and thymol. *International Journal of Pharmaceutics* 154: 67–77.
51. Chen, Y.L., et al. 2011. Transformation of cinnamic acid from trans- to cis-form raises a notable bactericidal and synergistic activity against multiple-drug resistant Mycobacterium tuberculosis. *European Journal of Pharmaceutical Sciences* 43 (3): 188–194.
52. Helander, I.M., et al. 1998. Characterization of the action of selected essential oil components on gram-negative bacteria. *Journal of Agricultural and Food Chemistry* 46: 3590–3595.
53. Pereira-da-Silva, L., et al. 2002. Osmolality of solutions, emulsions and drugs that may have a high osmolality: Aspects of their use in neonatal care. *The Journal of Maternal-Fetal & Neonatal Medicine* 11 (5): 333–338.
54. Stanton, R.A., K.M. Gernert, J.H. Nettles, and R. Aneja. 2011. Drugs that target dynamic microtubules: A new molecular perspective. *Medicinal Research Reviews* 31 (3): 443–481.
55. Sgarbossa, A., et al. 2013. The effects of ferulic acid on β-amyloid fibrillar structures investigated through experimental and computational techniques. *Biochimica et Biophysica Acta* 1830 (4): 2924–2937.
56. Perkins, H.R. 1969. Specificity of combination between mucopeptide precursors and vancomycin or ristocetin. *The Biochemical Journal* 111 (2): 195–205.
57. Grases, F., J. Perelló, R.M. Prieto, B.M. Simonet, and J.J. Torres. 2004. Dietary myo-inositol hexaphosphate prevents dystrophic calcifications in soft tissues: A pilot study in Wistar rats. *Life Sciences* 75 (1): 11–19.
58. Pécora, J.D., L.F. Guimarães, and R.N. Savioli. 1992. Surface tension of several drugs used in endodontics. *Brazilian Dental Journal* 2 (2): 123–127.
59. Tyler, S.E.B. 2014. The work surfaces of morphogenesis: The role of the morphogenetic field. *Biological Theory* 9: 194–208.

60. Allemann, R.K., and N.S. Scrutton. 2009. *Quantum tunnelling in enzyme-catalysed reactions*. Cambridge: The Royal Society of Chemistry.
61. Sahu, S., S. Ghosh, D. Fujita, and A. Bandyopadhyay. 2014. Live visualizations of single isolated tubulin protein self-assembly via tunneling current: Effect of electromagnetic pumping during spontaneous growth of microtubule. *Scientific Reports* 4: 7303.
62. Craddock, T.J., D. Friesen, J. Mane, S. Hameroff, and J.A. Tuszynski. 2014. The feasibility of coherent energy transfer in microtubules. *Journal of The Royal Society Interface* 11 (100): 20140677.
63. Dinicola, S., et al. 2016. Inositol induces mesenchymal-epithelial reversion in breast cancer cells through cytoskeleton rearrangement. *Experimental Cell Research* 345 (1): 37–50.
64. Gregori-Puigjané, E., et al. 2012. Identifying mechanism-of-action targets for drugs and probes. *Proceedings of the National Academy of Sciences of the United States of America* 109 (28): 11178–11183.
65. FDA. 2018. Clinical trial endpoints for the approval of cancer drugs and biologics guidance for industry. See at: https://www.fda.gov/Drugs/GuidanceComplianceRegulatoryInformation/Guidances/default.htm
66. Hesslow, G. 1993. Do we need a concept of disease? *Theoretical Medicine and Bioethics* 14: 1–14.
67. Murphy, D. 2015. Concepts of disease and health. In *The Stanford Encyclopedia of Philosophy*, ed. Edward N. Zalta, Spring 2015 edn. https://plato.stanford.edu/archives/spr2015/entries/health-disease
68. Reznek, L. 1987. *The nature of disease*. New York: Routledge.
69. Szasz, T. 1987. *Insanity*. New York: Wiley.
70. Horwitz, A.V. 2002. *Creating mental illness*. Chicago: University of Chicago Press.
71. Kitcher, P. 1997. *The lives to come: The genetic revolution and human possibilities*. New York: Simon & Schuster.
72. Low, G. 1999. Thomas Sydenham: The English Hippocrates. *The Australian and New Zealand Journal of Surgery* 69 (4): 258–262.
73. Schneider, H.C., and T. Klabunde. 2013. Understanding drugs and diseases by systems biology? *Bioorganic & Medicinal Chemistry Letters* 23 (5): 1168–1176.
74. Longo, G., M. Montevil, C. Sonnenschein, and A.M. Soto. 2015. In search of principles for a theory of organisms. *Journal of Biosciences* 40 (5): 955–968.
75. Lemoine, M. 2013. Defining disease beyond conceptual analysis: An analysis of conceptual analysis in philosophy of medicine. *Theoretical Medicine and Bioethics* 34: 309–325.
76. Scannell, J.W., A. Blanckley, H. Boldon, and B. Warrington. 2012. Diagnosing the decline in pharmaceutical R&D efficiency. *Nature Reviews. Drug Discovery* 11 (3): 191–200.
77. Wade, D.T., and P.W. Halligan. 2004. Do biomedical models of illness make for good healthcare systems? *BMJ* 329 (7479): 1398–1401.
78. Kirmayer, L.J., A. Young, and J.M. Robbins. 1994. Symptom attribution in cultural perspective. *Canadian Journal of Psychiatry* 39: 584–595.
79. Moynihan, R., I. Heath, and D. Henry. 2002. Selling sickness: The pharmaceutical industry and disease mongering. *BMJ* 324: 886–891.
80. Smith, R. 2002. In search of 'non-disease'. *BMJ* 324: 883–885.
81. Scully, J.L. 2004. What is a disease. *EMBO Reports* 5 (7): 650–653.
82. Hofmann, B. 2005. Simplified models of the relationship between health and disease. *Theoretical Medicine and Bioethics* 26 (5): 355–377.
83. Porter, R. 1997. *The greatest benefit to mankind. A medical history of humanity from antiquity to the present*. London: Harper Collins.
84. Stefl, S., H. Nishi, M. Petukh, A.R. Panchenko, and E. Alexov. 2013. Molecular mechanisms of disease-causing missense mutations. *Journal of Molecular Biology* 425 (21): 3919–3936.
85. Loscalzo, J., and A.L. Barabasi. 2011. Systems biology and the future of medicine. *Wiley Interdisciplinary Reviews. Systems Biology and Medicine* 3 (6): 619–627.
86. van Ommen, G.J., E. Bakker, and J.T. den Dunnen. 1999. The human genome project and the future of diagnostics, treatment, and prevention. *Lancet* 354 (Suppl 1): SI5–SI10.

87. Kato, G.J., M.T. Gladwin, and M.H. Steinberg. 2007. Deconstructing sickle cell disease: Reappraisal of the role of hemolysis in the development of clinical subphenotypes. *Blood Reviews* 21: 37–47.
88. Loscalzo, J., I. Kohane, and A.L. Barabasi. 2007. Human disease classification in the postgenomic era: A complex systems approach to human pathobiology. *Molecular Systems Biology* 3: 124.
89. Roizman, B. 1982. The family herpesviridae: General description taxonomy, and classification. In *The Herpesviruses*. New York: Plenum Press.
90. Thorley-Lawson, D.A. 1988. Basic virological aspects of Epstein-Barr virus infection. *Seminars in Hematology* 25: 247.
91. Wright, D.H. 1978. Burkitt's lymphoma and infectious mononucleosis. In *The immunopathology of lymphoreticular neoplasms*, ed. J.J. Twomey and R.A. Good, 391–423. New York: Plenum Publishing Corporation.
92. Grömminger, S., J. Mautner, and G.W. Bornkamm. 2012. Burkitt lymphoma: The role of Epstein-Barr virus revisited. *British Journal of Haematology* 156 (6): 719–729.
93. Rezk, S.A., X. Zhao, and L.M. Weiss. 2018. Epstein-Barr virus (EBV)-associated lymphoid proliferations, a 2018 update. *Human Pathology* 79: 18–41.
94. Naseem, M., et al. 2018. Outlooks on Epstein-Barr virus associated gastric cancer. *Cancer Treatment Reviews* 66: 15–22.
95. Weiss, R.A. 2016. Tumour-inducing viruses. *British Journal of Hospital Medicine (London, England: 2005)* 77 (10): 565–568.
96. Mastria, G., V. Mancini, A. Viganò, and V. Di Piero. 2016. Alice in wonderland syndrome: a clinical and pathophysiological review. *BioMed Research International* 2016: 1–10.
97. Ascherio, A., and K.L. Munger. 2015. EBV and autoimmunity. *Current Topics in Microbiology and Immunology* 390: 365–385.
98. Nussinovitch, M., D. Prais, B. Volovitz, R. Shapiro, and J. Amir. 2003. Post-infectious acute cerebellar ataxia in children. *La Clinica Pediatrica* 42 (7): 581–584.
99. Giller, R.H., and C. Grose. 1989. Epstein-Barr virus: The hematologic and oncologic consequences of virus-host interaction. *Critical Reviews in Oncology/Hematology* 9 (2): 149–195.
100. Farber, H., and J. Loscalzo. 2004. Pulmonary hypertension. *The New England Journal of Medicine* 351: 1655–1665.
101. Seidman, J.C., and C. Seidman. 2001. The genetic basis for cardiomyopathy from mutation identification to mechanistic paradigms. *Cell* 104: 557–567.
102. Sams-Dodd, F. 2005. Target-based drug discovery: Is something wrong? *Drug Discovery Today* 10 (2): 139–147.
103. Kaelin, W.G. 2005. The concept of synthetic lethality in the context of anticancer therapy. *Nature Reviews. Cancer* 5: 689–698.
104. Martincorena, I., et al. 2015. High burden and pervasive positive selection of somatic mutations in normal human skin. *Science* 348 (6237): 880–886.
105. Kamb, A., S. Wee, and C. Lengauer. 2007. Why is cancer drug discovery so difficult? *Nature Reviews. Drug Discovery* 6 (2): 115–120.
106. Seoane, J. 2017. Cancer: Division hierarchy leads to cell heterogeneity. *Nature* 549 (7671): 164–166.
107. Li, R.X., and R. Zeng. 2009. Dynamic proteomics for investigating the response of individual cancer cells under drug action. *Expert Review of Proteomics* 6 (1): 19–21.
108. Baird, P. 2001. The human genome project, genetics and health. *Community Genetics* 4 (2): 77–80.
109. McClellan, J., and M.C. King. 2010. Genetic heterogeneity in human disease. *Cell* 141: 210–217.
110. Goh, K.I., et al. 2007. The human disease network. *Proceedings of the National Academy of Sciences of the United States of America* 104 (21): 8685–8690.
111. Drews, J. 2006. Case histories, magic bullets and the state of drug discovery. *Nature Reviews. Drug Discovery* 5: 635–640.

112. Pujol, A., R. Mosca, J. Farrés, and P. Aloy. 2010. Unveiling the role of network and systems biology in drug discovery. *Trends in Pharmacological Sciences* 31 (3): 115–123.
113. Hirsh, S.R., and D.R. Weinberger. 1995. *Schizophrenia*. Oxford: Blackwell Science Ltd.
114. Brunello, N., et al. 2002. The role of noradrenaline and selective noradrenaline reuptake inhibition in depression. *European Neuropsychopharmacology* 12: 461–475.
115. Albino, A.P., R. Le Strange, A.I. Oliff, M.E. Furth, and L.J. Old. 1984. Transforming ras genes from human melanoma: A manifestation of tumor heterogeneity? *Nature* 308: 69–72.
116. Plattner, R., et al. 1996. Loss of oncogenic ras expression does not correlate with loss of tumorigenicity in human cells. *Proceedings of the National Academy of Sciences of the United States of America* 93: 6665–6670.
117. Bizzarri, M., A. Cucina, F. Conti, and F. D'Anslemi. 2008. Beyond the oncogenic paradigm: Understanding complexity in cancerogenesis. *Acta Biotheoretica* 56: 173–196.
118. Swanton, C. 2012. Intratumor heterogeneity: Evolution through space and time. *Cancer Research* 72: 4875–4882.
119. Bhang, H.E., et al. 2015. Studying clonal dynamics in response to cancer therapy using high-complexity barcoding. *Nature Medicine* 21: 440–448.
120. Scannell, J.W., and J. Bosley. 2016. When quality beats quantity: Decision theory, drug discovery and the reproducibility crisis. *PLoS One* 11 (2): e0147215.
121. Bizzarri, M. 2017. Do new anticancer drugs really work? A serious concern *Organisms. Journal of Biological Sciences* 1 (1): 9–10.
122. Wise, P.H. 2016. Cancer drugs, survival, and ethics. *BMJ* 355: i5792.
123. Kim, C., and V. Prasad. 2016. Strength of validation for surrogate end points used in the US Food and Drug Administration's approval of oncology drugs. *Mayo Clinic Proceedings* 91: 713–725.
124. Hawkes, N. 2011. High cost of cancer treatment doesn't reflect benefits, say specialists. *BMJ* 343: d6220.
125. Giles, F.J., M. O'Dwyer, and R. Swords. 2009. Class effects of tyrosine kinase inhibitors in the treatment of chronic myeloid leukemia. *Leukemia* 23: 1697–1607.
126. Pearson, H. 2002. Surviving a knockout blow. *Nature* 415 (6867): 8–9.
127. Eberhardt, F. 2017. Introduction to the foundations of causal discovery. *International Journal of Data Science and Analytics* 3: 81–91.
128. Szymkowski, D.E. 2001. Too many targets, not enough target validation. *Drug Discovery Today* 6 (8): 397.
129. Coveney, P.V., E.R. Dougherty, and R.R. Highfield. 2016. Big data need big theory too. *Philosophical Transactions of the Royal Society A* 374: 20160153.
130. Miklos, G.L.G. 2005. The Human Cancer Genome Project – One more misstep in the war on cancer. *Nature Biotechnology* 23: 535–537.
131. Leppert, J., and C. Patel. 2015. Beyond the genome. *Nature* 518 (7539): 273.
132. Soto, A.M., and C. Sonnenschein. 2014. One hundred years of somatic mutation theory of carcinogenesis: Is it time to switch? *BioEssays* 36 (1): 118–120.
133. Prasad, V. 2016. Perspective: The precision-oncology illusion. *Nature* 537 (7619): S63.
134. Calude, C.S., and G. Longo. 2017. The deluge of spurious correlations in big data. *Foundations of Science* 22: 595.
135. Karpatne, A., et al. 2017. Theory-guided data science: A new paradigm for scientific discovery from data. *IEEE Transactions on Knowledge and Data Engineering* 29: 2318–2331.
136. Anderson C. The end of theory: The data deluge makes the scientific method obsolete. http://archive.wired.com/science/discoveries/magazine/16-07/pb_theory/
137. Mazzocchi, F. 2015. Could Big Data be the end of theory in science? A few remarks on the epistemology of data-driven science. *EMBO Reports* 16 (10): 1250–1255.
138. Ioannidis, J.P.A. 2005. Why most published research findings are false. *PLoS Medicine* 2: e124.
139. Macleod, M.R., et al. 2014. Biomedical research: Increasing value, reducing waste. *Lancet* 383: 101–104.

140. Baker, M. 2016. 1,500 scientists lift the lid on reproducibility. *Nature* 533: 452–454.
141. Bizzarri, M. 2018. Is biology in an existential crisis? A diagnostic analysis and perhaps… An effective treatment" (Editorial). *Organisms. Journal of Biological Sciences* 2 (2): 1
142. Tannock, I.F., and J.A. Hickman. 2016. Limits to personalized cancer medicine. *The New England Journal of Medicine* 375: 1289–1294.
143. Hopkins, A.L. 2008. Network pharmacology: The next paradigm in drug discovery. *Nature Chemical Biology* 4 (11): 682–690.
144. Zambrowicz, B.P., and A.T. Sands. 2004. Modeling drug action in the mouse with knockouts and RNA interference. *Drug Discovery Today: Targets* 3: 198–207.
145. Barabasi, A.L., and Z.N. Oltvai. 2004. Network biology: Understanding the cell's functional organization. *Nature Reviews. Genetics* 5: 101–113.
146. Wagner, A., and J. Wright. 2007. Alternative routes and mutational robustness in complex regulatory networks. *Biosystems* 88 (1–2): 163–172.
147. Rosenfeld, R. 2011. Biomolecular self-defense and futility of high-specificity therapeutic targeting. *Gene Regulation and Systems Biology* 5: 89–104.
148. Albert, R., H. Jeong, and A.L. Barabasi. 2000. Error and attack tolerance of complex networks. *Nature* 406: 378–382.
149. Hillenmeyer, M.E., et al. 2008. The chemical genomic portrait of yeast: Uncovering a phenotype for all genes. *Science* 320: 362–365.
150. Keith, C.T., A.A. Borisy, and B.R. Stockwel. 2005. Multicomponent therapeutics for networked systems. *Nature Reviews. Drug Discovery* 4: 71–78.
151. Csermely, P., T. Korcsmáros, H.J. Kiss, G. London, and R. Nussinov. 2013. Structure and dynamics of molecular networks: A novel paradigm of drug discovery: A comprehensive review. *Pharmacology & Therapeutics* 138 (3): 333–408.
152. Kolodkin, A., et al. 2012. Emergence of the silicon human and network targeting drugs. *European Journal of Pharmaceutical Sciences* 46: 190–197.
153. Kola, I., and J. Bell. 2011. A call to reform the taxonomy of human disease. *Nature Reviews. Drug Discovery* 10: 641–642.
154. Anderson, B., A. Nielsen, D. McKee, A. Jeffres, and B. Kligler. 2012. Acupuncture and heart rate variability: A systems level approach to understanding mechanism. *Explore (New York, N. Y.)* 8 (2): 99–106.
155. Schwartz, J.H. 2013. Emergence of shape. *Biological Theory* 8: 209–210.
156. Haken, H. 1977. *Synergetics*. Berlin/New York: Springer Verlag.
157. Tanaka, G., K. Tsumoto, S. Tsuji, and K. Aihara. 2008. Bifurcation analysis on a hybrid systems model of intermittent hormonal therapy for prostate cancer. *Physica D* 237: 2616–2627.
158. Venegas, J.G., et al. 2005. Self-organized patchiness in asthma as a prelude to catastrophic shifts. *Nature* 434: 777–782.
159. Liu, X., et al. 2017. Quantifying critical states of complex diseases using single-sample dynamic network biomarkers. *PLoS Computational Biology* 13 (7): e1005633.
160. Chen, L., R. Liu, Z.P. Liu, M. Li, and K. Aihara. 2012. Detecting early-warning signals for sudden deterioration of complex diseases by dynamical network biomarkers. *Scientific Reports* 2: 342.
161. Dorogovtsev, S.N., and J.F.F. Mendes. 2001. Scaling properties of scale-free evolving networks: Continuous approach. *Physical Review E* 63: 056125.
162. Huang, S. 2002. Rational drug discovery: What can we learn from regulatory networks? *Drug Discovery Today* 7 (20 Suppl): S163–S169.
163. Motter, A.E. 2004. Cascade control and defense in complex networks. *Physical Review Letters* 93 (9): 098701.
164. Agoston, V., P. Csermely, and S. Pongor. 2005. Multiple weak hits confuse complex systems: A transcriptional regulatory network as an example. *Physical Review E – Statistical, Nonlinear, and Soft Matter Physics* 71 (5 Pt 1): 051909.

165. Csermely, P., V. Agoston, and S. Pongor. 2005. The efficiency of multi-target drugs: The network approach might help drug design. *Trends in Pharmacological Sciences* 26 (4): 178–182.
166. Gavras, I., and T. Rosenthal. 2004. Combination therapy as first-line treatment for hypertension. *Current Hypertension Reports* 6 (4): 267–272.
167. Liebovitch, L.S., N. Tsinoremas, and A. Pandya. 2007. Developing Combinatorial Multi-Component Therapies (CMCT) of drugs that are more specific and have fewer side effects than traditional one drug therapies. *Nonlinear Biomedical Physics* 1 (1): 11.
168. Wagner, H. 2006. Multitarget therapy--the future of treatment for more than just functional dyspepsia. *Phytomedicine* 13 (Suppl 5): 122–129.
169. Khan, M., A. Maryam, J.I. Qazi, and T. Ma. 2015. Targeting apoptosis and multiple signaling pathways with Icariside II in cancer cells. *International Journal of Biological Sciences* 11 (9): 1100–1112.
170. Kiyohara, H., T. Matsumoto, and H. Yamada. 2004. Combination effects of herbs in a multi-herbal formula: Expression of Juzen-taiho-to's Immuno-modulatory activity on the intestinal immune system. *Evidence-based Complementary and Alternative Medicine* 1 (1): 83–91.
171. Cantelli-Forti, G., et al. 1994. Interaction of licorice on glyzyrrhizin pharmacokinetics. *Environmental Health Perspectives* 102 (Suppl. 2): 65–68.
172. Ahn, A.C., et al. 2010. Applying principles from complex systems to studying the efficacy of CAM therapies. *Journal of Alternative and Complementary Medicine* 16 (9): 1015–1022.
173. Koithan, M., I.R. Bell, K. Niemeyer, and D. Pincus. 2012. A complex systems science perspective for whole systems of complementary and alternative medicine research. *Forschende Komplementärmedizin* 19 (Suppl 1): 7–14.
174. Niemeyer, K., I.R. Bell, and M. Koithan. 2013. Traditional knowledge of Western herbal medicine and complex systems science. *Journal of Herbal Medicine* 3 (3): 112–119.
175. Kauffman, S. 1995. *At home in the universe. The search for the laws of self-organization and complexity*. Oxford: Oxford University Press.
176. Hosoya, E. 1988. Scientific reevaluation of Kampo prescriptions using modern technology. In *Resent advances in the pharmacology of Kampo (Japanese herbal) medicines*, ed. E. Hosoya and Y. Yamamura, 17–29. Tokyo: Excerpta Medica.
177. Williamson, E.M. 2001. Synergy and other interactions in phytomedicine. *Phytomedicine* 8: 401–409.
178. Berenbaum, M. 1989. What is synergy? *Pharmacological Reviews* 41: 93–141.
179. Gilbert, S.F., and S. Sarkar. 2000. Embracing complexity: Organicism for the 21st century. *Developmental Dynamics* 219: 1–9.
180. Capasso, A., and L. Sorrentino. 2005. Pharmacological studies on the sedative and hypnotic effect of Kava kava und Passiflora. *Phytomedicine* 12: 39–45.
181. Wagner, H., and B. Steinke. 2005. Natural products chemistry and phytomedicine in the 21th century: New developments and challenges. *Pure and Applied Chemistry* 77 (1): 1–6.
182. Wang, L., et al. 2008. Dissection of mechanisms of Chinese medicinal formula Realgar-Indigo naturalis as an effective treatment for promyelocytic leukemia. *Proceedings of the National Academy of Sciences of the United States of America* 105: 4826–4831.
183. Koropatkin, N.M., E.A. Cameron, and E.C. Martens. 2012. How glycan metabolism shapes the human gut microbiota. *Nature Reviews. Microbiology* 10: 323–335.
184. Lam, W., et al. 2010. The four-herb Chinese medicine PHY906 reduces chemotherapy-induced gastrointestinal toxicity. *Science Translational Medicine* 2: 45ra59.
185. Zhou, S.S., et al. 2016. Gut microbiota-involved mechanisms in enhancing systemic exposure of ginsenosides by coexisting polysaccharides in ginseng decoction. *Scientific Reports* 6: 22474.
186. Tremaroli, V., and F. Backhed. 2012. Functional interactions between the gut microbiota and host metabolism. *Nature* 489: 242–249.
187. Van Regenmortel, M.H.V. 2004. Biological complexity emerges from the ashes of genetic reductionism. *Journal of Molecular Recognition* 17: 145–148.

188. Kitano, H. 2007. A robustness-based approach to systems-oriented drug design. *Nature Reviews. Drug Discovery* 6 (3): 202–210.
189. Goodacre, R., et al. 2004. Metabolomics by numbers: Acquiring and understanding global metabolite data. *Trends in Biotechnology* 22: 245–252.
190. Gu, Y., et al. 2012. Plasma metabonomics study of rheumatoid arthritis and its Chinese medicine subtypes by using liquid chromatography and gas chromatography coupled with mass spectrometry. *Molecular BioSystems* 8 (5): 1535–1543.
191. Harrigan, G.G., and R. Goodacre. 2003. *Metabolic profiling: Its role in biomarker discovery and gene function analysis*. Boston: Kluwer Academic Publishers.
192. Urbanczyk-Wochniak, E., et al. 2003. Parallel analysis of transcript and metabolic profiles: A new approach in systems biology. *EMBO Reports* 4: 989–993.
193. Kell, D.B. 2006. Systems biology, metabolic modelling and metabolomics in drug discovery and development. *Drug Discovery Today* 11 (23–24): 1085–1092.
194. Telerman, A., and R. Amson. 2009. The molecular programme of tumour reversion: The steps beyond malignant transformation. *Nature Reviews. Cancer* 9: 206–216.
195. Hendrix, M.J., et al. 2007. Reprogramming metastatic tumour cells with embryonic microenvironments. *Nature Reviews. Cancer* 7: 246–255.
196. Telerman, A., R. Amson, and M.J. Hendrix. 2010. Tumor reversion holds promise. *Oncotarget* 1: 233–234.
197. Livraghi, T., et al. 2005. Treatment with stem cell differentiation stage factors of intermediate-advanced hepatocellular carcinoma: An open randomized clinical trial. *Oncology Research* 15: 399–408.
198. Bizzarri, M., A. Cucina, and S. Proietti. 2014. The tumor microenvironment as a target for anticancer treatment. *Oncobiology and Targets* 1: 3–11.
199. Allegrucci, C., et al. 2011. Epigenetic reprogramming of breast cancer cells with oocyte extracts. *Molecular Cancer* 10: 7.
200. Ferranti, F., et al. 2013. TCam-2 seminoma cells exposed to egg-derived microenvironment modify their shape, adhesive pattern and migratory behaviour: A molecular and morphometric analysis. *PLoS One* 8: e76192.
201. Proietti, S., et al. 2019. Active fraction from embryo fish extracts induces reversion of the malignant invasive phenotype in breast cancer through down-regulation of TCTP and modulation of E-cadherin/β-catenin pathway. *International Journal of Molecular Sciences* 20 (9): pii:E2151.
202. Bizzarri, M., et al. 2011. Embryonic morphogenetic field induces phenotypic reversion in cancer cells. *Current Pharmaceutical Biotechnology* 12: 243–253.
203. Safe, S., and R. Kasiappan. 2016. Natural products as mechanism-based anticancer agents: Sp transcription factors as targets. *Phytotherapy Research* 30 (11): 1723–1732.
204. Suzuki, T., and N. Miyata. 2006. Epigenetic control using natural products and synthetic molecules. *Current Medicinal Chemistry* 13 (8): 935–958.
205. Sharma, R. 2009. Nutraceuticals and nutraceutical supplementation criteria in cancer: A literature survey. *The Open Nutraceuticals Journal* 2: 92–106.
206. Yu, T. 2011. The discovery of Artemisinin (qinghaosu) and gifts from Chinese medicine. *Nature Medicine* 17 (10): 1217–1220.

Dynamical Aspects of Pharmacokinetic/Pharmacodynamic & Quantitative Systems Pharmacology Models

Ioannis Loisios-Konstantinidis, Panteleimon D. Mavroudis, and Panos Macheras

Introduction

Dynamical Aspects of Homeostasis and Disease

The realization that organisms maintain a dynamic equilibrium in face of constant internal or external stressors, is embedded in the term "*homeostasis*" and the observation of Claude Bernard that the purpose of diverse physiological mechanisms is to maintain a stable "*Milieu intérieur*" (interior milieu) against growing, aging, disease, operation, accident, stress etc. Successful adaptation to a constantly changing environment consists of a variety of body reactions that work towards counteracting the effect of natural development or internal/external stress to re-establish homeostasis [18, 28]. Long before the brilliant insights of Claude

Authors Ioannis Loisios-Konstantinidis, Panteleimon D. Mavroudis have equally contributed to this chapter.

I. Loisios-Konstantinidis
Faculty of Pharmacy, SMARTc – CRCM – INSERM UMR1068 – CNRS UMR7258 – AMU UM105, Marseille Cedex, France

P. D. Mavroudis
School of Pharmacy and Pharmaceutical Sciences, State University of New York at Buffalo, Buffalo, NY, USA

P. Macheras (✉)
School of Pharmacy and Pharmaceutical Sciences, State University of New York at Buffalo, Buffalo, NY, USA

Department of Pharmacy, National and Kapodistrian University of Athens, Athens, Greece

PharmaInformatics Unit, ATHENA Research Center, Athens, Greece
e-mail: macheras@pharm.uoa.gr

M. Bizzarri (ed.), *Approaching Complex Diseases*, Human Perspectives in Health Sciences and Technology 2, https://doi.org/10.1007/978-3-030-32857-3_2

Bernard, in classic era, *Heraclitus* with his famous quote "*everything flows*" suggested that all systems retain the intrinsic capacity to undergo constant changes rather than being static. *Empedocles* had also proposed that all matters consist by the dynamic opposition and alliance of basic elements, and *Hippocrates* argued that health results from a balanced relationship of elements, whereas disease is indicative of a disharmony among them. Evidently, the notion that homeostasis is a complex, dynamic equilibrium with multiple levels of organization and control is not new.

Physiological complexity in humans spans from molecular and cellular networks to tissue functions and organs [16]. One of the most important manifestations of intrinsic complexity is a nearly 24 h (*circadian*) temporal coordination of biological processes that enables body to anticipate daily changes and optimize fitness. The 2017 Nobel Prize in Physiology or Medicine to Jeffrey Hall, Michael Rosbash, and Michael Young for their work on organism's inner circadian clocks indicates the paramount value of circadian rhythms [11]. The circadian timing mechanism consists of cell-autonomous molecular clocks present in almost all cells of the body, that are entrained by periodic environmental cues the most pervasive of which is the light-dark cycles. The suprachiasmatic nucleus (SCN) of the brain responds to light/dark cycles by regulating systemic signals, such as body temperature, hormone secretion, and activity and then transmits the 24 h light-dark information to the periphery of the body and synchronize the function of peripheral tissues [10]. Loss of entrainment between SCN and peripheral oscillators has been linked with several diseases such as obesity, diabetes, cardiovascular disease and cancer [77, 82, 117]. This hierarchical organization of circadian rhythms along with the clinical outcomes resulting from their disruption, represents a characteristic example of how organism's well-being emerges from the system's characteristics rather than its isolated parts' behavior.

Physiological variability and multiscale organization confers advantages not only to the homeostatic function of the body, but also to body's response to external stimuli such as injury or infection. In response to a stressor, body mounts an inflammatory response aiming to resolve the effects of the stressor and restore homeostasis. The inflammatory response involves multiple levels of organization such as transcriptional activation of inflammatory genes in multiple cell types, autonomic neural signaling, secretion of hormones and production of hormone-like inflammatory mediators (cytokines/chemokines) [36, 85, 132]. These intercommunicating systems are designed to confer a time-restrained return to homeostasis. However, when anti-inflammatory mechanisms fail to adequately counterbalance pro-inflammatory activity, the body can reach a state of prolonged, unresolving systemic inflammation. This dysregulated inflammatory state can cause significant harm to the body, even in the absence of any exogenous stressor. The inherent complexity of the inflammatory response and its multi-level organization necessitate a systems biology view to study how the individual parts interact to ultimately produce a self-regulated restoration of homeostasis.

Despite the established appreciation of human body complexity, drugs even today aim to fix certain physiological parameters to nominal values on patient's behalf. Many times, the notion that organism's stability and well-being could emerge as a property of the network is missing, and incorporation of systems dynamics in drug development is often limited. With the advances of systems theory and data analysis, our era urges the use of systems understanding in pharmacology [142].

Dynamical Aspects in PK/PD and Quantitative Systems Pharmacology

Maintenance of homeostasis usually involves a convolution of positive and negative feedback loops at multiple levels of body's organization, that function as control systems to counteract changes of various signals (negative feedback) or enhance system's response to a certain stimulus (positive feedback). Since early 80's, the emerging literature indicating that chaotic functions can accommodate the abundance of positive and negative feedback controls in physiological systems, particularly in EEG and ECG analysis, propelled van Rossum & de Bie to put forward the involvement of nonlinear dynamics by suggesting that chaotic behaviours could underlie pharmacological mechanisms [108–110]. These authors were the first to introduce the concept of attractor of a dynamical system in pharmaceutical sciences i.e., geometric forms that characterize long-term behavior in the phase space of the dynamical system. Alternatively, an attractor can be defined as a set of numerical values towards which a system tends to evolve, for a wide variety of starting conditions of the system. These geometrical forms correspond to predictable systems (point, limit cycle and torus attractor), while the strange attractor corresponds to unpredictable motions i.e. chaotic systems (Figure 1).

In general, pharmacotherapy assumes a reasonable degree of predictability and that usually variability in the observed systems arises from pure randomness. However, biological systems, including human body, are complex dynamic systems with a large number of variables and regulating mechanisms which operate simultaneously. A relatively simple mathematical presentation, the logistic map for population growth, has been used to explain the route to chaos of the variable x_n assigning different values to the control parameter α, based on the following difference equation (Eq. 1):

$$x_{n+1} = f[x_n] = a \cdot x_n \cdot [1 - x_n] \tag{1}$$

An illustrative example of the use of logistic map to reveal if the observed irregular behavior of observations arise from noise or chaos is shown in Fig. 2.

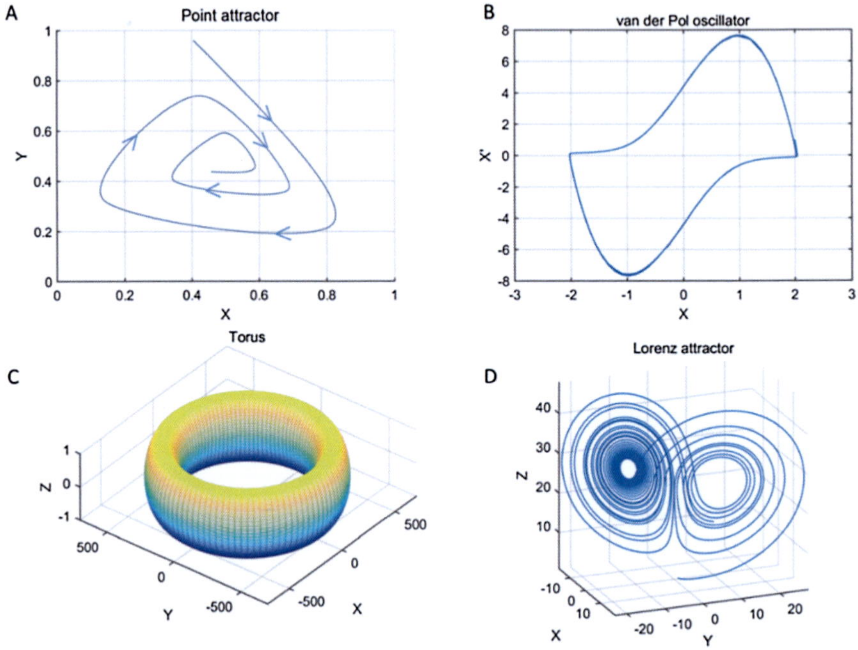

Fig. 1 A schematic representation of the various types of attractors. A: the point attractor. Regardless the initial conditions, the system ends up to the same steady state. B: Cycle attractor (van der Pol oscillator). The system always ends up doing a specific oscillation. C: A torus attractor. The torus is the two-dimensional (2D) equivalent of a circle. In fact, a circle can be called a 1-torus, the 2D torus can be called a 2-torus and there is also the 3-torus and generally the n-torus. The trajectory on a 2-torus is a 2D oscillation with the ratio of the frequencies of the two oscillations being non-rational. Because the trajectory never passes from the same point twice, in infinite time fills the entire surface of the torus. This type of trajectory is called quasiperiodic. Being an attractor, the torus attracts all trajectories to fall on its surface and follow the quasiperiodic behavior. D: Strange attractor (Lorenz attractor)

For a detailed mathematical analysis on the logistic map and a concise review on the principles of nonlinear dynamics including their applications in PD, the reader is referred to the publication by Dokoumetzidis et al. [29].

Danhof [25] classified dynamical systems into adaptive and non-adaptive. The former have changing system properties over time, particularly through emergence of self-organization, whereas the latter are characterized by functionally constant properties over time. The author suggested that the behaviors of complex biological systems are governed by the fundamental properties of hysteresis, non-linearity, individuality, variability, interdependency, convergence, resilience and multi-stationarity. Additional fundamental properties include emergence, robustness, self-organization and degeneracy [51, 141]. Multi-stationarity refers to the biological systems' phenomenon of existence of multiple, more or less stable, states and it is

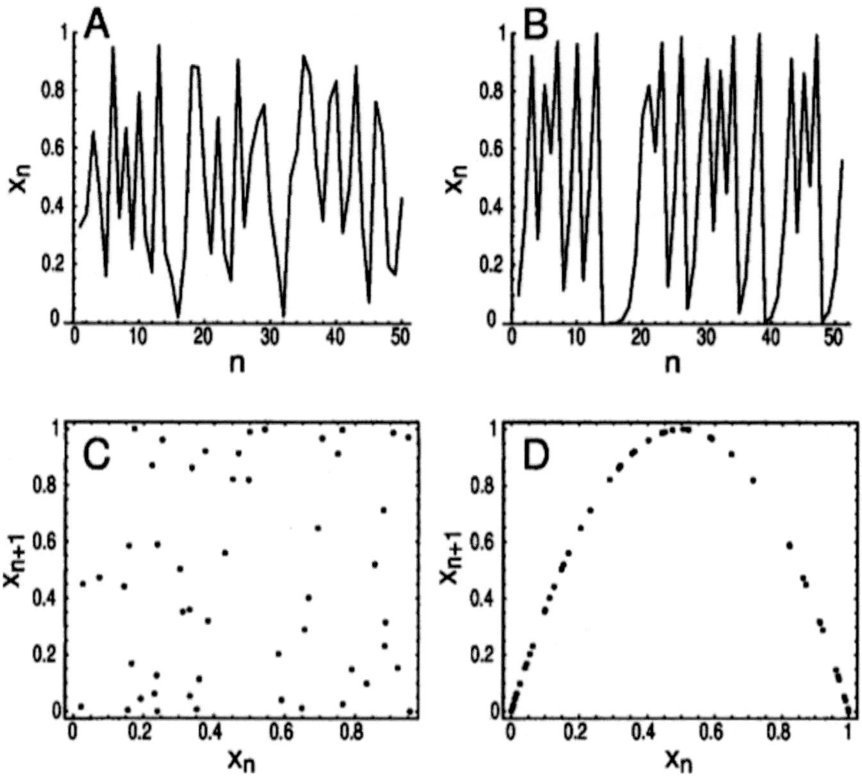

Fig. 2 (**A**) A series of uniformly distributed random numbers between 0 and 1. (**B**) Plot generated by the logistic map, a deterministic system of the form $x_n + 1 = 4x_n \cdot (1 - x_n)$. It is impossible to distinguish them visually (**A** or **B**). (**C** and **D**) The pseudophase plots of the two sequences of plots A and B, respectively. Each x_n is plotted against its consequent x_{n+1}. The random sequence (**A**) produces a pseudophase space of scattered points (**c**) showing that there is no correlation between successive points. On the contrary, the points of the deterministic sequence (**B**) lay in a well formed line (**D**). (Reproduced with permission from Ref. 29)

also characteristic of chaos theory. However, multi-stationarity is not a self-evident property. It is rarely immediately apparent through ordinary differential equations (ODEs)-based modeling and simulation, unless advanced mathematics techniques, like bifurcation analysis, are applied. Multi-stationarity can often confound the simulations, but the cause cannot be identified directly by visual inspection.

In pharmacokinetics/pharmacodynamics (PK/PD), simpler dynamics are associated with pharmacokinetic processes and this is the reason why pharmacokinetic studies, in general, are less variable than the pharmacodynamic ones [67]. Even when the system is simple, tools from dynamical systems theory can still be useful. When a system has only one variable, its behaviour can be studied by plotting the variable against its derivative. Such a plot is referred as phase plane (e.g. *dC*/*dt* vs. C in Michaelis-Menten kinetics). According to van Rossum and de Bie [109], the phase space of a pharmacokinetic system is ruled by a point attractor since the drug leaves the body and consequently the drug concentration in plasma tends to zero. On the other hand, in pharmacodynamics there are several examples of systems exhibiting nonlinear dynamical nature. Traditionally, pharmacodynamics has been based on the receptor occupancy theory without feedback, which leads to the classical direct and indirect Emax models. This simplified representation of a pharmacodynamic response is not always physiologically relevant and deviations from that can be anticipated when feedback mechanisms, induced by the ligand-receptor complex formation and functioning to maintain a basal ligand value, are considered.

Ever since, PK/PD models not only have been addressed by dynamic sub-models linking nonlinearly time-concentration-effect, but they have also evolved over time to state-of –the-art mechanism-based PK/PD, disease progression and quantitative systems pharmacology (QSP) models under the umbreall of systems biology. Such models can account for whole biochemical signaling pathways and biological networks rather than single transduction pathways implemented in classical PK/PD (e.g. turnover models). In addition, they are able to mathematically describe the sum of involved physiological processes and characterize the functional interactions within the biological network, which are of great importance for agents acting in multiple target and/or when homeostatic (feedback) mechanisms are operative in the network. Therefore, systems pharmacology models can be advantageous for illuminating irregular patterns of drug action (e.g. oscillatory behaviour). Apart from commonly mentioned issues of such complex models, regarding robustness, identifiability and granularity, in depth investigation of the system's dynamics is often neglected. In this chapter, specific examples from simple PK/PD to complex and large quantitative systems pharmacology models were carefully chosen and briefly presented to illustrate the importance of applying tools from the nonlinear dynamical systems theory in mathematical modeling and simulation. These examples are categorized based on the respective physiological systems including the cardiovascular, the central-nervous, the endocrine and the immune system.

Cardiovascular System

Historically, cardiac physiology has been extensively studied and numerous applications of nonlinear dynamics and chaos theory associated with the healthy and/or diseased cardiac function have been published [24, 45, 46, 59, 137]. Analyzing electrocardiogram (ECG) with either statistical (like spectral analysis) or dynamical (like phase space reconstruction) techniques has clearly indicated that heartbeat, blood pressure or in general signaling in the cardiovascular system is essentially irregular [46, 47, 124, 136, 138, 143]. In fact, the ECG was one of the first biological signals studied with such tools, where concepts from chaos theory have been applied to the analysis of its variability [24, 46, 52, 91, 92, 135].

There are a number of studies investigating the pharmacologic effect of drugs on the dynamics of cardiac physiology. These examples include, but are not limited to, the attempt to control cardiac chaos using ouabain, the induction of cellular chaos during quinidine toxicity and the effect of atropine and anticholinergic drug on cardiac inter-beat intervals and heart rate variability, respectively [43, 58, 114, 128]. De Brouwer et al. [6] have successfully applied concepts of nonlinear dynamics using a simple model based on coupled oscillators to describe the dynamics of agonist-induced vasomotion, where the route to chaos in the presence of a class IV antiarrhythmic drug, verapamil, was investigated.

Modeling of the spatial evolution dynamics of the cardiac electrical activity under the prism of chaos theory is very promising. In this effort, the cardiac tissue is considered to be an excitable medium, the electrical activity of which is described not only in time but also in space by reaction-diffusion partial differential equations [84]. This way, the system is able to produce spiral waves, which serve as precursors of the chaotic behavior. Such spiral waves signal the transition from the normal heart rate to tachycardia, while the chaotic regime, observed after spiral waves have broken up, corresponds to the transition to fibrillation. The latter transition is often characterized as electrical turbulence, due to its resemblance to the respective hydrodynamic phenomenon. Although such approaches have not yet directly implemented to PK/PD or QSP models and integration of pharmacologic response in the excitable media models remains challenging, they provide valuable information for the pharmacologic effect of antiarrhythmic drugs. Agents belonging to the antiarrhythmic classes I and III such as flecainide and moricizine have been proven to potentially increase sudden death rate caused by ventricular fibrillation [31]. Simulations of two- and three-dimentional (2D, 3D) cardiac tissue have been attempted. In this case, the 3D equivalent of spiral waves is scroll waves [42, 105, 125]. These models were able to differentiate between the antiarrhythmic action observed in a single cell system, lacking spatial evolution, and the proarrhythmic effect in a whole cardiac tissue system of either two or three spatial dimensions.

This new approach has given rise to evaluation of antiarrhythmic drugs based on the chaotic dynamics governing the transition from tachycardia to fibrillation [42, 105, 139]. This is also supported by experimental evidence [42]. These findings have indicated that the limitations of the classical approach of premature ventricular polarization suppression (i.e. initiation of tachycardia) might be associated with the failure to predict long-term efficacy of class I and III antiarrhythmic compounds [139]. However, sudden cardiac death, due to ventricular fibrillation, is divided into two main components: a) the initiation of tachycardia and b) its degeneration to fibrillation. As a result, a revised antiarrhythmic drugs' classification, incorporating both anti-tachycardiac with anti-fibrillary profiles, has been proposed.

Several implications of cardiovascular drugs such as the increased risk of bioequivalence failure of antiarrhythmic generics could be attributed to the chaotic behavior of the cardiac signaling. It is evident that the oversimplified approach of Emax model, adopted most of the times in PK/PD modeling of antiarrhythmic compounds, might be neither sufficient nor appropriate. Overall, further research to increase understanding of the high complexity degree of cardiac signaling, its association with disease and the effect of drugs acting on the cardiovascular system is required. Toward this direction, nonlinear dynamical systems analysis might be a very powerful, if not necessary, mathematical tool.

CNS-System

Evaluation of nonlinear dynamics in brain electrical activity has provided information about the underlying neuronal networks and brain disorders [81]. Most of studies applying tools from the nonlinear dynamical systems theory are based on experimental electroencephalogram (EEG) recordings and highlight the chaotic behavior of the brain electrical activity. Phase space reconstruction and fractality calculation of real time EEG recording are some of the techniques that have been used to assess the EEG variability [1]. These nonlinear systems tools demonstrate not only the underlying complexity of brain electrical activity, but also enrich the information obtained from classical techniques such as the Fourier analysis. In addition, they can be utilized to qualitatively distinguish between EEG recordings in different disease states like epileptic seizures, Alzheimer's and Parkinson's disease or schizophrenia [56, 57, 119, 123]. In the same context, reduction of the nonlinear structure of brain activity has been observed after administration of low doses of ethanol [33].

Usually, PK/PD studies of centrally acting drugs rely on some quantitative measures of EEG parameters [73]. Analysis of time series of EEG data in PD studies with CNS drugs using techniques of nonlinear dynamics are very limited. Some examples include the effect of pregnenolone sulfate, penicillin and lorazepam on the electrophysiological activity of brain [35, 54, 55, 65, 74, 107].

Modeling in the brain aims to rather illuminate the qualitative principles underlying the various phenomena, such as epileptic seizures, than to quantify and forecast them [64]. Accounting for changes in brain activity using tools from chaos theory can provide important information about the underlying dynamics and possibly reveal irregular dynamical behavior as source of the high variability observed in the PD parameters of CNS drugs. In this chapter, we focus on tools from chaos theory applied in different models of a well-studied CNS disease, the Parkinson's disease.

Parkinson's Disease

Parkinson's disease (PkD) is the second most common, after Alzheimer's, progressive neurodegenerative disease with substantial and growing socioeconomic burden and its incidence is expected to increase with life expectancy. Parkinson's is a complex multifactorial disease resulting from aging, genetic predisposition and exposure to environmental stimuli and is characterized by tremors and movement rigidity. Pathogenesis of PkD involves oxidative stress, aggregation of the a-synuclein (Asyn) protein and dysfunction of proteasomes as well as lysosomes. Both, positive and negative feedback motifs have been identified and all involve misfolding of Asyn. Even though the physiological proteolytic mechanisms are normally responsible for clearing misfolded proteins, misfolded Asyn is capable of partly inhibiting, through a double-negative feedback interaction, proteasomic and lysosomal function [70]. On the other hand, there are two double-positive feedback mechanisms, involving misfolded Asyn, that we put emphasis on: (1) increased cytosolic dopamine (DA) levels via permeabilization of DA-containing vehicles, which in turn induce the misfolding of native Asyn and (2) increased oxidative stress and as a result elevated levels of reactive oxygen and nitrogen species (ROS/RNS), due to mitochondrial damage. This leads Asyn to misfold even further. Of course, this is not an exhaustive list of the observed feedback mechanisms in Parkinson's disease, where several longer and more complex interaction pathways can take place, but outlines well identified pathways that have been utilized in mathematical models. Aside from the feedback mechanisms, the development and prognosis of Parkinson's disease is influenced by several other factors as well. This include, but are not limited to, increased brain concentrations of metal ions (like Fe^{2+}, Cu^{2+}), increased inflammation and age-related degenerative factors such as protein clearance and mitochondrial function attenuation.

Over the last decades, mechanism-based mathematical models of Parkinson's disease have been developed and improved with the increasing insight of brain physiology and experimental techniques. Recently, Bakshi et al. [2] carefully reviewed the mathematical biology models for Parkinson's disease. These mechanistic models were classified into three categories (i) Asyn aggregation, (ii) disease pathogenesis and (iii) pathology propagation models. The former category is typically ODE-based with the most parsimonious models and the greatest experimental support, whereas the latter has not gained much attention modeling-wise, apart from some limited attempts mainly by Kuznetsov and co-workers [61, 62, 126]. The pathogenesis models, in contrast, usually combine ODEs with network models, stochastic simulation algorithms, flux-balance analysis (FBA) and/or biochemical systems theory. In this chapter we focus on these models, since tools from the nonlinear dynamical systems theory have been more widely applied to their analysis.

Examples of pathogenesis models from three sub-categories: (a) reactive oxygen species (b) Ubiquitin proteasome pathway, chaperone-mediated autophagy and lysosomal clearance and (c) Dopamine metabolism models are summarized in this section.

Reactive Oxygen Species (ROS) Models

In 2009 a detailed brain energy metabolism model was proposed by Cloutier et al. [20] Glycolysis and mitochondrial energy metabolism in neurons and astrocytes were modelled and the predictions were compared with in vivo data from rat brains. The same authors in a later work updated their model by integrating the aging-related effect and by adding time-dependent reduction in the mitochondrial complex 1 efficiency. Using metabolomics, energy-related metabolites data from brain slices in healthy and PARK2 knockout mice were used to train the model. The authors suggested that regardless of genetic mutation, the brain cells were able to robustly maintain control of ATP levels and the observed reduction of those may not be related with Parkinson's pathology [97, 140]. However, brain energy metabolism models have several limitation since they are lacking spatial details such as diffusion and locus synaptic activity effects, in regards to capillaries. Recent research has shown that implementation of such effects could influence the prediction and that averaging spatially detailed models with ODEs might be appropriate only under certain parametrizations [12, 13].

The aforementioned energy metabolism model was modified to account for feedback between ROS and misfolded Asyn, Asyn aggregation and minimal description of its proteolytic clearance. The final model consisted of 33 ODEs and demonstrated a bistable-switch-like behavior with respect to various factors like mutation, environmental toxins and aging. The lower and higher steady states were characterized by low and increased ROS and misfolded Asyn levels, representing the healthy and disease states, respectively [22]. To get insight into the bistability and bifurcation behavior, a reduced model with only as variables the ROS and misfolded Asyn levels was implemented and was able to reproduce the full scale model [21]. The bistable-switching behavior in short-time scale has been also supported by in vivo experiments in paraquat-induced oxidative stress in rat brains [39]. In simulations of the model by Raichur et al. bistability was present, even though the authors did not comment on it at that time [121].

Ubiquitin Proteasome Pathway, Chaperone- Mediated Autophagy, and Lysosomal Clearance

Models for the ubiquitin proteasome pathway and the negative feedback involved between misfolded Asyn and proteasomes have been recently presented. However, in these cases mainly stochastic simulations to qualitatively test hypotheses and/or experimental results have been used [89, 100, 101]. Recent work has also focused on lysosomes and autophagosomes, but this is not subject of this section [37, 96, 145].

In this section, we discuss in more detail a theoretical model involving feedback between proteolytic pathways and Asyn aggregation. In this model, a minimal description of Asyn aggregation and its interaction with proteasomes was described using three ODEs [121]. As function of the ratio between Asyn fibrils and free proteasomes, a bifurcation behavior was observed. It was predicted that homeostasis could be maintained at lower, but not at higher ratios, where free proteasome levels oscillated with prolonged periods of low concentration. Parkinson's pathogenesis was assumed to be associated with extended proteasome depletion leading to accumulation of Asyn oligomers [121]. The authors concluded that the model was able to predict Parkinson's pathogenesis, even without explicitly modeling the proteasome inhibiting function of Asyn aggregates. In this case, similar to previously discussed ROS/NOS models, pathogenesis was predicted via systems undergoing bifurcation.

Dopamine (DA) Metabolism

Loss of functionality of dopaminergic neurons, which are involved in DA synthesis, storage, release and reuptake, causes DA depletion in brain. This results in increased DA cytoplasmic levels, which in turn lead to ROS generation and subsequently to Asyn misfolding. Using the biochemical systems theory (BST) formalism, Qi et al. developed a DA metabolism model to investigate the effects of key enzymes or transporters on DA homeostasis as well as the influence of rotenone and paraquat on DA metabolism [102–104]. Sass et al. [111] also adopted the BST approach with minimal kinetic information to understand the interplay between DA metabolism, Asyn and proteasomal/lysosomal athway. The authors showed that disruption of cellular pathways may result in Parkinson's disease.

Buchel et al. [9], using flux-balance analysis (FBA), in a model of the dopaminergic neuron, considered steady-state fluxes of several variables including DA, Asyn, ROS and proteasomal machinery. In a qualitative manner, it was shown that Parkinson's disease pathology can be associated with increased stressors levels (e.g. neurotoxins, Asyn). Tools from the nonlinear dynamical systems theory have been implemented in all the models discussed in this section. However, they all have several assumptions and limitations as well. The numerous variables and processes considered in BST and FBA models, given the limited kinetic information, has led the authors to use relative species concentrations and parametrization. In pathogenesis models the native Asyn concentrations used are of several orders of magnitude higher than the observed experimentally. This has a great impact for models exhibiting bifurcation or bistability, which are sensitive to native Asyn concentrations. In these models, recalibration of relevant parameter to reflect the physiological Asyn concentration range is mandatory. Furthermore, in models with bistability also other critical parameters may need to be revised. For example, whether the bistable behavior still remains at lower Asyn concentrations or not should be investigated. Quantitative comparisons of experimentally measured variables (such as misfolded Asyn) are necessary for calibration. Typically, the pathogenesis models only incorporate a single feedback system (e.g., ROS-Asyn feedback or Asyn-UPP feedback).

However, a dynamic model integrating multiple feedback loops to adequately capture the multifactorial nature of Parkinson's disease, provided that adequate parametrization is feasible, would be of interest. Such integrated models would not only provide further insight on the relative importance of various components involved in the pathogenesis, but they would also be helpful to compare effectiveness of different potential interventions and explore the synergistic value of combination therapies. System perturbation may be crucial to elucidate the link between disease biology and various aspects of the clinical manifestation. Last but not least, a multiscale model of Parkinson's, including molecular level and small timescale changes to neuronal connectivity to evolution of clinical UPDRS scores, would require an approach that integrates diverse modeling formalisms such as a combination of differential equation-based molecular level, agent-based cell-level model and network-based neural connectivity models [2].

Secretion and Regulation of Hormones

Pulsatility is a widely appreciated characteristic of hormone secretion. Since the early 70's, Hellman et al. observed the episodical secretion of cortisol in man [50]. It was soon revealed that pulsatility is inherent to physiological processes. Indeed, the highly regulated hormone secretion through extensive feedback mechanism by the central and autonomous nervous system, the biological rhythms (e.g. circadian clock) and the complex between-hormones interactions are only some of the reasons for the pulsatile behavior and the chaotic nature governing the endo-, para- and autocrine systems. In this context, it has been evident that implementing tools from the dynamical systems theory in the study of hormonal systems is not only useful, but also fundamental to increase our understanding. This has been addressed so far bidirectionally; using experimental as well as modeling and simulation approaches.

Experimental studies have been focusing on the phase space reconstruction technique. In this approach, the concentration-time plasma profiles of various hormones have been utilized to evaluate the dimensionality and emerge the chaotic nature of the underlying systems' dynamics. Such examples include prolactin, cortisol, the parathyroid and the growth hormones [53, 90, 98]. In all these studies, the phase plane reconstruction produced attractors with fractal dimensions showing evidence for the presence of nonlinear dynamics. Based on the concept of Lyapunov exponents, Pincus developed the Approximate Entropy algorithm (ApEn) method to quantify the hormone pulsatility [95]. This method has been successfully applied to pulsatility quantification and its betwwen-groups differences identification (e.g diseased vs. healthy, young vs. elderly) of several hormones such as cortisol, prolactin, insulin, testosterone, the adrenocorticotropic (ACTH), the growth (GH) and the luteinizing (LH) hormones [14, 15, 26, 38, 41, 66, 68, 69, 120, 130, 133]. Therefore, experimental evidence of pulsatility origin and the chaotic nature underlying the dynamics of hormonal systems is abundant. This should serve as a guide to advance the experimental research and encourage physiologically- sound, mechanistical modeling, accounting for the dynamical aspects on a proper level.

That said, several mathematical models have been published in literature attempting to provide insight on the hormonal behaviour (secretion and regulation) and/or the effect of drugs on it. In 1980, Smith et al. [120] developed a model to qualitatively describe the interaction between LH-releasing hormone (LHRH), LH and testosterone. This model was then further elaborated by Cartwright and Husain [14] incorporating time-delayed terms and exhibiting limit cycle solutions.

Additional improvements have been implemented by Liu and Deng [68] as well as by Das et al. [26]. Specific examples include models of the hypothalamic-pituitary-adrenal (HPA) axis for the system of corticotropin-releasing hormone (CRH), ACTH and cortisol and the beta- cells mass system for insulin/glucose interplay [66, 130]. A dynamic, with chaotic behaviour, model for hormonal systems coupled by negative feedbacks has been also proposed by Londergan and Peacock-Lopez [69].

In pharmacokinetics/pharmacodynamics, often the physiological hormonal secretion is perturbed by the drug's effect either as primary target system of action or as side effect and as a result many studies have considered the hormonal secretion along with the dominant PK/PD aspects. In this section, some typical examples are summarized, whereas models involving thorough analysis of the underlying nonlinear dynamics for cortisol and prolactin are described in detail separately. Chakraborty et al. investigated the effect of corticosteroids on circadian cortisol levels, [15] while Fattinger et al. [38] studied the impact of a LHRH antagonist on testosterone and LH. Further examples include the dopaminomimetic, calcimimetic effect on prolactin and parathyroid hormone, respectively; as well as the ipamorelin-mediated effect on GH [41, 44, 63].

Of course, this is not an exhaustive list of modeling examples. However, these studies are mentioned, because of a common feature that they are sharing and this is not other than a minimum oversimplified implementation of hormone secretion, giving a smooth hormone baseline. In these cases, only the most obvious characteristics of hormone secretion, like circadian rhythms, have been integrated. The underlying physiological pulsatility, since it is considered as noise, it is practically ignored or phenomenologically implemented through spline terms or Fourier harmonics, [15, 41] but not through modeling of its dynamical origin. Last but not least, cases where the pulsatility can be omitted have been also identified. This refers to the effect of ipamorelin on GH, where the baseline of the hormone is wisely set to zero due to the multifold amplification of GH levels post drug-administration [44].

Melatonin

There is multiple evidence showing that melatonin participates in a number of physiological processes including the regulation of body's circadian rhythms, sleep, mood, immune response, aging, and cancer [8]. Melatonin is secreted by the pineal gland of the brain in response to the light/dark signals emanating from the retina and sympathetic nervous system. The synthesis of melatonin is stimulated by darkness and inhibited by light.

One of the most widely used model to mathematically describe the circadian secretion of melatonin in the body has been published by Brown et al. [7] In this model, two-dimensional linear differential equations were formulated in order to analyze plasma melatonin levels in 18 normal healthy male subjects during a constant routine. The model includes two physiological compartments, namely pineal gland and plasma. Melatonin secretion to the pineal gland compartment was modelled through a step-wise function simulating pineal N-acetyltransferase (NAT) rise as a combination of two exponential processes with different time constants and plateau levels. Melatonin levels are then moved to the plasma compartment through a first order infusion and cleared from the plasma by a certain clearance rate. The model was successfully applied to describe melatonin physiological data and provided overall a more physiologically plausible estimate of the melatonin synthesis onset time. This model was further extended in a number of works in order to describe phase shifts observed upon exogenous melatonin administration [5, 122].

Depending on data availability and the question of interest, melatonin secretion and forward effects were simulated through various models. Sekula et al. [116] followed a statistical approach and fitted a linear beta-model to either healthy (control) or major depressive individuals shading light to probable differences on their onset, peak and offset times of melatonin secretion. In another study, Schwartz et al. [115] by using a simple model of two oscillators were able to represent melatonin secretion pattern and further predict amplitude changes in melatonin release under forced desynchronization. Finally, Scheff et al. modelled melatonin circadian secretion through a step-wise function by further considering its anti-inflammatory effects, which was ultimately included in a systems model of immune response [112, 113].

Cortisol

Over the last decades, as previously discussed, nonlinear dynamics have proved that deterministic systems even with few degrees of freedom can exhibit complex-random behaviors. Although only classical randomness is often involved in PK/PD, it has been shown variability can originate from the chaotic nature of the underlying dynamics. As indicated by numerous studies, hormone secretion is of chaotic nature and characterized by pulsatility [14, 26, 30, 68, 69, 90, 98, 120, 129, 130, 133].

In 2002, Dokoumetzidis et al. [30] described the erratic secretion of cortisol and its suppression by corticosteroids from a dynamical systems' perspective using a simple model. The model was relied on well-established concepts of hormonal erratic secretion and circadian rhythm [60]. Other factors controlling cortisol secretion have also been considered, but not explicitly modelled (e.g. negative feedback loops). The concentration of cortisol was described by a nonlinear time-delay differential equation with two terms, [48, 72, 84] particularly a first order term for disposition and elimination and a secretion rate term which adheres to the negative feedback mechanism and drives the pulsatile secretion as follows (Eq. 2):

$$\frac{dC}{dt} = k_1 \cdot \frac{a^n \cdot C_d}{a^n + C_d^n} - k_2 \cdot C \tag{2}$$

where C is the cortisol concentration, C_d is the value of C at time t-T, n is an exponent, k1 and k2 are the input and output rate constants, respectively. Phenomenological implementation of the circadian rhythm is achieved by the parameter alpha, which is a simple cosine function of a 24-h period, given by the following equation (Eq. 3):

$$a = A \cdot \cos\left[(t - f) \cdot \frac{2 \cdot \pi}{1440}\right] + B \tag{3}$$

where A and B are constants with concentration units, *f* is a constant with time units and t is time in min. A similar periodic function have been used by Rohatagi et al. [106] to describe the secretion rate of cortisol.

The time-delay equation (Eq. 2) has physical meaning because it is associated with cortisol auto-regulation and negative feedback loops and thus its own secretion. Since Eq. 2 has an infinite number of degrees of freedom, the authors constructed a pseudo-phase space for the system of Eqs. 2 and 3 using the model variables C(t), C (t-T/2), C(t-T) (Fig. 3) [4]. The attractor of the system was a strange attractor, shwoing that it had quite complex geometry. The use of three dimensions is also supported by the embedding dimension found by Ilias et al. [53]

The authors advocated that the extremely high variability often observed in experimental data (e.g. cortisol plasma concentration profiles) originates from

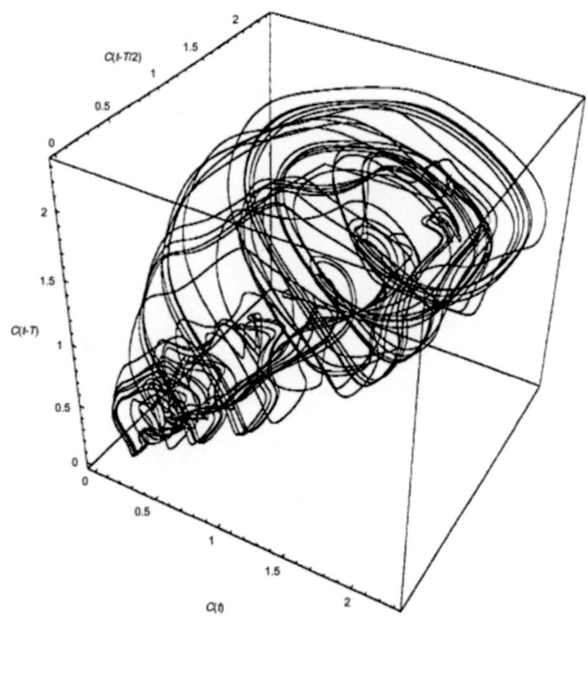

Fig. 3 A pseudo-phase space for the model of Eqs. 2 and 3 using the variables C(t), C(t-T/2) and C(t-T) expressed in mg 100 ml-1. The real phase space is of infinite dimension, however, trajectories may be considered to lie in a low dimensional space (attractor). The model parameters take the same values as in Fig. 1 and time runs for 10 days. (Reproduced from Ref. 30)

chaotic dynamics and not from pure randomness and therefore measures of central tendency should be questioned in these case and averaging might not be appropriate. Due to the erratic nature of the secretion, they suggested that an instantaneous, instead of average, secretion rate should be reported. It was also shown that the final simulated profiles with this model differed remarkably depending on the initial conditions and the sampling design.

Furthermore, the initial model was extended to allow for perturbation of the system due to external oral administration of corticosteroids. The pharmacokinetics of the external corticosteroids were assumed to be as simple as one-compartment disposition with first-order absorption, coupled with an effect compartment. Then, the effect-site concentration (C_e) was linked to the initial model so that it suppresses cortisol secretion (Eq. 4):

$$a' = a \cdot \left(1 - \frac{C_e}{C_{e50} + C_e}\right) \tag{4}$$

where C_{e50} is a coefficient that expresses the concentration of the drug when $\alpha' = \alpha/2$.

The final extended model was able to adequately capture the experimental cortisol plasma concentration profiles, generated after exogenous administration of fluticasone propionate, published by Chakraborty et al. [15] Based on the aforementioned discussion, the authors concluded that their analysis under the light of nonlinear dynamics provides mechanistical insight on cortisol secretion and that this approach should be more widely applied in PK/PD modeling, when supported by experimental and physiological evidence. They also highlighted that nonlinear dynamics construes a fundamentally new rational in PK/PD and that the clinical pharmacologists should be aware of the limitations of chaotic models for long-term predictions.

Interestingly, in recent studies Pilai et al. [93, 94] combined adaptive chaos synchronization and grid search to estimate physiological and pharmacological systems' parameters by exploring deterministic methods that are more appropriate and potentially perform better than classical numerical approaches (e.g. minimization of sum of squares, likelihood maximization). To illustrate these issues, they used the previously described cortisol model. They showed that chaos synchronization could help to avoid ending up in local minima, which is often observed with the gradient-based optimization algorithms. The hybrid adaptive chaos synchronization could be applied iteratively and was able to accurately estimate nonlinear parameters and track trajectories for a wide class of noisy chaotic systems. The authors concluded that their method could effectively estimate the parameters of the cortisol chaotic system and that its robustness against noise and data sampling rate effect may be of benefit for modeling nonlinear dynamical physiological systems.

Prolactin

Prolactin (PRL) is a polypeptide hormone with a primary role in the regulation of lactation in humans. It is predominantly secreted by specialized cells in the anterior pituitary gland, the lactotrophs [18]. In PK/PD, the response of prolactin to antipsychotic drugs has been studied by the classic precursor-pool model. Precursor-pool models have been used several times in the past to explain the tolerance and rebound effects induced by drug response [34, 49, 83, 118]. These are precursor-dependent indirect response models, which assume that the tolerance (or rebound) is the result of depletion (or accumulation) of finite pools of precursors that are responsible for the drug effect. The pool model has been applied to the PRL response after administration of antipsychotic drugs with the aim to explain the tolerance developed after repeated drug administration at narrowly spaced intervals. The original pool model by Movin-Osswald & Hammarlund-Udenaes was modified to account for the effect of remoxipride in rats [83]. By including a positive feedback (PF) component, Stevens et al. [127]. extended the already modified model, which had become nonlinear. Simulations, after using the model to investigate the effect of risperidone, predicted two separate baselines depending on dose. However, since no mathematical analysis of the underlying dynamics had been performed, such findings relied only on serendipitous discovery through simulations.

Recently, Bakshi et al. [3], using both the original and modified precursor-pool models for PRL, analyzed the systems' dynamics mathematically in an informative tutorial. The classic precursor-pool model consists only of two linear turnover equations in the dependent variables, and thus it has a unique steady state. However, the PF model it was shown that undergoes trans-critical bifurcation, meaning that the system changes from stable to unstable steady state or vice versa. The convergence to different steady states was also dose-dependent.

Through steady-state and phase-plane analysis they demonstrated that the nonlinearity of the model has resulted in multistationarity and that the higher of the two steady-states remains always stable In contrast, the lower one is stable only from below. In addition, under a parametric condition the desired steady state has been observed to be the higher and thus, always stable. However, in the parametrization by Stevens et al. the desired steady state is the lower one, which is unstable and reachable only from below. Simulations and plotting of the orbits in the phase plane illustrated that the parametric condition leading to the higher steady state is activated only by some orbits and this is the reason why the model exhibited convergence to two discrete steady states in response to different doses. Activation of the "if" condition depends not only on the relative timescale of drug clearance and the PRL endogenous elimination and secretion rate, but also on the accuracy of the numerical solver. Nevertheless, the authors concluded that even if the numerically solver was perfectly accurate, this irregular behavior would persist as structural property of the model.

Overall, the authors summarized the basic steps in dynamical systems analysis of ODE-based models, with the phase plane analysis being of particular interest for 2-ODE models allowing for comprehensive analysis of the underlying dynamics.

Immune System & Inflammatory Response

Immune response has triggered significant interest in systems-based approaches to understand the involved complexities and the individual interactions of a response [134]. In particular, systems modeling was applied to quantitatively evaluate not only the onset but also the resolution of the inflammatory response.

Modeling efforts on immune response vary from statistical and correlational to mechanism-based, deterministic and stochastic [19, 40, 79, 86, 113]. Clearly, inflammatory response encompasses a high level of interconnections through multiple levels of physiological organization and control. Chow et al. [17] investigated the acute inflammatory response in diverse shock states by implementing a highly detailed realization that incorporates individual cytokines, different types of immune cells, and other key physiological parameters. This model showed that different inflammatory outcomes can result from the same model with different initial states indicating that diverse inflammatory shock states share the same underlying mechanisms, even when cytokine-concentration data may be heterogeneous. The level of detail and complexity of such models can increase until eventually reaching the level of description of the host response [27, 88, 99, 131]. Similar, using high-throughput microarray mRNA measurements from peripheral blood cells, it was possible to develop semi-mechanistic models by linking the ligand (lipopolysaccharide, LPS) recognition by appropriate (TLR4) receptors to activation of inflammation-specific signaling cascades (NF-kB) which drive the peripheral release of pro- and anti-inflammatory cytokines [40, 87]. In a subsequent work it was demonstrated that an extended model with adeqaute signaling cascades was able to describe the complex phenomenon of endotoxin tolerance [144]. The inflammatory response induces the involvement of the neuroendocrine system, which modulates the release of anti-inflammatory hormones and neurotransmitters. As such, cortisol and epinephrine lead to anti-inflammatory downstream effects, cortisol through glucocorticoid receptor-mediated signaling and epinephrine through the stimulation of adrenergic receptors, leading to elevated intracellular cAMP concentrations. Leveraging established models of hormone activity to account for the effects of cortisol and epinephrine, allowed for investigation of cellular-level transcriptional responses to inflammation. The mechanisms through which hormone levels influence whether the system exhibits a healthy self-limited inflammatory response or a persistent chronic inflammatory state were also explored.

Circadian rhythms are of importance in the context of the inflammatory response since they impose patterns on a wide range of inflammatory mediators [23]. Meyer-Herman et al. [80] developed a mathematical model to evaluate the neuroendocrine-immune system interactions in rheumatoid arthritis. This model describes mainly the measured circadian responses of plasma levels of TNF, noradrenaline, and cortisol, making use of a set of ordinary differential equations. Based on their model, it was observed that treatment with glucocorticoids between 00: 00 and 02: 00 a.m. induced the strongest inhibitory effect on TNF secretion. In chronic inflammatory diseases, such as rheumatoid arthritis, where patients overexpressing inflammatory agents, significant reduction in pro-inflammatory mediators like the TNF is often a clinical target. Similarly, Scheff et al. [112] incorporated a multilevel mathematical modeling scope based on which they evaluated the interplay between inflammation and circadian rhythms. This model predicted that LPS administration during the night induces larger increase in inflammatory mediators and larger reduction in the heart rate variability (HRV) relative to administration in the morning. Finally, semi-mechanistic models were further explored to rationalize the potentially permissive-suppressive inflammatory effects of cortisol as manifestations of the balance between pro- and anti-inflammatory characteristics induced by circadian rhythms [75, 76, 78].

Conclusions & Future Directions

In this chapter, the presence of nonlinear dynamics and the chaotic nature of physiological systems is highlighted. Dynamical aspects in the analysis of the behaviour of PK/PD and QSP models are summarized with the aim to showcase the urge to implement toolz of nonlinear dynamics in it. Particularly, we have focused on dynamical aspects of: (a) the cardiovascular and (b) the central nervous, (c) the hormone secretion and regulation and (d) anti-inflammatory response, for which some case examples are analysed. At the same time, there are several other therapeutic areas where dynamic systems theory has been successfully applied and therefore this chapter should not be considered as an exhaustive literature review of all possible applications of nonlinear dynamics in PK/PD or QSP models. However, the reader is referred to Eftimie et al. [32] for a comprehensive review of mathematical models in immunology and to the book by Macheras and Iliadis [71] for a detailed explanation of fractal phenomena in biopharmaceutics, pharmacokinetics and pharmacodynamics.

From the above presentation, it is evident that significant progress has been made towards mechanism- and physiologically-based modeling. On the other hand, physiological systems are inherently nonlinear and chaotic. The chaotic behavior is often considered to be feature of healthy state, whereas periodic or non-chaotic states are associated with disease and pathological symptoms which can be attributed to a sudden qualitative change in the temporal pattern (e.g. bifurcation) [28]. This change can either be caused by endogenous factors or an external stimulus (e.g. drug administration) that alters one or more critical control parameters. In this context, integration of nonlinear dynamics might be also useful in the field of disease progression by introducing a new rationale for therapeutic strategies, which rather aims at restoring and maintaining the homeostasis than eliminiating the symptoms. This general concept has been introduced by Mackey and Milton [72, 81] and it is known as dynamic disease.

The notions of sensitivity from the initial conditions and the qualitatively different behavior for, even slightly, altered values of the control parameters, have a major impact and should be taken into account in modeling activities, especially when also experimentally supported. Advanced computational methods play a key role in simulating complex models, but usually offer only a limited picture of model behaviors, especially when the dynamics are rich in nonlinearities and counterintuitive behaviors. Such behaviors can be very difficult, if not impossible, to be revealed solely through simulations. It is noteworthy that even the simplest 2-ODE models can exhibit unpredictable behaviors and addition of one ODE explodes the range of possible behaviors, including chaotic ones. Systems pharmacology models are expected to be larger and even more complex and their dynamical behaviors are likely to hide an even greater degree of complexity.

In pharmacology, the dynamics of the underlying system is often only partially understood and models are a blend of biological, mechanical, physiological and pharmacological information accompanied by experimentally data. This might raise questions about the validity itself, the applicability, the parametrization, the granularity and the potential extrapolations of such models. Nowadays, many large biological, physiological, pharmacological or combination of those systems networks are formed from modules or motifs of smaller networks. Thorough understanding of the behavior of all the individual components is not only crucial, but also imperative, to increase confidence on their value, usefulness and performance. This also underlines the importance of using mathematical analysis to gain insight into model behavior.

In summary, mathematical techniques of dynamical systems' analysis allow for exploration of multistationarity, with the possibility of rejecting a priori models that are structurally unstable, and better understanding of the overall model behavior. Furthermore, such analysis can inform and guide experimentally testable hypotheses for verification/falsification of a model. At the same time, other mathematical techniques, such as quasi-steady state analysis and model reduction may be useful to reduce the model size and complement the dynamical systems' analysis.

Acknowledgments We wish to thank the Emeritus Prof. Athanasios Iliadis, Faculty of Pharmacy, Aix-Marseille University, Marseille, France for valuable discussions. ILK would like to particularly thank Prof. Iliadis for the continuous support and guidance throughout his research stay at the Department of Pharmacokinetics & Toxicokinetics, Aix-Marseille University, Marseille, France.

References

1. Accardo, A., M. Affinito, M. Carrozzi, and F. Bouquet. 1997. Use of the fractal dimension for the analysis of electroencephalographic time series. *Biological Cybernetics* 77: 339–350.
2. Bakshi, S., V. Chelliah, C. Chen, and P.H. van der Graaf. 2019. Mathematical biology models of Parkinson's disease. *CPT: Pharmacometrics & Systems Pharmacology* 8: 77–86.

3. Bakshi, S., E. de Lange, P. van der Graaf, M. Danhof, and L. Peletier. 2016. Understanding the behavior of systems pharmacology models using mathematical analysis of differential equations: Prolactin Modeling as a Case Study. *CPT: Pharmacometrics & Systems Pharmacology* 5: 339–351.
4. Bassingthwaighte, J.B., L.S. Liebovitch, and B.J. West. 1994. *Fractal physiology*. New York: Springer New York.
5. Breslow, E.R., A.J.K. Phillips, J.M. Huang, M.A. St. Hilaire, and E.B. Klerman. 2013. A mathematical model of the circadian phase-shifting effects of exogenous melatonin. *Journal of Biological Rhythms* 28: 79–89.
6. De Brouwer, S., D.H. Edwards, and T.M. Griffith. 1998. Simplification of the quasiperiodic route to chaos in agonist-induced vasomotion by iterative circle maps. *American Journal of Physiology. Heart and Circulatory Physiology* 274: H1315–H1326.
7. Brown, E.N., Y. Choe, T.L. Shanahan, and C.A. Czeisler. 1997. A mathematical model of diurnal variations in human plasma melatonin levels. *American Journal of Physiology-Endocrinology and Metabolism* 272: E506–E516.
8. Brzezinski, A. 1997. Melatonin in humans. *The New England Journal of Medicine* 336: 186–195.
9. Büchel, F., S. Saliger, A. Dräger, S. Hoffmann, C. Wrzodek, A. Zell, and P.J. Kahle. 2013. Parkinson's disease: Dopaminergic nerve cell model is consistent with experimental finding of increased extracellular transport of α-synuclein. *BMC Neuroscience* 14: 136.
10. Buijs, R.M., and A. Kalsbeek. 2001. Hypothalamic integration of central and peripheral clocks. *Nature Reviews. Neuroscience* 2: 521–526.
11. Callaway, E., and H. Ledford. 2017. Medicine Nobel awarded for work on circadian clocks. *Nature* 550: 18–18.
12. Calvetti, D., Y. Cheng, and E. Somersalo. 2015. A spatially distributed computational model of brain cellular metabolism. *Journal of Theoretical Biology* 376: 48–65.
13. Calvetti, D., Y. Cheng, and E. Somersalo 2016. Uncertainty quantification in flux balance analysis of spatially lumped and distributed models of neuron–astrocyte metabolism. *Journal of Mathematical Biology* 73: 1823–1849.
14. Cartwright, M., and M. Husain. 1986. A model for the control of testosterone secretion. *Journal of Theoretical Biology* 123: 239–250.
15. Chakraborty, A., W. Krzyzanski, and W.J. Jusko. 1999. Mathematical modeling of circadian cortisol concentrations using indirect response models: Comparison of several methods. *Journal of Pharmacokinetics and Biopharmaceutics* 27: 23–43.
16. Chauvet, G.A. 1993. Hierarchical functional organization of formal biological systems: a dynamical approach. I. The increase of complexity by self-association increases the domain of stability of a biological system. *Philosophical Transactions of the Royal Society of London. Series B, Biological Sciences* 339: 425–444.
17. Chow, C.C., G. Clermont, R. Kumar, C. Lagoa, Z. Tawadrous, D. Gallo, B. Betten, J. Bartels, G. Constantine, M.P. Fink, T.R. Billiar, and Y. Vodovotz. 2005. The acute inflammatory response in diverse shock states. *Shock* 24: 74–84.
18. Chrousos, G.P., and P.W. Gold. 1992. The concepts of stress and stress system disorders. Overview of physical and behavioral homeostasis. *JAMA* 267: 1244–1252.
19. Clermont, G., C.C. Chow, G.M. Constantine, Y. Vodovotz, and J. Bartels. 2004. Mathematical and statistical modeling of acute inflammation. In *Classification, clustering, and data mining applications*, 457–467. Berlin/Heidelberg: Springer.
20. Cloutier, M., F.B. Bolger, J.P. Lowry, and P. Wellstead. 2009. An integrative dynamic model of brain energy metabolism using in vivo neurochemical measurements. *Journal of Computational Neuroscience* 27: 391–414.
21. Cloutier, M., R. Middleton, and P. Wellstead. 2012. Feedback motif for the pathogenesis of Parkinson's disease. *IET Systems Biology* 6: 86–93.
22. Cloutier, M., and P. Wellstead. 2012. Dynamic modelling of protein and oxidative metabolisms simulates the pathogenesis of Parkinson's disease. *IET Systems Biology* 6: 65–72.

23. Coogan, A.N., and C.A. Wyse. 2008. Neuroimmunology of the circadian clock. *Brain Research* 1232: 104–112.
24. Costa, M.D., W.T. Schnettler, C. Amorim-Costa, J. Bernardes, A. Costa, A.L. Goldberger, and D. Ayres-De-Campos. 2014. Complexity-loss in fetal heart rate dynamics during labor as a potential biomarker of acidemia. *Early Human Development* 90: 67–71.
25. Danhof, M. 2016. Systems pharmacology – Towards the modeling of network interactions. *European Journal of Pharmaceutical Sciences* 94: 4–14.
26. Das, P., A.B. Roy, and A. Das. 1994. Stability and oscillations of a negative feedback delay model for the control of testosterone secretion. *Biosystems* 32: 61–69.
27. Daun, S., J. Rubin, Y. Vodovotz, A. Roy, R. Parker, and G. Clermont. 2008. An ensemble of models of the acute inflammatory response to bacterial lipopolysaccharide in rats: Results from parameter space reduction. *Journal of Theoretical Biology* 253: 843–853.
28. Davis, J.D., C.M. Kumbale, Q. Zhang, and E.O. Voit. 2019. Dynamical systems approaches to personalized medicine. *Current Opinion in Biotechnology* 58: 168–174.
29. Dokoumetzidis, A., A. Iliadis, and P. Macheras. 2001. Nonlinear dynamics and Chaos theory: Concepts and applications relevant to pharmacodynamics. *Pharmaceutical Research* 18: 415–426.
30. Dokoumetzidis, A., A. Iliadis, and P. Macheras. 2002. Nonlinear dynamics in clinical pharmacology: The paradigm of cortisol secretion and suppression. *British Journal of Clinical Pharmacology* 54: 21–29.
31. Echt, D.S., P.R. Liebson, L.B. Mitchell, R.W. Peters, D. Obias-Manno, A.H. Barker, D. Arensberg, A. Baker, L. Friedman, H.L. Greene, M.L. Huther, D.W. Richardson, and Investigators∗ the C. 1991. Mortality and morbidity in patients receiving Encainide, Flecainide, or Placebo. *The New England Journal of Medicine* 324: 781–788.
32. Eftimie, R., J.J. Gillard, and D.A. Cantrell. 2016. Mathematical models for immunology: Current state of the art and future research directions. *Bulletin of Mathematical Biology* 78: 2091–2134.
33. Ehlers, C.L., J. Havstad, D. Prichard, and J. Theiler. 1998. Low doses of ethanol reduce evidence for nonlinear structure in brain activity. *The Journal of Neuroscience* 18: 7474–7486.
34. Ekblad, E.B., and V. Licko. 1984. A model eliciting transient responses. *American Journal of Physiology. Regulatory, Integrative and Comparative Physiology* 246: R114–R121.
35. Elger, C.E., G. Widman, R. Andrzejak, J. Arnhold, P. David, and K. Lehnertz. 2000. Nonlinear EEG analysis and its potential role in epileptology. *Epilepsia* 41 (Suppl 3): S34–S38.
36. Epstein, F.H., and S. Reichlin. 1993. Neuroendocrine-immune interactions. *The New England Journal of Medicine* 329: 1246–1253.
37. Esteves, A.R., and S.M. Cardoso. 2017. LRRK2 at the crossroad between autophagy and microtubule trafficking. *The Neuroscientist* 23: 16–26.
38. Fattinger, K.E., D. Verotta, H.C. Porchet, A. Munafo, J.Y. Le Cotonnec, and L.B. Sheiner. 1996. Modeling a bivariate control system: LH and testosterone response to the GnRH antagonist antide. *American Journal of Physiology-Endocrinology and Metabolism* 271: E775–E787.
39. Finnerty, N.J., S.L. O'Riordan, J.P. Lowry, M. Cloutier, and P. Wellstead. 2013. Continuous real-time in vivo measurement of cerebral nitric oxide supports theoretical predictions of an irreversible switching in cerebral ROS after sufficient exposure to external toxins. *Journal of Parkinson's Disease* 3: 351–362.
40. Foteinou, P.T., S.E. Calvano, S.F. Lowry, and I.P. Androulakis. 2009. Modeling endotoxin-induced systemic inflammation using an indirect response approach. *Mathematical Biosciences* 217: 27–42.
41. Francheteau, P., J.-L. Steimer, C. Dubray, and D. Lavene. 1991. Mathematical model forin vivo pharmacodynamics integrating fluctuation of the response: Application to the prolactin suppressant effect of the dopaminomimetic drug DCN 203–922. *Journal of Pharmacokinetics and Biopharmaceutics* 19: 287–309.

42. Garfinkel, A., Y.-H. Kim, O. Voroshilovsky, Z. Qu, J.R. Kil, M.-H. Lee, H.S. Karagueuzian, J.N. Weiss, and P.-S. Chen. 2000. Preventing ventricular fibrillation by flattening cardiac restitution. *Proceedings of the National Academy of Sciences* 97: 6061–6066.
43. Garfinkel, A., M.L. Spano, W.L. Ditto, and J.N. Weiss. 1992. Controlling cardiac chaos. *Science* 257: 1230–1235.
44. Gobburu, J.V., H. Agersø, W.J. Jusko, and L. Ynddal. 1999. Pharmacokinetic-pharmacodynamic modeling of ipamorelin, a growth hormone releasing peptide, in human volunteers. *Pharmaceutical Research* 16: 1412–1416.
45. Goldberger, A.L. 1991. Is the normal heartbeat chaotic or homeostatic? *News in Physiological Sciences* 6: 87–91.
46. Goldberger, A.L., L.A. N Amaral, J.M. Hausdorff, P. Ch Ivanov, C. Peng, and H. Eugene Stanley. 2002. Self-organized complexity in the physical, biological, and social sciences. *Proceedings of the National Academy of Sciences* 99: 2466–2472.
47. Goldberger, A.L., R. Shabetai, V. Bhargava, B.J. West, and A.J. Mandell. 1984. Nonlinear dynamics, electrical alternans, and pericardial tamponade. *American Heart Journal* 107: 1297–1299.
48. Goodwin, B.C. 1965. Oscillatory behavior in enzymatic control processes. *Advances in Enzyme Regulation* 3: 425–437.
49. Hazra, A., W. Krzyzanski, and W.J. Jusko. 2006. Mathematical assessment of properties of precursor-dependent indirect pharmacodynamic response models 1. *Journal of Pharmacokinetics and Pharmacodynamics* 33: 683–717.
50. Hellman, L., F. Nakada, J. Curti, E.D. Weitzman, J. Kream, H. Roffwarg, S. Ellman, D.K. Fukushima, and T.F. Gallagher. 1970. Cortisol is secreted episodically by normal man. *The Journal of Clinical Endocrinology and Metabolism* 30: 411–422.
51. Holland, J.H. 2006. Studying complex adaptive systems. *Journal of Systems Science and Complexity* 19: 1–8.
52. Huikuri, H.V., T.H. Mäkikallio, and J. Perkiömäki. 2003. Measurement of heart rate variability by methods based on nonlinear dynamics. *Journal of Electrocardiology* 36 (Suppl): 95–99.
53. Ilias, I., A.N. Vgontzas, A. Provata, and G. Mastorakos. 2002. Complexity and non-linear description of diurnal cortisol and growth hormone secretory patterns before and after sleep deprivation. *Endocrine Regulations* 36: 63–72.
54. Isaacson, R.L., J.A. Varner, J.M. Baars, and D. De Wied. 1995. The effects of pregnenolone sulfate and ethylestrenol on retention of a passive avoidance task. *Brain Research* 689: 79–84.
55. Ishizuka, S., and H. Hayashi. 1998. Spontaneous epileptiform bursts and long-term potentiation in rat CA3 hippocampal slices induced by chaotic stimulation of mossy fibers. *Brain Research* 790: 108–114.
56. Jeong, J. 2004. EEG dynamics in patients with Alzheimer's disease. *Clinical Neurophysiology* 115: 1490–1505.
57. Jeong, J., D.J. Kim, J.H. Chae, S.Y. Kim, H.J. Ko, and I.H. Paik. 1998. Nonlinear analysis of the EEG of schizophrenics with optimal embedding dimension. *Medical Engineering & Physics* 20: 669–676.
58. Karagueuzian, H.S., B.Y. Kogan, S.S. Khan, T.A. Denton, W.J. Karplus, W.J. Mandel, and G.A. Diamond. 1992. Induction of cellular chaos during quinidine toxicity. Predictive power of nonlinear dynamic analysis for drug-induced proarrhythmia–a hypothesis. *Journal of Electrocardiology* 24 (Suppl): 91–96.
59. Kawczyk-Krupka, A., A. Sieroń, and M. Adamek. 1998. Chaotic dynamics in medicine. *Wiadomości Lekarskie* 51: 525–530.
60. Kraan, G.P.B., R.P.F. Dullaart, J.J. Pratt, B.G. Wolthers, N.M. Drayer, and R. de Bruin. 1998. The daily cortisol production reinvestigated in healthy men. The serum and urinary cortisol production rates are not significantly different. *The Journal of Clinical Endocrinology and Metabolism* 83: 1247–1252.
61. Kuznetsov, I.A., and A.V. Kuznetsov. 2016. What can trigger the onset of Parkinson's disease – A modeling study based on a compartmental model of α-synuclein transport and aggregation in neurons. *Mathematical Biosciences* 278: 22–29.

62. Kuznetsov, I.A., and A.V. Kuznetsov. 2016. Mathematical models of α-synuclein transport in axons. *Computer Methods in Biomechanics and Biomedical Engineering* 19: 515–526.
63. Lalonde, R., J. Gaudreault, D. Karhu, and T. Marriott. 1999. Mixed-effects modeling of the pharmacodynamic response to the calcimimetic agent R–568. *Clinical Pharmacology and Therapeutics* 65: 40–49.
64. Larter, R., B. Speelman, and R.M. Worth. 1999. A coupled ordinary differential equation lattice model for the simulation of epileptic seizures. *Chaos: An Interdisciplinary Journal of Nonlinear Science* 9: 795–804.
65. Lehnertz, K. 1999. Non-linear time series analysis of intracranial EEG recordings in patients with epilepsy – An overview. *International Journal of Psychophysiology* 34: 45–52.
66. Lenbury, Y., and P. Pacheenburawana. 1991. Modelling fluctuation phenomena in the plasma cortisol secretion system in normal man. *Biosystems* 26: 117–125.
67. Levy, G. 1998. Predicting effective drug concentrations for individual patients. *Clinical Pharmacokinetics* 34: 323–333.
68. Liu, B.Z., and G.M. Deng. 1991. An improved mathematical model of hormone secretion in the hypothalamo-pituitary-gonadal axis in man. *Journal of Theoretical Biology* 150: 51–58.
69. Londergan, C.H., and E. Peacock-López. 1998. Dynamic model of hormonal systems coupled by negative feedback. *Biophysical Chemistry* 73: 85–107.
70. Lotharius, J., and P. Brundin. 2002. Pathogenesis of parkinson's disease: Dopamine, vesicles and α-synuclein. *Nature Reviews. Neuroscience* 3: 932–942.
71. Macheras, P., and A. Iliadis. 2006. *Modeling in biopharmaceutics, pharmacokinetics, and pharmacodynamics homogeneous and heterogeneous approaches*. New York: Springer Science+Business Media, Inc.
72. Mackey, M., and L. Glass. 1977. Oscillation and chaos in physiological control systems. *Science (80-)* 197: 287–289.
73. Mandema, J.W., and M. Danhof. 1992. Electroencephalogram effect measures and relationships between pharmacokinetics and pharmacodynamics of centrally acting drugs. *Clinical Pharmacokinetics* 23: 191–215.
74. Di Mascio, M., G. Di Giovanni, V. Di Matteo, and E. Esposito. 1999. Decreased chaos of midbrain dopaminergic neurons after serotonin denervation. *Neuroscience* 92: 237–243.
75. Mavroudis, P.D., S.A. Corbett, S.E. Calvano, and I.P. Androulakis. 2014. Mathematical modeling of light-mediated HPA axis activity and downstream implications on the entrainment of peripheral clock genes. *Physiological Genomics* 46: 766–778.
76. Mavroudis, P.D., S.A. Corbett, S.E. Calvano, and I.P. Androulakis. 2015. Circadian characteristics of permissive and suppressive effects of cortisol and their role in homeostasis and the acute inflammatory response. *Mathematical Biosciences* 260: 54–64.
77. Mavroudis, P.D., J.D. Scheff, S.E. Calvano, and I.P. Androulakis. 2013. Systems biology of circadian-immune interactions. *Journal of Innate Immunity* 5: 153–162.
78. Mavroudis, P.D., J.D. Scheff, S.E. Calvano, S.F. Lowry, and I.P. Androulakis. 2012. Entrainment of peripheral clock genes by cortisol. *Physiological Genomics* 44: 607–621.
79. Mavroudis, P.D., J.D. Scheff, J.C. Doyle, Y. Vodovotz, and I.P. Androulakis. 2019. The impact of stochasticity and its control on a model of the inflammatory response. *Computation* 7: 3.
80. Meyer-Hermann, M., M.T. Figge, and R.H. Straub. 2009. Mathematical modeling of the circadian rhythm of key neuroendocrine-immune system players in rheumatoid arthritis: A systems biology approach. *Arthritis and Rheumatism* 60: 2585–2594.
81. Milton, J., and D. Black. 1995. Dynamic diseases in neurology and psychiatry. *Chaos: An Interdisciplinary Journal of Nonlinear Science* 5: 8–13.
82. Mormont, M.C., and F. Lévi. 1997. Circadian-system alterations during cancer processes: A review. *International Journal of Cancer* 70: 241–247.
83. Movin-Osswald, G., and M. Hammarlund-Udenaes. 1995. Prolactin release after remoxipride by an integrated pharmacokinetic-pharmacodynamic model with intra-and interindividual aspects. *The Journal of Pharmacology and Experimental Therapeutics* 274: 921–927.
84. Murray, J.D. 2002. *Mathematical biology. I. An introduction*. New York: Springer.
85. Nathan, C. 2002. Points of control in inflammation. *Nature* 420: 846–852.

86. Nguyen, T.T., S.E. Calvano, S.F. Lowry, and I.P. Androulakis. 2013. An agent-based model of cellular dynamics and circadian variability in human endotoxemia. *PLoS One* 8: e55550.
87. Nguyen, T.T., P.T. Foteinou, S.E. Calvano, S.F. Lowry, and I.P. Androulakis. 2011. Computational identification of transcriptional regulators in human endotoxemia. *PLoS One* 6: e18889.
88. Nieman, G., D. Brown, J. Sarkar, B. Kubiak, C. Ziraldo, J. Dutta-Moscato, C. Vieau, D. Barclay, L. Gatto, K. Maier, G. Constantine, T.R. Billiar, R. Zamora, Q. Mi, S. Chang, and Y. Vodovotz. 2012. A two-compartment mathematical model of endotoxin-induced inflammatory and physiologic alterations in swine. *Critical Care Medicine* 40: 1052–1063.
89. Ouzounoglou, E., D. Kalamatianos, E. Emmanouilidou, M. Xilouri, L. Stefanis, K. Vekrellis, and E.S. Manolakos. 2014. In silico modeling of the effects of alpha-synuclein oligomerization on dopaminergic neuronal homeostasis. *BMC Systems Biology* 8: 54.
90. Papavasiliou, S.S., T. Brue, P. Jaquet, and E. Castanas. 1995. Pattern of prolactin diurnal secretion in normal humans; Evidence for nonlinear dynamics. *Neuroendocrinology* 62: 444–453.
91. Peng, C.K., S. Havlin, J.M. Hausdorff, J.E. Mietus, H.E. Stanley, and A.L. Goldberger. 1995. Fractal mechanisms and heart rate dynamics. Long-range correlations and their breakdown with disease. *Journal of Electrocardiology* 28 (Suppl): 59–65.
92. Perkiömäki, J.S., T.H. Mäkikallio, and H.V. Huikuri. 2005. Fractal and complexity measures of heart rate variability. *Clinical and Experimental Hypertension* 27: 149–158.
93. Pillai, N., M. Craig, A. Dokoumetzidis, S.L. Schwartz, R. Bies, and I. Freedman. 2018. Chaos synchronization and Nelder-Mead search for parameter estimation in nonlinear pharmacological systems: Estimating tumor antigenicity in a model of immunotherapy. *Progress in Biophysics and Molecular Biology* 139: 23–30.
94. Pillai, N., S.L. Schwartz, T. Ho, A. Dokoumetzidis, R. Bies, and I. Freedman. 2019. Estimating parameters of nonlinear dynamic systems in pharmacology using chaos synchronization and grid search. *Journal of Pharmacokinetics and Pharmacodynamics* 46: 193–210.
95. Pincus, S.M. 1991. Approximate entropy as a measure of system complexity. *Proceedings of the National Academy of Sciences* 88: 2297–2301.
96. Plotegher, N., and M.R. Duchen. 2017. Crosstalk between lysosomes and mitochondria in Parkinson's disease. *Frontiers in Cell and Development Biology* 5: 110.
97. Poliquin, P.O., J. Chen, M. Cloutier, L.-É. Trudeau, and M. Jolicoeur. 2013. Metabolomics and in-Silico analysis reveal critical energy deregulations in animal models of Parkinson's disease. *PLoS One* 8: e69146.
98. Prank, K., H. Harms, M. Dammig, G. Brabant, F. Mitschke, and R.D. Hesch. 1994. Is there low-dimensional chaos in pulsatile secretion of parathyroid hormone in normal human subjects? *The American Journal of Physiology: Endocrinology and Metabolism* 266: E653–E658.
99. Prince, J.M., R.M. Levy, J. Bartels, A. Baratt, J.M. Kane, C. Lagoa, J. Rubin, J. Day, J. Wei, M.P. Fink, S.M. Goyert, G. Clermont, T.R. Billiar, Y. Vodovotz, and Y. Vodovotz. 2006. In silico and in vivo approach to elucidate the inflammatory complexity of CD14-deficient mice. *Molecular Medicine* 12: 88–96.
100. Proctor, C.J., P.J. Tangeman, and H.C. Ardley. 2010. Modelling the role of UCH-L1 on protein aggregation in age-related Neurodegeneration. *PLoS One* 5: e13175.
101. Proctor, C.J., M. Tsirigotis, and D.A. Gray. 2007. An in silico model of the ubiquitin-proteasome system that incorporates normal homeostasis and age-related decline. *BMC Systems Biology* 1: 17.
102. Qi, Z., G.W. Miller, and E.O. Voit. 2008. Computational systems analysis of dopamine metabolism. *PLoS One* 3: e2444.
103. Qi, Z., G.W. Miller, and E.O. Voit. 2009. Computational analysis of determinants of dopamine (DA) dysfunction in DA nerve terminals. *Synapse* 63: 1133–1142.
104. Qi, Z., G.W. Miller, and E.O. Voit. 2014. Rotenone and paraquat perturb dopamine metabolism: a computational analysis of pesticide toxicity. *Toxicology* 315: 92–101.
105. Qu, Z., J.N. Weiss, and A. Garfinkel. 1999. Cardiac electrical restitution properties and stability of reentrant spiral waves: a simulation study. *American Journal of Physiology. Heart and Circulatory Physiology* 276: H269–H283.

106. Rohatagi, S., A. Bye, A.E. Mackie, and H. Derendorf. 1996. Mathematical modeling of cortisol circadian rhythm and cortisol suppression. *European Journal of Pharmaceutical Sciences* 4: 341–350.
107. Röschke, J., and J.B. Aldenhoff. 1992. A Nonlinear approach to brain function: deterministic Chaos and sleep EEG. *Sleep* 15: 95–101.
108. van Rossum, J.M., and J.E.G.M. de Bie. 1989. Systems dynamics in clinical pharmacokinetics. *Clinical Pharmacokinetics* 17: 27–44.
109. ———. 1991. Chaos and illusion. *Trends in Pharmacological Sciences* 12: 379–383.
110. van Rossum, J.M., J.E.G.M. de Bie, G. van Lingen, and H.W.A. Teeuwen. 1989. Pharmacokinetics from a dynamical systems point of view. *Journal of Pharmacokinetics and Biopharmaceutics* 17: 365–392.
111. Sass, M.B., A.N. Lorenz, R.L. Green, and R.A. Coleman. 2009. A pragmatic approach to biochemical systems theory applied to an α-synuclein-based model of Parkinson's disease. *Journal of Neuroscience Methods* 178: 366–377.
112. Scheff, J.D., S.E. Calvano, S.F. Lowry, and I.P. Androulakis. 2010. Modeling the influence of circadian rhythms on the acute inflammatory response. *Journal of Theoretical Biology* 264: 1068–1076.
113. Scheff, J.D., P.D. Mavroudis, S.E. Calvano, S.F. Lowry, and I.P. Androulakis. 2011. Modeling autonomic regulation of cardiac function and heart rate variability in human endotoxemia. *Physiological Genomics* 43: 951–964.
114. Scheinin, H., A. Helminen, S. Huhtala, P. Grönroos, J.A. Bosch, T. Kuusela, J. Kanto, and T. Kaila. 1999. Spectral analysis of heart rate variability as a quantitative measure of parasympatholytic effect–integrated pharmacokinetics and pharmacodynamics of three anticholinergic drugs. *Therapeutic Drug Monitoring* 21: 141–151.
115. Schwartz, M.D., C. Wotus, T. Liu, W.O. Friesen, J. Borjigin, G.A. Oda, and H.O. de la Iglesia. 2009. Dissociation of circadian and light inhibition of melatonin release through forced desynchronization in the rat. *Proceedings of the National Academy of Sciences of the United States of America* 106: 17540–17545.
116. Sekula, L.K., J.F. Lucke, E.K. Heist, R.K. Czambel, and R.T. Rubin. 1997. Neuroendocrine aspects of primary endogenous depression. XV: Mathematical modeling of nocturnal melatonin secretion in major depressives and normal controls. *Psychiatry Research* 69: 143–153.
117. Sephton, S., and D. Spiegel. 2003. Circadian disruption in cancer: a neuroendocrine-immune pathway from stress to disease? *Brain, Behavior, and Immunity* 17: 321–328.
118. Sharma, A., W.F. Ebling, and W.J. Jusko. 1998. Precursor-dependent indirect pharmacodynamic response model for tolerance and rebound phenomena. *Journal of Pharmaceutical Sciences* 87: 1577–1584.
119. Silva, C., I.R. Pimentel, A. Andrade, J.P. Foreid, and E. Ducla-Soares. 1999. Correlation dimension maps of EEG from epileptic absences. *Brain Topography* 11: 201–209.
120. Smith, W.R. 1980. Hypothalamic regulation of pituitary secretion of luteinizing hormone. II. Feedback control of gonadotropin secretion. *Bulletin of Mathematical Biology* 42: 57–78.
121. Sneppen, K., L. Lizana, M.H. Jensen, S. Pigolotti, and D. Otzen. 2009. Modeling proteasome dynamics in Parkinson's disease. *Physical Biology* 6: 036005.
122. St Hilaire, M.A., C. Gronfier, J.M. Zeitzer, and E.B. Klerman. 2007. A physiologically based mathematical model of melatonin including ocular light suppression and interactions with the circadian pacemaker. *Journal of Pineal Research* 43: 294–304.
123. Stam, K.J., D.L. Tavy, B. Jelles, H.A. Achtereekte, J.P. Slaets, and R.W. Keunen. 1994. Non-linear dynamical analysis of multichannel EEG: Clinical applications in dementia and Parkinson's disease. *Brain Topography* 7: 141–150.
124. Stanley, H., L. Amaral, A. Goldberger, S. Havlin, P.C. Ivanov, and C. Peng. 1999. Statistical physics and physiology: Monofractal and multifractal approaches. *Physica A* 270: 309–324.
125. Starmer, C.F., D.N. Romashko, R.S. Reddy, Y.I. Zilberter, J. Starobin, A.O. Grant, and V.I. Krinsky. 1995. Proarrhythmic response to potassium channel blockade. *Circulation* 92: 595–605.

126. Stefano, G.B., N. Pilonis, R. Ptacek, J. Raboch, M. Vnukova, and R.M. Kream. 2018. Gut, microbiome, and brain regulatory axis: Relevance to neurodegenerative and psychiatric disorders. *Cellular and Molecular Neurobiology* 38: 1197–1206.
127. Stevens, J., B.A. Ploeger, M. Hammarlund-Udenaes, G. Osswald, and P.H. Van Der Graaf. 2012. Mechanism-based PK-PD model for the prolactin biological system response following an acute dopamine inhibition challenge: Quantitative extrapolation to humans. *Journal of Pharmacokinetics and Pharmacodynamics* 39: 463–477.
128. Sugihara, G., W. Allan, D. Sobel, and K.D. Allan. 1996. Nonlinear control of heart rate variability in human infants. *Proceedings of the National Academy of Sciences of the United States of America* 93: 2608–2613.
129. Tolić, I.M., E. Mosekilde, and J. Sturis. 2000. Modeling the insulin–glucose feedback system: the significance of pulsatile insulin secretion. *Journal of Theoretical Biology* 207: 361–375.
130. Topp, B., K. Promislow, G. Devries, R.M. Miura, and D.T. Finegood. 2000. A model of β -cell mass, insulin, and glucose kinetics: Pathways to diabetes. *Journal of Theoretical Biology* 206: 605–619.
131. Torres, A., T. Bentley, J. Bartels, J. Sarkar, D. Barclay, R. Namas, G. Constantine, R. Zamora, J.C. Puyana, and Y. Vodovotz. 2009. Mathematical modeling of posthemorrhage inflammation in mice: Studies using a novel, computer-controlled, closed-loop hemorrhage apparatus. *Shock* 32: 172–178.
132. Tracey, K.J. 2002. The inflammatory reflex. *Nature* 420: 853–859.
133. Veldhuis, J.D., and S.M. Pincus. 1998. Orderliness of hormone release patterns: A complementary measure to conventional pulsatile and circadian analyses. *European Journal of Endocrinology* 138: 358–362.
134. Vodovotz, Y. 2010. Translational systems biology of inflammation and healing. *Wound Repair and Regeneration* 18: 3–7.
135. Voss, A., S. Schulz, R. Schroeder, M. Baumert, and P. Caminal. 2009. Methods derived from nonlinear dynamics for analysing heart rate variability. *Philosophical Transactions. Series A, Mathematical, Physical, and Engineering Sciences* 367: 277–296.
136. Wagner, C.D., B. Nafz, and P.B. Persson. 1996. Chaos in blood pressure control. *Cardiovascular Research* 31: 380–387.
137. Wagner, C.D., and P.B. Persson. 1998. Chaos in the cardiovascular system: An update. *Cardiovascular Research* 40: 257–264.
138. Wagner, C.D., P.B. Persson, and P.B. Persson Nonlinear. 1995. Nonlinear chaotic dynamics of arterial pressure and renal blood flow. *American Journal of Physics* 268: H621–H627.
139. Weiss, J.N., A. Garfinkel, H.S. Karagueuzian, Z. Qu, and P.S. Chen. 1999. Chaos and the transition to ventricular fibrillation: A new approach to antiarrhythmic drug evaluation. *Circulation* 99: 2819–2826.
140. Wellstead, P., and M. Cloutier. 2011. An energy systems approach to Parkinson's disease. *Wiley Interdisciplinary Reviews. Systems Biology and Medicine* 3: 1–6.
141. Whitacre, J.M. 2010. Degeneracy: a link between evolvability, robustness and complexity in biological systems. *Theoretical Biology & Medical Modelling* 7: 6.
142. Wolkenhauer, O. 2001. Systems biology: The reincarnation of systems theory applied in biology? *Briefings in Bioinformatics* 2: 258–270.
143. Yambe, T., S. Nitta, T. Sonobe, S. Naganuma, S. Kobayashi, Y. Haga, M. Tanaka, T. Fukuju, M. Miura, and N. Sato. 1993. Identification of the deterministic chaos in cardiovascular dynamics by the use of the non-linear mathematics. *The Science Reports of the Research Institutes, Tohoku University Medicine* 39: 1–5.
144. Yang, Q., S.E. Calvano, S.F. Lowry, and I.P. Androulakis. 2011. A dual negative regulation model of Toll-like receptor 4 signaling for endotoxin preconditioning in human endotoxemia. *Mathematical Biosciences* 232: 151–163.
145. Zhang, J., M.L. Culp, J.G. Craver, and V. Darley-Usmar. 2018. Mitochondrial function and autophagy: Integrating proteotoxic, redox, and metabolic stress in Parkinson's disease. *Journal of Neurochemistry* 144: 691–709.

The Efficiency of Multi-target Drugs: A Network Approach

Lucas N. Alberca and Alan Talevi

Modern drug discovery paradigms have experienced considerable changes from the early XXth Century to the present. After initial discoveries based on serendipitous findings or exploitation of traditional folk knowledge, drug discovery efforts embraced systematic screening approximations (phenotypic or physiology-based screening, which at first relied on *in vivo* models but later progressed to *in vitro* primary screens) [1, 2]. From 1970s, though, target-driven drug discovery gained momentum until becoming the dominant paradigm for more than two decades. The interest in target- focused drug discovery has gone hand-in-hand with the development of pertinent technologies [1]: intensive enzymology research during 1950s and 1960s, extension of enzyme kinetics methodology to receptors in 1970s, technological advances on molecular biology in 1980s (e.g. recombinant DNA technology) and miniaturization and automation (high-throughput screening) in 1990s.

The premise behind target-oriented discovery is tempting: exquisitely selective ("clean") drugs ("magic bullets") potently inhibiting one and only one target of interest would be safer due to avoidance of off-target interactions. Such strategy is also attractive from a drug design perspective since it allows the definition of structure- based ("rational") drug discovery programs. After validation of a relevant, disease- modifying target, and provided that its structure is known or can be appropriately modelled, specific modulators displaying complementary features to the binding site/s may be investigated. Classic Quantitative Structure-Activity Relationships (QSAR) and the pharmacophore notion also arose in the context of a strong target-focused paradigm and are focused in (indirectly) mapping the ligand-target interaction [3, 4].

L. N. Alberca · A. Talevi (✉)
Laboratory of Bioactive Research and Development (LIDeB), Faculty of Exact Sciences, University of La Plata (UNLP), Buenos Aires, Argentina

Argentinean National Council of Scientific and Technical Research (CONICET), Buenos Aires, Argentina

M. Bizzarri (ed.), *Approaching Complex Diseases*, Human Perspectives in Health Sciences and Technology 2, https://doi.org/10.1007/978-3-030-32857-3_3

From a more pragmatic viewpoint, target-based screening platforms are faster, easier, and less costly to develop and run than phenotypic ones [5].

However, in many aspects, target-based drug discovery has failed to deliver. First, the highly selective agents proposed by such approximation might be too ambitious a goal to meet: drugs can produce surprising pharmacology not initially anticipated and even drugs developed under the "one target" paradigm often show a multi-target nature and off-target events [5]. Second, "the one disease, one gene, one drug" strategy might result over-reductionist to address complex conditions, such as mood disorders or neurodegenerative conditions [6, 7]. The physiological states of living organisms are often resilient to perturbation due to genetic redundancy and compensatory mechanisms [8], and a similar principle could apply for certain persistent physio-pathological states characterized by multiple (sometimes simultaneous, sometimes progressive) perturbations (in fact, it has been suggested that focus should shift from seeking disease-causing genes to disease-causing networks, see Chen et al. 2008 [9]). Several network structural motifs and statistical topological properties have been reported to be positively associated with the robustness of biological networks [10, 11]. The attention of the scientific community has been particularly drawn toward evidence that the organization of biological systems into a scale-free network (SFN) is a major contributor to biological robustness. Excessive attention to target-driven discovery has been signalled as a possible cause of declining productivity in the pharmaceutical industry [12], and a substantial proportion of recently approved drugs have been found using a phenotypic approach [13]. At last, drug resistance issues due to intrinsic or acquired alterations of the target, reactivation of the signalling pathway or bypass mechanisms are more likely to occur when using single-target therapies [14–16].

From the early 2000s, a complementary approach in drug discovery, the *network pharmacology* paradigm, has been introduced, in line with the systems biology philosophy. Such paradigm is expressed in the renewed interest in phenotypic screening [1], in the use of network topology analysis to propose new drug targets [17] and, prominently, in the increasing attention to polypharmacology [18] as a therapeutic strategy to treat complex conditions or prevent or solve drug-resistance issues. The notion of polypharmacology contemplates a single drug acting on multiple targets or multiple drugs acting on distinct targets. The former option is expected to deliver therapeutic options with simplified pharmacokinetics, reduced chances of drug interactions and improved treatment adherence. Polypharmacology of known drugs may also be exploited for drug repositioning programs.

The design of tailored multi-target agents may combine the advantages of network pharmacology with the rationality of structure- and computer-aided drug discovery. Under the light of this new paradigm, we will also challenge the validity of some common assumptions of target-driven discovery, and we will review some particular considerations that should be taken into account when designing this type of therapeutic agents.

The More Potent, the Better?

Under the single target-oriented paradigm, the search of high affinity ligands is doubtlessly one of the guiding principles behind drug discovery approaches. The general procedure in drug discovery involves the discovery of an initial hit (chemical start point for a drug discovery project) that typically displays potency in the high nM or low μM range, and which will be then subjected to hit-to-lead optimization to increase its potency to the low nM range [19]. The common assumption that potency should guide drug design was first challenged back in the early 1990s, when the pharmaceutical industry realized that seeking for more potent compounds was detrimental for the bioavailability and safety of the resulting drug candidates, leading to high ADMET-related attrition rates in clinical trials [20, 21]. The multiparametric nature of the drug discovery process is now well-accepted within the pharmaceutical community, which now prefers addressing the multiple determinant parameters of a (successful) drug in parallel rather than sequentially [3, 22]. Lusher and co-workers have cleverly compared the drug discovery and development process to solving a Rubik's cube, where each face represents a required parameter and changing one face will often negatively affect other; solving the puzzle might require sacrificing an already completed face, i.e. taking a local step back in favour of a global improvement [23]. In fact, it may be argued that the best practical solution will usually involve a compromise: accepting a local suboptimal result to approach a global optimal.

However, network theory (and a few empirical examples that we will briefly review) sheds light on more reasons not to overemphasize the importance of affinity and potency. The first question we will probably need to answer when deciding which molecular target/s we will pursue is whether we want to achieve a lethal effect (e.g. on a cancerous cell or a pathogen) or if we must adjust one or more molecular processes to restore a perturbed physiological state [17]. Central nodes, such as hubs or inter- modular overlaps and bridges (with emphasis on choke points/bottlenecks, i.e. nodes with high betweenness centrality) are highly efficient mediators of perturbations [24, 25] and may be considered as interesting targets when looking for deleterious effects (*central hit strategy*). In that case, potent impairment of such nodes mediated by high- affinity and high-efficacious drugs would be required. In any case, hitting several of these nodes simultaneously would decrease the occurrence of resistance issues; furthermore, even in that circumstances it has been shown that *partial knockout* or *attenuation* of a small number of targets (e.g. 3–5) may produce a larger effect than the complete knockout of a single target [26]. In contrast, the *network influence strategy* uses systems-level knowledge to find drug targets for complex disorders, which are presumed to have rigid networks, in order to restore well-functioning [25]. Targeting central nodes here may lead to side-effects and toxicity. The network influence strategy would then target nodes which are neither hubs nor central nodes but occupy strategic disease-relevant network positions able to influence central nodes. A recent retrospective study showed that drug targets are generally better spreaders of perturbations than non-target proteins, and that cancer-related and metabolic disease-related proteins were shown to have an average network distance to the related drug targets of 2.3 and ~5 network edges [27]. This agrees with the notion that rapidly proliferating cells, like those in cancer, are attacked at their central proteins, while differentiated cells, such as those involved in type-2 diabetes, are attacked at the neighbours of central proteins.

Efficient restoration of the healthy state might be accomplished by a simultaneous attack on many proteins, wherein the targeting efficiency on each of them may only be partial. Via low affinity binding (or, occasionally, partial agonists), multi-target drugs may avoid the frequent dual-trap of drug-resistance and toxicity [25, 28]. This has been clearly stated by Bianchi and co-workers when discussing innovative hypothesis for new, safer antiepileptic medications, as they stated that "the complexity of neural processes underlying seizure activity may be more amenable to multiple small perturbations than a single dominant mechanism" [29].

Precisely, a couple of interesting counterexamples to common thinking on the invariable importance of developing high affinity and high efficacy drugs can be found in the field of neurology. Memantine is one among few currently prescribed treatments of moderate-to-severe Alzheimer's disease and other types of dementia; it is prescribed when acetylcholinesterase inhibitors are not tolerated. It was the first N-methyl-D-aspartate receptors (NMDAR) antagonist approved for Alzheimer; several others had been developed based on the conventional paradigm but failed in advanced clinical trials in large part because of unacceptable side effects, including cognitive impairment and Olney's lesions [30, 31]. Memantine displays surprising low- affinity binding to NMDARs (high nM to low μM range), fast on/off kinetics, and almost no selectivity among NMDARs subtypes, being much better tolerated than high affinity compounds with similar pharmacologic profiles. It also shows uncompetitive antagonism on other receptors, including type 3 serotonin [32], dopamine D2 [33] and nicotinic [34] receptors.

Another interesting example is the antiepileptic drug levetiracetam, which has low affinity for its more frequently ascribed target, the synaptic vesicle protein SV2A, but also a complex pharmacology which includes low affinity for high voltage activated calcium channels [35]. Its recently approved high affinity, highly selective analogue, brivaracetam (clearly developed under the target-driven paradigm) has enhanced affinity and selectivity to SV2a (in the low nM range) [36]. It is intriguing whether brivaracetam would prove to be more efficacious and/or safer in the clinical scenario than levetiracetam. Meta-analysis studies and indirect comparisons have (very preliminarily) suggested brivaracetam is neither safer nor more efficacious than levetiracetam, and even that levetiracetam might be better tolerated [37, 38]. However, the methodologic approach to reach such conclusions has already been criticized and subjected to debate [39, 40] and we should wait for future studies to extract a more valid conclusion from the levetiracetam versus brivaracetam comparison.

A third example comes from the field of antipsychotics, where initial therapeutic approach involving inhibition of dopamine receptors as a primary target was associated with extrapyramidal motor symptoms. Side effects were considerably reduced by subsequent inclusion of anti-serotonergic activity, which also improved therapeutic efficacy [41]. Later, the modulation of efficacies on different targets involving partial and inverse agonism enabled functional selectivity on multiple receptors, improving response rates and tolerability [42].

Balancing Affinities

One of the more significant challenges of multi-target drugs involves attaining an optimal balance of their affinities to each of the pursued targets [43]; in other words, the potency ratio of the multiple ligand on its targets should be optimized, so that both targets are modulated to an appropriate (disease-modifying) degree in vivo at similar plasma or brain concentrations. The key here is the meaning of "an appropriate degree". In many available literature examples, the goal is to obtain in vitro activities within an order of magnitude of each other, under the assumption that this will lead to similar levels of receptor occupancy in vivo [44]. However, this may not necessarily be the case.

Have in mind that the extent of modulation of a drug on a given target and the impact of such modulation on the phenotype depend not only on the affinity of the drug to the target but on a number of parameters including the number of copies of the target in the disease-relevant tissue/cell, the levels of the target physiologic substrate, the affinity of the physiologic substrate to the target, the in vivo exposure of relevant tissues to the therapeutic agent, and the architecture of the metabolic pathway/s in which the target participate. Often, in vivo models of the disease will be required to find the optimal balance in potency on the targets of interest [43, 44]. Alternatively, mathematical flux models can be employed to predict the extent to which each target must be modulated for producing the desired therapeutic effect. Metabolic control analysis and kinetic models are valuable, sometimes overlooked tools to assist choosing appropriate drug targets and target combinations, and to decide on the potency requirements on the selected target/s.

The previous discussion is closely related to both the way potential new targets are proposed at present and the fact that drug candidates are not always exhaustively characterized from a kinetic point of view [45] (for instance, the quantitative metrics that are usually used to prioritize lead compounds, such as the IC50 values from assays on the purified target or cell-based assays, are not able to fully account for time-dependent changes in target engagement in the dynamic environment of the human body).

How are potential drug targets discovered today? There is a diversity of approaches, but genetic-driven target identification and validation has been gaining increasing attention in the pharmaceutical field, and a very substantial proportion of the drug discovery projects with direct genetic support to drug mechanism is currently active or successful at Phase II clinical trials [46]. Genome-wide association study findings have become a particularly powerful approach to identify intermediate phenotypes controlling key checkpoints in disease pathogenesis that can be modulated therapeutically. Differential expression of genes between diseased and healthy tissue is commonly taken as an indication of the specific targets of disease pathways [47] (although of course, association does not necessarily mean causality). Modern molecular approaches such as those based on small interfering RNA and antisense oligonucleotides may also be employed to reveal unknown targets open to therapeutic intervention. The phenotypical effects of gene downregulation (e.g. gene suppression) are frequently hold as evidence of essentiality to support novel drug targets [48].

However, as Saavedra and colleagues state in their recent, inspiring review on metabolic control analysis, essentiality assessment should be complemented with other considerations. Pharmacological inhibition of an enzyme or transporter that takes part in a metabolic pathway of interest seldom reaches the high levels of specific suppression achieved by the genetic strategies. Furthermore, there is normally an excess of enzyme/transporter activity in comparison to the pathway flux. In general, the enzymes or transporters with higher control on the pathway fluxes will be those with the lowest maximal velocities. Control refers to the degree of influence that each element has on the flux and the metabolite concentrations of a metabolic pathway. Partial modulation of a key, highly controlling ("leader") nodes in a metabolic pathway may have a substantial influence on the metabolic flux. In contrast, almost complete inhibition of a low influencing ("follower") node would be required to significantly affect the pathway, otherwise the effect would be negligible. The former discussion illustrates the idea that not every essential gene product requires the same degree of modulation to attain the pursued output, and that, with adequate parametrization, flux analysis can guide optimization of affinity balance for multi-target agents. For instance, Yang et al. performed flux analysis of the arachidonic acid metabolic network in human polymorphous leukocytes and predicted the most promising target combinations for synergistic effects [49].

Target Compatibility from a Chemical Perspective

One of the central issues in target selection for multi-target drugs is the chemical feasibility of a multi-target agent, which critically depends on the specificity requirements of the target combination. i.e. whether the targets are compatible in terms of ligand accommodation [43].

Conceptually, the development of tailored multi-target agents relies on the methodical combination of pharmacophores from single-target ligands, which may vary from conjugating distinct pharmacophores through a linker, to partially or fully overlapped pharmacophores [50]. Figure 1 illustrates the different options and provides examples of each of them. The conjugated pharmacophore approximation is the one that will usually pose more difficulties to accommodate the ligand in the different targets, whenever a stable linker is used to join both active scaffolds. The use of such linkers will often lead to compounds of considerable size and unfavourable pharmacokinetics (e.g., compounds that violate Lipinski's rules and have poor absorption and/or biodistribution). Whereas that issues may be solved by the used of cleavable linkers, this also limits some of the merits of the single multi-

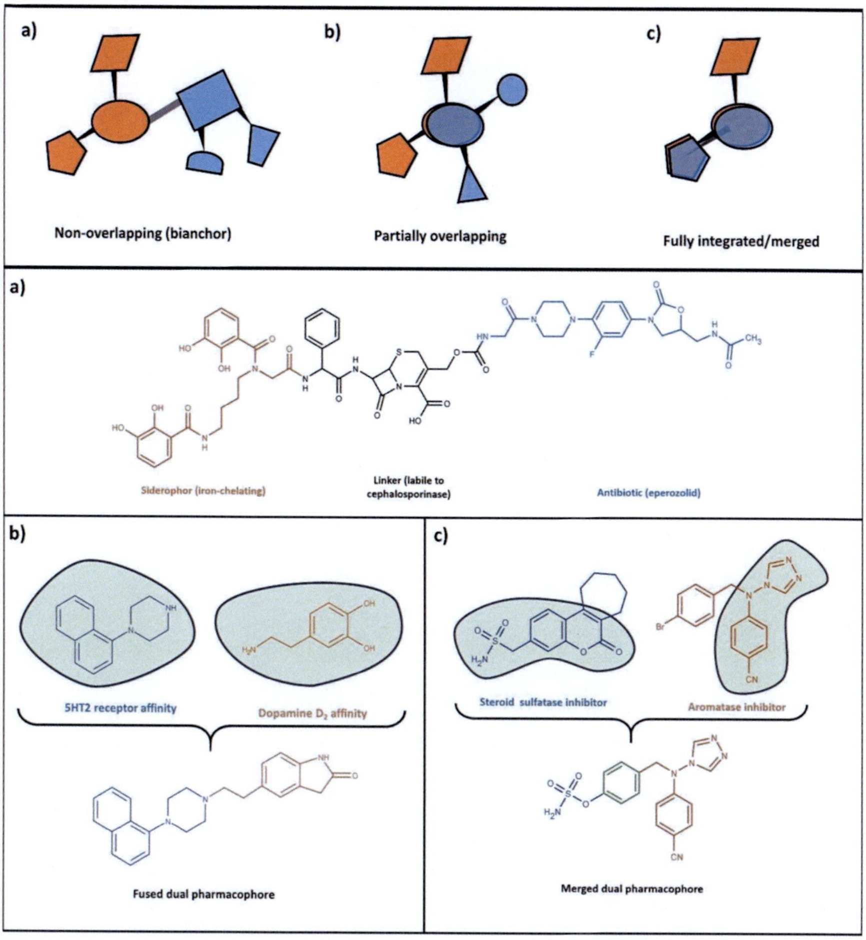

Fig. 1 Three different types of multi-target ligands, based on divergent (**a**) and partially and fully convergent pharmacophores (**b** and **c**, respectively). The examples were extracted from Morphy and Rankovic [60] and Proschak et al. [43]

target agent in comparison with combination therapies (e.g. reduced chances of drug interactions), although others are maintained (e.g. treatment adherence due to simplified posology). The fragment-based (conjugated pharmacophore) approach could also result in poor ligand efficiency metrics [51], which represent the binding efficiency per ligand atom. It might be speculated that, when engaged with one of the targets, only a part of this type of molecule will be interacting with such target, while the other part can become an obstacle for the binding event, reducing the binding efficiency (this is schematically represented in the awkward topology of the multi-blade key of Fig. 2, which resorts to the classic lock and key analogy to conceptualize multi-target agents). For non-cleavable linkers, length and geometry of the linker and attachment points to the pharmacophores will require special attention [52]; the position and characteristics of the linker may be guided by SAR

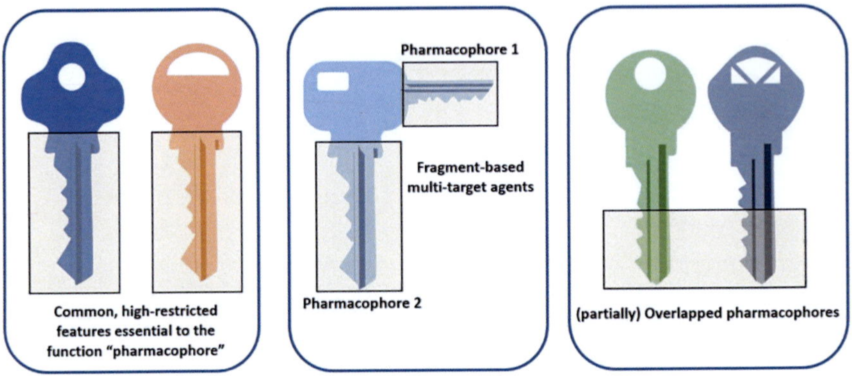

Fig. 2 Classic lock and key analogy adapted to multi-target agents

information on each separate pharmacophore. SAR on each individual target can also be systematically explored to identify substituent vectors that are tolerated by all targets of interest, as illustrated by a recent study on dual agents to counter non-alcoholic steatohepatitis [53]. It should be noted that the situation might once again resemble the Rubik's cube solution, as the extreme, best points of one SAR might exclude the best points of the other, and possibly a compromise solution should be selected.

The development of fused multiple-targeting agents emerging from partially or fully overlapping pharmacophores can be used to circumvent most of the limitations of conjugated multi-target agents. It can be intuitively realized that such approach requires the occurrence of similarities between the binding sites of the pursued targets. Whereas it may be thought that designing a dual-target ligand is easier for highly related targets (e.g. evolutionarily or functionally related, sharing substrate specificity or having a common chemical modulator), increasing evidence suggests that two unrelated targets might present binding site similarities [54, 55]. The identification of binding site similarities can be indirectly inferred from ligand similarities, as in the Similarity Ensemble Approach [56] or in a direct manner, using binding site comparison approximations [57].

It has also been suggested that chemical determinants of promiscuity might be tuned in favour of selective promiscuity, e.g. choosing chemotypes and scaffolds linked to a polypharmacology behaviour [58]. For instance, Hu and Bajorath have observed that specific chemical scaffolds seem to be related to particular sets of target families [59].

Some Considerations on Computational Approaches

Both structure- and ligand-based computational approximations may probe valuable to assist the rational discovery of multi-target agents. Strategies to obtain multiple ligands include virtual screening of compound libraries to identify potential binders simultaneously matching the structural requirements of all the targets of interest [61]; such in silico screen may include molecular docking, pharmacophore alignment or QSAR approaches. However, it has been discussed that, if looking for compounds for which affinity for a given pursued target is independent of affinity for the other ones (i.e. non-overlapping pharmacophores) the overall Positive Predictive Value (PPV, i.e. the probability of a predicted multi-target candidate confirming its activity against all predicted targets) will substantially decrease with the number of pursued targets, as the overall PPV would be the product of each individual PPV for a single target [12]. Accordingly, the use of overlapping pharmacophores would be also recommended from this perspective.

Alternatively, molecular docking and pharmacophore matching can also be used to design novel chemical multi-target entities.

In the particular case of using independent pharmacophores for each target, particular attention should be paid in this case to the definition of excluded volumes, in order to avoid clashes with the biological targets that lead to inactive compounds even though positive pharmacophore features are matched [62]. Alternatively, consensus pharmacophore models can also be developed to cover the common pharmacophore features of both targets of interest [63].

Conclusions

Network-based drug discovery integrates systems biology thinking, computational technology and omics data, providing a framework for the development of innovative therapeutic solutions. It challenges some deep-rooted concepts in the field of pharmaceutical research, e.g. the target-driven discovery and the notion that potent (nM) drug candidates should be pursued. Whereas these might still be valid concepts in some scenarios, they should be (and are being) revised, as strict adherence to existing paradigms could result in loosing valuable opportunities for improved therapeutics. The general idea here is not to discard previous ways of thinking, but to expand them and complement them with new ones, and to understand in which circumstances one might be more appropriate (i.e. lead to better results) than the others. In particular, the network strategy could be especially adequate when searching for therapeutic solutions for complex disorders (neurodegenerative conditions, mood disorders, epilepsies, metabolic diseases) and conditions prone to drug resistance development (i.e. cancer, infectious diseases).

Some of the more salient expressions of network pharmacology in the field of therapeutics involve the selection of target and target combinations that hold the greatest potential for pharmacological intervention (enhanced efficacy/safety balance); the renewed interest in phenotypic screening and; polypharmacology.

Polypharmacology is not a novel strategy (in fact, most herbal drugs and many old drugs discovered through phenotypic screening are multi-target agents); however, we are now in a position to approach it in a very rational manner, from choosing optimal target combinations to computer-aided drug design. As a matter of fact, tailored multi-target agents combine the best of the two precedent paradigms (phenotypic- and target-guided discovery), with certain specific concerns and peculiarities that have been overviewed throughout the chapter.

Network-pharmacology implies disregarding excessively reductionist approaches and realizing that we are not aiming at isolated targets, but we are targeting whole disease or physiologic pathways. Of particular note is the idea that essentiality, druggability and assayability are not the only points that define a potentially good drug target, but also the degree of control that it exerts on a given pathway in (dynamic) physiologic conditions.

References

1. Zheng, W., N. Thorne, and J. McKew. 2013. Phenotypic screens as a renewed approach for drug discovery. *Drug Discovery Today* 18 (21–22): 1067–1073.
2. Margineanu, D. 2014. Systems biology, complexity, and the impact on antiepileptic drug discovery. *Epilepsy & Behavior* 38: 131–142.
3. Cruz-Monteagudo, M., S. Schürer, E. Tejera, Y. Pérez-Castillo, J. Medina-Franco, A. Sánchez-Rodríguez, and F. Borges. 2017. Systemic QSAR and phenotypic virtual screening: chasing butterflies in drug discovery. *Drug Discovery Today* 22 (7): 994–1007.
4. Lo, Y., S. Rensi, W. Torng, and R. Altman. 2018. Machine learning in chemoinformatics and drug discovery. *Drug Discovery Today* 23 (8): 1538–1546.
5. Croston, G. 2017. The utility of target-based discovery. *Expert Opinion on Drug Discovery* 12 (5): 427–429.
6. Talevi, A. 2015. Multi-target pharmacology: Possibilities and limitations of the "skeleton key approach" from a medicinal chemist perspective. *Frontiers in Pharmacology* 6: 205.
7. Ramsay, R., M. Popovic-Nikolic, K. Nikolic, E. Uliassi, and M. Bolognesi. 2018. A perspective on multi-target drug discovery and design for complex diseases. *Clinical and Translational Medicine* 7 (1): 3.
8. Hopkins, A. 2008. Network pharmacology: The next paradigm in drug discovery. *Nature Chemical Biology* 4 (11): 682–690.
9. Chen, Y., J. Zhu, P. Lum, X. Yang, S. Pinto, D. MacNeil, C. Zhang, J. Lamb, S. Edwards, S. Sieberts, A. Leonardson, L. Castellini, S. Wang, M. Champy, B. Zhang, V. Emilsson, S. Doss, A. Ghazalpour, S. Horvath, T. Drake, A. Lusis, and E. Schadt. 2008. Variations in DNA elucidate molecular networks that cause disease. *Nature* 452 (7186): 429–435.
10. Aloy, P., and R. Russell. 2004. Taking the mystery out of biological networks. *EMBO Reports* 5 (4): 349–350.
11. Whitacre, J. 2012. Biological robustness: Paradigms, mechanisms, and systems principles. *Frontiers in Genetics* 3: 67.
12. Talevi, A. 2016. Tailored multi-target agents. Applications and design considerations. *Current Pharmaceutical Design* 22 (21): 3164–3170.
13. Swinney, D., and J. Anthony. 2011. How were new medicines discovered? *Nature Reviews. Drug Discovery* 10 (7): 507–519.
14. Groenendijk, F., and R. Bernards. 2014. Drug resistance to targeted therapies: Déjà vu all over again. *Molecular Oncology* 8 (6): 1067–1083.
15. Talevi, A., and B. Luis Enrique. 2013. On the development of new antiepileptic drugs for the treatment of Pharmacoresistant epilepsy: Different approaches to different hypothesis. In *Pharmacoresistance in epilepsy*, ed. L. Rocha and E. Cavalheiro, 1st ed., 207–224. New York: Springer.

16. Yao, J., and C. Rock. 2016. Resistance mechanisms and the future of bacterial Enoyl-acyl carrier protein reductase (FabI) antibiotics. *Cold Spring Harbor Perspectives in Medicine* 6 (3): a027045.
17. Pinto, J., R. Machado, J. Xavier, and M. Futschik. 2014. Targeting molecular networks for drug research. *Frontiers in Genetics* 5: 160.
18. Reddy, A., and S. Zhang. 2013. Polypharmacology: Drug discovery for the future. *Expert Review of Clinical Pharmacology* 6 (1): 41–47.
19. Hughes, J., S. Rees, S. Kalindjian, and K. Philpott. 2011. Principles of early drug discovery. *British Journal of Pharmacology* 162 (6): 1239–1249.
20. Kola, I., and J. Landis. 2004. Can the pharmaceutical industry reduce attrition rates? *Nature Reviews. Drug Discovery* 3 (8): 711–716.
21. Schuster, D., C. Laggner, and T. Langer. 2005. Why drugs fail – a study on side effects in new chemical entities. *Current Pharmaceutical Design* 11 (27): 3545–3559.
22. Talevi, A. 2018. Computer-aided drug design: An overview. *Methods in Molecular Biology* 1762: 1–19.
23. Lusher, S., R. McGuire, R. Azevedo, J. Boiten, R. van Schaik, and J. de Vlieg. 2011. A molecular informatics view on best practice in multi-parameter compound optimization. *Drug Discovery Today* 16 (13–14): 555–568.
24. Yu, H., P. Kim, E. Sprecher, V. Trifonov, and M. Gerstein. 2007. The importance of bottlenecks in protein networks: Correlation with gene essentiality and expression dynamics. *PLoS Computational Biology* 3 (4): e59.
25. Csermely, P., T. Korcsmáros, H. Kiss, G. London, and R. Nussinov. 2013. Structure and dynamics of molecular networks: A novel paradigm of drug discovery. *Pharmacology & Therapeutics* 138 (3): 333–408.
26. Ágoston, V., P. Csermely, and S. Pongor. 2005. Multiple weak hits confuse complex systems: A transcriptional regulatory network as an example. *Physical Review E* 71 (5): 051909.
27. Perez-Lopez, Á., K. Szalay, D. Türei, D. Módos, K. Lenti, T. Korcsmáros, and P. Csermely. 2015. Targets of drugs are generally and targets of drugs having side effects are specifically good spreaders of human interactome perturbations. *Scientific Reports* 5 (1): 10182.
28. Farkas, I., T. Korcsmaros, I. Kovacs, A. Mihalik, R. Palotai, G. Simko, K. Szalay, M. Szalay-Beko, T. Vellai, S. Wang, and P. Csermely. 2011. Network-based tools for the identification of novel drug targets. *Science Signaling* 4 (173): pt3.
29. Bianchi, M., J. Pathmanathan, and S. Cash. 2009. From ion channels to complex networks: Magic bullet versus magic shotgun approaches to anticonvulsant pharmacotherapy. *Medical Hypotheses* 72 (3): 297–305.
30. Lipton, S. 2006. Paradigm shift in neuroprotection by NMDA receptor blockade: Memantine and beyond. *Nature Reviews. Drug Discovery* 5 (2): 160–170.
31. Zheng, H., M. Fridkin, and M. Youdim. 2014. From single target to multitarget/network therapeutics in Alzheimer's therapy. *Pharmaceuticals* 7 (2): 113–135.
32. Rammes, G., R. Rupprecht, U. Ferrari, W. Zieglgänsberger, and C. Parsons. 2001. The N-methyl-d-aspartate receptor channel blockers memantine, MRZ 2/579 and other amino-alkyl-cyclohexanes antagonise 5-HT3 receptor currents in cultured HEK-293 and N1E-115 cell systems in a non-competitive manner. *Neuroscience Letters* 306 (1–2): 81–84.
33. Seeman, P., C. Caruso, and M. Lasaga. 2007. Memantine agonist action at dopamine D2High receptors. *Synapse* 62 (2): 149–153.
34. Aracava, Y., E. Pereira, A. Maelicke, and E. Albuquerque. 2005. Memantine blocks 7∗ nicotinic acetylcholine receptors more potently than N-methyl-D-aspartate receptors in rat hippocampal neurons. *The Journal of Pharmacology and Experimental Therapeutics* 312 (3): 1195–1205.
35. Klitgaard, H., A. Matagne, J. Nicolas, M. Gillard, Y. Lamberty, M. De Ryck, R. Kaminski, K. Leclercq, I. Niespodziany, C. Wolff, M. Wood, J. Hannestad, Kervyn, and B. Kenda. 2016. Brivaracetam: Rationale for discovery and preclinical profile of a selective SV2A ligand for epilepsy treatment. *Epilepsia* 57 (4): 538–548.

36. Gillard, M., B. Fuks, K. Leclercq, and A. Matagne. 2011. Binding characteristics of brivaracetam, a selective, high affinity SV2A ligand in rat, mouse and human brain: Relationship to anti-convulsant properties. *European Journal of Pharmacology* 664 (1–3): 36–44.
37. Zhang, L., S. Li, H. Li, and X. Zou. 2016. Levetiracetam vs. brivaracetam for adults with refractory focal seizures: A meta-analysis and indirect comparison. *Seizure* 39: 28–33.
38. Zhu, L., D. Chen, D. Xu, G. Tan, H. Wang, and L. Liu. 2017. Newer antiepileptic drugs compared to levetiracetam as adjunctive treatments for uncontrolled focal epilepsy: An indirect comparison. *Seizure* 51: 121–132.
39. Borghs, S., M. Charokopou, C. Brandt, and P. Klein. 2016. Response to Zhang et al.: Levetiracetam vs. brivaracetam for adults with refractory focal seizures: A meta- analysis and indirect comparison. *Seizure* 41: 182–183.
40. Lin, Z., and Z. Xiaoyi. 2016. Response to "Response to Zhang et al.: Levetiracetam vs. brivaracetam for adults with refractory focal seizures: A meta-analysis and indirect comparison". *Seizure* 41: 184–186.
41. Wong, E., F. Tarazi, and M. Shahid. 2010. The effectiveness of multi-target agents in schizophrenia and mood disorders: Relevance of receptor signature to clinical action. *Pharmacology & Therapeutics* 126 (2): 173–185.
42. Roth, B., D. Sheffler, and W. Kroeze. 2004. Magic shotguns versus magic bullets: Selectively non-selective drugs for mood disorders and schizophrenia. *Nature Reviews. Drug Discovery* 3 (4): 353–359.
43. Proschak, E., H. Stark, and D. Merk. 2019. Polypharmacology by design: A medicinal Chemist's perspective on multitargeting compounds. *Journal of Medicinal Chemistry* 62 (2): 420–444.
44. Morphy, R., and Z. Rankovic. 2005. Designed multiple ligands. An emerging drug discovery paradigm. *Journal of Medicinal Chemistry* 48 (21): 6523–6543.
45. Tonge, P. 2018. Drug–target kinetics in drug discovery. *ACS Chemical Neuroscience* 9 (1): 29–39.
46. Floris, M., S. Olla, D. Schlessinger, and F. Cucca. 2018. Genetic-driven Druggable target identification and validation. *Trends in Genetics* 34 (7): 558–570.
47. Cascante, M., L. Boros, B. Comin-Anduix, P. de Atauri, J. Centelles, and P. Lee. 2002. Metabolic control analysis in drug discovery and disease. *Nature Biotechnology* 20 (3): 243–249.
48. Saavedra, E., Z. Gonzalez-Chavez, R. Moreno-Sanchez, and P. Michels. 2018. Drug target selection for Trypanosoma cruzi metabolism by metabolic control analysis and kinetic modeling. *Current Medicinal Chemistry* 25: 1–19.
49. Yang, K., W. Ma, H. Liang, Q. Ouyang, C. Tang, and L. Lai. 2007. Dynamic simulations on the arachidonic acid metabolic network. *PLoS Computational Biology* 3 (3): e55.
50. Morphy, R., C. Kay, and Z. Rankovic. 2004. From magic bullets to designed multiple ligands. *Drug Discovery Today* 9 (15): 641–651.
51. Hopkins, A., G. Keserü, P. Leeson, D. Rees, and C. Reynolds. 2014. The role of ligand efficiency metrics in drug discovery. *Nature Reviews. Drug Discovery* 13 (2): 105–121.
52. Walles, M., A. Connor, and D. Hainzl. 2018. ADME and safety aspects of non- cleavable linkers in drug discovery and development. *Current Topics in Medicinal Chemistry* 17 (32): 3463–3475.
53. Schmidt, J., M. Rotter, T. Weiser, S. Wittmann, L. Weizel, A. Kaiser, J. Heering, T. Goebel, C. Angioni, M. Wurglics, A. Paulke, G. Geisslinger, A. Kahnt, D. Steinhilber, E. Proschak, and D. Merk. 2017. A dual modulator of Farnesoid X receptor and soluble epoxide hydrolase to counter nonalcoholic steatohepatitis. *Journal of Medicinal Chemistry* 60 (18): 7703–7724.
54. Haupt, V., S. Daminelli, and M. Schroeder. 2013. Drug promiscuity in PDB: Protein binding site similarity is key. *PLoS One* 8 (6): e65894.
55. Barelier, S., T. Sterling, M. O'Meara, and B. Shoichet. 2015. The recognition of identical ligands by unrelated proteins. *ACS Chemical Biology* 10 (12): 2772–2784.

56. Keiser, M., B. Roth, B. Armbruster, P. Ernsberger, J. Irwin, and B. Shoichet. 2007. Relating protein pharmacology by ligand chemistry. *Nature Biotechnology* 25 (2): 197–206.
57. Ehrt, C., T. Brinkjost, and O. Koch. 2016. Impact of binding site comparisons on medicinal chemistry and rational molecular design. *Journal of Medicinal Chemistry* 59 (9): 4121–4151.
58. Bellera, C.L., M.L. Sbaraglini, L.N. Alberca, J.I. Alice, and A. Talevi. 2019. In silico modeling of FDA-approved drugs for discovery of therapies against neglected diseases: A drug repurposing approach. In *In silico drug design*, ed. K. Roy, 1st ed., 625–648. Academic Press, London, UK.
59. Hu, Y., and J. Bajorath. 2010. Polypharmacology directed compound data mining: Identification of promiscuous chemotypes with different activity profiles and comparison to approved drugs. *Journal of Chemical Information and Modeling* 50 (12): 2112–2118.
60. Morphy, R., and Z. Rankovic. 2006. The physicochemical challenges of designing multiple ligands. *Journal of Medicinal Chemistry* 49 (16): 4961–4970.
61. Ma, X., Z. Shi, C. Tan, Y. Jiang, M. Go, B. Low, and Y. Chen. 2010. In-silico approaches to multi-target drug discovery. *Pharmaceutical Research* 27 (5): 739–749.
62. Handler, N. 2017. Pharmacophore generation for multiple ligands. In *Drug selectivity: An evolving concept in medicinal chemistry*, ed. N. Handler and H. Buschmann, 1st ed., 275–312. Weinheim: Wiley-VCH Verlag GmbH & Co.
63. Sindhu, T., and P. Srinivasan. 2015. Identification of potential dual agonists of FXR and TGR5 using e-pharmacophore based virtual screening. *Molecular BioSystems* 11 (5): 1305–1318.

Mining Complex Biomedical Literature for Actionable Knowledge on Rare Diseases

Vinicius M. Alves, Stephen J. Capuzzi, Nancy Baker, Eugene N. Muratov, Alexander Trospsha, and Anthony J. Hickey

Introduction

Since the Enlightenment, the approach to physical, chemical, and biological processes has involved the development of analytical methods that would use quantitative data to build simple integrated models leading to prediction of drug action [1]. This was a logical approach as it could capture a large quantity of data in a manner that was easy to store and transmit from one individual to the another. However, as scientific research has become increasingly prominent type of human activity, there has been a dramatic growth of scientific publications reporting the research outcomes in a form combining both textual description and numerical data

V. M. Alves · S. J. Capuzzi
UNC Catalyst for Rare Diseases, UNC Eshelman School of Pharmacy, University of North Carolina, Chapel Hill, NC, USA

Laboratory for Molecular Modeling, Division of Chemical Biology and Medicinal Chemistry, UNC Eshelman School of Pharmacy, University of North Carolina, Chapel Hill, NC, USA

N. Baker
ParlezChem, Hillsborough, NC, USA

E. N. Muratov
Laboratory for Molecular Modeling, Division of Chemical Biology and Medicinal Chemistry, UNC Eshelman School of Pharmacy, University of North Carolina, Chapel Hill, NC, USA

Department of Pharmaceutical Sciences, Federal University of Paraiba, Joao Pessoa, PB, Brazil

A. Trospsha
Laboratory for Molecular Modeling, Division of Chemical Biology and Medicinal Chemistry, UNC Eshelman School of Pharmacy, University of North Carolina, Chapel Hill, NC, USA

A. J. Hickey (✉)
UNC Catalyst for Rare Diseases, UNC Eshelman School of Pharmacy, University of North Carolina, Chapel Hill, NC, USA
e-mail: ahickey@unc.edu

M. Bizzarri (ed.), *Approaching Complex Diseases*, Human Perspectives in Health Sciences and Technology 2, https://doi.org/10.1007/978-3-030-32857-3_4

[2]. Most scientists do spend a lot of time thinking about the best verbal ways of describing and reasoning over their results and thus, a lot of useful information and knowledge could be obtained by reading the scientific literature. As important as reading is in the life of every scientist, the process of obtaining summative knowledge compiled from many publications is a non-scalable effort [3].

Fortunately, the advent of computer technology enables storing and efficient processing of large amounts of data, including textual sources. The analysis of this complex data allows mechanistic inferences to be drawn that promote novel hypotheses that illuminate fundamental natural phenomena. The importance of evolving computational methods that allow consideration of a wide range of data and their implications cannot be understated [4].

The current exponentially increasing cost and decades of inefficiency in drug discovery and development have resulted in a problematic situation with respect to pharmaceutical innovation and commercialization. The overwrought drug discovery pipeline may take up 15 years to develop a successful drug (considering hit-to-lead discovery/development, pre-clinical, and clinical studies), with an average cost estimated to be from $800 million to $1.5 billion [5]. This process is deemed as inadequate and unsustainable, especially as concerning its ability to provide a therapy for diseases that affect people in poor parts of the world, such as tropical diseases, as well as those affecting a limited number of patients, such as rare diseases, due to the potential resulting low revenue [6, 7]. A disruptive approach is required that can bring about revolutionary, not evolutionary, change equivalent to the changes that have occurred in the communications, electronics and financial industries over the last 25 years.

Rare diseases, which are defined as a condition that affects fewer than 200,000 people in the United States and 1 in 2000 people in the European Union, are particularly in need of disruptive and revolutionary drug discovery paradigms. Although, individually each rare disease affects a small portion of the total population only, their collective effect on the human population is substantial as there are over 7000 rare diseases that roughly affect 25–30 million people in the United States [8]. Alarmingly, very few patients can be treated with an approved medicine. Taken together, rare diseases represent a substantial burden on individuals, families, and whole economies [9, 10].

Developing a drug for a rare disease, on average, is half the cost of common diseases [11]. Still, considering the smaller amount of data and investment, any innovative approaches to treat these diseases will likewise be of value for drug discovery *writ large*. It is anticipated that once paradigms are developed for the rare diseases drug discovery, which have less financial benefit than more prevalent diseases, drug discovery efforts in general will become more efficient [12]. In addition to financial concerns and a limited patient populations, rare disease drug discovery also suffers from sparse and heterogeneous data [13], which hamper the ability to draw novel insights and treatment hypotheses. However, a growing number of rare diseases registries has been incorporate within in different databases [14, 15], such as Pharos [16] (https://pharos.nih.gov/), ClinVar [17, 18] (https://www.ncbi.nlm.nih.gov/clinvar/), the Online Mendelian Inheritance in Man (OMIM) (https://omim.org/), among others [7]. While efforts have been made to promote the sharing of information between multi-disciplinary collaborations [19], there is still need to curate and properly integrate all of this information [13, 20–22].

Computational approaches have emerged as a practical solution to accelerate drug discovery efforts and reduce costs [23, 24]. One promising approach, named Literature-Based-Discovery (LBD), seeks to unlock biological observations hidden within informational sources, such as published texts and manuscripts [25]. Since 1988, when the relationship between magnesium and migraine was discovered in the literature by Swanson [26], other treatment hypotheses have been generated for many diseases, such as Parkinson's disease [27], multiple sclerosis [28], and cancer [29]. This approach has been also used to elucidate adverse drug effects [25, 30]. As such, LBD is a powerful new technology in the drug discovery arsenal.

In this chapter, we aim to review the status of available biomedical data on PubMed and describe how mining complex drug-target-disease relationships within this database could contribute to finding new targets, new repurposed medications, and novel drug candidates for rare diseases. The intent of the following discussion is to focus on the impact of consideration of complexity in drug discovery and clinical data to allow new therapies to emerge that can be rapidly screened and progressed to clinical application. The overall approach described may likely be one component of a strategy that will regenerate pharmaceutical development and promote a rational approach to the pharmacological element of health care delivery.

Biomedical Knowledge Data in the Scientific Literature

Bioactivity data such as the outcome of *in vivo* and *in vitro* assays have been growing extensively in publicly available repositories such as ChEMBL [31, 32] (https://www.ebi.ac.uk/chembl/) and PubChem [33, 34] (http://pubchem.ncbi.nlm.nih.gov/). Despite the growth of these databases, the scientific literature remains the largest repository of untapped biomedical data [2]. The United States National Library of Medicine (NLM) journal citation database (MEDLINE) is the preeminent source of biomedical literature, with ~30 million citations [35]. This database can be accessed through PubMed, a search engine maintained by NLM at the National Institutes of Health (NIH). It is possible to retrieve reference for scientific articles stored in MEDLINE by querying specific terms named Medical Subject Headings (MeSH) [36], which are used to index and categorize publications stored in MEDLINE. MeSH terms encompass most drugs, targets, and diseases present in scientific publications and could potentially be used to accelerate drug discovery [37].

The major approach to manipulate knowledge stored in the literature is through natural language processing, a subfield of artificial intelligence that allows computers to understand, interpret, and manipulate human language [38]. For this purpose, many dictionary-based systems that recognize passages in the literature with ontological terms have been proposed and evaluated [39]. The SciLite Annotations platform (https://europepmc.org/Annotations) provides means to link research articles with biological data through text mining [40]. In a 2016 study, text mining on PubMed and social network analysis were integrated to analyze gene-gene interactions in order to identify new potential biomarkers for breast cancer [41]. More recently, text mining has been used to analyze gene-disease associations present in PubMed by integrating MeSH terms and co-occurrence methods [42].

Drug Repurposing

As discussed in the introduction, it may take a drug up to 15 years to reach the market [43]. This process usually includes discovery and development research, preclinical studies (*in vitro*/*in vivo* evaluation),0020and clinical research, divided in Phase I (safety and dose evaluation in healthy individuals), Phase II (efficacy and safety in small number of patients), Phase III (efficacy and safety in large number of patients), and Phase IV (post-market safety monitoring). During Phase II, approximately 90% of the compounds fail due to safety concerns and poor efficacy [44].

Drug repurposing, also known as repositioning or reprofiling, is a strategy to identify novel uses for approved or investigational drugs that are outside of the original therapeutic indication [45]. Recently, this approach has been a trending topic among researchers [46] and has attracted attention of companies due to the reduced cost associated with the low risk of failure, especially when safety evaluation has already been completed in preclinical and clinical trials [47]. Because repurposed drugs can skip safety evaluation during preclinical and Phase I studies, it is estimated that developing a repurposed drugs costs on average only $300 million over a 6.5 year period [48]. In addition to reduced cost and time, approximately 30% of repurposed drugs are approved, which can be seen as a market-oriented incentive to companies [45, 49]. For comparison, the typical approval rate for drugs entering clinical trials is 9.6% [50].

Repurposing studies very often are initiated after unexpected drug effects are observed during clinical trials or during pharmacovigilance upon their release on the market [51]. Many of the current repurposing studies have been initiated thanks to a serendipitous observation of unexpected drug effects upon clinical trials or following their release on the market. Prime examples of such discoveries are the stories of sildenafil (Viagra®) [52] and thalidomide [53, 54].

Recently, it has been shown in a bibliographic study [55] that more than 60% of all approved drugs or drug candidates (~35,000) have been tried in more than one disease, including 189 drugs that have been tried in >300 diseases each. Considering only approved drugs, more than 30% have been tested during their lifetime for at least one additional indication following their original approval [55]. Despite several success cases, drug repurposing still faces lack of financial support due to potentially low return, lower drug prices, and short patent duration [56, 57]. Nevertheless, this approach is still considered promising, especially for rare diseases [58]. Small grant programs to help develop drugs or treatments for rare diseases are usually available from rare disease foundations [59]. The National Organization for Rare Disorders (NORD) (http://rarediseases.org/) provides recommendations to such organizations.

Using Chemotext to Infer Novel Therapies and Targets

Biological insights about the etiology of diseases, such as causative protein mutations or aberrant pathway signaling, and the potential drug treatments of these diseases are stored primarily in the biomedical literature [2]. As such, there exists biomedically relevant relationships between drugs, biological targets, and diseases, which we call DTD triangles, that lie latent within published texts [3, 60]. Using text-mining approaches, therefore, these DTD triangles can be identified and extracted from the published biomedical literature [61].

Text-mining capabilities in conjunction with the wealth of text-based data stored within PubMed considerations led to the development of Chemotext [62], a computational algorithm which extracts MeSH terms describing "drugs", "targets", and "diseases" and generates DTD triangles. Chemotext is based on the frequency with which MeSH terms of interest co-occur in abstracts of papers annotated in PubMed. Chemotext is thus an extension of Swanson's ABC paradigm wherein "A" terms are drug (chemical) MeSH terms, "B" terms are target-associated MeSH terms, i.e., proteins and pathways, and "C" terms are disease MeSH terms (Fig. 1).

The underlying DTD triangle generation starts with the observation that the MeSH term of drug "A" co-occurs in the same articles as the MeSH term of target "B" while the MeSH term of disease "C" co-occurs in the same or additional articles with the same target B. Thus, if drug A and disease C have not been mentioned together in the same article, an "A–C" connection mediated though target B can be inferred, completing a DTD triangle. This analysis, enabled by the Chemotext approach, leads to the identification of a new possible therapeutic use of drug "A".

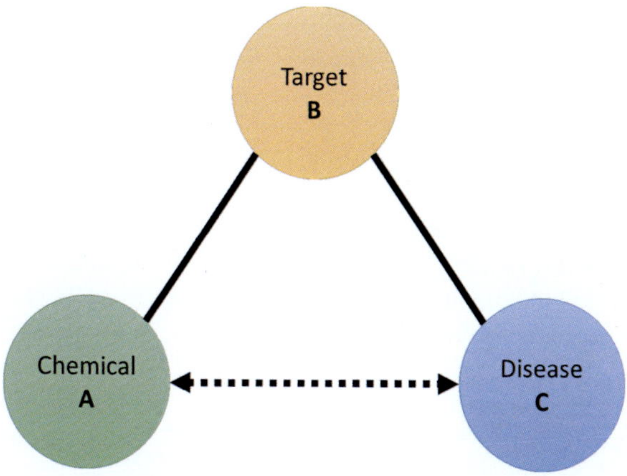

Fig. 1 Swanson's ABC paradigm incorporated in Chemotext. Chemical A is proposed to affect disease C since both terms are associated with target B. Solid lines (edges) indicate an actual text-based relationship, while dashed lines (edges) indicate proposed connections

The power and efficacy of Chemotext is demonstrated by elucidation of the antineoplastic agent imatinib as a potential drug repurposing candidate for the treatment of severe refractory asthma. Imatinib is an FDA-approved tyrosine kinase inhibitor that is used in the treatment chronic myeloid leukemia (CML). Imatinib inhibits the activity of KIT, which reduces bone marrow mast-cell numbers in patients with CML [63]. KIT is also present in lung mast cells and was hypothesized as a basis of the pathobiology of severe refractory asthma [64], which is characterized by an adverse response to traditional glucocorticoid asthma treatment [65]. Figure 2 shows how Chemotext can be used to link Imatinib (A), Proto-Oncogene Proteins c-kit (B), and asthma (C).

In 2017, a proof-of-principle trial demonstrated that imatinib reduced airway hyperresponsiveness, a physiological marker of severe asthma, as well as on airway mast-cell numbers and activation in patients with severe asthma. Since this publication had not yet been entered into the.

MEDLINE database, it was used a validation test of the Chemotext algorithm. Through co-occurrences of these MeSH terms in previously published studies, Chemotext was used to draw the interference between imatinib, KIT, and asthma, which constitutes a DTD triangle (Fig. 2). This case study demonstrates that Chemotext can identify drug repurposing candidates and targets through text-based inferences alone.

Mining Other Sources of Biomedical Data for Drug Repurposing

Mining literature data can afford rapid identification of all published studies that could confirm connections between drugs, their targets, underlying biological pathways, and diseases, including enabling new inferences of such connections [3, 60]. The elucidation of the mechanistic relationships between these connections is at the core of modern drug discovery research [61]. Currently, there are several databases with valuable information for drug discovery that could be connected to complete a DTD triangle. ChEMBL [31, 32] (https://www.ebi.ac.uk/chembl/) and PubChem [33, 34] (http://pubchem.ncbi.nlm.nih.gov/) contain many chemical–target ("A–B") and chemical–disease ("A– C") relationships. Other databases contain target–disease ("B–C") associations, such as ClinVar [17, 18] (https://www.ncbi.nlm.nih.gov/clinvar/), the Online Mendelian Inheritance in Man (OMIM) (https://omim.org/). Pharos [16] (https://pharos.nih.gov/), specifically, contains data on the whole DTD triangle for many diseases. Several databases are available containing parts of the triangle available for rare diseases, such as Malacards [66] (http://www.malacards.org/ the National Organization for Rare Disorders (NORD) [67] (https://rarediseases.org/), the Genetic and Rare Diseases Information Center (GARD) [68]

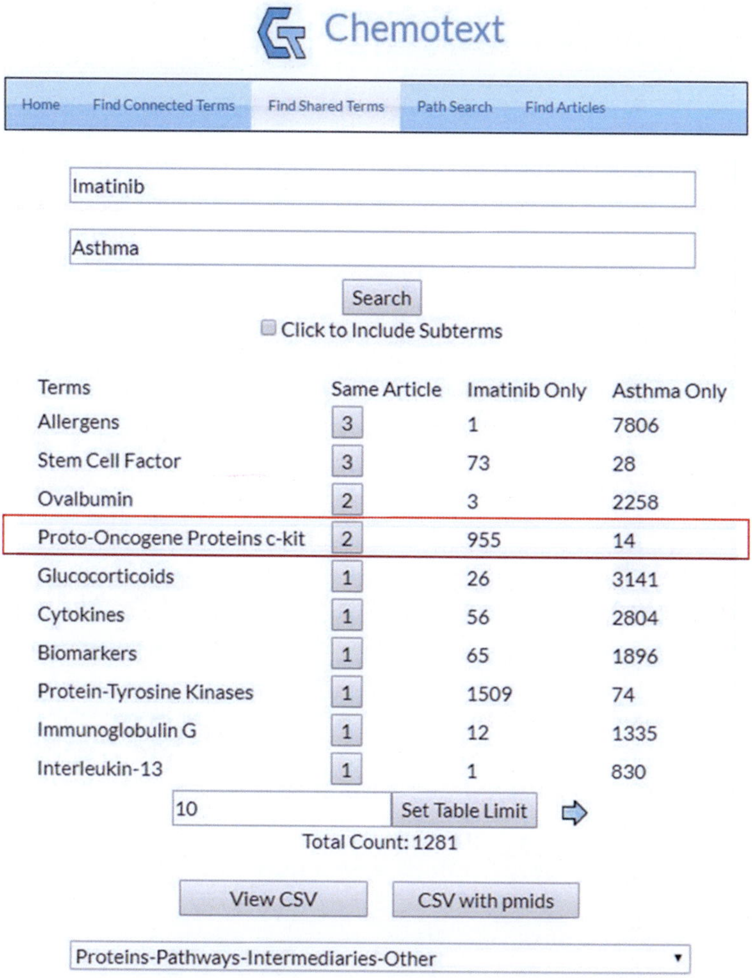

Fig. 2 Example showing how Chemotext connects Imatinib and Asthma with shared terms. In this example, query terms "Imatinib" and "Asthma" were searched in the Find Shared Terms module. The list of full associations was filtered by Proteins-Pathways-Intermediaries-Other. The MeSH term "Proto- Oncogene Proteins c-kit" was the fourth highest ranked shared term (two shared articles) selected as the potential biological target in the clinical outcome pathway

(https://rarediseases.info.nih.gov/), and the Infohub for Rare Diseases (https://rarediseases.oscar.ncsu.edu/).

Recently, NIH has launched the Biomedical Data Translator program (https://ncats.nih.gov/translator), which has integrated many data sources with multiple types of content, such as diseases, patient-reported outcomes, electronic health records, microbiome, proteins, genes, chemicals, among others. This massive project attempts to integrate currently available medical research data towards accelerated development of new treatments. The major challenge to establish valuable connections, as in any data science project, is proper curation of the data [13, 20–22]. To establish useful relationships between these sources of data, knowledge graphs have emerged as a practical solution. A knowledge graph is a network of entities that acquires and integrates information into an ontology and applies a reasoner to derive new knowledge [68]. A 2016 study has applied network-based modeling within to identify promising multi-target drugs for triple negative breast cancer [11]. More recently, a study has applied knowledge graphs to integrate different data sources on diseases and drugs to suggest the repurposing of 21 drugs for Autosomal Dominant Polycystic Kidney Disease (ADPKD) [68].

There has also been a growing interest in using social media to supplement established approaches for pharmacovigilance [69, 70]. The use of social media, also called "social listening", therefore, is a potential resource for repurposing. Social media has been recently used in public health to estimate trends of cholera outbreak in the after math of the 2010 earthquake in Haiti [71], seasonal influenza surveillance [72], and onset of mental illness [73]. As previously discussed, many repurposed drugs have been discovered through adverse side effects observed during clinical trials or pharmacovigilance. Many people have used social media to report adverse effects of their medications. Several studies analyzing adverse reactions on social media have been published recently [30, 74, 75], which makes social media a potential source of adverse effect data to be mined for repurposing.

Drug Repurposing and Bibliometric Analysis on Rare Diseases

Several repurposing stories for rare diseases have been reported in the recent years. For instance, metformin has been studied to treat idiopathic pulmonary fibrosis [76]. A recent study suggests that inhibitors of p110β, a catalytic subunit of the phosphoinositide 3-kinase (PI3K) gene family, commonly associated with cancer, might prevent cognitive and behavioral defects and become a promising disease-modifying strategy for fragile X syndrome and other brain disorders [77]. Fenfluramine, initially proposed as a an appetite suppressant and withdrawn from the market, has been submitted to the FDA for the treatment of Dravet syndrome [78].

Many computational approaches historically applied for drug discovery, such as quantitative structure-activity relationships (QSAR) modeling, similarity search, molecular docking, etc., have been successfully applied for drug repurposing as well [79]. Computational drug repurposing approaches have been widely applied to neglected tropical diseases [80–84], and, more recently, to rare diseases [58, 83]. The eMatchSite, a platform for compare drug-binding sites have been applied to propose the possibility to repurpose a steroidal aromatase inhibitor to treat Niemann-Pick disease type C [85]. A structure-based virtual screening approach has been applied to screen FDA approved drugs on ENGase, a potential target for the treatment of N-Glycanase (NGLY1) deficiency. The authors experimentally confirmed the activity of rabeprazole (IC50 = 4.4. μM) on ENGase as a promising treatment to patients suffering from NGLY1 deficiency [69].

Mining literature data allows the exploitation of opportunities to reposition known drugs interacting with proteins associated with diseases [3, 60]. The integration of data on drug-target-disease to form networks has become a valuable approach for computational drug repositioning research [86]. Recently, a study has used bioinformatics methods and bibliographic research to propose the repositioning of some drugs as potential competitors against idiopathic pulmonary fibrosis [87].

As of June of 2019, there are 244,911 references with the term "rare disease" through the text and 17,134 references with the term "rare disease" in the title or abstract. Here, we performed a brief bibliometric analysis on drug repurposing for rare diseases, similar to the one that was recently published by Baker et al. [55]. We mined PubMed using earlier text-mining work [37] to identify articles in PubMed where a chemical entity was described in terms of its therapeutic association with a rare disease. We determined this relationship by examining the MeSH annotations in a stepwise manner (described in the supplementary material online). All drug–disease combinations were extracted, along with the year the article was published, into a separate dataset. This set included citations with no abstract and those in languages other than English, as long as they were annotated, and the annotations met the criteria.

In our analysis, we found that only 1267 out of more than 7000 rare diseases have been studied in association with a chemical entity. It was known that only a small fraction of rare diseases has associated treatments, but our findings reveal there is still a major gap in research for rare diseases, since many of them have not been associated with any chemical entity as a potential treatment. These findings reinforce the need to expand research on the development of novel therapies for rare diseases. As one can see in Fig. 3, 6570 out of 12,376 chemicals (53%) have been associated with only one rare disease, while 4796 (38%) have been associated with two to ten diseases, 984 (7.0%) have been associated with eleven to 100 diseases, and 26 (0.20%) chemicals with more than 100 diseases.

We show in Table 1 the top 30 drugs that were tested for rare diseases. Sixteen out of 30 were among the top drugs most tested in the previous study [55]. As one can see, most of these drugs are used to suppress the immune system and/or to decrease inflammation, such as glucocorticoid medications (prednisone, prednisolone, dexamethasone, methylprednisolone, hydrocortisone, and cortisone), cancer chemotherapy agents (cyclophosphamide, bevacizumab, methotrexate), and medications used to prevent transplant rejections (sirolimus, rituximab, cyclosporine). The rare

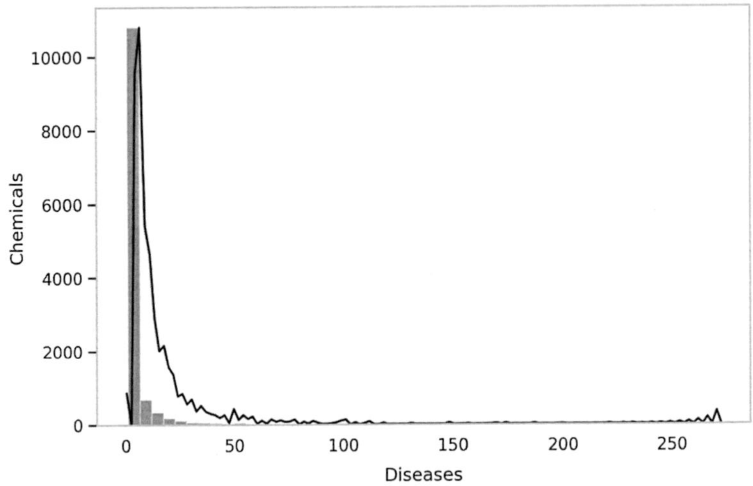

Fig. 3 Distribution of chemicals tested in rare diseases mined from PubMed

diseases with most publications and chemicals tested are presented in Table 2. Most of these diseases are rare forms of cancer, such as sarcoma, and neoplasm, and multiple forms of carcinoma, which explains why most of the most studied drugs present in Table 1 are suppressant of immune system, anti-inflammatory, and anti-cancer drugs. Surprisingly, none of the most studied drugs were used in some of the most studied diseases, such as malaria, tuberculosis, and Alzheimer.

Final Remarks

There is an urgent need for the development of treatments or cures for rare diseases. The complex biological systems and nature of drug discovery make iterative mechanistic strategies costly and inefficient. Current developments in database development, text mining, and machine learning tools allows efficient and inexpensive navigation through inferences to the identification of novel or repurposed drug candidates. The same principles can be employed to the traverse the complexity of drug delivery systems and biopharmaceutical principles that result in optimal drug disposition to achieve the desired therapeutic effect. In this manner, the development

Table 1 Top 30 drugs most tested in rare diseases with publications count

Chemicals	Rare diseases count	Publications count
Prednisolone	272	1627
Prednisone	233	1857
Dexamethasone	229	1598
Methylprednisolone	221	1162
Cyclophosphamide	199	2309
Cyclosporine	187	1376
Rituximab	174	2123
Methotrexate	170	1790
Interferon-alpha	167	3316
Immunoglobulin G	156	782
Sirolimus	149	802
Ascorbic Acid	148	504
Vitamin E	131	635
Infliximab	119	2467
Adrenocorticotropic Hormone	118	969
Tretinoin	112	776
Tacrolimus	112	416
Thalidomide	111	1320
Hydrocortisone	111	345
Aspirin	110	693
Heparin	106	396
Indomethacin	105	941
Curcumin	102	590
Bevacizumab	101	1271
Granulocyte Colony-Stimulating Factor	101	1253
Interferon-gamma	101	723
Cortisone	100	723
Acetylcysteine	100	315
Pentoxifylline	100	262

of novel pharmaceutical treatment options can focus on the generation of data suited to regulatory scrutiny and positive clinical outcomes without investment in the tangential iterative data generation that has historically been required to support statistical validation of the action, process, or clinical observations that surround the optimal approach.

Table 2 Top 30 rare diseases ranked by number of publications and chemicals tested

Rare disease	Publications	Chemicals
Adenocarcinoma	11811	2226
Liver Neoplasms	10773	2063
Carcinoma, Non-Small-Cell Lung	7527	1355
Carcinoma, Hepatocellular	6570	1647
Alzheimer Disease	5864	1500
Multiple Myeloma	5680	873
Malaria	5620	667
Urinary Bladder Neoplasms	5486	824
Leukemia, Myelogenous, Chronic, BCR-ABL Positive	5173	504
Leukemia, Myeloid, Acute	5066	897
Crohn Disease	4551	342
Kidney Neoplasms	4148	841
Tuberculosis	3876	535
Hemophilia A	3732	184
Carcinoma, Renal Cell	3193	627
Hodgkin Disease	2212	320
Cystic Fibrosis	2070	346
Cytomegalovirus Infections	2058	220
Neuroblastoma	2051	770
Osteosarcoma	2012	671
Myelodysplastic Syndromes	2010	235
Encephalomyelitis, Autoimmune, Experimental	1911	858
Anemia, Aplastic	1714	195
Sarcoma	1696	433
Spondylitis, Ankylosing	1667	171
Leprosy	1606	176
Fatty Liver	1603	651
Respiratory Distress Syndrome, Newborn	1530	180
Pulmonary Fibrosis	1459	574
Leishmaniasis, Visceral	1453	258

References

1. Hickey, A.J., and H.D.C. Smyth. 2011. *Pharmaco-complexity*. Boston: Springer US.
2. Hunter, L.E. 2017. Knowledge-based biomedical data science. *Data Science Journal* 1: 1–7. https://doi.org/10.3233/DS-170001.
3. Przybyła, P., M. Shardlow, S. Aubin, R. Bossy, R. Eckart de Castilho, S. Piperidis, J. McNaught, and S. Ananiadou. 2016. Text mining resources for the life sciences. *Database* 2016. https://doi.org/10.1093/database/baw145.
4. Pan, W., Z. Li, Y. Zhang, and C. Weng. 2018. The new hardware development trend and the challenges in data management and analysis. *Data Science and Engineering* 3: 263–276. https://doi.org/10.1007/s41019-018-0072-6.

5. DiMasi, J.A., H.G. Grabowski, and R.W. Hansen. 2016. Innovation in the pharmaceutical industry: New estimates of R&D costs. *Journal of Health Economics* 47: 20–33. https://doi.org/10.1016/J.JHEALECO.2016.01.012.
6. Baxter, K., E. Horn, N. Gal-Edd, K. Zonno, J. O'Leary, P.F. Terry, and S.F. Terry. 2013. An end to the myth: There is no drug development pipeline. *Science Translational Medicine* 5: 171cm1. https://doi.org/10.1126/scitranslmed.3003505.
7. Zhao, M., and D.-Q.Q. Wei. 2018. Rare diseases: Drug discovery and informatics resource. *Interdisciplinary Sciences: Computational Life Sciences* 10: 195–204. https://doi.org/10.1007/s12539-017-0270-3.
8. Valdez, R., L. Ouyang, and J. Bolen. 2016. Public health and rare diseases: oxymoron no more. *Preventing Chronic Disease* 13: 150491. https://doi.org/10.5888/pcd13.150491.
9. Kakkis, E.D., M. O'Donovan, G. Cox, M. Hayes, F. Goodsaid, P. Tandon, P. Furlong, S. Boynton, M. Bozic, M. Orfali, and M. Thornton. 2015. Recommendations for the development of rare disease drugs using the accelerated approval pathway and for qualifying biomarkers as primary endpoints. *Orphanet Journal of Rare Diseases* 10: 16. https://doi.org/10.1186/s13023-014-0195-4.
10. Angelis, A., D. Tordrup, and P. Kanavos. 2015. Socio-economic burden of rare diseases: A systematic review of cost of illness evidence. *Health Policy* 119: 964–979. https://doi.org/10.1016/j.healthpol.2014.12.016.
11. Vitali, F., L.D. Cohen, A. Demartini, A. Amato, V. Eterno, A. Zambelli, and R. Bellazzi. 2016. A network- based data integration approach to support drug repurposing and multi-target therapies in triple negative breast cancer. *PLoS One* 11: e0162407. https://doi.org/10.1371/journal.pone.0162407.
12. Ekins, S. 2017. Industrializing rare disease therapy discovery and development. *Nature Biotechnology* 35: 117–118. https://doi.org/10.1038/nbt.3787.
13. Roos M, López Martin E, Wilkinson MD (2017) Preparing data at the source to Foster interoperability across rare disease resources. In: de la Posada Paz M, Taruscio D, Groft S (eds) Rare diseases epidemiology: Update and overview. Advances in Experimental Medicine and Biology. Springer, Cham, pp 165–179
14. Kodra, Y., M. Posada de la Paz, A. Coi, M. Santoro, F. Bianchi, F. Ahmed, Y.R. Rubinstein, J. Weinbach, and D. Taruscio. 2017. Data quality in rare diseases registries. In *Advances in experimental medicine and biology*, 149–164. Cham: Springer.
15. Litterman, N.K., M. Rhee, D.C. Swinney, and S. Ekins. 2014. Collaboration for rare disease drug discovery research. *F1000Research* 3: 261. https://doi.org/10.12688/f1000research.5564.1.
16. Nguyen, D.T., S. Mathias, C. Bologa, S. Brunak, N. Fernandez, A. Gaulton, A. Hersey, J. Holmes, L.J. Jensen, A. Karlsson, G. Liu, A. Ma'ayan, G. Mandava, S. Mani, S. Mehta, J. Overington, J. Patel, A.D. Rouillard, S. Schürer, T. Sheils, A. Simeonov, L.A. Sklar, N. Southall, O. Ursu, D. Vidovic, A. Waller, J. Yang, A. Jadhav, T.I. Oprea, and R. Guha. 2017. Pharos: Collating protein information to shed light on the druggable genome. *Nucleic Acids Research* 45: D995–D1002. https://doi.org/10.1093/nar/gkw1072.
17. Landrum, M.J., J.M. Lee, M. Benson, G.R. Brown, C. Chao, S. Chitipiralla, B. Gu, J. Hart, D. Hoffman, W. Jang, K. Karapetyan, K. Katz, C. Liu, Z. Maddipatla, A. Malheiro, K. McDaniel, M. Ovetsky, G. Riley, G. Zhou, J.B. Holmes, B.L. Kattman, and D.R. Maglott. 2018. ClinVar: Improving access to variant interpretations and supporting evidence. *Nucleic Acids Research* 46: D1062–D1067. https://doi.org/10.1093/nar/gkx1153.
18. Landrum, M.J., J.M. Lee, G.R. Riley, W. Jang, W.S. Rubinstein, D.M. Church, and D.R. Maglott. 2014. ClinVar: Public archive of relationships among sequence variation and human phenotype. *Nucleic Acids Research* 42: D980–D985. https://doi.org/10.1093/nar/gkt1113.
19. Kaufmann, P., A.R. Pariser, and C. Austin. 2018. From scientific discovery to treatments for rare diseases – The view from the National Center for Advancing Translational Sciences –

Office of Rare Diseases Research. *Orphanet Journal of Rare Diseases* 13: 196. https://doi.org/10.1186/s13023-018-0936-x.
20. Fourches, D., E. Muratov, and A. Tropsha. 2010. Trust, but verify: On the importance of chemical structure curation in cheminformatics and QSAR modeling research. *Journal of Chemical Information and Modeling* 50: 1189–1204. https://doi.org/10.1021/ci100176x.
21. ———. 2016. Trust, but Verify II: A practical guide to chemogenomics data curation. *Journal of Chemical Information and Modeling* 56: 1243–1252. https://doi.org/10.1021/acs.jcim.6b00129.
22. ———. 2015. Curation of chemogenomics data. *Nature Chemical Biology* 11: 535–535. https://doi.org/10.1038/nchembio.1881.
23. Rognan, D. 2017. The impact of in silico screening in the discovery of novel and safer drug candidates. *Pharmacology & Therapeutics* 175: 47–66. https://doi.org/10.1016/j.pharmthera.2017.02.034.
24. Makhouri, F.R., and J.B. Ghasemi. 2019. Combating diseases with computational strategies used for drug design and discovery. *Current Topics in Medicinal Chemistry* 18: 2743–2773. https://doi.org/10.2174/1568026619666190121125106.
25. Henry, S., and B.T. McInnes. 2017. Literature based discovery: Models, methods, and trends. *Journal of Biomedical Informatics* 74: 20–32. https://doi.org/10.1016/j.jbi.2017.08.011.
26. Swanson, D.R. 1988. Migraine and magnesium: Eleven neglected connections. *Perspectives in Biology and Medicine* 31: 526–557.
27. Kostoff, R.N., and M.B. Briggs. 2008. Literature-Related Discovery (LRD): Potential treatments for Parkinson's disease. *Technological Forecasting and Social Change* 75: 226–238. https://doi.org/10.1016/j.techfore.2007.11.007.
28. Kostoff, R.N., M.B. Briggs, and T.J. Lyons. 2008. Literature-related discovery (LRD): Potential treatments for multiple sclerosis. *Technological Forecasting and Social Change* 75: 239–255. https://doi.org/10.1016/j.techfore.2007.11.002.
29. Choi, B.-K., T. Dayaram, N. Parikh, A.D. Wilkins, M. Nagarajan, I.B. Novikov, B.J. Bachman, S.Y. Jung, P.J. Haas, J.L. Labrie, C.R. Pickering, A.K. Adikesavan, S. Regenbogen, L. Kato, A. Lelescu, C.M. Buchovecky, H. Zhang, S.H. Bao, S. Boyer, G. Weber, K.L. Scott, Y. Chen, S. Spangler, L.A. Donehower, and O. Lichtarge. 2018. Literature-based automated discovery of tumor suppressor p53 phosphorylation and inhibition by NEK2. *Proceedings of the National Academy of Sciences* 115: 10666–10671. https://doi.org/10.1073/pnas.1806643115.
30. La, M.K., A. Sedykh, D. Fourches, E. Muratov, and A. Tropsha. 2018. Predicting adverse drug effects from literature- and database-mined assertions. *Drug Safety* 41: 1059–1072. https://doi.org/10.1007/s40264-018-0688-5.
31. Willighagen, E.L., A. Waagmeester, O. Spjuth, P. Ansell, A.J. Williams, V. Tkachenko, J. Hastings, B. Chen, and D.J. Wild. 2013. The ChEMBL database as linked open data. *Journal of Cheminformatics* 5: 23. https://doi.org/10.1186/1758-2946-5-23.
32. Gaulton, A., A. Hersey, M. Nowotka, A.P. Bento, J. Chambers, D. Mendez, P. Mutowo, F. Atkinson, L.J. Bellis, E. Cibrián-Uhalte, M. Davies, N. Dedman, A. Karlsson, M.P. Magariños, J.P. Overington, G. Papadatos, I. Smit, and A.R. Leach. 2017. The ChEMBL database in 2017. *Nucleic Acids Research* 45: D945–D954. https://doi.org/10.1093/nar/gkw1074.
33. Wang, Y., T. Suzek, J. Zhang, J. Wang, S. He, T. Cheng, B.A. Shoemaker, A. Gindulyte, and S.H. Bryant. 2014. PubChem BioAssay: 2014 update. *Nucleic Acids Research* 42: D1075–D1082. https://doi.org/10.1093/nar/gkt978.
34. Wang, Y., J. Xiao, T.O. Suzek, J. Zhang, J. Wang, Z. Zhou, L. Han, K. Karapetyan, S. Dracheva, B.A. Shoemaker, E. Bolton, A. Gindulyte, and S.H. Bryant. 2012. PubChem's BioAssay database. *Nucleic Acids Research* 40: D400–D412. https://doi.org/10.1093/nar/gkr1132.
35. Roberts, R.J. 2001. PubMed central: The GenBank of the published literature. *Proceedings of the National Academy of Sciences* 98: 381–382. https://doi.org/10.1073/pnas.98.2.381.

36. NLM. 2019. *Medical subject headings*. https://www.nlm.nih.gov/mesh/meshhome.html. Accessed 3 Jun 2019.
37. Baker, N.C., and B.M. Hemminger. 2010. Mining connections between chemicals, proteins, and diseases extracted from Medline annotations. *Journal of Biomedical Informatics* 43: 510–519. https://doi.org/10.1016/j.jbi.2010.03.008.
38. Kreimeyer, K., M. Foster, A. Pandey, N. Arya, G. Halford, S.F. Jones, R. Forshee, M. Walderhaug, and T. Botsis. 2017. Natural language processing systems for capturing and standardizing unstructured clinical information: A systematic review. *Journal of Biomedical Informatics* 73: 14–29. https://doi.org/10.1016/j.jbi.2017.07.012.
39. Funk, C., W. Baumgartner, B. Garcia, C. Roeder, M. Bada, K.B. Cohen, L.E. Hunter, and K. Verspoor. 2014. Large-scale biomedical concept recognition: An evaluation of current automatic annotators and their parameters. *BMC Bioinformatics* 15: 59. https://doi.org/10.1186/1471-2105-15-59.
40. Venkatesan, A., J.-H. Kim, F. Talo, M. Ide-Smith, J. Gobeill, J. Carter, R. Batista-Navarro, S. Ananiadou, P. Ruch, and J. McEntyre. 2016. SciLite: A platform for displaying text-mined annotations as a means to link research articles with biological data. *Wellcome Open Research* 1: 25. https://doi.org/10.12688/wellcomeopenres.10210.2.
41. Jurca, G., O. Addam, A. Aksac, S. Gao, T. Özyer, D. Demetrick, and R. Alhajj. 2016. Integrating text mining, data mining, and network analysis for identifying genetic breast cancer trends. *BMC Research Notes* 9: 236. https://doi.org/10.1186/s13104-016-2023-5.
42. Zhou, J., and B.-Q. Fu. 2018. The research on gene-disease association based on text-mining of PubMed. *BMC Bioinformatics* 19: 37. https://doi.org/10.1186/s12859-018-2048-y.
43. IFPMA. 2017. *The pharmaceutical industry and global health: Facts and figures*. https://www.ifpma.org/wp-content/uploads/2017/02/IFPMA-Facts-And-Figures-2017.pdf. Accessed 7 Jun 2019.
44. Arrowsmith, J. 2011. Trial watch: Phase II failures: 2008–2010. *Nature Reviews. Drug Discovery* 10: 328–329. https://doi.org/10.1038/nrd3439.
45. Ashburn, T.T., and K.B. Thor. 2004. Drug repositioning: Identifying and developing new uses for existing drugs. *Nature Reviews. Drug Discovery* 3: 673–683. https://doi.org/10.1038/nrd1468.
46. Langedijk, J., A.K. Mantel-Teeuwisse, D.S. Slijkerman, and M.-H.D.B. Schutjens. 2015. Drug repositioning and repurposing: Terminology and definitions in literature. *Drug Discovery Today* 20: 1027–1034. https://doi.org/10.1016/j.drudis.2015.05.001.
47. Cha, Y., T. Erez, I.J. Reynolds, D. Kumar, J. Ross, G. Koytiger, R. Kusko, B. Zeskind, S. Risso, E. Kagan, S. Papapetropoulos, I. Grossman, and D. Laifenfeld. 2018. Drug repurposing from the perspective of pharmaceutical companies. *British Journal of Pharmacology* 175: 168–180. https://doi.org/10.1111/bph.13798.
48. Nosengo, N. 2016. Can you teach old drugs new tricks? *Nature* 534: 314–316. https://doi.org/10.1038/534314a.
49. Hernandez, J.J., M. Pryszlak, L. Smith, C. Yanchus, N. Kurji, V.M. Shahani, and S.V. Molinski. 2017. Giving drugs a second chance: Overcoming regulatory and financial hurdles in repurposing approved drugs as cancer therapeutics. *Frontiers in Oncology* 7: 273. https://doi.org/10.3389/fonc.2017.00273.
50. Bio. 2016. *Clinical development success rates*. https://www.bio.org/sites/default/files/Clinical Development Success Rates 2006-2015 - BIO, Biomedtracker, Amplion 2016.pdf. Accessed 19 Jun 2019.
51. Pushpakom, S., F. Iorio, P.A. Eyers, K.J. Escott, S. Hopper, A. Wells, A. Doig, T. Guilliams, J. Latimer, C. McNamee, A. Norris, P. Sanseau, D. Cavalla, and M. Pirmohamed. 2019. Drug repurposing: Progress, challenges and recommendations. *Nature Reviews. Drug Discovery* 18: 41–58. https://doi.org/10.1038/nrd.2018.168.
52. Langtry, H.D., and A. Markham. 1999. Sildenafil. *Drugs* 57: 967–989. https://doi.org/10.2165/00003495-199957060-00015.

53. Franks, M.E., G.R. Macpherson, and W.D. Figg. 2004. Thalidomide. *Lancet* 363: 1802–1811. https://doi.org/10.1016/S0140-6736(04)16308-3.
54. NCI. 2006. *Thalidomide*. https://www.cancer.gov/about-cancer/treatment/drugs/thalidomide?redirect=true. Accessed 7 Jun 2019.
55. Baker, N.C., S. Ekins, A.J. Williams, and A. Tropsha. 2018. A bibliometric review of drug repurposing. *Drug Discovery Today* 23: 661–672. https://doi.org/10.1016/j.drudis.2018.01.018.
56. Novac, N. 2013. Challenges and opportunities of drug repositioning. *Trends in Pharmacological Sciences* 34: 267–272. https://doi.org/10.1016/j.tips.2013.03.004.
57. Ding, X. 2016. Drug repositioning needs a rethink. *Nature* 535: 355–355. https://doi.org/10.1038/535355d.
58. Delavan, B., R. Roberts, R. Huang, W. Bao, W. Tong, and Z. Liu. 2018. Computational drug repositioning for rare diseases in the era of precision medicine. *Drug Discovery Today* 23: 382–394. https://doi.org/10.1016/j.drudis.2017.10.009.
59. Sun, W., W. Zheng, and A. Simeonov. 2017. Drug discovery and development for rare genetic disorders. *American Journal of Medical Genetics. Part A* 173: 2307–2322. https://doi.org/10.1002/ajmg.a.38326.
60. Wei, C.-H., H.-Y. Kao, and Z. Lu. 2013. PubTator: A web-based text mining tool for assisting biocuration. *Nucleic Acids Research* 41: W518–W522. https://doi.org/10.1093/nar/gkt441.
61. Hughes, J.P., S. Rees, S.B. Kalindjian, and K.L. Philpott. 2011. Principles of early drug discovery. *British Journal of Pharmacology* 162: 1239–1249. https://doi.org/10.1111/j.1476-5381.2010.01127.x.
62. Capuzzi, S.J., T.E. Thornton, K. Liu, N. Baker, W.I. Lam, C. O'Banion, E.N. Muratov, D. Pozefsky, A. Tropsha, C.P. O'Banion, E.N. Muratov, D. Pozefsky, and A. Tropsha. 2018. Chemotext: A publicly available web server for mining drug–target–disease relationships in PubMed. *Journal of Chemical Information and Modeling* 58: 212–218. https://doi.org/10.1021/acs.jcim.7b00589.
63. Reichardt, P. 2018. The story of Imatinib in GIST – A journey through the development of a targeted therapy. *Oncology Research Treatment* 41: 472–477. https://doi.org/10.1159/000487511.
64. Fuehrer, N.E., A.M. Marchevsky, and J. Jagirdar. 2009. Presence of c-KIT-positive mast cells in obliterative bronchiolitis from diverse causes. *Archives of Pathology & Laboratory Medicine* 133: 1420–1425. https://doi.org/10.1043/1543-2165-133.9.1420.
65. Cahill, K.N., H.R. Katz, J. Cui, J. Lai, S. Kazani, A. Crosby-Thompson, D. Garofalo, M. Castro, N. Jarjour, E. DiMango, S. Erzurum, J.L. Trevor, K. Shenoy, V.M. Chinchilli, M.E. Wechsler, T.M. Laidlaw, J.A. Boyce, and E. Israel. 2017. KIT inhibition by Imatinib in patients with severe refractory asthma. *The New England Journal of Medicine* 376: 1911–1920. https://doi.org/10.1056/NEJMoa1613125.
66. Rappaport, N., M. Twik, I. Plaschkes, R. Nudel, T.I. Stein, J. Levitt, M. Gershoni, C.P. Morrey, M. Safran, and D. Lancet. 2017. MalaCards: An amalgamated human disease compendium with diverse clinical and genetic annotation and structured search. *Nucleic Acids Research* 45: D877–D887. https://doi.org/10.1093/nar/gkw1012.
67. Putkowski, S. 2010. The National Organization for Rare Disorders (NORD). *NASN School Nurse* 25: 38–41. https://doi.org/10.1177/1942602X09352796.
68. Lewis, J., M. Snyder, and H. Hyatt-Knorr. 2017. Marking 15 years of the genetic and rare diseases information center. *Translational Science of Rare Diseases* 2: 77–88. https://doi.org/10.3233/TRD-170011.
69. Bi, Y., M. Might, H. Vankayalapati, and B. Kuberan. 2017. Repurposing of proton pump inhibitors as first identified small molecule inhibitors of endo-β-N-acetylglucosaminidase (ENGase) for the treatment of NGLY1 deficiency, a rare genetic disease. *Bioorganic & Medicinal Chemistry Letters* 27: 2962–2966. https://doi.org/10.1016/j.bmcl.2017.05.010.
70. Tricco, A.C., W. Zarin, E. Lillie, S. Jeblee, R. Warren, P.A. Khan, R. Robson, B. Pham, G. Hirst, and S.E. Straus. 2018. Utility of social media and crowd-intelligence data for

pharmacovigilance: A scoping review. *BMC Medical Informatics and Decision Making* 18: 38. https://doi.org/10.1186/s12911-018-0621-y.
71. Chunara, R., J.R. Andrews, and J.S. Brownstein. 2012. Social and news media enable estimation of epidemiological patterns early in the 2010 Haitian cholera outbreak. *The American Journal of Tropical Medicine and Hygiene* 86: 39–45. https://doi.org/10.4269/ajtmh.2012.11-0597.
72. Kagashe, I., Z. Yan, and I. Suheryani. 2017. Enhancing seasonal influenza surveillance: Topic analysis of widely used medicinal drugs using twitter data. *Journal of Medical Internet Research* 19: e315. https://doi.org/10.2196/jmir.7393.
73. Reece, A.G., A.J. Reagan, K.L.M. Lix, P.S. Dodds, C.M. Danforth, and E.J. Langer. 2017. Forecasting the onset and course of mental illness with twitter data. *Scientific Reports* 7: 13006. https://doi.org/10.1038/s41598-017-12961-9.
74. Adrover, C., T. Bodnar, Z. Huang, A. Telenti, and M. Salathé. 2015. Identifying adverse effects of HIV drug treatment and associated sentiments using twitter. *JMIR Public Health and Surveillance* 1: e7. https://doi.org/10.2196/publichealth.4488.
75. MacKinlay, A., H. Aamer, and A.J. Yepes. 2017. Detection of adverse drug reactions using medical named entities on twitter. *AMIA Annual Symposium Proceedings. AMIA Symposium* 2017: 1215–1224.
76. Rangarajan, S., N.B. Bone, A.A. Zmijewska, S. Jiang, D.W. Park, K. Bernard, M.L. Locy, S. Ravi, J. Deshane, R.B. Mannon, E. Abraham, V. Darley-Usmar, V.J. Thannickal, and J.W. Zmijewski. 2018. Metformin reverses established lung fibrosis in a bleomycin model. *Nature Medicine* 24: 1121–1127. https://doi.org/10.1038/s41591-018-0087-6.
77. Gross, C., A. Banerjee, D. Tiwari, F. Longo, A.R. White, A.G. Allen, L.M. Schroeder-Carter, J.C. Krzeski, N.A. Elsayed, R. Puckett, E. Klann, R.A. Rivero, S.L. Gourley, and G.J. Bassell. 2019. Isoform-selective phosphoinositide 3-kinase inhibition ameliorates a broad range of fragile X syndrome-associated deficits in a mouse model. *Neuropsychopharmacology* 44: 324–333. https://doi.org/10.1038/s41386-018-0150-5.
78. Zogenix. 2019. *Zogenix submits new drug application to U.S.* Food & Drug Administration and Marketing authorization application to European Medicines Agency for FINTEPLA® for the treatment of Dravet syndrome – Zogenix, Inc. https://zogenixinc.gcs-web.com/news-releases/news-release-details/zogenix-submits-new-drug-application-us-food-drug-administration. Accessed 7 Jun 2019.
79. Vanhaelen, Q., P. Mamoshina, A.M. Aliper, A. Artemov, K. Lezhnina, I. Ozerov, I. Labat, and A. Zhavoronkov. 2017. Design of efficient computational workflows for in silico drug repurposing. *Drug Discovery Today* 22: 210–222. https://doi.org/10.1016/j.drudis.2016.09.019.
80. Ferreira, L.G., and A.D. Andricopulo. 2016. Drug repositioning approaches to parasitic diseases: A medicinal chemistry perspective. *Drug Discovery Today* 21: 1699–1710. https://doi.org/10.1016/j.drudis.2016.06.021.
81. Williams, K., E. Bilsland, A. Sparkes, W. Aubrey, M. Young, L.N. Soldatova, K. De Grave, J. Ramon, M. de Clare, W. Sirawaraporn, S.G. Oliver, and R.D. King. 2015. Cheaper faster drug development validated by the repositioning of drugs against neglected tropical diseases. *Journal of the Royal Society, Interface* 12: 20141289–20141289. https://doi.org/10.1098/rsif.2014.1289.
82. Alves, V.M., A. Golbraikh, S.J. Capuzzi, K. Liu, W.I. Lam, D.R. Korn, D. Pozefsky, C.H. Andrade, E.N. Muratov, and A. Tropsha. 2018. Multi-descriptor read across (MuDRA): A simple and transparent approach for developing accurate quantitative structure–activity relationship models. *Journal of Chemical Information and Modeling* 58: 1214–1223. https://doi.org/10.1021/acs.jcim.8b00124.
83. Ekins, S., A.J. Williams, M.D. Krasowski, and J.S. Freundlich. 2011. In silico repositioning of approved drugs for rare and neglected diseases. *Drug Discovery Today* 16: 298–310. https://doi.org/10.1016/j.drudis.2011.02.016.

84. Neves, B.J., R.C. Braga, J.C.B. Bezerra, P.V.L. Cravo, and C.H. Andrade. 2015. In silico repositioning chemogenomics strategy identifies new erugs with potential activity against multiple life stages of Schistosoma mansoni. *PLoS Neglected Tropical Diseases* 9: e3435. https://doi.org/10.1371/journal.pntd.0003435.
85. Govindaraj, R.G., M. Naderi, M. Singha, J. Lemoine, and M. Brylinski. 2018. Large-scale computational drug repositioning to find treatments for rare diseases. *NPJ Systems Biology and Applications* 4: 13. https://doi.org/10.1038/s41540-018-0050-7.
86. Sun, P., J. Guo, R. Winnenburg, and J. Baumbach. 2017. Drug repurposing by integrated literature mining and drug–gene–disease triangulation. *Drug Discovery Today* 22: 615–619. https://doi.org/10.1016/j.drudis.2016.10.008.
87. Karatzas, E., M.M. Bourdakou, G. Kolios, and G.M. Spyrou. 2017. Drug repurposing in idiopathic pulmonary fibrosis filtered by a bioinformatics-derived composite score. *Scientific Reports* 7: 12569. https://doi.org/10.1038/s41598-017-12849-8.

Big Data, Personalized Medicine and Network Pharmacology: Beyond the Current Paradigms

Alessandro Giuliani and Virginia Todde

A Reproducibility Crisis

The 2005 John Ioannidis paper '*Why most published research findings are false*' [1] was a stone in the pond of biomedical sciences. The paper provoked a great debate and, after few years, it was sufficiently clear that both 'ethical' (e.g. 'this is a consequence of misconduct of scientists that, pushed by the publish-or- perish dilemma, produce fake results') and 'technical' (e.g. 'there are big problems in the identification and labelling of cell lines') interpretations were completely inadequate to explain the crisis.

As often happens when in presence of a catastrophe, the first (mainly emotional) answers leave the place to a more rational evaluation. Inferential statistics methods were developed in the first half of the twentieth century with the intention to have an informal way to judge whether evidence was significant in the old-fashioned sense: i.e. worthy of a second look. In the twenty-first century, these methods were turned into a definitive test of truth [2]. Young and Kerr, in their paper [3], adopt a powerful expression to describe the reaction of the scientists in front of a $p < 0.05$ result: they behave '*as a deer caught in the headlight*'; freezing (like the deer) and thinking that 'magic' value indicates the presence of a real effect.

The 'p-value' idolatry is involved in the lack of reproducibility of biomedical sciences and, correctly, many authors advocated for a 'revolution' in the education in science so to give scientists a sufficiently solid statistical knowledge [4], in the same time leading scientific journals published 'declarations of correct statistical conduct' (e.g. [5]).

A. Giuliani (✉) · V. Todde
Istituto Superiore di Sanità, Roma, Italy
e-mail: alessandro.giuliani@iss.it

M. Bizzarri (ed.), *Approaching Complex Diseases*, Human Perspectives in Health Sciences and Technology 2, https://doi.org/10.1007/978-3-030-32857-3_5

This is for sure an important issue, but the question of why this lack of understanding was so pervasive, notwithstanding a relevant scientific literature dealing with both the importance of taking into consideration the size of an effect not focusing exclusively on the statistical significance [6, 7] and the related problems of chance correlation and overfitting [8], remains unanswered. The question is still more intriguing if we consider that many inferential statistics methods were developed within the realm of biological sciences [9].

We think that the sloppy [8] statistical treatment of results (while deprecable) is probably not the primary cause of information crisis, but only a symptom of the failure of one of the main (even if seldom overtly declared) pillars of biomedical science: i.e. that the 'causally relevant' phenomena are ever located at the molecular level.

This interpretation is consistent with Table 1 of the Ioannidis paper [1]: the positive predictive value (PPV) of biomedical investigations has a dramatic fall going from the organism level of correctly designed clinical trials (PPV = 0.85) to the molecular biology level of 'Discovery Oriented Exploratory Research' (PPV = 0.001).

These figures tell us of a lack of relevance of molecular level investigation with respect to the organism level outcome suggesting basic biological research is actually located at an inadequate (too detailed) noise- dominated level of investigation [8].

Where Interesting Events Do Happen

The increasing importance that molecular biology gained in the last 30 years, made the majority of biologists to think the 'ultimate explanations' must be looked for at the molecular level, being the paradigm of a biological explanation something like *'gene A provokes phenomenon (disease, phenotypic trait...) B by means of the pathway C'*. The existence of a single 'explanatory layer' is in sharp contrast with what we know about complex structured systems, where multi-layer causality is at work [9].

Ecology (the biology field with a most sensible use of quantitative tools) recognized since many years that the 'most microscopic' level of organization is not necessarily the place where 'the most relevant facts do happen'. On the contrary, the most fruitful scale of investigation is where 'non-trivial determinism is maximal' [10] i.e. the scale more 'rich' in meaningful correlations or, in ecological terms, the mesoscopic realm.

Non-trivial determinism can be defined in terms of prediction error as [10]:

$$\text{Prediction } r^2 = 1 - E^2/S^2$$

In the above formula, E is the mean prediction error and S the standard deviation. In the case of a simple linear regression in which a dependent variable Y should be predicted by an independent variable X, the non-trivial determinism corresponds to squared Pearson correlation. The formula can be extended to any other situation in which we wish to predict a system feature Y located at a hierarchical higher layer with respect to X, moreover both X and Y do not need to represent single variables but any suitable set of information at any definition scale.

Consequently, in the 'many Y'/'many X' case, the non-trivial determinism could correspond to the first canonical coefficient [11], while in the case of a binary diagnosis to the area below the ROC curve [12].

It is worth focusing on the specification 'non-trivial' attached to the word determinism. The statement '*Any protein is made up of 20 different amino-acid residues*' indicates a shared feature of the chemical composition of the protein molecules but, for the same fact it holds identical for all the proteins, it gives no information on the differences among protein structures. This should be obvious in the case of biology, where the relevance of a scientific statement stems from the ability of getting rid (e.g. by establishing a meaningful correlation) of the variance of the system at hand but, as often happens, 'obvious' statements tend to be overlooked.

The essence of any scientific enterprise is not the accumulation of as many details as possible on a given phenomenon, but the recognition of the very few crucial parameters that allow for a good predictive power. It is relatively immediate to understand that in real world (otherwise science should not be possible) incredibly complex microscopic models encompassing a huge number of details, end up into effective theories based upon few coarse-grained macroscopic parameters. Thus, while three-dimensional molecular liquids have huge microscopic complexity, in a certain regime (lengths and times large with respect to molecules vibration periods), their behavior is determined entirely by their viscosity and density [8].

This drastic collapse of information and subsequent simplification, stem from the presence of a correlation structure among the microscopic players that drastically limits the effective dimensionality of a system along few latent dimensions [8]: the discovery and quantification of such latent dimensions is the goal of multidimensional data analysis techniques like Principal Component Analysis [8, 13].

One the fathers of information science, Warren Weaver, in his fundamental "Science and Complexity" 1948 paper [14], proposed a three-class partition of science into: (1) Organized Simplicity, (2) Disorganized complexity and (3) Organized complexity that allows for a clear explanation of the issue.

The first class (simplicity) refers to the classical use of quantitative methods in science. Class 1 problems allow for extreme abstraction (e.g. a planet becomes a dimensionless 'material point'). This allows to write down differential equations predicting the behavior of the studied system relying on the stability in both space and time of the experimental (observational in the case of astronomy) results. The drastic reduction of the relevant properties down to very few basic features like mass and distance, allows for a straightforward appreciation of classical physics.

The consideration of mass and distances as the only relevant features of planets is not a 'collapse of information due to internal correlation of the system' but something completely different: is abstract thinking. In Class 1 problems, we can completely forget about the nature of the involved objects: this is why introductory Physics manuals use examples based upon balls, cars, stones, planets to explain the same laws of nature. This is a top-down approach made possible by the existence of (very few) general largely context independent laws. Hyper-reductionist approaches in Biology, in many cases, suppose that a gene (or a protein), if observed at the molecular basic level, display such a deterministic and unescapable character fuzzyfied at macro-level by the addition of external noise obscuring the natural simplicity of microscopic interactions. This is why molecular biology and biochemistry books are full of box-and-arrows diagrams.

This attitude has nothing to do with the 'bottom-up' recognition of coarse-grained parameters that is at the basis of Weaver class 2 problems.

Problems of disorganized complexity (class 2) allow for a greater generalization power (and similar accuracy) than class 1, by means of a very different style of reasoning. Here, the predictive power stems from the generation of macroscopic descriptors corresponding to gross averages on a transfinite number of atomic elements. Thermodynamics is the brightest example of this style of reasoning: emergent collective parameters like temperature or pressure allow for an accurate control of system behavior without the need to go into microscopic (noise-dominated) details.

Both the approaches must fulfill very stringent constraints. Class 1 approach asks for few involved elements interacting in a stable way, class 2 style needs a very large number of identical particles with only negligible (or very stable and invariant) interactions among them. Biological systems, only in very few cases do fulfill these constraints, so we step into Weaver's third class (organized complexity), here the emergent collective parameters ask for a less straightforward thinking.

Organized complexity arises when many (even if not so many as in class 2) non-identical elements each other interact with time-varying correlation strength. This provokes an extreme context dependence of the results so giving rise to the actual information crisis: the same experiment, performed in slightly different context, can give rise to opposite outcomes. Organized complexity is the 'middle kingdom' where life sciences reside that was recognized as the twenty-first century frontier of basic science [15].

Before going ahead, is worth reporting the original figure of the Weaver paper [14] sketching the three realms of science (Fig. 1):

Weaver claims that when dealing with complex organized systems, the focus of the investigation must shift from the detailed analysis of single elements to their wiring pattern.

The key point is to make the 'relevant parameters' to emerge from the consideration of the mutual correlations among the system descriptors.

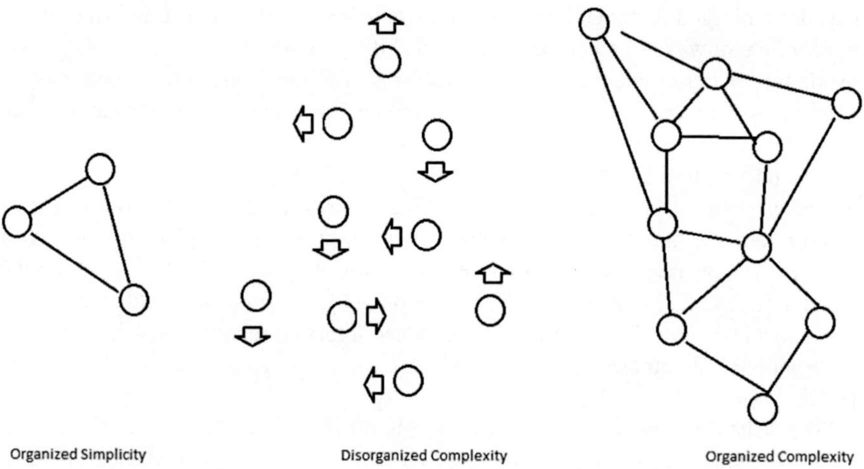

Fig. 1 Circles represent the elementary players, the lines their mutual relations. The arrows of the graph in the middle (disorganized complexity) correspond to the trajectories of the particles whose interactions are both random and contingent. The third panel depicts organized complexity: the interactions are both non- negligible and time varying

A Dangerous Trend

The clarity of the Weaver's message faded away by the action of a drastic terminological (and philosophical) revolution: it is not without consequences referring to quantitative approaches by the term 'Informatics' instead of 'Mathematics' (as actually was the case in the great majority of biological applications).

Bioinformatics revolution started with the need of generating (and storing) very long symbolic strings correspondent to the sequences of biopolymers (DNA, RNA and protein molecules). The analysis of symbolic strings is probably the 'most classical' problem of informatics dating back to the very birth of the discipline since Alan Turing seminal studies [16]. This act-of-birth influenced the development of the relation between Informatics and biological sciences. Bioinformatics tools are considered as purely technical devices (like a fridge or a spectrophotometer) helping the biologist to answer questions that arise from (largely qualitative) speculation. The 'quantitative step' is very stereotyped and reduces to a 'pattern matching' in which the problem is to find the 'maximal superposition' between a 'query' (e.g. an unknown biopolymer sequence found in a sample) and a 'target' (sequences of proteins whose physiological role is known). Even (apparently) much more sophisticated 'machine intelligence' approaches are part of the same class of applications.

This state of affairs transforms the information scientist into a 'servant' solving a practical problem with no relevant role in the emergence of new insights. The generation and interrogation of static (only growing for brutal addition of new data) repositories where to look for potentially useful hints for the problem at hand, constitutes the almost totality of the work. This approach, in the last two decades, enlarged its range from biopolymers to gene expression (transcriptomics), metabolic pathways (metabolomics), medical diagnoses but substantially stays well inside the 'pattern matching' class of problems. The maximization of 'non- trivial' determinism instead, asks for solutions allowing for a common explanation of an entire 'class' of phenomena and not only limited to a specific instance.

The information crisis exacerbated the 'Bioinformatics' stereotypy, as evident in nowadays 'Big Data' enchantment. The 'Big Data' approach starts from a correct assumption: the nowadays information crisis is an 'overfitting' crisis. When in presence of too many degrees of freedom (being they genes, proteins, metabolic reactions...) as consequence of the development of 'high-throughput' techniques allowing to measure thousands of different descriptors on relatively few independent observations, hypothesis-driven research based on few parameters is out of scale [1, 3].

The 'Big Data' proposal to overcome this problem is (roughly): '*Let's give up with theory-driven experimentations and let's look, without preconceived ideas, to the 'whole-thing'* (the development of various –omics makes this possible): the emerging correlations will allow for new ideas and findings spontaneously appear in a data-driven way' [17].

Notwithstanding the increasing funding of 'Big Data' initiatives, it is sufficiently clear that the pure enumeration of single relevant correlations across a huge number of variables only exacerbates the reproducibility crisis [8]. Pure data-driven approaches set forth by the 'Big Data' extremists claiming for the 'end of scientific method' (see for example [18]), risk to become the Heaven of chance correlations (see [19] for a very interesting critic to the pure informatics approach to science).

What we really need is to look for 'Universal Organization Principles' of complex systems (Weaver Class 3) moving from the single nodes to the wiring pattern level [10] so to discover general principles of organization [15] largely independent of microscopic details. In order to perform such a 'quantum leap' we could profitably make use of information science tools but only if we complement these tools with a sort of 'statistical mechanics of data' [20].

Statistical Mechanics of Data

As pointed out by Nicosia et al. [21]: "*Networks are the fabric of complex systems*". This is why different investigation fields – from protein science [22] to neuroscience [23] – build upon the consideration that shared organization rules should give rise to similar phenomenology, independently of the nature of the constituting elements. The quest for 'network laws' largely independent of the nature of the constituting nodes of the network, stems from the work of the Dutch electrical engineer Bernard Tellegen [24] that developed a sort of conservation principle of both potential and flux across a network analogous to Kirchoff's laws. The flux does not need to be an electrical current and the same holds for the potential. Any system modeled by a set of nodes linked by edges (being them metabolites linked by chemical reactions

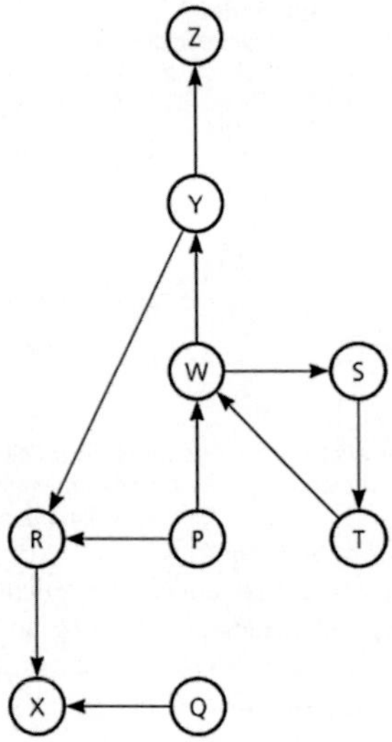

	P	Q	R	S	T	W	X	Y	Z
P	0	0	1	0	0	1	0	0	0
Q	0	0	0	0	0	0	1	0	0
R	0	0	0	0	0	0	1	0	0
S	0	0	0	0	1	0	0	0	0
T	0	0	0	0	0	1	0	0	0
W	0	0	0	1	0	0	0	1	0
X	0	0	0	0	0	0	0	0	0
Y	0	0	1	0	0	0	0	0	1
Z	0	0	0	0	0	0	0	0	0

Fig. 2 Left panel reports a complex network in the usual way of nodes (elements of the system) linked by edges (between elements interactions), right panel reports the identical information in the form of a binary adjacency matrix. In both cases we deal with the same information, the right panel representation is more practical for computational purposes

transforming one into the other or mutually interacting persons in an office. . .) has similar emerging properties. As aptly stressed in [25], the theorem opens the way to a sort of 'network thermodynamics', whose principles are strictly dependent from wiring architecture while largely independent of the constitutive laws governing the single elements.

Formalizing a given problem in terms of a graph (a mathematical graph is equivalent to a network expressed in terms of its adjacency matrix) allows for a thermodynamic-like approach (here focusing on relations and no more on means like in Weaver class 2) to be applied to complex systems.

We can roughly describe the network approach as the answer to the question "*What can we derive from the sole knowledge of the wiring diagram of a system?*"

An adjacency matrix (and consequently a complex network) can generate from any sensible correlation metrics applied to the elements of a system. A correlation matrix reporting the pairwise Pearson coefficients between continuous variables, Euclidean distances between discrete landmarks (e.g. amino- acid residues location in 3D protein structure, species abundance profiles) or the phase coherence of electrophysiological signals are only some examples of the situations that can profitably expressed as graphs (Fig. 2).

The most basic level of quantitative description of graphs (correspondent to descriptive statistics of random variables) is the computation of so called 'graph invariants' [26]. These invariants are relative to local (single nodes), global (entire network), and mesoscopic (clusters of nodes, optimal paths) levels.

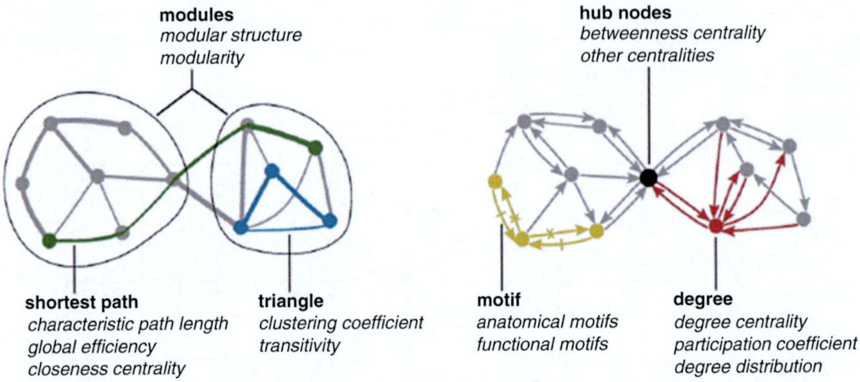

Fig. 3 Modules correspond to subset of nodes having much more links among them than with other nodes of the network. Measures of centrality (closeness, betweeness...) describe nodes in terms of the number of shortest paths traversing them. Shortest path is the characteristic metrics for networks: they correspond to the shortest distances (in terms of number of nodes/links to be traversed) for linking pairs of nodes

Thus, the "degree" (how many links are attached to a given node) is a local descriptor, the "average shortest path", corresponding to the average length of minimal paths connecting all the node pairs, is a mesoscopic feature, while the general connectivity of the network (density of links) is a global property [26] but all these descriptors are naturally inter-related.

Figure 3 reports an exemplar network structure with the indication of some relevant descriptors (graph invariants) of the wiring architecture: the values of local, mesoscopic and global descriptors depend on each other by the simple fact they are relative to the same network architecture. This creates a 'natural' microscopic-macroscopic link devoid of any theoretical assumption.

This very basic level of description allows for deriving useful biological information: e.g. protein structures can be formalized as graphs (protein contact networks) having amino-acid residues as nodes. A link is established between two i and j residues if $d(i,j) < R$, where $d(i,j)$ is the Euclidean distance between i and j and R corresponds to Van der Walls radius, the maximal distance the two residues can engage an effective relation (i.e. the maximal distance they can be considered in contact) [22]. Similar considerations hold true for metabolism, food webs, neural and social systems [27].

Using network invariants as such corresponds to the routine use of statistical indexes; a statement like '*Drug A significantly decreases average degree of gene expression network*' substitutes '*Drug A significantly decreases average expression of gene X*'. The only (but very important), difference between the two above statements is that, while the second statement refers to a microscopic level effect, the first one refers to a mesoscopic property.

Network Pharmacology and the 'Blessing of Dimensionality'

The acceptation of mesoscopic measures as proper biologically meaningful descriptions is a still open cultural problem of the biomedical sciences community, but we can safely affirm that the 'mesoscopic style' is gaining momentum. A very good example is the emerging field of 'Network Pharmacology' [26, 28–30].

For decades, the dominant paradigm of pharmacology was fully reductionist: the goal of pharmacological research was to look for the main molecular determinant of a given disease. This determinant, generally a protein molecule, was considered the "target" of the drug (receptor) and the candidate drugs were screened for their ability to bind selectively to the receptor [30]. The 'best binders' candidates entered into subsequent phases in which their efficacy was tested on animal models of the disease, and eventually go into clinical trials.

This strategy worked remarkably well for around 30 years then, almost abruptly, around the eighties of the last century, it entered a deep crisis provoking the apparent paradox of an exponential growth of basic knowledge going hand-in-hand with a drastic fall of newly marketed drugs. This crisis is the 'application' counterpart of the information crisis we described in the first paragraph: in [31] Overington and colleagues sketched an approximate estimate of 76% of drugs discovered in the last 20 years referring to receptor molecules discovered around the fifties. On the contrary, only the 6% bind to recently discovered targets, while for the remnants no reasonable hypothesis of mechanism of action does hold.

The promise of a "druggable genome" set forth by the completion of human genome sequencing with the consequent opening of a practically infinite horizon for the development of new drugs, failed: something very fundamental went wrong. The network pharmacology paradigm tries to overcome the above reductionist view posing that biological regulation, at any level, must be intended as a relational affair in which the observed action is the resultant behavior of an interacting network of agents.

The network paradigm changes the way we screen candidate drugs starting from the very first steps. In the reductionist paradigm, we look for molecules that selectively bind to the receptor and discard the candidates binding to a multiplicity of different receptors.

In the network paradigm, on the contrary, we prefer those candidate drugs weakly binding to a multitude of different receptors because they are more efficient "network modifiers" inducing a systemic effect pushing the system dynamics to another mode of functioning (attractor) [30, 32].

Simultaneous 'attacks' in different points of a network are more potentially 'disruptive' with respect to single target attack, in terms of outcome at the organism scale, for the simple reason that such an outcome happens at a macroscopic scale and thus has to do with a collective property of the underlying set of microscopic interactions. Such a general property is, in other words, an emergent feature, a sort of average (graph invariant) of the entire set of microscopic interactions, this allows for a drastic simplification of the studied phenomenon, a 'blessing of dimensionality' [20].

Brain is probably the most complex interaction network present in Nature; the interactions among brain elements span many organization layers (from single neurons to anatomical substructures) and physiological features (from electrical activity to glucose metabolism). It is thus out of scope to try to go down to the microscopic level in order to get rid of the fine mechanisms of complex pathologies like Alzheimer disease. On the contrary, the amount of variance explained by the first principal component (a sort of general amount of connectivity of the system) of brain metabolism (being the variables the metabolic rate of different brain areas evaluated by PET) scales almost perfectly (Pearson $r = 0.97$) with the transition from normal aging to Alzheimer disease [33]. This approach does not look into the subtleties of the wiring pattern of the brain networks, but focus on a thermodynamic (global) quantification of the system-as-a-whole that is both reliable and generalizable [8].

The 'Blessing of Dimensionality' is the secret of the success of statistical mechanics and implies the conscious acquiring of a physically motivated hypothesis on the phenomenon in study that is exactly the opposite attitude with respect to the Big Data philosophy advocating a pure brute force approach in facing high dimensionality problems.

Personalized Medicine: A Field of Battle

The Prasad 2016 paper [34] is only one of the many papers casting doubts on the possibility to develop 'personalized cures' on the basis of the specific genomic background of each single individual, the paper criticizes the possibility to profitably use a 'personal genomic signature' driving the cure. Some other authors do not agree with this view, and claim for the unescapable link between 'genomics' and disease cure (see for example [35]) that will undoubtedly push to success the project of precision (or personalized) medicine. Without entering into the battle from a biomedical perspective, here it is worth stressing some basic statistical/epistemological nodes of precision medicine that are particularly relevant to clarify the above described mesoscopic approach.

The term 'personal' stems from the irreducible singularity of each human being: the word comes from Latin (and Italian) word '*persona*' (English: person) that means *per-se-una* that we can translate as 'a unity in itself'. This implies that we can ever find one (or more) features that pertain only to that specific person.

Science relaxes this concept and, without asking for an impossible absolute identity of two persons (patients), limits the identity to the absence of a statistically significant departure of the specific patient as for any of a set of N relevant parameters. This is exactly what happens when we look at our clinical analyses score and we discover (with satisfaction) that all our haematological parameters are within their confidence intervals. Such intervals come from reference healthy individual populations that are taken as reference. Let us imagine a set of N variables (e.g. N cancer-relevant gene expression values) that define a hyper-volume (an N-dimensional area) of 'business-as-usual' correspondent to the use of standard therapy.

This 'standard zone' is defined by N-dimensional distribution of variables X1, X2, ...Xn. We assume that distributions are normal and each Xi variable independent from the others. This corresponds to imagine each variable as '*relevant per se*' given it is endowed by a singular and peculiar biological (clinical) meaning. Starting from a very simple statistical case (e.g. measuring the blood levels of a molecule, X1), we crudely divide the population in three (k) equal fractions, such that each individual is either below normal, normal or above normal. Thus, we set the "cut-off" for normal at P(1) = 1/3 (each tail is unhealthy). The probability to be "abnormal" with respect to X1 is then $1-(1/3) = 2/3$ (only the central interval is considered as 'normal').

Now if we consider two "health-dimensions", then the probability to be normal with respect to both, X1 and X2 is $P(1,2) = (1/3)^2$. For N-dimensions, the probability that a person is healthy with respect to all N health variables is minuscule:

$P(1...n) = (1/3)^n$. Hence, the probability to be "sick", defined by multiple clinical outcomes (or algorithm) as being in the abnormal tail fraction in at least one dimension would $(1 - (1/3))^n$. For n = 1000 this probability is near 0. Thus, anyone is practically certain to be labelled as "abnormal" in some respect. Considering the usual 95% confidence interval the figures become $(1 - (0.05))^n = 0.95^n$ that for n = 1000 (a number of dimensions in the range of the actual knowledge of the 'involved players' in cancer) we obtain a probability near ($P = 9.5^{999}$) (practically zero)to be in the 'standard zone'. If we consider as 'abnormal' even a single statistically significant departure from the normality range.

Even if we include some correction, (e.g. Bonferroni correction), the figures remain critical, increasing the stringency of significance threshold (i.e. passing from 0.05 to 0.005, 0.0005 etc.) does not solve the problem [36].

This is an 'extreme' way of reasoning. The fact that any relevant '-omics' refers to an highly correlated underlying system, decreases the probability that a 'peculiar' case is 'peculiar' in only one dimension and, in the same time, makes the 'standard zone' much more populated with respect to the extreme vacuum envisaged by plain statistical considerations. Notwithstanding that, the above example clearly indicates the need of a jump from the microscopic to the mesoscopic scale when in presence of high-dimensionality data sets. The root of information crisis described in the first paragraph lays in the use of formally correct statistical reasoning developed for few variables in the wrong context of high dimensionality [1, 3].

We need a completely different approach making high-dimensionality a blessing and not a curse [20].

The main determinant of the confidence interval length and consequently of the probability to define a single observation as 'abnormal' is the standard deviation of the reference population. Standard deviation (the square root of Variance) can be interpreted as the average of Euclidean distances of a population of N points from their centre of mass, so to their mean value. That is evident looking at the formula:

$$\text{Std.Dev.}(X) = \sqrt{\sum (Xi - M)^2 / N} \quad (1)$$

In (1) the summation extends to all the N statistical units, Xi indicates the value of the variable X relative to the i-th unit and M is the mean value (centre of mass) of the data set. The formula exactly corresponds to the average distance from the centre of mass (M) of a set of N points by the action of Euclidean distance in one dimension (the variable X). If we enlarge this concept to multivariate spaces and consider a set of N points around a common centre of mass in a p-dimensional space we can compute the distance from the centre of mass (mean) of each i-th observation according to the formula (2) with the summation extended from k = 1 to p being p the number of variables.;

$$\text{Dist}(i, M) = \sqrt{\sum (Xi, k - Mk)^2} \quad (2)$$

The average of the above distance values equates the standard deviation of the considered data set as for the multivariate space. If we perform the above computation on a reference 'healthy' population we will generate a 'healthy area' corresponding to the hyper-circle that keeps inside' a given percentage of observations (let's say the 95% of total population corresponding to a circle having the radius equal to 2SD).

Now we can easily appreciate how a Multivariate Confidence Interval looks like (Fig. 4): any observation that lies outside the circle is considered as 'abnormal'.

This implies that a point, in order to be 'out-of-the-circle' (i.e. aberrant with respect to the distribution or statistically significant) must have a distance from the centre much greater than the average distance of the reference population.

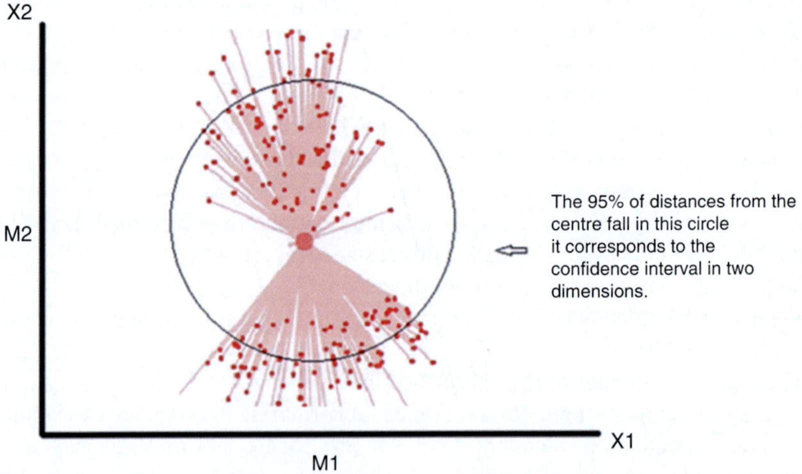

Fig. 4 Bi-dimensional visualization of multivariate Confidence Interval. The circle shows the 95% (CI) of distances from the centre

It is worth nothing that the paradoxical effect in which anything seems to be out of the norm in the case of multivariate distributions disappears by shifting to distances. In (2) a single 'anomalous' addendum (correspondent to a statistical significance for a single descriptor) cannot make the global distance from the centre to be anomalous as well, for the intuitive fact that, the higher the number of variables, the more 'diluted' is a single 'out-of-scale' result relative to a particular X variable.

In the case of multivariate normal distribution, the significance computations do not rely upon single t-test but on the Hotelling T-squared distribution. Variance scales with p (number of dimensions, degrees of freedom). This is exactly correspondent to the 'dilution factor' sketched above: the more variables taken into consideration, the higher the variance, the more demanding scoring a significant result. We are definitely out of the paradox territories taming the high dimensionality beast.

Thus, when dealing with multivariate distributions, we must think in terms of distances on the whole space and not in terms of a variable after the other in sequential order. This shift of focus deeply changes the nature of the concept of 'what is abnormal' that goes from '*An observation is 'out-of-norm' even if only one of its descriptors is outside the confidence interval*' to '*An observation is 'out-of-norm' if its distance from the centre-of-mass of a reference distribution is outside the confidence interval'*.

The shift to distance, while solving the paradoxical void of 'normal area' implicit in the classical 'variable- by-variable' is much more demanding in terms of biological interpretation. Why in single variable approach we must not to bother about the 'meaning' of each piece of information (it is perfectly sound for a physician taking into consideration the implications of a single out-of-norm index as the cholesterol level in the context of a given general frame of reference) this is not the case for distance approach. In multivariate spaces, by the action of distance computation, we are obliged to '*take all the packet of p variables as a whole*': we must pay a huge attention on the internal consistency of the pieces of information inside the packet. In order abnormality to emerge, it must have a widespread effect over the entire set of variables. This makes the issue of the correlation structure (and internal coherence) among descriptors the most important point.

The theme of between variables correlations has a clear counterpart in the geometrical perspective. In fact, any proper Euclidean metrics implies the axes are each other orthogonal, which means that the variables are each other independent or, in other words, they have a Pearson r equal to zero. Pearson r in fact is nothing else than the cosine of the angle defined by the two X1, X2 variable vectors. Thus, in order to compute properly defined Euclidean distances, the between variables correlation must be taken into consideration and eventually correct for the existing correlations.

This corresponds to shift from plain Euclidean distances to Mahalanobis distances that are Euclidean distances corrected for between variables correlations [37]. Practically, we could compute the Mahalanobis distance of a point (animal, human, cell culture, protein, etc.) from its reference distribution, and if significant, we are allowed to label it as 'abnormal' [38] and consequently decide (in the case of a patient) for a personalized treatment outside the standard protocol. The same result could be achieved by computing the Euclidean distance on the Principal Components full space where K components stand for k variables (principal components are each other orthogonal by construction so Mahalanobis is identical to plain Euclidean metrics). Furthermore, Principal Component Analysis can face the "dilution problem": if k (original dimensionality of the problem) is too high (so making it impossible to detect abnormal points), it is possible to limit the components to a number k1 << k.

Principal components correspond to the eigenvectors of the correlation (or covariance) matrix of the analysed variables, i.e. they are the image in light of the wiring architecture of the underlying interaction network [13]: this makes 'realistic' personalized medicine approaches to go back to the natural 'organized complexity' realm.

The crucial point to be stressed is that the success of the procedure is strictly dependent on the initial choice of the variables that involves a 'semantic' (human based) knowledge of the variable space.

The human 'subjective judgement' comes necessarily into play in terms of the interpretation of the components in terms of knowledge and expertise of the analyst: the k1 components to retain are not necessarily those endowed with the higher eigenvalues, but those that can be explained in scientific terms by means of the critical analysis of their loading pattern. This implies *a priori* knowledge of the physician that cannot be substituted by a purely data-driven approach that is necessarily strictly dependent of the considered data set.

Conclusions

The entanglement of 'content' and 'methodological' knowledge is the basic methodological novelty made necessary by the actual information crisis (and consequent lack of practical efficacy) of biomedical sciences. The classical separation of scientific enterprises into a linear sequence made of: 'hypothesis setting'- 'experimental methods' – 'data analysis'– 'hypothesis verification/falsification' is untenable in the high- throughput era. This by no means must be intended in terms of 'end-of-theory-driven-science' pretending the sole 'data analysis' step encompasses all the other segments of scientific work letting relevant results to naturally emerge by the brute force of computational power applied to massive data sets. On the contrary, all the different components remain alive and well but they are all present from the beginning to the end of a scientific enterprise. Each data analysis choice is strictly dependent from theoretical assumptions and each theoretical assumption is in turn influenced and modulated along the process by the emerging results. We hope to have demonstrated that the Big Data frame, when correctly interpreted, assigns a great relevance to theoretical work, but this work is not separated from the actual methodological choices to be adopted at each step of the analysis. It is not a case that the 'Results' and 'Discussion' sessions, usually distinct in scientific papers, collapse into a single 'Results and Discussion' section in data mining based investigations.

Curiously enough this style of reasoning is reminiscent of the traditional scientific work in which the same scientist built his/her investigation tools inspired by the theoretical work. Until the first half of the twentieth century a theoretical physicist like Enrico Fermi personally assembled his instrumentation inspired by his personal view of the physics problem he wanted to investigate [39].

This is not without consequences as for the kind of science education most fit for our times: we are again in need, resembling the title of Enrico Fermi biography [39] of 'know all scientists', while the nowadays hyper- specialization became detrimental for real science advancements. This 'know all' ideal can by no means take care of all the details of any specific field but asks for a drastic simplification of the different bodies of knowledge to get a shared minimal set of essential concepts. The necessary paradigm change starts with a fade from the prison of specialization.

References

1. Ioannidis, J.P. 2005. Why most published research findings are false. *PLoS Medicine* 2 (8): e1242.
2. Nuzzo, R. 2014. Scientific method: Statistical errors. *Nature* 506 (7487): 150.
3. Young, S., and A. Kerr. 2011. Deming, data and observational studies a process out of control and needing fixing. *Significance* 8 (3): 116–120.
4. Voosen, P. 2015, March 6. Amid a sea of false findings, the NIH tries reform. *The Chronicle of Higher education*
5. Munafò, Marcus R., et al. 2017. A manifesto for reproducible science. *Nature Human Behaviour* 1: 1–0021.
6. Kraemer, H.C., and D.J. Kupfer. 2006. Size of treatment effects and their importance to clinical research and practice. *Biological Psychiatry* 59 (11): 990–996.
7. Richardson, J.T. 1996. Measures of effect size. *Behavior Research Methods, Instruments, & Computers* 28 (1): 12–22.
8. Transtrum, M.K., B.B. Machta, K.S. Brown, B.C. Daniels, C.R. Myers, and J.P. Sethna. 2015. Perspective: Sloppiness and emergent theories in physics, biology, and beyond. *The Journal of Chemical Physics* 143 (1): 07B201_1.
9. Agresti, A., and C.A. Franklin. 2007. *Statistics: The art and science of learning from data*. Upper Saddle River: Pearson Prentice Hall.
10. Pascual, M., and S.A. Levin. 1999. From individuals to population densities: Searching for the intermediate scale of nontrivial determinism. *Ecology* 80 (7): 2225–2236.
11. Härdle, W., and L. Simar. 2007. Canonical correlation analysis. In *Applied multivariate statistical analysis*, 321–330. Berlin/Heidelberg: Springer.
12. Heagerty, P.J., and Y. Zheng. 2005. Survival model predictive accuracy and ROC curves. *Biometrics* 61 (1): 92–105.
13. Giuliani, A. 2017. The application of principal component analysis to drug discovery and biomedical data. *Drug Discovery Today* 22 (7): 1069–1076.
14. Weaver, W. 1948. Science and complexity. *American Scientist* 36: 536–549.
15. Laughlin, R.B., D. Pines, J. Schmalian, B.P. Stojković, and P. Wolynes. 2000. The middle way. *Proceedings of the National Academy of Sciences* 97 (1): 32–37.

16. Turing, A. M. 2006. Biological sequences and the exact string-matching problem. In *Introduction to computational biology*. Springer
17. Todde, V., and A. Giuliani. 2018. Big data. A briefing. *Annali dell'Istituto Superiore di Sanità* 54 (3): 174–175.
18. Anderson, C. 2008. The end of theory: The data deluge makes the scientific method obsolete. *Wired Magazine* 16 (7): 16–07.
19. Calude, C.S., and G. Longo. 2017. The deluge of spurious correlations in big data. *Foundations of Science* 22 (3): 595–612.
20. Gorban, A.N., and I.Y. Tyukin. 2018. Blessing of dimensionality: Mathematical foundations of the statistical physics of data. *Philosophical Transactions of the Royal Society A – Mathematical Physical and Engineering Sciences* 376 (2118): 20170237.
21. Nicosia, V., M. De Domenico, and V. Latora. 2014. Characteristic exponents of complex networks. *EPL (Europhysics Letters)* 106 (5): 58005.
22. Di Paola, L., M. De Ruvo, P. Paci, D. Santoni, and A. Giuliani. 2012. Protein contact networks: An emerging paradigm in chemistry. *Chemical Reviews* 113 (3): 1598–1613.
23. Hauser, T.U., V.G. Fiore, M. Moutoussis, and R.J. Dolan. 2016. Computational psychiatry of ADHD: Neural gain impairments across many levels of analysis. *Trends in Neurosciences* 39 (2): 63–73.
24. Tellegen, B. 1952. A general network theorem with application. *Phillips Research Reports* 7: 259–269.
25. Mickulecki, D. 2001. Network thermodynamics and complexity: A transition to relational systems theory. *Computers & Chemistry* 25: 369–391.
26. Csermely, P., et al. 2013. Structure and dynamics of molecular networks: A novel paradigm of drug discovery: A comprehensive review. *Pharmacology & Therapeutics* 138: 333–408.
27. Kohestani, H., and A. Giuliani. 2016. Organization principles of biological networks: An explorative study. *Biosystems* 141: 31–39.
28. Hopkins, A.L. 2008. Network pharmacology: The next paradigm in drug discovery. *Nature Chemical Biology* 4 (11): 682–690.
29. Ligeti, B., et al. 2015. A network-based target overlap score for characterizing drug combinations: High correlation with cancer clinical trial results. *PLoS One* 10 (6): e0129267.
30. Csermely, P., et al. 2005. The efficiency of multi-target drugs: The network approach might help drug design. *Trends in Pharmacological Sciences* 26: 178–182.
31. Overington, J.P., et al. 2006. How many drug targets are there? *Nature Reviews. Drug Discovery* 5 (12): 993–996.
32. Huang, S. 2009. Reprogramming cell fates: Reconciling rarity with robustness. *BioEssays* 31 (5): 546–560.
33. Pagani, M., et al. 2016. Predicting the transition from normal aging to Alzheimer's disease: A statistical mechanistic evaluation of FDG-PET data. *NeuroImage* 141: 282–290.
34. Prasad, V. 2016. Perspective: The precision-oncology illusion. *Nature* 537 (7619): S63.
35. Abrahams, E., and S.L. Eck. 2016. Molecular medicine: Precision oncology is not an illusion. *Nature* 539 (7629): 357.
36. Goh, W.W.B., and L. Wong. 2018. Dealing with confounders in omics analysis. *Trends in Biotechnology* 36 (5): 488–498.
37. Penny, K.I. 1996. Appropriate critical values when testing for a single multivariate outlier by using the Mahalanobis distance. *Applied Statistics* 45: 73–81.
38. De Sanctis, R., A. Viganò, A. Giuliani, A. Gronchi, A. De Paoli, P. Navarria, V. Quagliuolo, A. Santoro, and A. Colosimo. 2018. Unsupervised versus supervised identification of prognostic factors in patients with localized retroperitoneal sarcoma (RPS): a data clustering and the Mahalanobis distance approach. *Biomed Research International*.
39. Schwartz, David N. 2017. *The last man who knew everything: The Life and times of Enrico Fermi, father of the nuclear age*. New York: Hachette.

Epigenetic Control Using Small Molecules in Cancer

Tomohiro Kozako, Yukihiro Itoh, Shin-ichiro Honda, and Takayoshi Suzuki

Introduction

The control of gene expression sits at the core of biological processes, and epigenetic mechanisms, which are thought to primarily influence gene expression at the transcriptional level [68, 103, 162], are vital for appropriate differentiation and development (Fig. 1, Table 1) [176]. Epigenetics is defined as heritable alterations in gene expression that are not caused by changes in DNA sequence [21]. Epigenetic mechanisms include DNA methylation, post-translational modifications (PTMs) of histones, various RNA-mediated processes, gene imprinting, and other components of chromatin remodeling. Among them, the functional elements of the epigenomic machinery regarding DNA methylation and histone PTMs are divided into three categories [96, 165] (Table 1). These categories include "writers," which are enzymes that introduce functional groups such as methyl and acetyl groups into DNA or proteins (e.g., histone methyltransferases); "erasers" (also called "editors"), which are enzymes that remove the functional groups from the DNA or proteins (e.g., histone demethylases); and "readers," which are proteins or complexes that recognize the functional groups and interact specifically with the modified DNA or proteins (e.g., methyl-lysine binding proteins).

T. Kozako (✉) · S.-i. Honda
Department of Biochemistry, Faculty of Pharmaceutical Sciences, Fukuoka University, Fukuoka, Japan
e-mail: kozako@fukuoka-u.ac.jp

Y. Itoh
Graduate School of Medical Science, Kyoto Prefectural University of Medicine, Kyoto, Japan

T. Suzuki
Graduate School of Medical Science, Kyoto Prefectural University of Medicine, Kyoto, Japan

CREST, Japan Science and Technology Agency (JST), Saitama, Japan

M. Bizzarri (ed.), *Approaching Complex Diseases*, Human Perspectives in Health Sciences and Technology 2, https://doi.org/10.1007/978-3-030-32857-3_6

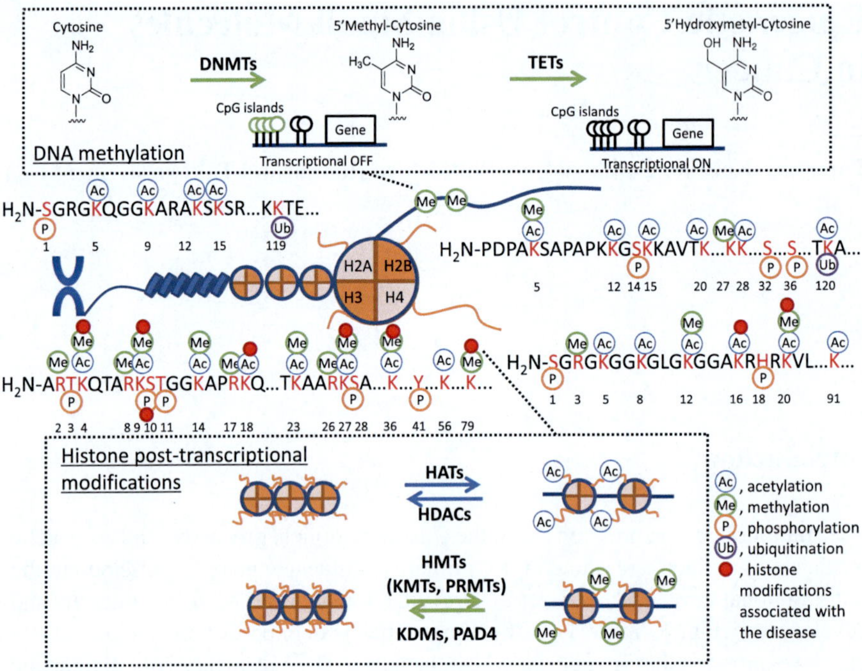

Fig. 1 Epigenetic mechanism. DNA methylation mechanism: Methylation of cytosine to 5-methylcytosine is catalyzed by DNMTs, through the methyl donor SAM, which is converted to SAH. Hypermethylation of CpG islands of promoter regions leads to transcriptional gene repression. Hydroxylation of 5mC–5hmC by TETs promotes transcriptional gene activation. Histone modifications: Covalent modifications on histones control the accessibility of DNA to transcription factors. The writers HATs and HMTs sign acetylated and methylated marks, using as co-factors acetyl-CoA and SAM, respectively. Acetylated and methylated marks can be removed by erasers, such as HDACs and KDMs, using as co-factors Zn^{2+} or NAD^{+} and FAD or Fe^{2+}/α-ketoglutarate, respectively. *DNMTs* DNA methyltransferases, *HATs* histone acetyltransferases, *HDACs* histone deacetylases, *HMTs* histone methyltransferases, *KDMs* histone demethylases, *KMTs* lysine methyltransferases, *PAD4* peptidylarginine deiminase 4, *PRMTs* protein arginine methyltransferases, *SAH* S-adenosylhomocysteine, *SAM* S-adenosylmethionine, *TETs* ten-eleven translocation family

Unfavorable epigenetic changes are often caused by loss of the balance between DNA methylation and histone PTMs, which are maintained by the "writers," "erasers," and "readers." Such epigenetic changes dramatically affect the expression of various genes that control cell behavior and function [30]. Unfavorable epigenetic changes, which typically occur in human cancers, are closely associated with the generation of malignant phenotypes. In addition, they are now known to cooperate with genetic changes to manipulate the phenotype of cancer [9]. The modulation of epigenetic mechanisms is expected to play a role in cancer therapies and compounds that control epigenetic modifications have promising anti-tumorigenic effects on malignancies [49]. In other words, the use of small molecules that modulate the epigenome in a specific manner is a viable approach for discovering cancer

Table 1 Epigenetic modifications

	Modification	Writers	Erasers	Readers (domain)	Function	References
DNA	Methylation	DNMT1, 3A, 3B	TET1–3, AID, TDG	MBD1, 2 (MBD domain), Kaiso (zinc finge domqin), UHRF1 (SRA domain)	Gene silencing	[30, 45, 56, 143, 183]
Histone	Acetylation	HATs: GNAT, GNAT family (Gcn5, PCAF and ELP3), p300/CBP (cAMP response element binding protein) family, MYST family (Tip60 and MYST 1–4)	HDAC1–11, SIRT1–7	BET family (BRD)	Transcriptional activation, DNA repair, condensation	[6, 58, 126, 150, 218]
	Lysine methylation	KMTs: SET domain-containing (G9, EZH2), Non-SET domain-containing (DOT1L)	KDMs: LSD1/2, JHDMs	HP1, L3MBTL1 (chromodomain), KDM5C (PHD fingers)	Transcription, DNA repair	[11, 30, 93, 94, 103, 126]
	Arginine methylation	PRMTs: Type I (PRMT1–4, 6 and 8), Type II (PRMT5 and 9, TypeIII (PRMT7)	PAD4	SMN, SPF30, TRD3 (tudor domain)	Transcription	

ADMA asymmetric NG,NG-di-methyl arginine, *AID* activation-induced cytidine deaminase, *BET* bromodomain and extra-terminal domain, *BRD* bromodomain, *CBP* cAMP response element binding protein, *DNMT* DNA-methyltransferase, *DOT1L* disruptor of telomeric silencing 1-like, *EZH* enhancer of zeste homolog, *Gcn5* general control non-repressible 5, *GNAT* Gcn5-related N-acetyltransferase, *HAT* histone acetyltransferase, *HDAC* histone deacetylase, *HP1* heterochromatin protein 1, *JMJD* Jumonji C (JmjC) domain-containing, *KDM* lysine demethylase, *KMT* lysine methyltransferase, *L3MBTL* loss of lethal(3)malignant brain tumor-like protein, *LSD* lysine-specific demethylase, *MBD* methylated CpG dinucleotide binding domain, *MYST* MOZ, YBF2/SAS3, SAS2, TIP60, *PAD4* peptidylarginine deiminase 4, *PCAF* P300/CBP-associated factor, *PHD* plant homeodomain, *PRMT* protein arginine methyltransferase, *SET* Su(var)3-9, Enhancer of Zeste, Trithorax, *SIRT* sirtuins, *SDMA* symmetric NG,NG'-di-methyl arginine, *SMN* survival motor neuron protein, *SPF30* splicing factor 30 kDa, *TET* ten-eleven translocation protein, *TDG* thymine DNA glycosylase, *TDRD3* tudor domain-containing protein 3, *UHRF* ubiquitin like with PHD and ring finger domain therapeutic agents [76, 147]. Indeed, a number of small-molecule modulators of DNA methylation, histoneacetylation, and histone methylation have been reported [22] (Tables 2, 3, and 4) and some of them are being not only pre-clinically tested but also clinically used/tested. This chapter describes epigenetics in cancer and small-molecule modulators against epigenetic enzymes for cancer therapy.

Epigenetic Mechanisms in Cancer

The existence of cancer stem cells with increased tumor-initiating capacities and self-renewal potential is a key issue in understanding cancer biology [73, 88]. Therefore, understanding the epigenetics of not only cancer cells but also cancer stem cells is a key issue in cancer biology. Cancer stem cells show genetic features similar to those of normal stem cells; there is a large proportion of genes with cancer-specific patterns of chromatin regulation in embryonic stem cells [142, 170, 205]. Among these genes, those that are unmethylated at promoter CpG islands are under the control of "bivalent chromatin" [16], which is characterized by the simultaneous presence of a repressive transcription mark (tri-methylated histone H3 lysine 27, H3K27me3) and an active transcription mark (tri-methylated histone H3 lysine 4, H3K4me3). In the bivalent state, the genes have low expression and are stable at the transcriptional level; switching to an active state with predominant promoter H3K4me3 or to a suppressed state with predominant H3K27me3 generally occurs during differentiation [16]. In addition to the bivalent state, promoter DNA hypermethylation negatively controls gene expression and modulates the balance between the maintenance of self-replication and differentiation [47]. This epigenetic mechanism works correctly in embryonic and adult stem cells, but not in cancer stem cells.

Human cancer cells are generated by both the activation and the inactivation of cancer-associated genes that regulate cellular processes including cell division, cell death, and cell migration from one part of the body to another [9]. Some oncogenes are activated during oncogenesis. In contrast, some tumor-suppressor genes are inactivated so that they are no longer able to prevent oncogenesis. The switching of such gene activation or inactivation occurs in a heritable manner via epigenetic alterations, rather than by mutation in the DNA sequence [9]. Epigenetic alterations include (i) global DNA hypomethylation and site-specific hypermethylation of CpG sites (CpG islands) at gene promoters, (ii) changes in histone PTMs caused by aberrations of histone-modifying enzymes such as histone acetyltransferases/deacetylases and methyltransferases/demethylases, (iii) alteration of "reader" proteins' ability to read histone marks and their binding to chromatin, (iv) alterations in nucleosome remodeling or histone exchange, and (v) changes in regulatory microRNA expression patterns [48, 64, 70]. In addition, epigenetic changes often cause mutations in genes; such mutations are frequently observed in genes that modify the epigenome [8]. Additionally, methylated cytosines in CpG islands of DNA are frequently mutated to thymines by spontaneous

Table 2 Inhibitors targeting epigenetic writers

Modification	Category	Compounds	Target	Development stage	Cancer type	References
DNA	**Methyltransferase inhibitor**					
Methylation	Nucleoside analugue	Azacytidine (5-azacytidine)	DNMT1	Approved	MDS, AML	[42]
		Decitabine (5-aza-2-deoxycytidine)	DNMT1	Approved	MDS, AML	[42]
		Zebularine	DNMT1	Preclinical	Solid tumors	[138]
		Duadecitabine (SGI-110)	DNMT1	Clinical	MDS, AML, ovarian cancer, HCC	[100]
		RX-3117	DNMT1	Clinical	Pancreatic cancer	[168]
		CP-4200	DNMT1	Preclinical	MDS	[26]
	Non-nucleoside analugue	RG108	DNMT1	Preclinical	Hematologic malignancies	[25]
		Procaine	DNMT1	Clinical	Solid tumors	[197]
		Procainamide	DNMT1	Clinical	Solid tumors	[171]
		Hydralazine	DNMT1, DNMT3A/B	Clinical	Solid tumors	[69]
		SGI-1027	DNMT1, DNMT3A/B	Preclinical	Solid tumors	[41]
		Nanaomycin A	DNMT3B	Preclinical	Lung, colon, leukemia	[110, 111]
		MG98	DNMT1	Clinical	MRCC	[4]
Histone	**Acetyltransferase inhibitor**	C646	p300	Preclinical	Solid tumors, AML	[203]
Acetylation		Isothiazolones	PCAF, p300	Preclinical	Colon cancer	[179]
		PU139, PU141	Gcn5, PCAF, p300	Preclinical	Neuroblastoma	[61]
		EML425	p300	Preclinical	Hematologic malignancies	[132]
		6-alkylsalicylates	Tip60	Preclinical	Cervical	[66]
		TH1834	Tip60	Preclinical	Breast cancer	[62]
		ICG-001	p300/ β-catenin	Preclinical	Colon carcinoma	[51]

(continued)

Table 2 (continued)

Modification	Category	Compounds	Target	Development stage	Cancer type	References
Histone	**Lysine methylytransferase inhibitor**	DZNep (3-deazaneplanocin A)	EZH2	Preclinical	Bresast, colon, prostate cancers	[184]
Methylation		EI1	EZH2	Preclinical	DLBCL	[153]
		CPI-1205	EZH2	Clinical	DLBCL	[194]
		GSK126	EZH2	Clinical	DLBCL	[130]
		Tazemetostat (EPZ6438)	EZH2	Clinical	DLBCL	[112]
		BIX-01294	G9a	Preclinical	Neuroblastoma, leukemia	[109]
		UNC0638	G9a	Preclinical	AML, breast cancer	[1]
		AZ-505	SMYD2	Preclinical	Polycystic kidney	[122]
		LLY-507	SMYD2	Preclinical	Esophageal, liver cancers	[140]
		Pinometostat (EPZ-5676)	DOT1L	Clinical	Hematologic malignancies	[204]
		SYC-522	DOT1L	Preclinical	Hematologic malignancies	[124]
	Arginine methylytransferase inhibitor	GSK3235025 (EPZ015666)	PRMT5	Preclinical	MCL	[33]
		GSK3326595 (EPZ015938)	PRMT5	Clinical	Solid tumors, NHL	[99]
		LLY-283	PRMT5	Preclinical	Solid tumors	[24]
		AMI-408	PRMT1	Preclinical	AML	[35]
		DB75	PRMT1	Preclinical	Hematologic malignancies	[212]
		TP-064	PRMT4 (CARM1)	Preclinical	MM	[139]
		GSK3359088 (EZM2302)	PRMT4 (CARM1)	Preclinical	MM	[44]

AML acute myeloid leukemia, *DLBCL* diffuse large B-cell lymphoma, *HCC* hepatocellular carcinoma, *MCL* mantle cell lymphoma, *MRCC* metastatic renal cell carcinoma, *NHL* non Hodgkin's lymphoma, *MM* multiple myeloma

Table 3 Inhibitors targeting epigenetic erasers

	Modification	Category	Compaunds	Target	Development stage	Cancer type	References
Histone	Deacetylation	**Deacethylase inhibitor**					
		Hydroxamates	Vorinostat (SAHA)	Class I, II, IV, KDM4E	Approved	CTCL	[128]
			Belinostat	Class I, II, IV	Approved	PTCL	[119]
			Abexinostat	Class I, II,	Clinical	HL, non-HL, CLL	[54]
			Givinostat	Class I, II,	Clinical	CLL, MM, HL	[60]
			Panobinostat	Class I, II, IV	Approved	MM	[213]
			Pracinostat	Class I, II, IV	Approved	AML, prostate cancer	[63]
			Resminostat	Class I, II,	Clinical	Colorectal cancer, HCC, HL	[27]
			Quisinostat	Class I, II,	Clinical	Solid tumors, lymphoma, CTCL	[195]
		Cyclic peptide	Romidepsin	Class I	Approved	CTCL, PTCL	[17]
		Fatty acids	Phenylbutyrate	Class I, II, IV	Clinical	Solid tumors	[34]
			Valproic acid	Class I, II	Approved	Advanced cancer	[20]
		Benzamides	Mocetinostat	Class I, IV	Clinical	Hematologic and solid tumors	[219]
			Etinostat	HDAC1–3	Clinical	NSCLC, solid tumors	[46]
		Sirtuin inhibitors	Suramin	SIRT1, 2 and 5	Clinical	Solid tumors	[202]
			EX-527	SIRT1	Preclinical	Solid tumors	[102]
			Sirtinol	SIRT1, 2	Preclinical	Hematologic and solid tumors	[104]
			Splitomicin	Sir2	Preclinical	Breast cancer	[10]
			Cambinol	SIRT1, 2	Preclinical	Hematologic and solid tumors	[79]
			Salermide	SIRT1, 2	Preclinical	Hematologic and solid tumors	[115]
			Tenovin-6	SIRT1, 2, 3	Preclinical	Hematologic and solid tumors	[114]
			AGK2	SIRT2	Preclinical	Glioma	[102]
			NCO-01/04	SIRT1 , 2	Preclinical	Hematologic and solid tumors	[106]

(continued)

Table 3 (continued)

	Modification	Category	Compaunds	Target	Development stage	Cancer type	References
			NCO-90/141	SIRT2	Preclinical	Hematologic malignancies	[107]
			KPM-2	SIRT2	Preclinical	Breast cancer	[131]
			SirReal2	SIRT2	Preclinical	Gastric cancer cells	[148]
			TM	SIRT2	Preclinical	Breast cancer	[97]
	Demethylation	**Lysine demethylase inhibitor**					
		KDM1	Tranylcypromine	LSD1	Clinical	AML, MDS	[223]
			Phenelzine	LSD1	Preclinical	Lung, prostate cancers	[152]
			Pargyline	LSD1	Preclinical	Prostate cancer	[201]
			ORY-1001	LSD1	Clinical	AML	[127]
			GSK2879552	LSD1	Clinical	AML, relapsed/refractory small-cell lung carcinoma	[223]
			IMG7289	LSD1	Clinical	AML, MDS	[141]
			Bizine	LSD1	Preclinical	Lung, prostate cancers	[152]
			PG11144	LSD1	Preclinical	Breast cancer	[224]
			Namoline	LSD1	Preclinical	Prostate cancer	[206]
			HCI-2509	LSD1	Preclinical	Prostate cancer	[71]
			GSK354, GSK690	LSD1	Clinical	AML	[127]
			Domatinostat (4SC-202)	LSD1, HDAC Class I	Clinical	Hematologic malignancies	[198]
		JHDMs	IOX1	JmjC domain	Preclinical	Cervical carcinoma	[83]
			GSK-J1, GSK-J4	KDM6A/B	Preclinical	to be determined	[78]
			EPT-1013182	KDM5B	Preclinical	to be determined	[74]

CLL chronic lymphocytic leukemia, *CTCL* cutaneous T-cell lymphoma, *HCC* hepatocellular carcinoma, *HL* Hodgkin's lymphoma, *MM* multiple myeloma, *NSCLC* non-small cell lung cancer, *PTCL* peripheral T-cell lymphoma

Table 4 Inhibitors targeting epigenetic readers

	Modification	Category	Compaunds	Target	Development stage	Cancer type	References
Histone	Acetylation	**BET bromodomain**	JQ1	BET proteins	Preclinical	Prostate cancer, AML with MLL translocations, MM, NUT midline carcinoma	[57]
			I-BET762 (GSK525762)	BET proteins	Clinical	Hematologic malignancies, NUT midline carcinoma, solid tumors	[222]
			ABBV-075 (Mivebresib)	BRD2, 4, BRDT	Clinical	Solid tumors	[28]
			BAY1238097	BET proteins	Clinical	Hematologic and solid tumors	[15]
			CPI-0610	BET proteins	Clinical	AML, MDS, MM, lymphoma	[174]
			FT-1101	BET proteins	Clinical	AML, MDS, NHL	[43]
			I-BET151	BET proteins	Clinical	MM	[31]
			INCB054329	BET proteins	Clinical	Advanced malignancies	[180]
			OTX015 (MK-8628)	BRD2, 3, 4	Clinical	AML	[14]
			TEN-010	BET proteins	Clinical	NUT midline carcinomas, solid tumors	[211]
			dBET1	BET proteins	Preclinical	Hematologic malignancies	[91]
		Non-BET bromodomain	BI-7273 and BI-9564	BRD9	Preclinical	AML	[129]
	Methylation	**MBT domain**	UNC280	L3MBTL1	Preclinical	to be determined	[80]
			UNC669	L3MBTL1	Preclinical	to be determined	[81]

(continued)

Table 4 (continued)

	Modification	Category	Compaunds	Target	Development stage	Cancer type	References
			UNC926	L3MBTL1	Preclinical	to be determined	[81]
			UNC1215	L3MBTL3	Preclinical	to be determined	[93]
			UNC1679	L3MBTL3	Preclinical	to be determined	[93]
		Chromodomain	MS37452	CBX7	Preclinical	to be determined	[157]
		PHD finger domain	Amiodarone	JARID1A	Preclinical	to be determined	[133]
			CF16	Pygo 2	Preclinical	to be determined	[199]

AML acute myeloid leukemia, *MDS* myelodysplastic syndrome, *MLL* mixed lineage leukemia, *MM* multiple myeloma, *NHL* non Hodgkin's lymphoma, *NUT* nuclear protein in testis hydrolytic deamination [9]. Hyper-methylation of CpG sites induces gene mutations because CpG site methylation increases the binding of some chemical carcinogens to DNA [216]. Thus, the epigenome contains key information regarding how epigenetic changes could be involved in cancer.

Against this background, any abnormalities in epigenetic mechanisms potentially lead to the development of human cancer. The well-studied epigenetic alterations associated with neoplastic phenotypes are variations in DNA methylation, and alterations in histone protein structure through PTMs and histone variants [22]. Focusing on DNA methylation, histone acetylation, and histone methylation, the following sections highlight the use of epigenetic regulation and its modulators as tools for cancer treatment.

DNA Modification

Adenine, thymine, cytosine, and guanine are the key nitrogenous bases that are found in eukaryotic organisms. These bases usually comprise the majority of sequences found in eukaryotic DNA. Cytosine methylation occurs in regions with a high frequency of CpG islands, which mostly reside at promoter regions, and is strongly implicated in transcriptional silencing [47, 177]. This silencing is induced by deregulation of the epigenetic machinery at several different levels; for example, it involves inappropriate methylation of cytosine residues in DNA CpG sequence motifs that govern gene expression (Fig. 1, Table 1).

The change in DNA methylation pattern was the first identified epigenetic alteration in cancer [56]. DNA methylation changes such as increases in methylation of CpG islands and overall decreases in global DNA methylation have been observed in cancer cells. For instance, CpG methylation inactivates the promoter activity of the human retinoblastoma tumor-suppressor gene [143]. Hypermethylation of the promoter region of *TIMP3*, which is a negative regulator of angiogenesis for tumor metastasis, also leads to cancer growth [158]. Thus, DNA methylation is deeply associated with cancer initiation and progression [56].

DNA Methylation

DNA Methyltransferase Inhibitors

The "writers" DNA-methyltransferase 1 (DNMT1), DNMT3A, and DNMT3B convert cytosine into 5-methylcytosine (5mC) using a methyl donor, S-adenosyl-L-methionine (SAM; Fig. 1, Table 1) [18, 19, 144]. DNMT1 preferably methylates the hemi-methylated state of DNA during replication, whereas DNMT3A and

DNMT3B are essential for both hemi- and unmethylated DNA, which means that they are *de novo* methyltransferases [145].

DNA methylation due to DNMT overactivation in the CpG islands of tumor-suppressor genes leads to the silencing of gene expression. Because DNMT inhibition may lead to the restoration of tumor-suppressor gene activity, DNMT inhibitors are interesting therapeutic agents (Table 2). There are two broad classes of DNMT inhibitors: nucleoside and non-nucleoside analogs. The nucleoside analogs 5-azacytidine (azacytidine) and 5-aza-2-deoxycytidine (decitabine), which were initially designed as antimetabolites, have been approved by the Food and Drug Administration (FDA; United States Department of Health and Human Services) for use in the management of myelodysplastic syndrome (MDS), acute myeloid leukemia (AML), and chronic myelomonocytic leukemia [38, 42]. After these compounds are incorporated into DNA during the S phase of the cell cycle, they are recognized by the DNMTs. They then act as suicide inhibitors that form a covalent complex with DNMTs and trigger proteasomal degradation of the enzymes [67, 167]. They also have antitumor effects via apoptosis or the differentiation of cancer stem-like cells accompanied by reduced genome-wide promoter DNA methylation [191]. Ongoing studies have led to the development of diverse new nucleoside inhibitors [13]. The toxicity of the older nucleoside analogs was addressed by the development of zebularine, which is a more effective nucleoside DNMT inhibitor [138]. The dinucleotide analog guadecitabine (SGI-110, S110) is a new hypomethylating agent that is derived from decitabine and is a promising candidate for the treatment of MDS and AML [215]; it is currently undergoing a phase III clinical trial following a phase II clinical trial for the treatment of AML [100]. RX-3117 is another nucleoside analog with the ability to enhance DNMT1 degradation that showed anti-cancer effects [36, 168] in a clinical trial in combination with nabpacitaxel (ClinicalTrials.gov Identifier: NCT03189914). Other investigational drugs such as an elaidic acid ester of azacitidine, which works as a prodrug (CP-4200), and thiocytidine analogs (4′-thio-2′-deoxycytidine and 5-aza-4′-thio-2′-deoxycytidine) inhibit DNMT1 in cancer cell lines and animal models of cancer [26, 188].

Non-nucleoside small-molecule DNMT inhibitors are also being developed. They can block DNMTs without being incorporated into the DNA [175]. These inhibitors include the phthalimide RG108 [25], procaine [197] and its amide analog procainamide [171], hydralazine [69], the quinoline-based derivative SGI-1027 [41], nanaomycin A [110], and natural products such as epigallocatechin-3-gallate (EGCG) [55] (Table 2). RG108, which was developed as a DNMT inhibitor through *in silico* drug design, induced E-cadherin expression in promyelocytic leukemia cells both alone and in combination with histone deacetylase (HDAC) inhibitors. It was also able to protect retinal pigment epithelial cells from oxidative stress by upregulating methylated silenced genes involved in producing antioxidant enzymes [169, 190]. Moreover, the antiarrhythmic drug procainamide and its ester analog procainamide (procaine) were able to interrupt the hemimethylase activity of DNMT1 without much alteration of DNMT3A and DNMT3B activity [118].

Procainamide also restored the expression of tumor-suppressor genes such as p16INK4a, RAR-β, and GSTP1 [156]. Furthermore, hydralazine, which is used for managing hypertension, reduced DNMT1, DNMT3A, and DNMT3B mRNA production. It was revealed that hydralazine has the potential to reduce malignant growth through epigenetic alterations in prostate cancer cells [69]. SGI-1027 is a lipophilic quinoline derivative that strongly inhibits DNMT3A but not human DNMT1; because of its basic properties, it weakly binds to AT-rich regions of DNA [161]. A quinone-containing antibiotic, nanaomycin A, which was isolated from a *Streptomyces* strain, exhibits selectivity for DNMT3B and was able to reactivate silenced tumor-suppressor genes [111]. Polyphenols have been broadly studied because of their DNMT inhibitory activity [29]. The flavan-3-ol EGCG contained in green tea directly inhibits the DNMT1 enzyme [214]. However, all examined non-nucleoside compounds have shown limited potency. In addition to small- molecule DNMT inhibitors, a second-generation DNMT1 inhibitor, MG98, which is an antisense oligonucleotide designed to bind with the 3′ untranslated region of DNMT1 mRNA to disturb its transcription, specifically inhibits DNMT1 without altering DNMT3 expression [3, 4].

DNA Demethylation Inhibitors

DNA demethylation is mediated by the sequential reactions that are catalyzed by several epigenetic enzymes (Fig. 1). Ten-eleven translocation protein (TET) methylcytosine dioxygenases (TET1, 2, and 3), which are named after a recurrent chromosomal translocation, t(10;11)(q22;q23), oxidatively convert 5mC to 5-hydroxymethyl (5hmC), 5- formyl (5fC), and 5-carbocarboxylcytosine (5caC) [183]. These modified cytosines are converted to unmodified ones by epigenetic enzymes such as activation- induced cytidine deaminase (AID) and thymine DNA glycosylase (TDG) [39, 90]. Therefore, TETs, AID, and TDG work as "erasers" for DNA methylation (Table 1). In many cancers, increased methylation is observed in promoter CpG islands of normally unmethylated genes, especially tumor-suppressor genes such as *p73*, *CDKN2A*, *MLH1*, *BRCA1*, and *VHL* [53, 98]. This suggests that the "writers" and "erasers" that control DNA methylation work normally in normal cells, but that the balance between them is lost in cancer cells.

No reports have addressed potent TET, AID, and TDG protein inhibitors in cancer, although vitamin C was found to potentiate TET activity [85]. R-2-Hydroxyglutarate (R-2HG) and its stereoisomer S-2HG, which are produced by the cancer-associated mutants IDH1 and IDH2, have been reported to inhibit the activity of TET proteins: S-2HG is often more effective [86, 220]. However, their inhibitory activity and TET-selectivity are too low to be tested in cancer cells. There is thus a need for the development of highly potent and selective small molecules modulating DNA methylation "erasers."

MDB Protein Inhibitors

DNA methylation directly silences gene expression by interfering with the binding of various transcription factors and indirectly by enrolling DNA methylation "readers" such as methylated CpG dinucleotide binding domain (MBD) proteins [45] (Table 1). The MBD protein family, Kaiso protein family, and SET- and RING finger-associated (SRA) domain family specifically induce interactions between methylated DNA methylation and histone modifications to promote a composite regulatory program [75, 192]. The ubiquitin-like with PHD and RING finger domain 1 (UHRF1) is a co-factor that cooperates with DNMT1 throughout the S phase of the cell cycle and interacts with hemi-methylated DNA using an SRA domain [154]. These proteins play crucial roles in establishing epigenetic gene regulation regarding DNA methylation. However, there are no known compounds that target these proteins. Thus far, a time-resolved fluorescence resonance energy transfer (FRET)-based assay system has been developed to identify small-molecule inhibitors of MBD2, which acts in multi-protein complexes containing histone-modifying enzymes to directly assemble repressive chromatin; however, no candidate MBD2 inhibitors have been reported [209].

Histone Modifications

Histone modifications alter DNA–histone and histone–histone binding interactions, and convert transcriptionally active to transcriptionally inactive chromatin [6]. Various PTMs at histones (lysine: acetylation, methylation, ubiquitination, sumoylation, crotonylation, butyrylation, and propionylation; arginine: methylation, citrullination, and ADP-ribosylation; serine, tyrosine, and threonine: phosphorylation) are regulated by many epigenetic enzymes [126, 134] (Fig. 1). The enzymes introduce or remove covalent attachments at specific histone residues [2, 155]. Representative examples are as follows: "writers": acetyltransferases (HATs) and histone methyltransferases (HMTs); and "erasers": histone deacetylases (HDACs) and histone lysine demethylases (KDMs) [150]. In addition, "reader" proteins such as bromodomain (BRD) and methyl-lysine binding (MKB) proteins recognize acetylated and methylated histones, respectively; they also recruit additional chromatin modifiers or remodeling enzymes (Table 1). In this section, we describe inhibitors of the "writers," "erasers," and "readers" for histone acetylation and methylation.

Histone Acetylation

Histone Acetyltransferase Inhibitors

HATs, "writers" for histone acetylation, produce acetylated marks using acetyl-CoA as an acetyl group donor (Fig. 1, Table 1). HAT-mediated acetylation of the ε-amino group of lysine residues in histone tails facilitates gene transcription by loosening chromatin compaction or enhancing the recruitment of transcriptional activators [6]. Histone acetylation such as acetylation of H3K56 is associated with the pathological activation of tumorigenesis [123, 207]. HATs interact with various proteins and form protein complexes that catalytically control histone acetylation. HATs are classified into two categories based on their cellular location [6]: type A HATs are located in the nucleus and type B HATs are located in the cytoplasm. HATs in the nucleus play crucial roles in transcriptional activation and are classified based on structural homology and catalytic mechanism [59, 117]. The major families of HATs involved in chromatin remodeling are the Gcn5-related N-acetyltransferase (GNAT) family (Gcn5, PCAF, and ELP3); the cAMP response element binding protein (p300/CBP) family; and the MOZ, YBF2/SAS3, SAS2, and TIP60 (MYST) family (Tip60 and MYST 1–4) [82].

Small-molecule HAT inhibitors have been identified and are classified as bi-substrate inhibitors, synthetic small molecules, natural products and their derivatives, or protein– protein interaction inhibitors [147] (Table 2). Bi-substrate inhibitors mimic the two HAT substrates, acetyl-CoA and peptide resembling a lysine: they are conjugates of an acetyl-CoA-mimicking moiety and a lysine-containing peptide connected via a linker [203]. The small-molecule p300 inhibitor C646, which was discovered by *in silico* screening, is competitive with acetyl-CoA and noncompetitive with the histone lysine substrate [146]. A recent study demonstrated that C646 inhibited the proliferation of prostate cancer and melanoma cells, induced cell cycle arrest in AML1-ETO-positive AML cells, and sensitized lung cancer cells to irradiation [203]. Virtual analysis and high-throughput screening identified isothiazolones as p300/CBP-associated factors (PCAFs) and p300 HAT inhibitors [179]. It was also revealed that these inhibitors killed microorganisms and reduced the proliferation of colon cancer cells. PU139 and PU141, which are pyridoisothiazolone derivatives, inhibit Gcn5, PCAF, and p300/CBP in several neoplastic cell lines and have anti-tumor activity *in vivo* [61]. Natural products such as garcinol, anacardic acid, and curcumin have been shown to be HAT inhibitors [203]. A benzylidene barbituric acid derivative (EML425) was developed from garcinol by structure-based drug design and showed improved selectivity for p300/CBP [132]. 6-Alkylsalicylates, which are anacardic acid analogs, have been developed as selective Tip60 inhibitors [66]. A pentamidine derivative, TH1834, also showed improved selectivity for Tip60 [62]. Histone lysine acetylation is involved in protein–protein interactions via BRDs [166]. In this context, HAT inhibitors that disrupt protein–protein interactions have been developed. ICG-001 is one of the protein–protein interaction inhibitors that target the interaction between CBP and β- catenin in colon carcinoma [51]. Although numerous HAT inhibitors have been developed, there are no reports on their use as clinical candidates [203]. The HAT inhibitors listed in Table 1 are at the preclinical stage.

Histone Deacetylase Inhibitors

HDACs remove the acetyl group from lysine residues of histones using the co-factor Zn^{2+} or nicotinamide adenine dinucleotide (NAD^{+}), which promotes a condensed chromatin status and consequent repression of gene transcription [218] (Fig. 1, Table 1). HDACs are divided into four classes according to structural and mechanistic similarities: zinc-dependent classes I (HDAC1–3, 8), II (HDAC4–7, 9, 10), and IV (HDAC11), and NAD^{+}-dependent class III (sirtuin family) [172].

HDAC-mediated histone deacetylation is associated with tumorigenesis via the transcriptional repression of tumor-suppressor genes [137]. Mutation and aberrant expression of HDACs are linked to oncogenic events [9, 135]. Thus, HDAC inhibition is involved in determining the fate of cancer cells (Table 3).

Both the Zn^{2+}-dependent (classes I, II, and IV) and NAD^{+}-dependent (class III) HDACs are considered validated drug targets for cancer treatment. Trichostatin A (TSA), a dienohydroxamic acid derivative isolated from *Streptomyces hygroscopicus*, was the first reported specific HDAC (class I/II) inhibitor [217]. TSA has some potential as an anti-cancer drug to promote the expression of apoptosis-related genes, although it has not been used in clinical trials because of its rapid metabolic inactivation [196]. The hydroxamic acid derivative vorinostat (SAHA) [159], which is an HDAC (class I/II) inhibitor, was the first HDAC inhibitor approved by the FDA [128]. Vorinostat is used for third-line therapy for cutaneous T-cell lymphoma and is currently being clinically tested as an adjuvant treatment of colorectal cancer in combination with hydroxychloroquine against the tyrosine kinase inhibitor regorafenib (NCT02316340) and in advanced solid tumors (NCT01023737). The combination of vorinostat with immunotherapy using the checkpoint inhibitor pembrolizumab (a therapeutic antibody that blocks the PD-1, programmed cell death protein 1) was also tested in patients with advanced NSCLC (NCT02638090). Moreover, the proteasome inhibitor bortezomib holds promise for therapeutic use in combination with vorinostat in NSCLC (NCT02211755). Thus, combination therapy with HDAC inhibitors may hold promise regarding its anticancer effects [181].

Belinostat, a pan-HDAC inhibitor [119], has also been approved to treat peripheral T-cell lymphomas by the FDA. Other hydroxamic acid derivatives that have been approved and/or are in various phases of development include abexinostat, givinostat, panobinostat, pracinostat, resminostat, and quisinostat [13, 27, 54, 60, 195]. Panobinostat was approved by the FDA for treating multiple myeloma, and is to be taken in combination with bortezomib and dexamethasone [213]. Pracinostat, a hydroxamic acid analog, was approved by the FDA and was designated as a breakthrough therapy in combination with azacitidine for elderly patients with AML [63]. Aside from hydroxamic acid derivatives, romidepsin was also approved by the FDA and is a cyclic tetrapeptide that shows strong inhibitory activity against HDAC1–3 and HDAC8 [17]. Short-chain fatty acids including phenylbutyrate and valproic acid are another class of HDAC inhibitors, but their HDAC-inhibitory activities are low [20, 34]. The benzamides mocetinostat (MGCD0103) [219] and etinostat (MS-275) [46] are currently being evaluated in combination with classical chemotherapy or other targeted drugs for the treatment of refractory mesothelioma, melanoma, lymphoma, MDS, and other cancers [65, 181].

Sirtuins (SIRT1–7) are NAD^+-dependent deacetylases or mono-[ADP-ribosyl] transferases that play essential roles in genome stability, cellular metabolism, DNA repair, chromosomal stability, longevity, and cancer development [89]. SIRT1–3, 6, and 7 can remove not only acetyl groups but also hydrophobic acyl groups, whereas SIRT5 exclusively removes negatively charged acyl groups [185]. Small-molecule agents targeted to sirtuins also have therapeutic potential for cancer [105, 182]. Suramin, which targets SIRT1, 2, and 5, has been used in clinical trials for lung cancer, breast cancer, prostate cancer, bladder cancer, and autism [202]. EX-527, which targets SIRT1, has been used as a treatment in Huntington's disease phase I/II clinical trials and suppressed cell growth and caused G1- phase arrest *in vitro* [102]; this established EX-527 as a clinical drug candidate. In addition, several preclinical compounds that act as sirtuin inhibitors have been discovered and characterized. In preclinical trials, sirtuin inhibitors such as sirtinol [104], splitomicin [10], cambinol [79], salermide [115], tenovin-6 [114], AGK2 [102], NCO- 01/04 [106], NCO-90/141 [107], KPM-2 [131], SirReal2 [148], and TM [97] were determined to be candidate anticancer agents for solid malignancies, such as breast cancer, glioma, AML, and adult T-cell leukemia/lymphoma.

BRD Protein Inhibitors

BRD proteins are known as "readers" that specifically recognize acetylated lysines on histone tails [166]. Based on sequence/structural similarity, BRD proteins have been divided into eight families (families I–VIII) [58]. Most of the HATs in the nucleus contain BRD as a catalytic component [95]. The BRD-containing HATs [lysine (K) acetyltransferase A, KAT2A, Gcn5; KAT2B, PCAF; KAT3A, CBP; KAT3B, p300] play important roles in the epigenetic landscape and cancer development.

Small molecules that target specific BRDs should help clarify the biological functions of BRD proteins and show promise as anti-cancer agents (Table 4). Small- molecule inhibitors of BRD proteins have demonstrated efficacy both preclinically and clinically [149]. Some potent selective inhibitors of bromodomain and extra-terminal domain (BET) proteins (BRD2, BRD3, BRD4, and bromodomain testis-specific protein) are in development for treating cancer: JQ1 (a thienotriazolodiazepine) is under preclinical trials and its analogs (benzodiazepines) are under clinical trials [37, 57]. These inhibitors decreased the expression of c-myc protein and ERK1/2 phosphorylation levels, prevented the proliferation of pancreatic cancer cells *in vitro*, and reduced protein levels of IL-6, phosphorylated ERK1/2, and phosphorylated STAT-3 *in vivo* [116]. JQ1, which has a short half-life, is not a candidate for clinical development. However, a JQ1 analog, I-BET762 (GSK525762), which has good potency and pharmacokinetic properties, is currently in clinical trials for the treatment of nuclear protein in testis (NUT) midline carcinoma, breast, lung, and other cancers [221, 222]. The efficacy of BET inhibitors (e.g., ABBV- 075 [28], BAY1238097 [15], CPI-0610 [174], FT-1101 [43], I-BET151 [31], INCB054329 [180], OTX015 [14], and TEN-010 [210]) in preclinical cancer models and RVX-208 [151] in cardiovascular disease provided the rationale for using them in a multitude of ongoing human clinical trials. These include trials for patients with hematological malignancies, BRD4–NUT-expressing NUT midline carcinoma, and various solid tumors [210]. Thus, BET family inhibitors have been extensively studied in recent years.

Inhibitors of other BRD proteins have also been reported. For example, I-BRD9 was identified through structure-based design, and is a non-BET inhibitor that is selective for BRD9 [187]. BI-7273 and BI-9564, which have pyridinone-like scaffolds, have also been determined to be selective for BRD9 and their anti-tumor activity was observed *in vivo* [129]. Identification of non-BET inhibitors that target BRD proteins with HAT activity can help elucidate their potential for treating cancer. Non- BET inhibitors target the HAT–BRD interaction and prevent the BRD proteins from binding to acetylated lysines. HAT–BRD interaction inhibitors have been developed for KAT3A (p300/CBP family) and PCAF (GNAT family). Representative examples include the natural product ischemin, a set of cyclic peptides, and small-molecule N^1-aryl-propane- 1,3-diamine derivatives [203]. SGC-CBP30 and I-CBP112 are also known as BRD inhibitors that exhibit selectivity towards CREBBP/p300 BRD [77]. Furthermore, bromosporine has been reported to act as a broad-spectrum BRD inhibitor [149].

More comprehensive information can be obtained from previous reviews [149, 210]. In addition to the BRD inhibitors, recently, small molecules that induce the degradation of BET proteins have been developed: dBET1 is one representative example [91]. This research should also contribute to the development of clinically used BRD modulators.

Histone Methylation

Histone Methyltransferase Inhibitors

Histone methylation occurs at lysine and arginine residues. There are three types of methylation state for each of lysine and arginine: for lysine, they include mono- (me1), di- (me2), and tri-methylation (me3), whereas for arginine, they include N^G-mono-methyl arginine, asymmetric N^G,N^G-di-methyl arginine (ADMA), and symmetric N^G,N^G-di- methylarginine (SDMA) [11] (Table 1). Furthermore, N-terminal histone tails undergo methylation at specific histone 17 lysine residues and seven arginine residues (Fig. 1). The multiple methylation states are associated with chromatin modifications and their functions [103]: these states are associated with transcriptional activation or repression based on the location of the lysine or arginine residues [6, 7, 12, 50, 164, 178] (Table 5). For example, H3K9me1 and H3K27me1 are abundant at active gene promoters, whereas H3K9me3 and H3K27me3 are associated with transcriptionally repressed gene promoters. Histone arginine methylation marks can also be activating (H4R3me2a, H3R2me2s, H3R17me2a, and H3R26me2a) or repressive (H3R2me2a, H3R8me2a, H3R8me2s, and H4R3me2s). These methylations are SAM-dependently catalyzed by HMTs as "writers" for histone methylation: 28 lysine methyltransferases (KMTs) including suppressor of variegation 3–9, enhancer of zeste, trithorax (SET) domain-containing or non-SET domain-containing methyltransferases, and 11 protein arginine methyltransferases (PRMTs) [13, 208]. Thus, histone KMTs are generally subdivided into SET domain-containing and non-SET domain-containing methyltransferases [189]. SET domain-containing proteins include mixed lineage leukemia 1 (MLL1)/KMT2A, SET and myeloid-Nervy-DEAF1 (MYND) domain containing protein 3 (SMYD3), suppressor of variegation 3–9 homologs 1 and 2 (SUV39H1 and 2)/KMT1A/B, G9a/KMT1C, enhancer of zeste homolog 2 (EZH2)/KMT6A, and nuclear receptor-binding SET domain protein 2 (NSD2)/MMSET/WHSC1. Non-SET domain-containing methyltransferases include disruptor of telomeric silencing 1-like (DOT1L). Alternatively, histone PRMTs generally belong to three classes: type I PRMTs (PRMT1, PRMT2, PRMT3, PRMT4/CARM1, PRMT6, and PRMT8), which produce ADMA; type II PRMTs (PRMT5 and PRMT9), which produce SDMA; and type III (PRMT7), which produce N^G-mono-methyl arginine over SDMA [13, 225]. PRMT10 and PRMT11 are putative proteins with homology to PRMT7 and PRMT9, respectively ([108] #1346). They have no methyltransferase activity, and their role is currently unknown. Tudor domain-containing proteins, such as the survival motor neuron protein (SMN), splicing factor 30 kDa (SPF30), and tudor domain-containing protein 3 (TDRD3), recognize these methylated arginines [40, 92]. Abnormal methylation of histone lysines and arginines is linked to carcinogenesis [147].

Small molecules that selectively inhibit oncogenic KMTs are promising therapeutic agents for cancer treatment [101] (Table 2). The EZH2 mutation represses the H3K27me3 mark and is associated with poor prognosis in diffuse large B-cell lymphoma (DLBCL), and breast and prostate cancer [136]. In contrast, functional loss of EZH2 due to mutations has been reported in myeloid malignancies and T-cell acute lymphoblastic leukemia [52]. Therefore, EZH2 may have either oncogenic or tumor-suppressor properties depending on the cellular context. An EZH2 inhibitor, DZNep (3-deazaneplanocin A), binds to the SAM-binding site of EZH2 and causes EZH2-degradation [184]. EI1 [153], CPI-1205 [194] (NCT02395601), and GSK126 [130] (NCT02082977), which are selective SAM- competitive EZH2 inhibitors, inhibit the proliferation of mutant DLBCL cell lines and the growth of these cells in xenografted mice. Other small-molecule KMT inhibitors such as tazemetostat

Table 5 Histone methylation sites and transcriptional activation

	Arginine residues					Lysine residues						
Modification	H3R2	H3R8	H3R17	H3R26	H4R3	H2BK5	H3K4	H3K9	H3K27	H3K36	H3K79	H4K20
Mono-methylation						+	+	+	+		+	DNA repair
Di-methylation	–	–	+	+	SMDA: – ADMA: +		+	–	–		+	DNA repair
Tri-methylation						–	+	–	–	Elongation	+	–

+: Activation
–: Repression

(EPZ6438) have been developed as potential anticancer therapeutics [112]. Additionally, G9a-specific inhibitors, BIX-01294 [109] and UNC0638 [1], have potential as anti-cancer agents. An SMYD2-selective inhibitor, AZ-505, delays cyst growth in a mouse model of polycystic kidney disease [122]. LLY-507, a selective inhibitor of SMYD2, also inhibits the proliferation of several cancer cell lines [140]. Alternatively, DOT1L, which catalyzes H3K79 methylation, is a promising target and has crucial roles in MLL-rearranged leukemias. The DOT1L inhibitor pinometostat (EPZ-5676) inhibits tumor growth with MLL rearrangements in association with decreases in H3K79me levels at homeobox genes [204] (NCT01684150). Another DOT1L inhibitor, SYC-522, also increases the sensitivity of MLL-rearranged leukemia cells to chemotherapeutics [124].

PRMTs are also expected to become therapeutic targets for anticancer strategies because PRMT overexpression has been observed in a variety of cancers [11]. The PRMT5 inhibitor GSK3235025 (EPZ015666) shows antitumor activity in xenografts of mantle cell lymphoma [33]. An improved PRMT5 inhibitor, GSK3326595 (formerly EPZ015938), has also been examined in clinical trials (solid tumors and non-Hodgkin's lymphoma) [99] (NCT02783300). LLY-283 is a specific inhibitor of PRMT5 with antitumor activity [24]. In addition, the PRMT1 inhibitor AMI-408 suppresses the transformative functions of MLL– GAS7 and MOZ–transcriptional intermediary factor 2 (TIF2) fusions in AML models [35]. A PRMT1-specific inhibitor, DB75, also inhibits cell proliferation in leukemia cell lines [212]. Selective CARM1 (PRMT4) inhibitors, TP-064 and GSK3359088 (EZM2302), were also shown to have beneficial effects on multiple myeloma in preclinical study [44, 139]. More comprehensive information on this issue can be obtained from a previous review [99].

Histone Demethylase Inhibitors

Methylation of lysines and arginines on histone tails is a dynamic modification. Methylated markers are removed by "erasers" such as KDMs and peptidylarginine deiminase 4 (PAD4) [6, 200] (Fig. 1, Table 1). Histone KDMs generally belong to two classes. The first class includes lysine-specific demethylase 1 (LSD1, also known as KDM1A) and LSD2 (KDM1B), which are flavin- dependent amine oxidase domain-containing enzymes. Their substrates are limited to mono- and dimethylated lysines. The second class includes Jumonji C (JmjC) domain-containing (JMJD) protein histone demethylases, which are Fe(II)- and 2-oxoglutarate (2- OG, alpha-ketoglutarate)-dependent enzymes (KDM2–8). KDM2–8 remove methyl groups from all three lysine methylation states [23, 189]. Because H3K4me2 is an essential chromatin mark associated with promoters of active genes, oxidative demethylation of H3K4me1 and H3K4me2 by LSD1 is associated with transcriptional repression [5]. Hyper-demethylation of this mark by highly expressed LSD1 prevents the expression of tumor-suppressor genes that are important in human cancer [87]. Thus, these KDMs play crucial roles in development, differentiation, and carcinogenesis [173]. Alternatively, the mechanisms underlying the demethylation of methylated arginines is unknown: PAD4, which is associated with transcription and chromatin decondensation, undergoes conversion from mono-methyl arginine into citrulline [121]. Recent reports have suggested that JMJD6 catalyzes both arginine demethylation of histone H3/H4 residues and lysyl hydroxylation [32, 113].

Dysregulation or mutations of KDMs have been reported in various cancers, which suggests that these enzymes are promising targets for anticancer therapies [193] (Table 3). Tranylcypromine, phenelzine, and pargyline are the initial compounds that were reported to inhibit LSD1 [152, 201, 223]. Because LSD1 has a C-terminal amine oxidase-like domain related to monoaminoxidases (MAOs), these compounds also inhibit MAOs by irreversibly binding to the FAD co-factor. Therefore, they have some problems with low LSD1-selectivity. Based on this background, LSD1-selective inhibitors have been developed as follows. ORY-1001 and GSK2879552, which are analogs of the MAO inhibitor tranylcypromine, show highly potent and selective LSD1-inhibitory activity. They have entered clinical trials [127, 223]. IMG-7289, which is a mimic of a highly potent LSD1-selective inhibitor, NCD38 [141], is also being clinically tested for the treatment of AML and MDS. The phenelzine analog bizine also inhibits LNCaP and H460 cell growth [152]. Moreover, a polyamine derivative (PG11144) [224], a reversible LSD1 inhibitor (namoline) [206], an irreversible LSD1 inhibitor (HCI-2509) [71], GSK354, and GSK690 have also been reported to inhibit LSD1 function in cancer cell lines [127]. Interestingly, the small molecule domatinostat (4SC-202) was determined to have dual functions that facilitate LSD1 and HDAC inhibition in clinical trials for patients with advanced hematological malignancies [198].

Various JmjC–KDM inhibitors have been reported to be candidate anticancer agents, including hydroxyquinoline analogs, metal-chelating hydroxamic acid scaffold, cyclic peptides, catechol molecules, and flavonoid analogs [72]. IOX1, an 8-hydroxyquinoline derivative, inhibits many KDM isoforms [83]. The hydroxamic acid derivative HDAC inhibitor SAHA (vorinostat) demonstrated KDM4E inhibition [163]. N-oxalylglycine, which is an amide analog of 2-OG, is a weaker inhibitor of KDM4 [84]. GSK-J1 is a competitive inhibitor with 2-OG, but not with the substrate. GSK-J1 and its prodrug GSK-J4 inhibit KDM6A (UTX) and KDM6B (JMJD3), but display less activity against KDM5A and 5B [78]. KDM6 inhibitors are of interest as anticancer agents because KDM6s are associated with cancers such as AML, multiple myeloma (MM), bladder cancer, T-ALL, and metastatic prostate cancer, in which KDM6s are mutated or overexpressed [120, 193]. A KDM5B-selective inhibitor, EPT-1013182, showed antiproliferative effects on many cancer cell lines and inhibited growth in MM xenograft models [74]. Although there are very few promising reports regarding its clinical use, some flavonoid- and catechol-type molecules, such as myricetin, epigallocatechin gallate, and caffeic acid, were also found to be JmjC KDM inhibitors [125]. More comprehensive information about histone lysine demethylase inhibitors can be obtained from previous reviews [13, 22, 23, 189].

MKB Protein Inhibitors

The "readers" of the histone code, which bind mono-, di-, or tri-methylated histone lysines, are important components in the epigenetic regulation of gene expression [80, 94]. The "readers" of methyl-lysine residues are composed of various proteins, including specialized domains that facilitate recognition of these modifications. The "readers" for methyl-lysine include ATRX-DNMT3-DNMT3L, ankyrin, chromodomain, chromobarrel, double chromodomain, malignant brain tumor (MBT) domain, tudor domain, tandem tudor domain, Pro-Trp-Trp-Pro, plant homeodomain (PHD), WD-40, bromo-adjacent homology, and zinc finger CW domain [13] (Table 1). Recognition of methyl-lysine marks by MBT domains leads to the compaction of chromatin and a repressed transcriptional state. Four general classes of protein folds/domains that bind methyl lysine marks have been identified: ankyrin repeats, WD-40 repeat domains, PHD fingers, and royal family proteins [186]. More comprehensive information regarding MKB proteins can be obtained from a previous review [186].

The loss of lethal (3) MBT-like protein 1 (L3MBTL1) and L3MBTL3 has been shown to contribute to tumorigenesis [160, 211]. This report prompted researchers to find MKB protein inhibitors. Although there has not been so much progress beyond preclinical studies to date (Table 4), several MBT inhibitors have been reported. For example, UNC280 [80], UNC669 [81], and UNC926 [81] inhibit L3MBTL1 peptide binding. Two other MBT inhibitors, UNC1215 and UNC1679, have been reported to exhibit selective inhibition for L3MBTL3 [93]. Alternatively, chromodomains such as chromobox proteins play crucial roles in tumorigenesis. A chromobox 7 inhibitor, MS37452, derepresses transcription of the polycomb repressive complex target gene *p16/CDKN2A* in prostate cancer cells [157]. Moreover, a recent study reported two PHD finger inhibitors, amiodarone and CF16 [133, 199]. Thus, targeting epigenetic "readers" can be a useful strategy to antagonize the effect of aberrant histone methylation profiles in cancer, although their effectiveness as clinical candidates has not been reported.

Conclusion

Recently, many molecular-targeted cancer drugs have been approved by regulatory authorities and have improved the lives of many patients. However, the discovery and development of new targeted drugs, particularly in late clinical trials, are unsatisfactorily slow and they have high failure rates; ultimately, the development of drug resistance remains an issue. Substantial evidence has emerged that epigenetic mechanisms, including DNA methylation and histone modifications, play important roles in cancer development and onset. The abundant genetic variation that occurs in epigenetic regulatory complexes and proteins provides many basic targets for epigenetic drug discovery for cancer treatment.

In this chapter, we mainly focused on small-molecule modulators of cancer-related epigenetic mechanisms as potential cancer treatment targets. As we mentioned above, the development of effective inhibitors of DNA methyltransferases and histone deacetylases has already been successful to some degree. Epigenetic modulators of other targets such as "reader" proteins and chromatin remodeling complexes are also expected to be suitable candidates for cancer therapy. Additionally, combining epigenetic drugs with chemotherapeutic agents will broaden cancer treatment options. We hope that a better understanding of the epigenetic mechanisms in cancer collapse will lead to a more mechanistically based rationale for using specific epigenetic inhibitors for cancer therapy.

Acknowledgments This work was supported in part by the Central Research Institute of Fukuoka University (Grant No. 186005 to T.K.). We thank Edanz (www.edanzediting.co.jp) for editing the English text of a draft of this manuscript.

References

1. Agarwal, P., and S.P. Jackson. 2016. G9a inhibition potentiates the anti-tumour activity of DNA double-strand break inducing agents by impairing DNA repair independent of p53 status. *Cancer Letters* 380: 467–475.
2. Allfrey, V.G., R. Faulkner, and A.E. Mirsky. 1964. Acetylation and methylation of histones and their possible role in the regulation of RNA synthesis. *Proceedings of the National Academy of Sciences of the United States of America* 51: 786–794.
3. Amato, R.J. 2007. Inhibition of DNA methylation by antisense oligonucleotide MG98 as cancer therapy. *Clinical Genitourinary Cancer* 5: 422–426.
4. Amato, R.J., J. Stephenson, S. Hotte, J. Nemunaitis, K. Belanger, G. Reid, and R.E. Martell. 2012. MG98, a second-generation DNMT1 inhibitor, in the treatment of advanced renal cell carcinoma. *Cancer Investigation* 30: 415–421.
5. Amente, S., L. Lania, and B. Majello. 2013. The histone LSD1 demethylase in stemness and cancer transcription programs. *Biochimica et Biophysica Acta* 1829: 981–986.
6. Bannister, A.J., and T. Kouzarides. 2011. Regulation of chromatin by histone modifications. *Cell Research* 21: 381–395.
7. Barski, A., S. Cuddapah, K. Cui, T.Y. Roh, D.E. Schones, Z. Wang, G. Wei, I. Chepelev, and K. Zhao. 2007. High-resolution profiling of histone methylations in the human genome. *Cell* 129: 823–837.
8. Baylin, S.B., and P.A. Jones. 2011. A decade of exploring the cancer epigenome – biological and translational implications. *Nature Reviews. Cancer* 11: 726–734.
9. ———. 2016. Epigenetic determinants of cancer. *Cold Spring Harbor Perspectives in Biology* 8 (9): a019505.
10. Bedalov, A., T. Gatbonton, W.P. Irvine, D.E. Gottschling, and J.A. Simon. 2001. Identification of a small molecule inhibitor of Sir2p. *Proceedings of the National Academy of Sciences of the United States of America* 98: 15113–15118.
11. Bedford, M.T., and S.G. Clarke. 2009. Protein arginine methylation in mammals: Who, what and why. *Molecular Cell* 33: 1–13.
12. Benevolenskaya, E.V. 2007. Histone H3K4 demethylases are essential in development and differentiation. *Biochemistry and Cell Biology* 85: 435–443.
13. Bennett, R.L., and J.D. Licht. 2018. Targeting epigenetics in cancer. *Annual Review of Pharmacology and Toxicology* 58: 187–207.

14. Berenguer-Daize, C., L. Astorgues-Xerri, E. Odore, M. Cayol, E. Cvitkovic, K. Noel, M. Bekradda, S. Mackenzie, K. Rezai, F. Lokiec, M.E. Riveiro, and L. Ouafik. 2016. OTX015 (MK-8628), a novel BET inhibitor, displays in vitro and in vivo antitumor effects alone and in combination with conventional therapies in glioblastoma models. *International Journal of Cancer* 139: 2047–2055.
15. Bernasconi, E., E. Gaudio, P. Lejeune, C. Tarantelli, L. Cascione, I. Kwee, F. Spriano, A. Rinaldi, A.A. Mensah, E. Chung, A. Stathis, S. Siegel, N. Schmees, M. Ocker, E. Zucca, B. Haendler, and F. Bertoni. 2017. Preclinical evaluation of the BET bromodomain inhibitor BAY 1238097 for the treatment of lymphoma. *British Journal of Haematology* 178: 936–948.
16. Bernstein, B.E., T.S. Mikkelsen, X. Xie, M. Kamal, D.J. Huebert, J. Cuff, B. Fry, A. Meissner, M. Wernig, K. Plath, R. Jaenisch, A. Wagschal, R. Feil, S.L. Schreiber, and E.S. Lander. 2006. A bivalent chromatin structure marks key developmental genes in embryonic stem cells. *Cell* 125: 315–326.
17. Bertino, E.M., and G.A. Otterson. 2011. Romidepsin: A novel histone deacetylase inhibitor for cancer. *Expert Opinion on Investigational Drugs* 20: 1151–1158.
18. Bestor, T.H. 2000. The DNA methyltransferases of mammals. *Human Molecular Genetics* 9: 2395–2402.
19. Bestor, T., A. Laudano, R. Mattaliano, and V. Ingram. 1988. Cloning and sequencing of a cDNA encoding DNA methyltransferase of mouse cells. The carboxyl-terminal domain of the mammalian enzymes is related to bacterial restriction methyltransferases. *Journal of Molecular Biology* 203: 971–983.
20. Bilen, M.A., S. Fu, G.S. Falchook, C.S. Ng, J.J. Wheler, M. Abdelrahim, B. Erguvan-Dogan, D.S. Hong, A.M. Tsimberidou, R. Kurzrock, and A. Naing. 2015. Phase I trial of valproic acid and lenalidomide in patients with advanced cancer. *Cancer Chemotherapy and Pharmacology* 75: 869–874.
21. Bird, A. 2007. Perceptions of epigenetics. *Nature* 447: 396–398.
22. Biswas, S., and C.M. Rao. 2017. Epigenetics in cancer: Fundamentals and beyond. *Pharmacology & Therapeutics* 173: 118–134.
23. ———. 2018. Epigenetic tools (the writers, the readers and the erasers) and their implications in cancer therapy. *European Journal of Pharmacology* 837: 8–24.
24. Bonday, Z.Q., G.S. Cortez, M.J. Grogan, S. Antonysamy, K. Weichert, W.P. Bocchinfuso, F. Li, S. Kennedy, B. Li, M.M. Mader, C.H. Arrowsmith, P.J. Brown, M.S. Eram, M.M. Szewczyk, D. Barsyte-Lovejoy, M. Vedadi, E. Guccione, and R.M. Campbell. 2018. LLY-283, a potent and selective inhibitor of arginine methyltransferase 5, PRMT5, with antitumor activity. *ACS Medicinal Chemistry Letters* 9: 612–617.
25. Brueckner, B., R. Garcia Boy, P. Siedlecki, T. Musch, H.C. Kliem, P. Zielenkiewicz, S. Suhai, M. Wiessler, and F. Lyko. 2005. Epigenetic reactivation of tumor suppressor genes by a novel small-molecule inhibitor of human DNA methyltransferases. *Cancer Research* 65: 6305–6311.
26. Brueckner, B., M. Rius, M.R. Markelova, I. Fichtner, P.A. Hals, M.L. Sandvold, and F. Lyko. 2010. Delivery of 5-azacytidine to human cancer cells by elaidic acid esterification increases therapeutic drug efficacy. *Molecular Cancer Therapeutics* 9: 1256–1264.
27. Brunetto, A.T., J.E. Ang, R. Lal, D. Olmos, L.R. Molife, R. Kristeleit, A. Parker, I. Casamayor, M. Olaleye, A. Mais, B. Hauns, V. Strobel, B. Hentsch, and J.S. De Bono. 2013. First-in-human, pharmacokinetic and pharmacodynamic phase I study of Resminostat, an oral histone deacetylase inhibitor, in patients with advanced solid tumors. *Clinical Cancer Research* 19: 5494–5504.
28. Bui, M.H., X. Lin, D.H. Albert, L. Li, L.T. Lam, E.J. Faivre, S.E. Warder, X. Huang, D. Wilcox, C.K. Donawho, G.S. Sheppard, L. Wang, S. Fidanze, J.K. Pratt, D. Liu, L. Hasvold, T. Uziel, X. Lu, F. Kohlhapp, G. Fang, S.W. Elmore, S.H. Rosenberg, K.F. Mcdaniel, W.M. Kati, and Y. Shen. 2017. Preclinical characterization of BET family bromodomain inhibitor ABBV-075 suggests combination therapeutic strategies. *Cancer Research* 77: 2976–2989.

29. Busch, C., M. Burkard, C. Leischner, U.M. Lauer, J. Frank, and S. Venturelli. 2015. Epigenetic activities of flavonoids in the prevention and treatment of cancer. *Clinical Epigenetics* 7: 64.
30. Cedar, H., and Y. Bergman. 2009. Linking DNA methylation and histone modification: Patterns and paradigms. *Nature Reviews. Genetics* 10: 295–304.
31. Chaidos, A., V. Caputo, K. Gouvedenou, B. Liu, I. Marigo, M.S. Chaudhry, A. Rotolo, D.F. Tough, N.N. Smithers, A.K. Bassil, T.D. Chapman, N.R. Harker, O. Barbash, P. Tummino, N. Al-Mahdi, A.C. Haynes, L. Cutler, B. Le, A. Rahemtulla, I. Roberts, M. Kleijnen, J.J. Witherington, N.J. Parr, R.K. Prinjha, and A. Karadimitris. 2014. Potent antimyeloma activity of the novel bromodomain inhibitors I-BET151 and I-BET762. *Blood* 123: 697–705.
32. Chang, B., Y. Chen, Y. Zhao, and R.K. Bruick. 2007. JMJD6 is a histone arginine demethylase. *Science* 318: 444–447.
33. Chan-Penebre, E., K.G. Kuplast, C.R. Majer, P.A. Boriack-Sjodin, T.J. Wigle, L.D. Johnston, N. Rioux, M.J. Munchhof, L. Jin, S.L. Jacques, K.A. West, T. Lingaraj, K. Stickland, S.A. Ribich, A. Raimondi, M.P. Scott, N.J. Waters, R.M. Pollock, J.J. Smith, O. Barbash, M. Pappalardi, T.F. Ho, K. Nurse, K.P. Oza, K.T. Gallagher, R. Kruger, M.P. Moyer, R.A. Copeland, R. Chesworth, and K.W. Duncan. 2015. A selective inhibitor of PRMT5 with in vivo and in vitro potency in MCL models. *Nature Chemical Biology* 11: 432–437.
34. Chen, J.S., D.V. Faller, and R.A. Spanjaard. 2003. Short-chain fatty acid inhibitors of histone deacetylases: Promising anticancer therapeutics? *Current Cancer Drug Targets* 3: 219–236.
35. Cheung, N., T.K. Fung, B.B. Zeisig, K. Holmes, J.K. Rane, K.A. Mowen, M.G. Finn, B. Lenhard, L.C. Chan, and C.W. So. 2016. Targeting aberrant epigenetic networks mediated by PRMT1 and KDM4C in acute myeloid leukemia. *Cancer Cell* 29: 32–48.
36. Choi, W.J., H.J. Chung, G. Chandra, V. Alexander, L.X. Zhao, H.W. Lee, A. Nayak, M.S. Majik, H.O. Kim, J.H. Kim, Y.B. Lee, C.H. Ahn, S.K. Lee, and L.S. Jeong. 2012. Fluorocyclopentenyl-cytosine with broad spectrum and potent antitumor activity. *Journal of Medicinal Chemistry* 55: 4521–4525.
37. Chung, C.W., H. Coste, J.H. White, O. Mirguet, J. Wilde, R.L. Gosmini, C. Delves, S.M. Magny, R. Woodward, S.A. Hughes, E.V. Boursier, H. Flynn, A.M. Bouillot, P. Bamborough, J.M. Brusq, F.J. Gellibert, E.J. Jones, A.M. Riou, P. Homes, S.L. Martin, I.J. Uings, J. Toum, C.A. Clement, A.B. Boullay, R.L. Grimley, F.M. Blandel, R.K. Prinjha, K. Lee, J. Kirilovsky, and E. Nicodeme. 2011. Discovery and characterization of small molecule inhibitors of the BET family bromodomains. *Journal of Medicinal Chemistry* 54: 3827–3838.
38. Constantinides, P.G., P.A. Jones, and W. Gevers. 1977. Functional striated muscle cells from non-myoblast precursors following 5-azacytidine treatment. *Nature* 267: 364–366.
39. Cortellino, S., J. Xu, M. Sannai, R. Moore, E. Caretti, A. Cigliano, M. Le Coz, K. Devarajan, A. Wessels, D. Soprano, L.K. Abramowitz, M.S. Bartolomei, F. Rambow, M.R. Bassi, T. Bruno, M. Fanciulli, C. Renner, A.J. Klein-Szanto, Y. Matsumoto, D. Kobi, I. Davidson, C. Alberti, L. Larue, and A. Bellacosa. 2011. Thymine DNA glycosylase is essential for active DNA demethylation by linked deamination-base excision repair. *Cell* 146: 67–79.
40. Cote, J., and S. Richard. 2005. Tudor domains bind symmetrical dimethylated arginines. *The Journal of Biological Chemistry* 280: 28476–28483.
41. Datta, J., K. Ghoshal, W.A. Denny, S.A. Gamage, D.G. Brooke, P. Phiasivongsa, S. Redkar, and S.T. Jacob. 2009. A new class of quinoline-based DNA hypomethylating agents reactivates tumor suppressor genes by blocking DNA methyltransferase 1 activity and inducing its degradation. *Cancer Research* 69: 4277–4285.
42. Derissen, E.J., J.H. Beijnen, and J.H. Schellens. 2013. Concise drug review: Azacitidine and decitabine. *The Oncologist* 18: 619–624.
43. Doroshow, D.B., J.P. Eder, and P.M. Lorusso. 2017. BET inhibitors: A novel epigenetic approach. *Annals of Oncology* 28: 1776–1787.

44. Drew, A.E., O. Moradei, S.L. Jacques, N. Rioux, A.P. Boriack-Sjodin, C. Allain, M.P. Scott, L. Jin, A. Raimondi, J.L. Handler, H.M. Ott, R.G. Kruger, M.T. Mccabe, C. Sneeringer, T. Riera, G. Shapiro, N.J. Waters, L.H. Mitchell, K.W. Duncan, M.P. Moyer, R.A. Copeland, J. Smith, R. Chesworth, and S.A. Ribich. 2017. Identification of a CARM1 inhibitor with potent in vitro and in vivo activity in preclinical models of multiple myeloma. *Scientific Reports* 7: 17993.
45. Du, Q., P.L. Luu, C. Stirzaker, and S.J. Clark. 2015. Methyl-CpG-binding domain proteins: Readers of the epigenome. *Epigenomics* 7: 1051–1073.
46. Duque-Afonso, J., A. Yalcin, T. Berg, M. Abdelkarim, O. Heidenreich, and M. Lubbert. 2011. The HDAC class I-specific inhibitor entinostat (MS-275) effectively relieves epigenetic silencing of the LAT2 gene mediated by AML1/ETO. *Oncogene* 30: 3062–3072.
47. Easwaran, H., S.E. Johnstone, L. Van Neste, J. Ohm, T. Mosbruger, Q. Wang, M.J. Aryee, P. Joyce, N. Ahuja, D. Weisenberger, E. Collisson, J. Zhu, S. Yegnasubramanian, W. Matsui, and S.B. Baylin. 2012. A DNA hypermethylation module for the stem/progenitor cell signature of cancer. *Genome Research* 22: 837–849.
48. Easwaran, H., H.C. Tsai, and S.B. Baylin. 2014. Cancer epigenetics: Tumor heterogeneity, plasticity of stem-like states, and drug resistance. *Molecular Cell* 54: 716–727.
49. Egger, G., G. Liang, A. Aparicio, and P.A. Jones. 2004. Epigenetics in human disease and prospects for epigenetic therapy. *Nature* 429: 457–463.
50. Eissenberg, J.C., and A. Shilatifard. 2010. Histone H3 lysine 4 (H3K4) methylation in development and differentiation. *Developmental Biology* 339: 240–249.
51. Emami, K.H., C. Nguyen, H. Ma, D.H. Kim, K.W. Jeong, M. Eguchi, R.T. Moon, J.L. Teo, H.Y. Kim, S.H. Moon, J.R. Ha, and M. Kahn. 2004. A small molecule inhibitor of beta-catenin/CREB-binding protein transcription [corrected]. *Proceedings of the National Academy of Sciences of the United States of America* 101: 12682–12687.
52. Ernst, T., A.J. Chase, J. Score, C.E. Hidalgo-Curtis, C. Bryant, A.V. Jones, K. Waghorn, K. Zoi, F.M. Ross, A. Reiter, A. Hochhaus, H.G. Drexler, A. Duncombe, F. Cervantes, D. Oscier, J. Boultwood, F.H. Grand, and N.C. Cross. 2010. Inactivating mutations of the histone methyltransferase gene EZH2 in myeloid disorders. *Nature Genetics* 42: 722–726.
53. Esteller, M. 2007. Cancer epigenomics: DNA methylomes and histone-modification maps. *Nature Reviews. Genetics* 8: 286–298.
54. Evens, A.M., S. Balasubramanian, J.M. Vose, W. Harb, L.I. Gordon, R. Langdon, J. Sprague, M. Sirisawad, C. Mani, J. Yue, Y. Luan, S. Horton, T. Graef, and N.L. Bartlett. 2016. A phase I/II multicenter, open-label study of the oral histone deacetylase inhibitor abexinostat in relapsed/refractory lymphoma. *Clinical Cancer Research* 22: 1059–1066.
55. Fang, M.Z., Y. Wang, N. Ai, Z. Hou, Y. Sun, H. Lu, W. Welsh, and C.S. Yang. 2003. Tea polyphenol (-)-epigallocatechin-3-gallate inhibits DNA methyltransferase and reactivates methylation-silenced genes in cancer cell lines. *Cancer Research* 63: 7563–7570.
56. Feinberg, A.P., and B. Vogelstein. 1983. Hypomethylation distinguishes genes of some human cancers from their normal counterparts. *Nature* 301: 89–92.
57. Filippakopoulos, P., J. Qi, S. Picaud, Y. Shen, W.B. Smith, O. Fedorov, E.M. Morse, T. Keates, T.T. Hickman, I. Felletar, M. Philpott, S. Munro, M.R. Mckeown, Y. Wang, A.L. Christie, N. West, M.J. Cameron, B. Schwartz, T.D. Heightman, N. La Thangue, C.A. French, O. Wiest, A.L. Kung, S. Knapp, and J.E. Bradner. 2010. Selective inhibition of BET bromodomains. *Nature* 468: 1067–1073.
58. Filippakopoulos, P., S. Picaud, M. Mangos, T. Keates, J.P. Lambert, D. Barsyte-Lovejoy, I. Felletar, R. Volkmer, S. Muller, T. Pawson, A.C. Gingras, C.H. Arrowsmith, and S. Knapp. 2012. Histone recognition and large-scale structural analysis of the human bromodomain family. *Cell* 149: 214–231.
59. Furdas, S.D., S. Kannan, W. Sippl, and M. Jung. 2012. Small molecule inhibitors of histone acetyltransferases as epigenetic tools and drug candidates. *Archiv der Pharmazie (Weinheim)* 345: 7–21.

60. Furlan, A., V. Monzani, L.L. Reznikov, F. Leoni, G. Fossati, D. Modena, P. Mascagni, and C.A. Dinarello. 2011. Pharmacokinetics, safety and inducible cytokine responses during a phase 1 trial of the oral histone deacetylase inhibitor ITF2357 (givinostat). *Molecular Medicine* 17: 353–362.
61. Gajer, J.M., S.D. Furdas, A. Grunder, M. Gothwal, U. Heinicke, K. Keller, F. Colland, S. Fulda, H.L. Pahl, I. Fichtner, W. Sippl, and M. Jung. 2015. Histone acetyltransferase inhibitors block neuroblastoma cell growth in vivo. *Oncogene* 4: e137.
62. Gao, C., E. Bourke, M. Scobie, M.A. Famme, T. Koolmeister, T. Helleday, L.A. Eriksson, N.F. Lowndes, and J.A. Brown. 2014. Rational design and validation of a Tip60 histone acetyltransferase inhibitor. *Scientific Reports* 4: 5372.
63. Garcia-Manero, G., E. Atallah, S.K. Khaled, M. Arellano, M.M. Patnaik, T.A. Butler, C. Ashby, and B.C. Medeiros. 2015. Final results from a phase 2 study of pracinostat in combination with azacitidine in elderly patients with acute myeloid leukemia (AML). *Blood* 126 (3): –453.
64. Garraway, L.A., and E.S. Lander. 2013. Lessons from the cancer genome. *Cell* 153: 17–37.
65. Gatzka, M.V. 2018. Targeted tumor therapy remixed-an update on the use of small-molecule drugs in combination therapies. *Cancers (Basel)* 10 (6).
66. Ghizzoni, M., J. Wu, T. Gao, H.J. Haisma, F.J. Dekker, and Y. George Zheng. 2012. 6-alkylsalicylates are selective Tip60 inhibitors and target the acetyl-CoA binding site. *European Journal of Medicinal Chemistry* 47: 337–344.
67. Ghoshal, K., J. Datta, S. Majumder, S. Bai, H. Kutay, T. Motiwala, and S.T. Jacob. 2005. 5-Aza-deoxycytidine induces selective degradation of DNA methyltransferase 1 by a proteasomal pathway that requires the KEN box, bromo-adjacent homology domain, and nuclear localization signal. *Molecular and Cellular Biology* 25: 4727–4741.
68. Gibney, E.R., and C.M. Nolan. 2010. Epigenetics and gene expression. *Heredity (Edinburgh)* 105: 4–13.
69. Graca, I., E.J. Sousa, P. Costa-Pinheiro, F.Q. Vieira, J. Torres-Ferreira, M.G. Martins, R. Henrique, and C. Jeronimo. 2014. Anti-neoplastic properties of hydralazine in prostate cancer. *Oncotarget* 5: 5950–5964.
70. Gronbaek, K., C. Hother, and P.A. Jones. 2007. Epigenetic changes in cancer. *APMIS* 115: 1039–1059.
71. Gupta, S., K. Doyle, T.L. Mosbruger, A. Butterfield, A. Weston, A. Ast, M. Kaadige, A. Verma, and S. Sharma. 2018. Reversible LSD1 inhibition with HCI-2509 induces the p53 gene expression signature and disrupts the MYCN signature in high-risk neuroblastoma cells. *Oncotarget* 9: 9907–9924.
72. Hamada, S., T. Suzuki, K. Mino, K. Koseki, F. Oehme, I. Flamme, H. Ozasa, Y. Itoh, D. Ogasawara, H. Komaarashi, A. Kato, H. Tsumoto, H. Nakagawa, M. Hasegawa, R. Sasaki, T. Mizukami, and N. Miyata. 2010. Design, synthesis, enzyme-inhibitory activity, and effect on human cancer cells of a novel series of jumonji domain-containing protein 2 histone demethylase inhibitors. *Journal of Medicinal Chemistry* 53: 5629–5638.
73. Hamburger, A.W., and S.E. Salmon. 1977. Primary bioassay of human tumor stem cells. *Science* 197: 461–463.
74. Hancock, R.L., K. Dunne, L.J. Walport, E. Flashman, and A. Kawamura. 2015. Epigenetic regulation by histone demethylases in hypoxia. *Epigenomics* 7: 791–811.
75. Hashimoto, H., P.M. Vertino, and X. Cheng. 2010. Molecular coupling of DNA methylation and histone methylation. *Epigenomics* 2: 657–669.
76. Hauser, A.T., D. Robaa, and M. Jung. 2018. Epigenetic small molecule modulators of histone and DNA methylation. *Current Opinion in Chemical Biology* 45: 73–85.
77. Hay, D.A., O. Fedorov, S. Martin, D.C. Singleton, C. Tallant, C. Wells, S. Picaud, M. Philpott, O.P. Monteiro, C.M. Rogers, S.J. Conway, T.P. Rooney, A. Tumber, C. Yapp, P. Filippakopoulos, M.E. Bunnage, S. Muller, S. Knapp, C.J. Schofield, and P.E. Brennan. 2014. Discovery and optimization of small-molecule ligands for the CBP/p300 bromodomains. *Journal of the American Chemical Society* 136: 9308–9319.

78. Heinemann, B., J.M. Nielsen, H.R. Hudlebusch, M.J. Lees, D.V. Larsen, T. Boesen, M. Labelle, L.O. Gerlach, P. Birk, and K. Helin. 2014. Inhibition of demethylases by GSK-J1/J4. *Nature* 514: E1–E2.
79. Heltweg, B., T. Gatbonton, A.D. Schuler, J. Posakony, H. Li, S. Goehle, R. Kollipara, R.A. Depinho, Y. Gu, J.A. Simon, and A. Bedalov. 2006. Antitumor activity of a small-molecule inhibitor of human silent information regulator 2 enzymes. *Cancer Research* 66: 4368–4377.
80. Herold, J.M., T.J. Wigle, J.L. Norris, R. Lam, V.K. Korboukh, C. Gao, L.A. Ingerman, D.B. Kireev, G. Senisterra, M. Vedadi, A. Tripathy, P.J. Brown, C.H. Arrowsmith, J. Jin, W.P. Janzen, and S.V. Frye. 2011. Small-molecule ligands of methyl-lysine binding proteins. *Journal of Medicinal Chemistry* 54: 2504–2511.
81. Herold, J.M., L.I. James, V.K. Korboukh, C. Gao, K.E. Coil, D.J. Bua, J.L. Norris, D.B. Kireev, P.J. Brown, J. Jin, W.P. Janzen, O. Gozani, and S.V. Frye. 2012. Structure-activity relationships of methyl-lysine reader antagonists. *Medchemcomm* 3: 45–51.
82. Hodawadekar, S.C., and R. Marmorstein. 2007. Chemistry of acetyl transfer by histone modifying enzymes: Structure, mechanism and implications for effector design. *Oncogene* 26: 5528–5540.
83. Hopkinson, R.J., A. Tumber, C. Yapp, R. Chowdhury, W. Aik, K.H. Che, X.S. Li, J.B.L. Kristensen, O.N.F. King, M.C. Chan, K.K. Yeoh, H. Choi, L.J. Walport, C.C. Thinnes, J.T. Bush, C. Lejeune, A.M. Rydzik, N.R. Rose, E.A. Bagg, M.A. Mcdonough, T. Krojer, W.W. Yue, S.S. Ng, L. Olsen, P.E. Brennan, U. Oppermann, S. Muller-Knapp, R.J. Klose, P.J. Ratcliffe, C.J. Schofield, and A. Kawamura. 2013a. 5-carboxy-8-hydroxyquinoline is a broad spectrum 2-oxoglutarate oxygenase inhibitor which causes iron translocation. *Chemical Science* 4: 3110–3117.
84. Hopkinson, R.J., A. Tumber, C. Yapp, R. Chowdhury, W. Aik, K.H. Che, X.S. Li, J.B.L. Kristensen, O.N.F. King, M.C. Chan, K.K. Yeoh, H. Choi, L.J. Walport, C.C. Thinnes, J.T. Bush, C. Lejeune, A.M. Rydzik, N.R. Rose, E.A. Bagg, M.A. Mcdonough, T.J. Krojer, W.W. Yue, S.S. Ng, L. Olsen, P.E. Brennan, U. Oppermann, S. Muller, R.J. Klose, P.J. Ratcliffe, C.J. Schofield, and A. Kawamura. 2013b. 5-Carboxy-8-hydroxyquinoline is a broad spectrum 2-oxoglutarate oxygenase inhibitor which causes iron translocation. *Chemical Science* 4: 3110–3117.
85. Hore, T.A., F. Von Meyenn, M. Ravichandran, M. Bachman, G. Ficz, D. Oxley, F. Santos, S. Balasubramanian, T.P. Jurkowski, and W. Reik. 2016. Retinol and ascorbate drive erasure of epigenetic memory and enhance reprogramming to naive pluripotency by complementary mechanisms. *Proceedings of the National Academy of Sciences of the United States of America* 113: 12202–12207.
86. Huang, Y., and A. Rao. 2014. Connections between TET proteins and aberrant DNA modification in cancer. *Trends in Genetics* 30: 464–474.
87. Huang, Y., Greene, E., Murray Stewart, T., Goodwin, A. C., Baylin, S. B., Woster, P. M. & Casero, R. A., JR. (2007) Inhibition of lysine-specific demethylase 1 by polyamine analogues results in reexpression of aberrantly silenced genes. *Proceedings of the National Academy of Sciences of the United States of America,* 104**,** 8023-8028.
88. Huntly, B.J., and D.G. Gilliland. 2005. Leukaemia stem cells and the evolution of cancer-stem-cell research. *Nature Reviews. Cancer* 5: 311–321.
89. Imai, S., C.M. Armstrong, M. Kaeberlein, and L. Guarente. 2000. Transcriptional silencing and longevity protein Sir2 is an NAD-dependent histone deacetylase. *Nature* 403: 795–800.
90. Ito, S., A.C. D'alessio, O.V. Taranova, K. Hong, L.C. Sowers, and Y. Zhang. 2010. Role of Tet proteins in 5mC to 5hmC conversion, ES-cell self-renewal and inner cell mass specification. *Nature* 466: 1129–1133.
91. Itoh, Y. 2018. Chemical protein degradation approach and its application to epigenetic targets. *Chemical Record* 18: 1681–1700.
92. Itoh, Y., T. Suzuki, and N. Miyata. 2013. Small-molecular modulators of cancer-associated epigenetic mechanisms. *Molecular BioSystems* 9: 873–896.

93. James, L.I., D. Barsyte-Lovejoy, N. Zhong, L. Krichevsky, V.K. Korboukh, J.M. Herold, C.J. Macnevin, J.L. Norris, C.A. Sagum, W. Tempel, E. Marcon, H. Guo, C. Gao, X.P. Huang, S. Duan, A. Emili, J.F. Greenblatt, D.B. Kireev, J. Jin, W.P. Janzen, P.J. Brown, M.T. Bedford, C.H. Arrowsmith, and S.V. Frye. 2013a. Discovery of a chemical probe for the L3MBTL3 methyllysine reader domain. *Nature Chemical Biology* 9: 184–191.
94. James, L.I., V.K. Korboukh, L. Krichevsky, B.M. Baughman, J.M. Herold, J.L. Norris, J. Jin, D.B. Kireev, W.P. Janzen, C.H. Arrowsmith, and S.V. Frye. 2013b. Small-molecule ligands of methyl-lysine binding proteins: Optimization of selectivity for L3MBTL3. *Journal of Medicinal Chemistry* 56: 7358–7371.
95. Jeanmougin, F., Wurtz, J. M., LE Douarin, B., Chambon, P. & Losson, R. (1997) The bromodomain revisited. *Trends in Biochemical Sciences,* 22**,** 151-153.
96. Jenuwein, T., and C.D. Allis. 2001. Translating the histone code. *Science* 293: 1074–1080.
97. Jing, H., J. Hu, B. He, Y.L. Negron Abril, J. Stupinski, K. Weiser, M. Carbonaro, Y.L. Chiang, T. Southard, P. Giannakakou, R.S. Weiss, and H. Lin. 2016. A SIRT2-selective inhibitor promotes c-Myc oncoprotein degradation and exhibits broad anticancer activity. *Cancer Cell* 29: 767–768.
98. Jones, P.A., and S.B. Baylin. 2007. The epigenomics of cancer. *Cell* 128: 683–692.
99. Kaniskan, H.U., M.L. Martini, and J. Jin. 2018. Inhibitors of protein methyltransferases and demethylases. *Chemical Reviews* 118: 989–1068.
100. Kantarjian, H.M., G.J. Roboz, P.L. Kropf, K.W.L. Yee, C.L. O'Connell, R. Tibes, K.J. Walsh, N.A. Podoltsev, E.A. Griffiths, E. Jabbour, G. Garcia-Manero, D. Rizzieri, W. Stock, M.R. Savona, T.L. Rosenblat, J.G. Berdeja, F. Ravandi, E.P. Rock, Y. Hao, M. Azab, and J.J. Issa. 2017. Guadecitabine (SGI-110) in treatment-naive patients with acute myeloid leukaemia: Phase 2 results from a multicentre, randomised, phase 1/2 trial. *The Lancet Oncology* 18: 1317–1326.
101. Kim, K.H., and C.W. Roberts. 2016. Targeting EZH2 in cancer. *Nature Medicine* 22: 128–134.
102. Kim, H.W., S.A. Kim, and S.G. Ahn. 2016. Sirtuin inhibitors, EX527 and AGK2, suppress cell migration by inhibiting HSF1 protein stability. *Oncology Reports* 35: 235–242.
103. Kouzarides, T. 2007. Chromatin modifications and their function. *Cell* 128: 693–705.
104. Kozako, T., A. Aikawa, T. Shoji, T. Fujimoto, M. Yoshimitsu, S. Shirasawa, H. Tanaka, S. Honda, H. Shimeno, N. Arima, and S. Soeda. 2012. High expression of the longevity gene product SIRT1 and apoptosis induction by sirtinol in adult T-cell leukemia cells. *International Journal of Cancer* 131: 2044–2055.
105. Kozako, T., T. Suzuki, M. Yoshimitsu, N. Arima, S.I. Honda, and S. Soeda. 2014. Anticancer agents targeted to sirtuins. *Molecules* 19: 20295–20313.
106. Kozako, T., T. Suzuki, M. Yoshimitsu, Y. Uchida, A. Kuroki, A. Aikawa, S. Honda, N. Arima, and S. Soeda. 2015. Novel small-molecule SIRT1 inhibitors induce cell death in adult T-cell leukaemia cells. *Scientific Reports* 5: 11345.
107. Kozako, T., P. Mellini, T. Ohsugi, A. Aikawa, Y.I. Uchida, S.I. Honda, and T. Suzuki. 2018. Novel small molecule SIRT2 inhibitors induce cell death in leukemic cell lines. *BMC Cancer* 18: 791.
108. Krause, C.D., Z.H. Yang, Y.S. Kim, J.H. Lee, J.R. Cook, and S. Pestka. 2007. Protein arginine methyltransferases: evolution and assessment of their pharmacological and therapeutic potential. *Pharmacology & Therapeutics* 113: 50–87.
109. Kubicek, S., R.J. O'Sullivan, E.M. August, E.R. Hickey, Q. Zhang, M.L. Teodoro, S. Rea, K. Mechtler, J.A. Kowalski, C.A. Homon, T.A. Kelly, and T. Jenuwein. 2007. Reversal of H3K9me2 by a small-molecule inhibitor for the G9a histone methyltransferase. *Molecular Cell* 25: 473–481.
110. Kuck, D., T. Caulfield, F. Lyko, and J.L. Medina-Franco. 2010a. Nanaomycin A selectively inhibits DNMT3B and reactivates silenced tumor suppressor genes in human cancer cells. *Molecular Cancer Therapeutics* 9: 3015–3023.

111. Kuck, D., N. Singh, F. Lyko, and J.L. Medina-Franco. 2010b. Novel and selective DNA methyltransferase inhibitors: docking-based virtual screening and experimental evaluation. *Bioorganic & Medicinal Chemistry* 18: 822–829.
112. Kuntz, K.W., J.E. Campbell, H. Keilhack, R.M. Pollock, S.K. Knutson, M. Porter-Scott, V.M. Richon, C.J. Sneeringer, T.J. Wigle, C.J. Allain, C.R. Majer, M.P. Moyer, R.A. Copeland, and R. Chesworth. 2016. The importance of being me: Magic methyls, methyltransferase inhibitors, and the discovery of tazemetostat. *Journal of Medicinal Chemistry* 59: 1556–1564.
113. Kwok, J., M. O'shea, D.A. Hume, and A. Lengeling. 2017. Jmjd6, a JmjC dioxygenase with many interaction partners and pleiotropic functions. *Frontiers in Genetics* 8: 32.
114. Lain, S., J.J. Hollick, J. Campbell, O.D. Staples, M. Higgins, M. Aoubala, A. Mccarthy, V. Appleyard, K.E. Murray, L. Baker, A. Thompson, J. Mathers, S.J. Holland, M.J. Stark, G. Pass, J. Woods, D.P. Lane, and N.J. Westwood. 2008. Discovery, in vivo activity, and mechanism of action of a small-molecule p53 activator. *Cancer Cell* 13: 454–463.
115. Lara, E., A. Mai, V. Calvanese, L. Altucci, P. Lopez-Nieva, M.L. Martinez-Chantar, M. Varela-Rey, D. Rotili, A. Nebbioso, S. Ropero, G. Montoya, J. Oyarzabal, S. Velasco, M. Serrano, M. Witt, A. Villar-Garea, A. Imhof, J.M. Mato, M. Esteller, and M.F. Fraga. 2009. Salermide, a sirtuin inhibitor with a strong cancer-specific proapoptotic effect. *Oncogene* 28: 781–791.
116. Leal, A.S., C.R. Williams, D.B. Royce, P.A. Pioli, M.B. Sporn, and K.T. Liby. 2017. Bromodomain inhibitors, JQ1 and I-BET 762, as potential therapies for pancreatic cancer. *Cancer Letters* 394: 76–87.
117. Lee, K.K., and J.L. Workman. 2007. Histone acetyltransferase complexes: One size doesn't fit all. *Nature Reviews. Molecular Cell Biology* 8: 284–295.
118. Lee, B.H., S. Yegnasubramanian, X. Lin, and W.G. Nelson. 2005. Procainamide is a specific inhibitor of DNA methyltransferase 1. *The Journal of Biological Chemistry* 280: 40749–40756.
119. Lee, H.Z., V.E. Kwitkowski, P.L. del Valle, M.S. Ricci, H. Saber, B.A. Habtemariam, J. Bullock, E. Bloomquist, Y. Li Shen, X.H. Chen, J. Brown, N. Mehrotra, S. Dorff, R. Charlab, R.C. Kane, E. Kaminskas, R. Justice, A.T. Farrell, and R. Pazdur. 2015. FDA approval: Belinostat for the treatment of patients with relapsed or refractory peripheral T-cell lymphoma. *Clinical Cancer Research* 21: 2666–2670.
120. Ler, L.D., S. Ghosh, X. Chai, A.A. Thike, H.L. Heng, E.Y. Siew, S. Dey, L.K. Koh, J.Q. Lim, W.K. Lim, S.S. Myint, J.L. Loh, P. Ong, X.X. Sam, D. Huang, T. Lim, P.H. Tan, S. Nagarajan, C.W. Cheng, H. Ho, L.G. Ng, J. Yuen, P.H. Lin, C.K. Chuang, Y.H. Chang, W.H. Weng, S.G. Rozen, P. Tan, C.L. Creasy, S.T. Pang, M.T. Mccabe, S.L. Poon, and B.T. Teh. 2017. Loss of tumor suppressor KDM6A amplifies PRC2-regulated transcriptional repression in bladder cancer and can be targeted through inhibition of EZH2. *Science Translational Medicine* 9, 378.
121. Li, P., J. Hu, and Y. Wang. 2012. Methods for analyzing histone citrullination in chromatin structure and gene regulation. *Methods in Molecular Biology* 809: 473–488.
122. Li, L.X., L.X. Fan, J.X. Zhou, J.J. Grantham, J.P. Calvet, J. Sage, and X. Li. 2017. Lysine methyltransferase SMYD2 promotes cyst growth in autosomal dominant polycystic kidney disease. *The Journal of Clinical Investigation* 127: 2751–2764.
123. Liu, B.L., J.X. Cheng, X. Zhang, R. Wang, W. Zhang, H. Lin, X. Xiao, S. Cai, X.Y. Chen, and H. Cheng. 2010. Global histone modification patterns as prognostic markers to classify glioma patients. *Cancer Epidemiology, Biomarkers & Prevention* 19: 2888–2896.
124. Liu, W., L. Deng, Y. Song, and M. Redell. 2014. DOT1L inhibition sensitizes MLL-rearranged AML to chemotherapy. *PLoS One* 9: e98270.
125. Lohse, B., A.L. Nielsen, J.B. Kristensen, C. Helgstrand, P.A. Cloos, L. Olsen, M. Gajhede, R.P. Clausen, and J.L. Kristensen. 2011. Targeting histone lysine demethylases by truncating the histone 3 tail to obtain selective substrate-based inhibitors. *Angewandte Chemie (International Ed. in English)* 50: 9100–9103.

126. Luger, K., and T.J. Richmond. 1998. The histone tails of the nucleosome. *Current Opinion in Genetics & Development* 8: 140–146.
127. Maes, T., E. Carceller, J. Salas, A. Ortega, and C. Buesa. 2015. Advances in the development of histone lysine demethylase inhibitors. *Current Opinion in Pharmacology* 23: 52–60.
128. Mann, B.S., J.R. Johnson, M.H. Cohen, R. Justice, and R. Pazdur. 2007. FDA approval summary: Vorinostat for treatment of advanced primary cutaneous T-cell lymphoma. *The Oncologist* 12: 1247–1252.
129. Martin, L.J., M. Koegl, G. Bader, X.L. Cockcroft, O. Fedorov, D. Fiegen, T. Gerstberger, M.H. Hofmann, A.F. Hohmann, D. Kessler, S. Knapp, P. Knesl, S. Kornigg, S. Muller, H. Nar, C. Rogers, K. Rumpel, O. Schaaf, S. Steurer, C. Tallant, C.R. Vakoc, M. Zeeb, A. Zoephel, M. Pearson, G. Boehmelt, and D. Mcconnell. 2016. Structure-based design of an in vivo active selective BRD9 inhibitor. *Journal of Medicinal Chemistry* 59: 4462–4475.
130. Mccabe, M.T., H.M. Ott, G. Ganji, S. Korenchuk, C. Thompson, G.S. Van Aller, Y. Liu, A.P. Graves, A. Della Pietra 3rd, E. Diaz, L.V. Lafrance, M. Mellinger, C. Duquenne, X. Tian, R.G. Kruger, C.F. Mchugh, M. Brandt, W.H. Miller, D. Dhanak, S.K. Verma, P.J. Tummino, and C.L. Creasy. 2012. EZH2 inhibition as a therapeutic strategy for lymphoma with EZH2-activating mutations. *Nature* 492: 108–112.
131. Mellini, P., Y. Itoh, H. Tsumoto, Y. Li, M. Suzuki, N. Tokuda, T. Kakizawa, Y. Miura, J. Takeuchi, M. Lahtela-Kakkonen, and T. Suzuki. 2017. Potent mechanism-based sirtuin-2-selective inhibition by an in situ-generated occupant of the substrate-binding site, "selectivity pocket" and NAD(+)-binding site. *Chemical Science* 8: 6400–6408.
132. Milite, C., A. Feoli, K. Sasaki, V. La Pietra, A.L. Balzano, L. Marinelli, A. Mai, E. Novellino, S. Castellano, A. Tosco, and G. Sbardella. 2015. A novel cell-permeable, selective, and noncompetitive inhibitor of KAT3 histone acetyltransferases from a combined molecular pruning/classical isosterism approach. *Journal of Medicinal Chemistry* 58: 2779–2798.
133. Miller, T.C., T.J. Rutherford, K. Birchall, J. Chugh, M. Fiedler, and M. Bienz. 2014. Competitive binding of a benzimidazole to the histone-binding pocket of the Pygo PHD finger. *ACS Chemical Biology* 9: 2864–2874.
134. Miranda-Goncalves, V., A. Lameirinhas, R. Henrique, and C. Jeronimo. 2018. Metabolism and epigenetic interplay in cancer: Regulation and putative therapeutic targets. *Frontiers in Genetics* 9: 427.
135. Miremadi, A., M.Z. Oestergaard, P.D. Pharoah, and C. Caldas. 2007. Cancer genetics of epigenetic genes. *Human Molecular Genetics* 16 (Spec No 1): R28–R49.
136. Morin, R.D., N.A. Johnson, T.M. Severson, A.J. Mungall, J. An, R. Goya, J.E. Paul, M. Boyle, B.W. Woolcock, F. Kuchenbauer, D. Yap, R.K. Humphries, O.L. Griffith, S. Shah, H. Zhu, M. Kimbara, P. Shashkin, J.F. Charlot, M. Tcherpakov, R. Corbett, A. Tam, R. Varhol, D. Smailus, M. Moksa, Y. Zhao, A. Delaney, H. Qian, I. Birol, J. Schein, R. Moore, R. Holt, D.E. Horsman, J.M. Connors, S. Jones, S. Aparicio, M. Hirst, R.D. Gascoyne, and M.A. Marra. 2010. Somatic mutations altering EZH2 (Tyr641) in follicular and diffuse large B-cell lymphomas of germinal-center origin. *Nature Genetics* 42: 181–185.
137. Nakagawa, M., Y. Oda, T. Eguchi, S. Aishima, T. Yao, F. Hosoi, Y. Basaki, M. Ono, M. Kuwano, M. Tanaka, and M. Tsuneyoshi. 2007. Expression profile of class I histone deacetylases in human cancer tissues. *Oncology Reports* 18: 769–774.
138. Nakamura, K., K. Nakabayashi, K. Htet Aung, K. Aizawa, N. Hori, J. Yamauchi, K. Hata, and A. Tanoue. 2015. DNA methyltransferase inhibitor zebularine induces human cholangiocarcinoma cell death through alteration of DNA methylation status. *PLoS One* 10: e0120545.
139. Nakayama, K., M.M. Szewczyk, C. Dela Sena, H. Wu, A. Dong, H. Zeng, F. Li, R.F. De Freitas, M.S. Eram, M. Schapira, Y. Baba, M. Kunitomo, D.R. Cary, M. Tawada, A. Ohashi, Y. Imaeda, K.S. Saikatendu, C.E. Grimshaw, M. Vedadi, C.H. Arrowsmith, D. Barsyte-Lovejoy, A. Kiba, D. Tomita, and P.J. Brown. 2018. TP-064, a potent and selective small molecule inhibitor of PRMT4 for multiple myeloma. *Oncotarget* 9: 18480–18493.

140. Nguyen, H., A. Allali-Hassani, S. Antonysamy, S. Chang, L.H. Chen, C. Curtis, S. Emtage, L. Fan, T. Gheyi, F. Li, S. Liu, J.R. Martin, D. Mendel, J.B. Olsen, L. Pelletier, T. Shatseva, S. Wu, F.F. Zhang, C.H. Arrowsmith, P.J. Brown, R.M. Campbell, B.A. Garcia, D. Barsyte-Lovejoy, M. Mader, and M. Vedadi. 2015. LLY-507, a cell-active, potent, and selective inhibitor of protein-lysine methyltransferase SMYD2. *The Journal of Biological Chemistry* 290: 13641–13653.
141. Ogasawara, D., Y. Itoh, H. Tsumoto, T. Kakizawa, K. Mino, K. Fukuhara, H. Nakagawa, M. Hasegawa, R. Sasaki, T. Mizukami, N. Miyata, and T. Suzuki. 2013. Lysine-specific demethylase 1-selective inactivators: Protein-targeted drug delivery mechanism. *Angewandte Chemie (International Ed. in English)* 52: 8620–8624.
142. Ohm, J.E., K.M. Mcgarvey, X. Yu, L. Cheng, K.E. Schuebel, L. Cope, H.P. Mohammad, W. Chen, V.C. Daniel, W. Yu, D.M. Berman, T. Jenuwein, K. Pruitt, S.J. Sharkis, D.N. Watkins, J.G. Herman, and S.B. Baylin. 2007. A stem cell-like chromatin pattern may predispose tumor suppressor genes to DNA hypermethylation and heritable silencing. *Nature Genetics* 39: 237–242.
143. Ohtani-Fujita, N., T. Fujita, A. Aoike, N.E. Osifchin, P.D. Robbins, and T. Sakai. 1993. CpG methylation inactivates the promoter activity of the human retinoblastoma tumor-suppressor gene. *Oncogene* 8: 1063–1067.
144. Okano, M., S. Xie, and E. Li. 1998. Cloning and characterization of a family of novel mammalian DNA (cytosine-5) methyltransferases. *Nature Genetics* 19: 219–220.
145. Okano, M., D.W. Bell, D.A. Haber, and E. Li. 1999. DNA methyltransferases Dnmt3a and Dnmt3b are essential for de novo methylation and mammalian development. *Cell* 99: 247–257.
146. Ott, M., and E. Verdin. 2010. HAT trick: p300, small molecule, inhibitor. *Chemistry & Biology* 17: 417–418.
147. Pachaiyappan, B., and P.M. Woster. 2014. Design of small molecule epigenetic modulators. *Bioorganic & Medicinal Chemistry Letters* 24: 21–32.
148. Peng, L., Z. Xu, Z. Liu, Y. Wei, H. Sun, Z. Li, X. Zhao, and C. Gao. 2015. An iron-based green approach to 1-h production of single-layer graphene oxide. *Nature Communications* 6: 5716.
149. Perez-Salvia, M., and M. Esteller. 2017. Bromodomain inhibitors and cancer therapy: From structures to applications. *Epigenetics* 12: 323–339.
150. Peserico, A., and C. Simone. 2011. Physical and functional HAT/HDAC interplay regulates protein acetylation balance. *Journal of Biomedicine & Biotechnology* 2011: 371832.
151. Picaud, S., C. Wells, I. Felletar, D. Brotherton, S. Martin, P. Savitsky, B. Diez-Dacal, M. Philpott, C. Bountra, H. Lingard, O. Fedorov, S. Muller, P.E. Brennan, S. Knapp, and P. Filippakopoulos. 2013. RVX- 208, an inhibitor of BET transcriptional regulators with selectivity for the second bromodomain. *Proceedings of the National Academy of Sciences of the United States of America* 110: 19754–19759.
152. Prusevich, P., J.H. Kalin, S.A. Ming, M. Basso, J. Givens, X. Li, J. Hu, M.S. Taylor, A.M. Cieniewicz, P.Y. Hsiao, R. Huang, H. Roberson, N. Adejola, L.B. Avery, R.A. Casero Jr., S.D. Taverna, J. Qian, A.J. Tackett, R.R. Ratan, O.G. Mcdonald, A.P. Feinberg, and P.A. Cole. 2014. A selective phenelzine analogue inhibitor of histone demethylase LSD1. *ACS Chemical Biology* 9: 1284–1293.
153. Qi, W., H. Chan, L. Teng, L. Li, S. Chuai, R. Zhang, J. Zeng, M. Li, H. Fan, Y. Lin, J. Gu, O. Ardayfio, J.H. Zhang, X. Yan, J. Fang, Y. Mi, M. Zhang, T. Zhou, G. Feng, Z. Chen, G. Li, T. Yang, K. Zhao, X. Liu, Z. Yu, C.X. Lu, P. Atadja, and E. Li. 2012. Selective inhibition of Ezh2 by a small molecule inhibitor blocks tumor cells proliferation. *Proceedings of the National Academy of Sciences of the United States of America* 109: 21360–21365.
154. Qin, W., P. Wolf, N. Liu, S. Link, M. Smets, F. La Mastra, I. Forne, G. Pichler, D. Horl, K. Fellinger, F. Spada, I.M. Bonapace, A. Imhof, H. Harz, and H. Leonhardt. 2015. DNA methylation requires a DNMT1 ubiquitin interacting motif (UIM) and histone ubiquitination. *Cell Research* 25: 911–929.

155. Ramakrishnan, V. 1997. Histone structure and the organization of the nucleosome. *Annual Review of Biophysics and Biomolecular Structure* 26: 83–112.
156. Ren, J., B.N. Singh, Q. Huang, Z. Li, Y. Gao, P. Mishra, Y.L. Hwa, J. Li, S.C. Dowdy, and S.W. Jiang. 2011. DNA hypermethylation as a chemotherapy target. *Cellular Signalling* 23: 1082–1093.
157. Ren, C., K. Morohashi, A.N. Plotnikov, J. Jakoncic, S.G. Smith, J. Li, L. Zeng, Y. Rodriguez, V. Stojanoff, M. Walsh, and M.M. Zhou. 2015. Small- molecule modulators of methyl-lysine binding for the CBX7 chromodomain. *Chemistry & Biology* 22: 161–168.
158. Rhee, I., K.E. Bachman, B.H. Park, K.W. Jair, R.W. Yen, K.E. Schuebel, H. Cui, A.P. Feinberg, C. Lengauer, K.W. Kinzler, S.B. Baylin, and B. Vogelstein. 2002. DNMT1 and DNMT3b cooperate to silence genes in human cancer cells. *Nature* 416: 552–556.
159. Richon, V.M., S. Emiliani, E. Verdin, Y. Webb, R. Breslow, R.A. Rifkind, and P.A. Marks. 1998. A class of hybrid polar inducers of transformed cell differentiation inhibits histone deacetylases. *Proceedings of the National Academy of Sciences of the United States of America* 95: 3003–3007.
160. Richter, C., K. Oktaba, J. Steinmann, J. Muller, and J.A. Knoblich. 2011. The tumour suppressor L(3)mbt inhibits neuroepithelial proliferation and acts on insulator elements. *Nature Cell Biology* 13: 1029–1039.
161. Rilova, E., A. Erdmann, C. Gros, V. Masson, Y. Aussagues, V. Poughon-Cassabois, A. Rajavelu, A. Jeltsch, Y. Menon, N. Novosad, J.M. Gregoire, S. Vispe, P. Schambel, F. Ausseil, F. Sautel, P.B. Arimondo, and F. Cantagrel. 2014. Design, synthesis and biological evaluation of 4-amino-N- (4-aminophenyl)benzamide analogues of quinoline-based SGI-1027 as inhibitors of DNA methylation. *ChemMedChem* 9: 590–601.
162. Rodriguez-Paredes, M., and M. Esteller. 2011. Cancer epigenetics reaches mainstream oncology. *Nature Medicine* 17: 330–339.
163. Rose, N.R., S.S. Ng, J. Mecinovic, B.M. Lienard, S.H. Bello, Z. Sun, M.A. Mcdonough, U. Oppermann, and C.J. Schofield. 2008. Inhibitor scaffolds for 2-oxoglutarate-dependent histone lysine demethylases. *Journal of Medicinal Chemistry* 51: 7053–7056.
164. Rosenfeld, J.A., Z. Wang, D.E. Schones, K. Zhao, R. Desalle, and M.Q. Zhang. 2009. Determination of enriched histone modifications in non-genic portions of the human genome. *BMC Genomics* 10: 143.
165. Ruthenburg, A.J., C.D. Allis, and J. Wysocka. 2007. Methylation of lysine 4 on histone H3: Intricacy of writing and reading a single epigenetic mark. *Molecular Cell* 25: 15–30.
166. Sanchez, R., and M.M. Zhou. 2009. The role of human bromodomains in chromatin biology and gene transcription. *Current Opinion in Drug Discovery & Development* 12: 659–665.
167. Santi, D.V., A. Norment, and C.E. Garrett. 1984. Covalent bond formation between a DNA-cytosine methyltransferase and DNA containing 5-azacytosine. *Proceedings of the National Academy of Sciences of the United States of America* 81: 6993–6997.
168. Sarkisjan, D., J.R. Julsing, K. Smid, D. de Klerk, A.B. van Kuilenburg, R. Meinsma, Y.B. Lee, D.J. Kim, and G.J. Peters. 2016. The cytidine analog fluorocyclopentenylcytosine (RX-3117) is activated by uridine-cytidine kinase 2. *PLoS One* 11: e0162901.
169. Savickiene, J., G. Treigyte, A. Jazdauskaite, V.V. Borutinskaite, and R. Navakauskiene. 2012. DNA methyltransferase inhibitor RG108 and histone deacetylase inhibitors cooperate to enhance NB4 cell differentiation and E-cadherin re- expression by chromatin remodelling. *Cell Biology International* 36: 1067–1078.
170. Schlesinger, Y., R. Straussman, I. Keshet, S. Farkash, M. Hecht, J. Zimmerman, E. Eden, Z. Yakhini, E. Ben-Shushan, B.E. Reubinoff, Y. Bergman, I. Simon, and H. Cedar. 2007. Polycomb-mediated methylation on Lys27 of histone H3 pre-marks genes for de novo methylation in cancer. *Nature Genetics* 39: 232–236.
171. Segura-Pacheco, B., C. Trejo-Becerril, E. Perez-Cardenas, L. Taja-Chayeb, I. Mariscal, A. Chavez, C. Acuna, A.M. Salazar, M. Lizano, and A. Duenas-Gonzalez. 2003. Reactivation of tumor suppressor genes by the cardiovascular drugs hydralazine and procainamide and their potential use in cancer therapy. *Clinical Cancer Research* 9: 1596–1603.

172. Seto, E., and M. Yoshida. 2014. Erasers of histone acetylation: The histone deacetylase enzymes. *Cold Spring Harbor Perspectives in Biology* 6: a018713.
173. Shi, Y. 2007. Histone lysine demethylases: Emerging roles in development, physiology and disease. *Nature Reviews. Genetics* 8: 829–833.
174. Siu, K.T., H. Eda, L. Santo, J. Ramachandran, M. Koulnis, J. Mertz, R.J. Sims, M. Cooper, and N.S. Raje. 2015. Effect of the BET inhibitor, Cpi-0610, alone and in combination with lenalidomide in multiple myeloma. *Blood* 126 (23): 4255.
175. Song, S.H., S.W. Han, and Y.J. Bang. 2011. Epigenetic-based therapies in cancer: Progress to date. *Drugs* 71: 2391–2403.
176. Sproul, D., N. Gilbert, and W.A. Bickmore. 2005. The role of chromatin structure in regulating the expression of clustered genes. *Nature Reviews. Genetics* 6: 775–781.
177. Srinivasan, P.R., and E. Borek. 1964. Species variation of the RNA methylases. *Biochemistry* 3: 616–619.
178. Steger, D.J., M.I. Lefterova, L. Ying, A.J. Stonestrom, M. Schupp, D. Zhuo, A.L. Vakoc, J.E. Kim, J. Chen, M.A. Lazar, G.A. Blobel, and C.R. Vakoc. 2008. DOT1L/KMT4 recruitment and H3K79 methylation are ubiquitously coupled with gene transcription in mammalian cells. *Molecular and Cellular Biology* 28: 2825–2839.
179. Stimson, L., M.G. Rowlands, Y.M. Newbatt, N.F. Smith, F.I. Raynaud, P. Rogers, V. Bavetsias, S. Gorsuch, M. Jarman, A. Bannister, T. Kouzarides, E. Mcdonald, P. Workman, and G.W. Aherne. 2005. Isothiazolones as inhibitors of PCAF and p300 histone acetyltransferase activity. *Molecular Cancer Therapeutics* 4: 1521–1532.
180. Stubbs, M., R. Collins, A. Volgina, M.K. Liu, M. Favata, M. Rupar, X. Wen, R. Sparks, T. Maduskuie, M. Covington, T. Burn, B. Ruggeri, A.P. Combs, W.Q. Yao, R. Huber, G. Hollis, P. Scherle, and P.C.C. Liu. 2016. Activity of the BET inhibitor INCB054329 in models of lymphoma. *Cancer Research* 76.
181. Suraweera, A., K.J. O'Byrne, and D.J. Richard. 2018. Combination therapy with histone deacetylase inhibitors (HDACi) for the treatment of cancer: Achieving the full therapeutic potential of HDACi. *Frontiers in Oncology* 8: 92.
182. Suzuki, T., M.N. Khan, H. Sawada, E. Imai, Y. Itoh, K. Yamatsuta, N. Tokuda, J. Takeuchi, T. Seko, H. Nakagawa, and N. Miyata. 2012. Design, synthesis, and biological activity of a novel series of human sirtuin-2-selective inhibitors. *Journal of Medicinal Chemistry* 55: 5760–5773.
183. Tahiliani, M., K.P. Koh, Y. Shen, W.A. Pastor, H. Bandukwala, Y. Brudno, S. Agarwal, L.M. Iyer, D.R. Liu, L. Aravind, and A. Rao. 2009. Conversion of 5-methylcytosine to 5-hydroxymethylcytosine in mammalian DNA by MLL partner TET1. *Science* 324: 930–935.
184. Tan, J., X. Yang, L. Zhuang, X. Jiang, W. Chen, P.L. Lee, R.K. Karuturi, P.B. Tan, E.T. Liu, and Q. Yu. 2007. Pharmacologic disruption of Polycomb-repressive complex 2-mediated gene repression selectively induces apoptosis in cancer cells. *Genes & Development* 21: 1050–1063.
185. Tanabe, K., J. Liu, D. Kato, H. Kurumizaka, K. Yamatsugu, M. Kanai, and S.A. Kawashima. 2018. LC-MS/MS-based quantitative study of the acyl group- and site-selectivity of human sirtuins to acylated nucleosomes. *Scientific Reports* 8: 2656.
186. Teske, K.A., and M.K. Hadden. 2017. Methyllysine binding domains: Structural insight and small molecule probe development. *European Journal of Medicinal Chemistry* 136: 14–35.
187. Theodoulou, N.H., P. Bamborough, A.J. Bannister, I. Becher, R.A. Bit, K.H. Che, C.W. Chung, A. Dittmann, G. Drewes, D.H. Drewry, L. Gordon, P. Grandi, M. Leveridge, M. Lindon, A.M. Michon, J. Molnar, S.C. Robson, N.C. Tomkinson, T. Kouzarides, R.K. Prinjha, and P.G. Humphreys. 2016. Discovery of I-BRD9, a selective cell active chemical probe for bromodomain containing protein 9 inhibition. *Journal of Medicinal Chemistry* 59: 1425–1439.
188. Thottassery, J.V., V. Sambandam, P.W. Allan, J.A. Maddry, Y.Y. Maxuitenko, K. Tiwari, M. Hollingshead, and W.B. Parker. 2014. Novel DNA methyltransferase-1 (DNMT1)

depleting anticancer nucleosides, 4′-thio-2′-deoxycytidine and 5-aza-4′-thio-2′-deoxycytidine. *Cancer Chemotherapy and Pharmacology* 74: 291–302.

189. Tian, X., S. Zhang, H.M. Liu, Y.B. Zhang, C.A. Blair, D. Mercola, P. Sassone-Corsi, and X. Zi. 2013. Histone lysine-specific methyltransferases and demethylases in carcinogenesis: New targets for cancer therapy and prevention. *Current Cancer Drug Targets* 13: 558–579.
190. Tokarz, P., K. Kaarniranta, and J. Blasiak. 2016. Inhibition of DNA methyltransferase or histone deacetylase protects retinal pigment epithelial cells from DNA damage induced by oxidative stress by the stimulation of antioxidant enzymes. *European Journal of Pharmacology* 776: 167–175.
191. Tsai, H.C., H. Li, L. Van Neste, Y. Cai, C. Robert, F.V. Rassool, J.J. Shin, K.M. Harbom, R. Beaty, E. Pappou, J. Harris, R.W. Yen, N. Ahuja, M.V. Brock, V. Stearns, D. Feller-Kopman, L.B. Yarmus, Y.C. Lin, A.L. Welm, J.P. Issa, I. Minn, W. Matsui, Y.Y. Jang, S.J. Sharkis, S.B. Baylin, and C.A. Zahnow. 2012. Transient low doses of DNA-demethylating agents exert durable antitumor effects on hematological and epithelial tumor cells. *Cancer Cell* 21: 430–446.
192. Unoki, M., T. Nishidate, and Y. Nakamura. 2004. ICBP90, an E2F-1 target, recruits HDAC1 and binds to methyl-CpG through its SRA domain. *Oncogene* 23: 7601–7610.
193. Van Haaften, G., G.L. Dalgliesh, H. Davies, L. Chen, G. Bignell, C. Greenman, S. Edkins, C. Hardy, S. O'Meara, J. Teague, A. Butler, J. Hinton, C. Latimer, J. Andrews, S. Barthorpe, D. Beare, G. Buck, P.J. Campbell, J. Cole, S. Forbes, M. Jia, D. Jones, C.Y. Kok, C. Leroy, M.L. Lin, D.J. Mcbride, M. Maddison, S. Maquire, K. Mclay, A. Menzies, T. Mironenko, L. Mulderrig, L. Mudie, E. Pleasance, R. Shepherd, R. Smith, L. Stebbings, P. Stephens, G. Tang, P.S. Tarpey, R. Turner, K. Turrell, J. Varian, S. West, S. Widaa, P. Wray, V.P. Collins, K. Ichimura, S. Law, J. Wong, S.T. Yuen, S.Y. Leung, G. Tonon, R.A. Depinho, Y.T. Tai, K.C. Anderson, R.J. Kahnoski, A. Massie, S.K. Khoo, B.T. Teh, M.R. Stratton, and P.A. Futreal. 2009. Somatic mutations of the histone H3K27 demethylase gene UTX in human cancer. *Nature Genetics* 41: 521–523.
194. Vaswani, R.G., V.S. Gehling, L.A. Dakin, A.S. Cook, C.G. Nasveschuk, M. Duplessis, P. Iyer, S. Balasubramanian, F. Zhao, A.C. Good, R. Campbell, C. Lee, N. Cantone, R.T. Cummings, E. Normant, S.F. Bellon, B.K. Albrecht, J.C. Harmange, P. Trojer, J.E. Audia, Y. Zhang, N. Justin, S. Chen, J.R. Wilson, and S.J. Gamblin. 2016. Identification of (R)-N-((4-Methoxy-6-methyl-2-oxo-1,2-dihydropyridin-3-yl)methyl)-2- methyl-1-(1-(1 -(2,2,2-trifluoroethyl)piperidin-4-yl)ethyl)-1H-indole-3-carboxamide (CPI- 1205), a potent and selective inhibitor of histone methyltransferase EZH2, suitable for phase I clinical trials for B-cell lymphomas. *Journal of Medicinal Chemistry* 59: 9928–9941.
195. Venugopal, B., R. Baird, R.S. Kristeleit, R. Plummer, R. Cowan, A. Stewart, N. Fourneau, P. Hellemans, Y. Elsayed, S. Mcclue, J.W. Smit, A. Forslund, C. Phelps, J. Camm, T.R. Evans, J.S. DE Bono, and U. Banerji. 2013. A phase I study of quisinostat (JNJ-26481585), an oral hydroxamate histone deacetylase inhibitor with evidence of target modulation and antitumor activity, in patients with advanced solid tumors. *Clinical Cancer Research* 19: 4262–4272.
196. Vigushin, D.M., S. Ali, P.E. Pace, N. Mirsaidi, K. Ito, I. Adcock, and R.C. Coombes. 2001. Trichostatin A is a histone deacetylase inhibitor with potent antitumor activity against breast cancer in vivo. *Clinical Cancer Research* 7: 971–976.
197. Villar-Garea, A., M.F. Fraga, J. Espada, and M. Esteller. 2003. Procaine is a DNA-demethylating agent with growth-inhibitory effects in human cancer cells. *Cancer Research* 63: 4984–4989.
198. Von Tresckow, B., C. Sayehli, W.E. Aulitzky, M.E. Goebeler, M. Schwab, E. Braz, B. Krauss, R. Krauss, F. Hermann, R. Bartz, and A. Engert. 2018. Phase I study of domatinostat (4SC-202), a class I histone deacetylase inhibitor in patients with advanced hematological malignancies. *European Journal of Haematology* 102 (2): 163–173.

199. Wagner, E.K., N. Nath, R. Flemming, J.B. Feltenberger, and J.M. Denu. 2012. Identification and characterization of small molecule inhibitors of a plant homeodomain finger. *Biochemistry* 51: 8293–8306.
200. Wang, Y., J. Wysocka, J. Sayegh, Y.H. Lee, J.R. Perlin, L. Leonelli, L.S. Sonbuchner, C.H. Mcdonald, R.G. Cook, Y. Dou, R.G. Roeder, S. Clarke, M.R. Stallcup, C.D. Allis, and S.A. Coonrod. 2004. Human PAD4 regulates histone arginine methylation levels via demethylimination. *Science* 306: 279–283.
201. Wang, M., X. Liu, J. Guo, X. Weng, G. Jiang, Z. Wang, and L. He. 2015. Inhibition of LSD1 by Pargyline inhibited process of EMT and delayed progression of prostate cancer in vivo. *Biochemical and Biophysical Research Communications* 467: 310–315.
202. Wang, Y., J. He, M. Liao, M. Hu, W. Li, H. Ouyang, X. Wang, T. Ye, Y. Zhang, and L. Ouyang. 2019. An overview of Sirtuins as potential therapeutic target: Structure, function and modulators. *European Journal of Medicinal Chemistry* 161: 48–77.
203. Wapenaar, H., and F.J. Dekker. 2016. Histone acetyltransferases: Challenges in targeting bi-substrate enzymes. *Clinical Epigenetics* 8: 59.
204. Waters, N.J., S.R. Daigle, B.N. Rehlaender, A. Basavapathruni, C.T. Campbell, T.B. Jensen, B.F. Truitt, E.J. Olhava, R.M. Pollock, K.A. Stickland, and A. Dovletoglou. 2015. Exploring drug delivery for the DOT1L inhibitor pinometostat (EPZ-5676): Subcutaneous administration as an alternative to continuous IV infusion, in the pursuit of an epigenetic target. *Journal of Controlled Release* 220: 758–765.
205. Widschwendter, M., H. Fiegl, D. Egle, E. Mueller-Holzner, G. Spizzo, C. Marth, D.J. Weisenberger, M. Campan, J. Young, I. Jacobs, and P.W. Laird. 2007. Epigenetic stem cell signature in cancer. *Nature Genetics* 39: 157–158.
206. Willmann, D., S. Lim, S. Wetzel, E. Metzger, A. Jandausch, W. Wilk, M. Jung, I. Forne, A. Imhof, A. Janzer, J. Kirfel, H. Waldmann, R. Schule, and R. Buettner. 2012. Impairment of prostate cancer cell growth by a selective and reversible lysine-specific demethylase 1 inhibitor. *International Journal of Cancer* 131: 2704–2709.
207. Wilting, R.H., and J.H. Dannenberg. 2012. Epigenetic mechanisms in tumorigenesis, tumor cell heterogeneity and drug resistance. *Drug Resistance Updates* 15: 21–38.
208. Wolf, S.S. 2009. The protein arginine methyltransferase family: An update about function, new perspectives and the physiological role in humans. *Cellular and Molecular Life Sciences* 66: 2109–2121.
209. Wyhs, N., D. Walker, H. Giovinazzo, S. Yegnasubramanian, and W.G. Nelson. 2014. Time-resolved fluorescence resonance energy transfer assay for discovery of small-molecule inhibitors of methyl-CpG binding domain protein 2. *Journal of Biomolecular Screening* 19: 1060–1069.
210. Xu, Y., and C.R. Vakoc. 2017. Targeting cancer cells with BET bromodomain inhibitors. *Cold Spring Harbor Perspectives in Medicine* 7 (7).
211. Xu, T., S.S. Park, B.D. Giaimo, D. Hall, F. Ferrante, D.M. Ho, K. Hori, L. Anhezini, I. Ertl, M. Bartkuhn, H. Zhang, E. Milon, K. Ha, K.P. Conlon, R. Kuick, B. Govindarajoo, Y. Zhang, Y. Sun, Y. Dou, V. Basrur, K.S. Elenitoba-Johnson, A.I. Nesvizhskii, J. Ceron, C.Y. Lee, T. Borggrefe, R.A. Kovall, and J.F. Rual. 2017. RBPJ/CBF1 interacts with L3MBTL3/MBT1 to promote repression of Notch signaling via histone demethylase KDM1A/LSD1. *The EMBO Journal* 36: 3232–3249.
212. Yan, L., C. Yan, K. Qian, H. Su, S.A. Kofsky-Wofford, W.C. Lee, X. Zhao, M.C. Ho, I. Ivanov, and Y.G. Zheng. 2014. Diamidine compounds for selective inhibition of protein arginine methyltransferase 1. *Journal of Medicinal Chemistry* 57: 2611–2622.
213. Yee, A.J., and N.S. Raje. 2018. Panobinostat and multiple myeloma in 2018. *The Oncologist* 23: 516–517.
214. Yiannakopoulou, E.C. 2015. Targeting DNA methylation with green tea catechins. *Pharmacology* 95: 111–116.

215. Yoo, C.B., S. Jeong, G. Egger, G. Liang, P. Phiasivongsa, C. Tang, S. Redkar, and P.A. Jones. 2007. Delivery of 5-aza-2′-deoxycytidine to cells using oligodeoxynucleotides. *Cancer Research* 67: 6400–6408.
216. Yoon, J.H., L.E. Smith, Z. Feng, M. Tang, C.S. Lee, and G.P. Pfeifer. 2001. Methylated CpG dinucleotides are the preferential targets for G-to-T transversion mutations induced by benzo [a]pyrene diol epoxide in mammalian cells: Similarities with the p53 mutation spectrum in smoking-associated lung cancers. *Cancer Research* 61: 7110–7117.
217. Yoshida, M., M. Kijima, M. Akita, and T. Beppu. 1990. Potent and specific inhibition of mammalian histone deacetylase both in vivo and in vitro by trichostatin A. *The Journal of Biological Chemistry* 265: 17174–17179.
218. Yoshida, M., N. Kudo, S. Kosono, and A. Ito. 2017. Chemical and structural biology of protein lysine deacetylases. *Proceedings of the Japan Academy. Series B, Physical and Biological Sciences* 93: 297–321.
219. Younes, A., Y. Oki, R.G. Bociek, J. Kuruvilla, M. Fanale, S. Neelapu, A. Copeland, D. Buglio, A. Galal, J. Besterman, Z. Li, M. Drouin, T. Patterson, M.R. Ward, J.K. Paulus, Y. Ji, L.J. Medeiros, and R.E. Martell. 2011. Mocetinostat for relapsed classical Hodgkin's lymphoma: An open-label, single-arm, phase 2 trial. *The Lancet Oncology* 12: 1222–1228.
220. Zhang, J., S. Zhang, Y. Wang, H. Cheng, L. Hao, Y. Zhai, Z. Zhang, X. An, X. Ma, X. Zhang, Z. Li, and B. Tang. 2017. Effect of TET inhibitor on bovine parthenogenetic embryo development. *PLoS One* 12: e0189542.
221. Zhang, D., A.S. Leal, S. Carapellucci, K. Zydeck, M.B. Sporn, and K.T. Liby. 2018. Chemoprevention of preclinical breast and lung cancer with the bromodomain inhibitor I-BET 762. *Cancer Prevention Research (Philadelphia, Pa.)* 11: 143–156.
222. Zhao, Y., C.Y. Yang, and S. Wang. 2013. The making of I-BET762, a BET bromodomain inhibitor now in clinical development. *Journal of Medicinal Chemistry* 56: 7498–7500.
223. Zheng, Y.C., B. Yu, G.Z. Jiang, X.J. Feng, P.X. He, X.Y. Chu, W. Zhao, and H.M. Liu. 2016. Irreversible LSD1 inhibitors: Application of tranylcypromine and its derivatives in cancer treatment. *Current Topics in Medicinal Chemistry* 16: 2179–2188.
224. Zhu, Q.S., Y. Huang, L.J. Marton, P.M. Woster, N.E. Davidson, and R.A. Casero. 2012. Polyamine analogs modulate gene expression by inhibiting lysine-specific demethylase 1 (LSD1) and altering chromatin structure in human breast cancer cells. *Amino Acids* 42: 887–898.
225. Zurita-Lopez, C.I., T. Sandberg, R. Kelly, and S.G. Clarke. 2012. Human protein arginine methyltransferase 7 (PRMT7) is a type III enzyme forming omega-NG- monomethylated arginine residues. *The Journal of Biological Chemistry* 287: 7859–7870.

Multiscale Modelling of Cancer: Micro-, Meso- and Macro-scales of Growth and Spread

Mark A. J. Chaplain

Introduction

Cancer is a complex, dynamic disease with underlying processes occurring over the full range of biological scales from genetic, through proteomic, cellular, tissue, organ, to organism and sometimes even the whole population level. The first detectable (palpable) symptoms are almost always macroscopic, but mechanisms are also present *a priori* at the cellular level and these in turn originate from changes/mutations in the individual's DNA. Perhaps one of the most difficult questions of modern medicine is how to intervene and manipulate the complex system of the patient's body to affect changes in dynamics which can bring it back from a state of disease to either full remission or stabilisation. Given the complexity of the system a chance to answer that question should be sought by complementing the traditional clinical methods with mathematical and computational modelling and simulations. However, while developing predictive models one of the most important key aspect of the disease to be considered, if not the key aspect, is its multi-scale character.

In one of the most influential cancer papers of the last two decades, Hanahan and Weinberg [26] defined what they termed to be the six hallmarks of cancer: (i) sustaining proliferative signalling; (ii) evading growth suppressors; (iii) activating invasion and metastasis; (iv) enabling replicative immortality; (v) inducing angiogenesis; (vi) resisting cell death. More recently the authors [27] updated this list to also include two other emerging hallmarks: (i) deregulating cellular energetics; (ii) avoiding immune destruction, and two enabling characteristics (i) genome instability and mutation; and (ii) tumour promoting inflammation. These hallmarks are linked with phenotypic traits that give cancer cells an evolutionary advantage over healthy cells. Furthermore, in [27] Hanahan and Weinberg described four main types

M. A. J. Chaplain (✉)
School of Mathematics and Statistics, University of St Andrews, St Andrews, Scotland
e-mail: majc@st-andrews.ac.uk

M. Bizzarri (ed.), *Approaching Complex Diseases*, Human Perspectives in Health Sciences and Technology 2, https://doi.org/10.1007/978-3-030-32857-3_7

of intracellular circuit (signal transduction pathway) regulating the operation of cells: (i) proliferation circuits; (ii) viability circuits; (iii) motility circuits; and (iv) cytostasis and differentiation circuits. The failure or dysregulation of these four circuits jointly make up the characteristic phenotype of cancer cells, corresponding directly with four of the hallmarks given above. In contrast to healthy cells that carefully control the production of specific growth and proliferative signals, cancer cells have an abnormal progression through the cell cycle and divide rapidly. Equally they have much higher viability compared to normal cells; resisting cell death, avoiding immune destruction, genome instability and mutation make cancer cells somewhat "immortal". The outcome is the formation of macroscopic structures such as solid tumours that can be observed clinically. Despite enormous progress full understanding of these processes is difficult because we are dealing with a complex interplay between various subprocesses occurring with different dynamics at different spatial scales.

One of the most dangerous properties of malignant tumours is their ability to invade surrounding tissues and to metastasize. The invasion or infiltration of surrounding tissue by cancer cells can impair the tissue or organ function. However, a more dangerous aspect of invasion is the infiltration of blood and lymph vessels. When cancer cells penetrate the vessel lumen they may migrate with blood or lymph to distant sites in the body to form new tumours, i.e. metastases. It is worth mentioning that angiogenesis also contributes; through the formation of new blood vessels within the tumour it facilitates the migration of tumour cells. Metastasis of cancer makes patient's treatment very difficult. It prevents the effective resection of the primary tumour, as new outbreaks cause recurrence of the disease. There are many mechanisms that enable cancer cells invasion and metastasis, together making the motility circuit. One can mention here the frequently occurring over-expression of genes encoding extracellular matrix-degrading enzymes such as matrix metalloproteinases (MMPs). However, perhaps the most characteristic change is the loss of the functionality of the protein E-cadherin, which is the main molecule responsible for binding between epithelial cells.

While it is clear that there are many different, inter-connected temporal and spatial scales that are important during the development of any tumour, within these there are three clearly demarcated "fundamental scales" linked to each other which, when considered together, go to make up understanding the complex phenomenon that is cancer: the intra-cellular scale, the cellular scale and the tissue scale. At the level of intracellular processes we must include within the description complicated but essential phenomena such as signal transduction cascades, gene regulatory networks or cell cycle regulation. Doing so aids our understanding of the differences at the intracellular level between normal and transformed cells and therefore improves the efficiency of anti-cancer cell-cycle-dependent drugs. Another challenge while modelling intracellular processes is to understand how the three-dimensional structure of DNA and chromatin affects gene expression within signalling pathways crucial for the disease development. Although it is known that cancer is most often caused by the accumulation of mutations in genes involved in cell cycle regulation and apoptosis, another important issue is how the disease progression is influenced by structural or epigenetic changes within the cell nucleus.

At the level of cellular colonies and tissue there are two main approaches towards modelling complex biological processes like cancer: continuum and discrete. Continuum methods, that are derived from principles of continuum mechanics, have proved to be very useful in modelling phenomena at the tissue scale such as general tumour growth. However, one of the main drawbacks of continuum modelling is the difficulty in representing individual cell properties. Including these and intracellular processes in multi-scale phenomena such as cancer is becoming more and more important as experimental data across multiple scales becomes available. Alternative approaches rely on an individual-based description of a single cell. The main advantage of such methods is related to the relative simplicity of transmitting detailed biological processes into dynamics and development of cell populations and tissue. The main disadvantage is the computational cost which increases rapidly with the number of simulated cells. However the problem of high computational complexity can be addressed by selecting appropriate algorithms and by efficient implementation on high performance computing (HPC) systems.

Further milestones related to cancer modelling will be adapting the models for specific cancer types and specific patients. The latter means not only the acquisition of biochemical parameters but also the acquisition of medical image data for individual patients. This will be a definite step towards personalised medicine, which has a chance to completely reform our approach to the patient and his treatment. Already today imaging studies are of great importance in diagnosis and planning surgical procedures. However, especially for treatment of non-resectable tumours, such imaging studies could also be important in selecting the appropriate treatment or monitoring the disease dynamics.

In this chapter we provide a brief overview of current cancer modelling ("multi-scale mathematical oncology") at the three different scales previously mentioned – intra-cellular, cellular and tissue – drawing on recent work by Sturrock et al. [49, 50], Szymańska et al. [51] and Domschke et al. [15]. In Sect. 2 we discuss the modelling of intracellular dynamics, specifically gene regulatory networks (GRNs). In particular we focus on the canonical transcription factor – Hes1. In Sect. 3 we focus on the cell scale, in particular investigating cell-cell/cell-matrix dynamics using an individual-based (or agent based) force-based model. In Sect. 4 we model cancer cell invasion at the tissue or macroscale using a system of nonlinear, nonlocal partial differential equations. This system explicitly accounts for cell-cell adhesion and also cell-matrix adhesion through a non-local term developed originally by Armstrong et al. [7] and then originally applied to cancer invasion modelling by Gerisch and Chaplain [19]. In the final Discussion Sect. 5, we provide directions for future research by combining modelling at all scales and highlight recent work on modelling cancer treatment (chemotherapy and radiotherapy) and the metastatic spread of cancer.

The Microscale: Gene Regulatory Networks and Transcription Factors

At the heart of cellular function and communication lies segments of DNA (genes) and their associated gene regulatory networks (GRNs). A GRN can be defined as a collection of genes in a cell which interact with each other indirectly through their RNA and protein products. GRNs are vital to intracellular signal transduction and indirectly control many important cellular functions such as cell division, apoptosis and adhesion. One specific class of GRN involves proteins called transcription factors, which alter the transcription rate of genes in response to intra- or extracellular cues. Transcription factors may reduce or increase the transcription rate of a given gene, respectively inhibiting or promoting its production. If the inhibition (or promotion) is directed towards the transcription factor's own gene, either directly or indirectly, there is negative (or positive) feedback. Negative feedback loops are an important component of many gene networks and are found within a wide range of biological processes e.g. inflammation, meiosis, apoptosis and the heat shock response [32]. Mechanically speaking, systems which contain negative feedback should (and in fact are known to) exhibit oscillations in the levels of the molecules involved. Furthermore, in many biological processes, it is the oscillatory expression which is of particular importance.

Mathematical modelling of GRNs began some 50 years ago with the papers of [21, 23], in which a negative feedback model for a simple, single mRNA-protein feedback system was proposed. However, while GRNs are known to exhibit periodic fluctuations in mRNA and protein concentrations (e.g. the results for the Hes1 system, cf. [29]), these early models, which were restricted to purely temporal ODEs, could not produce oscillatory behaviour. Subsequently, discrete delay ODE models were proposed, which although reproducing the oscillatory dynamics, neglect the spatial structure of the cell (nucleus/cytoplasm). The first spatial models (for theoretical intracellular systems) were proposed in the 1970s by Glass and co-workers [20, 47] and similarly in the 1980s by Mahaffy and co-workers [8, 36, 37]. The inclusion of spatial terms (rather than, for example, a delay in a system of ODEs) is capable of producing the experimentally observed oscillations [31, 33–35, 49, 50, 52]. Moreover, in addition to the computational results of the previous papers, the work of Chaplain et al. [9] has rigorously proved, for the Hes1 system, that molecular diffusion causes oscillations.

The Hes1 System

The Hes1 protein may be viewed as the simplest transcription factor i.e. the Hes1 protein downregulates its own hes1 mRNA production, making it the canonical negative feedback system. This system (as well as the more complex p53-Mdmd2 system) was considered as a spatial problem in [49, 50]. Figure 1 shows a schematic

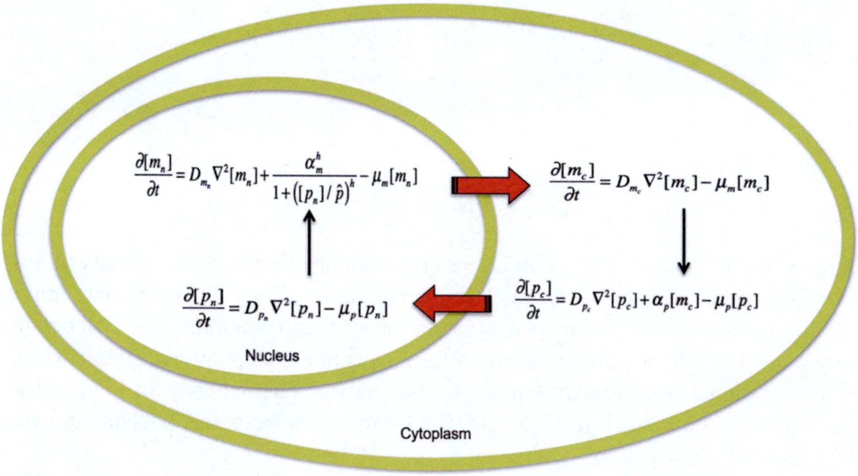

Fig. 1 Schematic diagram showing the Hes1 gene regulatory network. Hes1 protein is produced in the cytoplasm (translation), diffuses through the cytoplasm, across the nuclear membrane and into the nucleus where it down-regulates hes1 mRNA. The hes1 mRNA is produced in the nucleus (transcription), diffuses, crosses the nuclear membrane into the cytoplasm and is then translated into Hes1 protein. The equations show the reaction-diffusion events for each molecule and the red arrows denoted passage across the nuclear membrane

diagram of the Hes1 system, with Hes1 protein being produced in the cytoplasm (protein synthesis or translation), diffusing through the cytoplasm, across the nuclear membrane and into the nucleus where it down-regulates hes1 mRNA production (transcription). The hes1 mRNA itself can then diffuse to the nuclear membrane, move across the membrane and into the cytoplasm where it diffuses and is translated in to Hes1 protein in the ribosomes (translation).

These processes of molecular diffusion, protein production (translation) and mRNA production (transcription), along with the downregulation of mRNA can be modelled by the system of PDEs (where $m(\mathbf{x}, t)$, $p(\mathbf{x}, t)$ are the concentrations of hes1 mRNA and Hes1 protein respectively, with subscript n denoting the nucleus, and subscript c denoting the cytoplasm) as follows:

$$\frac{\partial [m_n]}{\partial t} = D_{m_n} \nabla^2 [m_n] + \underbrace{\frac{\alpha_m}{1 + ([p_n]/\widehat{p})^h}}_{transcription} - \mu_m [m_n], \tag{1}$$

$$\frac{\partial [m_c]}{\partial t} = D_{m_c} \nabla^2 [m_c] - \mu_m [m_c], \tag{2}$$

$$\frac{\partial [p_c]}{\partial t} = D_{p_c} \nabla^2 [p_c] + \underbrace{\alpha_p [m_c]}_{synthesis} - \mu_p [p_c], \tag{3}$$

$$\frac{\partial [p_n]}{\partial t} = D_{p_n} \nabla^2 [p_n] - \mu_p [p_n], \tag{4}$$

along with zero-flux boundary conditions at the cell membrane and continuity of flux boundary conditions across the nuclear membrane (cf. Fig. 1). Appropriate initial conditions for each molecular species closes the system mathematically. Full details are provided in [49, 50]. Figure 2 shows the results of a computational simulation of the above model in a domain similar to that shown schematically in Fig. 1. The oscillations in both hes1 mRNA and Hes1 protein in both the nucleus and the cytoplasm are clearly seen.

The computational results obtained in [49, 50] indicate that the molecular diffusion plays a major role in generating and controlling the oscillations. This numerical observation was complemented by a full analysis of a 1-dimensional caricature model of the system in [9]. A (nondimensionalised) one dimensional gene regulatory network model (i.e. a caricature of the Hes1 system) was considered on a 1-dimensional spatial domain shown in Fig. 3, with governing equations given by:

$$\frac{\partial m}{\partial t} = D \frac{\partial^2 m}{\partial x^2} + \alpha_m f(p) \delta^{\varepsilon}_{x_M}(x) - \mu m \qquad \text{in } (0,T) \times (0,1), \tag{5}$$

$$\frac{\partial p}{\partial t} = D \frac{\partial^2 p}{\partial x^2} + \alpha_p g(x) m - \mu p \qquad \text{in } (0,T) \times (0,1), \tag{6}$$

with boundary and initial conditions:

$$\frac{\partial m(t,0)}{\partial x} = \frac{\partial m(t,1)}{\partial x} = 0, \qquad \frac{\partial p(t,0)}{\partial x} = \frac{\partial p(t,1)}{\partial x} = 0 \qquad \text{in } (0,T),$$

$$m(0,x) = m_0(x), \qquad p(0,x) = p_0(x) \qquad \text{in } (0,1),$$

where D, α_m, α_p, μ_m and μ_p are positive constants (the diffusion coefficient, transcription rate, translation rate and decay rates of hes1 mRNA and Hes1 protein respectively). Full details can be found in the papers of Sturrock et al. [49, 50] and [9]. Here l denotes the position of the nuclear membrane and therefore the domain is partitioned into two distinct regions, $(0, l)$ the cell nucleus and $(l, 1)$ the cell cytoplasm, for some $l \in (0, 1)$. The point $x_M \in (0, l)$ is the position of the centre of the gene site and by $\delta^{\varepsilon}_{x_M}$ we denote the Dirac approximation of the δ-distribution located at x_M, with $\varepsilon > 0$ a small parameter and $\delta^{\varepsilon}_{x_M}$ has compact support (cf. Fig. 3).

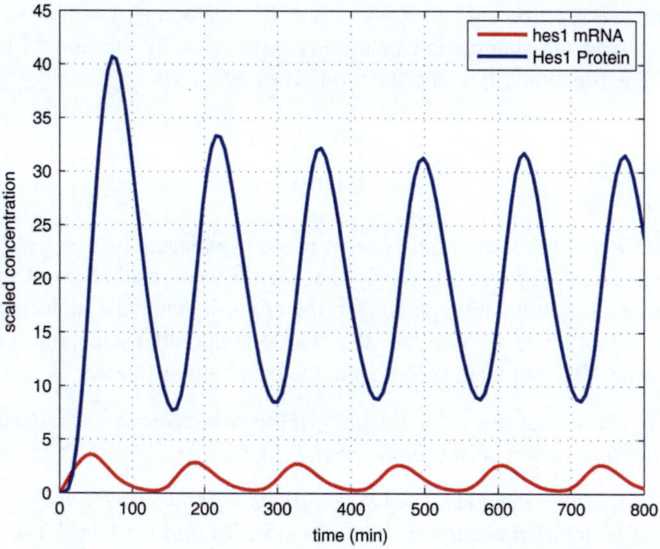

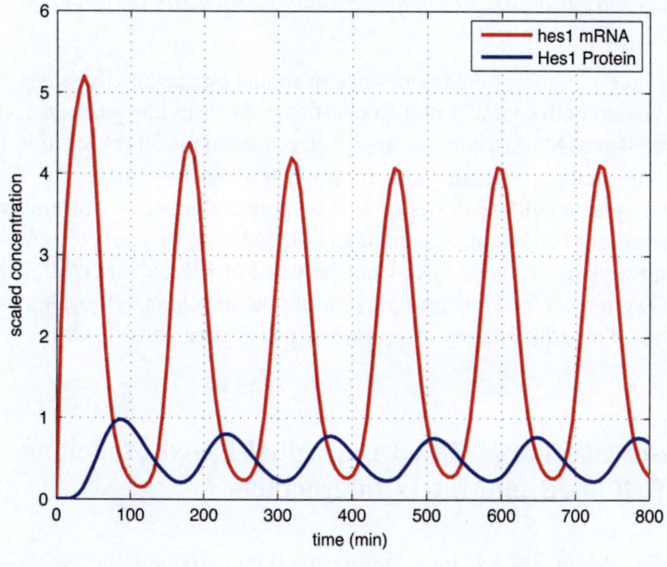

Fig. 2 Plots showing the oscillations in both molecular species (hes1 mRNA, Hes1 protein) in both the nucleus and the cytoplasm

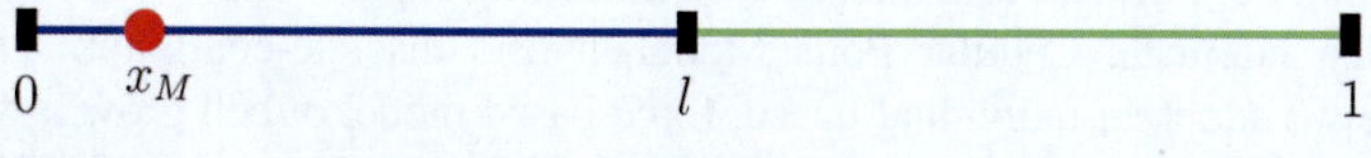

Fig. 3 1-dimensional spatial domain for the caricature Hes1 model. The blue region denotes the nucleus (0,l), while the green region denotes the cytoplasm (l,1). The location x_M (red circle) denotes the location of the gene site where the Hes1 protein binds and down-regulates hes1 mRNA

The nonlinear reaction term $f : \mathbb{R} \to \mathbb{R}$ is a Hill function $f(p) = 1/(1 + p^h)$, with $h \geq 2$, modelling the suppression of mRNA production by the protein (negative feedback). The function g is a step function given by

$$g(x) = \begin{cases} 0, \text{if } x < l, \\ 1, \text{if } x \geq l, \end{cases}$$

since the process of translation only occurs in the cytoplasm.

Chaplain et al. [9] proved rigorously that the diffusion coefficient of the molecules acts as a Hopf bifurcation parameter, therefore showing that molecular movement alone is sufficient to generate the (spatio-temporal) oscillations i.e. space influences time. The two main theorems in the paper are as follows:

Theorem 1 *There exist two critical values of the parameter D, i.e.* D_1^c *and* D_2^c *for which a Hopf bifurcation occurs in the model* (5), (6).

Theorem 2 *At both critical values of the bifurcation parameter* D_1^c *and* D_2^c *a supercritical Hopf bifurcation occurs in the system* (5), (6) *and the families of periodic orbits bifurcating from the stationary solution at each Hopf bifurcation point are stable.*

Further investigation of the importance of spatial aspects in GRNs has examined the Hes1 system both spatially and stochastically [48]. In this paper, a continuous-time discrete-space Markov process approach is used to model the reaction-diffusion kinetics. Since cell populations are naturally heterogenous, a stochastic description with spatial aspects built in allows us to incorporate a variety of differences and to look for emergent behaviour. The approach of [48] can be applied to model other natural pathways or synthetic GRNs cf. the work of Macnamara et al. [33–35], in particular key molecules known to play an important role in cell-cycle control and apoptosis and the inflammatory response *viz.* p53-Mdm2 and NF-κB.

The Mesoscale: Force-Based Individual-Based Modelling of Cell-Cell and Cell-Matrix Interactions

While the model of the previous section highlighted (stochastic) spatio-temporal models of intracellular pathways, in this section, we will focus on a model of cancer growth at the individual cell level developed by Szymańska et al. [51]. There are now a number of different individual-based modelling approaches that one can adopt cf. cellular automata, Cellular Potts Model, hybrid discrete-continuum [1, 5, 6, 14]. Here we adopt an individual-based, force-based model of cell growth which is driven by forces acting upon the cell, and is based upon the model of [42]. More recently this approach has been extended and implemented on a massively parallel system (IBM Blue Gene/Q system) allowing hybrid high performance simulations to describe, for example, tumour growth in its early clinical stage. Details of the implementation can be found in [11–13]. Adopting this approach, each cell is modelled as an isotropic elastic object capable of migration and division and parameterise it by cell-kinetic, biophysical and cell-biological parameters that can be experimentally measured, from both *in vitro* and *in vivo* experiments [10, 24, 30, 39, 40, 43, 45, 46, 54]. We assume that an individual cell ci in isolation is spherical and characterise the cell shape by its radius R. The position of the cell in 3D space is described by the Cartesian coordinates $(x_{c_i}, y_{c_i}, z_{c_i})$ of its centre.

Regarding cell kinetics, we assume that the cell-cycle is divided into four phases, i.e. mitosis – M-phase, followed by G1-, S-, and G2-phases, after which mitosis occurs again. During a complete cell-cycle, the cell must accurately duplicate its DNA once during S-phase and distribute an identical set of chromosomes equally to two progeny cells during M-phase. M-phase consists of two major events: the division of the nucleus called mitosis and subsequent cytoplasmic division called cytokinesis. G1-phase is an interval between mitosis and the initiation of nuclear DNA replication. It provides additional time for a cell to grow and to replicate its cytoplasmic organelles. G2-phase is again an interval between the completion of nuclear DNA replication and mitosis. Over the course of both the G1- and G2-phases, the cell checks the internal and external environment to ensure that the conditions are suitable and complete preparation for entry into either S-phase or M-phase. When DNA is damaged cell cycle is arrested in G1 or G2 so that the cell can repair DNA damages prior to its duplication or before cell division.

Cell cycle events must occur in a precise order, which should be maintained, even when one of the steps takes longer than usual. For instance, this means that cell division cannot start before DNA replication is complete. Similarly, when DNA is damaged the cell cycle is arrested so that the cell can repair the damage. This is possible because the cell is equipped with molecular mechanisms that can stop the cycle at various checkpoints. Two main checkpoints are located within the G1- and G2-phases. The G1 checkpoint allows the cell to check whether its environment is conducive to divisions and whether its DNA is damaged. If environmental conditions make cell division impossible, instead of entering S-phase a cell can enter a resting state – G0-phase, where it remains until conditions improve and it continues the cell cycle. The G2 checkpoint ensures that the cell has no DNA damage, and DNA replication will be completed before the beginning of mitosis [2].

Interactions between cells are modelled by taking into account the repulsive and attractive forces between cells. Upon compression, i.e. with decreasing distance $d_{c_ic_j}$ between the centres of two adjacent cells, c_i and c_j, of radii, r_{c_i} and r_{c_j}, both their surface contact area and the number of adhesive contacts increase, resulting in an attractive interaction. We assume that adhesive forces are proportional to ρ_m, which is the density of the surface adhesion molecules in the contact zone (which we assume is given for particular cell type), k_B, which is the Boltzmann constant, T, which denotes temperature and $D_{c_ic_j}$, which measures the contact between cells c_i and c_j and is calculated as the volume of the common area of intersecting spheres representing those cells. Experiments suggest that cells only have a small compressibility – the Poisson numbers are close to 0.5, [3, 38]. In this instance, both the limited deformability and the limited compressibility give rise to a repulsive interaction. Repulsive forces are inversely proportional to the term E_{c_i,c_j}, which is calculated form Young moduli, E_{c_i} and E_{c_j}, and Poisson ratios, ν_{c_i} and ν_{c_j}. The precise formula is given by:

$$E_{c_i,c_j} = \frac{3}{4}\left(\frac{1-\nu_{c_i}^2}{E_{c_i}} + \frac{1-\nu_{c_j}^2}{E_{c_j}}\right). \tag{7}$$

We model the combination of the repulsive and attractive energy contributions by a modified Hertz-model [18, 44] which has the advantage that both the interaction energy and the force can be represented as an analytical expression [16]. Inertia terms are neglected due to the high friction of cells with their environment, and we also do not consider the existence of any memory term as in [18].

$$V_{c_ic_j} = \underbrace{(r_{c_i} + r_{c_j} - d_{c_ic_j})^{\frac{5}{2}} \frac{1}{5E_{c_ic_j}} \sqrt{\frac{r_{c_i}r_{c_j}}{r_{c_i} + r_{c_j}}}}_{\text{repulsive interactions}} + \underbrace{\rho_m D_{c_ic_j} 25 k_B T}_{\text{adhesion}}. \tag{8}$$

Cells require access to oxygen from the circulatory system in order to grow and survive. It is well known that cancer cells grow preferentially around blood vessels. Those tumour cells that are located more than about 0.2 mm away from blood vessels were found to be quiescent, while others even farther away were found to be necrotic. This threshold of approximately 0.2 mm represents the distance that oxygen can effectively diffuse through living tissue [53]. Because of the low redox potential and high activation energy that occurs in living organisms, reactions involving molecular oxygen occur only in mitochondria. Therefore, we assume that the loss of oxygen in the tissue takes place only due to its consumption by the cells. The general equation governing the external oxygen concentration $Q(\mathbf{x}, t)$ in the cells' environment may be written:

$$\frac{\partial}{\partial t} Q(\mathbf{x}, t) = D_Q \nabla^2 Q(\mathbf{x}, t) - G(\mathbf{x}, t) + H(\mathbf{x}, t), \tag{9}$$

where D_Q, is the oxygen diffusion coefficient. The function $G(\mathbf{x}, t)$ models the oxygen uptake by cells and the function $H(\mathbf{x}, t)$ models the production of oxygen by vessels. Both of these functions are computed in each time step of the simulation from the current spatial organisation of cells and vessels through interpolation. The force associated with a given cell, c_i, is then given by the expression:

$$F_{c_i} = \underbrace{\nabla V_{c_i}}_{\text{inter-cellular interactions}} + \underbrace{\lambda \nabla Q(x, t)}_{\text{chemotaxis}} \tag{10}$$

where λ is a measure of a cell's chemotactic sensitivity to the oxygen concentration and V_{c_i} is given by

$$V_{c_i} = \sum_{c_j \in B_{\epsilon_{c_i}}} V_{c_i c_j} \tag{11}$$

with $B_{\in_{c_i}}(c_i)$ a sphere (i.e. a ball in $\mathbb{R}^3$) centred on $(x_{c_i}, y_{c_i}, z_{c_i})$, radius $\in_{c_i}$, denoting the maximum inter-cellular interaction region.

Summing all the forces between the cells and assuming a frictional force/drag force proportional to a cell's velocity and then applying Newton's Second Law of motion allows us to integrate a Langevin-type equation to give the spatial location of the cells over time. The direct use of equations of motion for the cells permits one to include more easily the limiting case of very small (or no) noise and is more intuitive. In this approach cells move under the influence of forces and a random contribution to the locomotion which results from the local exploration of space.

Solving the oxygen concentration (which is a global field) together with the individual-based particle system of up to 10^9 cells is a challenging task in the context of parallel processing. First of all, it requires the use of appropriate data structures to optimize the computations of interactions between lattice-free cells. In our approach, the main data structure that stores information about cells is an octal tree. We assume that the domain of simulation is a 3D cube. The cells are arranged in a tree based on the position of their centers. The tree is built recursively starting from the whole domain of simulation, which corresponds to the root of the tree. Subsequently, the cubes are divided recursively into 8 equal cubes with edges reduced by a factor of a half. This procedure is repeated until in the cube under consideration there is only one cell centre. Full details are provided in [51].

Model Application

Figures 4 and 5 show the results of applying the individual-based model to the scenario of a solid tumour (a tumour cord) growing around a central blood vessel

Fig. 4 Plot showing the growing tumour cord around a central blood vessel at times 300, 400, 700 and 1300 h. As the tumour cord grows, cells further away from the vessel become necrotic (black). At the final time of 1300 h, there is a total of around 10,000 cells

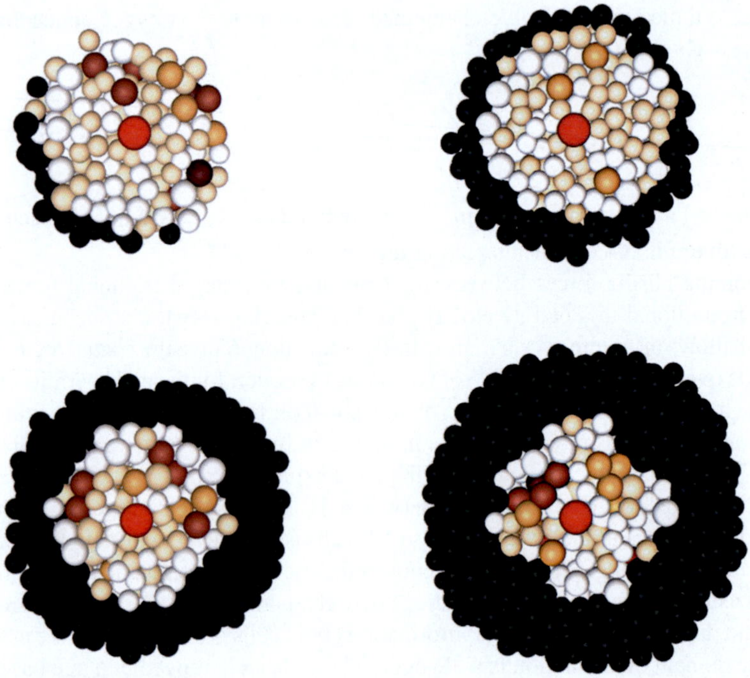

Fig. 5 Plot showing cross-sections of the growing tumour cord around a central blood vessel at times 300, 400, 700 and 1300 h. As the tumour cord grows, cells further away from the vessel become necrotic (black). At the final time of 1300 h, there is a total of around 10,000 cells

which supplies oxygen to the surrounding tissue and cancer cells. The figures show the development of the tumour cord over time and the formation of necrotic (dead) cells towards the outer boundary of the cord, since these are furthest from the blood vessel and the source of oxygen. More detailed simulations can be found in [51], while applications of the approach to avascular solid tumours can be found in Cytowski and Szymańska [11–13].

The Macroscale: Cancer Invasion and Metastasis

This section considers a macroscale model of cancer invasion based focusing on the role of cancer cell adhesion – both cell-cell and cell-matrix. The underlying basis for the model was developed by Armstrong et al. [7] who considered a model for cell sorting, and then developed by Gerisch and Chaplain in [19] as a model for cancer invasion. This approach was further developed more recently by Domschke et al [15] and it is this model that we present here. The variables in the model are cancer cells (density $c(t, x)$), extracellular matrix, ECM, (density $\mathrm{v}(t, x)$) and matrix degrading enzymes, MDE, (concentration $m(t, x)$). The model considers several populations of cancer cells $c_1(t, x)$, $c_2(t, x) \dots c_n(t, x)$ which are written $\mathbf{c} = (c_1, c_2, \dots c_n)^{\mathrm{T}}$.

The evolution of the cancer cell sub-population densities is driven by cell random motility, cell-cell and cell-matrix adhesion-mediated directed migration, proliferation, and mutations between the cancer cell sub-populations. This can be expressed as

$$\frac{\partial \mathbf{c}}{\partial t} = \nabla \cdot [\mathbf{D}\nabla \mathbf{c} - \operatorname{diag}(\mathbf{c})\mathcal{A}(t, x, \mathbf{u}(t, \cdot))] + P(t, \mathbf{u})\mathbf{c} + \mathbf{M}(\mathbf{u})\mathbf{c}. \tag{12}$$

Here, the diagonal matrix $\mathbf{D} = \operatorname{diag}(D_{1,1}, \dots, D_{1,n}) \in \mathbb{R}^{n,n}$ contains the random motility coefficients $D_{1,\mathrm{i}} > 0$ of the cancer cell sub-populations. In this work we assume that those are constants.

Adhesion-mediated directed cancer cell migration is represented using the non-local operator

$$\mathcal{A}(t, x, \mathbf{u}(t, \cdot)) := \begin{bmatrix} \mathcal{A}_1(t, x, \mathbf{u}(t, \cdot))^{\mathrm{T}} \\ \mathcal{A}_2(t, x, \mathbf{u}(t, \cdot))^{\mathrm{T}} \\ \vdots \\ \mathcal{A}_n(t, x, \mathbf{u}(t, \cdot))^{\mathrm{T}} \end{bmatrix} \in \mathbb{R}^{n,p}$$

which maps (t, x) together with the space-dependent function $\mathbf{u}(t, \cdot)$, that is $\mathbf{c}(t, \cdot)$ and $\mathrm{v}(t, \cdot)$, to an $n \times p$ matrix depending on (t, x). Row i in that matrix, i.e. $\mathcal{A}_i\,(t, x, \mathbf{u}(t, \cdot))^T$, represents the velocity of directed cancer cell migration of sub-population i which is induced by cell-cell and cell-matrix adhesion properties of cancer cells and ECM. Here cell-cell adhesion refers to adhesion between cells of sub-population i itself, *self-adhesion*, as well as between cells of sub-population i and sub-population $j \neq i$, *cross-adhesion*. The velocity for sub-population i is defined by the following vector-valued integral, cf. [7] or [19],

$$\mathcal{A}_i(t, x, \mathbf{u}(t, \cdot)) = \frac{1}{R} \int_{B(0,R)} \mathrm{n}(y) \cdot \Omega_i\left(\|y\|_2\right) \cdot g_i(t, \mathbf{u}(t, x + y))\, \mathrm{d}y. \tag{13a}$$

Here, $R > 0$ is the *sensing radius*, $B(0, R) \subset \mathbb{R}^p$ is the ball of radius R centred at zero, and for $x \in$ d the set $x + B(0, R)$ is the *sensing region* at x. Note that for points $x \in$ d, which are so close to the boundary of d such that $x + B(0, R) \cap d) \not\subset d$, the integral in Eq. (13a) is not yet well-defined; we resolve this issue when discussing the boundary conditions for our model in the end of this section. For $y \in B(0, R)$, the unit vector pointing from x to x + y, is denoted by $\mathrm{n}(y)$, i.e.

$$n(y) := \begin{cases} y/\|y\|_2 & \text{if } y \neq 0 \\ 0 \in \mathbb{R}^p & \text{otherwise} \end{cases}. \tag{13b}$$

Furthermore, $\Omega_i(r)$, with $r = \|y\|_2$, is the *radial dependency function* for sub-population i. It characterizes the relative importance of points at distance r from x for adhesion-mediated cell migration. This function is non-negative and normalised such that

$$1 = \int_{B(0,R)} \Omega_i(\|y\|_2)\,\mathrm{d}y. \tag{13c}$$

Finally, the function $g_i(t, \mathbf{u})$ is the i-th component of

$$\mathbf{g}(t, \mathbf{u}) \equiv \mathbf{g}(t, \mathbf{c}, v) = [\mathbf{S}_{\mathbf{cc}}(t)\mathbf{c} + \mathbf{S}_{\mathbf{cv}}(t)\mathbf{1}v] \cdot (1 - \rho(\mathbf{u}))^+. \tag{13d}$$

In the above, $\mathbf{1} \in \mathbb{R}^n$ is the all-one vector, $\mathbf{Scv}(t) \in \mathbb{R}^{n,n}$ is the diagonal matrix containing the non-negative cell-matrix adhesion coefficients of all cancer cell sub-populations with the ECM, and $\mathbf{Scc}(t) \in \mathbb{R}^{n,n}$ represents the matrix containing the non-negative cell-cell adhesion coefficients. Note that these matrices may have coefficients depending explicitly on time. We introduce the additional notation that $S_{c_i c_j} := (\mathbf{Scc})_{i,\, j}$ is the self-adhesion coefficient of sub-population i if $i = j$ and the cross-adhesion coefficient between sub-populations i and j if $j \neq i$. Furthermore $(\mathbf{Scv})i, i =: S_{c_{i,v}}$. In the usual notation, the positive part of an expression is denoted by $(\cdot)^+ := \max\{0, \cdot\}$ and the factor $(1 - \rho(\mathbf{u}))^+$ models an inhibition of migration due to volume filling effects, see e.g. [28].

Cancer cells mutate and thus change membership from one cancer cell sub-population to another one. This gives rise to the structured-population model with n cancer cell sub-populations as considered here. The matrix $\mathbf{M}(\mathbf{u}(t, x)) \in \mathbb{R}^{n,n}$, multiplied by $\mathbf{c}$, represents the effect of mutations in (12). As in the case of the proliferation term, the factor $\mathbf{c}$ makes explicit that cells of sub-population i may mutate only if they already exist. Since mutations of cells of sub-population i correspond to a loss of cells in that sub-population and mutations of cells into cells of sub-population i correspond to a gain of cells in that sub-population, the diagonal elements of $\mathbf{M}$ must be non-positive and the off-diagonal elements of $\mathbf{M}$ must be non-negative. Furthermore, in order to ensure conservation of cell mass, we require that the column sums of $\mathbf{M}$ equal zero, i.e.

$$\sum_{i=1}^{n} M_{ij} = 0, \qquad \text{for} \quad j = 1, 2, \ldots, n. \tag{14}$$

Different additional structural conditions may apply to the matrix $\mathbf{M}$. For instance, if we assume that the cancer cell sub-populations are ordered such that mutations occur only towards sub-populations with a larger index, then matrix $\mathbf{M}$ is a lower triangular matrix, or, if we even assume that mutations occur only towards the sub-population with the next larger index, then $\mathbf{M}$ is even a lower bidiagonal matrix.

The evolution of the ECM density is governed by MDE-mediated matrix degradation as well as ECM remodelling. This is expressed as

$$\frac{\partial v}{\partial t} = -\gamma m v + \psi(t, \mathbf{u}), \tag{15}$$

where $\psi(t, \mathbf{u})$ represents the ECM remodelling law, and γ is the rate constant of ECM degradation due to the presence of MDEs. We require that $v = 0$ implies that $\psi(t, \mathbf{u}) \geq 0$ as this will ensure the non-negativity of the ECM density.

Finally, the evolution of the MDE concentration is determined by molecular diffusion of the enzymes, by natural decay, and by the release of MDEs by the cancer cell sub-populations into the tumour microenvironment. Hence we obtain

$$\frac{\partial m}{\partial t} = \nabla \cdot [D_3 \nabla m] + \alpha^{\top} \mathbf{c} - \lambda m. \tag{16}$$

In the above equation, D_3 is the positive MDE diffusion constant, $\boldsymbol{\alpha} \in \mathbb{R}^n$ is the non-negative vector of MDE release rates of the cancer cell sub-populations, and λ is the non-negative decay constant.

Model Application

Using the general formulation (12), (13a), (13b), (13c), (13d), (14), (15), and (16), we apply this framework to model the scenario of two cancer cell populations (phenotypes), c_1 and c_2, one of which may mutate into the other. We envisage a scenario where, as time develops, some of the cancer cells of type c_1 mutate to a more aggressive population c_2, leading to an increase of tumour malignancy [15]. In a different investigation, focused on the uPA system, [4] considered two cancer cells sub-populations within the context of a local-haptotaxis tumour cell movement model. The model for the two cancer cell populations, including mutation from one to the other, secretion of MDEs and interaction with the ECM is as follows:

$$\begin{aligned}
\frac{\partial c_1}{\partial t} &= \nabla \cdot [D_{11} \nabla c_1 - \mathcal{A}_1(u(t,\cdot)) c_1] + \mu_{11} c_1 (1 - c_1 - c_2 - v) - \delta_{c_1} F(t, v), \\
\frac{\partial c_2}{\partial t} &= \nabla \cdot [D_{12} \nabla c_2 - \mathcal{A}_2(u(t,\cdot)) c_2] + \mu_{12} c_2 (1 - c_1 - c_2 - v) - \delta_{c_1} F(t, v), \\
\frac{\partial v}{\partial t} &= -\gamma m v + \mu_2 (1 - c_1 - c_2 - v)^{+}, \\
\frac{\partial m}{\partial t} &= \nabla \cdot [D_3 \nabla m] + \alpha_1 c_1 + \alpha_2 c_2 - \lambda m,
\end{aligned} \tag{17}$$

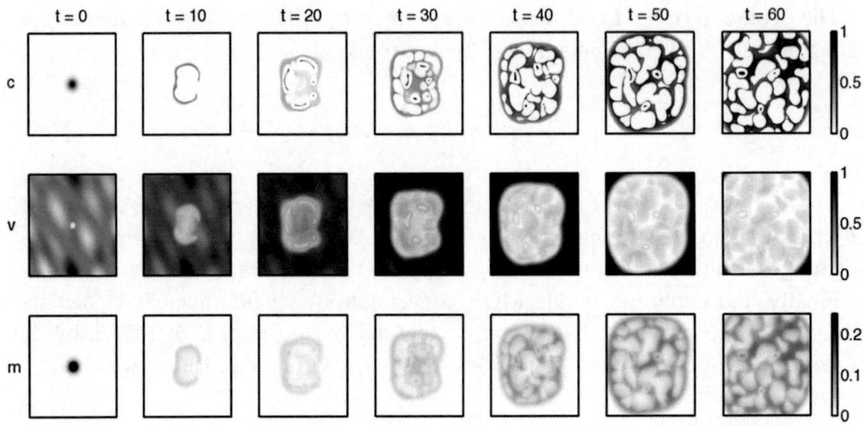

Fig. 6 Plots showing the cancer cell densities in the top row (black: $c1$, red: $c2$), ECM density v in the centre row, and the MDE concentration m in the bottom row at $t = 0$ (IC) and $t = 10, 20, \ldots,$ 60 obtained from a simulation of model (17) with ECM reproduction rate $\mu_2 = 0.05$

with the conversion or mutation function

$$F(t, v) = H(t - t_{12}) \cdot H(v - v_{min}).$$

Here, H denotes the Heaviside function, t_{12} the time when the conversion from population 1 to population 2 starts and v_{min} is the minimal ECM density that is needed for a conversion to take place [4]. We assume that matrix remodelling process takes place while the locally available volume is not entirely occupied, i.e., as long as $1 - c_1 - c_2 - \mathrm{v} > 0$. Full details can be found in Domschke et al. [15].

Figure 6 shows the result of a simulation with the following parameter values (cf. [15, 19]). The cell-cell and cell-matrix adhesion parameters of both cancer cell sub-populations are kept constant and, from (13d), these are defined as

$$\mathbf{S_{cc}} = \begin{pmatrix} 0.5 & 0 \\ 0 & 0.3 \end{pmatrix} \qquad \mathbf{S_{cv}} = \begin{pmatrix} 0.3 & 0 \\ 0 & 0.5 \end{pmatrix}.$$

The remaining parameter values are as follows:

$$\begin{array}{lll} \gamma = 10 & \mu_2 = 0.05 & D_3 = 10^{-3} \\ \lambda = 0.5 & R = 0.1 & \end{array}$$

along with

$$\begin{array}{llll} c_1: & D_{11} = 10^{-4} & \mu_{11} = 0.1 & \alpha_1 = 0.1 \\ c_2: & D_{12} = 10^{-4} & \mu_{12} = 0.25 & \alpha_2 = 0.1 \end{array}$$

and

$$\delta = 0.3 \qquad t_{12} = 10 \qquad v_{min} = 0.3.$$

From the plots in Fig. 6, we see that after time $t_{12} = 10$, the second population c_2 begins to emerges in the overall tumour cell density profile. Since the cell-cell cross-adhesion parameter $S_{c_1c_2}$ is zero, the two sub-populations do not mix. However, even though we have constant adhesive properties, the two cancer cell sub-populations form together with the (continuously remodelling) ECM a strongly heterogeneous pattern, reminiscent of invasion patterns observed clinically in lung and oesophageal cancers. A range of rich, heterogeneous spatio-temporal dynamics can be obtained by varying key parameters of the model such as the cell-cell and cell-matrix adhesive strengths in the matrices $\mathbf{S_{cc}}$ and $\mathbf{S_{cv}}$. Full details can be found in [15].

Discussion and Future Directions

In this chapter we have presented an overview of cancer modelling at various different important scales (intracellular, cellular and tissue) and focussing on key aspects (cf. hallmarks) of cancer – control of proliferation and differentiation, growth around blood vessels, local spread and invasion. While the modelling of GRNs (transcription factors) and cancer invasion was focussed on the relevant single scale (intracellular and tissue), the individual-based modelling in Sect. 3, with its inclusion of a basic cell-cycle in each cancer cell and an external oxygen field was genuinely multiscale in structure.

It is such multiscale modelling that holds out the best possibility for the development of optimal, individualised patient-based therapy in the future. Such a multiscale approach for modelling potential optimal treatment strategies (chemotherapy and radiotherapy) has already been explored by Powathil and co-workers [25, 41], while the very recent work of Franssen et al. [17] has, for the first time, developed a framework to model the metastatic spread of cancer from the primary tumour to secondary sites in the body. Since it is metastatic spread which is responsible for around 90% of deaths from cancer [22, 26], developing and clinically implementing predictive multiscale mathematical and computational models may well become an important part of cancer treatment in the years to come.

References

1. Alarcon, T., H. Byrne, and P. Maini. 2003. A cellular automaton model for tumour growth in inhomogeneous environment. *Journal of Theoretical Biology* 225: 257–274.
2. Alberts, B., D. Bray, K. Hopkin, A. Johnson, J. Lewis, M. Raff, K. Roberts, and P. Walter, eds. 2010. *Essential cell biology*. New York/London: Garland Publishing, Inc.
3. Alcaraz, J.L., M. Buscemi, X. Grabulosa, B. Trepat, R. Fabry, D. Farre, and D. Navajas. 2003. Microrheology of human lung epithelial cells measured by atomic force. *Biophysical Journal* 84: 2071–2079.
4. Andasari, V., A. Gerisch, G. Lolas, A. South, and M. Chaplain. 2011. Mathematical modeling of cancer cell invasion of tissue: Biological insight from mathematical analysis and computational simulation. *Journal of Mathematical Biology* 63 (1): 141–171.
5. Andasari, V., R. Roper, M.H. Swat, and M.A.J. Chaplain. 2012. Integrating intracellular dynamics using CompuCell3D and Bionetsolver: Applications to multiscale modelling of cancer cell growth and invasion. *PLoS ONE* 7 (3): e33726.
6. Anderson, A.R.A., and M.A.J. Chaplain. 1998. Continuous and discrete mathematical models of tumour-induced angiogenesis. *Bulletin of Mathematical Biology* 60: 857–899.
7. Armstrong, N.J., K.J. Painter, and J.A. Sherratt. 2006. A continuum approach to modelling cell–cell adhesion. *Journal of Theoretical Biology* 243 (1): 98–113.
8. Busenberg, S., and J.M. Mahaffy. 1985. Interaction of spatial diffusion and delays in models of genetic control by repression. *Journal of Mathematical Biology* 22: 313–333.
9. Chaplain, M.A.J., M. Ptashnyk, and M. Sturrock. 2015. Hopf bifurcation in a gene regulatory network model: Molecular movement causes oscillations. *Mathematical Models and Methods in Applied Sciences* 25 (6): 1179–1215.
10. Chu, Y.S., W.A. Thomas, O. Eder, E. Pincet, J.P. Thiery, and S. Dufour. 2004. Force measurements in e-cadherin–mediated cell doublets reveal rapid adhesion strengthened by actin cytoskeleton remodeling through rac and cdc42. *The Journal of Cell Biology* 167: 1183–1194.
11. Cytowski, M., and Z. Szymańska. 2014. Large scale parallel simulations of 3-d cell colony dynamics. *IEEE Computational Science and Engineering* 16 (5): 86–95.
12. ———. 2015. Enabling large scale individual-based modelling through high performance computing. *ITM Web of Conferences* 5: 00014.
13. ———. 2015. Large scale parallel simulations of 3-d cell colony dynamics. ii. coupling with continuous description of cellular environment. *IEEE Computational Science and Engineering* 17 (5): 44–48.
14. D'Antonio, G., P. Macklin, and L. Preziosi. 2013. An agent-based model for elasto-plastic mechanical interactions between cells, basement membrane and extracellular matrix. *Mathematical Biosciences and Engineering* 10: 75–101.
15. Domschke, P., D. Trucu, A. Gerisch, and M.A.J. Chaplain. 2014. Mathematical modelling of cancer invasion: Implications of cell adhesion variability for tumour infiltrative growth patterns. *Journal of Theoretical Biology* 361: 41–60.
16. Drasdo, D., and S. Höhme. 2005. A single-cell-based model of tumor growth in vitro: Monolayers and spheroids. *Physical Biology* 2: 133–147.
17. Franssen, L.C., T. Lorenzi, A.E.F. Burgess, and M.A.J. Chaplain. 2019. A mathematical framework for modelling the metastatic spread of cancer. *Bulletin of Mathematical Biology* 81: 1965–2010.
18. Galle, J., M. Loeffler, and D. Drasdo. 2005. Modelling the effect of deregulated proliferation and apoptosis on the growth dynamics of epithelial cell populations in vitro. *Biophysical Journal* 88: 62–75.
19. Gerisch, A., and M. Chaplain. 2008. Mathematical modelling of cancer cell invasion of tissue: Local and non-local models and the effect of adhesion. *Journal of Theoretical Biology* 250 (4): 684–704.

20. Glass, L., and S.A. Kauffman. 1970. Co-operative components, spatial localization and oscillatory cellular dynamics. *Journal of Theoretical Biology* 34: 219–237.
21. Goodwin, B.C. 1965. Oscillatory behaviour in enzymatic control processes. *Advances in Enzyme Regulation* 3: 425–428.
22. GP, G.P.G., and J. Massagué. 2006. Cancer metastasis: Building a framework. *Cell* 127 (4): 679–695.
23. Griffith, J.S. 1968. Mathematics of cellular control processes. i. negative feedback to one gene. *Journal of Theoretical Biology* 20: 202–208.
24. Gumbiner, B.M. 2005. Regulation of cadherin-mediated adhesion in morphogenesis. *Nature Reviews. Molecular Cell Biology* 6: 622–634.
25. Hamis, S., G.G. Powathil, and M.A.J. Chaplain. 2019. Blackboard to bedside: A mathematical modeling bottom-up approach toward personalized cancer treatments. *JCO Clinical Cancer Informatics* (3): 1–11. https://doi.org/10.1200/CCI.18.00068.
26. Hanahan, D., and R.A. Weinberg. 2000. The hallmarks of cancer. *Cell* 100: 57–70.
27. ———. 2011. Hallmarks of cancer: The next generation. *Cell* 144: 646–674.
28. Hillen, T., and K. Painter. 2001. Global existence for a parabolic chemotaxis model with prevention of overcrowding. *Advances in Applied Mathematics* 26 (4): 280–301.
29. Hirata, H., S. Yoshiura, T. Ohtsuka, Y. Bessho, T. Harada, K. Yoshikawa, and R. Kageyama. 2002. Oscillatory expression of the bHLH factor Hes1 regulated by a negative feedback loop. *Science* 298: 840–843.
30. Jagiella, N., B. Müller, M. Müller, I.E. Vignon-Clementel, and D. Drasdo. 2016. Inferring growth control mechanisms in growing multi-cellular spheroids of nsclc cells from spatial-temporal image data. *PLoS Computational Biology* 12 (2): e1004412.
31. Lachowicz, M., M. Parisot, and Z. Szymańska. 2016. Intracellular protein dynamics as a mathematical problem. *Discrete and Continuous Dynamical Systems. Series B* 21: 2551–2566.
32. Lahav, G., N. Rosenfeld, A. Sigal, N. Geva-Zatorsky, A.J. Levine, M.B. Elowitz, and U. Alon. 2004. Dynamics of the p53-Mdm2 feedback loop in individual cells. *Nature Genetics* 36: 147–150.
33. Macnamara, C.K., and M.A.J. Chaplain. 2016. Diffusion driven oscillations in gene regulatory networks. *Journal of Theoretical Biology* 407: 51–70.
34. ———. 2017. Spatio-temporal models of synthetic genetic oscillators. *Mathematical Biosciences and Engineering* 14: 249–262.
35. Macnamara, C.K., E.I. Mitchell, and M.A.J. Chaplain. 2019. Spatial-stochastic modelling of synthetic gene regulatory networks. *Journal of Theoretical Biology* 468: 27–44.
36. Mahaffy, J.M. 1988. Genetic control models with diffusion and delays. *Mathematical Biosciences* 90: 519–533.
37. Mahaffy, J.M., and C.V. Pao. 1984. Models of genetic control by repression with time delays and spatial effects. *Journal of Mathematical Biology* 20: 39–57.
38. Mahaffy, R.E., C.K. Shih, F.C. McKintosh, and J. Kaes. 2000. Scanning probe-based frequency-dependent microrheology of polymer gels and biological cells. *Physical Review Letters* 85: 880–883.
39. Miron-Mendoza, M., V. Koppaka, C. Zhou, and W.M. Petroll. 2013. Techniques for assessing 3-d cell-matrix mechanical interactions in vitro and in vivo. *Experimental Cell Research* 319: 2470–2480.
40. Näthke, I.S., L. Hinck, and W.J. Nelson. 1995. The cadherin/catenin complex: Connections to multiple cellular processes involved in cell adhesion, proliferation and morphogenesis. *Seminars in Developmental Biology* 6: 89–95.
41. Powathil, G.G., D.J. Adamson, and M.A.J. Chaplain. 2013. Towards predicting the response of a solid tumour to chemotherapy and radiotherapy treatments: Clinical insights from a computational model. *PLoS Computational Biology* 9 (7): e1003120. https://doi.org/10.1371/journal.pcbi.1003120.

42. Ramis-Conde, I., D. Drasdo, A.R.A. Anderson, and M.A.J. Chaplain. 2008. Modelling the the influence of the E-cadherin-β-catenin pathway in cancer cell invasion: A multi-scale approach. *Biophysical Journal* 95: 155–165.
43. Ritchie, T., W. Zhou, E. McKinstry, M. Hosch, Y. Zhang, I.S. Näthke, and J.F. Engelhardt. 2001. Developmental expression of catenins and associated proteins during submucosal gland morphogenesis in the airway. *Experimental Lung Research* 27: 121–141.
44. Schaller, G., Meyer-Hermann, M.: Multicellular tumor spheroid in an off-lattice Voronoi-Delaunay cell model. Physical Review E 71, 051910–1–051910–16 (2005)
45. Schlüter, D.K., I. Ramis-Conde, and M.A.J. Chaplain. 2012. Computational modeling of single cell migration: The leading role of extracellular matrix fibers. *Biophysical Journal* 103: 1141–1151.
46. ———. 2015. Multi-scale modelling of the dynamics of cell colonies: Insights into cell-adhesion forces and cancer invasion from in silico simulations. *Journal of the Royal Society, Interface* 12: 20141080.
47. Shymko, R.M., and L. Glass. 1974. Spatial switching in chemical reactions with heterogeneous catalysis. *The Journal of Chemical Physics* 60: 835–841.
48. Sturrock, M., A. Hellander, A. Matzavinos, and M.A.J. Chaplain. 2013. Spatial stochastic modelling of the hes1 gene regulatory network: Intrinsic noise can explain heterogeneity in embryonic stem cell differentiation. *Journal of the Royal Society, Interface* 10: 20120988.
49. Sturrock, M., A.J. Terry, D.P. Xirodimas, A.M. Thompson, and M.A.J. Chaplain. 2011. Spatio-temporal modelling of the Hes1 and p53-Mdm2 intracellular signalling pathways. *Journal of Theoretical Biology* 273: 15–31.
50. ———. 2012. Influence of the nuclear membrane, active transport, and cell shape on the Hes1 and p53-Mdm2 pathways: Insights from spatio-temporal modelling. *Bulletin of Mathematical Biology* 74: 1531–1579.
51. Szymańska, Z., M. Cytowski, E.I. Mitchell, C.K. Macnamara, and M.A.J. Chaplain. 2018. Computational modelling of cancer development and growth: Modelling at multiple scales and multiscale modelling. *Bulletin of Mathematical Biology* 80: 1366–1403.
52. Szymańska, Z., M. Parisot, and M. Lachowicz. 2014. Mathematical modeling of the intracellular protein dynamics: The importance of active transport along microtubules. *Journal of Theoretical Biology* 363: 118–128.
53. Weinberg, R.A. 2007. *The biology of cancer*. New York: Garland Science.
54. Zaman, M.H., L.M. Trapani, A.L. Sieminski, D. MacKellar, H. Gong, R.D. Kamm, A. Wells, D.A. Lauffenburger, and P. Matsudaira. 2006. Migration of tumor cells in 3d matrices is governed by matrix stiffness along with cell-matrix adhesion and proteolysis. *PNAS* 103: 10889–10894.

Precision Oncology vs Phenotypic Approaches in the Management of Cancer: A Case for the Postmitotic State

Armando Aranda-Anzaldo and Myrna A. R. Dent

Precision Oncology and the Somatic Mutation Theory of Cancer

Cancer is a major concern for human societies. The statistics worldwide indicate that 1/4 to 1/3 of contemporary humans that live beyond their third decade shall be diagnosed with cancer at some time in their lives [13, 57, 99]. On the other hand, Precision Medicine aims to use multiple types of data to classify patients into groups that in principle will most likely respond to a given treatment. Within this framework, the so-called Precision Oncology (PO) approach is currently using data from next-generation genome sequencing of tumors to select therapy for oncologic patients independently of the type of cancer as traditionally defined on the basis of anatomy and histology. Therefore, in PO therapy is determined according to the profile of mutations in putative cancer driver genes found in the genome of the tumor sample sequenced. The chosen drugs for therapy are usually small molecules or antibodies designed to inhibit specific oncogenic mutations (as manifested in the target proteins) or to target key regulatory nodes in biochemical and cell-signaling pathways that supposedly drive tumorigenesis or underlie cancer vulnerabilities. Hence, the whole rationale of the PO effort rests on assuming the somatic mutation theory (SMT) of cancer causation as a proved fact.

In a nutshell, the SMT understands cancer as a genetic disease caused by induced or spontaneous mutations in cellular genes, leading to a transformed cellular phenotype. The standard narrative suggests that the resulting founder mutated cell somehow acquires the property of unlimited proliferation (immortalization) and from this point onwards, further mutations create variation within the resulting transformed

A. Aranda-Anzaldo (✉) · M. A. R. Dent
Laboratorio de Biología Molecular y Neurociencias, Facultad de Medicina, Universidad Autónoma del Estado de México, Toluca, Mexico
e-mail: aaa@uaemex.mx

M. Bizzarri (ed.), *Approaching Complex Diseases*, Human Perspectives in Health Sciences and Technology 2, https://doi.org/10.1007/978-3-030-32857-3_8

cell population so that cellular clones with varied phenotypes arise (clonal heterogeneity) and start competing among them for resources and space, evolving in a Darwinian fashion towards further maximization of their spreading and survival capacities (invasivity and metastasis) at the expense of their competitors, including the normal, non-transformed cells of the host [76, 180]. This theory became mainstream from about 1982 [208] and it has remained so despite that varied long-standing evidence from research on cancer is not consistent with the SMT, such as the fact that several well-established carcinogens are not mutagenic ([17, 40, 107, 162, 163] p.32). Yet the most serious counterargument for the SMT is Peto's paradox: the incidence of cancer does not correlate with the number of cells in the organism [143, 145]. This paradox points out that if the probability of carcinogenesis from mutations (either spontaneous or induced) is constant across cells then huge, long-lived animals (elephants, whales) should not exist since their large number of cells (equivalent to tens or hundreds of human body equivalents) imply that they should be overwhelmed by an almost universal cancer incidence. Despite that several theoretical mechanisms and experimental findings have been proposed for explaining away the paradox [1, 35] the paradox remains [34, 142].

Within the context of the SMT it is assumed that a driver gene mutation confers directly or indirectly a selective growth advantage to the cell affected while a passenger mutation has no direct or indirect effect on the selective growth advantage of the cell [203]. The big question then is how to discriminate objectively between both kinds of mutations since the degree of success of PO depends on the ability to verify the driving role of a targeted mutation/alteration in tumor development. However, there is no unified method or approach for identifying driver mutations in tumor genomic profiles and in any case the evidence shows that the presence of passenger mutations is overwhelming [166, 203]. It has been shown that the number of somatic mutations in tumors of self-renewing tissues is positively correlated with the age of the patient at diagnosis [189]. More recently it was also shown that even in healthy epithelial tissue from the skin and esophagus it is common the presence of mutations in cancer-associated genes that correlate with the formation of cell clones in situ [118, 119, 214], suggesting that that such mutations may promote cell proliferation but not necessarily carcinogenesis. For example, the gene NOTCH1, that when mutated is usually classified as a cancer-driving gene, is the most frequently mutated gene in healthy esophageal tissue and paradoxically at a higher frequency than in samples of esophageal cancer [119, 214]. Such a finding casts a serious doubt on the notion that common gene alterations in tumor samples have a cancer-promoting role and this is worrisome considering that such is the most common approach for identifying putative cancer-driver genes [158]. Moreover, the evidence also shows that the overall number of mutations, the number of mutations in cancer-associated genes and the size of the corresponding cellular clones are greater in samples from healthy older people when compared to those of healthy younger people [119, 214]. This finding seriously undermines the SMT because classical epidemiological studies on the correlation between age and cancer incidence suggested that some six rate-limiting steps occur in the road to cancer [16, 133] and such rate-limiting steps where interpreted, within the SMT, as mutations in tumor-driver genes. This necessarily implies that in older individuals, that have already accumulated somatic mutations in potentially cancer-driving genes (as shown in: [118, 119, 189, 214]), the induction of cancer by known carcinogens should be easier (either at a lower dose or at a shorter exposure) than in younger individuals. However, the experimental evidence in animal models of carcinogenesis and epidemiological studies in humans, clearly indicate that age itself do not materially affects the process of carcinogenesis [56, 143–145] suggesting that the rate-limiting steps in cancer development do not really correspond to genetic mutations.

Nevertheless, it has been suggested that some 3–8 somatic mutations are required to drive cancer [90, 188, 203] but in fact identifying such a set of driver mutations in any given tumor is a very difficult biological and computational problem. In cancer genomics the putative driver genes are divided in two groups: oncogenes and tumor suppressor genes (TSG), depending on whether the evidence indicates that they positively or negatively regulate the growth and proliferation of cells. Genomic sequencing may identify a large number of genetic variants but determining which of these are causally related to disease is quite a challenge. The observation that relatively few mutations occur in a significantly recurrent fashion across tumors [117] holds, despite the development of sophisticated statistical tools for evaluating the significance of mutations. Indeed, in most cancer datasets there is a large set of genes with infrequent mutations, where the putative drivers are statistically indistinguishable from passenger mutations [70, 90]. Moreover, because of the high variance in background mutation rates in different cancers, very large numbers of tumor samples would need to be sequenced for detecting putative drivers occurring at slightly above the high background mutation rate [86]. The problem becomes more acute when considering that usually there is no well-established mechanistic explanation of how the molecular changes may relate to the disease phenotype. Therefore, even though a given mutation in a given gene may have been confirmed as a cancer-driver mutation in an animal model, it is unwarranted the assumption that any other non-synonymous mutation in the same gene necessarily has functional consequences. Indeed, there is a gulf between finding a mutation in a cancer genome and proving that such a mutation is able to drive cancer development [219]. In its current version (GRCh38.COSMICv87) the Cancer Gene Census (CGC) from the Catalogue of Somatic Mutations in Cancer (COSMIC) includes 576 Tier 1 genes that in principle contain mutations causally implicated in cancer. It is remarkable that ~12% of such genes are ambiguously classified as oncogene/TSG indicating that the role of such genes in cancer depends on the biological context. Thus, it is unjustified to assume that the new variants that might be found in putative cancer driver genes may have an impact on cellular phenotype and as such on cancer development in absence of experimental characterization that proves or at least suggests the mechanism of action.

Precision Oncology an Unrealizable Medical Utopia

Given that the biomedical mainstream views cancer as a genetic disease, PO has largely focused on the determination of genetic biomarkers and clinical trials to test whether targeting these genetic alterations in cancer can either cure the disease or at least prolong survival. However, genomic data from patients with advanced cancer suggest that less than 10% have "actionable" mutations susceptible of drug-based intervention [29, 146]. The range of genomic mutations varies widely among different cancer types, roughly from 0.28 to 8.15 per genomic megabase [95] and, as previously mentioned, some of the most common putative driver mutations found in several types of cancer are also present in normal cells [117, 119, 214] or are found paradoxically at a higher frequency in normal tissue or benign lesions unlikely to become cancer [96, 119, 214]. On the other hand, the results of a randomized trial of PO that compared whether genetic profiling coupled to pathway directed therapy improved the outcome relative to the therapy chosen by expert opinion, found no progression-free survival difference between both strategies [105].

Genomic profiling provides information regarding genetic mutations, amplifications, deletions and epigenetic modifications in DNA such as cytosine methylation. However, there are limitations of using genomics as a single platform for biomarker identification given for example, that some common cancer types, such as prostate cancer, have very few or even no recurrent mutations detected [90, 203]. In addition, genomic profiling does not provide information regarding the activity of actual protein products supposedly mediating oncogenic or tumor suppressor gene functions. In other words, as previously mentioned, genetic variations in oncogenes and/or TSG do not necessarily predict activation of the corresponding biological pathway, and vice versa: cancer driver pathways can be active without the presence of mutations [38]. Moreover, intra-tumor heterogeneity (ITH); namely, differences in the molecular makeup of tumor cells within individual patients, is a major source of uncertainty for interpreting the results of tumor genome profiling. This uncertainty results from the fact that sequencing of genomic DNA from a bulk tumor sample may highlight the putative driver mutations in the dominant tumor clone and yet there may be other tumor cell populations that do not carry such mutations and so, that in principle may survive the therapeutic onslaught designed according to the gross results of tumor genomic profiling. Indeed, evidence for overwhelming ITH has been established in solid tumors by deep sequencing of several hundred microbiopsies from the same tumor and the results do not support the notion that Darwinian positive selection of the fittest cellular clones drives tumor evolution [108]. In vitro studies with a large set of established cancer cell lines also indicate that cancer cells are genetically heterogeneous and their genomes are continuously changing in a spontaneous and unpredictable fashion [23]. ITH manifests as differences in genetic, epigenetic and signaling networks of individual tumor cells but also is coupled with heterogeneity within the stromal compartment [126]. While ITH is influenced by the inherent tumor genetic makeup, epigenetic states (influenced by the location of tumor cells), as well as factors from the microenvironment are also equally important in determining the cellular states that drive the primary tumor or its metastatic lesions [168].

Despite the hype surrounding PO, the fact is that this approach has only improved the clinical outcomes of a picked if not anecdotic minority of patients [29, 146]. Indeed, the designed or highly defined small molecules or antibodies used in PO-based treatments, almost universally fail to eradicate the tumor that usually relapses within a year and recurs in a more malignant fashion [80, 88, 169]. Hence, sequential therapy with further inhibitors that specifically address emerging mutations within the original target [94] or drugs targeting newly established putative driver mutations [21] have been used in an attempt to overcome resistance. However, multiple resistance mechanisms (genetic and non-genetic) can act in concert to confer resistance to targeted therapy and so management of resistant disease remains dismal and short-lived, partly because multiple resistance mutations (in addition to other mechanisms) can occur simultaneously, suggesting that targeting a single pathway in a tumor is in most cases, not sufficient to achieve a sustained response.

Considering the arguments above together with the fact that no genetic alterations have been found to correlate with well-characterized predictive cancer biomarkers such as the expression of estrogen receptor or androgen receptor, in breast or prostate cancer respectively, some have suggested that in order to overcome these challenges a more comprehensive approach is necessary, requiring the simultaneous characterization of the genome, epigenome, transcriptome, proteome, and metabolome of tumors and their surrounding stroma; as these are all crucial parameters for defining the cellular phenotypes involved in cancer pathogenesis, as well as in characterizing responsiveness to therapy [169]. However, these parameters are dynamic and may change in response to external stimuli and as such are expected to show spatial heterogeneity. Therefore, the analysis of multiple biopsies and longitudinal follow-up of patients would need to be performed to predict the initial responses to therapy and to identify putative mechanisms of drug resistance. Thus, a serious limiting factor in performing multiple "omics" approaches and functional testing is tissue availability. Altogether these considerations highlight the potential high cost and the actual impracticality of the PO scheme

Phenotype-Driven Therapeutic Cancer Research as an Alternative to Precision Oncology

Drug development for Precision Medicine based on precision drug design for targeting specific cellular molecules is facing many challenges, including high failure rates in clinical trials that have been reflected in higher R & D costs. This situation is reinvigorating traditional phenotypic drug discovery [134]. Phenotypic screening for drug discovery identifies substances that alter the phenotype of an organism or a cell in a desired fashion and historically this has been the basis for discovering new drugs. Implicitly, this phenotypic screening assumes a "black box" approach so that only after discovery of the appropriate compound or drug there would be attempts for identifying the biological targets and action mechanisms involved. However, some of the most successful drugs in history such as acetylsalicylic acid (aspirin) have been medically used on a regular basis long before there was a hint of their actual mechanisms of action [197]. Moreover, given the pleiotropic effects of several well-established drugs, they can be repositioned or repurposed for treating other medical conditions as exemplified again by aspirin that currently is used for the treatment of myocardial infarction and for preventing its recurrence [18].

In humans ~ 90% of cancer is epithelial in origin and yet the skin, breast, prostate, lung and colon concentrate ~ 65% of the total incidence of human cancer [52]. Given the current statistics that one in every three humans shall be diagnosed with cancer worldwide, the question has been asked of why the other two do not get cancer [99]. The same can be asked about whether certain cell types cannot be the seat of cancer or whether there is a cellular phenotype that do not supports carcinogenesis. It is common knowledge that cancer may only arise from cells with proliferating potential ([163], p. 10). Thus adult organisms whose tissues are mostly constituted by postmitotic cells cannot develop cancer. Such is the case of *Drosophila melanogaster* in which treatment with known carcinogens shows that this fly may only develop cancer before the larva stage [71, 82]. The brain is rarely the seat of carcinogenesis but it is quite remarkable that neurons, the cells that according to current estimates constitute half of the brain cell population in humans [85] are not susceptible to carcinogenesis since so far there is not a single bona fide report of a brain tumor derived from cortical neurons [151]. The standard explanation for this is a circular argument: neoplastic transformation may only occur in cells endowed with a proliferating potential and given that neurons are terminally differentiated, postmitotic, cells they cannot be the origin of cancer. This pseudo-explanation is just a statement about a matter of fact. However, since the postmitotic state is a naturally occurring cellular phenotype that is an absolute tumor suppressor it would be reasonable to find ways for inducing such a state in tumor cells for preventing or curtailing tumor progression at least in the most common types of cancer. From this perspective it is helpful to have an understanding of the cellular basis of the postmitotic state.

Cell Differentiation and the Postmitotic State

It is widely known the negative correlation between cell differentiation and proliferating potential. Postmitotic cells are considered to be terminally differentiated cells that have permanently lost the capacity to divide ([163], p. 128). In mammals some of the most important tissues such a brain, heart, skeletal muscle, include a large proportion of postmitotic cells (neurons, cardiomiocytes, myotubes). Such cells are usually unable to re- enter the cell cycle yet they are able to remain alive and functional for long periods of time (actually decades in the case of humans).

However, terminal differentiation correlates with but it is not necessarily the same as the postmitotic state. For example, in neurons there is evidence that terminal selector transcription factors are necessary for initiating and maintaining the expression of genes associated with terminal differentiation and the postembryonic removal of such terminal selectors causes the loss of differentiated functional properties in neurons [64, 87]. Nevertheless, the postmitotic condition persists in neurons despite the loss of functional markers of terminal differentiation, suggesting that it do not depends on genetic factors or functions associated with terminal differentiation. Moreover, the fact that differentiated cells can be reprogrammed to a pluripotent state by the artificial over-expression of defined transcription factors [183] indicates the reversible nature of the differentiated state but the evidence also shows that the more differentiated a cell is, the more difficult it is to reprogram [139]. In the case of neurons, it has been reported that induced pluripotent cells (iPSCs) can be obtained from postnatal day 7 cortical neurons from mice. However, the forced ectopic over-expression of the classical four transcription factors used for this purpose was not sufficient for neuron reprogramming and so it was also necessary to inhibit the expression of the p53 tumor suppressor gene and the over-expression of REST (a negative regulator of neuronal genes that it is usually expressed in non-neuron cells) in order to achieve measurable neuron reprogramming [220]. Since the reprogramming strategy requires the in vitro stationary culture of neurons for several days and P7 is the latest time point at which efficient ex vivo neuronal culture is possible, the authors of this report acknowledge that it would be necessary to perform the same experiments in cultures of adult neurons (so far not technically feasible) so as to confirm that terminally differentiated neurons can be reprogrammed and to discard the possibility that the resulting iPSCs came from contaminating non-neuronal cells.

In mammalian embryos when the neuronal precursors at the ventricular and subventricular zone of the neural tube divide, the resulting neurons that migrate into the cortical plate become postmitotic for the rest of their lifespan [149]. Neurogenesis may occur in specific but limited regions of the postnatal mammalian brain (such as the hippocampus) and in the human neurogenesis drops to undetectable levels in adults [177]. Moreover, the new neurons always arise from precursor cells but never by division from terminally differentiated neurons [27, 150, 151]. Currently, embryo cloning by nuclear transfer is the gold standard for testing the capacity of a differentiated cell nucleus for achieving karyokinesis (and so for evaluating the reversibility or not of the postmitotic state). The successful cloning of mice using nuclei from postmitotic olfactory sensory neurons has been achieved and in principle demonstrates that the mechanisms causing the irreversible mitotic arrest in neurons can be reverted in the environment of the egg [60]. However, attempts to clone mice with nuclei from adult cortical neurons were completely unsuccessful [204]. Indeed, additional studies have shown that the developmental potency of neuronal nuclei is dramatically reduced as the cells migrate from the ventricular zone (where neural precursors reside) to the pial zone (where the neurons are already postmitotic) of the developing cerebral cortex [213]. Therefore, the available evidence shows that postmitotic condition of cortical neurons is non-reversible.

Cellular Senescence and the Postmitotic State

Terminal differentiation associated with the postmitotic state in vivo has been thought to be similar to the replicative senescence characterized by a permanent cellular arrest observed in normal mammalian cells after serial passages in culture [75]. On the other hand, classical evidence suggested that the serial division of fibroblasts in culture constitutes a differentiation process leading to terminally differentiated cells [116]. However, it has been shown that postmitotic cells may enter a senescent state associated with cellular stress independently of their postmitotic condition [165]. Indeed, as it has been pointed out, the indiscriminate application of the term "cellular senescence" to a broad range of processes that result in stable cellular quiescence has led to misunderstand the postmitotic state as a hallmark of cellular senescence [112] but such an assumption is unwarranted given the fact that fully differentiated, postmitotic cells such as neurons and cardiomyocytes are already present in the tissues of healthy newborns and such cells may survive for decades in a completely functional state without manifesting the morphological and functional features [103] associated with cellular senescence. Therefore, it must be clarified that "cellular senescence" refers to at least three different phenomena involving stable cellular arrest: (1) classical replicative senescence (RS) as described originally by Hayflick and Moorehead [83] in serial cultures of normal cells that results in a permanent cellular arrest as a function of cellular divisions. In human cells RS has been linked to the gradual attrition of the chromosomal telomeres as a result of serial rounds of DNA synthesis [171]. (2) Stress induced premature senescence (SIPS or STASIS) that occurs independently of the number of cell divisions and it is induced by a number of factors that cause cellular stress such as hipoxia, hiperoxia, DNA damage or the over-expression of oncogenes [171]. (3) Spontaneous replicative senescence (SRS) that occurs independently of stress or cell divisions as shown hereunder.

Spontaneous Replicative Senescence (SRS) and the Postmitotic State

Studies with clonal populations of normal mammalian cells in culture show that cells lose their proliferative potential in a stochastic non-reversible fashion independently of their previous number of cell divisions, reaching a SRS which is a *de facto* postmitotic state [175, 176]. The presence of large differences in the proliferative potential of pairs of daughter cells also indicates that there is a stochastic component in the cellular proliferative potential [221]. Vigorously dividing cell cultures are nevertheless heterogeneous and always contain a percentage of growth- arrested cells; this percentage progressively increases until all cells in the population become postmitotic [176, 222]. Indeed, diploid cells in culture lose their proliferative potential in a stochastic, age-independent fashion that satisfies an exponential decay distribution for the overall proliferative potential of the culture [175, 223]. This phenomenon is consistent with a one-hit model (first order kinetics) suggesting that cells in culture reach the postmitotic state entirely by chance. Indeed, careful studies have shown that the replicative lifespan of fibroblasts in culture has no correlation with donor age since the proliferative potential of cell lines derived from the same individual at different ages show no correlation with age, suggesting that in vivo the loss of proliferative potential occurs as a mosaic where cells with low and high proliferative potential co-exist at any time [47].

The stochastic loss of proliferative potential is not the consequence of cellular stressors acting upon the cells in culture, thus it is different from SIPS which depends on the action of specific gene products and so it can be reverted by specific mutations [171, 184], also it is independent of chromosome telomere attrition as it occurs in cells from rodents that posses very lengthy telomeres that do not shorten with each cell division since telomerase is continuously active in their somatic cells [113, 140]. Moreover, this stochastic loss of cellular proliferative potential also occurs in vivo as instantiated by the case of normal rat liver hepatocytes which are usually quiescent but retain a remarkable proliferative potential so that in a young rat after a 2/3 partial hepatectomy (PH) ~95% of the remaining hepatocytes re-enter the cell cycle and achieve complete liver regeneration within 7 days [128]. However, in healthy but older animals the fraction of remaining hepatocytes able to re-enter the cell cycle after PH is severely reduced to ~ 30% and so liver regeneration takes much longer [167]. This loss of proliferating potential correlates with the increase of ploidy in hepatocytes [173, 174] that suggests the impossibility of performing karyokinesis and mitosis despite the ability to achieve DNA synthesis in the polyploid cells.

Polyploidy, Aneuploidy and the Postmitotic State

A typical feature of normal tissues enriched with postmitotic cells is the presence of polyploid and aneuploid cells [28, 49], phenomenon also observed in cultures of normal cells reaching the average limit of their proliferative potential [111, 121, 172]. In most tissues advanced polyploidy is considered an indicator of terminal differentiation [78, 173]. In the developing brain about a third of the neural mitotic progenitor cells display aneuploidy and there is a significant presence of postmitotic aneuploid neurons in the mature brain that functionally integrate into the normal brain [98, 155, 156, 209]. Moreover, there is evidence that the generation of aneuploid cells is intrinsic to normal brain and liver development [59, 155]. Interestingly polyploidy precedes the appearance of aneuploidy in both normal and transformed tissues [37, 58, 179]. The presence of polyploid neurons in the postnatal brain indicates that such cells able to replicate their genome have been unable to perform karyokinesis and as such mitosis thus resulting in cells with polyploid nuclei [48]. Thus, true polyploidy implies uncoupling between DNA synthesis and mitosis since there is an essential failure of karyokinesis. The selective loss of polyploid cells during liver regeneration confirms that they are not usually permissive for renewed cell-cycling [174]. Evidence has been provided for limited cardiomyocyte renewal in postnatal mammals, the new cardiomyocytes arise from pre-exiting cardiomyocytes and not from precursor cells [26]. Yet only diploid cardiomyocytes are capable of completing mitosis resulting in new cells while polyploid cardiomyocytes may synthesize DNA but cannot complete cell division thus not contributing to heart renewal or repair [170]. Experimental correction of liver failure by transplantation of hepatocytes into livers of suitable mice show that polyploid donor hepatocytes may proliferate so as to repopulate the damaged liver but in the process most of them undergo ploidy reversal by a single step mechanism and the majority of the donor-derived cells become aneuploid, [58]. Interestingly in cultures from late-passage fibroblasts corresponding to the stage when such cells perform their last mitoses before becoming stably postmitotic, there is a chaotic partitioning of DNA between daughter cells indicating widespread aneuploidy in the resulting postmitotic cells [111].

The Postmitotic State and the Cell Cycle

Given the stable postmitotic condition of cortical neurons it is remarkable that they preserve the whole molecular machinery necessary for DNA synthesis. This machinery as well as their molecular regulators can be reactivated by cellular stress or experimental manipulation [63, 102, 138]. Yet in most cases reentry of postmitotic neurons into the cell cycle leads to cell death by apoptosis [45, 217]. Interestingly, CDK5, a peculiar cyclin-dependent kinase very active in postmitotic neurons, is a powerful suppressor of the cell cycle thus acting as a neuroprotector by limiting the possibility of cell-cycle related neuronal death [217, 218]. Also the anaphase-promoting complex/cyclosome (APC/C)-Cdh1 complex actively suppresses the aberrant cell cycle reentry and related death of neurons [199]. Thus in neurons molecular "guardians" actively suppress the reentry into the cell cycle so as to avoid unwanted neuronal death. Indeed, reentry of neurons into the cell cycle has been observed in a number of pathological conditions and so it is considered as a hallmark of neurodegeneration or brain injury [32, 50, 53, 205].

In other cell types it has been suggested that the stable non-proliferative state depends on the continued activity of specific gene products [135]. For example, the myotubes of the striated muscle are terminally differentiated and postmitotic and yet the experimental elimination of inhibitors of the cell cycle leads to reentry of such cells into the cell cycle. However, after replicating DNA the myotubes either die or remain permanently blocked in G2 but never successfully divide [136] putting in evidence their non-reversible postmitotic condition.

Although there are published claims for the existence of neuronal tumors [141] the fact is that the neuronal origin of such tumors is highly debatable. A survey of a large number of brain tumors in children found some evidence of neuronal differentiation in about a fifth of them with an spectrum going from ganglionic tumors difficult to distinguish from cortical dysplasia, to neuroectodermal tumors and in most cases the tumors were thought to be either congenital tumors or developmental malformations that do not qualify as cancers [22]. On the other hand, the so-called mutinodular and vacuolating neuronal tumors, of which just a handful have been described worldwide, are more likely malformative lesions than true neoplasms that originate from a progenitor neuro-glial cell types showing aberrant maturation [186]. Primary Brain Tumors (PBT) are a heterogeneous group of malignancies but gliomas are the most common with glioblastoma multiforme accounting for 50% of PBT [74]. Since the discovery of neural stem cells (NSC) in the postnatal brain the expert opinion favors the notion that PBT result from the oncogenic transformation of NSC residing in the subventricular zone, rather than from lineage committed progenitors or de-differentiated astrocytes [55, 74]. There is an isolated report suggesting the tumorous transformation of mature neurons in transgenic mice by ad hoc experimental manipulations. However, the resulting tumors are always gliomas with no remaining evidence of the neuronal phenotype [66]. This report tried to revive the claim that neuronal de-differentiation into progenitor cells is a possible road for tumorigenesis in the brain, but considering previous warnings on the incorrect labeling and identification of neurons in brain tissue [150] there is scope for doubting whether the identification of the artificially transformed neurons was truly unambiguous. This perception is reinforced by the retraction of another paper that claimed the in vitro transformation of retinal ganglion neurons resulting in a transformed neuronal cell line [101] as it was found that the cell line was really derived from induced retinal tumors that arose from neuroectodermal precursors in transgenic mice [100]. Indeed, a previous report had already shown that the tumor- inducing procedure caused photoreceptor degeneration and death instead of tumorous growth when used in postmitotic photoreceptor cells [3] thus conforming to the principle that reentry of postmitotic neurons into the cell cycle is lethal [50].

The Postmitotic State and the SMT

The SMT implies that sets of spontaneous or induced mutations in cellular genes (some of them activating but other inactivating mutations) are enough for causing the abnormal, non- regulated cell proliferation that is considered the fundamental hallmark of cancer [81]. There is hard evidence of high levels of insertional retrotransposition [19, 131] and copy number variations (CNV) in adult human brain neurons [125]. Indeed, genomic deletions are quite common in neurons and a significant fraction displays highly aberrant genomes with multiple alterations [125], indicating that somatic mosaicism is widespread among neurons and this also implies that neurons are prone to large-scale mutational events. However, neurons remain immune to carcinogenic transformation. Thus considering that there are billions of mammalian brains and given that the neuronal genome undergoes varied mutagenic events, it is quite remarkable that nobody has documented the spontaneous proliferation of adult mammalian neurons, since all biological states or processes controlled by genes and gene products can be inhibited, reverted, or bypassed by spontaneous or induced mutations in the genes involved. Therefore, the standing evidence indicates that the postmitotic state of neurons is not dependent on the action of specific genes. This situation suggests a non-genetic basis for the postmitotic state, thus it is worth considering the relationship between the structural organization in the cell nucleus and the postmitotic state.

The Nuclear Higher Order Structure (NHOS)

For a long time it has been known that during the cellular interphase, the nuclear DNA of metazoans is organized in negatively supercoiled loops anchored to a compartment or substructure commonly known as the nuclear matrix [24, 44, 154, 202]. Negative DNA supercoiling results spontaneously from the anchoring of DNA loops to the NM, as this is thermodynamically favorable [130]. The nuclear matrix (NM) is operationally defined as the residual nuclear substructure that results from extracting the nucleus with non-ionic detergents, high salt and DNase [62, 132, 192]. Originally considered as a sort of cage or stable framework [25, 84, 216], the current view is that the NM is a rather dynamic and diffuse compartment, which nevertheless is involved in the structural and functional organization of the cell nucleus [192]. A common set of ~ 300 proteins defines the core NM proteome [62] but there is also evidence of a large set of tissue- specific NM proteins [51, 181]. The interactions between nuclear DNA and the NM can be divided into stable-structural that resist extraction with high salt and transient perhaps functional that cannot resist such an extraction [61, 122, 153]. DNA interacts with the NM through segments of relatively short length ($\leq$ 200 bp) known as matrix attachment/associated regions or MARs [41, 224] and those segments in the subset corresponding to the structural interactions are known as true loop attachment regions or LARs [153]. However, there is no specific sequence or set of sequences that predetermines the MAR/LARs and the current evidence suggest that the recognition between DNA and NM proteins is mediated by the local topology of the DNA so that the NM proteins show non-overlapping affinity for single-stranded [79] supercoiled [97] or linear double-stranded DNA, the last property is very likely the distinctive feature of the LARs bound in situ and thus anchoring the genome to the NM [11, 69].

The set of structural DNA-NM interactions defines a nuclear higher-order structure (NHOS) that can be studied in substructures known as nucleoids obtained by extracting the nucleus for NM but without using DNase. These substructures consist of the whole set of naked but supercoiled DNA loops bound to the NM [44] and show that the structural interactions DNA-NM are of higher stability than those between DNA and chromatin proteins that are completely removed by the nucleoid-extraction procedure [202]. The naked supercoiled loops bound to the NM can be induced to unwind or rewind by mechanical force using the DNA intercalating agent

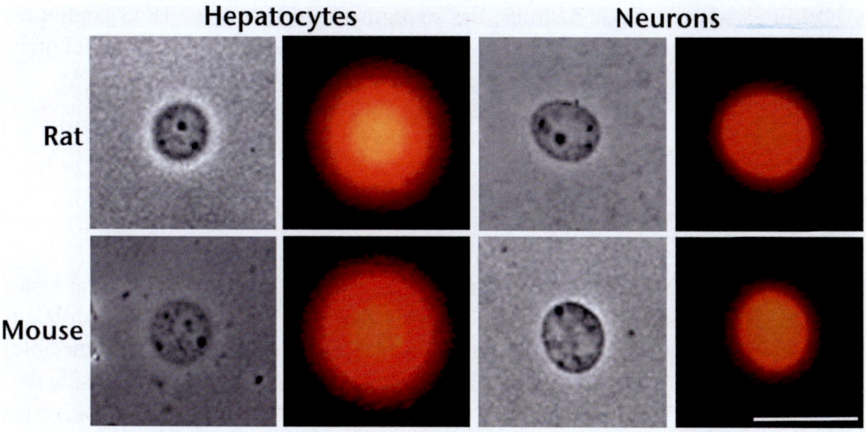

Fig. 1 The average size of the DNA loops bound to the nuclear matrix (NM) is cell- type specific. Representative phase contrast micrographs showing the morphology and typical size of the NM from nucleoids of primary hepatocytes and cortical neurons from two closely related mammals (rat and mouse). The fluorescence micrographs show the DNA halos surrounding the corresponding NM that result from the unwinding of the naked supercoiled DNA loops anchored to the NM induced by ethidium bromide that intercalates between the rungs of base pairs in DNA, thus acting as a molecular lever. White bar = 15 μm. The estimated average DNA loop size from base to tip is 14.1 ± 2.3 μm (41.5 kb) for rat hepatocytes and 13.0 ± 1.3 μm (38.4 kb) for mouse hepatocytes. For neurons the average DNA loops size is 8.24 ± 0.88 μm (24.23 kb) for rat and 8.15 ± 0.86 μm (23.97 kb) for mouse. These values correspond to half-length of the whole DNA loop. Photographs kindly provided by Evangelina Silva-Santiago, Ph.D

ethidium bromide (EB). This approach permits the evaluation of the integrity of DNA and the status of DNA supercoiling [9, 12, 14, 161, 225] as well as of the average size and distribution of the DNA loops [92, 154]. Moreover, it also permits the evaluation of the strength and stability of the DNA-NM interactions [123]. Remarkably, nucleoids of primary cells from different tissues show differential average DNA loop size (Fig. 1) and differential stability of the DNA-NM interactions. For example, rat nucleoids from newborn hepatocytes or from naive B- lymphocytes are completely disaggregated after the DNA halo expansion induced by EB [123, 190]. Yet as the hepatocytes mature there is a significant reduction in the average size of the DNA loops (which means a larger number of stable DNA attachments to the NM) coupled to an increase of the NM density due to the increasing abundance of some of its major components such as lamin A. This results in the overall strengthening of the NHOS in aged hepatocytes which becomes resistant to destabilization by EB [123] but this goes in hand with a significant increase in polyploid hepatocytes that correlates with the loss of the proliferating potential of such cells [73, 167].

Remarkably, the nucleoids from postmitotic cortical neurons display a shorter average DNA-loop size as well as high resistance to destabilization by mechanical force when compared to hepatocyte nucleoids, and this is observed even in neuronal nucleoids of very early postnatal stages [4]. However, despite the high stability of the

NHOS in early postmitotic neurons, the strengthening of such a NHOS continues during postnatal life coupled to both the increase in the proportion of DNA anchored to the NM and the abundance of NM constituents [5].

The NHOS and Structural Tensegrity

The fact that mechanical force directly applied to the nuclear DNA is able to disaggregate the NM [123] strongly suggests that DNA has a structural role in sustaining the integrity of the NM compartment. Cells are high-wired systems able to transduce mechanical information [8, 206]. In the nuclear envelope and inside the nucleus lamins can be assembled into a meshwork that is actually very sensitive to mechanical forces [20]. Cellular progression towards terminal differentiation correlates with a relative increase of NM constituents able to interact with DNA, such as the lamins A/C that are not expressed in immature cells but actively expressed in differentiated cells [43, 160]. Moreover, in differentiated cells A-type nuclear lamins increase their expression with age and this correlates with the strengthening of the NM [91, 123] besides establishing a larger binding surface for furthering DNA-NM interactions.

Tensegrity is an architectural-engineering principle for building lightweight but highly resistant and resilient structures. In tensegrity structures discontinuous compression elements balance the force generated by continuous tension elements reaching a structural equilibrium that is largely independent of gravity [68]. Such structures present isometric tension or pre-stress that leads to a configuration that minimizes the stored elastic energy of the whole structural system [68, 89]. The interphase nuclear organization in which the telomeric DNA of chromosomes is anchored to the NM component corresponding to the peripheral nuclear lamina [109] while the rest of the chromosomal DNA is multiply anchored to nodes of internal NM proteins, constitutes a structural system based on tensegrity [11, 15].

Thus, the DNA-NM interactions create a structural whole in which discontinuous compression elements (NM proteins) and continuous tension elements (chromosomal DNA) interact constituting a highly stable overall structure (Fig. 2) and the trend for increasing the number of DNA-NM interactions results in a highly stable cell nucleus. Indeed, the stiffness of the cell nucleus directly correlates with the stage of cellular differentiation so that embryonic stem cells have a highly deformable nucleus while knocking down lamin A/C expression in differentiated cells leads to nuclear deformability similar to that of adult stem cells [137]. This correlates with the biophysical evidence that the NHOS of cells with high proliferating potential is less stable than that from cells with reduced or null proliferating potential, thus requiring lesser energy input for achieving its complete destabilization [4, 5, 123].

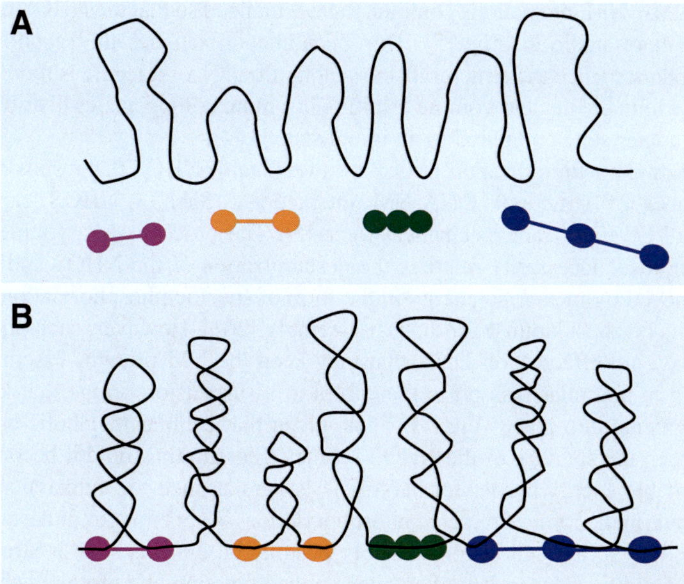

Fig. 2 Tensegrity in the cell nucleus. The principle of structural tensegrity implies that discontinuous compression elements and continuous tension elements interact constituting a highly stable overall structure. (**a**) in the cell nucleus nuclear matrix (NM) proteins (colored dots and beams) function as compression elements while the chromosomal DNA loops (black line) provide the continuous tension elements. (**b**) the fastening of the DNA loops to the NM proteins results in isometric pre-stress and so in DNA supercoiling. Therefore, chromosomal DNA interconnects hubs/nodes of NM proteins. Moreover, during the interphase the chromosomal telomeres are stably anchored to the peripheral inner lamina component of the NM [109]. This general pattern of DNA-NM interactions results in a higher-order structure that permeates the whole nucleus. (Illustration from Aranda-Anzaldo et al. [15])

A Structural Basis for the Postmitotic State

Spontaneous looping dissipates structural stress along chromosome DNA but attachment to the NM is necessary for stabilizing such loops against thermal agitation otherwise a long DNA may break into pieces [15]. This process obeys thermodynamic constraints and so the number of DNA-NM interactions should increase on average with time, provided there is any remaining structural stress to be dissipated [10]. The topological organization of the NHOS based on a selective use of a limited set of potential MAR/LARs, as it occurs in nuclei from newborn hepatocytes [123], is highly asymmetrical and the trend for physical systems is towards reducing asymmetries in time. A configuration in which most potential MAR/LARs become attached to the NM, resulting in shorter, more numerous and more stable DNA loops, is a symmetrical structural attractor [15]. Indeed, even in postmitotic cells the potential DNA-NM interactions continue to be actualized so that the NHOS becomes more and more stable in time [5]. This phenomenon satisfies the second law of thermodynamics since the structural stress along the DNA molecule is more evenly dispersed within the nuclear volume resulting in a more homogeneous distribution of energy in a later state compared to an initial state [104].

Thus mature postmitotic neurons display a very stable NHOS that results from the high number of structural DNA-NM interactions. Such a NHOS cannot be destabilized by significant mechanical force [5]. The process of karyokinesis that precedes mitosis necessarily requires the destabilization of the NHOS and in vivo this is achieved by metabolic energy in the form of specific phosphorylation of NM components (such as lamins) for their disassembly [201]. However, such phosphorylations have no effect upon DNA that may keep the NM proteins assembled by performing in a similar fashion to the cables in a suspension bridge that keep the deck's segments into place (Fig. 3). Thus, given that cellular metabolic energy is limited, when the stability of the NHOS reaches a certain threshold it becomes and insurmountable energy barrier for karyokinesis (and as such for mitosis), since the energy input (in the form of targeted phosphorylation) is not enough for destabilizing the NHOS and so the cell becomes stably postmitotic but this state is structurally defined and so it is independent of any gene-coded function, therefore not able to be reverted by somatic mutations. Moreover, this structural postmitotic condition is achieved even in cells from postnatal tissues still endowed with proliferating potential, as shown by the case of hepatocytes that spontaneously lose their proliferating potential as they reach the threshold of high NHOS stability in a stochastic fashion [123, 167] and this is consistent with a thermodynamically driven process.

NHOS and DNA Replication

Currently there is compelling evidence that the structural DNA loops attached to the NM correspond to the actual subdivision of the genome into units of replication (replicons) [30, 46, 106, 153, 159, 202] and that the NM is an organizing compartment for DNA replication [6, 207, 210]. Moreover, the experimental evidence indicates that in vivo DNA replication occurs by a reeling mechanism in which loop DNA moves or is moved to the replication factories organized upon the NM [159, 226]. The evidence also suggests that the spatial distribution of the replication complexes somehow depends on the NHOS [6, 30, 46, 210] so that in a replication focus only adjacent replicons belonging to the same chromosomal region may be simultaneously replicated [124]. As previously mentioned negative DNA supercoiling results spontaneously from the anchoring of DNA loops to the NM [130] and given that in metazoans there are no specific DNA sequences that determine the replication origins it is remarkable that the proteins of the origin recognition complex (ORC) bind to negatively supercoiled DNA [157] which is a strong feature of DNA subjected to torsional stress nearby to the LARs. Moreover,

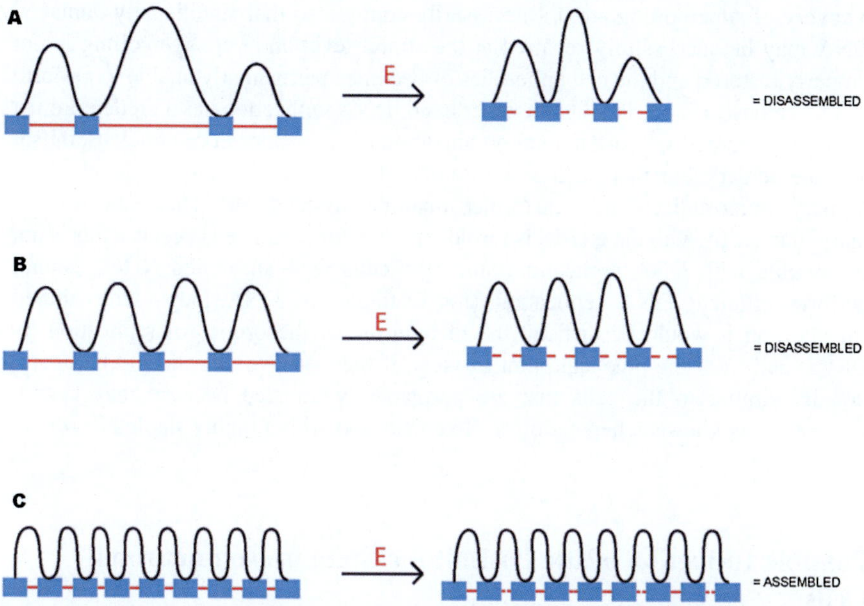

Fig. 3 The nuclear higher order structure (NHOS) and the control of karyokinesis and mitosis. Thermodynamic, entropic forces, favor the looping of chromosomal DNA (black lines) as this is a highly probable conformation for a fiber-like polymer. DNA loops become stabilized by anchoring to the proteinaceous constituents (blue rectangles) of the nuclear matrix (NM). Looping dissipates the intrinsic structural stress of DNA and so it preserves the integrity of the chromosomal DNA [11, 15]. The NM elements are also connected by energy-labile links (red lines). (**a**) in embryonic, non-differentiated cells with high proliferating potential, the DNA loops per unit length of chromosomal DNA are relatively few and quite heterogeneous in size. This correlates with the relatively low abundance of major NM components. During mitosis, metabolic energy (E) is used for rupturing the links (by phosphorylation of molecular targets) resulting in the efficient disassembly of the NHOS necessary for karyokinesis and so for mitosis. (**b**) cell differentiation correlates with quantitative changes in NM components and so the DNA loops per unit length of chromosomal DNA are more numerous and more homogeneous in size. Nevertheless, the available metabolic energy is still enough for destabilizing the NHOS as a pre-condition for achieving mitosis. (**c**) in terminally differentiated cells there is a further increase in the abundance of major NM constituents, this results in a higher number of DNA loops per unit length of chromosomal DNA and the loop size becomes quite homogeneous. In this case the metabolic energy available is not enough for destabilizing the NHOS, because the anchored DNA loops bridge the NM constituents despite the breakup of targeted molecular links. The highly stable NHOS becomes an insurmountable energy barrier for karyokinesis and mitosis. (Illustration from Aranda-Anzaldo and Dent [228])

the evidence available indicates that mammalian origins of replication are not determined by the nucleotide sequence but by 3D topological features of DNA [36, 127, 198]. Thus, a finely tuned degree of negative supercoiling is necessary for the recognition of the potential origins of replication and the initiation of DNA replication [185]. Indeed, the cutting or nicking of DNA induces the loss of supercoiling. Hence correct repair of damaged DNA requires the appropriate recovery of negative supercoiling. However, the repair of DNA strand integrity and the recovery of supercoiling are not necessarily coupled so that significantly damaged DNA may be successfully repaired at the strand level and yet supercoiling is not properly restored and the cell either dies or becomes permanently unable to perform DNA synthesis [9, 12, 14]. On the other hand, DNA replication has a predetermined but highly ordered spatio-temporal organization in the metazoan cell nucleus, thus it is quite remarkable that such spatio-temporal organization remains identical in primary, immortalized and transformed mammalian cells [54]. This indicates that major tampering with the NHOS is forbidden even for tumor cells because this is not compatible with DNA replication, since the cells with an altered NHOS cannot perform efficient DNA replication (for example, in DNA loops with altered supercoiling it would be difficult the recognition of the origins of replication by ORCs) and from the practical point of view, if they survive, they would be operationally similar to the cells that are permanently arrested because they cannot perform karyokinesis when endowed (like neurons) with a highly stable NHOS.

Possible Induction of the Postmitotic State in Non-neuronal Cells

For the last 40 years there has been scattered but consistent evidence that supports the possibility of inducing a postmitotic state in different types of cells. Dimethyl sulfoxide (DMSO) a polar aprotic solvent miscible with water and able to dissolve a large range of polar and nonpolar molecules has been widely used in cell biology as a cryoprotective agent [215]. In cells all molecules function in aqueous solution, the Hofmeister effect is the effect that solutes have on the structure and behavior of interfacial water. Co-solvents (in water) that improve the stability of the interactions between water molecules (known as kosmotropes) indirectly stabilize the intramolecular interactions in macromolecules such as proteins. On the other hand, co-solvents that disrupt the structure of water (chaotropes) also destabilize solute aggregates or the internal structure of macromolecules. When mixed with water at low concentrations ($\leq 2\%$) DMSO is a kosmotrope while at high concentrations ($> 15\%$) is a chaotrope [215]. State of the art techniques have shown that low dose DMSO induces significant structural changes in proteins and nucleic acids that may undoubtedly affect or modify their functions [193]. Indeed, proteins endowed with a high internal stability are poorly interactive with other molecules while structurally unstable proteins promiscuously interact with other molecules [194]. Therefore depending on its concentration DMSO may affect the structure and function of cellular macromolecules and as such the cellular state, as shown by the fact that DMSO reduces cell proliferation and cytokine production in cultures of normal peripheral blood lymphocytes [2].

A large number of classical studies reported that DMSO at low concentrations may induce the differentiation of cells from a wide variety of tumor derived cell lines [178]. Also DMSO improves the capacity of pluripotent stem cells for directed differentiation into multiple lineages [39]. However, although DMSO may induce some tumor cells along specific differentiation pathways [42, 67, 93] there are also several reports in which DMSO-treated tumor cells become "normalized" in the sense that they regain contact inhibition and reduced saturation density resulting in stable growth arrest, while becoming flattened and elongated even though the resulting morphology do not corresponds to the normal appearance of the original cell type [77, 120, 200]. Moreover, DMSO is commonly used in the biotechnology industry as a cytostatic since stable growth arrest enhances the production of recombinant proteins from mammalian cells [182]. The evidence suggests that the "differentiation" induced by DMSO is an all or nothing event so that cells stochastically shift to the "differentiated" or non-proliferating state. However, continued treatment with DMSO for several weeks results in a homogeneously growth-arrested population of cells and the effect remains after the removal of DMSO from the culture medium [120, 147, 191]. Interestingly, DMSO modifies the transition temperature of DNA and induces its unwinding thus it has a direct effect on the topology of DNA including supercoiling [227]. This property has been exploited for improving the performance of PCR DNA amplification by reducing the formation of unwanted secondary DNA structures that may inhibit the amplification reaction [65]. It has been shown that DMSO induces a thorough rearrangement of the NHOS in tumor cells and this correlates with their apparent "normalization" that includes a dramatic change in their cellular and nuclear morphology (Fig. 4) and stable growth arrest [120].

Despite the evidence that the effects of DMSO are non-specific and depend on the structural alteration of major biological macromolecules [7, 193, 215], further "inducers of differentiation" have been synthesized but starting from the rather misguided notion that there must be specific cellular targets or receptor sites for such compounds. DMSO and the synthetic "inducers of differentiation" are currently mislabeled as histone deacetylase inhibitors (HDACi) given their effect on such also mislabeled enzymes, since histone deacetylases are not specific for histones and have hundreds of other types of proteins as substrates [114, 115]. This HDACi mislabeling is more a marketing strategy for fitting these compounds within the fashionable field of chromatin epigenetics that deals with the characterization of transient biochemical modifications to chromatin proteins that correlate with modifications in gene expression [148]. However, this marketing of DMSO and other compounds as HDACi neglects the fact that their inhibitory effect upon their putative target enzymes is usually transient and yet the phenotypical effects on cells may become permanent [120, 147]. Methyl sulfone, a naturally occurring molecule that is present in the milk of grass-eating cows and that is closely related to DMSO, thus sharing its chemical and biological properties, is also able to induce "normalized" phenotypes that include the recovery of contact inhibition and permanent proliferative arrest in cells from metastatic tumors [33].

The synthetic suberoylanilide hydroxamic acid (SAHA), which is the most potent HDACi authorized for clinical use so far, induces either apoptosis or polyploidy (the last one coupled to stable cell proliferation arrest) in the treated cells and this occurs despite the fact that only a relative minority of genes are affected in their expression

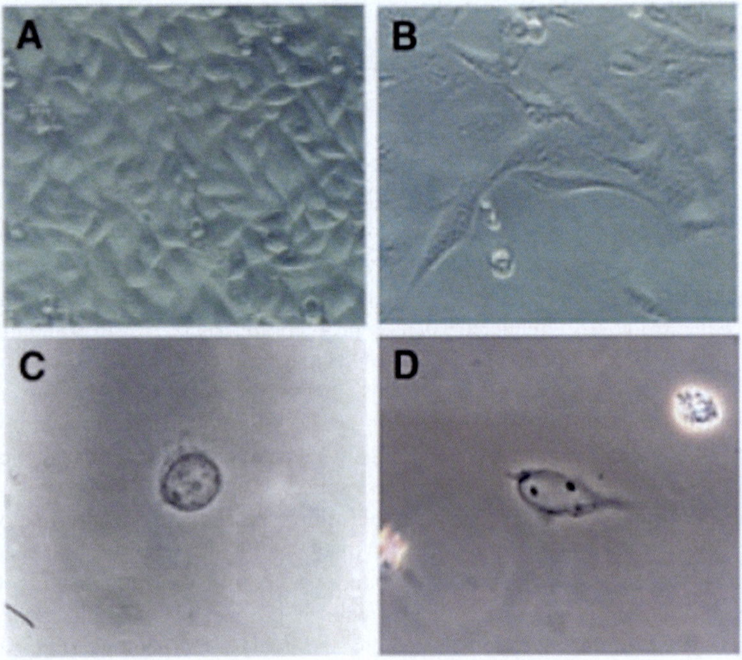

Fig. 4 The effect of dimethyl sulfoxide (DMSO) on the phenotype of tumor cells. (**a**) typical morphology of proliferating, non-contact inhibited HeLa cells. (**b**) after 6 weeks of treatment with 2% DMSO HeLa cells are growth arrested, contact-inhibited and display a flattened and elongated morphology. (**c**) phase contrast micrograph showing the typical round morphology of the nuclear matrix (NM) from a HeLa cell. (**d**) phase contrast micrograph showing the typical elongated morphology of the NM from a DMSO-treated, growth-arrested HeLa cell. The phenotypical changes persist after removal of DMSO from the medium in the chronically treated cultures. (Illustration from Martínez-Ramos et al. [120])

by SAHA [114, 211]. However, there are reports of tumor cells resistant to the pro-apoptotic effects of SAHA. For example, high intrinsic Bcl-2 expression in the prostate cancer cell line PC3 makes them resistant to SAHA-induced apoptosis [212] and yet the same cells are completely susceptible to the growth-suppressive effects of SAHA [31]. The traditional goal in cancer chemotherapy has been the killing of the tumor cells [80, 88] therefore resistance to induction of apoptosis may be considered a failure for the drug in question [129]. However, if the treatment of tumor cells leads to stable growth arrest the result in vivo would be tumor dormancy. A clinical trial of SAHA in patients with metastatic breast cancer showed that SAHA did not meet the standing Response Evaluation Criteria in Solid Tumors but nevertheless, 29% of patients treated experienced clinical benefit as shown by having stable disease and a median time to disease progression of 8.5 months without significant toxicity [110]. The serendipitous finding that SAHA may inhibit HDACs, led to a shift in focus about understanding its mechanisms of action and this also applied to DMSO [114]. Thus it was lost from sight the fact that classical inducers of "differentiation" such as DMSO or dimethylformamide [178] are polar aprotic solvents with the capacity to modify the structure of biological macromolecules [193] without necessarily having specific targets for their obvious effects (Fig. 4). This pleiotropic or multi-target activity makes difficult the appearance of true resistance to the overall effect of such compounds.

DMSO is widely used in industry and biological studies as solvent for water-insoluble compounds and it is commonly used as a vehicle for drug therapy. Also it has been used in several human therapeutic situations [164]. Toxicological studies in vivo and in vitro classify DMSO as an unusually nontoxic organic solvent and the side effects associated with exposure to DMSO are usually mild [72] and certainly minor when compared to the typically serious side effects of the specific inhibitors of mitosis used in the chemotherapy of cancer [187]. DMSO is not mutagenic and it is not listed as carcinogen despite the fact that DMSO and its metabolites, dimethyl sulfide and methyl sulfone, have been widely reported in a variety of foods [72].

For 40 years DMSO has been used as the mainstay therapy for interstitial cystitis/bladder pain syndrome in humans using frequent direct instillation to the bladder of 50–70% aqueous DMSO without evidence of toxic effects [152]. It has been shown that exposure of epithelial cells to 0.5% DMSO is more than enough for producing significant changes in DNA topology and protein structure [193]. 0.5% DMSO roughly equals 0.4g/kg in a 70 kg person and this dose is tenfold lower than the reported LD50 for monkeys either when injected in the blood stream or by mouth [72].

Conclusion

The fact that normal adult brain neurons present widespread mutations and chromosomal alterations without becoming cancerous indicates that mutations per se are not enough for transforming cells to a neoplastic state in a non-experimental setting. This also implies that the non-reversible postmitotic state of neurons is not genetically determined or dependent on genetic activities, otherwise mutations may be able to revert such a state. The evidence available suggests that the postmitotic state of neurons is consequence of the high structural stability of their NHOS that becomes an energy barrier for karyokinesis and mitosis. The evidence also suggests that it is possible to induce a postmitotic condition, by exposure to small molecules such as DMSO, in cells from epithelial tissues that are the most frequent targets of carcinogenesis. The interpretation of experimental results always depends on the assumed theoretical framework, therefore it is necessary to reconsider and to reevaluate the effects of DMSO and related compounds on cell differentiation and/or normalization in both the laboratory and clinical settings, given that so far the results obtained have been mostly interpreted from the perspective of the quite limited and inconsistent effects of such compounds on gene expression while neglecting the evidence that DMSO and related compounds may have an effect on the structural properties of proteins and nucleic acids and as such on the NHOS. A current alternative for cancer therapy is "induced tumor dormancy" which aims at prolonging the survival of patients together with preserving a good quality of life [195, 196]. DMSO and similar agents may fit into the aforementioned therapeutic approach, since by inducing a stable cellular growth arrest, either because the treatment induces a high stability of the NHOS or a global perturbation of the NHOS that disables the initiation of DNA replication, the resulting de facto postmitotic state would be an effective suppressor of cancer progression.

References

1. Abegglen, L.M., A.F. Caulin, A. Chan, K. Lee, R. Robinson, M.S. Campbell, W.K. Kiso, D.L. Schmitt, P.J. Waddell, S. Bhaskara, S.T. Jensen, C.C. Maley, and J.D. Schiffman. 2015. Potential mechanisms for cancer resistance in elephants and comparative cellular response to DNA damage in humans. *JAMA* 314: 1850–1860.
2. Abreu Costa, L., M.E. Fernandes Ottoni, M.G. dos Santos, A. Batista Meireles, V. Gomes de Almeida, W. de Fátima Pereira, B. Alves de Avelar Freitas, and G.E.A. Brito-Melo. 2017. Dimethyl sulfoxide (DMSO) decreases cell proliferation and TNF-alpha, IFN-gamma, and IL-2 cytokines production in cultures of peripheral blood lymphocytes. *Molecules* 22: pii E1789.
3. Al-Ubaidi, M.R., J.G. Hollyfield, P.A. Overbeek, and W. Baehr. 1992. Photoreceptor degeneration induced by the expression of simian virus 40 large tumor antigen in the retina of transgenic mice. *Proceedings of the National Academy of Sciences of the United States of America* 89: 1194–1198.
4. Alva-Medina, J., M.A.R. Dent, and A. Aranda-Anzaldo. 2010. Aged and post-mitotic cells share a very stable higher-order structure in the cell nucleus in vivo. *Biogerontology* 11: 703–716.
5. Alva-Medina, J., A. Maya-Mendoza, M.A.R. Dent, and A. Aranda-Anzaldo. 2011. Continued stabilization of the nuclear higher-order structure of post-mitotic neurons in vivo. *PLoS ONE* 6 (6): e21360.
6. Anachkova, B., V. Djeliova, and G. Russev. 2005. Nuclear matrix support of DNA replication. *Journal of Cellular Biochemistry* 96: 951–961.
7. Arakawa, T., Y. Kita, and S.N. Timasheff. 2007. Protein precipitation and denaturation by dimethyl sulfoxide. *Biophysical Chemistry* 131: 62–70.
8. Aranda-Anzaldo, A. 1989. On the role of chromatin higher-order structure and mechanical interactions in the regulation of gene expression. *Speculations in Science and Technology* 12: 163–176.
9. ———. 1992. Altered chromatin higher-order structure in cells infected by herpes simplex type 1. *Archives of Virology* 124: 245–253.
10. ———. 2009. A structural basis for cellular senescence. *Aging* 1: 598–607.
11. ———. 2016. The interphase mammalian chromosome as a structural system based on tensegrity. *Journal of Theoretical Biology* 393: 51–59.
12. Aranda-Anzaldo, A., and M.A.R. Dent. 1997. Loss of DNA supercoiling and organization in cells infected by herpes simplex virus type 1. *Research in Virology* 148: 397–408.
13. ———. 2003. Developmental noise, ageing and cancer. *Mechanisms of Ageing and Development* 124: 711–720.
14. Aranda-Anzaldo, A., F. Orozco-Velasco, E. García-Villa, and P. Gariglio. 1999. p53 is a rate-limiting factor in the repair of higher-order DNA structure. *Biochimica et Biophysica Acta* 1446: 181–192.

15. Aranda-Anzaldo, A., M.A.R. Dent, and A. Gómez-Martínez. 2014. The higher-order structure in the cell nucleus as the structural basis of the postmitotic state. *Progress in Biophysics and Molecular Biology* 114: 137–145.
16. Armitage, P., and R. Doll. 1954. The age distribution of cancer and a multi-stage theory of carcinogenesis. *British Journal of Cancer* 8: 1–12.
17. Autian, J., R. Singh, J.E. Turner, G.W.C. Hung, L.J. Nunez, and W.H. Lawrence. 1975. Carcinogenesis from polyurethans. *Cancer Research* 35: 1591–1596.
18. Baigent, C., L. Blackwell, R. Collins, J. Emberson, J. Godwin, R. Peto, J. Buring, C. Hennekens, P. Kearney, T. Meade, C. Patrono, M.C. Roncaglioni, and A. Zanchetti. 2009. Aspirin in the primary and secondary prevention of vascular disease: Collaborative meta-analysis of individual participant data from randomised trials. *Lancet* 373: 1849–1860.
19. Baillie, J.K., M.W. Barnett, K.R. Upton, D.J. Gerhardt, T.A. Richmond, F. De Sapio, P.M. Brennan, P. Rizzu, S. Smith, M. Fell, R.T. Talbot, S. Gustincich, T.C. Freeman, J.S. Mattick, D.A. Hume, P. Heutink, P. Carninci, J.A. Jeddeloh, and G.J. Faulkner. 2011. Somatic retrotransposition alters the genetic landscape of the human brain. *Nature* 479: 534–537.
20. Barboro, P., C. D'Arrigo, E. Repaci, E. Patrone, and C. Balbi. 2010. Organization of the lamin scaffold in the internal nuclear matrix of normal and transformed hepatocytes. *Experimental Cell Research* 316: 992–1001.
21. Bardelli, A., S. Corso, A. Bertotti, S. Hobor, E. Valtorta, G. Siravegna, A. Sartore-Bianchi, E. Scala, A. Cassingena, D. Zecchin, M. Apicella, G. Migliardi, F. Galimi, C. Lauricella, C. Zanon, T. Perera, S. Veronese, G. Corti, A. Amatu, M. Gambacorta, L.A. Diaz Jr., M. Sausen, V.E. Velculescu, P. Comoglio, L. Trusolino, F. Di Nicolantonio, S. Giordano, and S. Siena. 2013. Amplification of the MET receptor drives resistance to anti-EFGR therapies in colorectal cancer. *Cancer Discovery* 3: 658–673.
22. Becker, L.E. 1995. Central neuronal tumors in childhood: Relationship to dysplasia. *Journal of Neuro-Oncology* 24: 13–19.
23. Ben-David, U., B. Siranosian, G. Ha, H. Tang, Y. Oren, K. Hinohara, C.A. Strathdee, J. Dempster, N.J. Lyons, R. Burns, A. Nag, G. Kugener, B. Cimini, P. Tsvetkov, Y.E. Maruvka, R. O'Rourke, A. Garrity, A.A. Tubelli, P. Bandopadhayay, A. Tsherniak, F. Vazquez, B. Wong, C. Birger, M. Ghandi, A.R. Thorner, J.A. Bittker, M. Meyerson, G. Getz, R. Beroukhim, and T.R. Golub. 2018. Genetic and transcriptional evolution alters cancer cell line drug response. *Nature* 560: 325–330.
24. Benyajati, C., and A. Worcel. 1976. Isolation, characterization and structure of the folded interphase genome of Drosophila melanogaster. *Cell* 9: 393–407.
25. Berezney, R., and D.S. Coffey. 1974. Identification of a nuclear protein matrix. *Biochemical and Biophysical Research Communications* 60: 1410–1417.
26. Bergmann, O., R.D. Bhardwaj, S. Bernard, S. Zdunek, F. Barnabé-Heider, S. Walsh, J. Zupicich, K. Alkass, B.A. Buchholz, H. Druid, S. Jovinge, and J. Frisén. 2009. Evidence for cardiomyocyte renewal in humans. *Science* 324: 98–102.
27. Bhardwaj, R.D., M.A. Curtis, K.L. Spalding, B.A. Buchholz, D. Fink, T. Björk-Eriksson, C. Nordborg, F.H. Gage, H. Druid, P.S. Eriksson, and J. Frisén. 2006. Neocortical neurogenesis in humans is restricted to development. *Proceedings of the National Academy of Sciences of the United States of America* 103: 12564–12568.
28. Biesterfeld, S., K. Gerres, G. Fischer-Wein, and A. Böcking. 1994. Polyploidy in non-neoplastic tissues. *Journal of Clinical Pathology* 47: 38–42.
29. Brock, A., and S. Huang. 2017. Precision oncology: Between vaguely right and precisely wrong. *Cancer Research* 77: 6473–6479.
30. Buongiorno-Nardelli, M., G. Micheli, M.T. Carri, and M. Marilley. 1982. A relationship between replicon size and supercoiled loop domains in the eukaryotic genome. *Nature* 298: 100–102.
31. Butler, L.M., D.B. Agus, H.I. Scher, B. Higgins, A. Rose, C. Cordon-Cardo, H.T. Thaler, R.A. Rifkind, P.A. Marks, and V.M. Richon. 2000. Suberoylanilide hydroxamic acid, an

inhibitor of histone deacetylase, suppresses the growth of prostate cancer cells in vitro and in vivo. *Cancer Research* 60: 5165–5170.
32. Byrnes, K.R., and A.I. Faden. 2007. Role of cell cycle proteins in CNS injury. *Neurochemical Research* 32: 1799–1807.
33. Caron, J.M., M. Bannon, L. Rosshirt, J. Luis, L. Monteagudo, J.M. Caron, and G.M. Sternstein. 2010. Methyl sulfone induces loss of metastatic properties and reemergence of normal phenotypes in metastatic Cloudman S-91 (M3) murine melanoma cell line. *PLoS ONE* 5: e11788.
34. Casola, C. 2016. TP53 gene and cancer resistance in elephants. *JAMA* 315: 1788–1789.
35. Caulin, A.F., T.A. Graham, L.-S. Wang, and C. Maley. 2015. Solutions to Peto's paradox revealed by mathematical modelling and cross-species cancer gene analysis. *Philosophical Transactions of the Royal Society B* 36: 20140222.
36. Cayrou, C., P. Coulombe, and M. Méchali. 2010. Programming DNA replication origins and chromosome organization. *Chromosome Research* 18: 137–145.
37. Celton-Morizur, S., and C. Desdouets. 2010. Ployploidization of liver cells. *Advances in Experimental Medicine and Biology* 676: 123–135.
38. Chen, J.C., M.J. Alvarez, F. Talos, H. Dhruv, G.E. Rieckhof, A. Iyer, K.L. Diefes, K. Aldape, M. Berens, M.M. Shen, and A. Califano. 2014. Identification of causal genetic drivers of human disease through systems-level analysis of regulatory networks. *Cell* 159: 402–414.
39. Chetty, S., F. Walton Pagliuca, C. Honore, A. Kweudjeu, A. Rezania, and D.A. Melton. 2013. A simple tool to improve pluripotent stem cell differentiation. *Nature Methods* 10: 553–556.
40. Clark, W.H. 1995. The nature of cancer: Morphogenesis and progressive (self)- disorganization in neoplastic development and progression. *Acta Oncologica* 34: 3–21.
41. Cockerill, P.N., and W.T. Garrard. 1986. Chromosomal loop anchorage of the kappa immunoglobulin gene occurs next to the enhancer in a region containing toposiomerase II sites. *Cell* 31: 273–282.
42. Collins, J.M., and K.A. Foster. 1983. Differentiation of promyelocytic (HL-60) cells into mature granulocytes: Mitochondrial-specific rhodamine 123 fluorescence. *The Journal of Cell Biology* 96: 94–99.
43. Constantinescu, D., H.L. Gray, P.J. Sammak, G.P. Schatten, and A.B. Csoka. 2006. Lamin A/C is a marker of mouse and human embryonic stem cell differentiation. *Stem Cells* 24: 177–185.
44. Cook, P.R., I. Brazell, and E. Jost. 1976. Characterization of nuclear structures containing superhelical DNA. *Journal of Cell Science* 22: 303–324.
45. Copani, A., D. Uberti, M.A. Sortino, V. Bruno, F. Nicoletti, and M. Memo. 2001. Activation of cell- cycle-associated proteins in neuronal death: A mandatory or dispensable path? *Trends in Neurosciences* 24: 25–31.
46. Courbet, S., S. Gay, N. Arnoult, G. Wronka, M. Anglana, O. Brison, and M. Debatisse. 2008. Replication fork movement sets chromatin loop size and origin choice in mammalian cells. *Nature* 455: 557–560.
47. Cristofalo, V.J., R.G. Allen, R.J. Pignolo, B.J. Martin, and J.C. Beck. 1998. Relationship between donor age and the replicative lifespan of human cells in culture. A reevaluation. *Proceedings of the National Academy of Sciences of the United States of America* 95: 10614–10619.
48. Currais, A., T. Hortobágyi, and S. Soriano. 2009. The neuronal cell cycle as a mechanism of pathogenesis in Alzheimer's disease. *Aging* 1: 363–371.
49. Del Monte, U. 2006. The puzzle of ploidy of Purkinje neurons. *Cerebellum* 5: 23–26.
50. Demir, O., S. Singh, L. Klimaschewski, and I.A. Kurnaz. 2009. From birth till death: Neurogenesis, cell cycle and neurodegeneration. *The Anatomical Record* 292: 1953–1961.
51. Dent, M.A.R., E. Segura-Anaya, J. Alva-Medina, and A. Aranda-Anzaldo. 2010. NeuN/Fox3 is an intrinsic component of the neuronal nuclear matrix. *FEBS Letters* 584: 2767–2771.
52. DePinho, R.A. 2000. The age of cancer. *Nature* 408: 248–254.

53. Di Giovanni, S., V. Movsesyan, F. Ahmed, I. Cernak, S. Schinelli, B. Stoica, and A.I. Faden. 2005. Cell cycle inhibition provides neuroprotection and reduces glial proliferation and scar formation after traumatic brain injury. *Proceedings of the National Academy of Sciences of the United States of America* 102: 8333–8838.
54. Dimitrova, D.S., and R. Berezney. 2002. The spatio-temporal organization of DNA replication sites is identical in primary, immortalized and transformed mammalian cells. *Journal of Cell Science* 115: 4037–4051.
55. Dirks, P.B. 2008. Brain tumor stem cells: The undercurrents of human brain cancer and their relationship to neural stem cells. *Philosophical Transactions of the Royal Society of London. Series B, Biological Sciences* 363: 139–152.
56. Doll, R., and R. Peto. 1981. The causes of cancer: Quantitative estimates of avoidable risks of cancer in the United States today. *Journal of the National Cancer Institute* 66: 1193–1308.
57. Duesberg, P., R. Li, A. Fabarius, and R. Hehlmann. 2005. The chromosomal basis of cancer. *Cellular Oncology* 27: 293–318.
58. Duncan, A.W., M.H. Taylor, R.D. Hickey, A.E. Hanlon Newell, M.L. Lenzi, S.B. Olson, M.J. Finegold, and M. Grompe. 2010. Ploidy conveyor of mature hepatocytes as a source of genetic variation. *Nature* 467: 707–710.
59. Duncan, A.W., A.E. Hanlon Newell, L. Smith, E.M. Wilson, S.B. Olson, M.J. Thaver, S.C. Strom, and M. Grompe. 2012. Frequent aneuploidy among normal human hepatocytes. *Gastroneterology* 142: 25–28.
60. Eggan, K., K. Baldwin, M. Tackett, J. Osborne, J. Gogos, A. Chess, R. Axel, and R. Jaenisch. 2004. Mice cloned from olfactory sensory neurons. *Nature* 428: 44–49.
61. Elcock, L.S., and J.M. Bridger. 2008. Exploring the effects of a dysfunctional nuclear matrix. *Biochemical Society Transactions* 36: 1378–1383.
62. Engelke, R., J. Riede, J. Hegermann, A. Wuerch, S. Eimer, J. Dengjel, and G. Mittler. 2014. The quantitative nuclear matrix proteome as a biochemical snapshot of nuclear organization. *Journal of Proteome Research* 13: 3940–3956.
63. Feddersen, R.M., R. Ehlensfeldt, W.S. Yunis, H.B. Clark, and H.T. Orr. 1992. Disrupted cerebellar cortical development and progressive degeneration of Purkinje cells in SV40 T antigen transgenic mice. *Neuron* 9: 955–966.
64. Flames, N., and O. Hobert. 2011. Transcriptional control of the terminal fate of monoaminergic neurons. *Annual Review of Neuroscience* 34: 153–184.
65. Frackman, S., G. Kobs, D. Simpson, and D. Storts. 1998. Betaine and DMSO: Enhancing agents for PCR. *Promega Notes* 65: 27.
66. Friedmann-Morvinski, D., E.A. Bushong, E. Ke, Y. Soda, T. Marumoto, O. Singer, M.H. Ellisman, and I.M. Verma. 2012. Dedifferentiation of neurons and astrocytes by oncogenes can induce gliomas in mice. *Science* 338: 1080–1084.
67. Friend, C., and H.A. Freedman. 1978. Effects and possible mechanism of action of dimethylsulfoxide on Friend cell differentiation. *Biochemical Pharmacology* 27: 1309–1313.
68. Galli, C., S. Guizazardi, G. Passeri, G.M. Macaluso, and R. Scandroglio. 2005. Life on the wire: On tensegrity and force balance in cells. *Acta Biologica et Medica* 76: 5–12.
69. García-Vilchis, D., and A. Aranda-Anzaldo. 2017. DNA length modulates the affinity of fragments of genomic DNA for the nuclear matrix in vitro. *Journal of Cellular Biochemistry* 118: 4487–4497.
70. Garraway, L.A., and E.S. Lander. 2013. Lessons from the cancer genome. *Cell* 153: 17–37.
71. Gateff, E. 1978. Malignant neoplasms of genetic origin in Drosophila melanogaster. *Science* 200: 1448–1459.
72. Gaylord Chemical. 2014. Dimethyl sulfoxide, health and safety information. *Bulletin 106*, June 2014. Retrieved at: https://www.gaylordchemical.com/wp- content/uploads/2015/07/GC-Literature-106.pdf.
73. Gentric, G., S. Celton-Morizur, and C. Desdouets. 2012. Polyploidy and liver proliferation. *Clinics and Research in Hepatology and Gastroenterology* 36: 29–34.

74. Germano, I., V. Swiss, and P. Casaccia. 2010. Primary brain tumors, neural stem cell, and brain tumor cancer cells: Where is the link? *Neuropharmacology* 58: 903–910.
75. Goldstein, S. 1990. Replicative senescence: The human fibroblast comes of age. *Science* 249: 1129–1133.
76. Greaves, M., and C.C. Maley. 2012. Clonal evolution in cancer. *Nature* 481: 306–313.
77. Grunt, T.W., C. Somay, M. Pavelka, A. Ellinger, E. Dittirch, and C. Dittrich. 1991. The effects of dimethyl sulfoxide and retinoic acid on the cell growth and the phenotype of ovarian cancer cells. *Journal of Cell Science* 100: 657–666.
78. Gupta, S. 2000. Hepatic polyploidy and liver growth control. *Seminars in Cancer Biology* 10: 161–171.
79. Hakes, D.J., and R. Berezney. 1991. DNA binding properties of the nuclear matrix and individual nuclear matrix proteins. *The Journal of Biological Chemistry* 266: 1113–11140.
80. Hanahan, D. 2014. Rethinking the war on cancer. *Lancet* 383: 558–563.
81. Hanahan, D., and R.A. Weinberg. 2011. Hallmarks of cancer: The next generation. *Cell* 144: 646–674.
82. Hartung, E.W. 1942. The effects of roentgen radiation on tumor incidence in Drosophila melanogaster. *Cancer Research* 2: 837–840.
83. Hayflick, L., and P.S. Moorhead. 1961. The serial cultivation of human diploid cell strains. *Experimental Cell Research* 25: 585–621.
84. He, D.C., J.A. Nickerson, and S. Penman. 1990. Core filaments of the nuclear matrix. *The Journal of Cell Biology* 110: 569–580.
85. Herculano-Houzel, S. 2009. The human brain in numbers: A linearly scaled-up primate brain. *Frontiers in Human Neuroscience* 3: 31.
86. Hofree, M., J.P. Shen, H. Carter, A. Gross, and T. Ideker. 2013. Network-based stratification of tumor mutations. *Nature Methods* 10: 1108–1115.
87. Holmberg, J., and T. Perlmann. 2012. Maintaining differentiated cellular identity. *Nature Reviews. Genetics* 13: 429–439.
88. Huang, S. 2014. The war on cancer: Lessons from the war on terror. *Frontiers in Oncology* (4): 293.
89. Ingber, D.E., N. Wang, and D. Stamenovic. 2014. Tensegrity, cellular biophysics, and the mechanics of living systems. *Reports on Progress in Physics* 77: 046603.
90. Iranzo, J., I. Martincorena, and E.V. Koonin. 2018. Cancer-mutation network and the number and specificity of driver mutations. *Proceedings of the National Academy of Sciences of the United States of America* 115: E6010–E6019.
91. Ivanović-Matić, S., S. Dinic, M. Vujosevic, and G. Poznanovic. 2000. The protein composition of the hepatocyte nuclear matrix is differentiation-stage specific. *IUBMB Life* 49: 511–517.
92. Jackson, D.A., P. Dickinson, and P.R. Cook. 1990. The size of chromatin loops in HeLa cells. *The EMBO Journal* 9: 567–571.
93. Jasmin, Spray D.C., A.C. Campos de Carvalho, and R. Mendez-Otero. 2010. Chemical induction of cardiac differentiation in P19 embryonal carcinoma stem cells. *Stem Cells and Development* 19: 403–411.
94. Jia, Y., C.-H. Yun, E. Park, D. Ercan, M. Manuia, J. Juarez, C. Xu, K. Rhee, T. Chen, H. Zhang, S. Palakurthi, J. Jang, G. Lelais, M. DiDonato, B. Bursulaya, P.Y. Michellys, R. Epple, T.H. Marsilje, M. McNeill, W. Lu, J. Harris, S. Bender, K.K. Wong, P.A. Jänne, and M.J. Eck. 2016. Overcoming EGFR (T790M) and EGFR (C797S) resistance with mutant-selective allosteric inhibitors. *Nature* 534: 129–132.
95. Kandoth, C., M.D. Mclellan, F. Vandin, K. Ye, B. Niu, C. Lu, M. Xie, Q. Zhang, J.F. McMichael, M.A. Wyczalkowski, M.D.M. Leiserson, C.A. Miller, J.S. Welch, M.J. Walter, M.C. Wendl, T.J. Ley, R.K. Wilson, B.J. Raphael, and L. Ding. 2013. Mutational landscape and significance across 12 major cancer types. *Nature* 502: 333–339.
96. Kato, S., S.M. Lippman, K.T. Flaherty, and R. Kuzrock. 2016. The conundrum of genetic "drivers" in benign conditions. *Journal of the National Cancer Institute* 108: djw036.

97. Kay, V., and J. Bode. 1994. Binding specificity of a nuclear scaffold: Supercoiled, single-stranded, and scaffold-attached-region DNA. *Biochemistry* 33: 367–374.
98. Kingsbury, M.A., B. Friedman, M.J. McConnell, S.K. Rehen, A.H. Yang, D. Kaushal, and J. Chun. 2005. Aneuploid neurons are functionally active and integrated into brain circuitry. *Proceedings of the National Academy of Sciences* 102: 6143–6147.
99. Klein, G. 2009. Towards a genetics of cancer resistance. *Proceedings of the National Academy of Sciences of the United States of America* 106: 859–863.
100. Krishnamoorty, R.R., A.F. Clark, D. Daudt, J.K. Vishwanatha, and T. Yorio. 2013. A forensic path to RGC-5 cell line identification: Lessons learned. *Investigative Ophthalmology & Visual Science* 54: 5712–5719.
101. Krishnamoorty, R.R., P. Agarwal, G. Prasanna, et al. 2014. Retraction notice to "characterization of a transformed rat retinal ganglion cell line. *Brain Research* 1544: 62
102. Kuan, C.Y., A.J. Schloemer, A. Lu, K.A. Burns, W.L. Weng, M.T. Williams, K.I. Strauss, C.V. Vorhees, R.A. Flavell, R.J. Davis, F.R. Sharp, and P. Rakic. 2004. Hypoxia-ischemia induces DNA synthesis without cell proliferation in dying neurons in adult rodent brain. *The Journal of Neuroscience* 24: 10763–10772.
103. Kuilman, T., C. Michaloglou, W.J. Mooi, and D.S. Peeper. 2010. The essence of senescence. *Genes & Development* 24: 2463–2479.
104. Lambert, F. 2002. Disorder, a cracked crutch for supporting entropy discussions. *Journal of Chemical Education* 79: 187–192.
105. Le Tourneau, C., J.P. Delord, A. Goncalves, et al. 2015. Molecularly targeted therapy based on tumour molecular profiling versus conventional therapy for advanced cancer (SHIVA): A multicentre, open-label, proof-of-concept, randomised, controlled phase 2 trial. *The Lancet Oncology* 16: 1324–1334.
106. Lemaitre, J.M., E. Danis, P. Pasero, Y. Vassetzky, and M. Méchali. 2005. Mitotic remodeling of the replicon and chromosome structure. *Cell* 123: 787–801.
107. Li, R., G. Yerganian, P. Duesberg, A. Kraemer, A. Willer, C. Rausch, and R. Hehlmann. 1997. Aneuploidy correlates 100% with chemical transformation of Chinese hamster cells. *Proceedings of the National Academy of Sciences of the United States of America* 94: 14506–14511.
108. Ling, S., Z. Hu, Z. Yang, Li Y, P. Lin, K. Chen, L. Dong, L. Cao, Y. Tao, L. Hao, Q. Chen, Q. Gong, D. Wu, W. Li, W. Zhao, X. Tian, C. Hao, E.A. Hungate, D.V. Catenacci, R.R. Hudson, W.H. Li, X. Lu, and C.I. Wu. 2015. Extremely high genetic diversity in a single tumor points to prevalence of non-Darwinian cell evolution. *Proceedings of the National Academy of Sciences of the United States of America* 112: E6496–E6505.
109. Luderus, M.E., B. van Steensel, L. Chong, O.C. Sibon, F.F. Cremers, and T. de Lange. 1996. Structure, subnuclear localization and nuclear matrix association of the mammalian telomeric complex. *The Journal of Cell Biology* 135: 867–881.
110. Luu, T.H., R.J. Morgan, L. Leong, D. Lim, M. McNamara, J. Portnow, P. Frankel, D.D. Smith, J.H. Doroshow, C. Wong, A. Aparicio, D.R. Gandara, and G. Somlo. 2008. A phase II trial of vorinostat (suberoylanilide hydroxamic acid) in metastatic breast cancer: A California Cancer Consortium study. *Clinical Cancer Research* 14: 7138–7142.
111. Macieira-Coelho, A. 1994. Chaos in DNA partition during the last mitoses of the proliferative life-span of human fibroblasts. *FEBS Letters* 358: 126–128.
112. ———. 2010. Cancers and the concept of cell senescence. *Biogerontology* 11: 211–227.
113. ———. 2011. Cell division and aging of the organism. *Biogerontology* 12: 508–515.
114. Marks, P.A., and R. Breslow. 2007. Dimehtyl sulfoxide to vorinostat: Development of this histone acetylase inhibitor as an anticancer drug. *Nature Biotechnology* 25: 84–90.
115. Marks, P.A., and W.S. Xu. 2009. Histone deacetylase inhibitors: Potential in cancer therapy. *Journal of Cellular Biochemistry* 107: 600–608.
116. Martin, G.M., C.A. Sprague, T.H. Norwood, and W.R. Pendergrass. 1974. Clonal selection, attenuation and differentiation in an in vitro model of hyperplasia. *The American Journal of Pathology* 74: 137–154.

117. Martincorena, I., and P.J. Campbell. 2015. Somatic mutation in cancer and normal cells. *Science* 349: 1483–1489.
118. Martincorena, I., A. Roshan, M. Gerstung, P. Ellis, P. Van Loo, S. McLaren, D.C. Wedge, A. Fullam, L.B. Alexandrov, J.M. Tubio, L. Stebbings, A. Menzies, S. Widaa, M.R. Stratton, P.H. Jones, and P.J. Campbell. 2015. Tumor evolution. High burden and pervasive positive selection of somatic mutations in normal human skin. *Science* 348: 880–886.
119. Martincorena, I., J.C. Fowler, A. Wabik, A.R.J. Lawson, F. Abascal, Hall MWJ, A. Cagan, K. Murai, K. Mahbubani, M.R. Stratton, R.C. Fitzgerald, P.A. Handford, P.J. Campbell, K. Saeb-Parsy, and P.H. Jones. 2018. Somatic mutant clones colonize the human esophagus with age. *Science* 362: 911–917.
120. Martínez-Ramos, I., A. Maya-Mendoza, P. Gariglio, and A. Aranda-Anzaldo. 2005. A global but stable change in HeLa cell morphology induces reorganization of DNA structural loop domains within the cell nucleus. *Journal of Cellular Biochemistry* 96: 79–88.
121. Matsumura, T. 1980. Multinucleation and polyploidization of aging human cells in culture. *Advances in Experimental Medicine and Biology* 129: 31–38.
122. Maya-Mendoza, A., R. Hernández-Muñoz, P. Gariglio, and A. Aranda-Anzaldo. 2003. Gene positional changes relative to the nuclear substructure correlate with the proliferating status of hepatocytes during liver regeneration. *Nucleic Acids Research* 31: 6168–6179.
123. ———. 2005. Natural ageing in the rat liver correlates with progressive stabilisation of DNA-nuclear matrix interactions and withdrawal of genes from the nuclear substructure. *Mechanisms of Ageing and Development* 126: 767–782.
124. Maya-Mendoza, A., P. Olivares-Chauvet, A. Shaw, and D.A. Jackson. 2010. S phase progression in human cells is dictated by the genetic continuity of DNA foci. *PLoS Genetics* 6 (4): e1000900.
125. McConnell, M., M.R. Lindberg, K.J. Brennand, J.C. Piper, T. Voet, C. Cowing-Zitron, S. Shumilina, R.S. Lasken, J.R. Vermeesch, I.M. Hall, and F.H. Gage. 2013. Mosaic copy number variation in human neurons. *Science* 342: 632–637.
126. McGranahan, N., and C. Swanton. 2017. Clonal heterogeneity and tumor evolution: Past, present, and the future. *Cell* 168: 613–628.
127. Méchali, M. 2010. Eukaryotic DNA replication origins: Many choices for appropriate answers. *Nature Reviews. Molecular Cell Biology* 11: 728–738.
128. Michalopoulos, G.K., and M.C. DeFrances. 1997. Liver regeneration. *Science* 276: 60–66.
129. Min, H.-Y., S.-C. Lee, J.K. Woo, H.J. Jung, K.H. Park, H.M. Jeong, S.Y. Hyun, J. Cho, W. Lee, J.E. Park, S.J. Kwon, H.J. Lee, X. Ni, Y.K. Shin, F.M. Johnson, M. Duvic, and H.Y. Lee. 2017. Essential role of DNA methyltransferase 1-mediated transcription of insulin-like growth factor 2 in resistance to histone deacetylase inhibitors. *Clinical Cancer Research* 23: 1299–1311.
130. Mirkin, S.M. 2001, May. DNA topology: Fundamentals. In *eLS*. Chichester: Wiley. http://www.els.net. https://doi.org/10.1038/npg.els.0001038.
131. Muotri, A.R., V.T. Chu, M.C. Marchetto, W. Deng, J.V. Moran, and F.H. Gage. 2005. Somatic mosaicism in neuronal precursor cells mediated by L1 retrotransposition. *Nature* 435: 903–910.
132. Nickerson, J.A. 2001. Experimental observations of a nuclear matrix. *Journal of Cell Science* 114: 463–474.
133. Nordling, C.O. 1953. A new theory on cancer-inducing mechanism. *British Journal of Cancer* 7: 68–72.
134. Nussinov, R., H. Jang, C.-J. Tsai, and F. Cheng. 2019. Precision medicine review: Rare driver mutations and their biophysical classification. *Biophysical Reviews* 11: 5–19.
135. Pajalunga, D., A. Mazzola, A.M. Salzano, M.G. Biferi, G. De Luca, and M. Crescenzi. 2007. Critical requirement for cell cycle inhibitors in sustaining nonproliferative states. *The Journal of Cell Biology* 176: 807–818.

136. Pajalunga, D., E.M.R. Puggioni, A. Mazzola, V. Leva, A. Montecucco, and M. Crescenzi. 2010. DNA replication is intrinsically hindered in terminally differentiated myotubes. *PLoS ONE* 5: e11559.
137. Pajerowski, J.D., K.N. Dahl, F.L. Zhong, P.J. Sammak, and D.E. Discher. 2007. Physical plasticity of the nucleus in stem cell differentiation. *Proceedings of the National Academy of Sciences of the United States of America* 104: 15619–15624.
138. Park, D.S., A. Obeidat, A. Giovanni, and L.A. Greene. 2000. Cell cycle regulators in neuronal death evoked by exocitotoxic stress: Implications for neurodegeneration and its treatment. *Neurobiology of Aging* 21: 771–781.
139. Pasque, V., J. Jullien, K. Miyamoto, R.P. Halley-Stott, and J.B. Gurdon. 2011. Epigenetic factors influencing resistance to nuclear reprogramming. *Trends in Genetics* 27: 516–525.
140. Patil, C.K., I. Saira Mian, and J. Campisi. 2005. The thorny path linking cellular senescence to organismal aging. *Mechanisms of Ageing and Development* 126: 1040–1045.
141. Pearson, J., M. Milstoc, J. Harris, G. Budzilovich, and I. Feign. 1976. Anaplastic neuronal tumors of brain. *Cancer* 38: 1424–1437.
142. Perez, R.P., and T. Komiya. 2016. TP53 and cancer resistance in elephants. *JAMA* 315: 1789–1790.
143. Peto, R. 2015. Quantitative implications of the approximate irrelevance of mammalian body size and lifespan to lifelong cancer risk. *Philosophical Transactions of the Royal Society of London. Series B, Biological Sciences* 370: pii 20150198.
144. Peto, R., and R. Doll. 1997. There is no such thing as aging: Old age is associated with disease, but does not cause it. *BMJ [British Medical Journal]* 315: 1030–1032.
145. Peto, R., F.J.C. Roe, P.N. Lee, L. Levy, and J. Clack. 1975. Cancer and ageing in mice and men. *British Journal of Cancer* 32: 411–426.
146. Prasad, V. 2016. The precision-oncology illusion. *Nature* 537: S63.
147. Preisler, H.D., and M. Giladi. 1975. Differentiation of erythroleukemic cells in vitro: Irreversible induction by dimehtyl sulfoxide (DMSO). *Journal of Cellular Physiology* 85: 537–546.
148. Ptashne, M. 2013. Epigenetics: Core misconcept. *Proceedings of the National Academy of Sciences of the United States of America* 110: 7101–7103.
149. Rakic, P. 1974. Neurons in rhesus visual cortex: Systematic relation between time of origin and eventual disposition. *Science* 183: 425–427.
150. ———. 2002. Neurogenesis in adult primate neocortex: An evaluation of the evidence. *Nature Reviews. Neuroscience* 3: 65–71.
151. ———. 2006. Neuroscience. No more cortical neurons for you. *Science* 313: 928–929.
152. Rawls, W.F., L. Cox, and E.S. Rovner. 2017. Dimethyl sulfoxide (DMSO) as intravesical therapy for interstitial cystitis/bladder pain syndrome: A review. *Neurology and Urodynamics* 36: 1677–1684.
153. Razin, S.V. 2001. The nuclear matrix and chromosomal DNA loops: Is there any correlation between partitioning of the genome into loops and functional domains? *Cellular & Molecular Biology Letters* 6: 59–69.
154. Razin, S.V., I.I. Gromova, and O.V. Iarovaia. 1995. Specificity and functional significance of DNA interactions with the nuclear matrix: New approaches to clarify the old questions. *International Review of Cytology* 162B: 405–448.
155. Rehen, S.K., M.J. McConnell, D. Kaushal, M.A. Kingsbury, A.H. Yang, and J. Chun. 2001. Chromosomal variation in neurons of the developing and adult mammalian nervous system. *Proceedings of the National Academy of Sciences* 98: 13361–13366.
156. Rehen, S.K., Y.C. Yung, M.P. McCreight, D. Kaushal, A.H. Yang, B.S. Almeida, M.A. Kingsbury, K.M. Cabral, M.J. McConnell, B. Anliker, M. Fontanoz, and J. Chun. 2005. Constitutional aneuploidy in the normal human brain. *The Journal of Neuroscience* 25: 2176–2180.
157. Remus, D., E.L. Beall, and M.R. Botchan. 2004. DNA topology, not DNA sequence, is a critical determinant for Drosophila ORC-DNA binding. *The EMBO Journal* 23: 897–907.

158. Repana, D., J. Nulsen, L. Dressler, M. Bortolomeazzi, S. Kuppili Venkata, A. Tourna, A. Yakovleva, T. Palmieri, and F.D. Cicarelli. 2019. The network of cancer genes (NCG): A comprehensive catalogue of known and candidate cancer genes from cancer sequencing screens. *Genome Biology* 20: 1.
159. Rivera-Mulia, J.C., R. Hernández-Muñoz, F. Martínez, and A. Aranda-Anzaldo. 2011. DNA moves sequentially towards the nuclear matrix during DNA replication in vivo. *BMC Cell Biology* 12: 3.
160. Röber, R.A., H. Sauter, K. Weber, and M. Osborn. 1990. Cells of the immune and hematopoietic system of the mouse lack lamins A/C: Distinction versus other somatic cells. *Journal of Cell Science* 95: 587–598.
161. Roti-Roti, J.L., W.D. Wright, and Y.C. Taylor. 1993. DNA loop structure and radiation response. *Advances in Radiation Biology* 17: 227–259.
162. Rubin, H. 1980. Is somatic mutation the major mechanism of malignant transformation? *Journal of the National Cancer Institute* 64: 995–1000.
163. Ruddon, R.W. 2007. *Cancer biology*. 4th ed. Oxford: Oxford University Press.
164. Santos, N.C., J. Figueira-Coelho, J. Martins-Silva, and C. Saldanha. 2003. Multidisciplinary utilization of dimethyl sulfoxide: Pharmacological, cellular, and molecular aspects. *Biochemical Pharmacology* 65: 1035–1041.
165. Sapieha, P., and F.A. Mallette. 2018. Cellular senescence in postmitotic cells: Beyond growth arrest. *Trends in Cell Biology* 28: 595–607.
166. Satgé, D. 2013, September. Analysis of somatic mutations in cancer tissues challenges the somatic mutation theory of cancer. In: *eLS*. Chichester: Wiley. http://www.els.net. https://doi.org/10.1002/9780470015902.a0024465.
167. Schmucker, D.L., and H. Sanchez. 2011. Liver regeneration and aging: A current perspective. *Current Gerontology and Geriatrics Research* 2011: 526379.
168. Senft, D., and Z.A. Ronai. 2016. Adaptive stress responses during tumor metastasis and dormancy. *Trends in Cancer* 2: 429–442.
169. Senft, D., M.D.M. Leiserson, E. Ruppin, and Z.A. Ronai. 2017. Precision oncology: The road ahead. *Trends in Molecular Medicine* 23: 874–898.
170. Senyo, S.E., M.L. Steinhauser, C.L. Pizzimenti, V.K. Yank, L. Cai, M. Wang, T.-D. Wu, J.-L. Guerquin- Kern, C.-P. Lechene, and R.T. Lee. 2013. Mammalian heart renewal by pre-existing cardiomyocytes. *Nature* 493: 433–436.
171. Shay, J.W., and W.E. Wright. 2005. Senescence and immortalization: Role of telomeres and telomerase. *Carcinogenesis* 26: 867–874.
172. Sherwood, S.W., D. Rush, J.L. Ellsworth, and R.T. Schimke. 1988. Defining cellular senescence in IMR-90 cells: A flow cytometric analysis. *Proceedings of the National Academy of Sciences of the United States of America* 85: 9086–9090.
173. Sigal, S.H., S. Gupta, D.F. Gebhard Jr., P. Holst, D. Neufeld, and L.M. Reid. 1995. Evidence for a terminal differentiation process in the rat liver. *Differentiation* 59: 35–42.
174. Sigal, S.H., P. Rajvanshi, G.R. Gorla, R.P. Sokhl, R. Saxena, D.R. Gebhard, L.M. Reid, and S. Gupta. 1999. Partial hepatectomy-induced polyploidy attenuates hepatocyte replication and activates cell aging events. *The American Journal of Physiology* 276: G1260–G1272.
175. Smith, J.R., and L. Hayflick. 1974. Variation in the life-span of clones derived from human diploid cell strains. *The Journal of Cell Biology* 62: 48–53.
176. Smith, J.R., and R.G. Whitney. 1980. Intraclonal variation in proliferative potential of human diploid fibroblasts: Stochastic mechanism for cellular aging. *Science* 207: 82–84.
177. Sorrells, S.F., M.F. Paredes, A. Cebrian-Silla, Sandoval K, D. Qi, K.W. Kelley, D. James, S. Mayer, J. Chang, K.I. Auguste, E.F. Chang, A.J. Gutierrez, A.R. Kriegstein, G.W. Mathern, M.C. Oldham, E.J. Huang, J.M. Garcia-Verdugo, Z. Yang, and A. Alvarez-Buylla. 2018. Human hippocampal neurogenesis drops sharply in children to undetectable levels in adults. *Nature* 555: 377–381.
178. Spremulli, E.N., and D.L. Dexter. 1984. Polar solvents: A novel class of antineoplastic agents. *Journal of Clinical Oncology* 2: 227–241.

179. Storchova, Z., and D. Pellman. 2004. From polyploidy to aneuploidy, genome instability and cancer. *Nature Reviews. Molecular Cell Biology* 5: 45–54.
180. Stratton, M.R., P.J. Campbell, and P.A. Futreal. 2009. The cancer genome. *Nature* 458: 719–724.
181. Stuurman, N., A.M.L. Meijne, A.J. van Der Pol, L. de Jong, R. van Driel, and J. van Renswoude. 1990. The nuclear matrix from cells of different origin. *The Journal of Biological Chemistry* 265: 5460–5465.
182. Sunley, K., and M. Butler. 2010. Strategies for the enhancement of recombinant protein production from mammalian cells by growth arrest. *Biotechnology Advances* 28: 385–394.
183. Takahashi, K., and S. Yamanaka. 2013. Induced pluripotent stem cells in medicine and biology. *Development* 140: 2457–2461.
184. Takahashi, A., N. Ohtani, and E. Hara. 2007. Irreversibility of cellular senescence: Dual roles of $p16^{ink4a}$/Rb-pathway in cell cycle control. *Cell Division* 2: 10.
185. Takahashi, S., S. Motooka, S. Kawasaki, H. Kurita, T. Mizuno, S.I. Matsuura, F. Hanaoka, A. Mizuno, M. Oshige, and S. Katsura. 2018. Direct single-molecule observations of DNA unwinding by SV40 large tumor antigen under a negative DNA supercoil state. *Journal of Biomolecular Structure & Dynamics* 36: 32–44.
186. Thom, M., J. Liu, A. Bongaarts, R.J. Reinten, B. Paradiso, H.R. Jäger, C. Reeves, A. Somani, S. An, D. Marsdon, A. McEvoy, A. Miserocchi, L. Thorne, F. Newman, S. Bucur, M. Honavar, T. Jacques, and E. Aronica. 2018. Multinodular and vacuolating neuronal tumors in epilepsy: Dysplasia or neoplasia? *Brain Pathology* 28: 155–171.
187. Tischer, J., and F. Gergely. 2019. Anti-mitotic therapies in cancer. *The Journal of Cell Biology* 218: 10–11.
188. Tomasetti, C., L. Marchionni, M.A. Nowak, G. Parmigiani, and B. Vogelstein. 2015. Only three driver mutations are required for the development of lung and colorectal cancers. *Proceedings of the National Academy of Sciences of the United States of America* 112: 118–123.
189. Tomasseti, C., B. Vogelstein, and G. Parmigiani. 2013. Half or more of the somatic mutations in cancers of self-renewing tissues originate prior to tumor initiaion. *Proceedings of the National Academy of Sciences of the United States of America* 110: 1999–2004.
190. Trevilla-García, C., and A. Aranda-Anzaldo. 2011. Cell-type-specific organization of nuclear DNA into structural looped domains. *Journal of Cellular Biochemistry* 112: 531–540.
191. Tsiftsoglou, A.S., and A.C. Sartorelli. 1979. Dimethyl sulfoxide-induced differentiation of Friend erytroleukemia cells in absence of cytokinesis. *Cancer Research* 39: 4058–4063.
192. Tsutsui, K.M., K. Sano, and K. Tsutsui. 2005. Dynamic view of the nuclear matrix. *Acta Medica Okayama* 59: 113–120.
193. Tunçer, S., R. Gurbanov, I. Sheraj, E. Solel, O. Esenturk, and S. Banerjee. 2018. Low dose dimethyl sulfoxide driven gross molecular changes have the potential to interfere with various cellular processes. *Scientific Reports* 8: 14828.
194. Turoverov, K., I.M. Kuznetsova, and V.N. Uversky. 2010. The protein kingdom extended: Ordered and intrinsically disordered proteins, their folding, supramolecular complex formation, and aggregation. *Progress in Biophysics and Molecular Biology* 102: 73–84.
195. Uhr, J.W., and K. Pantel. 2011. Controversy in clinical cancer dormancy. *Proceedings of the National Academy of Sciences of the United States of America* 108: 12396–12400.
196. Uhr, J.W., R.H. Scheuermann, N.E. Street, and E.S. Vitetta. 1997. Cancer dormancy: Opportunities for new therapeutic approaches. *Nature Medicine* 3: 505–509.
197. Vane, J.R. 1971. Inhibition of prostaglandin synthesis as a mechanism of action for aspirin-like drugs. *Nature: New Biology* 231: 232–235.
198. Vashee, S., C. Cvetic, W. Lu, P. Simancek, T.J. Kelly, and J.C. Walter. 2004. Sequence-independent DNA binding and replication initiation by the human origin recognition complex. *Genes & Development* 17: 1894–1908.
199. Veas-Pérez de Tudela, M., C. Maestre, M. Delgado-Esteban, J.P. Bolaños, and A. Almeida. 2015. Cdk5-mediated inhibition of APC/C-Cdh1 switches on the cyclin D1-Cdk4-pRb pathway causing aberrant S-phase entry of postmitotic neurons. *Scientific Reports* 5: 18180.

200. Viza, D., A. Aranda-Anzaldo, C. Zompett, and J.M. Vich. 1991. Dimethyl sulfoxide inhibits human immunodeficiency virus production in vitro. *Intervirology* 32: 59–64.
201. Vlcek, S., T. Dechat, and R. Foisner. 2001. Nuclear envelope and nuclear matrix; interactions and dynamics. *Cellular and Molecular Life Sciences* 58: 1758–1765.
202. Vogelstein, B., D.M. Pardoll, and D.S. Coffey. 1980. Supercoiled loops and eukaryotic DNA replication. *Cell* 22: 79–85.
203. Vogelstein, B., N. Papadopoulos, V.E. Velculescu, S. Zhou, L.A. Diaz Jr., and K.W. Kinzler. 2013. Cancer genome landscapes. *Science* 339: 1546–1558.
204. Wakayama, T., A.C.F. Perry, M. Zuccotti, K.R. Johnson, and R. Yanagimachi. 1998. Full-term development of mice from enucleated oocytes injected with cumulus cell nuclei. *Nature* 394: 369–374.
205. Wang, W., B. Bu, M. Xie, M. Zhang, Z. Yu, and D. Tao. 2009. Neural cell cycle dysregulation and central nervous system diseases. *Progress in Neurobiology* 89: 1–17.
206. Wang, N., J.D. Tytell, and D.E. Ingber. 2009. Mechanotransduction at a distance: Mechanically coupling the extracellular matrix and the nucleus. *Nature Reviews. Molecular Cell Biology* 10: 75–82.
207. Wei, X., J. Samarabandu, R.S. Devdhar, A.J. Siegel, R. Acharya, and R. Berezney. 1998. Segregation of transcription and replication sites into higher order domains. *Science* 281: 1502–1506.
208. Weinberg, R.A. 2014. Coming full circle-from endless complexity to simplicity and back again. *Cell* 157: 267–271.
209. Westra, J.W., R.R. Rivera, D.M. Bushman, Y.C. Yung, S.E. Peterson, S. Barral, and J. Chun. 2010. Neuronal DNA content variation (DCV) with regional and individual differences in the human brain. *Journal of Comparative Neurology* 518: 3981–4000.
210. Wilson, R.H., and D. Coverley. 2013. Relationship between DNA replication and the nuclear matrix. *Genes to Cells* 18: 17–31.
211. Xu, W.-S., G. Perez, L. Ngo, C.-Y. Gui, and P.A. Marks. 2005. Induction of polyploidy by histone acetylase inhibitor: A pathway for antitumor effects. *Cancer Research* 65: 7832–7839.
212. Xu, W., L. Ngo, G. Perez, M. Dokmanovic, and P.A. Marks. 2006. Intrinsic apoptotic and thioredoxin pathways in human prostate cancer cell response to histone deacetylase inhibitor. *Proceedings of the National Academy of Sciences of the United States of America* 103: 15540–15545.
213. Yamazaki, Y., H. Makino, K. Hamaguchi-Hamada, S. Hamada, H. Sugino, E. Kawase, T. Miyata, M. Ogawa, R. Yanagimachi, and T. Yagi. 2001. Assessment of the developmental totipotency of neural cells in the cerebral cortex of mouse embryo by nuclear transfer. *Proceedings of the National Academy of Sciences of the United States of America* 98: 14022–14026.
214. Yokoyama, A., N. Kakiuchi, T. Yoshizato, Y. Nannya, H. Suzuki, Y. Takeuchi, Y. Shiozawa, Y. Sato, K. Aoki, S.K. Kim, Y. Fujii, K. Yoshida, K. Kataoka, M.M. Nakagawa, Y. Inoue, T. Hirano, Y. Shiraishi, K. Chiba, H. Tanaka, M. Sanada, Y. Nishikawa, Y. Amanuma, S. Ohashi, I. Aoyama, T. Horimatsu, S. Miyamoto, S. Tsunoda, Y. Sakai, M. Narahara, Brown JB, Sato Y, G. Sawada, K. Mimori, S. Minamiguchi, H. Haga, H. Seno, S. Miyano, H. Makishima, M. Muto, and S. Ogawa. 2019. Age-related remodelling of oesophageal epithelia by mutated cancer drivers. *Nature* 565: 312–317.
215. Yu, Z.-W., and P.J. Quinn. 1994. Dimethyl sulphoxide: A review of its applications in cell biology. *Bioscience Reports* 14: 259–281.
216. Zbarsky, I.B. 1998. On the history of nuclear matrix manifestation. *Cell Research* 8: 99–103
217. Zhang, J., and K. Herrup. 2008. Cdk5 and the non-catalytic arrest of the neuronal cell cycle. *Cell Cycle* 7: 3487–3490.
218. Zhang, J., H. Li, O. Yabut, H. Fitzpatrick, G. D'Arcangelo, and K. Herrup. 2010. Cdk5 suppresses the neuronal cell cycle by disrupting the E2F1-DP1 complex. *The Journal of Neuroscience* 30: 5219–5228.

219. González-Sánchez, Juan Carlos, Francesco Raimondi, and Robert B. Russell. 2018. Cancer genetics meets biomolecular mechanism—Bridging an age-old gulf. *FEBS Letters* 592 (4): 463–474.
220. Kim, Jongpil, Christopher J. Lengner, Oktay Kirak, Jacob Hanna, John P. Cassady, Michael A. Lodato, Su Wu, Dina A. Faddah, Eveline J. Steine, Qing Gao, Fu Dongdong, Meelad Dawlaty, and Rudolf Jaenisch. 2011. Reprogramming of postnatal neurons into induced pluripotent stem cells by defined factors. *Stem Cells* 29 (6): 992–1000.
221. Jones, Richard B., Ronald G. Whitney, and James R. Smith. 1985. Intramitotic variation in proliferative potential: Stochastic events in cellular aging. *Mechanisms of Ageing and Development* 29 (2): 143–149.
222. Cristofalo, V.J., and B.B. Sharf. 1973. Cellular senescence and DNA synthesis. Thymidine incorporation as a measure of population age in human diploid cells. *Experimental Cell Research* 76: 419–427.
223. Merz, George S., and John D. Ross. 1969. Viability of human diploid cells as a function ofin vitro age. *Journal of Cellular Physiology* 74 (3): 219–221.
224. Jackson, D.A., P.R. Cook, and S.B. Patel. 1984. Attachment of repeated sequences to the nuclear cage. *Nucleic Acids Research* 12 (17): 6709–6726.
225. Cook, P.R., and I.A. Brazell. 1976. Conformational constraints in nuclear DNA. *Journal of Cell Science* 22: 287–302.
226. Cook, P.R. 1999. The Organization of Replication and Transcription. *Science* 284 (5421): 1790–1795.
227. Lee, C.H., H. Mizusawa, and T. Kakefuda. 1981. Unwinding of double-stranded DNA helix by dehydration. *Proceedings of the National Academy of Sciences* 78 (5): 2838–2842.
228. Aranda-Anzaldo, Armando, and Myrna A.R. Dent. 2017. Why cortical neurons cannot divide, and why do they usually die in the attempt? *Journal of Neuroscience Research* 95 (4): 921–929.

Migrastatics – Anti-metastatic Drugs Targeting Cancer Cell Invasion

Aneta Gandalovičová, Daniel Rosel, and Jan Brábek

Migratory Phenotype of Cancer Cells

One of the characteristics of life is movement. Accordingly, cells have the intrinsic ability to migrate. However, complex multicellular organisms have developed mechanisms that ensure temporal and spatial regulation of cell movement in order to form complex multi-cellular structures, such as organs made of specialized tissues. In human, cell migration is preserved in specific types of cells, such as immune cells, and largely regulated to maintain homeostasis. Unsurprisingly, dysregulated migration of cells disrupts tissue integrity and is associated with various diseases, including the deadliest – metastatic cancer [1–4]. While characteristics such as uncontrolled growth, apoptosis evasion or increased angiogenesis are shared by both benign and metastatic cancer, only the latter is able to invade from the primary site to secondary sites and establish secondary tumors, i.e. metastases, which are the cause of death in over 90% of all cancer associated deaths [5, 6].

Overall, the ability to migrate is enabled by re-structuralizing the complex network of the cytoskeleton, composed of actin, microtubules and intermediate filaments. This process is organized by many signaling pathways converging on cytoskeleton regulatory proteins, which directly drive the structural changes [7] (Fig. 1). Interestingly, the same set of proteins that maintains epithelial polarity also drives the polarization in migrating cells by interacting with cytoskeletal proteins [8]. Underlying force generation necessary for cell body translocation is

A. Gandalovičová · D. Rosel · J. Brábek (✉)
Department of Cell Biology, Charles University, Prague, Czech Republic

Biotechnology and Biomedicine Centre of the Academy of Sciences and Charles University (BIOCEV), Vestec u Prahy, Czech Republic
e-mail: brabek@natur.cuni.cz

M. Bizzarri (ed.), *Approaching Complex Diseases*, Human Perspectives in Health Sciences and Technology 2, https://doi.org/10.1007/978-3-030-32857-3_9

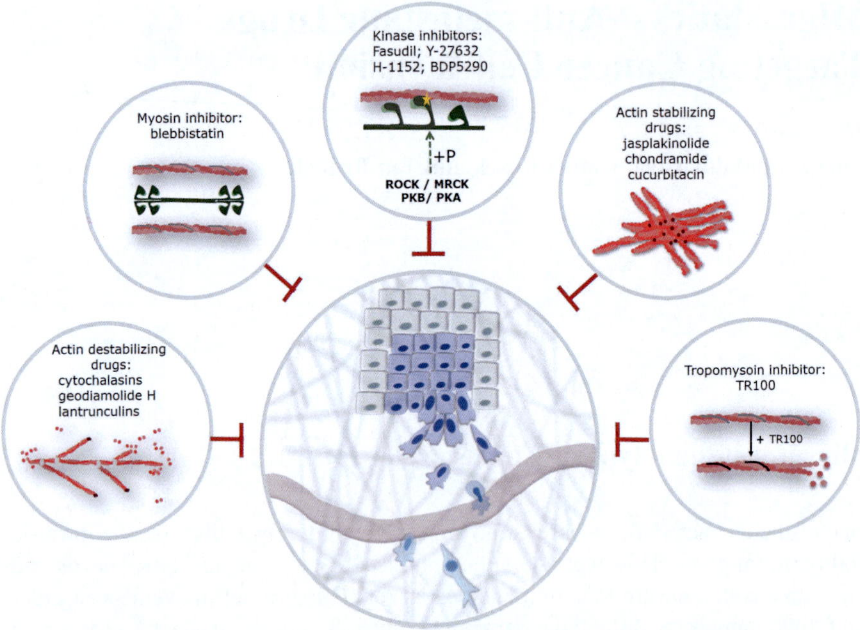

Fig. 1 Overview of candidate migrastatics. Agents targeting actin organization (actin stabilizing drugs, actin destabilizing drugs), tropomyosin (TR100) and actomyosin contractility (kinase inhibitors, blebbistatin) are suitable migrastatic candidates by affecting actin polymerization and actomyosin contractility, which are processes underlying cancer cell motility. In result, migrastatic drugs inhibit dissemination of metastatic cells (blue) from the primary tumor (grey)

polymerization and contraction of actin, which are processes regulated by Rho, Rac and Cdc42 proteins from the family of Rho GTPases [7, 9, 10]. Rac and Cdc42 signaling regulates activation of many actin associated proteins, such as Wiskott-Aldrich Syndrome proteins (e.g. WASP, N-WASP and WAVE1/2), that drive polymerization of actin leading to the formation of actin-based protrusions. On the other hand, Rho activation is the key signaling event in regulation of actomyosin contractility, which is propelled by phosphorylation of myosin light chain (MLC) [11] (Fig. 2). Rho activates Rho-kinase (ROCK), which directly phosphorylates MLC [12]. Additionally, ROCK along with myotonic dystrophy kinase-related Cdc42-binding kinase (MRCK) deactivate myosin light chain phosphatase (MLCP), a negative regulator of MLC phosphorylation [13, 14].

Various migration modes have been described, ranging from collective to single cell invasion, which differ in their requirements for actin protrusive forces and/or actomyosin contractility, resulting in diverse strategies of movement through the extracellular matrix (ECM). During collective migration cells move as a group of cells, which can be organized into clusters, sheets or strands that hold together by cell-cell adhesions. This type of migration is common during embryonic development or wound healing [15]. Cells at the leading front, referred to as leader cells, are

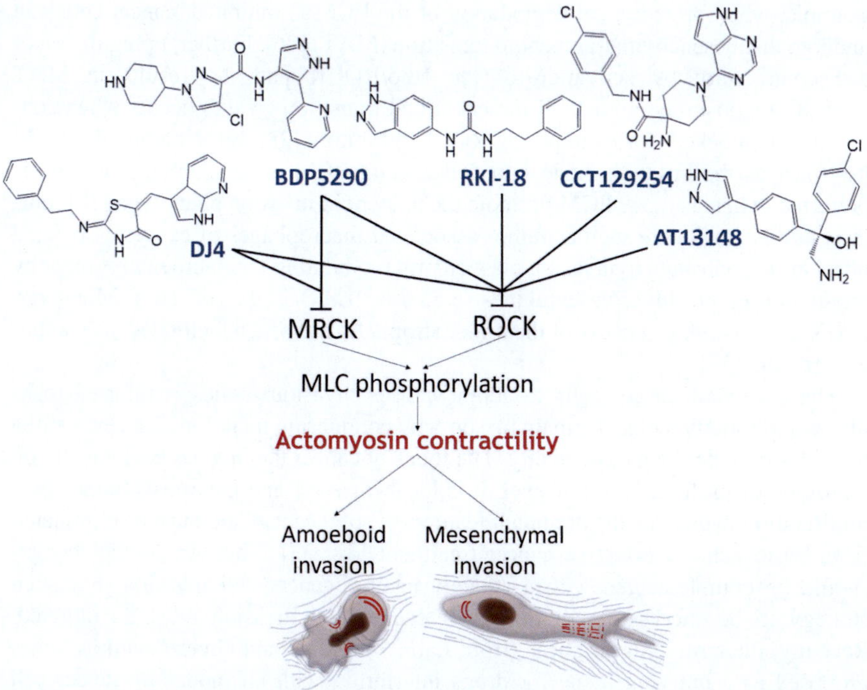

Fig. 2 Actomyosin contractility as a suitable target for migrastatics. The activity of ROCK and MRCK kinases results in increased phosphorylation of myosin light chain, which activates actomyosin contractility. Drugs targeting ROCK or MRCK are candidates for efficient migrastatics as they act to inhibit actomyosin contractility, which is necessary for of all cell invasion modes

responsible for proteolytical degradation of the extracellular matrix (ECM), which provides space for the group of cells to move forward [16, 17]. Unlike collective migration, single- cell migration does not require cell-cell adhesion. Cells invading as single cells adopt the protease- dependent mesenchymal invasion mode, protease-independent amoeboid mode, or various intermediate phenotypes. Mesenchymally migrating cells, such as fibroblasts, are characterized by elongated cell shape and numerous actin-based protrusions, and due to their proteolytical activity form tunnels through the ECM [18, 19]. Cells utilizing the amoeboid invasion mode enhance actomyosin contractility to dynamically deform their cell body, enabling them to push through pre-existing pores in the ECM. They are typically round with numerous membrane blebs that form due to high pressure inside the cells [20, 21]. A typical example of amoeboid migrating cells are leukocytes [22].

Apart from physiological significance of the migration modes, they are utilized by cancer cells during metastasis. In this manner, cancer cells have evolved to be "masters" of invasion, as they can utilize all the above-mentioned modes, including many transient phenotypes. More importantly, they are able to switch among the modes in order to most efficiently adapt to surrounding conditions [23–25]. For example, when proteolytical degradation of the ECM is inhibited, cancer cells can undergo the mesenchymal-amoeboid transition (MAT) [26]. Further, upregulation of cell contractility by activation of the Rho/ROCK pathway results in MAT [27, 28]. Opposingly, amoeboid-mesenchymal transition (AMT) occurs when contractility is blocked, for example by ROCK inhibitors [28, 29]. Notably, the ECM itself affects the invasion mode – in dense, rigid ECM mesenchymal migration is preferred, whereas loose ECM promote the amoeboid invasion phenotype [30]. Further, non-cancer cells, such as tumor associated macrophages or cancer-associated fibroblasts, contribute to invasion plasticity by modulating the microenvironment by production of pro-invasive cytokines (e.g. IL6, IL8, VEGF, LIF etc.). Moreover, CAFs realign collagen fibers of the tumor stroma ECM, which facilitates invasion of cancer cells [31].

The ability of cancer cells to adopt various invasion modes is referred to as invasion plasticity and is the main reason why conquering metastatic cancer is still a mere desire rather than close reality. The focus of cancer therapy has been mostly on inhibition of proliferation; however, it is local invasion and metastasis, rather than proliferation, which are the dominant features of solid cancer and must be eliminated in order to achieve effective cancer treatment [32, 33]. Thus, cytostatic therapy should be complemented with a strategy against cancer cell invasion. For such strategy to be successful, all modes of cancer cell invasion must be targeted. Recently, the term "migrastatics" (from Latin "migrare" and Greek "statikos") was proposed as a unifying name for drugs interfering with all modes of cancer cell invasion and consequently their ability to metastasize.

Due to the large variability of the invasion modes and even larger spectra of regulatory pathways, it is the common mechanism shared by all invasion modes which must be targeted. This requirement is fulfilled only by ultimate downstream effectors of cell migration, which are actin polymerization and actomyosin contractility. Unlike upstream regulators, which can be by-passed and compensated, there is no feasible way for a migrating cell to overcome inhibition of actin polymerization or contractility.

Migrastatic Drugs

Potential migrastatic candidates include drugs destabilizing actin cytoskeleton, such as cytochalasins, lantrunculins and Geodiamolide H. Further included are drugs stabilizing actin cytoskeleton, for example Jasplakinolide, chondramides and cucurbitacin E. Another class of candidates for migrastatic agents are those targeting contractility, such as TR-100, a tropomyosin inhibitor or blebbistatin, a non-muscle myosin II inhibitor. Many actomyosin targeting drugs are kinase inhibitors, for example ML-7 and ML-9 that inhibit MLCK or BDP5290, which inhibits MRCK. ROCK targeting inhibitors include fasudil, Y-27632, H-1152, Wf-536, RKI-1447 and RKI-18. DJ4 inhibits both ROCK and MRCK (Fig. 2). All the above mentioned have proven to decrease cancer cell invasion *in vitro* on model cancer cell lines.

Additionally, in case of TR-100, ML-7 and fasudil *in vivo* data exist. For a complete set of references see [33].

Importantly, ROCK/PKA/PKB multi-kinase inhibitors have been demonstrated to abolish both amoeboid and mesenchymal invasion both *in vitro* and *in vivo* [34]. CCT129254 was shown to reduce invasiveness and metastatic ability of melanoma cells *in vivo*. Another tested compound, AT13148, showed similar abilities, however with increased toxicity [34]. Notably, AT13148 is currently being tested in clinical development for oncological applications [35]. As such, ROCK/PKA/PKB multi-kinase inhibitors represent the most promising migrastatic candidates up to date (Fig. 2).

Apart from direct inhibition of cancer cell motility, migrastatic drugs could further decrease cancer cell invasion by affecting signaling within the tumor stroma and organization of the surrounding ECM. Various non-cancer cells that support tumor growth and invasion reside in the tumor stroma, such as tumor-associated macrophages, which produce cytokines known to promote invasive behavior of cancer cells, and also non-cancer cells, such as cancer-associated fibroblasts [36]. These signaling circuits are largely dependent on actomyosin contractility, which makes them possible targets of migrastatic drugs [37]. For example, direct macrophage contact with cancer cells activates RhoA signaling [38]. Vice versa, amoeboid cells with high myosin II activity recruit monocytes and stimulate their differentiation into TAMs [39]. Importantly, actomyosin contractility is essential for the activity of cancer-associated fibroblasts, which remodel the ECM by aligning collagen fibers. This increases ECM stiffness, which is known to facilitate cancer cell invasion [40]. Taken together, inhibition of Rho/ROCK signaling by migrastatics could disrupt the complex pro-malignant interactions within the tumor stroma and in reduce cancer cell dissemination.

As with all drugs, the main concern with migrastatics lies in possible toxicity [41]. After all, it was already mentioned here that the migration modes are employed also by non-cancer cells; and interfering with actin polymerization and contractility can be expected to affect both normal and cancer cells. Nevertheless, it can be anticipated, based on published examples, that concentrations targeting cancer cells, but not healthy cells, can be determined using suitable *in vivo* models. Moreover, the recent advancements in drug delivery to neoplastic tissues may disclose further potential of migrastatics when targeted directedly [42].

Migrastatic Therapy, Challenges and Advantages

It is important to stress that migrastatic therapy is not intended to replace the conventional anti- proliferative strategy. Rather, administering migrastatic agents should suitably complement cytotoxic therapy and increase its benefit. Due to aberrant cytokinesis caused by migrastatics, the tumor cells may be more responsive to DNA-modifying drugs, such alkylators or nucleoside analogs [43]. Also, it has already been shown that combination of actin-binding drugs with mitosis targeting

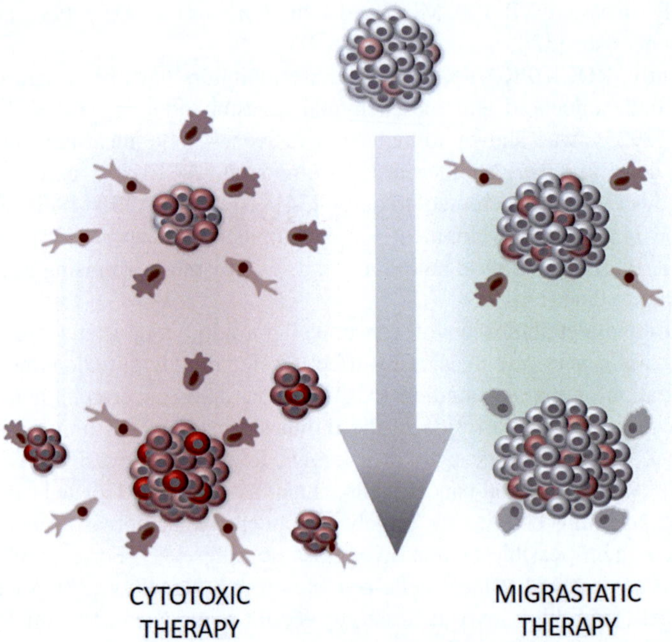

Fig. 3 Comparison of cytotoxic therapy and migrastatic therapy. Cytotoxic therapy reduces cell proliferation and thus tumor size. However, this is accompanied by accumulation of drug- resistant clones (pink and red) due to their proliferative advantage. Also, cell invasion is not primarily targeted, and metastases can be formed. On the other hand, migrastatic therapy is not expected to decrease tumor size, but impair cancer cell invasion and in result reduce secondary tumors. Moreover, drug-resistant clones do not gain proliferative advantages, decreasing the risk of gaining drug-resistant tumors. Altogether, combining anti-proliferative cytotoxic therapy and anti-invasive migrastatic therapy may offer novel therapeutic possibilities

microtubule drugs may result in reinforced inhibition of proliferation, as already documented in case of cytochalasin B and vincristine [44]. In fact, migrastatics themselves may have anti-proliferative effects. For example, inhibition of ROCK kinases decreases proliferation and loss of both ROCK isoforms blocks tumor formation in mice [45].

Moreover, during conventional cytotoxic treatment, tumor cells are prone to undergo Darwinian selection of drug-resistant clones, and these will inevitably overgrow the susceptible population [46] (Fig. 3). Additionally, this can result in treatment-induced metastasis [47]. However, it is anticipated that resistance to migrastatics will not provide cells with proliferative advantage, as in the case of acquired resistance to cytotoxic drugs. Overall, suitable combination or sequential administration of cytotoxic drugs and migrastatics may be especially effective in inhibiting cancer cell dissemination.

Administration of migrastatic drugs should be long term, which is associated with higher risk of adverse effects. The effective dose of migrastatics for metastasis prevention could be, however, lower than what is necessary to stop cell to migrate.

It was shown that the effectivity of cancer cell dissemination is very low – despite large number of cells escaping the primary tumor only very small number of cells successfully give rise secondary tumors [48]. The invasive phenotype of cancer cells is required any many steps of metastatic cascade including tissue invasion on primary site, extravasation, intravasation and tissue invasion on secondary site. Therefore, the effect of slowing down cancer cell migration on metastases formation is potentially multiplied by several folds.

In summary, migrastatic therapy provides a novel approach to anti-metastatic therapy. By targeting the basic processes of cell migration, migrastatics target and inhibit all modes of cell invasion. In combination with conventional treatment they allow synergistic impairment of tumor cells. Hopefully, migrastatics will provide a fundamental shift in the treatment of metastatic tumors.

References

1. Wedlich, D. 2006. *Cell migration in development and disease*. New Jersey: Wiley.
2. Biname, F., G. Pawlak, P. Roux, and U. Hibner. 2010. What makes cells move: Requirements and obstacles for spontaneous cell motility. *Molecular BioSystems* 6: 648–661.
3. Ridley, A.J., M.A. Schwartz, K. Burridge, R.A. Firtel, M.H. Ginsberg, G. Borisy, et al. 2003. Cell migration: Integrating signals from front to back. *Science* 302: 1704–1709.
4. Micuda, S., D. Rosel, A. Ryska, and J. Brabek. 2010. ROCK inhibitors as emerging therapeutic candidates for sarcomas. *Current Cancer Drug Targets (Bentham Science Publishers)* 10: 127–134.
5. Lazebnik, Y. 2010. What are the hallmarks of cancer? *Nature Reviews Cancer (Nature Publishing Group)* 10: 232–233.
6. Hanahan, D., and R.A. Weinberg. 2011. Hallmarks of cancer: The next generation. *Cell (Elsevier Inc.)* 144: 646–674.
7. Fife, C.M., J.A. McCarroll, and M. Kavallaris. 2014. Movers and shakers: Cell cytoskeleton in cancer metastasis. *British Journal of Pharmacology (England)* 171: 5507–5523.
8. Gandalovičová, A., T. Vomastek, D. Rosel, and J. Brábek. 2016. Cell polarity signaling in the plasticity of cancer cell invasiveness. *Oncotarget* 7 (18): 25022–25049.
9. Lauffenburger, D., and F. Horwitz. 1996. Cell migration: A physically integrated molecular process. *Cell* 84: 359–369.
10. Mitchison, T.J., and L.P. Cramer. 1996. Actin-based cell motility and cell locomotion. *Cell* 84: 371–379.
11. Spiering, D., and L. Hodgson. 2011. Dynamics of the Rho-family small GTPases in actin regulation and motility. *Cell Adhesion & Migration* 5: 170–180.
12. Amano, M., M. Ito, K. Kimura, Y. Fukata, K. Chihara, T. Nakano, et al. 1996. Phosphorylation and activation of myosin by Rho-associated kinase (Rho-kinase). *The Journal of Biological Chemistry* 271: 20246–20249.
13. Kimura, K., M. Ito, M. Amano, K. Chihara, Y. Fukata, M. Nakafuku, et al. 1996. Regulation of myosin phosphatase by Rho and Rho-associated kinase (Rho-kinase). *Science* 273: 245–248.
14. Wilkinson, S., H.F. Paterson, and C.J. Marshall. 2005. Cdc42-MRCK and Rho-ROCK signalling cooperate in myosin phosphorylation and cell invasion. *Nature Cell Biology* 7: 255–261.
15. Friedl, P., and D. Gilmour. 2009. Collective cell migration in morphogenesis, regeneration and cancer. *Nature Reviews. Molecular Cell Biology (Nature Publishing Group)* 10: 445–457.
16. Friedl, P., J. Locker, E. Sahai, and J.E. Segall. 2012. Classifying collective cancer cell invasion. *Nature Cell Biology* 14: 777–783.

17. Haeger, A., K. Wolf, M.M. Zegers, and P. Friedl. 2015. Collective cell migration: Guidance principles and hierarchies. *Trends in Cell Biology* 25: 556–566.
18. Friedl, P., and K. Wolf. 2008. Tube travel: The role of proteases in individual and collective cancer cell invasion. *Cancer Research* 68: 7247–7249.
19. Tolde, O., D. Rosel, R. Janostiak, P. Vesely, and J. Brabek. 2012. Dynamics and morphology of focal adhesions in complex 3D environment. *Folia Biologica (Praha)* 58: 177–184.
20. Sabeh, F., R. Shimizu-Hirota, and S.J. Weiss. 2009. Protease-dependent versus-independent cancer cell invasion programs: Three-dimensional amoeboid movement revisited. *The Journal of Cell Biology* 185: 11–19.
21. Charras, G., and E. Paluch. 2008. Blebs lead the way: How to migrate without lamellipodia. *Nature Reviews. Molecular Cell Biology* 9: 730–736.
22. Friedl, P., S. Borgmann, and E.B. Brocker. 2001. Amoeboid leukocyte crawling through extracellular matrix: Lessons from the Dictyostelium paradigm of cell movement. *Journal of Leukocyte Biology* 70: 491–509.
23. Yilmaz, M., and G. Christofori. 2010. Mechanisms of motility in metastasizing cells. *Molecular Cancer Research* 8: 629–642.
24. Friedl, P., and K. Wolf. 2010. Plasticity of cell migration: A multiscale tuning model. *The Journal of Cell Biology* 188: 11–19.
25. Panková, K., D. Rösel, M. Novotný, and J. Brábek. 2010. The molecular mechanisms of transition between mesenchymal and amoeboid invasiveness in tumor cells. *Cellular and Molecular Life Sciences* 67: 63–71.
26. Wolf, K., I. Mazo, H. Leung, K. Engelke, U.H. von Andrian, E.I. Deryugina, et al. 2003. Compensation mechanism in tumor cell migration: Mesenchymal-amoeboid transition after blocking of pericellular proteolysis. *The Journal of Cell Biology* 160: 267–277.
27. Rösel, D., J. Brábek, O. Tolde, C.T. Mierke, D.P. Zitterbart, C. Raupach, et al. 2008. Up-regulation of Rho/ROCK signaling in sarcoma cells drives invasion and increased generation of protrusive forces. *Molecular Cancer Research* 6: 1410–1420.
28. Sahai, E., and C.J. Marshall. 2003. Differing modes of tumour cell invasion have distinct requirements for Rho/ROCK signalling and extracellular proteolysis. *Nature Cell Biology* 5: 711–719.
29. Sanz-Moreno, V., G. Gadea, J. Ahn, H. Paterson, P. Marra, S. Pinner, et al. 2008. Rac activation and inactivation control plasticity of tumor cell movement. *Cell* 135: 510–523.
30. Van Goethem, E., R. Poincloux, F. Gauffre, I. Maridonneau-Parini, and V. Le Cabec. 2010. Matrix architecture dictates three-dimensional migration modes of human macrophages: Differential involvement of proteases and podosome-like structures. *Journal of Immunology* 184: 1049–1061.
31. Malik, R., P.I. Lelkes, and E. Cukierman. 2015. Biomechanical and biochemical remodeling of stromal extracellular matrix in cancer. *Trends in Biotechnology* 33: 230–236.
32. Steeg, P.S. 2016. Targeting metastasis. *Nature Reviews. Cancer* 16: 201–218.
33. andalovičová, A., D. Rosel, M. Fernandes, P. Veselý, P. Heneberg, V. Čermák, et al. 2017. Migrastatics—Anti-metastatic and anti-invasion drugs: Promises and challenges. *Trends in Cancer* 3: 391–406.
34. Sadok, A., A. McCarthy, J. Caldwell, I. Collins, M.D. Garrett, M. Yeo, et al. 2015. Rho kinase inhibitors block melanoma cell migration and inhibit metastasis. *Cancer Research* 75: 2272–2284.
35. Feng, Y., P.V. LoGrasso, O. Defert, and R. Li. 2016. Rho kinase (ROCK) inhibitors and their therapeutic potential. *Journal of Medicinal Chemistry* 59: 2269–2300.
36. Noy, R., and J.W. Pollard. 2016. Tumor-associated macrophages: From mechanisms to therapy. *Immunity* 41: 49–61.
37. Rodriguez-Hernandez, I., G. Cantelli, F. Bruce, and V. Sanz-Moreno. 2016. Rho, ROCK and actomyosin contractility in metastasis as drug targets. *F1000Research* 5: 783. F1000 Faculty Rev.

38. Roh-Johnson, M., J.J. Bravo-Cordero, A. Patsialou, V.P. Sharma, P. Guo, H. Liu, et al. 2014. Macrophage contact induces RhoA GTPase signaling to trigger tumor cell intravasation. *Oncogene* 33: 4203–4212.
39. Georgouli, M., C. Herraiz, E. Crosas-Molist, B. Fanshawe, O. Maiques, A. Perdrix, et al. 2019. Regional activation of myosin II in cancer cells drives tumor progression via a secretory cross-talk with the immune microenvironment. *Cell* 176: 757–774.e23.
40. Samuel, M.S., J.I. Lopez, E.J. McGhee, D.R. Croft, D. Strachan, P. Timpson, et al. 2011. Actomyosin- mediated cellular tension drives increased tissue stiffness and beta-catenin activation to induce epidermal hyperplasia and tumor growth. *Cancer Cell (United States)* 19: 776–791.
41. Gewirtz, D.A., M.L. Bristol, and J.C. Yalowich. 2010. Toxicity issues in cancer drug development. *Current Opinion in Investigational Drugs (England)*: 612–614.
42. Rosenblum, D., N. Joshi, W. Tao, J.M. Karp, and D. Peer. 2018. Progress and challenges towards targeted delivery of cancer therapeutics. *Nature Communications* 9: 1410.
43. Trendowski, M. 2014. Exploiting the cytoskeletal filaments of neoplastic cells to potentiate a novel therapeutic approach. *Biochimica et Biophysica Acta* 1846: 599–616.
44. Kolber, M.A., and P. Hill. 1992. Vincristine potentiates cytochalasin B-induced DNA fragmentation in vitro. *Cancer Chemotherapy and Pharmacology (Germany)* 30: 286–290.
45. Kumper, S., F.K. Mardakheh, A. McCarthy, M. Yeo, G.W. Stamp, A. Paul, et al. 2016. Rho-associated kinase (ROCK) function is essential for cell cycle progression, senescence and tumorigenesis. *eLife* 5: e12994.
46. Rosel, Daniel, Michael Fernandes, Victoria Sanz-Moreno, and Jan Brábek. 2019. Migrastatics: Redirecting R&D in solid Cancer towards metastasis? *Trends in Cancer* 5 (12): 755–756.
47. Ebos, J.M.L. 2015. Prodding the beast: Assessing the impact of treatment-induced metastasis. *Cancer Research* 75: 3427–3435.
48. Luzzi, K.J., I.C. MacDonald, E.E. Schmidt, N. Kerkvliet, V.L. Morris, A.F. Chambers, et al. 1998. Multistep nature of metastatic inefficiency: Dormancy of solitary cells after successful extravasation and limited survival of early micrometastases. *The American Journal of Pathology* 153: 865–873.

Critical Steps in Epithelial-Mesenchymal Transition as Target for Cancer Treatment

Evgeny V. Denisov, Mohit Kumar Jolly, Vitaly P. Shubin, Alexey S. Tsukanov, and Nadezhda V. Cherdyntseva

Introduction

Epithelial-mesenchymal transition (EMT) is a fundamental process that underlies many physiological and pathological conditions. In normal physiology, EMT is critically needed for the generation of new tissue types during embryogenesis and tissue regeneration during wound healing. In pathological states, particularly cancer, EMT results in the acquisition of a mobile mesenchymal phenotype in epithelial cancer cells. This phenotype allows cells to migrate, resist apoptosis and aging, and also release lytic enzymes to destroy the extracellular matrix (ECM) and the basement membrane [1–3]. A number of studies have shown that EMT is involved in the metastatic cascade including invasion, intra- and extravasation, and the establishment of micrometastasis as well as in resistance to radio-, chemo-, and immunotherapy [1, 4, 5].

Considering the crucial role of EMT in cancer, the modulation of this process is of great clinical interest. Substantial efforts have been devoted to the development of potent therapeutics that could inhibit EMT or cause its reversal called a mesenchymal-epithelial transition (MET) [6–9]. Many studies have been conducted to evaluate if targeting of the tumor microenvironment (TME) may suppress EMT [9–11]. Recent reports have raised the question that EMT can induce a hybrid epithelial/mesenchymal phenotype when cancer cells possess tumorigenicity [12, 13]. Nevertheless, EMT can be used as an instrument to transdifferentiate

E. V. Denisov (✉) · N. V. Cherdyntseva
Cancer Research Institute, Tomsk National Research Medical Center, Tomsk, Russia
e-mail: d_evgeniy@oncology.tomsk.ru

M. K. Jolly
Centre for BioSystems Science and Engineering, Indian Institute of Science, Bangalore, India

V. P. Shubin · A. S. Tsukanov
State Scientific Centre of Coloproctology, Moscow, Russia

M. Bizzarri (ed.), *Approaching Complex Diseases*, Human Perspectives in Health Sciences and Technology 2, https://doi.org/10.1007/978-3-030-32857-3_10

tumor cells to other cells [14]. Below we provide the detailed information about the molecular nature of EMT in cancer and potential approaches for targeting epithelial-mesenchymal plasticity.

EMT Signaling in Cancer

The initiation and completion of the EMT process specify a variety of molecular mechanisms and regulatory pathways at the transcriptional, post-transcriptional, translational and post-translational levels [15]. Among the large number of molecular mechanisms that contribute to EMT, transcription factors (EMT-TFs) play a major role. The EMT-TF family, leading to a decrease in E-cadherin expression, includes proteins: SNAI1 and SNAI2 (SLUG), TWIST1/2, ZEB1/2, and others [15]. MicroRNAs (miRNAs) play a significant role in the metastasis of tumors via the regulation of EMT and MET [16]. They directly affect EMT-TFs or signaling pathways acting as tumor suppressors or oncogenes [16]. The most significant miRNAs belong to the family of miR-200 (targets: ZEB and WNT/β-catenin), miR-34 (SNAI1, β-catenin, LRP6, LEF1, and AXIN2), miR-580 (TWIST1/2), etc. that control the plasticity between epithelial and mesenchymal states [17–19].

Several signaling pathways can trigger the EMT process [15]. However, the TGF-β, WNT, Notch, and Hedgehog pathways are one of the well-studied signaling pathways involved in EMT. The TGF-β pathway is involved in the regulation of cell growth, differentiation, adhesion, migration, and cell death [20, 21]. The term TGF-β covers a large family of secreted polypeptides with different functions [22] that is represented by three major isoforms: TGF-β1, TGF-β2, and TGF-β3. These proteins bind to transmembrane receptors, TGF-β type I and type II (TβRI and TβRII), which transmit signals further to SMAD proteins (SMADs 1, 2, 3, 5, and 8) [23, 24]. SMAD proteins form complexes with the "common-partner SMAD" (co-SMAD), which accumulates in the nucleus and regulates the transcription of various target genes assisted with a number of transcription factors and transcription modulators [20, 25]. In addition, TGF-β can activate SMAD-independent signaling pathways such as PI3K/AKT and MAPK [26, 27]. TGF-β plays an important role in several stages of cancer development: growth modulation at an early stage of cancer initiation, enhanced formation of ECM and fibrosis in the TME, promotion of EMT and/or metastasis, and immunosuppression [28]. Interestingly, TGF-β can act as a tumor suppressor and a tumor promoter depending on the type of cancer and the stage of its progression [29]. Various miRNAs targeting the TGF-β pathway have been identified in many types of cancers: miR-21, miR-17/92 cluster, miR-106b, miR-211, and miR-590 [28].

The role of the E-cadherin/β-catenin complex in the formation of cell junctions is well known [30]. At the same time, β-catenin is a key protein in the WNT signaling pathway. The main event in this pathway is the accumulation of β-catenin in the cytoplasm followed by its translocation and activation in the nucleus [30]. Further, β-catenin interacts with DNA-binding proteins TCF/LEF, Legless, and Pygopus [30, 31]. These interactions lead to the regulation of the expression of *MMP7*, *Myc*, *CCND1*, *CD44*, *TWIST*, and *SNAI* genes [32, 33]. If WNT signaling is not activated, β-catenin is phosphorylated by GSK-3β and CKI-α and degraded in proteasome [34]. The WNT pathway players are often mutated. Mutations in the *CTNNB1* (β-catenin) gene are common in many cancers [35], and hereditary mutations of the *APC* gene are found in familial adenomatous polyposis and colorectal cancer [36]. The inhibition of WNT/β-catenin pathway in EMT is caused by the following miRNAs: the miR-200 family, miR-429, miR-29c, etc., whereas the activation – miR-30, miR-181, miR-374a, etc. [17, 37–39].

Notch1 is a type I transmembrane receptor that activates transcription of target genes when interacting with Notch ligands [40]. The active form of Notch1 released by proteolysis from the full-length Notch1 is transferred to the cell nucleus and interacts with a DNA-binding protein to assemble a transcriptional complex that activates downstream target genes. The Notch signaling pathway plays an important role in cell differentiation, proliferation, apoptosis, adhesion, migration, angiogenesis, and oncogenesis and is a major inducer of EMT in a number of epithelial cancers [41, 42]. The high expression of Notch1 has been found to correlate to the progression of breast cancer [43], gastric cancer [44], colorectal cancer [45], glioblastoma [46], and squamous cell carcinoma of the oral cavity [47]. On the other hand, Notch1 acts as a tumor suppressor for prostate cancer [48], liver cancer [49], and pancreatic cancer [50]. There are several microRNAs targeting Notch/EMT pathway: miR-139-5p, miR-34a, miR-200 family, miR-9, miR-34c, and some others [51, 52].

Activation of the Hedgehog signaling pathway is implicated in carcinogenesis in different organs (lungs, ovaries, colon, etc.) and capable to trigger the EMT program [53, 54]. The Hedgehog pathway includes three proteins for signal transmission: Sonic Hedgehog (Shh), Indian Hedgehog (Ihh), and Desert Hedgehog (Dhh). These ligands transmit a signal by binding to Patched (PTCH) 12-transmembrane domain receptor that induces nuclear localization of the Gli family of transcription factors (GLI1, GLI2, and GLI3) via activation of Smoothened (Smo) protein [55]. Oncogenic activation of the Hedgehog pathway is usually caused by activating mutations in the *SMO* gene or inactivating mutations in the *PTCH1* gene [55, 56]. As other pathways, Hedgehog signaling is regulated by miRNAs: miR-378, miR-200, miR-431, etc. [57–59].

Thus, a huge number of mechanisms regulate EMT and lead to the appearance of migratory and invasive characteristics in cancer cells. Impact on the molecular targets of the EMT signaling might give us the ability to slow down, stop or reverse EMT and accordingly prevent cancer progression.

Strategies for Targeting EMT

Suppression of EMT

As mentioned above, EMT is triggered by various EMT-TFs the expression of which is induced by different signaling pathways [15]. Many of the EMT-TFs act synergistically and use common pathways. Nevertheless, several studies show that inhibition of a single transcription factor is sufficient to suppress EMT [15]. EMT can be inhibited in two ways: direct and indirect [60]. The direct way includes the inactivation of the EMT signaling pathways mainly via targeting upstream (ligands, receptors, adaptor proteins, kinases, etc.) and downstream (transcription factors) components. The indirect suppression of EMT is aimed at the modification of the activity of non-coding RNAs (ncRNAs) and other molecular units and proteins that modulate EMT.

Inhibition of Upstream EMT Pathway Components TGF-β shows a predominant role in the induction of EMT under both physiological and pathological conditions [61]. The TGF-β pathway can be modulated using three main approaches: (1) suppression of TGF-β production by different compounds; (2) blocking ligand-receptor interaction by TGF-β-specific antibodies or soluble TGF-β decoy receptors (traps); (3) inhibition of the kinase activity of TGF-β receptors by small-molecule inhibitors or aptamers that bind the downstream SMAD signaling proteins [62]. For example, TGF-β2-specific antisense oligodeoxynucleotides AP 12009 were effective in the treatment of patients with recurrent or refractory malignant (high-grade) glioma [63]. Fresolimumab, monoclonal antibodies against TGF-β, showed antitumor activity in melanoma and renal cell carcinoma and is under clinical trials for other cancers [64]. TGF-β type I receptor kinase (ALK5) inhibitors suppressed EMT and decreased breast cancer metastasis [65–68]. Tranilast, an antiallergic and antifibrotic agent, inhibited phospho-SMAD2 generation, blocked TGF-β signaling and EMT and reduced the growth (>50%) and metastasis (>90%) of breast carcinoma in mice [69]. Metallofullerenol nanoparticles were shown to lead to EMT blocking via inhibition of TGF-β signaling, elimination of cancer stem cells (CSCs), and abrogation of breast tumor initiation and metastasis (Table 1) [70]. Plant-derived agents, such as nobiletin, oridonin, and resveratrol, were effective in suppression of EMT via TGF-β signaling and decrease in metastasis in different cancers (Table 2) [71–73]. Other chemical agents that target various components of the TGF-β pathway including drugs under clinical trials are comprehensively described in the recent reviews [8, 62].

Multiple studies show that targeting WNT/β-catenin, Notch, and Hedgehog pathways represent a promising therapeutic strategy in inhibition of epithelial-mesenchymal plasticity and cancer progression. Moreover, targeting the crosstalk network of these pathways could potentially increase the efficacy of anti-cancer therapy [89]. A number of chemically-synthesized inhibitors of WNT, Notch, and Hedgehog signaling have been developed and are under clinical trials

Table 1 Chemically synthesized compounds that are effective in EMT inhibition and metastasis suppression via modulation of EMT pathways and transcription factors

Compound	Description	Target (pathway)	Cancer	References
Ki26894 EW-7195, EW-7197 IN-1130	TGF-β type I receptor kinase (ALK5) inhibitors breast	TGF-β/SMAD	Breast	[65] [66, 67] [68]
Tranilast	Antiallergic and antifibrotic agent	TGF-β/SMAD		[69]
Fresolimumab	TGF-β inhibitor	TGF-β/SMAD	Breast, lung, renal, melanoma	[64]
Gd@C82(OH) 22	Gd-metallofullerenol nanoparticles	TGF-β/SMAD	Breast	[70]
Ormeloxifene	Estrogen receptor modulator	WNT/β-catenin	Prostate	[74]
LGK974 Vantictumab	Porcupine (PORCN) inhibitor Frizzled-7 (FZD7) inhibitor	WNT/β-catenin	Colorectal, head and neck Lung, pancreatic, breast	[75]
PF-03084014 MK-0752	γ-Secretase inhibitor	Notch	Liver Breast	[76] [77]
GDC-0449 Vismodegib	Antagonist of G protein coupled receptor Smoothened (SMO) inhibitor	Hedgehog	Lung Basal cell	[78] [79]
HS-173	PI3K inhibitor	PI3K/AKT/mTOR	Pancreatic	[80]
Bosutinib Selumetinib	Src inhibitor MEK inhibitor	MAPK	Thyroid Breast	[81] [82]
BI 5700	IκB kinase 2 inhibitor	NF-κB	Breast, colon	[83]
Palbociclib	CDK4/6 inhibitor	c-Jun/COX-2	Breast	[84]
ECO/siβ3	Lipid ECO-based nanoparticles with β3 integrin siRNA	β3 integrin	Breast	[85]
Nimotuzumab	Humanized recombinant IgG1	SNAI1, vimentin, MMP9	Cholangiocarcinoma	[86]
Tranylcypromine	LSD1 inhibitor	SNAI2 (SLUG)	Breast	[87]
Panobinostat	Pan-deacetylase inhibitor	ZEB1/2	Breast	[88]

[8, 53]. However, it is unknown whether the anti-cancer effect of many of these inhibitors is related to EMT blocking because these pathways are also implicated in the control of other cellular processes. Nevertheless, several pharmacological agents

Table 2 Natural compounds that are effective in metastasis suppression via EMT inhibition

Compound	Source	Target (pathway)	Cancer	References
Arenobufagin	*Bufo arenarum*	β-catenin	Prostate	[90]
Emodin	Rhubarb, aloes, and others	β-catenin, AKT	Head and neck	[91]
		ILK/AKT/mTOR, AKT/GSK-3β/β-catenin	Breast	[92, 93]
		miR-1271	Pancreatic	[94]
Ginsenoside	Ginseng	EGFR, MAPK, NF-κB	Lung	[95]
Rg3		HIF-1α	Ovarian	[96]
Fisetin	Many fruits and vegetables	PTEN/AKT/GSK-3β	Breast	[97]
Nobiletin	*Citrus depressa*	TGF-β/SMAD3	Lung	[71]
Oleanolic acid	Various foods and plants	Nitric oxide synthase	Liver	[98]
Oridonin	*Rabdosia rubescens*	PI3K/AKT/GSK-3β	Melanoma	[99]
		TGF-β/SMAD2/3	Osteosarcoma	[72]
		WNT/β-catenin	Pancreatic	[100]
Phloroglucinol	Brown algae	PI3K/AKT, RAS/RAF-1/ERK	Breast	[101]
Quercetin	Widely distributed bioflavonoid	SNAI1/2	Lung	[102]
Resveratrol	Red grape	NF-κB	Melanoma	[103]
		TGF-β/SMAD	Colorectal	[73]
Withaferin A	*Withania somnifera*	Vimentin	Breast, lung	[104–107]

including natural compounds (Table 2) and targeted therapeutics have been shown to suppress cancer progression via EMT inhibition. Ormeloxifene, a clinically approved selective estrogen receptor modulator, has been found to inhibit growth and metastatic potential of prostate cancer cells in a xenograft mouse model via suppression of the WNT/β-catenin pathway [74]. Notch inhibitor (γ-secretase inhibitor), PF-03084014, blocked lung metastasis of hepatocellular carcinoma in a patient-derived xenograft via suppression of Notch1 activity, decrease in STAT3 activation and phosphorylation of the Akt signaling pathway, and reduction of EMT [76]. The small interfering RNA (siRNA)-mediated or pharmacological inhibition of Hedgehog signaling by GDC-0449 inhibitor abrogated resistance of lung cancer cells to erlotinib and cisplatin through modulation of EMT-regulating miRNAs [78]. EMT inhibition and a decrease in cancer metastasis were also found when the activity of PI3K/AKT/mTOR, MAPK, NF-κB, c-Jun/COX-2, and integrin pathways was suppressed using natural compounds and newly synthesized agents (Table 1).

Inhibition of EMT-TFs Suppression of the activity of EMT-TFs is another approach to inhibit EMT. *In vitro* studies have revealed that targeting EMT-TFs could significantly inhibit tumor growth, retard tumor metastasis, and reverse drug resistance. Several pharmacological agents showed high efficacy to suppress SNAI-induced EMT [108]. A humanized recombinant IgG1, Nimotuzumab, that has been approved for treatment of different cancers inhibited cholangiocarcinoma cell metastasis via suppression of SNAI, vimentin and MMP9 expression [86]. Cobalt (III)-DNA conjugates were found to inhibit SNAI and invasive characteristics of breast cancer cells [109, 110]. Different natural compounds were capable to suppress EMT, migration, and invasion of cancer cells through targeting SNAI as well as SLUG and LEF-1 [102, 111]. Moreover, quercetin, a widely distributed bioflavonoid, showed significant activity in inhibition of lung cancer metastasis through downregulating SNAI-dependent Akt activation and SNAI-independent ADAM9 expression pathways in a xenograft model (Table 2) [102]. Nevertheless, many more SNAI-inhibiting strategies need to be developed especially focused on the degradation of SNAI and prevention of its transport to the nucleus [108].

Knockdown of SLUG via RNA interference significantly suppressed lung and colorectal cancer cell invasion and metastasis [112, 113]. SLUG silencing or the treatment with tranylcypromine, an inhibitor of LSD1 that blocks its interaction with SLUG, markedly inhibited migration and invasion of triple-negative breast cancer cells and their metastatic spread in mice (Table 1) [87]. Nucleolin aptamer-siRNA chimeras were able to significantly knock down the expression of SLUG and neuropilin 1 (NRP1) in lung cancer cell lines and suppress tumor growth, invasiveness, circulating tumor cell amount, and angiogenesis in a xenograft model without affecting liver and kidney functions [114].

The inactivation of TWIST by siRNAs or chemotherapeutic approaches has proved successful, and several antagonists of TWIST signaling have been identified [115]. TWIST1 suppression by shRNA lentiviral constructs resulted in inhibition of tumorigenicity and invasion in glioma cells [116, 117]. TWIST knockout diminished the expression of other EMT-TFs (SNAI, SLUG, and ZEB2) in breast tumors. Mice with TWIST knockout tumors did not have circulating tumor cells and developed very little lung metastasis [118]. Knockdown of TWIST also suppressed the EMT phenotype, inhibited the migratory ability and increased the chemosensitivity of liver cancer cells [119]. The harmala alkaloid, harmine, inhibited multiple TWIST1 functions, including single-cell dissemination and proliferation of lung cancer cells *in vitro* and suppressed tumor growth in a xenograft model without toxicity [120]. Genetic silencing of *TWIST1* or treatment with harmine resulted in growth inhibition and apoptosis in *EGFR*-mutant lung cancer cells and increased sensitivity to EGFR tyrosine kinase inhibitors [121].

Targeting ZEB1/2 may block EMT, invasion, and metastasis, decrease drug resistance, increase apoptotic potential, and promote tumor immunity [122, 123]. Inhibition of ZEB1 was able to abrogate the HIF-1α-induced EMT and colorectal cell invasion [124]. Depletion of ZEB1 decreased invasion of ovarian cancer cells and made them more sensitive to paclitaxel [125]. Panobinostat, pan-deacetylase inhibitor, demonstrated high efficiency in the suppression of triple-negative breast cancer metastasis via inhibition of ZEB1/2 (Table 1) [88]. The HDAC class I-specific inhibitor, mocetinostat, reduced activity of ZEB1 and its targets and sensitized pancreatic cancer cells to chemotherapy

Table 3 Non-coding RNAs involved in EMT activation or inhibition

ncRNAs	Targets, functions	EMT	References
lncRNAs			
ATB	ZEB1 upregulation	Activation	[131]
Dreh	Vimentin downregulation	Inhibition	[132]
HOTAIR	SNAI upregulation	Activation	[133]
Hh	TWIST, hedgehog pathway activation	Activation	[134]
H19	ZEB1/2, binds with miRNA200a	Activation	[135]
AOC4P	Vimentin downregulation	Inhibition	[136]
miRNAs			
34a	Suppression of Notch1, SNAI1, and SMAD	Inhibition	[137, 138]
148a	Suppression of β-catenin, cyclin D1, and MMP9	Inhibition	[139]
200-family	Suppression of ZEB1 and ZEB2	Inhibition	[140]
206	NRP1, SMAD2, TGF-β inhibition	Inhibition	[141]
137	SNAI1 suppression	Inhibition	[142]
205	ASPP2 suppression	Inhibition	[143]
124	ZEB2 suppression	Inhibition	[144]
655	ZEB1 suppression	Inhibition	[145]

[126]. Naringin, a citrus bioflavonoid, inhibited proliferation and invasion and induced apoptosis in human osteosarcoma cells by downregulating ZEB1 [127].

EMT Inhibition via Targeting ncRNAs Non-coding RNAs (ncRNAs) represent a highly diverse group of molecules that are not translated to proteins but show various regulatory functions. NcRNAs play an important role in the regulation of gene expression at the level of transcription, processing, and translation. There are several types of ncRNAs: long ncRNAs (lncRNAs), miRNAs, small interfering RNAs (siRNAs), and piwi-interacting RNAs (piRNAs) [128]. An increasing number of studies show that ncRNAs are involved in the regulation (activation/suppression) of EMT (Table 3) in embryonic development, cellular homeostasis, and pathological conditions, especially in malignant diseases [129, 130].

Targeting lncRNAs is a promising strategy for suppression of EMT, cancer cell invasion, and metastasis. Several studies showed that lncRNA MALAT1 is a critical regulator of the metastatic phenotype particularly via EMT induction in different cancers [146–148]. Antisense oligonucleotides blocking MALAT1 expression have been found to prevent lung cancer metastasis in a mouse xenograft model [148]. Another lncRNA, SNHG15, promotes breast and colon cancer progression by sponging tumor suppressor miR-211-3p and stabilizing transcription factor SLUG, respectively [149, 150]. SNHG15 knockdown by siRNAs impaired breast cancer cell invasion and metastasis to lungs [149]. The inhibition of EMT and suppression of cancer cell migration and invasion were also found when other lncRNAs such as SPRY4-IT1 and lnc-ATB were downregulated [151–153].

Reconstitution of miRNAs that block EMT is another attractive approach to attenuate cancer progression. MiR-875-5p was shown to downregulated in prostate cancer [154] and works as a tumor suppressor in colorectal carcinoma [155]. MiR-875-5p re-expression using miRNA mimics counteracted EMT in prostate cancer cell lines through downregulation of ZEB1 [154] and inhibited invasiveness in colorectal cancer cells [155] through underexpression of MMP-7 and -9 and in hepatocellular carcinoma cells via deficiency of AEG-1 expression [156]. The most promising results have been achieved with reconstitution of miR-34a and miR-770 expression in nasopharyngeal carcinoma [138] and triple-negative breast cancer cells [157], respectively. In particular, their overexpression not only inhibited EMT and invasion but also suppressed metastasis *in vivo* [138, 157]. The similar effects on cancer cell invasion and metastasis were obtained in restoring expression of other miRNAs [158–160].

However, EMT inhibition by targeting ncRNAs is challenged due to several reasons: the necessity of considering the balance between miRNAs and lncRNAs, inefficient systemic transport of RNA-based therapeutics to the target cells, and host resistance by serum nucleases and innate immune system [161, 162]. Despite these challenges, the first clinical trials are being conducted [162]. In addition, some of these problems can be overcome by using natural compounds that target ncRNAs. Emodin, an anthraquinone, isolated from rhizomes of Rhubarb, aloes, and other plants, was found to inhibit EMT, invasion, and metastasis of pancreatic cancer cells by up-regulating miR-1271 (Table 2) [94]. Corylin, a flavonoid extracted from the nuts of *Psoralea corylifolia* L. (Fabaceae), inhibited EMT, migration, and invasion of liver cancer cells of via downregulation of tumor suppressor lncRNA GAS5 [163]. Pharmacological agents such as artesunate and propofol were also shown to modulate expression of ncRNAs and suppress EMT in liver, colorectal and lung cancer cells [164–167].

EMT Suppression Through Targeting Other Regulatory Proteins Many other proteins are involved in the program of EMT. Most of them play a secondary role in EMT or their expression is altered when this process was already initiated. However, downregulation or knockdown of such proteins was turned out to have a significant effect on EMT. Inhibition of Src, a central mediator in multiple cancer signaling pathways, prevented invasion and lung metastasis of thyroid cancer cells in a mouse model. This effect was mediated by downregulation of MAPK signaling pathways and inhibition of EMT [81]. Inhibition of Brd4, a mammalian bromodomain protein that binds to acetylated chromatin, changed the expression of several EMT proteins and suppressed colorectal cancer metastasis to the liver in an animal model [168]. The significant decrease in metastasis via EMT blocking was found with pharmacological inhibitors of ALK5, MEK, and CDK4/6 proteins (Table 1).

Targeting exosomes can be another potential approach for blocking of the EMT program. Lung cancer cells were found to produce exosomes with ZEB1 mRNA that can be transferred to other tumor cells and activate EMT [169, 170]. Such EMT-inducing exosomes were turned out to be secreted by only mesenchymal tumor cells. The inactivation of exosomes with EMT mediators is probably a critical point and can be taken into account for developing anti-EMT therapy [170]. Several studies have demonstrated that SDF-1/CXCR4 plays a significant role in cell migration and metastases via induction of EMT [171, 172]. The suppression of CXCR4 by CRISPR/Cas9 led to a decrease in mesenchymal gene expression, an increase in epithelial markers, and suppression of invasiveness in liver cancer cells. Moreover, CXCR4 knockdown increased sensitivity to cisplatin [173].

Epithelial-mesenchymal plasticity can be suppressed by natural compounds that target EMT-related proteins. Oleanolic acid that is widely spread in food, medicinal herbs and other plants both alone and in combination with regorafenib (angiogenic agent) inhibited EMT in liver cancer cells via induction of nitric oxide synthase (iNOS) and suppression of EMT-related proteins and attenuated invasion and lung metastasis [98]. Withaferin A, the most abundant component in the root of *Withania somnifera* (Ashwagandha), directly binds vimentin and inhibits its assembly [174]. Several studies showed that withaferin A suppresses EMT in breast and lung cancer cell lines and prevents metastasis in animal models (Table 2) [104–107]. The promising results were reported with inhibitors of ONECUT2, newly discovered transcription factor, and restoring of the tumor-suppressor *Nkx2.8* gene expression in blocking EMT and reducing ovarian and urothelial carcinoma cell invasion, respectively [175, 176].

Thus, many studies show that EMT can be successfully inhibited using different natural plant- and animal-derived compounds and chemically-synthesized agents. However, most of them lack the information regarding the effect of EMT suppression on the cancer growth, invasion, and most importantly metastasis. Only a small number of studies demonstrated a significant inhibitory effect on cancer progression through EMT blocking (Tables 1 and 2).

EMT Reversal (MET)

The conversion of mesenchymal tumor cells to epithelial tumor cells through MET represents another approach for countering EMT-associated invasiveness, metastasis, and therapeutic resistance. Most of the chemical compounds that are mentioned above as suppressors of the EMT program can be considered as drugs restoring epithelial characteristics in tumor cells. Indeed, many studies that report EMT inhibition using various strategies have declared not only downregulation of EMT-TFs, but also upregulation of epithelial adhesive molecules (E-cadherin, EpCAM, etc.). However, it is not understood whether the restoration of the expression of the classic epithelial markers may indicate the complete reversal of EMT and the acquisition of the epithelial phenotype. Regaining epithelial expression can be an indicator of partially reversed EMT as shown previously with the withdrawal of TGF-α/EGF, knockdown of SNAI, and re-expression of several miRNAs in breast cancer cells [177–180]. The epithelial phenotype is characterized by intercellular adhesion and cohesive interactions, apical-basal cell polarity and distribution of the organelles and cytoskeleton components, lack of mobility [181]. Tumor epithelial cells also show a high clonogenicity and capacity to grow [182, 183]. Ideally, the simultaneous existence of all these traits can indicate that tumor cells underwent a complete MET and demonstrate the differentiated epithelial phenotype. However, it is not always feasible in experimental practice. Nevertheless, several studies have reported a complete reversal of EMT in various cell lines using chemical inhibitors and genetic approaches.

The complete reversal of EMT has been demonstrated to require re-establishing both epithelial gene transcription and cell structural components. The combination of the inhibitors of TGF-β type I receptor kinase and ROCK protein resulted in the elimination of mesenchymal gene expression and F-actin stress fibers and restoring epithelial cadherin activity in renal tubular epithelial cells. The similar effect was obtained when the ROCK inhibitor was combined with ZEB1/2 knockdown [184]. Interestingly, individual inhibition of signaling proteins (p38 MAPK, MEK1, JNK, and ROCK) that are implicated in the TGF-β-induced EMT was not able to reverse the actin stress fiber morphology and induce a complete MET [184]. Similarly, another study found that ZEB1 and SYDE1 involved in cytoskeletal remodeling are functionally relevant in EMT reversal. Ovarian cancer cells, in which SYDE1 and ZEB1 were downregulated by siRNAs or tyrosine kinase inhibitor, nintedanib, showed induction of E-cadherin expression and colony compaction. Moreover, SYDE1-silenced cells demonstrated increased anoikis that is known to be suppressed in EMT [183]. The fact that EMT can be completely reversed through downregulation of EMT-inducing proteins and cytoskeleton remodeling has been demonstrated in the treatment of breast cancer cells with inositol, a cyclohexane polyol [185].

Other studies showed that targeting annexins can reverse EMT. Annexins, Ca^{2+}/phospholipid-binding proteins, play an important role in cell cycle, exocytosis, and apoptosis. Annexin A1 re-expression but not blockade of TGF-β pathway was found to reverse completely EMT in immortalized tumorigenic mammary cells and abolish metastasis. The complete reversal of Ras/TGFb-induced EMT was also induced by the antibiotic salinomycin [186]. Knockdown of annexin 2 reversed the EMT phenotype and gefitinib resistance of lung cancer cells induced by cancer-associated fibroblasts (CAFs). The combination of c-met and IGF-1R inhibitors suppressed annexin 2 and reduced CAFs-induced EMT and chemoresistance [187].

A partial EMT is suggested to be similar to stemness and provides high plasticity for cancer cells through the generation of hybrid (epithelial/mesenchymal, E/M) cell phenotypes [12, 188]. The recent study showed that the expression of telomerase reverse transcriptase (hTERT), the core component of telomerase, and the mesenchymal phenotype of CSCs are mutually exclusive. The loss of the mesenchymal state represses TERT expression. Knockdown of hTERT results in the MET in CSCs and loss of aggressive properties such as chemoresistance and tumorsphere formation [189]. These findings suggest that hTERT may be a potential target to induce EMT reversal in tumor cells namely CSCs and to suppress EMT/stemness-related cancer aggressiveness and progression.

Despite these encouraging results, therapy focused on a complete reversal of EMT may have undesirable effects by promoting the growth of micrometastases.

Because it is well known that MET is critically needed for the metastatic colonization of distant organs, namely the development of solid metastases from micrometastases [1, 4].

Modulation of Tumor Microenvironment

As cancer is considered to be successfully reprogrammed by modifying its dynamical relationship with microenvironment, the cell-stroma interactions are identified as targets for pharmacological intervention. This approach bears huge implications from both a fundamental and clinical perspective because it may provide a novel anticancer strategy focused on mimicking or activating the tumor reversion pathway [9, 190, 191].

Increasing evidence shows that TME plays a key role in the development of different tumors, including initiation, promotion, and metastatic spreading [192–196]. The main "players" of TME are presented by endothelial, inflammatory and immune cells, fibroblasts, ECM components and soluble factors located near the tumor cells [10]. Tumor-infiltrating inflammatory cells are mobilized and recruited by tumor-derived factors, as well as by mediators secreted by various types of host cellular components which both contribute to the TME. Tumor-associated macrophages (TAMs) are derived from monocytic precursors and show various functional programs providing either antitumor or tumor supporting capacity [197, 198]. Principally there are two main subpopulations of macrophages in the TME: classically activated M1 type exerting antitumor activity and M2 alternatively activated macrophages with protumor role [197, 198]. Protumor M2-TAMs facilitate tumor initiation, progression, and metastasis through the secretion of proteolytic molecules to promote ECM remodeling and cytokines and growth factors to stimulate tumor cell proliferation, migration, and invasion. In addition, M2-TAMs interfere with the antitumor functions of other immune cells providing immunosuppressive effect and tumor promotion [191, 193, 198]. However, in fact, there are many more polarization statuses of macrophages that can acquire distinct phenotypes to support tumor cell proliferation, neoangiogenesis, immune suppression or metastasis [195, 196, 199, 200]. CAFs, a substantial component of tumor stroma, play an important role in supporting the proliferative and invasive behavior of cancer cells through cell-cell interaction or extracellular signaling molecules.

Tumor cell ability to metastasis via EMT is well known to be initiated by specific signals from microenvironment suggesting that modulation of TME is a successful strategy in anti-cancer therapy [4, 201]. ECM structural proteins were found to induce EMT via activation of EMT-TFs. Collagen-I binds its receptor DDR2 resulting in activation SRC/ERK2 signaling and stabilization of SNAI1 in breast cancer cells that in turn upregulates MT1-MMP and collagen-I and promotes invasion [202]. Fibronectin also exhibits the ability to induce SNAI1 expression in tumor cells partly through binding to integrin receptors [203]. Increased collagen expression has been found to be associated with the highest propensity to develop distant metastases in triple-negative breast cancer patients having evidence of central fibrosis. Pirfenidone, an anti-fibrotic agent as well as a TGF-β antagonist, has inhibitory effects on cell viability of CAFs, collagen production and TGF-β signaling and suppresses tumor growth and lung metastasis synergistically in combination with doxorubicin [204].

The release of interleukins by immune cells, endothelial cells and fibroblasts contribute to EMT. IL-6 promotes EMT in head and neck cancer cells and correlates with increased TWIST1 and SNAI1 expression [205]. IL-6-TWIST1 positive feedback loop induces EMT in breast cancer cells [206]. CAFs facilitate tumor cells to undergo EMT through secretion of cytokines such as IL-6 and TCF21 [5, 6]. TGF-β, the best-characterized EMT inducer, is abundantly secreted not only by tumor cells but also by TAMs, CAFs, and platelets. M2-TAM-derived TGF-β1 results in the enhancement of the stemness and migration abilities of glioma cells via the activation of SMAD2/3 pathway and upregulation of the expression of SOX4 and SOX2 [193]. Overall, T helper 2-type inflammation in the TME has been suggested to favor EMT, invasion, and metastasis [201].

Induction of T helper 1-specific immune response and inhibition of Th2 responses as well as anti-inflammatory treatment to suppress inflammation in the TME is believed to be an effective strategy for suppression of cancer progression [201]. In a murine tumor, ablation of TAMs using legumain (asparaginyl endopeptidase)-activated prodrugs resulted in tumor growth and metastasis inhibition accompanied by a decrease in angiogenesis, release of circulating tumor cells and myeloid immune suppressor Gr-1+/CD11b + cells [207]. Photoimmunotherapy by TAM-targeted probe that represents conjugate of a monoclonal anti-CD206 antibody with a near-infrared phthalocyanine dye was effective to suppress the growth of sorafenib-resistant breast tumors and to inhibit lung metastasis [208]. Macrophages are the main cells that provide apoptotic cell clearance by phagocytosis to maintain tissue homeostasis. Treatment of macrophages with UV-irradiated apoptotic cancer cells was found to produce PTEN-containing exosomes that are taken up by recipient cancer cells and inhibit cancer progression and lung metastasis via suppression of SNAI1/2, ZEB1/2, and TWIST1 activity, EMT, migration, and invasion. A single injection of apoptotic cancer cells inhibited lung metastasis in syngeneic immunocompetent mice with enhanced PPARγ/PTEN signaling both in TAMs and in tumor cells [209]. Exosomal regulation of EMT was also observed after photodynamic therapy of head and neck squamous cell carcinoma patients. It turned out that in co-incubation with cancer cells, exosomes obtained on a week or more after treatment restored the epithelial phenotype and inhibited proliferation, migration, and invasion [210].

TME is also a factor that mediates EMT-driven drug resistance. The involvement of TAMs in the tumor response to cancer therapy is intensively studied. Many preclinical studies have revealed that the response of tumors to radiotherapy and chemotherapy can be improved by depleting TAMs from tumors or by prohibiting their polarization to an M2 phenotype. Chemotherapy is known to induce the accumulation of TAMs preferentially polarized into M2 type with high expression of the angiopoietin receptor, Tie2, [211] that is driven both by tumor hypoxia [212] and the production of CSF-1 and its ligand IL-34 [213, 214]. Blockade of the CSF-1/CSF-1R axis was found to lead to the selective killing of the M2 macrophages and antitumor immunity activation [215]. In preclinical studies, the suppression of CXCL12/CXCR4 and CCL2/CCR2 pathways individually which promote macrophage infiltration into tumors after therapy demonstrates significant improvement in the response to anti-cancer therapy. However, the abrogation of the CCL2-CCR2 pathway has not been successful in clinical trials. Further studies are required to highlight the clinical significance of TAM targeting combined with radiation or chemotherapy using inhibitors of the SDF-1/CXCR4 or CSF-1/CSF-1R pathways [216].

Hypoxia is another important feature of TME that promotes cancer cells to undergo EMT and become resistant to chemotherapy. Specifically, hypoxia can promote EMT via hypoxia-inducible factor-1α (HIF-1α) [217]. HIF-1α was found to increase SNAI1 protein stability leading to suppression of E-cadherin in ovarian carcinoma [218]. HIF-1α also induces TWIST1 expression by binding directly to the TWIST1 promoter [219]. In addition, HIF-1α cooperates with inflammatory cytokines to promote EMT. For example, HIF-1α together with TGF-β promotes SNAI1 nuclear translocation to induce EMT through the suppression of estrogen receptor β [220]. Also, HIF-1α enhances the expression of TWIST1 by up-regulating TNFα, IL-6, and TGF-β [221]. Hypoxia together with the WNT/β-catenin signaling can promote SNAI1 stability by inhibiting GSK-3β [222]. Activation of HIF-1α under hypoxic condition promotes EMT in hepatocellular carcinoma cells and induces drug resistance by increasing the expression of the *ABCB1* gene encoding the MDR1 protein [223]. Knock-down of HIF-1α reverses the EMT phenotype and abolishes the drug-resistant phenotype of liver cancer cells under hypoxia [6]. Taken together, HIF-1α represents a promising strategy to target hypoxia signaling in cancer.

Different strategies to target hypoxic cancer cells and/or HIFs include hypoxia-activated prodrugs and inhibition of HIF mRNA or protein expression, dimerization, DNA binding capacity and transcriptional activity [10, 11]. Several drugs that target hypoxia have been developed and are under clinical trials [224]. In contrast to standard targeted drugs, which are able to induce the chemoresistance in most treated tumors, targeting the hypoxic phenotype is considered to be a more general approach to eradicate malignant cells. Indeed, HIF inhibitors are likely to target multiple important carcinogenetic processes. Importantly, dysregulated HIF-2 is common in various tumor types regardless of their genetic and molecular diversity, and HIF-2 inhibition could potentially overcome drug resistance. A recent achievement in clinical immunotherapy and evidence into how HIFs regulate the tumor immune response suggests that combined immunotherapy and HIF inhibition is likely to be a powerful therapeutic approach [11].

Recently, it has been found that EMT correlates with expression of the immunosuppressive immune checkpoint molecules, PD-L1, PD-1, CTLA-4, etc., across a range of tumor types. Immune checkpoint inhibitors and/or EMT inhibition/reversion are being considered to overcome immunosuppression [225]. However, further studies are required to explore the presence of EMT as a marker of efficacy for immune checkpoint therapy.

Thus, tumor microenvironment greatly contributes to drug resistance and cancer progression through triggering EMT mainly in response to metabolic stress and antitumor treatment. Discovery of the intrinsic patterns of tumor-microenvironment cross-talk and mechanistic understanding of EMT-related progression and treatment resistance could identify the range of targetable processes or signaling pathways to achieve a better patient outcome.

Stimulating EMT Process Toward the Terminal Mesenchymal Stage

Recently accumulated evidence has suggested that EMT is not an "all-or-none" or binary process, instead, cells can stably acquire one or more hybrid (E/M) phenotypes that can combine various epithelial and mesenchymal traits such as adhesion and migration [3]. EMT is a high-dimensional and nonlinear process, involving changes in multiple interconnected properties such as molecular markers (cadherin switch), cell-cell adhesion, basement membrane remodeling, cell individualization and migration, apicobasal polarity, etc. [226] Thus, different EMT-inducing factors may be pushing different cells to different extents on these different axes, and hence EMT program driven by two different inducers may overlap to varying extents in this high-dimensional space. This complexity enables the existence of multiple hybrid E/M phenotypes with the different repertoire of morphological and molecular traits [227].

The hybrid E/M phenotype(s) has(have) been largely considered as 'metastable' or 'intermediate' to the terminal endpoint mesenchymal phenotype for cells en route EMT [228]. However, recent *in vitro*, *in vivo*, and *in silico* evidence suggests that the hybrid E/M phenotype(s) may be the end point of a transition, i.e. cells undergoing EMT or MET need not complete their transition to the other end of the spectrum, instead can stably acquire one or more hybrid E/M phenotypes [227, 229]. Importantly, this hybrid E/M phenotype has been proposed as the 'fittest' phenotype for metastasis [230] due to its enhanced tumor-initiation potential/stemness, increased drug resistance traits, augmented adaptive plasticity, and hiked propensity to give rise to clusters of circulating tumor cells – the major drivers of metastasis [231].

Owing to the binary paradigm for investigating EMT, a mesenchymal phenotype was initially reported to be enriched in stemness in breast cancer [232]; EMT was shown to enrich for $CD44^{+}/CD24^{-}$ – the canonical markers that can enrich for identifying CSCs. Similar observations were made in other carcinomas; however, another study argued that EMT suppresses tumor-initiating ability in prostate cancer [233]. Furthermore, in breast cancer, another sub-set of CSCs was reported – $ALDH^{+}$ – which was reportedly epithelial, hence confounding the connection between EMT and stemness [234]. Mathematical modeling efforts to help resolve this conundrum mapped the dynamical traits of regulatory networks for EMT (miR-200/ZEB/SNAI) and stemness (LIN28/let-7/OCT4), and predicted that a hybrid E/M phenotype was most likely to be associated with stemness instead of those on extreme ends of the epithelial-mesenchymal spectrum [235]; in other words, the 'stemness window' was most likely situated midway on 'EMT axis'.

Follow-up experiments that categorized cells into three phenotypes instead of two – epithelial ($CD24^+/CD44^-$), mesenchymal ($CD24^-/CD44^+$) and hybrid E/M ($CD24^+/CD44^+$) – indeed observed a 10-times more mammosphere forming potential *in vitro* for hybrid E/M cells as compared to either epithelial or mesenchymal population, highlighting the increased tumor-initiation potential of hybrid E/M phenotype [188]. Similarly, $CD24^+/CD44^+$ cells were shown to form more aggressive tumors *in vivo* [236]. More recently, a stable hybrid E/M phenotype was proposed to be essential for tumorigenicity of basal breast cancer cells *in vivo* [229] where the "extremely epithelial" and "extremely mesenchymal" populations led to substantial loss of tumorigenicity. Even a co-culture of these "extreme" cells could not lead to tumor formation, suggesting that the maximum stemness resides at the cells midway on the EMT spectrum. Thus, driving cells out of a hybrid E/M phenotype to a "locked" mesenchymal state may be a promising mechanism to reduce metastasis and tumor aggressiveness. This proposition is reminiscent of reports suggesting that disseminated cells that get "locked" in a mesenchymal phenotype and do not undergo MET may not be able to eventually contribute much to colonization [237].

How can cells be pushed out of a hybrid E/M phenotype? Recent studies have identified various 'phenotypic stability factors' (PSFs) for a hybrid E/M phenotype, whose knockdown can drive the cells to a completely mesenchymal phenotype *in vitro*. For instance, H1975 cells that can stably maintain a hybrid E/M phenotype over 2 months tend to move collectively switch to individual migration and lose E-cadherin completely upon knockdown of PSFs such as OVOL2, GRHL2, and NUMB [12, 13]. These PSFs act as 'molecular brakes' on EMT, and prevent "cells that have gained partial plasticity" from undergoing a complete EMT, thus maintaining them in hybrid E/M phenotype [238]; their functional implications in maintaining collective cell migration has also been reported *in vivo* in developmental contexts; their effect on mediating tumor-initiation potential *in vivo* needs to be further investigated. Similarly, knockdown of JAG1 – a ligand of cell-cell communication pathway Notch signaling that has been implicated in forming clusters of circulating tumor cells and in maintaining hybrid E/M phenotype [239, 240] – led to reduced tumor-initiation potential, thus bolstering the association between hybrid E/M phenotype and stemness [241]. Finally, a recent study that demonstrated reduced metastatic potential by turning breast cancer cells into adipocytes argued that the efficiency of this reprogramming is enhanced if cells are in a hybrid E/M phenotype instead of a fully mesenchymal one [14]. Put together, these results indicate that pushing cells out of a hybrid E/M phenotype may restrict cellular fitness and control metastatic potential.

Conclusion

Many studies *in vitro* and *in vivo* show that targeting EMT through its inhibition or reversal allow to suppress significantly tumor growth and metastasis and to increase sensitivity to chemo- and targeted therapy and can be an effective strategy in the treatment of different cancers. However, this therapeutic approach is challenging due to the involvement of EMT in normal physiological processes. EMT suppression may counteract wound healing and tissue regeneration as well as stem cell renewal in patients with cancer. On the other hand, reversal of EMT (induction of MET) may promote the metastatic process through the activation of the switch from micrometastasis to macrometastasis. A potential "safe" way to overcome EMT-associated cancer progression can be targeting oncogenic mutations that lead to the induction of EMT. For example, small-molecule inhibitors have been developed to target isocitrate dehydrogenases (IDH1/2) which mutations occur in diverse tumor types and disrupt normal epithelial morphology through EMT induction [242, 243]. Another strategy to inhibit EMT without potential "toxic" effects on normal tissues can be reprogramming the tumor microenvironment from protumor to antitumor state or complete suppression of inflammation [201]. A novel promising approach to prevent metastasis through targeting EMT can be modulation of the activity of phenotypic stability factors of a hybrid E/M phenotype to drive cancer cells to a mesenchymal phenotype or other cell types [12–14]. However, further studies should verify whether EMT-related transdifferentiation of cancer cells is an effective and safe strategy in anti-cancer treatment.

Acknowledgments The study was supported by the Russian Science Foundation (grant #19-75-30016).

References

1. Kalluri, R., and R.A. Weinberg. 2009. The basics of epithelial-mesenchymal transition. *The Journal of Clinical Investigation* 119 (6): 1420–1428.
2. Lamouille, S., J. Xu, and R. Derynck. 2014. Molecular mechanisms of epithelial-mesenchymal transition. *Nature Reviews. Molecular Cell Biology* 15 (3): 178–196.
3. Nieto, M.A., R.Y. Huang, R.A. Jackson, and J.P. Thiery. 2016. EMT: 2016. *Cell* 166 (1): 21–45.
4. Tsai, J.H., and J. Yang. 2013. Epithelial-mesenchymal plasticity in carcinoma metastasis. *Genes & Development* 27 (20): 2192–2206.
5. Sui, H., L. Zhu, W. Deng, and Q. Li. 2014. Epithelial-mesenchymal transition and drug resistance: Role, molecular mechanisms, and therapeutic strategies. *Oncology Research and Treatment* 37 (10): 584–589.
6. Du, B., and J.S. Shim. 2016. Targeting epithelial-mesenchymal transition (EMT) to overcome drug resistance in cancer. *Molecules* 21 (7): E965.
7. Voon, D.C., R.Y. Huang, R.A. Jackson, and J.P. Thiery. 2017. The EMT spectrum and therapeutic opportunities. *Molecular Oncology* 11 (7): 878–891.

8. Cho, E.S., H.E. Kang, N.H. Kim, and J.I. Yook. 2019. Therapeutic implications of cancer epithelial-mesenchymal transition (EMT). *Archives of Pharmacal Research* 42 (1): 14–24.
9. Bizzarri, M., A. Cucina, and S. Proietti. 2017. Tumor reversion: mesenchymal-epithelial transition as a critical step in managing the tumor-microenvironment cross-talk. *Current Pharmaceutical Design* 23 (32): 4705–4715.
10. Jung, H.Y., L. Fattet, and J. Yang. 2015. Molecular pathways: linking tumor microenvironment to epithelial-mesenchymal transition in metastasis. *Clinical Cancer Research* 21 (5): 962–968.
11. Wigerup, C., S. Pahlman, and D. Bexell. 2016. Therapeutic targeting of hypoxia and hypoxia-inducible factors in cancer. *Pharmacology & Therapeutics* 164: 152–169.
12. Jolly, M.K., S.C. Tripathi, D. Jia, S.M. Mooney, M. Celiktas, S.M. Hanash, S.A. Mani, K.J. Pienta, E. Ben-Jacob, and H. Levine. 2016. Stability of the hybrid epithelial/mesenchymal phenotype. *Oncotarget* 7 (19): 27067–27084.
13. Bocci, F., M.K. Jolly, S.C. Tripathi, M. Aguilar, S.M. Hanash, H. Levine, and J.N. Onuchic. 2017. Numb prevents a complete epithelial-mesenchymal transition by modulating Notch signalling. *Journal of the Royal Society Interface* 14 (136): 20170512.
14. Ishay-Ronen, D., M. Diepenbruck, R.K.R. Kalathur, N. Sugiyama, S. Tiede, R. Ivanek, G. Bantug, M.F. Morini, J. Wang, C. Hess, and G. Christofori. 2019. Gain fat-lose metastasis: Converting invasive breast cancer cells into adipocytes inhibits cancer metastasis. *Cancer Cell* 35 (1): 17–32.e16.
15. Gonzalez, D.M., and D. Medici. 2014. Signaling mechanisms of the epithelial-mesenchymal transition. *Science Signaling* 7 (344): re8.
16. Moyret-Lalle, C., E. Ruiz, and A. Puisieux. 2014. Epithelial-mesenchymal transition transcription factors and miRNAs: "Plastic surgeons" of breast cancer. *World Journal of Clinical Oncology* 5 (3): 311–322.
17. Cong, N., P. Du, A. Zhang, F. Shen, J. Su, P. Pu, T. Wang, J. Zjang, C. Kang, and Q. Zhang. 2013. Downregulated microRNA-200a promotes EMT and tumor growth through the wnt/beta-catenin pathway by targeting the E-cadherin repressors ZEB1/ZEB2 in gastric adenocarcinoma. *Oncology Reports* 29 (4): 1579–1587.
18. Guo, F., B.C. Parker Kerrigan, D. Yang, L. Hu, I. Shmulevich, A.K. Sood, F. Xue, and W. Zhang. 2014. Post-transcriptional regulatory network of epithelial-to-mesenchymal and mesenchymal-to-epithelial transitions. *Journal of Hematology & Oncology* 7: 19.
19. Tian, Y., Q. Pan, Y. Shang, R. Zhu, J. Ye, Y. Liu, X. Zhong, S. Li, Y. He, L. Chen, J. Zhao, W. Chen, Z. Peng, and R. Wang. 2014. MicroRNA-200 (miR-200) cluster regulation by achaete scute-like 2 (Ascl2): impact on the epithelial-mesenchymal transition in colon cancer cells. *The Journal of Biological Chemistry* 289 (52): 36101–36115.
20. David, C.J., and J. Massague. 2018. Contextual determinants of TGFbeta action in development, immunity and cancer. *Nature Reviews. Molecular Cell Biology* 19 (7): 419–435.
21. Morikawa, M., R. Derynck, and K. Miyazono. 2016. TGF-beta and the TGF-beta family: Context-dependent roles in cell and tissue physiology. *Cold Spring Harbor Perspectives in Biology* 8 (5): a021873.
22. Moustakas, A., and C.H. Heldin. 2009. The regulation of TGFbeta signal transduction. *Development* 136 (22): 3699–3714.
23. Derynck, R., B.P. Muthusamy, and K.Y. Saeteurn. 2014. Signaling pathway cooperation in TGF-beta-induced epithelial-mesenchymal transition. *Current Opinion in Cell Biology* 31: 56–66.
24. Miyazawa, K., M. Shinozaki, T. Hara, T. Furuya, and K. Miyazono. 2002. Two major Smad pathways in TGF-beta superfamily signalling. *Genes to Cells* 7 (12): 1191–1204.
25. Massague, J. 2012. TGFbeta signalling in context. *Nature Reviews. Molecular Cell Biology* 13 (10): 616–630.
26. Moustakas, A., and C.H. Heldin. 2005. Non-Smad TGF-beta signals. *Journal of Cell Science* 118 (Pt 16): 3573–3584.

27. Zhang, Y.E. 2017. Non-Smad signaling pathways of the TGF-beta family. *Cold Spring Harbor Perspectives in Biology* 9 (2): a022129.
28. Suzuki, H.I. 2018. MicroRNA control of TGF-beta signaling. *International Journal of Molecular Sciences* 19 (7): E1901.
29. Bierie, B., and H.L. Moses. 2006. Tumour microenvironment: TGFbeta: The molecular Jekyll and Hyde of cancer. *Nature Reviews Cancer* 6 (7): 506–520.
30. Heuberger, J., and W. Birchmeier. 2010. Interplay of cadherin-mediated cell adhesion and canonical Wnt signaling. *Cold Spring Harbor Perspectives in Biology* 2 (2): a002915.
31. Nusse, R. 2005. Wnt signaling in disease and in development. *Cell Research* 15 (1): 28–32.
32. Guo, Y., L. Xiao, L. Sun, and F. Liu. 2012. Wnt/beta-catenin signaling: A promising new target for fibrosis diseases. *Physiological Research* 61 (4): 337–346.
33. Yoshida, G.J., and H. Saya. 2014. Inversed relationship between CD44 variant and c-Myc due to oxidative stress-induced canonical Wnt activation. *Biochemical and Biophysical Research Communications* 443 (2): 622–627.
34. Clevers, H., and R. Nusse. 2012. Wnt/beta-catenin signaling and disease. *Cell* 149 (6): 1192–1205.
35. Rosenbluh, J., X. Wang, and W.C. Hahn. 2014. Genomic insights into WNT/beta-catenin signaling. *Trends in Pharmacological Sciences* 35 (2): 103–109.
36. Tsukanov, A.S., N.I. Pospekhova, V.P. Shubin, A.M. Kuzminov, V.N. Kashnikov, S.A. Frolov, and Y.A. Shelygin. 2017. Mutations in the APC gene in Russian patients with classic form of familial adenomatous polyposis. *Russian Journal of Genetics* 53 (3): 369–375.
37. Taylor, M.A., K. Sossey-Alaoui, C.L. Thompson, D. Danielpour, and W.P. Schiemann. 2013. TGF-beta upregulates miR-181a expression to promote breast cancer metastasis. *The Journal of Clinical Investigation* 123 (1): 150–163.
38. Cai, J., H. Guan, L. Fang, Y. Yang, X. Zhu, J. Yuan, J. Wu, and M. Li. 2013. MicroRNA-374a activates Wnt/beta-catenin signaling to promote breast cancer metastasis. *The Journal of Clinical Investigation* 123 (2): 566–579.
39. Ghahhari, N.M., and S. Babashah. 2015. Interplay between microRNAs and WNT/beta-catenin signalling pathway regulates epithelial-mesenchymal transition in cancer. *European Journal of Cancer* 51 (12): 1638–1649.
40. Fortini, M.E. 2009. Notch signaling: The core pathway and its posttranslational regulation. *Developmental Cell* 16 (5): 633–647.
41. Brabletz, S., K. Bajdak, S. Meidhof, U. Burk, G. Niedermann, E. Firat, U. Wellner, A. Dimmler, G. Faller, J. Schubert, and T. Brabletz. 2011. The ZEB1/miR-200 feedback loop controls Notch signalling in cancer cells. *The EMBO Journal* 30 (4): 770–782.
42. Bolos, V., J. Grego-Bessa, and J.L. de la Pompa. 2007. Notch signaling in development and cancer. *Endocrine Reviews* 28 (3): 339–363.
43. Reedijk, M., S. Odorcic, L. Chang, H. Zhang, N. Miller, D.R. McCready, G. Lockwood, and S.E. Egan. 2005. High-level coexpression of JAG1 and NOTCH1 is observed in human breast cancer and is associated with poor overall survival. *Cancer Research* 65 (18): 8530–8537.
44. Zhang, H., X. Wang, J. Xu, and Y. Sun. 2014. Notch1 activation is a poor prognostic factor in patients with gastric cancer. *British Journal of Cancer* 110 (9): 2283–2290.
45. Reedijk, M., S. Odorcic, H. Zhang, R. Chetty, C. Tennert, B.C. Dickson, G. Lockwood, S. Gallinger, and S.E. Egan. 2008. Activation of Notch signaling in human colon adenocarcinoma. *International Journal of Oncology* 33 (6): 1223–1229.
46. Kanamori, M., T. Kawaguchi, J.M. Nigro, B.G. Feuerstein, M.S. Berger, L. Miele, and R.O. Pieper. 2007. Contribution of Notch signaling activation to human glioblastoma multiforme. *Journal of Neurosurgery* 106 (3): 417–427.
47. Joo, Y.H., C.K. Jung, M.S. Kim, and D.I. Sun. 2009. Relationship between vascular endothelial growth factor and Notch1 expression and lymphatic metastasis in tongue cancer. *Otolaryngology and Head and Neck Surgery* 140 (4): 512–518.

48. Whelan, J.T., A. Kellogg, B.M. Shewchuk, K. Hewan-Lowe, and F.E. Bertrand. 2009. Notch-1 signaling is lost in prostate adenocarcinoma and promotes PTEN gene expression. *Journal of Cellular Biochemistry* 107 (5): 992–1001.
49. Wang, M., L. Xue, Q. Cao, Y. Lin, Y. Ding, P. Yang, and L. Che. 2009. Expression of Notch1, Jagged1 and beta-catenin and their clinicopathological significance in hepatocellular carcinoma. *Neoplasma* 56 (6): 533–541.
50. Mullendore, M.E., J.B. Koorstra, Y.M. Li, G.J. Offerhaus, X. Fan, C.M. Henderson, W. Matsui, C.G. Eberhart, A. Maitra, and G. Feldmann. 2009. Ligand-dependent Notch signaling is involved in tumor initiation and tumor maintenance in pancreatic cancer. *Clinical Cancer Research* 15 (7): 2291–2301.
51. Li, J., Q. Li, L. Lin, R. Wang, L. Chen, W. Du, C. Jiang, and R. Li. 2018. Targeting the Notch1 oncogene by miR-139-5p inhibits glioma metastasis and epithelial-mesenchymal transition (EMT). *BMC Neurology* 18 (1): 133.
52. Zhang, Y., B. Xu, and X.P. Zhang. 2018. Effects of miRNAs on functions of breast cancer stem cells and treatment of breast cancer. *OncoTargets and Therapy* 11: 4263–4270.
53. Salaritabar, A., I. Berindan-Neagoe, B. Darvish, F. Hadjiakhoondi, A. Manayi, K.P. Devi, D. Barreca, I.E. Orhan, I. Suntar, A.A. Farooqi, D. Gulei, S.F. Nabavi, A. Sureda, M. Daglia, A.R. Dehpour, S.M. Nabavi, and S. Shirooie. 2019. Targeting Hedgehog signaling pathway: Paving the road for cancer therapy. *Pharmacological Research* 141: 466–480.
54. Katoh, Y., and M. Katoh. 2008. Hedgehog signaling, epithelial-to-mesenchymal transition and miRNA (review). *International Journal of Molecular Medicine* 22 (3): 271–275.
55. Walter, K., N. Omura, S.M. Hong, M. Griffith, A. Vincent, M. Borges, and M. Goggins. 2010. Overexpression of smoothened activates the sonic hedgehog signaling pathway in pancreatic cancer-associated fibroblasts. *Clinical Cancer Research* 16 (6): 1781–1789.
56. Jiang, J., and C.C. Hui. 2008. Hedgehog signaling in development and cancer. *Developmental Cell* 15 (6): 801–812.
57. Kim, J., J. Hyun, S. Wang, C. Lee, and Y. Jung. 2018. MicroRNA-378 is involved in hedgehog-driven epithelial-to-mesenchymal transition in hepatocytes of regenerating liver. *Cell Death & Disease* 9 (7): 721.
58. Yu, F., Y. Zheng, W. Hong, B. Chen, P. Dong, and J. Zheng. 2015. MicroRNA200a suppresses epithelialtomesenchymal transition in rat hepatic stellate cells via GLI family zinc finger 2. *Molecular Medicine Reports* 12 (6): 8121–8128.
59. Liu, Y., L. Li, Z. Liu, Q. Yuan, and X. Lu. 2018. Downregulation of MiR-431 expression associated with lymph node metastasis and promotes cell invasion in papillary thyroid carcinoma. *Cancer Biomarkers* 22 (4): 727–732.
60. Santamaria, P.G., G. Moreno-Bueno, F. Portillo, and A. Cano. 2017. EMT: Present and future in clinical oncology. *Molecular Oncology* 11 (7): 718–738.
61. Zhang, J., X.J. Tian, and J. Xing. 2016. Signal transduction pathways of EMT induced by TGF-beta, SHH, and WNT and their crosstalks. *Journal of Clinical Medicine* 5 (4): E41.
62. Haque, S., and J.C. Morris. 2017. Transforming growth factor-beta: A therapeutic target for cancer. *Human Vaccines & Immunotherapeutics* 13 (8): 1741–1750.
63. Hau, P., P. Jachimczak, R. Schlingensiepen, F. Schulmeyer, T. Jauch, A. Steinbrecher, A. Brawanski, M. Proescholdt, J. Schlaier, J. Buchroithner, J. Pichler, G. Wurm, M. Mehdorn, R. Strege, G. Schuierer, V. Villarrubia, F. Fellner, O. Jansen, T. Straube, V. Nohria, M. Goldbrunner, M. Kunst, S. Schmaus, G. Stauder, U. Bogdahn, and K.H. Schlingensiepen. 2007. Inhibition of TGF-beta2 with AP 12009 in recurrent malignant gliomas: From preclinical to phase I/II studies. *Oligonucleotides* 17 (2): 201–212.
64. de Gramont, A., S. Faivre, and E. Raymond. 2017. Novel TGF-beta inhibitors ready for prime time in onco-immunology. *Oncoimmunology* 6 (1): e1257453.
65. Ehata, S., A. Hanyu, M. Fujime, Y. Katsuno, E. Fukunaga, K. Goto, Y. Ishikawa, K. Nomura, H. Yokoo, T. Shimizu, E. Ogata, K. Miyazono, K. Shimizu, and T. Imamura. 2007. Ki26894, a novel transforming growth factor-beta type I receptor kinase inhibitor, inhibits in vitro

invasion and in vivo bone metastasis of a human breast cancer cell line. *Cancer Science* 98 (1): 127–133.

66. Park, C.Y., J.Y. Son, C.H. Jin, J.S. Nam, D.K. Kim, and Y.Y. Sheen. 2011. EW-7195, a novel inhibitor of ALK5 kinase inhibits EMT and breast cancer metastasis to lung. *European Journal of Cancer* 47 (17): 2642–2653.
67. Son, J.Y., S.Y. Park, S.J. Kim, S.J. Lee, S.A. Park, M.J. Kim, S.W. Kim, D.K. Kim, J.S. Nam, and Y.Y. Sheen. 2014. EW-7197, a novel ALK-5 kinase inhibitor, potently inhibits breast to lung metastasis. *Molecular Cancer Therapeutics* 13 (7): 1704–1716.
68. Park, C.Y., K.N. Min, J.Y. Son, S.Y. Park, J.S. Nam, D.K. Kim, and Y.Y. Sheen. 2014. An novel inhibitor of TGF-beta type I receptor, IN-1130, blocks breast cancer lung metastasis through inhibition of epithelial-mesenchymal transition. *Cancer Letters* 351 (1): 72–80.
69. Chakrabarti, R., V. Subramaniam, S. Abdalla, S. Jothy, and G.J. Prud'homme. 2009. Tranilast inhibits the growth and metastasis of mammary carcinoma. *Anti-Cancer Drugs* 20 (5): 334–345.
70. Liu, Y., C. Chen, P. Qian, X. Lu, B. Sun, X. Zhang, L. Wang, X. Gao, H. Li, Z. Chen, J. Tang, W. Zhang, J. Dong, R. Bai, P.E. Lobie, Q. Wu, S. Liu, H. Zhang, F. Zhao, M.S. Wicha, T. Zhu, and Y. Zhao. 2015. Gd-metallofullerenol nanomaterial as non-toxic breast cancer stem cell-specific inhibitor. *Nature Communications* 6: 5988.
71. Da, C., Y. Liu, Y. Zhan, K. Liu, and R. Wang. 2016. Nobiletin inhibits epithelial-mesenchymal transition of human non-small cell lung cancer cells by antagonizing the TGF-beta1/Smad3 signaling pathway. *Oncology Reports* 35 (5): 2767–2774.
72. Sun, Y., X. Jiang, Y. Lu, J. Zhu, L. Yu, B. Ma, and Q. Zhang. 2018. Oridonin prevents epithelial-mesenchymal transition and TGF-beta1-induced epithelial-mesenchymal transition by inhibiting TGF-beta1/Smad2/3 in osteosarcoma. *Chemico-Biological Interactions* 296: 57–64.
73. Ji, Q., X. Liu, Z. Han, L. Zhou, H. Sui, L. Yan, H. Jiang, J. Ren, J. Cai, and Q. Li. 2015. Resveratrol suppresses epithelial-to-mesenchymal transition in colorectal cancer through TGF-beta1/Smads signaling pathway mediated Snail/E-cadherin expression. *BMC Cancer* 15: 97.
74. Hafeez, B.B., A. Ganju, M. Sikander, V.K. Kashyap, Z.B. Hafeez, N. Chauhan, S. Malik, A.E. Massey, M.K. Tripathi, F.T. Halaweish, N. Zafar, M.M. Singh, M.M. Yallapu, S.C. Chauhan, and M. Jaggi. 2017. Ormeloxifene suppresses prostate tumor growth and metastatic phenotypes via inhibition of oncogenic beta-catenin signaling and EMT progression. *Molecular Cancer Therapeutics* 16 (10): 2267–2280.
75. Zhan, T., N. Rindtorff, and M. Boutros. 2017. Wnt signaling in cancer. *Oncogene* 36 (11): 1461–1473.
76. Wu, C.X., A. Xu, C.C. Zhang, P. Olson, L. Chen, T.K. Lee, T.T. Cheung, C.M. Lo, and X.Q. Wang. 2017. Notch inhibitor PF-03084014 inhibits hepatocellular carcinoma growth and metastasis via suppression of cancer stemness due to reduced activation of Notch1-Stat3. *Molecular Cancer Therapeutics* 16 (8): 1531–1543.
77. Venkatesh, V., R. Nataraj, G.S. Thangaraj, M. Karthikeyan, A. Gnanasekaran, S.B. Kaginelli, G. Kuppanna, C.G. Kallappa, and K.M. Basalingappa. 2018. Targeting Notch signalling pathway of cancer stem cells. *Stem Cell Investigation* 5: 5.
78. Ahmad, A., M.Y. Maitah, K.R. Ginnebaugh, Y. Li, B. Bao, S.M. Gadgeel, and F.H. Sarkar. 2013. Inhibition of Hedgehog signaling sensitizes NSCLC cells to standard therapies through modulation of EMT-regulating miRNAs. *Journal of Hematology & Oncology* 6 (1): 77.
79. Yang, X., and M.S. Dinehart. 2017. Triple hedgehog pathway inhibition for basal cell carcinoma. *The Journal of Clinical and Aesthetic Dermatology* 10 (4): 47–49.
80. Rumman, M., K.H. Jung, Z. Fang, H.H. Yan, M.K. Son, S.J. Kim, J. Kim, J.H. Park, J.H. Lim, S. Hong, and S.S. Hong. 2016. HS-173, a novel PI3K inhibitor suppresses EMT and metastasis in pancreatic cancer. *Oncotarget* 7 (47): 78029–78047.

81. Kim, W.G., C.J. Guigon, L. Fozzatti, J.W. Park, C. Lu, M.C. Willingham, and S.Y. Cheng. 2012. SKI-606, an Src inhibitor, reduces tumor growth, invasion, and distant metastasis in a mouse model of thyroid cancer. *Clinical Cancer Research* 18 (5): 1281–1290.
82. Bartholomeusz, C., X. Xie, M.K. Pitner, K. Kondo, A. Dadbin, J. Lee, H. Saso, P.D. Smith, K.N. Dalby, and N.T. Ueno. 2015. MEK inhibitor selumetinib (AZD6244; ARRY-142886) prevents lung metastasis in a triple-negative breast cancer xenograft model. *Molecular Cancer Therapeutics* 14 (12): 2773–2781.
83. Huber, M.A., H.J. Maier, M. Alacakaptan, E. Wiedemann, J. Braunger, G. Boehmelt, J.B. Madwed, E.R. Young, D.R. Marshall, H. Pehamberger, T. Wirth, N. Kraut, and H. Beug. 2010. BI 5700, a selective chemical inhibitor of IkappaB kinase 2, specifically suppresses epithelial-mesenchymal transition and metastasis in mouse models of tumor progression. *Genes & Cancer* 1 (2): 101–114.
84. Qin, G., F. Xu, T. Qin, Q. Zheng, D. Shi, W. Xia, Y. Tian, Y. Tang, J. Wang, X. Xiao, W. Deng, and S. Wang. 2015. Palbociclib inhibits epithelial-mesenchymal transition and metastasis in breast cancer via c-Jun/COX-2 signaling pathway. *Oncotarget* 6 (39): 41794–41808.
85. Parvani, J.G., M.D. Gujrati, M.A. Mack, W.P. Schiemann, and Z.R. Lu. 2015. Silencing beta3 integrin by targeted ECO/siRNA nanoparticles inhibits EMT and metastasis of triple-negative breast cancer. *Cancer Research* 75 (11): 2316–2325.
86. Padthaisong, S., M. Thanee, A. Techasen, N. Namwat, P. Yongvanit, A. Liwatthakun, K. Hankla, S. Sangkhamanon, and W. Loilome. 2017. Nimotuzumab inhibits cholangiocarcinoma cell metastasis via suppression of the epithelial-mesenchymal transition process. *Anticancer Research* 37 (7): 3591–3597.
87. Ferrari-Amorotti, G., C. Chiodoni, F. Shen, S. Cattelani, A.R. Soliera, G. Manzotti, G. Grisendi, M. Dominici, F. Rivasi, M.P. Colombo, A. Fatatis, and B. Calabretta. 2014. Suppression of invasion and metastasis of triple-negative breast cancer lines by pharmacological or genetic inhibition of slug activity. *Neoplasia* 16 (12): 1047–1058.
88. Rhodes, L.V., C.R. Tate, H.C. Segar, H.E. Burks, T.B. Phamduy, V. Hoang, S. Elliott, D. Gilliam, F.N. Pounder, M. Anbalagan, D.B. Chrisey, B.G. Rowan, M.E. Burow, and B.M. Collins-Burow. 2014. Suppression of triple-negative breast cancer metastasis by pan-DAC inhibitor panobinostat via inhibition of ZEB family of EMT master regulators. *Breast Cancer Research and Treatment* 145 (3): 593–604.
89. Chatterjee, S., and P.C. Sil. 2019. Targeting the crosstalks of Wnt pathway with Hedgehog and Notch for cancer therapy. *Pharmacological Research* 142: 251–261.
90. Chen, L., W. Mai, M. Chen, J. Hu, Z. Zhuo, X. Lei, L. Deng, J. Liu, N. Yao, M. Huang, Y. Peng, W. Ye, and D. Zhang. 2017. Arenobufagin inhibits prostate cancer epithelial-mesenchymal transition and metastasis by down-regulating beta-catenin. *Pharmacological Research* 123: 130–142.
91. Way, T.D., J.T. Huang, C.H. Chou, C.H. Huang, M.H. Yang, and C.T. Ho. 2014. Emodin represses TWIST1-induced epithelial-mesenchymal transitions in head and neck squamous cell carcinoma cells by inhibiting the beta-catenin and Akt pathways. *European Journal of Cancer* 50 (2): 366–378.
92. Ma, J.W., C.M. Hung, Y.C. Lin, C.T. Ho, J.Y. Kao, and T.D. Way. 2016. Aloe-emodin inhibits HER-2 expression through the downregulation of Y-box binding protein-1 in HER-2-overexpressing human breast cancer cells. *Oncotarget* 7 (37): 58915–58930.
93. Song, X., X. Zhou, Y. Qin, J. Yang, Y. Wang, Z. Sun, K. Yu, S. Zhang, and S. Liu. 2018. Emodin inhibits epithelialmesenchymal transition and metastasis of triple negative breast cancer via antagonism of CCchemokine ligand 5 secreted from adipocytes. *International Journal of Molecular Medicine* 42 (1): 579–588.
94. Li, N., C. Wang, P. Zhang, and S. You. 2018. Emodin inhibits pancreatic cancer EMT and invasion by upregulating microRNA1271. *Molecular Medicine Reports* 18 (3): 3366–3374.

95. Tian, L., D. Shen, X. Li, X. Shan, X. Wang, Q. Yan, and J. Liu. 2016. Ginsenoside Rg3 inhibits epithelial-mesenchymal transition (EMT) and invasion of lung cancer by down-regulating FUT4. *Oncotarget* 7 (2): 1619–1632.
96. Liu, T., L. Zhao, Y. Zhang, W. Chen, D. Liu, H. Hou, L. Ding, and X. Li. 2014. Ginsenoside 20(S)-Rg3 targets HIF-1alpha to block hypoxia-induced epithelial-mesenchymal transition in ovarian cancer cells. *PLoS One* 9 (9): e103887.
97. Li, J., X. Gong, R. Jiang, D. Lin, T. Zhou, A. Zhang, H. Li, X. Zhang, J. Wan, G. Kuang, and H. Li. 2018. Fisetin inhibited growth and metastasis of triple-negative breast cancer by reversing epithelial-to-mesenchymal transition via PTEN/Akt/GSK3beta signal pathway. *Frontiers in Pharmacology* 9: 772.
98. Wang, H., W. Zhong, J. Zhao, H. Zhang, Q. Zhang, Y. Liang, S. Chen, H. Liu, S. Zong, Y. Tian, H. Zhou, T. Sun, Y. Liu, and C. Yang. 2019. Oleanolic acid inhibits epithelial-mesenchymal transition of hepatocellular carcinoma by promoting iNOS dimerization. *Molecular Cancer Therapeutics* 18 (1): 62–74.
99. Li, C.Y., Q. Wang, S. Shen, X.L. Wei, and G.X. Li. 2018. Oridonin inhibits migration, invasion, adhesion and TGF-beta1-induced epithelial-mesenchymal transition of melanoma cells by inhibiting the activity of PI3K/Akt/GSK-3beta signaling pathway. *Oncology Letters* 15 (1): 1362–1372.
100. Liu, Q.Q., K. Chen, Q. Ye, X.H. Jiang, and Y.W. Sun. 2016. Oridonin inhibits pancreatic cancer cell migration and epithelial-mesenchymal transition by suppressing Wnt/beta-catenin signaling pathway. *Cancer Cell International* 16: 57.
101. Kim, R.K., Y. Suh, K.C. Yoo, Y.H. Cui, E. Hwang, H.J. Kim, J.S. Kang, M.J. Kim, Y.Y. Lee, and S.J. Lee. 2015. Phloroglucinol suppresses metastatic ability of breast cancer cells by inhibition of epithelial-mesenchymal cell transition. *Cancer Science* 106 (1): 94–101.
102. Chang, J.H., S.L. Lai, W.S. Chen, W.Y. Hung, J.M. Chow, M. Hsiao, W.J. Lee, and M.H. Chien. 2017. Quercetin suppresses the metastatic ability of lung cancer through inhibiting Snail-dependent Akt activation and Snail-independent ADAM9 expression pathways. *Biochimica et Biophysica Acta, Molecular Cell Research* 1864 (10): 1746–1758.
103. Chen, M.C., W.W. Chang, Y.D. Kuan, S.T. Lin, H.C. Hsu, and C.H. Lee. 2012. Resveratrol inhibits LPS-induced epithelial-mesenchymal transition in mouse melanoma model. *Innate Immunity* 18 (5): 685–693.
104. Lee, J., E.R. Hahm, A.I. Marcus, and S.V. Singh. 2015. Withaferin A inhibits experimental epithelial-mesenchymal transition in MCF-10A cells and suppresses vimentin protein level in vivo in breast tumors. *Molecular Carcinogenesis* 54 (6): 417–429.
105. Kyakulaga, A.H., F. Aqil, R. Munagala, and R.C. Gupta. 2018. Withaferin A inhibits epithelial to mesenchymal transition in non-small cell lung cancer cells. *Scientific Reports* 8 (1): 15737.
106. Thaiparambil, J.T., L. Bender, T. Ganesh, E. Kline, P. Patel, Y. Liu, M. Tighiouart, P.M. Vertino, R.D. Harvey, A. Garcia, and A.I. Marcus. 2011. Withaferin A inhibits breast cancer invasion and metastasis at sub-cytotoxic doses by inducing vimentin disassembly and serine 56 phosphorylation. *International Journal of Cancer* 129 (11): 2744–2755.
107. Yang, Z., A. Garcia, S. Xu, D.R. Powell, P.M. Vertino, S. Singh, and A.I. Marcus. 2013. Withania somnifera root extract inhibits mammary cancer metastasis and epithelial to mesenchymal transition. *PLoS One* 8 (9): e75069.
108. Kaufhold, S., and B. Bonavida. 2014. Central role of Snail1 in the regulation of EMT and resistance in cancer: A target for therapeutic intervention. *Journal of Experimental & Clinical Cancer Research* 33: 62.
109. Harney, A.S., T.J. Meade, and C. LaBonne. 2012. Targeted inactivation of Snail family EMT regulatory factors by a Co(III)-Ebox conjugate. *PLoS One* 7 (2): e32318.
110. Vistain, L.F., N. Yamamoto, R. Rathore, P. Cha, and T.J. Meade. 2015. Targeted inhibition of Snail activity in breast cancer cells by using a Co(III) -Ebox conjugate. *Chembiochem* 16 (14): 2065–2072.

111. Finetti, F., A. Moglia, I. Schiavo, S. Donnini, G.N. Berta, F. Di Scipio, A. Perrelli, C. Fornelli, L. Trabalzini, and S.F. Retta. 2018. Yeast-derived recombinant avenanthramides inhibit proliferation, migration and epithelial mesenchymal transition of colon cancer cells. *Nutrients* 10 (9): E1159.
112. Wang, Y.P., M.Z. Wang, Y.R. Luo, Y. Shen, and Z.X. Wei. 2012. Lentivirus-mediated shRNA interference targeting SLUG inhibits lung cancer growth and metastasis. *Asian Pacific Journal of Cancer Prevention* 13 (10): 4947–4951.
113. Qian, J., H. Liu, W. Chen, K. Wen, W. Lu, C. Huang, and Z. Fu. 2013. Knockdown of Slug by RNAi inhibits the proliferation and invasion of HCT116 colorectal cancer cells. *Molecular Medicine Reports* 8 (4): 1055–1059.
114. Lai, W.Y., W.Y. Wang, Y.C. Chang, C.J. Chang, P.C. Yang, and K. Peck. 2014. Synergistic inhibition of lung cancer cell invasion, tumor growth and angiogenesis using aptamer-siRNA chimeras. *Biomaterials* 35 (9): 2905–2914.
115. Khan, M.A., H.C. Chen, D. Zhang, and J. Fu. 2013. Twist: A molecular target in cancer therapeutics. *Tumour Biology* 34 (5): 2497–2506.
116. Mikheeva, S.A., A.M. Mikheev, A. Petit, R. Beyer, R.G. Oxford, L. Khorasani, J.P. Maxwell, C.A. Glackin, H. Wakimoto, I. Gonzalez-Herrero, I. Sanchez-Garcia, J.R. Silber, P.J. Horner, and R.C. Rostomily. 2010. TWIST1 promotes invasion through mesenchymal change in human glioblastoma. *Molecular Cancer* 9: 194.
117. Mikheev, A.M., S.A. Mikheeva, L.J. Severs, C.C. Funk, L. Huang, J.L. McFaline-Figueroa, J. Schwensen, C. Trapnell, N.D. Price, S. Wong, and R.C. Rostomily. 2018. Targeting TWIST1 through loss of function inhibits tumorigenicity of human glioblastoma. *Molecular Oncology* 12 (7): 1188–1202.
118. Xu, Y., D.K. Lee, Z. Feng, Y. Xu, W. Bu, Y. Li, L. Liao, and J. Xu. 2017. Breast tumor cell-specific knockout of Twist1 inhibits cancer cell plasticity, dissemination, and lung metastasis in mice. *Proceedings of the National Academy of Sciences of the United States of America* 114 (43): 11494–11499.
119. Li, R., C. Wu, H. Liang, Y. Zhao, C. Lin, X. Zhang, and C. Ye. 2018. Knockdown of TWIST enhances the cytotoxicity of chemotherapeutic drugs in doxorubicin-resistant HepG2 cells by suppressing MDR1 and EMT. *International Journal of Oncology* 53 (4): 1763–1773.
120. Yochum, Z.A., J. Cades, L. Mazzacurati, N.M. Neumann, S.K. Khetarpal, S. Chatterjee, H. Wang, M.A. Attar, E.H. Huang, S.N. Chatley, K. Nugent, A. Somasundaram, J.A. Engh, A.J. Ewald, Y.J. Cho, C.M. Rudin, P.T. Tran, and T.F. Burns. 2017. A first-in-class TWIST1 inhibitor with activity in oncogene-driven lung cancer. *Molecular Cancer Research* 15 (12): 1764–1776.
121. Yochum, Z.A., J. Cades, H. Wang, S. Chatterjee, B.W. Simons, J.P. O'Brien, S.K. Khetarpal, G. Lemtiri-Chlieh, K.V. Myers, E.H. Huang, C.M. Rudin, P.T. Tran, and T.F. Burns. 2019. Targeting the EMT transcription factor TWIST1 overcomes resistance to EGFR inhibitors in EGFR-mutant non-small-cell lung cancer. *Oncogene* 38 (5): 656–670.
122. Zhang, Y., L. Xu, A.Q. Li, and X.Z. Han. 2019. The roles of ZEB1 in tumorigenic progression and epigenetic modifications. *Biomedicine & Pharmacotherapy* 110: 400–408.
123. Fardi, M., M. Alivand, B. Baradaran, M. Farshdousti Hagh, and S. Solali. 2019. The crucial role of ZEB2: from development to epithelial-to-mesenchymal transition and cancer complexity. *Journal of Cellular Physiology* 234: 14783–14799.
124. Zhang, W., X. Shi, Y. Peng, M. Wu, P. Zhang, R. Xie, Y. Wu, Q. Yan, S. Liu, and J. Wang. 2015. HIF-1alpha promotes epithelial-mesenchymal transition and metastasis through direct regulation of ZEB1 in colorectal cancer. *PLoS One* 10 (6): e0129603.
125. Sakata, J., F. Utsumi, S. Suzuki, K. Niimi, E. Yamamoto, K. Shibata, T. Senga, F. Kikkawa, and H. Kajiyama. 2017. Inhibition of ZEB1 leads to inversion of metastatic characteristics and restoration of paclitaxel sensitivity of chronic chemoresistant ovarian carcinoma cells. *Oncotarget* 8 (59): 99482–99494.
126. Meidhof, S., S. Brabletz, W. Lehmann, B.T. Preca, K. Mock, M. Ruh, J. Schuler, M. Berthold, A. Weber, U. Burk, M. Lubbert, M. Puhr, Z. Culig, U. Wellner, T. Keck, P. Bronsert,

S. Kusters, U.T. Hopt, M.P. Stemmler, and T. Brabletz. 2015. ZEB1-associated drug resistance in cancer cells is reversed by the class I HDAC inhibitor mocetinostat. *EMBO Molecular Medicine* 7 (6): 831–847.
127. Ming, H., Q. Chuang, W. Jiashi, L. Bin, W. Guangbin, and J. Xianglu. 2018. Naringin targets Zeb1 to suppress osteosarcoma cell proliferation and metastasis. *Aging (Albany NY)* 10 (12): 4141–4151.
128. Cech, T.R., and J.A. Steitz. 2014. The noncoding RNA revolution-trashing old rules to forge new ones. *Cell* 157 (1): 77–94.
129. Xu, Q., F. Deng, Y. Qin, Z. Zhao, Z. Wu, Z. Xing, A. Ji, and Q.J. Wang. 2016. Long non-coding RNA regulation of epithelial-mesenchymal transition in cancer metastasis. *Cell Death & Disease* 7 (6): e2254.
130. Exposito-Villen, A., E.A. Aránega, and D. Franco. 2018. Functional role of non-coding RNAs during epithelial-to-mesenchymal transition. *Noncoding RNA* 4 (2): E14.
131. Shi, S.J., L.J. Wang, B. Yu, Y.H. Li, Y. Jin, and X.Z. Bai. 2015. LncRNA-ATB promotes trastuzumab resistance and invasion-metastasis cascade in breast cancer. *Oncotarget* 6 (13): 11652–11663.
132. Huang, J.F., Y.J. Guo, C.X. Zhao, S.X. Yuan, Y. Wang, G.N. Tang, W.P. Zhou, and S.H. Sun. 2013. Hepatitis B virus X protein (HBx)-related long noncoding RNA (lncRNA) down-regulated expression by HBx (Dreh) inhibits hepatocellular carcinoma metastasis by targeting the intermediate filament protein vimentin. *Hepatology* 57 (5): 1882–1892.
133. Xu, Z.Y., Q.M. Yu, Y.A. Du, L.T. Yang, R.Z. Dong, L. Huang, P.F. Yu, and X.D. Cheng. 2013. Knockdown of long non-coding RNA HOTAIR suppresses tumor invasion and reverses epithelial-mesenchymal transition in gastric cancer. *International Journal of Biological Sciences* 9 (6): 587–597.
134. Zhou, M., Y. Hou, G. Yang, H. Zhang, G. Tu, Y.E. Du, S. Wen, L. Xu, X. Tang, S. Tang, L. Yang, X. Cui, and M. Liu. 2016. LncRNA-Hh strengthen cancer stem cells generation in Twist-positive breast cancer via activation of Hedgehog signaling pathway. *Stem Cells* 34 (1): 55–66.
135. Liang, W.C., W.M. Fu, C.W. Wong, Y. Wang, W.M. Wang, G.X. Hu, L. Zhang, L.J. Xiao, D.C. Wan, J.F. Zhang, and M.M. Waye. 2015. The lncRNA H19 promotes epithelial to mesenchymal transition by functioning as miRNA sponges in colorectal cancer. *Oncotarget* 6 (26): 22513–22525.
136. Wang, T.H., Y.S. Lin, Y. Chen, C.T. Yeh, Y.L. Huang, T.H. Hsieh, T.M. Shieh, C. Hsueh, and T.C. Chen. 2015. Long non-coding RNA AOC4P suppresses hepatocellular carcinoma metastasis by enhancing vimentin degradation and inhibiting epithelial-mesenchymal transition. *Oncotarget* 6 (27): 23342–23357.
137. Tang, Y., Y. Tang, and Y.S. Cheng. 2017. miR-34a inhibits pancreatic cancer progression through Snail1-mediated epithelial-mesenchymal transition and the Notch signaling pathway. *Scientific Reports* 7: 38232.
138. Huang, G., M.Y. Du, H. Zhu, N. Zhang, Z.W. Lu, L.X. Qian, W. Zhang, X. Tian, X. He, and L. Yin. 2018. MiRNA-34a reversed TGF-beta-induced epithelial-mesenchymal transition via suppression of SMAD4 in NPC cells. *Biomedicine & Pharmacotherapy* 106: 217–224.
139. Peng, L., Z. Liu, J. Xiao, Y. Tu, Z. Wan, H. Xiong, Y. Li, and W. Xiao. 2017. MicroRNA-148a suppresses epithelial-mesenchymal transition and invasion of pancreatic cancer cells by targeting Wnt10b and inhibiting the Wnt/beta-catenin signaling pathway. *Oncology Reports* 38 (1): 301–308.
140. Shelygin, Y.A., V.P. Shubin, S.A. Frolov, S.I. Achkasov, O.I. Sushkov, A.S. Tsukanov, V.N. Kashnikov, and N.I. Pospekhova. 2015. The analysis of microRNAs miR-200C and miR-145 expression in colorectal cancer of different molecular subtypes. *Doklady. Biochemistry and Biophysics* 463: 243–246.
141. Yin, K., W. Yin, Y. Wang, L. Zhou, Y. Liu, G. Yang, J. Wang, and J. Lu. 2016. MiR-206 suppresses epithelial mesenchymal transition by targeting TGF-beta signaling in estrogen receptor positive breast cancer cells. *Oncotarget* 7 (17): 24537–24548.

142. Dong, P., Y. Xiong, H. Watari, S.J. Hanley, Y. Konno, K. Ihira, T. Yamada, M. Kudo, J. Yue, and N. Sakuragi. 2016. MiR-137 and miR-34a directly target Snail and inhibit EMT, invasion and sphere-forming ability of ovarian cancer cells. *Journal of Experimental & Clinical Cancer Research* 35 (1): 132.
143. Wang, X., M. Yu, K. Zhao, M. He, W. Ge, Y. Sun, and Y. Wang. 2016. Upregulation of MiR-205 under hypoxia promotes epithelial-mesenchymal transition by targeting ASPP2. *Cell Death & Disease* 7 (12): e2517.
144. Ji, H., M. Sang, F. Liu, N. Ai, and C. Geng. 2019. miR-124 regulates EMT based on ZEB2 target to inhibit invasion and metastasis in triple-negative breast cancer. *Pathology, Research and Practice* 215 (4): 697–704.
145. Harazono, Y., T. Muramatsu, H. Endo, N. Uzawa, T. Kawano, K. Harada, J. Inazawa, and K. Kozaki. 2013. miR-655 Is an EMT-suppressive microRNA targeting ZEB1 and TGFBR2. *PLoS One* 8 (5): e62757.
146. Ying, L., Q. Chen, Y. Wang, Z. Zhou, Y. Huang, and F. Qiu. 2012. Upregulated MALAT-1 contributes to bladder cancer cell migration by inducing epithelial-to-mesenchymal transition. *Molecular BioSystems* 8 (9): 2289–2294.
147. Shen, L., L. Chen, Y. Wang, X. Jiang, H. Xia, and Z. Zhuang. 2015. Long noncoding RNA MALAT1 promotes brain metastasis by inducing epithelial-mesenchymal transition in lung cancer. *Journal of Neuro-Oncology* 121 (1): 101–108.
148. Gutschner, T., M. Hammerle, M. Eissmann, J. Hsu, Y. Kim, G. Hung, A. Revenko, G. Arun, M. Stentrup, M. Gross, M. Zornig, A.R. MacLeod, D.L. Spector, and S. Diederichs. 2013. The noncoding RNA MALAT1 is a critical regulator of the metastasis phenotype of lung cancer cells. *Cancer Research* 73 (3): 1180–1189.
149. Kong, Q., and M. Qiu. 2018. Long noncoding RNA SNHG15 promotes human breast cancer proliferation, migration and invasion by sponging miR-211-3p. *Biochemical and Biophysical Research Communications* 495 (2): 1594–1600.
150. Jiang, H., T. Li, Y. Qu, X. Wang, B. Li, J. Song, X. Sun, Y. Tang, J. Wan, Y. Yu, J. Zhan, and H. Zhang. 2018. Long non-coding RNA SNHG15 interacts with and stabilizes transcription factor Slug and promotes colon cancer progression. *Cancer Letters* 425: 78–87.
151. Liu, H., Z. Lv, and E. Guo. 2015. Knockdown of long noncoding RNA SPRY4-IT1 suppresses glioma cell proliferation, metastasis and epithelial-mesenchymal transition. *International Journal of Clinical and Experimental Pathology* 8 (8): 9140–9146.
152. Li, R.H., M. Chen, J. Liu, C.C. Shao, C.P. Guo, X.L. Wei, Y.C. Li, W.H. Huang, and G.J. Zhang. 2018. Long noncoding RNA ATB promotes the epithelial-mesenchymal transition by upregulating the miR-200c/Twist1 axe and predicts poor prognosis in breast cancer. *Cell Death & Disease* 9 (12): 1171.
153. Zhang, Y., J. Li, S. Jia, Y. Wang, Y. Kang, and W. Zhang. 2018. Down-regulation of lncRNA-ATB inhibits epithelial-mesenchymal transition of breast cancer cells by increasing miR-141-3p expression. *Biochemistry and Cell Biology* 97 (2): 193–200.
154. El Bezawy, R., D. Cominetti, N. Fenderico, V. Zuco, G.L. Beretta, M. Dugo, N. Arrighetti, C. Stucchi, T. Rancati, R. Valdagni, N. Zaffaroni, and P. Gandellini. 2017. miR-875-5p counteracts epithelial-to-mesenchymal transition and enhances radiation response in prostate cancer through repression of the EGFR-ZEB1 axis. *Cancer Letters* 395: 53–62.
155. Zhang, T., X. Cai, Q. Li, P. Xue, Z. Chen, X. Dong, and Y. Xue. 2016. Hsa-miR-875-5p exerts tumor suppressor function through down-regulation of EGFR in colorectal carcinoma (CRC). *Oncotarget* 7 (27): 42225–42240.
156. Hu, C., S. Cui, J. Zheng, T. Yin, J. Lv, J. Long, W. Zhang, X. Wang, S. Sheng, H. Zhang, Y. Sun, H. Wang, and C. Li. 2018. MiR-875-5p inhibits hepatocellular carcinoma cell proliferation and migration by repressing astrocyte elevated gene-1 (AEG-1) expression. *Translational Cancer Research* 7 (1): 158–169.
157. Li, Y., Y. Liang, Y. Sang, X. Song, H. Zhang, Y. Liu, L. Jiang, and Q. Yang. 2018. MiR-770 suppresses the chemo-resistance and metastasis of triple negative breast cancer via direct targeting of STMN1. *Cell Death & Disease* 9 (1): 14.

158. Qiu, H., F. Chen, and M. Chen. 2019. MicroRNA-138 negatively regulates the hypoxia-inducible factor 1α to suppress melanoma growth and metastasis. *Biology Open* 8: bio042937.
159. Kong, X., J. Zhang, J. Li, J. Shao, and L. Fang. 2018. MiR-130a-3p inhibits migration and invasion by regulating RAB5B in human breast cancer stem cell-like cells. *Biochemical and Biophysical Research Communications* 501 (2): 486–493.
160. Shu, S., X. Liu, M. Xu, X. Gao, J. Fan, H. Liu, and R. Li. 2018. MicroRNA-424 regulates epithelial-mesenchymal transition of endometrial carcinoma by directly targeting insulin-like growth factor 1 receptor. *Journal of Cellular Biochemistry* 120: 2171-2179.
161. Slaby, O., R. Laga, and O. Sedlacek. 2017. Therapeutic targeting of non-coding RNAs in cancer. *The Biochemical Journal* 474 (24): 4219–4251.
162. Zaravinos, A. 2015. The regulatory role of microRNAs in EMT and cancer. *Journal of Oncology* 2015: 865816.
163. Chen, C.Y., C.C. Chen, T.M. Shieh, C. Hsueh, S.H. Wang, Y.L. Leu, J.H. Lian, and T.H. Wang. 2018. Corylin suppresses hepatocellular carcinoma progression via the inhibition of epithelial-mesenchymal transition, mediated by long noncoding RNA GAS5. *International Journal of Molecular Sciences* 19 (2): E380.
164. Jing, W., H. Dong, M. Min, Z. Runpeng, X. Xuewei, C. Ru, X. Yingru, N. Shengfa, T. Baoxian, Y. Jinbo, H. Weidong, and Z. Rongbo. 2019. Dependence of artesunate on long noncoding RNA-RP11 to inhibit epithelial-mesenchymal transition of hepatocellular carcinoma. *Journal of Cellular Biochemistry* 120 (4): 6026–6034.
165. Xu, K., W. Tao, and Z. Su. 2018. Propofol prevents IL-13-induced epithelial-mesenchymal transition in human colorectal cancer cells. *Cell Biology International* 42 (8): 985–993.
166. Liu, W.Z., and N. Liu. 2018. Propofol inhibits lung cancer A549 cell growth and epithelial-mesenchymal transition process by upregulation of microRNA-1284. *Oncology Research* 27 (1): 1–8.
167. Liu, Z., J. Zhang, G. Hong, J. Quan, L. Zhang, and M. Yu. 2016. Propofol inhibits growth and invasion of pancreatic cancer cells through regulation of the miR-21/Slug signaling pathway. *American Journal of Translational Research* 8 (10): 4120–4133.
168. Hu, Y., J. Zhou, F. Ye, H. Xiong, L. Peng, Z. Zheng, F. Xu, M. Cui, C. Wei, X. Wang, Z. Wang, H. Zhu, P. Lee, M. Zhou, B. Jiang, and D.Y. Zhang. 2015. BRD4 inhibitor inhibits colorectal cancer growth and metastasis. *International Journal of Molecular Sciences* 16 (1): 1928–1948.
169. Shimada, Y., and J.D. Minna. 2017. Exosome mediated phenotypic changes in lung cancer pathophysiology. *Translational Cancer Research* 6 (Suppl 6): S1040–s1042.
170. Lobb, R.J., R. van Amerongen, A. Wiegmans, S. Ham, J.E. Larsen, and A. Moller. 2017. Exosomes derived from mesenchymal non-small cell lung cancer cells promote chemoresistance. *International Journal of Cancer* 141 (3): 614–620.
171. Hu, T.H., Y. Yao, S. Yu, L.L. Han, W.J. Wang, H. Guo, T. Tian, Z.P. Ruan, X.M. Kang, J. Wang, S.H. Wang, and K.J. Nan. 2014. SDF-1/CXCR4 promotes epithelial-mesenchymal transition and progression of colorectal cancer by activation of the Wnt/beta-catenin signaling pathway. *Cancer Letters* 354 (2): 417–426.
172. Li, X., Q. Ma, Q. Xu, H. Liu, J. Lei, W. Duan, K. Bhat, F. Wang, E. Wu, and Z. Wang. 2012. SDF-1/CXCR4 signaling induces pancreatic cancer cell invasion and epithelial-mesenchymal transition in vitro through non-canonical activation of Hedgehog pathway. *Cancer Letters* 322 (2): 169–176.
173. Wang, X., W. Zhang, Y. Ding, X. Guo, Y. Yuan, and D. Li. 2017. CRISPR/Cas9-mediated genome engineering of CXCR4 decreases the malignancy of hepatocellular carcinoma cells in vitro and in vivo. *Oncology Reports* 37 (6): 3565–3571.
174. Bargagna-Mohan, P., A. Hamza, Y.E. Kim, Y. Khuan Abby Ho, N. Mor-Vaknin, N. Wendschlag, J. Liu, R.M. Evans, D.M. Markovitz, C.G. Zhan, K.B. Kim, and R. Mohan. 2007. The tumor inhibitor and antiangiogenic agent withaferin A targets the intermediate filament protein vimentin. *Chemistry & Biology* 14 (6): 623–634.

175. Lu, T., B. Wu, Y. Yu, W. Zhu, S. Zhang, Y. Zhang, J. Guo, and N. Deng. 2018. Blockade of ONECUT2 expression in ovarian cancer inhibited tumor cell proliferation, migration, invasion and angiogenesis. *Cancer Science* 109 (7): 2221–2234.
176. Yu, C., Z. Liu, Q. Chen, Y. Li, L. Jiang, Z. Zhang, and F. Zhou. 2018. Nkx2.8 inhibits epithelial-mesenchymal transition in bladder urothelial carcinoma via transcriptional repression of Twist1. *Cancer Research* 78 (5): 1241–1252.
177. Zhang, H., K. Cai, J. Wang, X. Wang, K. Cheng, F. Shi, L. Jiang, Y. Zhang, and J. Dou. 2014. MiR-7, inhibited indirectly by lincRNA HOTAIR, directly inhibits SETDB1 and reverses the EMT of breast cancer stem cells by downregulating the STAT3 pathway. *Stem Cells* 32 (11): 2858–2868.
178. Ward, A., A. Balwierz, J.D. Zhang, M. Kublbeck, Y. Pawitan, T. Hielscher, S. Wiemann, and O. Sahin. 2013. Re-expression of microRNA-375 reverses both tamoxifen resistance and accompanying EMT-like properties in breast cancer. *Oncogene* 32 (9): 1173–1182.
179. Cattan, N., N. Rochet, C. Mazeau, E. Zanghellini, B. Mari, C. Chauzy, H. Stora de Novion, J. Amiel, J.L. Lagrange, B. Rossi, and J. Gioanni. 2001. Establishment of two new human bladder carcinoma cell lines, CAL 29 and CAL 185. Comparative study of cell scattering and epithelial to mesenchyme transition induced by growth factors. *British Journal of Cancer* 85 (9): 1412–1417.
180. Zhao, R., L. Gong, L. Li, L. Guo, D. Zhu, Z. Wu, and Q. Zhou. 2013. nm23-H1 is a negative regulator of TGF-beta1-dependent induction of epithelial-mesenchymal transition. *Experimental Cell Research* 319 (5): 740–749.
181. Larue, L., and A. Bellacosa. 2005. Epithelial-mesenchymal transition in development and cancer: Role of phosphatidylinositol 3′ kinase/AKT pathways. *Oncogene* 24 (50): 7443–7454.
182. Malfettone, A., J. Soukupova, E. Bertran, E. Crosas-Molist, R. Lastra, J. Fernando, P. Koudelkova, B. Rani, A. Fabra, T. Serrano, E. Ramos, W. Mikulits, G. Giannelli, and I. Fabregat. 2017. Transforming growth factor-beta-induced plasticity causes a migratory stemness phenotype in hepatocellular carcinoma. *Cancer Letters* 392: 39–50.
183. Huang, R.Y., K.T. Kuay, T.Z. Tan, M. Asad, H.M. Tang, A.H. Ng, J. Ye, V.Y. Chung, and J.P. Thiery. 2015. Functional relevance of a six mesenchymal gene signature in epithelial-mesenchymal transition (EMT) reversal by the triple angiokinase inhibitor, nintedanib (BIBF1120). *Oncotarget* 6 (26): 22098–22113.
184. Das, S., B.N. Becker, F.M. Hoffmann, and J.E. Mertz. 2009. Complete reversal of epithelial to mesenchymal transition requires inhibition of both ZEB expression and the Rho pathway. *BMC Cell Biology* 10: 94.
185. Dinicola, S., G. Fabrizi, M.G. Masiello, S. Proietti, A. Palombo, M. Minini, A.H. Harrath, S.H. Alwasel, G. Ricci, A. Catizone, A. Cucina, and M. Bizzarri. 2016. Inositol induces mesenchymal-epithelial reversion in breast cancer cells through cytoskeleton rearrangement. *Experimental Cell Research* 345 (1): 37–50.
186. Maschler, S., C.A. Gebeshuber, E.M. Wiedemann, M. Alacakaptan, M. Schreiber, I. Custic, and H. Beug. 2010. Annexin A1 attenuates EMT and metastatic potential in breast cancer. *EMBO Molecular Medicine* 2 (10): 401–414.
187. Yi, Y., S. Zeng, Z. Wang, M. Wu, Y. Ma, X. Ye, B. Zhang, and H. Liu. 2018. Cancer-associated fibroblasts promote epithelial-mesenchymal transition and EGFR-TKI resistance of non-small cell lung cancers via HGF/IGF-1/ANXA2 signaling. *Biochimica et Biophysica Acta - Molecular Basis of Disease* 1864 (3): 793–803.
188. Grosse-Wilde, A., A. Fouquier d'Herouel, E. McIntosh, G. Ertaylan, A. Skupin, R.E. Kuestner, A. del Sol, K.A. Walters, and S. Huang. 2015. Stemness of the hybrid epithelial/mesenchymal state in breast cancer and its association with poor survival. *PLoS One* 10 (5): e0126522.
189. El-Badawy, A., N.I. Ghoneim, M.A. Nasr, H. Elkhenany, T.A. Ahmed, S.M. Ahmed, and N. El-Badri. 2018. Telomerase reverse transcriptase coordinates with the epithelial-to-mesenchymal transition through a feedback loop to define properties of breast cancer stem cells. *Biology Open* 7 (7): bio034181.

190. Kzhyshkowska, J., M. Bizzarri, R. Apte, and N. Cherdyntseva. 2017. Editorial: Targeting of cancer cells and tumor microenvironment: Perspectives for personalized therapy. *Current Pharmaceutical Design* 23 (32): 4703–4704.
191. Stakheyeva, M., V. Riabov, I. Mitrofanova, N. Litviakov, E. Choynzonov, N. Cherdyntseva, and J. Kzhyshkowska. 2017. Role of the immune component of tumor microenvironment in the efficiency of cancer treatment: Perspectives for the personalized therapy. *Current Pharmaceutical Design* 23 (32): 4807–4826.
192. Sielska, M., P. Przanowski, B. Wylot, K. Gabrusiewicz, M. Maleszewska, M. Kijewska, M. Zawadzka, J. Kucharska, K. Vinnakota, H. Kettenmann, K. Kotulska, W. Grajkowska, and B. Kaminska. 2013. Distinct roles of CSF family cytokines in macrophage infiltration and activation in glioma progression and injury response. *The Journal of Pathology* 230 (3): 310–321.
193. Liu, Z., W. Kuang, Q. Zhou, and Y. Zhang. 2018. TGF-beta1 secreted by M2 phenotype macrophages enhances the stemness and migration of glioma cells via the SMAD2/3 signalling pathway. *International Journal of Molecular Medicine* 42 (6): 3395–3403.
194. Litviakov, N., M. Tsyganov, I. Larionova, M. Ibragimova, I. Deryusheva, P. Kazantseva, E. Slonimskaya, I. Frolova, E. Choinzonov, N. Cherdyntseva, and J. Kzhyshkowska. 2018. Expression of M2 macrophage markers YKL-39 and CCL18 in breast cancer is associated with the effect of neoadjuvant chemotherapy. *Cancer Chemotherapy and Pharmacology* 82 (1): 99–109.
195. Mitrofanova, I., M. Zavyalova, V. Riabov, N. Cherdyntseva, and J. Kzhyshkowska. 2018. The effect of neoadjuvant chemotherapy on the correlation of tumor-associated macrophages with CD31 and LYVE-1. *Immunobiology* 223 (6–7): 449–459.
196. Buldakov, M., M. Zavyalova, N. Krakhmal, N. Telegina, S. Vtorushin, I. Mitrofanova, V. Riabov, S. Yin, B. Song, N. Cherdyntseva, and J. Kzhyshkowska. 2017. CD68+, but not stabilin-1+ tumor associated macrophages in gaps of ductal tumor structures negatively correlate with the lymphatic metastasis in human breast cancer. *Immunobiology* 222 (1): 31–38.
197. Mantovani, A., and D.L. Longo. 2018. Macrophage checkpoint blockade in cancer – back to the future. *The New England Journal of Medicine* 379 (18): 1777–1779.
198. Mantovani, A., A. Ponzetta, A. Inforzato, and S. Jaillon. 2019. Innate immunity, inflammation and tumour progression: Double-edged swords. *Journal of Internal Medicine* 285: 524–532.
199. Hamilton, T.A., C. Zhao, P.G. Pavicic Jr., and S. Datta. 2014. Myeloid colony-stimulating factors as regulators of macrophage polarization. *Frontiers in Immunology* 5: 554.
200. Liu, T., I. Larionova, N. Litviakov, V. Riabov, M. Zavyalova, M. Tsyganov, M. Buldakov, B. Song, K. Moganti, P. Kazantseva, E. Slonimskaya, E. Kremmer, A. Flatley, H. Kluter, N. Cherdyntseva, and J. Kzhyshkowska. 2018. Tumor-associated macrophages in human breast cancer produce new monocyte attracting and pro-angiogenic factor YKL-39 indicative for increased metastasis after neoadjuvant chemotherapy. *Oncoimmunology* 7 (6): e1436922.
201. Perelmuter, V.M., L.A. Tashireva, V.N. Manskikh, E.V. Denisov, O.E. Savelieva, E.V. Kaygorodova, and M.V. Zavyalova. 2018. Heterogeneity and plasticity of immune inflammatory responses in the tumor microenvironment: Their role in the antitumor effect and tumor aggressiveness. *Biology Bulletin Reviews* 8 (5): 431–448.
202. Zhang, K., C.A. Corsa, S.M. Ponik, J.L. Prior, D. Piwnica-Worms, K.W. Eliceiri, P.J. Keely, and G.D. Longmore. 2013. The collagen receptor discoidin domain receptor 2 stabilizes SNAIL1 to facilitate breast cancer metastasis. *Nature Cell Biology* 15 (6): 677–687.
203. Park, J., and J.E. Schwarzbauer. 2014. Mammary epithelial cell interactions with fibronectin stimulate epithelial-mesenchymal transition. *Oncogene* 33 (13): 1649–1657.
204. Takai, K., A. Le, V.M. Weaver, and Z. Werb. 2016. Targeting the cancer-associated fibroblasts as a treatment in triple-negative breast cancer. *Oncotarget* 7 (50): 82889–82901.
205. Su, Y.W., T.X. Xie, D. Sano, and J.N. Myers. 2011. IL-6 stabilizes Twist and enhances tumor cell motility in head and neck cancer cells through activation of casein kinase 2. *PLoS One* 6 (4): e19412.

206. Sullivan, N.J., A.K. Sasser, A.E. Axel, F. Vesuna, V. Raman, N. Ramirez, T.M. Oberyszyn, and B.M. Hall. 2009. Interleukin-6 induces an epithelial-mesenchymal transition phenotype in human breast cancer cells. *Oncogene* 28 (33): 2940–2947.
207. Lin, Y., C. Wei, Y. Liu, Y. Qiu, C. Liu, and F. Guo. 2013. Selective ablation of tumor-associated macrophages suppresses metastasis and angiogenesis. *Cancer Science* 104 (9): 1217–1225.
208. Zhang, C., L. Gao, Y. Cai, H. Liu, D. Gao, J. Lai, B. Jia, F. Wang, and Z. Liu. 2016. Inhibition of tumor growth and metastasis by photoimmunotherapy targeting tumor-associated macrophage in a sorafenib-resistant tumor model. *Biomaterials* 84: 1–12.
209. Kim, Y.B., Y.H. Ahn, J.H. Jung, Y.J. Lee, J.H. Lee, and J.L. Kang. 2019. Programming of macrophages by UV-irradiated apoptotic cancer cells inhibits cancer progression and lung metastasis. *Cellular and Molecular Immunology* 16: 851–867.
210. Theodoraki, M.N., S.S. Yerneni, C. Brunner, J. Theodorakis, T.K. Hoffmann, and T.L. Whiteside. 2018. Plasma-derived exosomes reverse epithelial-to-mesenchymal transition after photodynamic therapy of patients with head and neck cancer. *Oncoscience* 5 (3–4): 75–87.
211. Hughes, R., B.Z. Qian, C. Rowan, M. Muthana, I. Keklikoglou, O.C. Olson, S. Tazzyman, S. Danson, C. Addison, M. Clemons, A.M. Gonzalez-Angulo, J.A. Joyce, M. De Palma, J.W. Pollard, and C.E. Lewis. 2015. Perivascular M2 macrophages stimulate tumor relapse after chemotherapy. *Cancer Research* 75 (17): 3479–3491.
212. Chen, L., J. Li, F. Wang, C. Dai, F. Wu, X. Liu, T. Li, R. Glauben, Y. Zhang, G. Nie, Y. He, and Z. Qin. 2016. Tie2 expression on macrophages is required for blood vessel reconstruction and tumor relapse after chemotherapy. *Cancer Research* 76 (23): 6828–6838.
213. Stafford, J.H., T. Hirai, L. Deng, S.B. Chernikova, K. Urata, B.L. West, and J.M. Brown. 2016. Colony stimulating factor 1 receptor inhibition delays recurrence of glioblastoma after radiation by altering myeloid cell recruitment and polarization. *Neuro-Oncology* 18 (6): 797–806.
214. Escamilla, J., S. Schokrpur, C. Liu, S.J. Priceman, D. Moughon, Z. Jiang, F. Pouliot, C. Magyar, J.L. Sung, J. Xu, G. Deng, B.L. West, G. Bollag, Y. Fradet, L. Lacombe, M.E. Jung, J. Huang, and L. Wu. 2015. CSF1 receptor targeting in prostate cancer reverses macrophage-mediated resistance to androgen blockade therapy. *Cancer Research* 75 (6): 950–962.
215. Zhu, Y., B.L. Knolhoff, M.A. Meyer, T.M. Nywening, B.L. West, J. Luo, A. Wang-Gillam, S.P. Goedegebuure, D.C. Linehan, and D.G. DeNardo. 2014. CSF1/CSF1R blockade reprograms tumor-infiltrating macrophages and improves response to T-cell checkpoint immunotherapy in pancreatic cancer models. *Cancer Research* 74 (18): 5057–5069.
216. Brown, J.M., L. Recht, and S. Strober. 2017. The promise of targeting macrophages in cancer therapy. *Clinical Cancer Research* 23 (13): 3241–3250.
217. Gort, E.H., A.J. Groot, E. van der Wall, P.J. van Diest, and M.A. Vooijs. 2008. Hypoxic regulation of metastasis via hypoxia-inducible factors. *Current Molecular Medicine* 8 (1): 60–67.
218. Imai, T., A. Horiuchi, C. Wang, K. Oka, S. Ohira, T. Nikaido, and I. Konishi. 2003. Hypoxia attenuates the expression of E-cadherin via up-regulation of SNAIL in ovarian carcinoma cells. *The American Journal of Pathology* 163 (4): 1437–1447.
219. Yang, M.H., M.Z. Wu, S.H. Chiou, P.M. Chen, S.Y. Chang, C.J. Liu, S.C. Teng, and K.J. Wu. 2008. Direct regulation of TWIST by HIF-1alpha promotes metastasis. *Nature Cell Biology* 10 (3): 295–305.
220. Mak, P., I. Leav, B. Pursell, D. Bae, X. Yang, C.A. Taglienti, L.M. Gouvin, V.M. Sharma, and A.M. Mercurio. 2010. ERbeta impedes prostate cancer EMT by destabilizing HIF-1alpha and inhibiting VEGF-mediated snail nuclear localization: Implications for Gleason grading. *Cancer Cell* 17 (4): 319–332.

221. Kim, H.J., J.W. Park, Y.S. Cho, C.H. Cho, J.S. Kim, H.W. Shin, D.H. Chung, S.J. Kim, and Y.S. Chun. 2013. Pathogenic role of HIF-1alpha in prostate hyperplasia in the presence of chronic inflammation. *Biochimica et Biophysica Acta* 1832 (1): 183–194.
222. Cannito, S., E. Novo, A. Compagnone, L. Valfre di Bonzo, C. Busletta, E. Zamara, C. Paternostro, D. Povero, A. Bandino, F. Bozzo, C. Cravanzola, V. Bravoco, S. Colombatto, and M. Parola. 2008. Redox mechanisms switch on hypoxia-dependent epithelial-mesenchymal transition in cancer cells. *Carcinogenesis* 29 (12): 2267–2278.
223. Jiao, M., and K.J. Nan. 2012. Activation of PI3 kinase/Akt/HIF-1alpha pathway contributes to hypoxia-induced epithelial-mesenchymal transition and chemoresistance in hepatocellular carcinoma. *International Journal of Oncology* 40 (2): 461–468.
224. Baran, N., and M. Konopleva. 2017. Molecular pathways: Hypoxia-activated prodrugs in cancer therapy. *Clinical Cancer Research* 23 (10): 2382–2390.
225. Redfern, A.D., L.J. Spalding, and E.W. Thompson. 2018. The Kraken Wakes: Induced EMT as a driver of tumour aggression and poor outcome. *Clinical & Experimental Metastasis* 35 (4): 285–308.
226. Jolly, M.K., K.E. Ware, S. Gilja, J.A. Somarelli, and H. Levine. 2017. EMT and MET: Necessary or permissive for metastasis? *Molecular Oncology* 11 (7): 755–769.
227. Jolly, M.K., J.A. Somarelli, M. Sheth, A. Biddle, S.C. Tripathi, A.J. Armstrong, S.M. Hanash, S.A. Bapat, A. Rangarajan, and H. Levine. 2019. Hybrid epithelial/mesenchymal phenotypes promote metastasis and therapy resistance across carcinomas. *Pharmacology & Therapeutics* 194: 161–184.
228. Lee, J.M., S. Dedhar, R. Kalluri, and E.W. Thompson. 2006. The epithelial-mesenchymal transition: New insights in signaling, development, and disease. *The Journal of Cell Biology* 172 (7): 973–981.
229. Kroger, C., A. Afeyan, J. Mraz, E.N. Eaton, F. Reinhardt, Y.L. Khodor, P. Thiru, B. Bierie, X. Ye, C.B. Burge, and R.A. Weinberg. 2019. Acquisition of a hybrid E/M state is essential for tumorigenicity of basal breast cancer cells. *Proceedings of the National Academy of Sciences of the United States of America* 116 (15): 7353–7362.
230. Jolly, M.K., S.A. Mani, and H. Levine. 2018. Hybrid epithelial/mesenchymal phenotype(s): The 'fittest' for metastasis? *Biochimica Et Biophysica Acta. Reviews on Cancer* 1870 (2): 151–157.
231. Aceto, N., A. Bardia, D.T. Miyamoto, M.C. Donaldson, B.S. Wittner, J.A. Spencer, M. Yu, A. Pely, A. Engstrom, H. Zhu, B.W. Brannigan, R. Kapur, S.L. Stott, T. Shioda, S. Ramaswamy, D.T. Ting, C.P. Lin, M. Toner, D.A. Haber, and S. Maheswaran. 2014. Circulating tumor cell clusters are oligoclonal precursors of breast cancer metastasis. *Cell* 158 (5): 1110–1122.
232. Mani, S.A., W. Guo, M.J. Liao, E.N. Eaton, A. Ayyanan, A.Y. Zhou, M. Brooks, F. Reinhard, C.C. Zhang, M. Shipitsin, L.L. Campbell, K. Polyak, C. Brisken, J. Yang, and R.A. Weinberg. 2008. The epithelial-mesenchymal transition generates cells with properties of stem cells. *Cell* 133 (4): 704–715.
233. Celia-Terrassa, T., O. Meca-Cortes, F. Mateo, A. Martinez de Paz, N. Rubio, A. Arnal-Estape, B.J. Ell, R. Bermudo, A. Diaz, M. Guerra-Rebollo, J.J. Lozano, C. Estaras, C. Ulloa, D. Alvarez-Simon, J. Mila, R. Vilella, R. Paciucci, M. Martinez-Balbas, A.G. de Herreros, R.R. Gomis, Y. Kang, J. Blanco, P.L. Fernandez, and T.M. Thomson. 2012. Epithelial-mesenchymal transition can suppress major attributes of human epithelial tumor-initiating cells. *The Journal of Clinical Investigation* 122 (5): 1849–1868.
234. Liu, S., Y. Cong, D. Wang, Y. Sun, L. Deng, Y. Liu, R. Martin-Trevino, L. Shang, S.P. McDermott, M.D. Landis, S. Hong, A. Adams, R. D'Angelo, C. Ginestier, E. Charafe-Jauffret, S.G. Clouthier, D. Birnbaum, S.T. Wong, M. Zhan, J.C. Chang, and M.S. Wicha. 2014. Breast cancer stem cells transition between epithelial and mesenchymal states reflective of their normal counterparts. *Stem Cell Reports* 2 (1): 78–91.

235. Jolly, M.K., B. Huang, M. Lu, S.A. Mani, H. Levine, and E. Ben-Jacob. 2014. Towards elucidating the connection between epithelial-mesenchymal transitions and stemness. *Journal of the Royal Society Interface* 11 (101): 20140962.
236. Goldman, A., B. Majumder, A. Dhawan, S. Ravi, D. Goldman, M. Kohandel, P.K. Majumder, and S. Sengupta. 2015. Temporally sequenced anticancer drugs overcome adaptive resistance by targeting a vulnerable chemotherapy-induced phenotypic transition. *Nature Communications* 6: 6139.
237. Tsai, J.H., J.L. Donaher, D.A. Murphy, S. Chau, and J. Yang. 2012. Spatiotemporal regulation of epithelial-mesenchymal transition is essential for squamous cell carcinoma metastasis. *Cancer Cell* 22 (6): 725–736.
238. Watanabe, K., A. Villarreal-Ponce, P. Sun, M.L. Salmans, M. Fallahi, B. Andersen, and X. Dai. 2014. Mammary morphogenesis and regeneration require the inhibition of EMT at terminal end buds by Ovol2 transcriptional repressor. *Developmental Cell* 29 (1): 59–74.
239. Boareto, M., M.K. Jolly, A. Goldman, M. Pietila, S.A. Mani, S. Sengupta, E. Ben-Jacob, H. Levine, and J.N. Onuchic. 2016. Notch-Jagged signalling can give rise to clusters of cells exhibiting a hybrid epithelial/mesenchymal phenotype. *J R Soc Interface* 13 (118): 20151106.
240. Cheung, K.J., V. Padmanaban, V. Silvestri, K. Schipper, J.D. Cohen, A.N. Fairchild, M.A. Gorin, J.E. Verdone, K.J. Pienta, J.S. Bader, and A.J. Ewald. 2016. Polyclonal breast cancer metastases arise from collective dissemination of keratin 14-expressing tumor cell clusters. *Proceedings of the National Academy of Sciences of the United States of America* 113 (7): E854–E863.
241. Bocci, F., L. Gearhart-Serna, M. Boareto, M. Ribeiro, E. Ben-Jacob, G.R. Devi, H. Levine, J.N. Onuchic, and M.K. Jolly. 2019. Toward understanding cancer stem cell heterogeneity in the tumor microenvironment. *Proceedings of the National Academy of Sciences of the United States of America* 116 (1): 148–157.
242. Grassian, A.R., F. Lin, R. Barrett, Y. Liu, W. Jiang, M. Korpal, H. Astley, D. Gitterman, T. Henley, R. Howes, J. Levell, J.M. Korn, and R. Pagliarini. 2012. Isocitrate dehydrogenase (IDH) mutations promote a reversible ZEB1/microRNA (miR)-200-dependent epithelial-mesenchymal transition (EMT). *The Journal of Biological Chemistry* 287 (50): 42180–42194.
243. Waitkus, M.S., B.H. Diplas, and H. Yan. 2018. Biological role and therapeutic potential of IDH mutations in cancer. *Cancer Cell* 34 (2): 186–195.

Targeting the Tumor-Associated Macrophages for 'Normalizing' Cancer

Julia Kzhyshkowska, Evgenia Grigoryeva, and Irina Larionova

Introduction

Despite significant progress in cancer diagnostics and development of novel therapeutic approaches, successful treatment of cancer is still a challenge and may require personalized therapeutic approaches. Cancer cells in solid tumors are surrounded by cellular and molecular microenvironment that actively involved in tumor development. Local tumor microenvironment (TME) consisting of immune cells, surrounding blood vessels, fibroblasts, the extracellular matrix (ECM) strongly modulates responses to treatment [1].

Immune microenvironment provides primary tumor growth, activating invasion and metastasis, inducing angiogenesis, tumor-promoting inflammation, immune suppression and resistance to chemo- and radiotherapy [1–3]. Numerous studies identified tumor-associated macrophages (TAMs) as key cells of immune component in TME. During treatment TAMs mediate tumor revascularization, resistance to chemotherapy, tumor re-growth, suppression of cytotoxic T cell immunity, activation of anti-apoptotic program in tumor cells [4–7]. These properties make TAMs

J. Kzhyshkowska (✉)
Institute of Transfusion Medicine and Immunology, Medical Faculty Mannheim, University of Heidelberg, Mannheim, Germany

German Red Cross Blood Service Baden-Württemberg – Hessen, Mannheim, Germany

Laboratory of translational cellular and molecular biomedicine, National Research Tomsk State University, Tomsk, Russia
e-mail: julia.kzhyshkowska@medma.uni-heidelberg.de

E. Grigoryeva · I. Larionova
Laboratory of translational cellular and molecular biomedicine, National Research Tomsk State University, Tomsk, Russia

Cancer Research Institute, Tomsk National Research Medical Center of the Russian Academy of Sciences, Tomsk, Russia

M. Bizzarri (ed.), *Approaching Complex Diseases*, Human Perspectives in Health Sciences and Technology 2, https://doi.org/10.1007/978-3-030-32857-3_11

attractive targets for immunomodulatory cancer therapy. Moreover, plasticity of TAM phenotype creates the possibility for directional re-programming/conversion of this population to exert anti-tumor activity [8, 9]. The main advantage of targeting the tumor microenvironment is the genetic stability of non-tumor cells, in contrast to tumor cells that are often highly unstable and can rapidly accumulate adaptive mutations resulting in drug resistance. Conventional chemotherapy is not sufficient to eliminate tumor-supporting TAMs, so several strategies have been developed with the aim of manipulating macrophages [4]. Clinical and experimental studies indicated that re-programming of immune components in tumor can be achieved by conventional or metronomic chemotherapy, TAM targeting and immunotherapy [10–12]. Functional "re-orientation" of macrophages into the antitumor phenotype triggers a cascade of events resulting in the inhibiting tumor growth, blocking tumor cell metastatic potential and creating equilibrium between cancer cell and immune component of TME that leads to suppression of tumor progression. The main approaches of targeting of TAMs can be identified, such as inhibitors of the recruitment of monocytes, inductors of apoptosis of TAMs and the most perspective and beneficial approach based on reprogramming of macrophages.

Origin of Macrophages

Monocytes are precursors of the cells of the mononuclear phagocytic system, which includes monocytes, macrophages and dendritic cells (DC). It is known that the DC population can also directly differentiate from the DC precursor [13]. Phagocytic system provides on the one hand, the removal of microorganisms by phagocytic activity of the cells, and on the other hand, the subsequent induction of adaptive immune responses mediated by T-cells. The cells of the macrophage system have fundamental function in tissue repair during inflammation and in maintaining tissue homeostasis by regulating the formation of vascular network [14–16].

As soon as monocytes are attracted from bone marrow (BM) to the circulation, they can migrate to various tissues and differentiate into macrophages, and subsequently become macrophages with their own phenotypic and functional characteristics depending on the tissue. According to the specific anatomic site they are differently called: alveolar macrophages in the lungs, microglia in the central nervous system (CNS), Kupffer cells in the liver etc.

Blood monocytes are reasonably considered to be precursors of tissue macrophages, since the main fraction of macrophages is derived from circulating blood monocytes. It is important to note that there are tissue resident macrophages which originate from monocytes, and their self-renewal, distribution, origin, and reproduction are not fully understood. In addition to monocytes, the yolk sac and fetal liver were found to be two additional sources of macrophages carrying the colony-stimulating factor-1 receptor (CSF-1R) [17, 18]. The presence of macrophages in the yolk sac of the mouse was reported on the 9th day of pregnancy, indicating that part of the macrophages can exist there before promonocytes and before the development of monocytes [19]. Hematopoiesis in the fetal liver initially seeded by hematopoietic progenitors from the yolk sac and subsequently from the hematogenic endothelium of the aorto-gonadal-mesonephros region of the embryo. During embryogenesis the fetal liver is the source of definitive hematopoiesis generating circulating monocytes. The post-natal formation of bone is followed by the fetal liver hematopoiesis reduction and the replacement by BM hematopoiesis [20].

In mouse model it has been demonstrated that microglia can be functionally maintained independently of the BM progenitor cells [21]. There is also experimental evidence confirming that microglia consists of hematopoietic cells. Inhibition of monocytes in the CNS prevents autoimmune encephalitis, that confirms the origin of macrophages from blood monocytes or other sources [22]. It is not clear how the population of resident macrophages is constantly self-renewing. Colony stimulating factor 1 (CSF-1) has been proposed as a protein regulating the amount of tissue-resident macrophages [23]; however, interleukin-4 (IL-4) has been shown to play a major role in the local proliferation of macrophages in parasitic infections. In this case, IL-4- mediated proliferation is independent of CSF-1 [24].

Diversity and Plasticity of Macrophages

Macrophages are heterogeneous cells with high plasticity which represent various phenotypes in response to different signals in microenvironment (for example, bacterial infections, tissue damage, tumor development). Macrophages are polarized towards the classically activated or "M1" phenotype using Th1 lymphocyte cytokines (IFN-γ) and/or activation of Toll-like receptors when interacting with bacterial components (for example, lipopolysaccharides) [20, 25]. Therefore, M1 macrophages are involved into the responses to pathogens and are able to increase the level of pro-inflammatory cytokines, such as IL-12, IL-8, IL-6 and tumor necrosis factor α (TNF-alpha), increase the expression of molecules of the main histocompatibility complex of class II (MHC II), generate reactive oxygen species (ROS) and intermediate nitrogen products and stimulate cell death [26]. In response to IL-4, IL-10 and IL-13, macrophages are polarized towards an alternatively activated or "M2" phenotype involved in the Th2 type reactions, which include dampening of inflammation and wound healing, angiogenesis, immunosuppression, and tissue remodeling [20, 27]. They are characterized by a high level of expression of scavenger receptors, mannose and galactose receptors, activation of arginase, production of IL-10, vascular endothelial growth factor (VEGF) and matrix metalloproteinases (MMPs) and effective phagocytic activity [26, 27].

Macrophage secreted factors are indicative for the chronic inflammation. Macrophages control the process of resolution of the inflammation and the healing phase. During the healing phase, macrophages suppress inflammatory activities of other immune cells by releasing tolerogenic cytokines, clear the tissue from unwanted-self components and apoptotic cells, induce and support angiogenesis to supply healing tissue with oxygen and nutrition, and stimulate reconstitution of normal tissue composition [28]. In the process of chronic inflammation M1- polarized macrophages mediate tissue damage and initiate inflammatory responses. Mechanisms of the tissue repair are accompanied by infiltrating M2 macrophages [26].

Numerous studies confirmed that type of macrophage polarization is defined by the set of secreted and surface markers. Molecular profiles of the macrophage subpopulations have been verified by various methods including RT-PCR, Western blotting, flow cytometry, immunofluorescence/confocal microscopy as well as ELISA. Our previous results demonstrated that IL-4 is a major driver for extracellular matrix (ECM) remodeling activity of M2 macrophage by the releasing matrix metalloproteinases (MMPs), tissue transglutaminase, CCL18, and other ECM components [29] (Fig. 1). The major effect of glucocorticoids is the enhancement of endocytic and phagocytic activity in macrophages toward unwanted-self components that are essential both for the resolution of inflammation and maintenance of

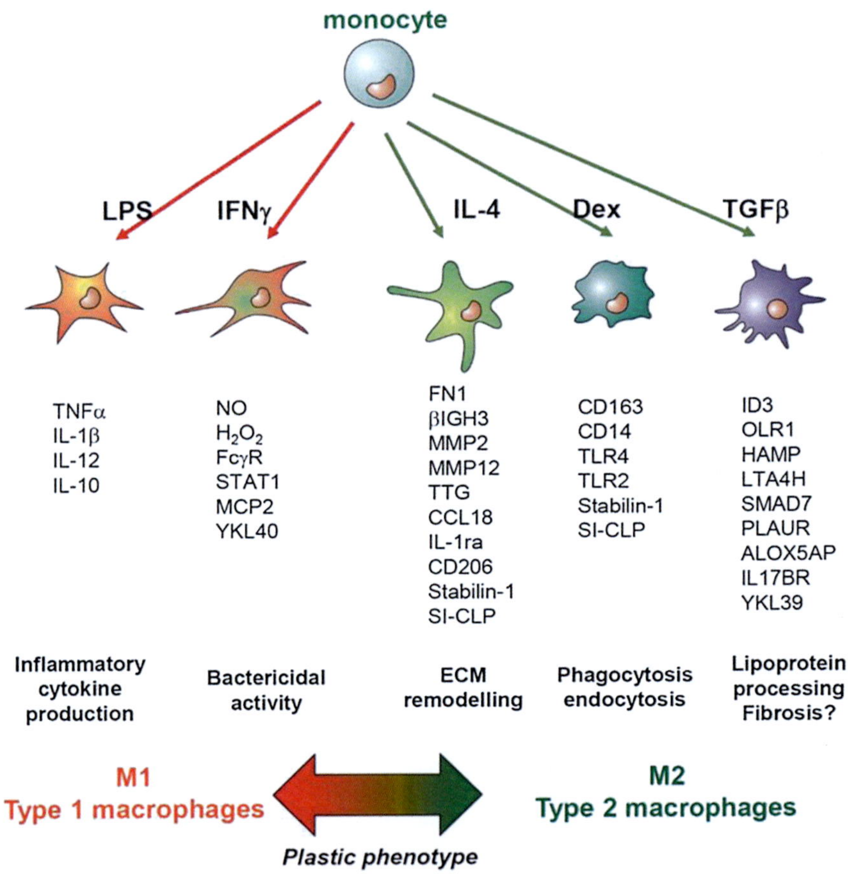

Fig. 1 Molecular signatures and functional subtypes of human ex vivo generated monocyte-derived macrophages. (Copyright from Kzhyshkowska et al. [28])

homeostatic tissue balance [28–31]. LPS was identified to induce inflammatory cytokine production in macrophages while IFNgamma activates strong microbicidal and tumoricidal activity through the high production of reactive nitrogen and oxygen intermediates that are grouped asM1-driving stimuli [26].

High plasticity of macrophages, i.e. the ability to reprogram functions under the influence of different external factors indicates their significance as therapeutic targets, the impact on which can significantly modulate the conditions of the pathological microenvironment. Diversity of macrophages allows selective targeting of specific subsets by using single of multiple surface markers.

Tumor-Associated Macrophages

Tumors synthesize the factors which re-polarize resident macrophages or, more likely, attract new monocytes that differentiate into tumor-promoting cells. Tumor infiltration with monocytes is mediated by a number of chemokines (such as CCL2, CCL5 and CXCL12), CSF-1, as well as components of the complement [32]. Circulating monocytes are recruited to the tumor site and programmed by tumor derived factors into tumor-associated macrophages (TAMs), which have tumor-supportive M2 phenotype [26, 33]. The polarization of macrophages requires IL-4, produced by CD4+ T-lymphocytes and/or tumor cells [34], and growth factors of tumor origin, for example, CSF-1 [35] and GM-CSF [36].

A significant number of experimental and clinical data indicate a supportive role for macrophages in cancer development [33, 37]. In solid tumors, TAMs can promote primary tumor growth, induce angiogenesis, lymphangiogenesis, stromal remodeling, metastasis and a suppression of immunity [38, 39]. TAMs express molecules that directly affect cancer cell proliferation, including epidermal growth factor (EGF), members of the fibroblast growth factor (FGF) family, and transforming growth factor beta (TGFβ) [37, 40]. TAMs produce proangiogenic growth factors, such as vascular endothelial growth factor A (VEGF-A), VEGF-C, tumor necrosis factor α (TNFα), FGF, thymidine phosphorylase (TP), urokinase plasminogen activator (uPA), adrenomedullin (ADM), and semaphorin 4D (Sema4D) [37, 39]. TAMs also produce several factors that are responsible for the induction of lymphangiogenesis, including VEGF-C, VEGF-D, VEGF-A, MMP2, MMP9, CXCL8 and many others [37, 39, 40]. The ability of TAMs to support tumor cell invasion and metastasis is mediated by the regulation of ECM by releasing plasmin, uPA, matrix metalloproteinases (MMPs), cathepsin B, CCL18 and TGF-β1 [38, 39]. TAMs have an important role in the recruitment and activation of Treg cells and the suppression of effector T cells in tumor microenvironment (TME), by secreting a number of cytokines and chemokines, such as CCL3, CCL4, CCL5, CCL22, TGFβ, and IL10 [33, 41]. TAMs resemble M2 macrophages with enhanced clearance function and express the scavenger receptor CD163, mannose receptor MRC1/CD206, macrophage scavenger receptor I and (CD204) and stabilin-1 [20, 42].

Biomarkers (surface, intracellular) of TAMs in tumor tissues are indicative for the prognosis, the effect of therapy and metastasis status. Diversity of macrophages allows to find the association of TAMs with tumor progression by using single or multiple surface markers. Moreover, there is much experimental clinical evidence which indicates a dual role of TAMs in tumor progression and survival. Thus, the tumor-supporting functions of CD68+ TAMs have been demonstrated for breast cancer, bladder cancer, esophageal cancer, stomach cancer, human Ewing sarcoma, myxoid liposarcoma [43–49] while anti-tumor role of CD68+ macrophages was found in non-small-cell lung cancer (NSCLC), ovarian cancer, colorectal cancer and breast cancer [42, 47, 50, 51] (Table 1).

An increased density of CD163+ was indicative for tumor progression in ovarian cancer, gastric cancer, lung cancer, breast cancer, leiomyosarcoma [52–56]. Similarly, a positive correlation between CD206+ macrophage infiltration and poor survival rates was found in gastric cancer, ovarian cancer, renal cell carcinoma, and hepatocellular carcinoma [57–60] (Table 1).

Macrophage progenitor cells are one of the important components of a metastatic niche [70–72]. The concept of "metastatic niches" proposed by D. Lyden [73] suggests the formation of a cluster of cells of BM origin, which are located in the place of the future development of metastasis before the tumor cells penetrate into it. The metastatic niche provides the homing and survival of the metastatic tumor cell. TAMs can co-migrate with tumor cells, promote tumor cell invasiveness and optimization of local microenvironment to suppress local immune reactions against metastatic cells [74]. Clinical manifestation of metastasis is possible only with the development of angiogenesis and inflammation in the site of metastasis [74]. Thus, the growth of primary tumor, epithelial-mesenchymal transition, underlying invasive growth and metastasis, the formation of metastatic niches, the growth of a secondary metastatic tumor occurs with the participation of macrophages, which play a key role in these processes. In this regard, the possibility to manage functional state of macrophages is promising way for the treatment of primary tumors, and for the prevention and treatment of metastatic tumors.

Macrophages interact with cancer cells and other cells of TME and control tissue remodeling not only by expressing various growth factors, cytokines and components of ECM but also through the internalization and degradation of these factors. This clearance function is particularly effective in alternatively activated macrophages (M2) [75]. Thus, clearance of secreted protein acidic and rich in cysteine (SPARC), a soluble component of the extracellular matrix, was shown to inhibit angiogenesis by modulating the expression of vascular endothelial growth factor (VEGF) and MMPs [76] in alternatively activated macrophages in humans [77]. The multifunctional scavenger receptor stabilin-1 that plays important role in the clearance of "unwanted" self-substances [61] is responsible for the uptake of SPARC and its targeting for degradation in lysosomes. High number of stabilin-1+ TAMs were found in metastasizing primary human breast tumors, and stabilin-1+ TAM have been shown to support tumor growth in a mammary adenocarcinoma mouse model [61].

Table 1 The association of the major macrophage markers and cancer prognosis

Markers	Expression in cancer	Bad prognosis	Good prognosis
CD68 (Scavenger receptor)	Breast cancer	Decreased disease-free survival Enhanced invasion increased number of lymphatic and distant metastasis In tumor stroma positively correlated with tumor size [48, 49]	In gaps of ductal tumor structures negatively correlate with the lymphatic metastasis in human breast cancer [42]
	Human Ewing Sarcoma	Decreased survival [43]	
	Bladder cancer	Decreased survival [44]	
	Gastric Cancer	Enhanced metastasis [45]	
	Myxoid liposarcoma	Decreased overall survival [46]	
	Non-small cell lung cancer (NSCLC)		Higher 5-year survival [47]
	Colorectal cancer		Longer survival [50]
CD163 (Scavenger receptor)	Leiomyosarcoma	Decreased overall survival and disease-specific survival [52]	
	Urinary bladder cancer		Longer survival [53]
	Gastric cancer	Tumor invasion and poor prognosis [54]	
	Breast cancer	Higher grade, larger tumor size, Ki67 positivity, estrogen receptor negativity, progesterone receptor negativity triple-negative/basal-like breast cancer [55]	
	Ovarian cancer	Poor prognosis [56]	
CD206 (Mannose receptor)	Renal cell carcinoma Gastric cancer Ovarian cancer Hepatocellular carcinoma	Decreased survival [57–60]	

(continued)

Table 1 (continued)

Markers	Expression in cancer	Bad prognosis	Good prognosis
Stabilin-1 (scavenger receptor)	Breast cancer	Satbilin-1 expression is enhanced on early stages of human breast cancer breast cancer growth is suppressed in stabilin-1 knockout mice (mouse model) [61]	
	Melanoma	Stabilin-1+ TAMs found in all pathological stages of melanoma [62]	
	Bladder cancer		Stabilin-1+macrophages were not associated with BC mortality, but a high Stabilin-1+ vessel count, by contrast, was associated with improved survival in univariate models in the TUR-BT cohort [44]
LYVE-1 (Hyaluronan receptor)	Melanoma		LYVE-1+macrophages decreases in melanoma metastasis [63]
	Breast cancer	Associated with poor outcome [64]	
YKL-40 (Chitinase-like protein)	Glioblastoma Breast cancer Colon cancer Lung cancer Prostate cancer Bladder cancer Stomach cancer Endometrial cancer Esophageal squamous cell carcinoma Liver cancer Pancreas cancer	Poor outcome or decreased disease- free survival [65–68]	
	Head and neck cancer (circulation level)		
YKL-39 (Chitinase-like protein)	Breast cancer	Elevated levels of YKL-39 after neoadjuvant chemotherapy are predictive for high metastatic potential and tumor resistance to the treatment [69]	

Macrophages serve as a major source of human chitinase-like proteins (CLPs), which belong to the family of Glyco_18 domain containing proteins and in humans include YKL-39, YKL-40 and SI-CLP [78, 79]. Biological activities of CLPs related to tumor progression include chemotactic activity, growth factor activity, stimulation of angiogenesis [79]. YKL-39 was identified by us to combine monocytes chemotactic and pro-angiogenic activities [69]. We showed that the expression of YKL-39 positively correlated with high metastasis rate and poor response to neoadjuvant chemotherapy in patients with breast cancer. Elevated levels of circulating YKL-40 are related to poor outcome or short disease-free survival in many cancers in humans, including glioblastoma, melanoma, ovarian, breast, colon, lung, and prostate cancers [65–68]. Moreover, an elevated serum levels of YKL-40 is used as an independent prognostic biomarker [68].

A number of studies demonstrated that TAMs may contribute to resistance to therapy and facilitate tumor progression via several mechanisms, including the macrophage-induced suppression of T cell immunity, the maintenance of tumor cell survival and the stimulation of tumor revascularization [4, 80–82]. Alternatively activated M2 macrophages can mediate chemotherapeutic resistance by secreting growth factors and inhibiting cell death signaling pathways in tumor cells, protecting them from the cytotoxic effects of chemotherapy [82]. For example, the infiltration of CD68+ and CD163+ macrophages in esophageal cancer tissue after neoadjuvant chemotherapy, significantly correlated with tumor depth, lymphatic and blood vessel invasion, and poor prognosis [83]. Chemotherapeutic treatment of mouse Lewis lung carcinoma model and mouse model of breast cancer metastasis resulted in a significant increase in the number of CD206+ TAMs, accumulating mostly in the vascularized regions of tumors that caused tumor revascularization and relapse [84]. In breast cancer accumulation of macrophages protecting tumors was also found after neoadjuvant chemotherapy [85].

These data indicated that chemotherapy is not sufficient to eliminate tumor-supporting macrophages. The combination of chemotherapy and depletion of M2-like TAMs or their reprogramming into the M1-like phenotype can enhance the efficacy of treatment and can be an efficient way to suppress tumor reccurance. An increasing number of studies have focused on complex therapeutic approaches in cancer treatment, including not only chemotherapy regimens but also immunotherapy, designed to activate immune response to increase the efficacy of CTLs against cancers, immune checkpoint blockade therapy, inhibiting immune suppressor molecules, and others approaches that discussed below.

Targeting Tumor-Associated Macrophages

The main advantage of targeting the tumor microenvironment is the genetic stability of non-tumor cells, in contrast to tumor cells that are often highly unstable and can rapidly accumulate adaptive mutations resulting in drug resistance. Conventional chemotherapy is not sufficient to eliminate tumor-supporting TAMs, so several strategies have been developed with the aim of manipulating macrophages, and their re-programming and depletion can be considered the most promising approaches. (Table 2).

Table 2 The most investigated compounds for macrophage immunomodulation and their clinical trials

Compound	Target	Clinical Phase	Tumor type	References
Inhibition of recruitment				
CNTO 888 (carlumab)	CCL2	Phase 2	Metastatic castration-resistant prostate cancer	[86]
CNTO 888 + docetaxel, gemcitabine, paclitaxel/carboplatin, pegylated liposomal doxorubicin		Phase 1b	Solid tumors (pancreas, NSCLC, prostate, other)	[87]
CNTO 888 (carlumab)		Phase 1	Solid tumors (colorectal, ovarian, prostate, other)	[88]
PF-6309	CCR2	Phase 1b/2	Pancreatic ductal adenocarcinoma	[89]
Cabiralizumab + nivolumab	CSF-1R	Phase 1a/b	Pancreatic ductal adenocarcinoma	[89]
AMG 820		Phase 1	Solid tumors (colorectal, non-small cell lung cancer, ovarian, other)	[91]
RG7155 (Emactuzumab)		Phase 1	Diffuse-Type Giant Cell Tumor	[92]
PLX-3397 (Pexidartinib)		Phase 1/2	Peripheral nerve sheath tumor	[93]
PLX-3397 (Pexidartinib)		Phase 2	Melanoma	NCT02071940
PLX-3397 (Pexidartinib)		Phase 1	Tenosynovial giant cell tumor	[94]
BLZ945		Phase 1/2	Solid tumors (glioblastoma, pancreatic cancer and triple negative breast cancer)	NCT02829723
Interference with survival; induction of apoptosis				
Trabectedin (ET743, Yondelis)	Caspase-8-dependent apoptosis in macrophages	Phase 2	Metastatic or inoperable soft tissue sarcomas	[95]
Clodronate	Mitochondria-mediated apoptosis	Phase 3	Breast Cancer	NCT00009945
Repolarization: Inhibition of pro-tumor activity				
Curcumin	STAT3	Phase 2	Pancreatic cancer	NCT00094445
Gefitinib	STAT6	Phase 4	Non-small cell lung cancer	NCT03264794

(continued)

Table 2 (continued)

Compound	Target	Clinical Phase	Tumor type	References
Repolarization: Stimulation of anti -tumor activity				
CP-870,893 monoclonal antibody	CD40	Phase 1	Pancreatic ductal adenocarcinoma	[96]
Resiquimod (R848)	TLR7/8	Phase 1	Cutaneous T-cell lymphoma	[97]
Motolimod + pegylated liposomal doxorubicin	TLR8	Phase 2	Ovarian cancer	[98]
Hu5F9-G4 (Humanised anti-CD47 antibody) + cetuximab	CD47	Phase 1/2	Colorectal cancer, head and neck cancer	NCT02953782
Hu5F9-G4 (Humanised anti-CD47 antibody)		Phase 1	Solid tumors	NCT02216409
CC-90002 (Humanised anti-CD47 antibody)		Phase 1	Solid and hemato-logic cancers	NCT02367196
TTI-621 (SIRPα-Fc fusion protein)	SIRPα	Phase 1	Solid and hemato-logic cancers	NCT02663518
ALX148 (High-affinity SIRPα variant)		Phase 1	Solid tumors	NCT03013218

Inhibition of the Recruitment of Monocytes to the Tumor Mass

The recruitment of TAMs can be inhibited in several ways (Fig. 2). Firstly, by using blocking antibodies against CCR2 receptor or its ligand CCL2, the key chemokine that regulates migration and infiltration of monocytes/macrophages. It has been shown to block of the CCL2/CCR2 axis decreased macrophage infiltration and reduced tumor growth [99]. Inhibition of CCL2 with specific antibodies reduced tumor growth of glioma, colon, prostate cancers, and melanoma in animal models [100, 101]. CCL2 directly interacts with CCR2 on the endothelial cell surface, leading to increased vessel formation and angiogenesis [102]. Clinical trials have been carried out with the use of the anti-CCL2 antibody (clinical trials NCT00537368, NCT00992186, and NCT01204996) but the result was rather contradictory. In the Phase I clinical trial, administration of anti-CCL2 monoclonal antibody, carlumab (CNTO 888), was well tolerated and showed significant efficacy in patients with solid tumors. However, a Phase 2 study of carlumab in patients with metastatic castration-resistant prostate cancer (CRPC) showed that antibody did not block the CCL-2/CCR-2 axis and did not demonstrate antitumor activity while using as a single agent in metastatic CRPC [86]. Similar results were observed in the study of Brana et al. [87] while conducted a first-in-human phase 1b study of carlumab with one of four chemotherapy regimens (docetaxel, gemcitabine, paclitaxel + carboplatin, and pegylated liposomal doxorubicin HCl [PLD]). Combination

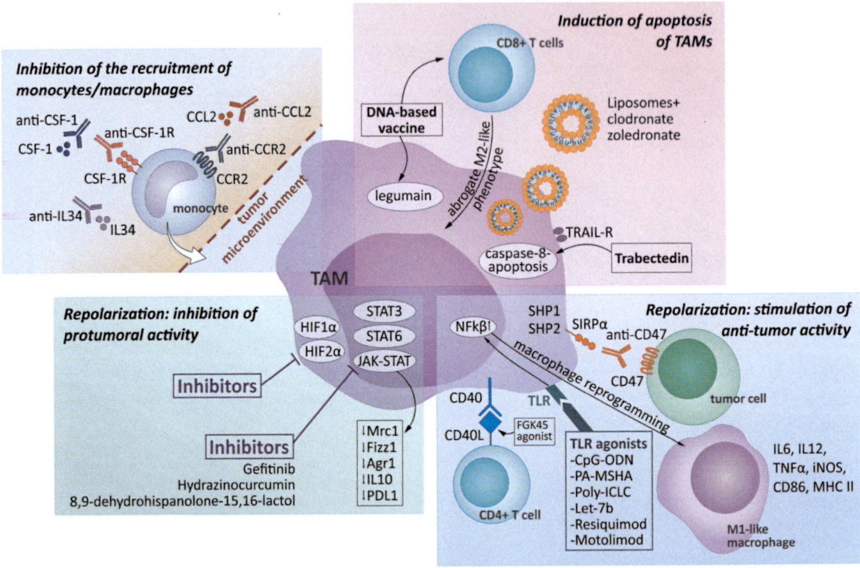

Fig. 2 Immunomodulatory strategies for targeting tumor-associated macrophages (TAMs). The recruitment of TAMs can be inhibited in several ways: antibodies against chemokines that regulate migration and infiltration of monocytes/macrophages and its receptors, such as CCR2 and CCL2, CSF-1/CSF1R axis, IL34. Induction of apoptosis of TAMs may be achieved by using novel legumain-based DNA vaccine, that stimulates CD8+ T cells and abrogates M2-like TAMs. Liposomes loaded with bisphosphonates were proved to decrease the numbers of monocytes and macrophage. Trabectedin activates caspase-8-dependent apoptosis in macrophages carried decoy TRAIL receptors. Using inhibitors of STAT signaling and hypoxia inducible factors, HIF-1α and HIF-2α, provides inhibition of pro-tumor activity of TAMs. Stimulation of anti-tumor activity may occur during using CD40 agonists. Another way of inducing M1 polarization is stimulation of TAMs by different TLR agonists. New approach for TAM polarization associated with targeting the CD47-SIRPα axis

treatment with carlumab had no clinically relevant pharmacokinetic effect [87]. According to Sandhu et al. [88], carlumab was well tolerated with evidence of transient CCL2 suppression and preliminary antitumor activity [95]. Further clinical research is needed to clarify that mechanisms and patterns found in animal models are comparable to human pathological processes. An alternative way to inhibit CCL2/CCR2 axis is blocking chemokine receptor CCR2. CCR-2 kinase inhibitor, PF-0416309 (PF-6309) in combination with FOLFIRINOX chemotherapy (oxaliplatin, irinotecan plus leucovorin and fluorouracil) is currently being studied in a Phase 1b/2 trial for pancreatic cancer and did not result in additional toxicity [89]. CCR2 blockade demonstrated a reduction in the TAM infiltration and indicated the endogenous anti-tumor immune response. More research is needed to fully understand the potential of CCR2 inhibition in cancer and identify potential immunotherapy combinations. MLN1202 is a humanized monoclonal antibody with high specificity to CCR2 is under clinical investigation (clinical trial NCT01015560).

Another approach based on gene silencing method involved complexing siRNAs to TAT cell penetrating peptides (Ca-TAT) through non-covalent calcium cross-linking currently being developed. Ca-TAT/siRNA complexes penetrated 3D collagen cultures of breast cancer cells and inhibited CCL2 expression more effectively than conventional antibody neutralization [103].

Reduction of TAMs may also be obtained by inhibition of recruitment of their precursors. CSF-1 receptor (CSF-1R) is expressed by most of the cells of the monocytic lineage and supports proliferation, differentiation and cell survival. CSF-1 also stimulates the chemotactic activity of monocytes and macrophages and is associated with poor overall survival in human triple-negative breast cancer [104]. Antagonists or antibodies to CSF-1/CSF1R have been developed and tested in various preclinical models (e.g., cervical cancer, pancreatic cancer, and glioblastoma) in combination with chemotherapy, radiation therapy, and checkpoint inhibitors. These studies reliably demonstrate their ability to deplete pool of immunosuppressive macrophages and to increase the CD8/CD4 ratio in tumors [92]. Cabiralizumab, a humanized IgG4 monoclonal antibody, binds to CSF-1R and blocks its signaling followed by reducing TAMs and promoting a proinflammatory microenvironment. In a phase 1a/b clinical study combination of cabiralizumab with nivolumab was tolerable and showed tumor immune modulation and prolonged clinical benefit in heavily pretreated patients with advanced pancreatic ductal adenocarcinoma [90, 105]. AMG 820 is human CSF1R antibody that inhibits binding of the ligands CSF1 and IL34 and subsequent ligand-mediated receptor activation. This first-in-human phase I study evaluated the safety pharmacokinetics, pharmacodynamics, and antitumor activity of AMG 820 [91]. RG7155 (Emactuzumab) is a humanized monoclonal antibody that blocks CSF1R activation, is under the phase I clinical trials. Administration of RG7155 to patients led to intense reductions of CSF-1R+CD163+ macrophages in tumor tissues, that was manifested in objective clinical responses in diffuse-type giant cell tumor (Dt-GCT) patients [92]. Pexidartinib (PLX-3397), another CSF1R inhibitor was developed as a selective FLT3 inhibitor for hematological malignancies, but it functions as an inhibitor of CSF-1 receptor-associated kinases. There is data confirmed that pexidartinib increased antitumor immune responses when combined with radiotherapy in glioblastoma models [106]. Several phase 1/2 clinical trials examine the efficiency of PLX397 in the treatment of different tumors (clinical trials NCT02584647, NCT02071940, NCT02472275, NCT02452424, NCT02371369, NCT01596751, NCT02401815, NCT01349049, NCT02734433, NCT01790503, NCT01525602, NCT01042379, and NCT02777710). BLZ 945, a potent, selective and brain-penetrant CSF-1R inhibitor, showing more than 1000-fold selectivity against its closest receptor tyrosine kinase homologs. In glioma-bearing mice, BLZ945 blocks tumor progression and significantly improves survival via CSF-1R inhibition. BLZ945 also inhibits orthotopic tumor growth of patient-derived glioblastoma spheres and proneural glioblastoma cell lines in vivo [107].

Interference with Survival; Induction of Apoptosis of TAMs

Inducing apoptosis of TAMs appears to be an effective immunotherapeutic approach for tumors. Lewēn S. et al. [108] hypothesized that suppression of TAMs can be achieved in mouse model with minigene vaccines against murine MHC class I antigen epitopes of Legumain which has been observed to be highly expressed in several types of tumors and may play a pivotal role in carcinogenesis [108]. A legumain-based DNA vaccine stimulates CD8+ T cells and selectively abrogates M2-like TAMs (Fig. 2) in mice with metastatic breast, colon and lung cancers, thereby increasing survival rate and regression of metastasis and angiogenesis [109]. According to Guo P. et al. [110] gastric cancer patients with Legumain-positive localized tumors had lower 5-year overall survival than those with Legumain-negative tumors [110].

The most radical way to inhibit TAMs activity in tumors is their depletion by target compounds that are toxic specifically to macrophages. This approach allows to abrogate the network of signals supporting tumor growth and progression. Nano-technology approaches were developed to deliver bisphosphonates, such as clodronate or zoledronate, to tumors resulting in better antitumoral effect, impaired angiogenesis and decreased metastasis (Fig. 2) [111]. The intratumoral injection of alendronate conjugated with glucomannan into sarcoma-bearing mice was used to target the mannose receptors in TAMs, resulting in their effective depletion via apoptosis [112]. Hattori and colleagues have discovered the effectiveness of folate-decorated zoledronic acid-encapsulating liposomes in inhibiting tumor angiogenesis and tumor growth in a murine model of colon adenocarcinoma. Despite of the compound was able to induce selective cytotoxicity in vitro via the folate receptor, a severe toxicity limited its use in vivo [113]. More recently, nitrogen-containing bisphosphonates were found to be also effective. For example, risedronate was able to bind to micro-calcifications, which were subsequently taken up by macrophages, but not by tumor cells [114]. Furthermore combination of docetaxel with risedronate might be effective for the treatment of angiosarcoma by targeting immunosuppressive cells such as M2 macrophages [115]. This combination of docetaxel with risedronate significantly decreased the production of CCL18, which was previously reported to correlate with the severity and prognosis in cancer patients. Clodronate-loaded liposomes (clondlip) were proved to decrease the numbers of monocytes and macrophages, that correlated with a decrease in tumor growth in different animal models [116]. Recently, an anti-macrophage activity of liposomal clodronate has been confirmed in a mouse model ofhepatocarcinoma [117].

Trabectedin (ET743, Yondelis) is a natural alkaloid derived initially from the Caribbean tunicate, which was originally recognized for its ability to induce cell cycle arrest and death, was found to cause a partial depletion of circulating monocytes and TAMs in cancer patients [95, 118]. Trabectedin is proved to have a strong antitumor properties and obtained full marketing approval from the European Commission in 2015 for the treatment of ovarian cancer and soft-tissue sarcomas and gained the United States Food and Drug Administration (US FDA) approval for the treatment of unresectable or metastatic liposarcoma or leiomyosarcoma [95]. Currently, a number of clinical trials of trabectedin alone and in combination with other drugs are underway. Preclinical studies showed that trabectedin strongly inhibited tumor growth by inducing double-strand breaks in DNA and interrupting the cell cycle. Trabectedin activates caspase-8-dependent apoptosis selectively in monocytes but not in neutrophils and lymphocytes due to the differential expression of signaling and decoy TRAIL receptors (Fig. 2) [118]. Trabectedin was shown to significantly decrease the production of cytokines and chemokines produced by TAMs and supporting tumor growth such as CCL-2, CXCL-8, IL-6, and VEGF [119]. Germano et al. [119] showed that trabectedin strongly decreased tumor growth and the number of TAMs in tumor tissue in vivo.

Also, targeted cell death of TAMs can be induced by immunotoxins which are chimeric proteins consisting of two parts. First one is a binding domain, which is commonly an antibody and the second one is a toxic domain, which is an enzyme usually derived from bacteria or plants. After binding of the antibody to a target cell antigen, the immunotoxin is internalized followed by endosomal processing and releasing the toxin into the cytoplasm, where it induces cell death. Combined molecule of anti-CD64 antibody and ricin showed efficacy for depleting the activated synovial macrophages in rheumatoid arthritis (RA) in vitro [120]. Anti-CD64-ricin immunotoxin also inhibited production of TNFalpha and IL-1beta, and cartilage-degrading activity of RA synovial tissue explants. Nagai T. et al. [121] found that folate receptor beta (FR beta) was expressed on macrophages in human glioblastomas. Recombinant immunotoxin consisting of immunoglobulin heavy and light chain Fv portions of an anti-mouse FR beta monoclonal antibody and Pseudomonas exotoxin A was produced for targeting FR beta-expressing TAMs [121] and showed the apoptosis of such macrophages and suppression in NO and VEGF levels in them.

There are several specific bacteria which can target the macrophage population and induce macrophage apoptosis. For example, a single injection of an attenuated strain of Shigella resulted in caspase-1 dependent apoptosis in TAMs followed by a 74% reduction in transgenic tumors of MMTV-HER-2 mice. TAM depletion was sustained and associated with complete tumor regression [122]. In addition, certain bacteria resided in macrophages, such as Listeria monocytogenes, Chlamydia psittaci and Legionella pneumophila, are also being considered for TAM-targeted immunotherapy [123].

Repolarization: Inhibition of Pro-tumor Activity

Despite promising results of preclinical and clinical trials of TAM-targeted approaches such as macrophage depletion or inhibition of TAM recruitment, an alternative strategy was also developed. Reprogramming TAMs towards a proinflammatory and antitumoral phenotype has become an attractive strategy in immunotherapy. Large-scale transcriptome studies performed on alternative activated macrophages have identified the key genes and signaling pathways that play a critical role in alternative way of macrophage polarization. For example, a role of the myeloid-specific Src family kinase member HCK as a key regulator of alternative polarization of M2- monocytes has been described. Aberrant activation of HCK plays role as a tumor cell-intrinsic oncogene triggering hematological malignancies. Furthermore, high HCK levels correlate with reduced survival in colorectal cancer patients. Accordingly, pharmacological inhibition or genetic reduction of Hck activity suppresses alternative activation of TAMs and as a result growth of colon cancer xenografts [124].

Two members of STAT protein family, STAT3 and STAT6, play an important role in tumor-promoting macrophage polarization. The STAT3 phosphorylation inhibitor hydrazinocurcumin converts TAMs to an M1-like phenotype to suppress angiogenesis and tumor progression in breast cancer [125]. Another small-molecule inhibitor of STAT3 significantly reduced M2-polarization in patients with malignant glioma [126]. TAMs from STAT6 deficient mice displayed M1-like phenotype and enhanced antitumor activity [127]. The study of L. Jiménez-García [128] revealed that 8,9-dehydrohispanolone-15,16-lactol significantly inhibited IL-4- or IL-13-stimulated M2 macrophage activation, following by reduced expression of M2 markers, via suppressing the JAK-STAT signaling pathway [128]. STAT6 was shown to be a target molecule for gefitinib, an EGFR tyrosine kinase inhibitor, which has been approved to the treatment of patients with non-small cell lung cancer metastasis [129]. Gefitinib inhibits IL-13- induced phosphorylation of STAT6, which is a crucial signaling pathway in M2-like polarization [130]. Gefitinib also decreases the mRNA levels of M2 genes including Mrc1, Fizz1, Arg1, Ym1, and IL-10 (Fig. 2).

Hypoxia is a common feature of solid tumors; it has been proposed as a trigger of monocyte migration into tumor site and macrophage repolarization [131]. The most hypoxic areas of solid tumors characterized by an accumulation of the M2-like macrophages [132]. M. M. Leblond et al. [133] evaluated the impact of hypoxia on macrophages tropism and polarization in human glioblastoma in vivo [133] and concluded that circulating monocytes enter the tumor as M0-macrophages and then acquire M2 phenotype with decreased oxygen levels. Factors responsible for the hypoxia-dependent macrophage polarization could be the hypoxia inducible factors, HIF-1α and HIF-2α. N. Takeda et al. [134] showed that inducible NO synthase gene and the arginase 1 gene in M2 polarized macrophages are specifically regulated by HIFs. This study demonstrated that the HIF-α isoforms can be differentially activated: HIF-1α is induced by Th1 cytokines in M1 polarization, whereas HIF-2α is induced by Th2 cytokines during an M2 response [134]. Hypoxia-induced HIF-1α also drives expression of the immune checkpoint protein PD-L1, that is essential for immune suppression and evasion [135]. Overexpression of HIF-1α or HIF-2α is associated with poor clinical outcomes in cancer patients, however it depends on the types of cancer [136]. Hypoxia and activation of the HIF pathway support resistance to chemotherapy, radiation therapy and targeted therapy in various types of cancer [137]. Several drugs targeting HIF mRNA expression, protein synthesis, and dimerization are now investigated in the ongoing clinical trials, summarized in the recent review of Fallah J. et al. (Fig. 2) [137].

Repolarization: Stimulation of Anti-tumor Activity

It is known that strong activation of the M1 phenotype in macrophages requires two consecutive signals. Firstly, in response to IFNγ the expression of TLR significantly increased triggering signal (for example LPS binding initiates a maximal cytotoxic macrophage response). CD40 agonists can also serve as a priming signal leading to the up-regulation of TLR. CD40, a tumor necrosis factor α receptor superfamily member, is widely expressed on cells, such as B cells, dendritic cells and macrophages, as well as endothelial cells and tumor cells [138]. After binding of CD40 with its ligand CD154 on the cell surface, antigen-presenting cells start to secrete proinflammatory cytokines and up-regulate costimulatory molecules CD80 and CD86 that are required for costimulation via CD28 on CD8+ T-cells [139]. In mouse models, the FGK.45 antibody was used to activate CD40 via cross-linking with FcγRIIB [140]. CD40 agonists inhibit tumor growth in animal models, and promote clinical responses in patients with advanced solid tumors [96].

The most common methods of inducing M1 polarization is stimulation by TLR agonists (Fig. 2). This binding activates NF-κB proteins by degradation of inhibitor molecule IκB and allows its translocation into the nucleus where it promotes transcription of proinflammatory cytokine genes [141]. There are different TLR agonists, such aspolyI:C, CpG-oligodeoxynucleotide (CpG-ODN), TLR ligands, anti-IL10R, which could switch M2-like TAMs into M1-like cells with enhanced anti-tumor activity [142]. L. Yang et al. [143] found that treatment of CD163+ macrophages from malignant pleural effusion with immune adjuvant PA-MSHA (pseudomonas aeruginosa – mannose-sensitive hemagglutinin) reeducates CD163+ TAMs to M1 macrophages through TLR4-mediated pathway [143]. Using the immunostimulant Poly-ICLC on murine tumor model resulted in inhibition of tumor growth by activation of TLR3 signaling pathway [144].

In the study of A. Vidyarthi et al. [145] the administration of TLR-3 L in the murine tumor switched the M2 macrophages to M1-phenotype and inhibited the tumor growth [145]. They observed an increased production of IL6, IL12, TNFα, iNOS, an enhanced antigen uptake and improved T-cell priming. Polarization to M1 macrophage was mediated by IFN-αβ signaling pathway. TLR7 stimulation is another attractive target for TAM repolarization as it activates both NFκβ and IRF7-signalling in macrophages.

Let-7b is a synthetic miRNA mimic which interacts with TLR7 and switches TAMs to proinflammatory phenotype. Furthermore, Let-7b directly suppresses IL10 production in CD4+ T cells preventing M2 polarization [146]. Mannose receptor C type 1-targeted nanoparticles, loaded with Let-7b, displayed potent TAM repolarization in vivo, characterized by an increased expression of CD86, MHC II, IL12 and iNOS and a decrease in ARG1 and IL10. Furthermore, Let-7b nanoparticles induced massive infiltration of CD8+ T cells, reduced primary tumor volume by up to 70% and significantly increase survival in a mouse model of breast cancer [147].

Resiquimod (R848) is a synthetic TLR7/8 agonist and was recently reported to potently re-educate TAMs. C. Rodell et al. [148] demonstrated that resiquimod is a potent driver of the M1 phenotype in vitro and that the treatment with R848-loaded β-cyclodextrin nanoparticles (CDNP-R848) results in efficient drug delivery to TAMs in vivo. As a monotherapy, the administration of CDNP-R848 in multiple tumor models in mice altered the functional orientation of the tumor immune microenvironment towards an M1 phenotype, leading to reduced tumor growth [148].

Motolimod is a highly potent and selective TLR8 agonist and now is being evaluated in a randomized Phase 2 trial in patients with ovarian cancer. Despite of motolimod shows potent immune activation, even in the late-stage cancer patients its combination with standard therapy (carboplatin or cisplatin, fluorouracil, cetuximab) does not improve survival in squamous cell carcinoma of the head and neck [149]. Thus, recent data demonstrated that TLR-mediated TAM repolarization could be a perspective antitumour strategy.

Another way of stimulating M1 polarization is based on the direct administration of proinflammatory cytokines. CSF2 is a cytokine that regulates macrophage function by enhancing antigen presentation and immune responsiveness. T. D. Eubank [150] showed that administration of GM-CSF caused a switch from M2 to M1 phenotype in TAMs, characterized by increased iNOS expression and decreased IL4 and IL10 production [150]. Furthermore, their study revealed that GM-CSF stimulates monocytes to secrete VEGF receptor-1 (sVEGFR-1), which binds and inactivates VEGF that inhibits angiogenesis.

IL12 is a cytokine produced by macrophages and DCs and played an important role in cell-mediated immunity against intracellular infection primarily by T and NK cells [151]. In the study of Q. Wang et al. [152] IL-12-overexpressed monocytes co-cultured with hepatocellular carcinoma cells showed M1-like phenotype identified as CD197high IL-12high and CD206low IL-10low, while the expression of pro-tumorigenic cytokines MMP-9, TGF-β and VEGF-A were significantly suppressed. The macrophage polarization is mediated by downregulating the p-Stat3 and c-myc expression. The results of transwell invasion assay indicated that IL-12-overexpressing monocytes significantly reduced the invasiveness of co-cultured hepatocellular carcinoma cells [152].

New approach associated with CD47 molecule for TAM polarization was developed. CD47 expression on tumor cell membranes act as a “don’t eat me” signal for macrophages that makes its blocking a potential therapeutic approach for cancer treatment [153]. CD47 binding to signal-regulatory protein alpha (SIRPα) on macrophages lead to the signal cascade that inhibits macrophage-mediated tumor cell phagocytosis [154]. A number of different drugs targeting the CD47-SIRPα axis are evaluated in patients with solid tumors in clinical trials (http://clinicaltrials.gov identifiers: NCT02216409, NCT02890368, NCT02953782, and NCT03013218) (Fig. 2). The study of S. Gu [155] revealed that anti-CD47 blocking antibody promoted phagocytosis of endometrial cancer cells by macrophages [155]. Moreover, CD47 blockade inhibited the growth of the endometrial tumors in vivo and increased the infiltration of TME by macrophages with antitumor activity. Another recent study have demonstrated that the treatment of primary human glioblastoma cell lines with an anti-CD47 monoclonal antibody led to enhanced tumor-cell phagocytosis by M1 and M2 macrophages, wherein anti-CD47-induced phagocytosis by M1 was more prominent than for M2. Thus, disruption of the CD47/SIRPα axis using antibodies or recombinant proteins followed by stimulation of phagocytosis of cancer cells could be promising approach in cancer treatment.

Specific miRNAs for repolarizing M2 macrophages to M1 phenotype could be a perspective tool for antitumor treatment. However, the macrophage-targeted delivery of miRNA remains a challenge. Liu L. et al. [156] generated redox/pH dual-responsive hybrid polypeptide nanovectors, which consisted of self-crosslinked redox-responsive nanoparticles based on galactose-functionalized n-butylamine-poly(l-lysine)-b-poly(l-cysteine) polypeptides (GLC) coated with DCA-grafted sheddable PEG-PLL (sPEG) copolymers [156]. Encapsulation with sPEG/GLC nanovectors effectively facilitated macrophage-targeted miRNA delivery in the acidic area, but diminished miRNA uptake at neutral pH. Administration of miR155-loaded sPEG/GLC (sPEG/GLC/155) nanocomplexes resulted in100–400 fold increase in miR155 expression in TAMs both in vitro and in vivo. Furthermore, sPEG/GLC/155 effectively repolarized immunosuppressive TAMs to anti-tumor M1 macrophages through elevating M1 macrophage markers (IL-12, iNOS, MHC II) and suppressing M2 macrophage markers (Msr2 and Arg1) in TAMs.

In addition to target TAMs using chemo- and immunotherapies, it is also possible to influence the macrophage polarization with radiotherapy. In vitro, human unpolarized monocyte-derived macrophages switched toward M1-like macrophages after moderating doses of irradiation (2 Gy × 5) that was confirmed by upregulation of such M1 macrophage markers as HLA-DR and CD86 and downregulation of M2 markers CD163, CD206 and reduced production of IL-10. M1-mediated phagocytosis was also increased after moderating radiation irradiation [157].

In the study of H. Prakash [158] the irradiation of CD11b + peritoneal resident TAMs from 26-week-old RT5 mice (late stage macrophages) with 2 Gyresulted in enhancement of the constitutive levels of iNOS protein, as well as NO levels indicating M1 programming of late stage tumor macrophages by irradiation [158]. Interestingly that both low (0,5Gy) and high (20 Gy) doses of irradiation reprogrammed macrophages toward M2-like phenotype [159].

Conclusive Remarks

New therapeutic approaches, based on the combination of targeting TAMs with pharmacological molecules to deplete or re-educate TAMs, have the potential to abolish immunosuppressive mechanisms and activate anti-tumor immune responses in the tumor microenvironment. We combined the results of clinical trials that give an idea about the possibilities of using the macrophage-targeting treatment of cancer in a wide clinical practice. High plasticity of macrophages provides the possibility of their functional programming. The modern immunotherapeutic approaches of targeting tumor-infiltrating macrophages may include inhibition of the recruitment of monocytes to the tumor, activation of apoptosis in macrophages, repolarization, based on the inhibition of pro-tumor activity or stimulation of anti-tumor activity.

Despite macrophages are perspective candidates as targets in the treatment of cancer, it is necessary to take into account that macrophage-targeting therapy has a limitation, due to the phenotypic switch after the completion of the course of treatment. Further researches are needed to establish the role of different macrophage phenotypes in various types of tumors. The use of compounds polarizing macrophages in the right direction in combination with chemo- and radiotherapy may allow to treat cancer in a personalized manner, to provide greater progress in the treatment of tumors and ultimately lead to improved outcomes for cancer patients.

Acknowledgments This work was supported by grant of Russian Scientific Foundation №19-15-00151.

References

1. Whiteside, T.L. 2008. The tumor microenvironment and its role in promoting tumor growth. *Oncogene* 27 (45): 5904–5912. https://doi.org/10.1038/onc.2008.271.
2. Stakheyeva, M., V. Riabov, I. Mitrofanova, N. Litviakov, E. Choynzonov, N. Cherdyntseva, and J. Kzhyshkowska. 2017. Role of the immune component of tumor microenvironment in the efficiency of cancer treatment: Perspectives for the personalized therapy. *Current Pharmaceutical Design* 23 (32): 4807–4826. https://doi.org/10.2174/1381612823666170714161703.
3. Wang, M., J. Zhao, L. Zhang, F. Wei, Y. Lian, Y. Wu, Z. Gong, S. Zhang, J. Zhou, K. Cao, X. Li, W. Xiong, G. Li, Z. Zeng, and C. Guo. 2017. Role of tumor microenvironment in tumorigenesis. *Journal of Cancer* 8 (5): 761–773. https://doi.org/10.7150/jca.17648.
4. Larionova, I., N. Cherdyntseva, T. Liu, M. Patysheva, M. Rakina, and J. Kzhyshkowska. 2019. Interaction of tumor-associated macrophages and cancer chemotherapy. *Oncoimmunology* 8 (7): 1596004. https://doi.org/10.1080/2162402X.2019.1596004.
5. Hughes, R., B.Z. Qian, C. Rowan, M. Muthana, I. Keklikoglou, O.C. Olson, et al. 2015. Perivascular M2 macrophages stimulate tumor relapse after chemotherapy. *Cancer Research* 75 (17): 3479–3491. https://doi.org/10.1158/0008-5472.CAN-14-3587.
6. Ruffell, B., and L.M. Coussens. 2015. Macrophages and therapeutic resistance in cancer. *Cancer Cell* 27 (4): 462–472. https://doi.org/10.1016/j.ccell.2015.02.015.
7. Sugimura, K., H. Miyata, K. Tanaka, T. Takahashi, Y. Kurokawa, M. Yamasaki, et al. 2015. High infiltration of tumor-associated macrophages is associated with a poor response to chemotherapy and poorprognosis of patients undergoing neoadjuvant chemotherapy foresophageal cancer. *Journal of Surgical Oncology* 111 (6): 752–759. https://doi.org/10.1002/jso.23881.
8. Chang, W.J., Y. Du, X. Zhao, L.Y. Ma, and G.W. Cao. 2014. Inflammation-relatedfactors predicting prognosis of gastric cancer. *World Journal of Gastroenterology* 20 (16): 4586–4596. https://doi.org/10.3748/wjg.v20.i16.4586.
9. Lin, E.Y., and J.W. Pollard. 2007. Tumor-associated macrophages press theangiogenic switch in breast cancer. *Cancer Research* 67 (11): 5064–5066. https://doi.org/10.1158/0008-5472.CAN-07-0912.
10. Farkona, S., E.P. Diamandis, and I.M. Blasutig. 2016. Cancer immunotherapy: The beginning of the end of cancer? *BMC Medicine* 14: 73. https://doi.org/10.1186/s12916-016-0623-5.

11. Peng, D., I. Kryczek, N. Nagarsheth, L. Zhao, S. Wei, W. Wang, et al. 2015. Epigenetic silencing of TH1-type chemokines shapes tumour immunity and immunotherapy. *Nature* 527 (7577): 249–253. https://doi.org/10.1038/nature15520.
12. Weir, G.M., R.S. Liwski, and M. Mansour. 2011. Immune modulation by chemotherapy or immunotherapy to enhance cancer vaccines. *Cancers* 3 (3): 3114–3142. https://doi.org/10.3390/cancers3033114.
13. Collin, M., N. McGovern, and M. Haniffa. 2018. Human dendritic cell subsets: An update. *Immunology* 154 (1): 3–20. https://doi.org/10.1111/imm.12888.
14. Fantin, A., J.M. Vieira, G. Gestri, L. Denti, Q. Schwarz, S. Prykhozhij, et al. 2010. Tissue macrophages act as cellular chaperones for vascular anastomosis downstream of VEGF-mediated endothelial tip cell induction. *Blood* 116 (5): 829–840. https://doi.org/10.1182/blood-2009-12-257832.
15. Springall, R., L.M. Amezcua-Guerra, H. Gonzalez-Pacheco, J. Furuzawa-Carballeda, L. Gomez-Garcia, R. Marquez-Velasco, et al. 2013. Interferon-gamma increases the ratio of matrix metalloproteinase-9/tissue inhibitor of metalloproteinase-1 in peripheral monocytes from patients with coronary artery disease. *PLoS One* 8 (8): e72291. https://doi.org/10.1371/journal.pone.0072291.
16. Schaale, K., J. Brandenburg, A. Kispert, M. Leitges, S. Ehlers, and N. Reiling. 2013. Wnt6 is expressed in granulomatous lesions of Mycobacterium tuberculosis-infected mice and is involved I macrophage differentiation and proliferation. *Journal of Immunology* 191 (10): 5182–5195. https://doi.org/10.4049/jimmunol.1201819.
17. Murray, P.J., J.E. Allen, S.K. Biswas, E.A. Fisher, D.W. Gilroy, S. Goerdt, S. Gordon, J.A. Hamilton, L.B. Ivashkiv, T. Lawrence, et al. 2014. Macrophage activation and polarization: Nomenclature and experimental guidelines. *Immunity* 41 (1): 14–20. https://doi.org/10.1016/j.immuni.2014.06.008.
18. Jones, C.V., and S.D. Ricardo. 2013. Macrophages and CSF-1: Implications for development and beyond. *Organogenesis* 9 (4): 249–260. https://doi.org/10.4161/org.25676.
19. Takahashi, K., F. Yamamura, and M. Naito. 1989. Differentiation, maturation, and proliferation of macrophages in the mouse yolk sac: A light-microscopic, enzyme-cytochemical, immunohistochemical, and ultrastructural study. *Journal of Leukocyte Biology* 45 (2): 87–96. https://doi.org/10.1002/jlb.45.2.87.
20. Szebeni, G.J., C. Vizler, K. Kitajka, and L.G. Puskas. 2017. Inflammation and cancer: Extra- and intracellular determinants of tumor-associated macrophages as tumor promoters. *Mediators of Inflammation* 2017: 9294018. https://doi.org/10.1155/2017/9294018.
21. Ajami, B., J.L. Bennett, C. Krieger, W. Tetzlaff, and F.M. Rossi. 2007. Local self-renewal can sustain CNS microglia maintenance and function throughout adult life. *Nature Neuroscience* 10 (12): 1538–1543. https://doi.org/10.1038/nn2014.
22. Hess, D.C., T. Abe, W.D. Hill, A.M. Studdard, J. Carothers, M. Masuya, et al. 2004. Hematopoietic origin of microglial and perivascular cells in brain. *Experimental Neurology* 186 (2): 134–144. https://doi.org/10.1016/j.expneurol.2003.11.005.
23. Lee, A.W., Y. Mao, J.M. Penninger, and S. Yu. 2011. Gab2 promotes colony-stimulating factor 1-regulated macrophage expansion via alternate effectors at different stages of development. *Molecular and Cellular Biology* 31 (22): 4563–4581. https://doi.org/10.1128/MCB.05706-11.
24. Jenkins, S.J., D. Ruckerl, G.D. Thomas, J.P. Hewitson, S. Duncan, F. Brombacher, et al. 2013. IL-4 directly signals tissue-resident macrophages to proliferate beyond homeostatic levels controlled by CSF-1. *The Journal of Experimental Medicine* 210 (11): 2477–2491. https://doi.org/10.1084/jem.20121999.
25. Hamidzadeh, K., S.M. Christensen, E. Dalby, P. Chandrasekaran, and D.M. Mosser. 2017. Macrophages and the recovery from acute and chronic inflammation. *Annual Review of Physiology* 79: 567–592. https://doi.org/10.1146/annurev-physiol-022516-034348.
26. Sica, A., and A. Mantovani. 2012. Macrophage plasticity and polarization: in vivo veritas. *Journal of Clinical Investigation* 122: 787–795. https://doi.org/10.1172/JCI59643.

27. Gordon, S., and F.O. Martinez. 2010. Alternative activation of macrophages: Mechanism and functions. *Immunity* 32 (5): 593–604. https://doi.org/10.1016/j.immuni.2010.05.007.
28. Kzhyshkowska, J., A. Gudima, K. Moganti, A. Gratchev, and A. Orekhov. 2016. Perspectives for monocyte/macrophage-based diagnostics of chronic inflammation. *Transfusion Medicine and Hemotherapy* 43 (2): 66–77. https://doi.org/10.1159/000444943.
29. Gratchev, A., J. Kzhyshkowska, J. Utikal, and S. Goerdt. 2005. Interleukin-4 and dexamethasone counterregulate extracellular matrix remodelling and phagocytosis in type-2 macrophages. *Scandinavian Journal of Immunology* 61: 10–17. https://doi.org/10.1111/j.0300-9475.2005.01524.x.
30. Kzhyshkowska, J., G. Workman, M. Cardo-Vila, W. Arap, R. Pasqualini, A. Gratchev, L. Krusell, S. Goerdt, and E.H. Sage. 2006. Novel function of alternatively activated macrophages: Stabilin-1-mediated clearance of SPARC. *Journal of Immunology* 176: 5825–5832. https://doi.org/10.4049/jimmunol.176.10.5825.
31. Kzhyshkowska, J., A. Gratchev, C. Schmuttermaier, H. Brundiers, L. Krusell, S. Mamidi, J. Zhang, G. Workman, E.H. Sage, C. Anderle, P. Sedlmayr, and S. Goerdt. 2008. Alternatively activated macrophages regulate extracellular levels of the hormone placental lactogen via receptor-mediated uptake and transcytosis. *Journal of Immunology* 180: 3028–3037. https://doi.org/10.4049/jimmunol.180.5.3028.
32. Bonavita, E., S. Gentile, M. Rubino, V. Maina, R. Papait, P. Kunderfranco, C. Greco, F. Feruglio, M. Molgora, I. Laface, et al. 2015. PTX3 acts as an extrinsic oncosuppressor by regulating complement-dependent inflammation in cancer. *Cell* 160 (4): 700–714. https://doi.org/10.1016/j.cell.2015.01.004.
33. Qian, B.Z., and J.W. Pollard. 2010. Macrophage diversity enhances tumor progression and metastasis. *Cell* 141 (1): 39–51. https://doi.org/10.1016/j.cell.2010.03.014.
34. Coussens, L.M., L. Zitvogel, and A.K. Palucka. 2013. Neutralizing tumor-promoting chronic inflammation: A magic bullet? *Science* 339 (6117): 286–291. https://doi.org/10.1126/science.1232227.
35. Lin, E.Y., A.V. Nguyen, R.G. Russell, and J.W. Pollard. 2001. Colony-stimulating factor 1 promotes progression of mammary tumors to malignancy. *The Journal of Experimental Medicine* 193 (6): 727–740. https://doi.org/10.1084/jem.193.6.727.
36. Su, S., Q. Liu, J. Chen, J. Chen, F. Chen, C. He, D. Huang, W. Wu, L. Lin, W. Huang, et al. 2014. A positive feedback loop between mesenchymal-like cancer cells and macrophages is essential to breast cancer metastasis. *Cancer Cell* 25: 605–620. https://doi.org/10.1016/j.ccr.2014.03.021.
37. Riabov, V., A. Gudima, N. Wang, A. Mickley, A. Orekhov, and J. Kzhyshkowska. 2014. Role of tumor associated macrophages in tumor angiogenesis and lymphangiogenesis. *Frontiers in Physiology* 5 (March): 1–13. https://doi.org/10.3389/fphys.2014.00075.
38. Bögels, M., R. Braster, P.G. Nijland, N. Gül, W. van de Luijtgaarden, R.J. Fijneman, G.A. Meijer, C.R. Jimenez, R.H. Beelen, and M. van Egmond. 2012. Carcinoma origin dictates differential skewing of monocyte function. *OncoImmunology* 1 (6): 798–809. https://doi.org/10.4161/onci.20427.
39. Mantovani, A., T. Schioppa, C. Porta, P. Allavena, and A. Sica. 2006. Role of tumor-associated macrophages in tumor progression and invasion. *Cancer Metastasis Reviews* 25 (3): 315–322. https://doi.org/10.1007/s10555-006-9001-7.
40. Allavena, P., and A. Mantovani. 2012. Immunology in the clinic review series; focus on cancer: Tumour-associated macrophages: Undisputed stars of the inflammatory tumour microenvironment. *Clinical and Experimental Immunology* 167 (2): 195–205. https://doi.org/10.1111/j.1365-2249.2011.04515.x.
41. Noy, R., and J.W. Pollard. 2014. Tumor-associated macrophages: From mechanisms to therapy. *Immunity* 41 (1): 49–61. https://doi.org/10.1016/j.immuni.2014.06.010.
42. Buldakov, M., M. Zavyalova, N. Krakhmal, N. Telegina, S. Vtorushin, I. Mitrofanova, V. Riabov, S. Yin, B. Song, N. Cherdyntseva, et al. 2017. CD68+, but not stabilin-1+ tumor associated macrophages in gaps of ductal tumor structures negatively correlate with the lymphatic metastasis in human breast cancer. *Immunobiology* 222 (1): 31–38. https://doi.org/10.1016/j.imbio.2015.09.011.

43. Fujiwara, T., J. Fukushi, S. Yamamoto, Y. Matsumoto, N. Setsu, Y. Oda, H. Yamada, S. Okada, K. Watari, M. Ono, M. Kuwano, S. Kamura, K. Iida, Y. Okada, M. Koga, and Y. Iwamoto. 2011. Macrophage infiltration predicts a poor prognosis for human ewing sarcoma. *The American Journal of Pathology* 179 (3): 1157–1170. https://doi.org/10.1016/j.ajpath.2011.05.034.
44. Boström, M.M., H. Irjala, T. Mirtti, P. Taimen, T. Kauko, A. Ålgars, S. Jalkanen, and P.J. Boström. 2015. Tumor-associated macrophages provide significant prognostic information in urothelial bladder cancer. *PLoS One* 10 (7): e0133552. https://doi.org/10.1371/journal.pone.0133552.
45. Zhang, J., Y. Yan, Y. Yang, L. Wang, M. Li, J. Wang, X. Liu, X. Duan, and J. Wang. 2016. High infiltration of tumor-associated macrophages influences poor prognosis in human gastric cancer patients, associates with the phenomenon of EMT. *Medicine* 95 (6): 1–6. https://doi.org/10.1097/MD.0000000000002636.
46. Nabeshima, A., Y. Matsumoto, J. Fukushi, K. Iura, T. Matsunobu, M. Endo, T. Fujiwara, K. Iida, Y. Fujiwara, M. Hatano, N. Yokoyama, S. Fukushima, Y. Oda, and Y. Iwamoto. 2015. Tumour-associated macrophages correlate with poor prognosis in myxoid liposarcoma and promote cell motility and invasion via the HB-EGF-EGFR-PI3K/Akt pathways. *British Journal of Cancer* 112 (3): 547–555. https://doi.org/10.1038/bjc.2014.637.
47. Welsh, T.J., R.H. Green, D. Richardson, D.A. Waller, K.J. O'Byrne, and P. Bradding. 2005. Macrophage and mast-cell invasion of tumor cell islets confers a marked survival advantage in non-small-cell lung cancer. *Journal of Clinical Oncology* 23 (35): 8959–8967.
48. Mahmoud, S.M.A., A.H.S. Lee, E.C. Paish, R.D. Macmillan, I.O. Ellis, and A.R. Green. 2012. Tumour-infiltrating macrophages and clinical outcome in breast cancer. *Journal of Clinical Pathology* 65 (2): 159–163. https://doi.org/10.1136/jclinpath-2011-200355.
49. Yang, J., X. Li, X.P. Liu, and Y. Liu. 2015. The role of tumor-associated macrophages in breast carcinoma invasion and metastasis. *International Journal of Clinical and Experimental Pathology* 8 (6): 6656–6664.
50. Forssell, J., Å. Öberg, M.L. Henriksson, R. Stenling, A. Jung, and R. Palmqvist. 2007. High macrophage infiltration along the tumor front correlates with improved survival in colon cancer. *Clinical Cancer Research* 13 (5): 1472–1479. https://doi.org/10.1158/1078-0432.CCR-06-2073.
51. Mitrofanova, I., M. Zavyalova, N. Telegina, M. Buldakov, V. Riabov, N. Cherdyntseva, and J. Kzhyshkowska. 2017. Tumor-associated macrophages in human breast cancer parenchyma negatively correlate with lymphatic metastasis after neoadjuvant chemotherapy. *Immunobiology* 222 (1): 101–109. https://doi.org/10.1016/j.imbio.2016.08.001.
52. Kostine, M., I.H. Briaire-de Bruijn, A.H.G. Cleven, C. Vervat, W.E. Corver, M.W. Schilham, E. Van Beelen, H. van Boven, R.L. Haas, A. Italiano, A.M. Cleton-Jansen, and J.V.M.G. Bovée. 2017. Increased infiltration of M2-macrophages, T-cells and PD-L1 expression in high grade leiomyosarcomas supports immunotherapeutic strategies. *Oncoimmunology* 7 (2): e1386828. https://doi.org/10.1080/2162402X.2017.
53. Aljabery, F., H. Olsson, O. Gimm, S. Jahnson, and I. Shabo. 2018. M2-macrophage infiltration and macrophage traits of tumor cells in urinary bladder cancer. *Urologic Oncology: Seminars and Original Investigations* 36 (4): 159.e19–159.e26. https://doi.org/10.1016/j.urolonc.2017.11.020.
54. Cheng, Z., D. Zhang, B. Gong, P. Wang, and F. Liu. 2017. Research paper CD163 as a novel target gene of STAT3 is a potential therapeutic target for gastric cancer. *Oncotarget* 8 (50): 87244–87262.
55. Medrek, C., F. Pontén, K. Jirström, and K. Leandersson. 2012. The presence of tumor associated macrophages in tumor stroma as a prognostic marker for breast cancer patients. *BMC Cancer* 12: 306. https://doi.org/10.1186/1471-2407-12-306.
56. Lan, C., X. Huang, S. Lin, H. Huang, Q. Cai, T. Wan, J. Lu, and J. Liu. 2013. Expression of M2-polarized macrophages is associated with poor prognosis for advanced epithelial ovarian

cancer. *Technology in Cancer Research & Treatment* 12 (3): 259–267. https://doi.org/10.7785/tcrt.2012.500312.
57. Xu, L., Y. Zhu, L. Chen, H. An, W. Zhang, G. Wang, Z. Lin, and J. Xu. 2014. Prognostic value of diametrically polarized tumor-associated macrophages in renal cell carcinoma. *Annals of Surgical Oncology* 21 (9): 3142–3150. https://doi.org/10.1245/s10434-014-3601-1.
58. Liu, D.R., Q.L. Guan, M.T. Gao, L. Jiang, and H.X. Kang. 2017. Mannose receptor as a potential biomarker for gastric cancer: A pilot study. *The International Journal of Biological Markers* 32 (3): e278–e283.
59. Shu, Q.H., Y.S. Ge, H.X. Ma, X.Q. Gao, J.J. Pan, D. Liu, G.L. Xu, J.L. Ma, and W.D. Jia. 2016. Prognostic value of polarized macrophages in patients with hepatocellular carcinoma after curative resection. *Journal of Cellular and Molecular Medicine* 20 (6): 1024–1035. https://doi.org/10.1111/jcmm.12787.
60. Le Page, C., A. Marineau, P.K. Bonza, K. Rahimi, L. Cyr, I. Labouba, J. Madore, N. Delvoye, A.M. Mes-Masson, D.M. Provencher, et al. 2012. BTN3A2 expression in epithelial ovarian cancer is associated with higher tumor infiltrating T cells and a better prognosis. *PLoS One* 7 (6): e38541. https://doi.org/10.1371/journal.pone.0038541.
61. Riabov, V., S. Yin, B. Song, A. Avdic, K. Schledzewski, I. Ovsiy, A. Gratchev, M. Llopis Verdiell, C. Sticht, et al. 2016. Stabilin-1 is expressed in human breast cancer and supports tumor growth in mammary adenocarcinoma mouse model. *Oncotarget* 7 (21): 31097–31110. https://doi.org/10.18632/oncotarget.8857.
62. Schönhaar, K., K. Schledzewski, J. Michel, C. Dollt, C. Gkaniatsou, C. Géraud, J. Kzhyshkowska, S. Goerdt, and A. Schmieder. 2014. Expression of stabilin-1 in M2 macrophages in human granulomatous disease and melanocytic lesions. *International Journal of Clinical and Experimental Pathology* 7 (4): 1625–1634.
63. Dollt, C., K. Becker, J. Michel, S. Melchers, C.A. Weis, K. Schledzewski, A. Krewer, L. Kloss, C. Gebhardt, J. Utikal, and A. Schmieder. 2017. The shedded ectodomain of Lyve-1 expressed on M2-like tumor-associated macrophages inhibits melanoma cell proliferation. *Oncotarget* 8 (61): 103682–103692. https://doi.org/10.18632/oncotarget.21771.
64. Bono, P., V.M. Wasenius, P. Heikkilä, J. Lundin, D.G. Jackson, and H. Joensuu. 2004. High LYVE-1-positive lymphatic vessel numbers are associated with poor outcome in breast cancer. *Clinical Cancer Research* 10 (21): 7144–7149.
65. Bi, J., S.H. Lau, Z.L. Lv, D. Xie, W. Li, Y.R. Lai, Y.R. Lai, J.M. Zhong, H.Q. Wu, Q. Su, et al. 2009. Overexpression of YKL-40 is an independent prognostic marker in gastric cancer. *Human Pathology* 40 (12): 1790–1797. https://doi.org/10.1016/j.humpath.2009.07.005.
66. Thorn, A.P., S. Daugaard, L.H. Christensen, I.J. Christensen, and M.M. Petersen. 2016. YKL-40 protein in osteosarcoma tumor tissue. *APMIS* 124 (6): 453–461. https://doi.org/10.1111/apm.12524.
67. Johansen, J.S., L. Drivsholm, P.A. Price, and I.J. Christensen. 2004. High serum YKL-40 level in patients with small cell lung cancer is related to early death. *Lung Cancer* 46 (3): 333–340. https://doi.org/10.1016/j.lungcan.2004.05.010.
68. Cintin, C., J.S. Johansen, I.J. Christensen, P.A. Price, S. Sørensen, and H.J. Nielsen. 2002. High serum YKL-40 level after surgery for colorectal carcinoma is related to short survival. *Cancer* 95 (2): 267–274. https://doi.org/10.1002/cncr.10644.
69. Liu, T., I. Larionova, N. Litviakov, V. Riabov, M. Zavyalova, M. Tsyganov, M. Buldakov, B. Song, K. Moganti, P. Kazantseva, et al. 2018. Tumor-associated macrophages in human breast cancer produce new monocyte attracting and pro-angiogenic factor YKL-39 indicative for increased metastasis after neoadjuvant chemotherapy. *Oncoimmunology* 7 (6): e1436922. https://doi.org/10.1080/2162402X.2018.1436922.
70. Gao, D., N. Joshi, H. Choi, S. Ryu, M. Hahn, R. Catena, H. Sadik, P. Argani, P. Wagner, L.T. Vahdat, et al. 2012. Myeloid progenitor cells in the premetastatic lung promote metastases by inducing mesenchymal to epithelial transition. *Cancer Research* 72: 1384–1394. https://doi.org/10.1158/0008-5472.CAN-11-2905.

71. Malanchi, I., A. Santamaria-Martínez, E. Susanto, et al. 2011. Interactions between cancer stem cells and their niche govern metastatic colonization. *Nature* 481 (7379): 85–89. https://doi.org/10.1038/nature10694.
72. Barcellos-Hoff, M.H., D. Lyden, and T.C. Wang. 2013. The evolution of the cancer nicheduring multistage carcinogenesis. *Nature Reviews. Cancer* 13 (7): 511–518. https://doi.org/10.1038/nrc3536.
73. Peinado, H., S. Lavotshkin, and D. Lyden. 2011. The secreted factors responsible for premetastatic niche formation: Old sayings and new thoughts. *Seminars in Cancer Biology* 21 (2): 139–146. https://doi.org/10.1016/j.semcancer.2011.01.002.
74. Sanchez, L.R., L. Borriello, D. Entenberg, J.S. Condeelis, M.H. Oktay, and G.S. Karagiannis. 2019. The emerging roles of macrophages in cancer metastasis and response to chemotherapy. *Journal of Leukocyte Biology* 106 (2): 259–274. https://doi.org/10.1002/JLB.MR0218-056RR.
75. Kzhyshkowska, J., and L. Krusell. 2009. Cross-talk between endocytic clearance and secretion in macrophages. *Immunobiology* 214 (7): 576–593. https://doi.org/10.1016/j.imbio.2009.03.007.
76. Zhang, J.L., G.W. Chen, Y.C. Liu, P.Y. Wang, X. Wang, Y.L. Wan, et al. 2012. Secreted protein acidic and rich in cysteine (SPARC) suppresses angiogenesis by down-regulating the expression of VEGF and MMP-7 in gastric cancer. *PLoS One* 7 (9): e44618. https://doi.org/10.1371/journal.pone.0044618.
77. Kzhyshkowska, J., A. Gratchev, and S. Goerdt. 2006. Stabilin-1, a homeostatic scavenger receptor with multiple functions. *Journal of Cellular and Molecular Medicine* 10 (3): 635–649. https://doi.org/10.1111/j.1582-4934.2006.tb00425.x.
78. Larionova, I.V., T.N. Sevastyanova, A.A. Rakina, N.V. Cherdyntseva, and J.G. Kzhyshkowska. 2018. Chitinase-like proteins as promising markers in cancer patients. *Siberian Journal of Oncology* 17 (4): 99–105. https://doi.org/10.21294/1814-4861-2018-17-4-99-105.
79. Kzhyshkowska, J., S. Yin, T. Liu, V. Riabov, and I. Mitrofanova. 2016. Role of chitinase-like proteins in cancer. *Biological Chemistry* 397 (3): 231–247. https://doi.org/10.1515/hsz-2015-0269.
80. Dijkgraaf, E.M., M. Heusinkveld, B. Tummers, L.T.C. Vogelpoel, R. Goedemans, V. Jha, et al. 2013. Chemotherapy alters monocyte differentiation to favor generation of cancer-supporting M2 macrophages in the tumor microenvironment. *Cancer Research* 73 (8): 2480–2492. https://doi.org/10.1158/0008-5472.CAN-12-3542.
81. De Palma, M., and C.E. Lewis. 2011. Macrophages limit chemotherapy. *Cancer Discovery* 1 (1): 54–67. https://doi.org/10.1038/472303a.
82. Mantovani, A., and P. Allavena. 2015. The interaction of anticancer therapies with tumor-associated macrophages. *The Journal of Experimental Medicine* 212 (4): 435–445. https://doi.org/10.1084/jem.20150295.
83. Sugimura, K., H. Miyata, K. Tanaka, T. Takahashi, Y. Kurokawa, M. Yamasaki, K. Nakajima, S. Takiguchi, M. Mori, Y. Doki, et al. 2015. High infiltration of tumor-associated macrophages is associated with a poor response to chemotherapy and poor prognosis of patients undergoing neoadjuvant chemotherapy for esophageal cancer. *Journal of Surgical Oncology* 111 (6): 752–759. https://doi.org/10.1002/jso.23881.
84. Hughes, R., B.Z. Qian, C. Rowan, M. Muthana, I. Keklikoglou, O.C. Olson, S. Tazzyman, S. Danson, C. Addison, M. Clemons, et al. 2015. Perivascular M2 macrophages stimulate tumor relapse after chemotherapy. *Cancer Research* 75 (17): 3479–3491. https://doi.org/10.1158/0008-5472.CAN-14-3587.
85. Shree, T., O.C. Olson, B.T. Elie, J.C. Kester, A.L. Garfall, K. Simpson, K.M. Bell-McGuinn, E.C. Zabor, E. Brogi, and J.A. Joyce. 2011. Macrophages and cathepsin proteases blunt chemotherapeutic response in breast cancer. *Genes & Development* 25 (23): 2465–2479. https://doi.org/10.1101/gad.180331.111.

86. Pienta, K.J., J.P. Machiels, D. Schrijvers, et al. 2013. Phase 2 study of carlumab (CNTO 888), a human monoclonal antibody against CC-chemokine ligand 2 (CCL2), in metastatic castration-resistant prostate cancer. *Investigational New Drugs* 31 (3): 760–768. https://doi.org/10.1007/s10637-012-9869-8.
87. Brana, I., A. Calles, P.M. Lo Russo, et al. 2015. Carlumab, an anti-C-C chemokine ligand 2 monoclonal antibody, in combination with four chemotherapy regimens for the treatment of patients with solid tumors: An open-label, multicenter phase 1b study. *Targeted Oncology* 10 (1): 111–123. https://doi.org/10.1007/s11523-014-0320-2.
88. Sandhu, S.K., K. Papadopoulos, P.C. Fong, et al. 2013. A first-in-human, first-in-class, phase I study of carlumab (CNTO 888), a human monoclonal antibody against CC-chemokine ligand 2 in patients with solid tumors. *Cancer Chemotherapy and Pharmacology* 71 (4): 1041–1050. https://doi.org/10.1007/s00280-013-2099-8.
89. Nywening, T.M., A. Wang-Gillam, D.E. Sanford, et al. 2016. Targeting tumour-associated macrophages with CCR2 inhibition in combination with FOLFIRINOX in patients with borderline resectable and locally advanced pancreatic cancer: A single-centre, open-label, dose-finding, non-randomised, phase 1b trial. *The Lancet Oncology* 17 (5): 651–662. https://doi.org/10.1016/S1470-2045(16)00078-4.
90. Wainberg, Z.A., S.A. Piha-Paul, J. Luke, et al. 2018. First-in-human phase 1 dose escalation and expansion of a novel combination, anti–CSF-1 receptor (cabiralizumab) plus anti–PD-1 (nivolumab), in patients with advanced solid tumors. https://doi.org/10.13140/RG.2.2.28962.53443.
91. Papadopoulos, K.P., L. Gluck, L.P. Martin, et al. 2017. First-in-human study of AMG 820, a monoclonal anti-colony-stimulating factor 1 receptor antibody, in patients with advanced solid tumors. *Clinical Cancer Research* 23 (19): 5703–5710. https://doi.org/10.1158/1078-0432.CCR-16-3261.
92. Ries, C.H., M.A. Cannarile, S. Hoves, et al. 2014. Targeting tumor-associated macrophages with anti-CSF-1R antibody reveals a strategy for cancer therapy. *Cancer Cell* 25 (6): 846–859. https://doi.org/10.1016/j.ccr.2014.05.016.
93. Manji, G.A., P. Patwardhan, and S.M. Lee. 2017. Phase 1/2 study of combination therapy with pexidartinib and sirolimus to target tumor-associated macrophages in malignant peripheral nerve sheath tumors. *Journal of Clinical Oncology* 34 (15_suppl): TPS11070. https://doi.org/10.1200/JCO.2016.34.15_suppl.TPS11070.
94. Gelhorn, H.L., S. Tong, K. McQuarrie, et al. 2016. Patient-reported symptoms of tenosynovial giant cell tumors. *Clinical Therapeutics* 38 (4): 778–793. https://doi.org/10.1016/j.clinthera.2016.03.008.
95. Gordon, E.M., K.K. Sankhala, N. Chawla, et al. 2016. Trabectedin for soft tissue sarcoma: Current status and future perspectives. *Advances in Therapy* 33 (7): 1055–1071. https://doi.org/10.1007/s12325-016-0344-3.
96. Beatty, G.L., D.A. Torigian, E.G. Chiorean, et al. 2013. A phase I study of an agonist CD40 monoclonal antibody (CP-870,893) in combination with gemcitabine in patients with advanced pancreatic ductal adenocarcinoma. *Clinical Cancer Research* 19 (22): 6286–6295. https://doi.org/10.1158/1078-0432.CCR-13-1320.
97. Rook, A.H., J.M. Gelfand, M. Wysocka, et al. 2015. Topical resiquimod can induce disease regression and enhance T-cell effector functions in cutaneous T-cell lymphoma. *Blood* 126 (12): 1452–1461. https://doi.org/10.1182/blood-2015-02-630335.
98. Monk, B.J., M.F. Brady, and C. Aghajanian. 2017. A phase 2, randomized, double-blind, placebo-controlled study of chemo-immunotherapy combination using motolimod with pegylated liposomal doxorubicin in recurrent or persistent ovarian cancer: A Gynecologic Oncology Group partners study. *Annals of Oncology* 28 (5): 996–1004. https://doi.org/10.1093/annonc/mdx049.
99. Lim, S.Y., A.E. Yuzhalin, A.N. Gordon-Weeks, et al. 2016. Targeting the CCL2-CCR2 signaling axis in cancer metastasis. *Oncotarget* 7 (19): 28697–28710. https://doi.org/10.18632/oncotarget.7376.

100. Loberg, R.D., L.L. Day, J. Harwood, et al. 2006. CCL2 is a potent regulator of prostate cancer cell migration and proliferation. *Neoplasia* 8: 578–586.
101. Fridlender, Z.G., V. Kapoor, G. Buchlis, et al. 2011. Monocyte chemoattractant protein-1 blockade inhibits lung cancer tumor growth by altering macrophage phenotype and activating CD8+ cells. *American Journal of Respiratory Cell and Molecular Biology* 44 (2): 230–237. https://doi.org/10.1165/rcmb.2010-0080OC.
102. Salcedo, R., M.L. Ponce, H.A. Young, et al. 2000. Human endothelial cells express CCR2 and respond to MCP-1: direct role of MCP-1 in angiogenesis and tumor progression. *Blood* 96 (1): 34–40.
103. Fang, W.B., M. Yao, G. Brummer, et al. 2016. Targeted gene silencing of CCL2 inhibits triple negative breast cancer progression by blocking cancer stem cell renewal and M2 macrophage recruitment. *Oncotarget* 7 (31): 49349–49367. https://doi.org/10.18632/oncotarget.9885.
104. Hollmén, M., S. Karaman, S. Schwager, et al. 2015. G-CSF regulates macrophage phenotype and associates with poor overall survival in human triple-negative breast cancer. *Oncoimmunology* 5 (3): e1115177.
105. Carleton, M., J. Powers, and P. Phillips. 2018. Pharmacodynamics (PD) and genomic profiling of pts treated with cabiralizumab (cabira) + nivolumab (NIVO) provide evidence of on-target tumor immune modulations and support future clinical applications. *Journal of Clinical Oncology* 36 (15_suppl): 3020. https://doi.org/10.1200/JCO.2018.36.15_suppl.3020.
106. Stafford, J.H., T. Hirai, L. Deng, et al. 2016. Colony stimulating factor 1 receptor inhibition delays recurrence of glioblastoma after radiation by altering myeloid cell recruitment and polarization. *Neuro-Oncology* 18 (6): 797–806. https://doi.org/10.1093/neuonc/nov272.
107. Pyonteck, S.M., L. Akkari, A.J. Schuhmacher, et al. 2013. CSF-1R inhibition alters macrophage polarization and blocks glioma progression. *Nature Medicine* 19 (10): 1264–1272. https://doi.org/10.1038/nm.3337.
108. Lewēn, S., H. Zhou, H.D. Hu, et al. 2008. A Legumain-based minigene vaccine targets the tumor stroma and suppresses breast cancer growth and angiogenesis. *Cancer Immunology, Immunotherapy* 57 (4): 507–515.
109. Luo, Y., H. Zhou, J.J. Krueger, et al. 2006. Targeting tumor-associated macrophages as a novel strategy against breast cancer. *The Journal of Clinical Investigation* 116 (8): 2132–2141.
110. Guo, P., Z. Zhu, Z. Sun, et al. 2013. Expression of legumain correlates with prognosis and metastasis in gastric carcinoma. *PLoS One* 8 (9): e73090. https://doi.org/10.1371/journal.pone.0073090.
111. Andón, F.T., E. Digifico, A. Maeda, et al. 2017. Targeting tumor associated macrophages: The new challenge for nanomedicine. *Seminars in Immunology* 34: 103–113. https://doi.org/10.1016/j.smim.2017.09.004.
112. Zhan, X., L. Jia, Y. Niu, et al. 2014. Targeted depletion of tumour-associated macrophages by an alendronate-glucomannan conjugate for cancer immunotherapy. *Biomaterials* 35 (38): 10046–10057. https://doi.org/10.1016/j.biomaterials.2014.09.007.
113. Hattori, Y., J. Yamashita, C. Sakaida, et al. 2015. Evaluation of antitumor effect of zoledronic acid entrapped in folate-linked liposome for targeting to tumor-associated macrophages. *Journal of Liposome Research* 25 (2): 131–140. https://doi.org/10.3109/08982104.2014.954128.
114. Junankar, S., G. Shay, J. Jurczyluk, et al. 2015. Real-time intravital imaging establishes tumor-associated macrophages as the extraskeletal target of bisphosphonate action in cancer. *Cancer Discovery* 5 (1): 35–42. https://doi.org/10.1158/2159-8290.CD-14-0621.
115. Fujimura, T., Y. Kambayashi, S. Furudate, et al. 2013. Immunomodulatory effect of bisphosphonate risedronate sodium on CD163+ arginase 1+ M2 macrophages: The development of a possible supportive therapy for angiosarcoma. *Clinical & Developmental Immunology* 325412: 2013–2017. https://doi.org/10.1155/2013/325412.
116. Banciu, M., J.M. Metselaar, R.M. Schiffelers, et al. 2008. Antitumor activity of liposomal prednisolone phosphate depends on the presence of functional tumor-associated macrophages in tumor tissue. *Neoplasia* 10 (2): 108–117.

117. Piaggio, F., V. Kondylis, F. Pastorino, et al. 2016. A novel liposomal clodronate depletes tumor-associated macrophages in primary and metastatic melanoma: Anti-angiogenic and anti-tumor effects. *Journal of Controlled Release* 223: 165–177. https://doi.org/10.1016/j.jconrel.2015.12.037.
118. Germano, G., R. Frapolli, M. Simone, et al. 2010. Antitumor and anti-inflammatory effects of trabectedin on human myxoid liposarcoma cells. *Cancer Research* 70 (6): 2235–2244. https://doi.org/10.1158/0008-5472.CAN-09-2335.
119. Germano, G., R. Frapolli, C. Belgiovine, et al. 2013. Role of macrophage targeting in the antitumor activity of trabectedin. *Cancer Cell* 23 (2): 249–262. https://doi.org/10.1016/j.ccr.2013.01.008.
120. van Roon, J.A., A.J. van Vuuren, S. Wijngaarden, et al. 2003. Selective elimination of synovial inflammatory macrophages in rheumatoid arthritis by an Fcgamma receptor I-directed immunotoxin. *Arthritis and Rheumatism* 48 (5): 1229–1238.
121. Nagai, T., M. Tanaka, Y. Tsuneyoshi, and B. Xu. 2009. Targeting tumor-associated macrophages in an experimental glioma model with a recombinant immunotoxin to folate receptor beta. *Cancer Immunology, Immunotherapy* 58 (10): 1577–1586. https://doi.org/10.1007/s00262-009-0667-x.
122. Galmbacher, K., M. Heisig, C. Hotz, et al. 2010. Shigella mediated depletion of macrophages in a murine breast cancer model is associated with tumor regression. *PLoS One* 5 (3): e9572. https://doi.org/10.1371/journal.pone.0009572.
123. Weigert, A., D. Sekar, B. Brüne, et al. 2009. Tumor-associated macrophages as targets for tumor immunotherapy. *Immunotherapy* 1 (1): 83–95. https://doi.org/10.2217/1750743X.1.1.83.
124. Bhattacharjee, A., S. Pal, G.M. Feldman, and M.K. Cathcart. 2011. Hck is a key regulator of gene expression in alternatively activated human monocytes. *The Journal of Biological Chemistry* 286 (42): 36709–36723. https://doi.org/10.1074/jbc.M111.291492.
125. Zhang, X., W. Tian, X. Cai, et al. 2013. Hydrazinocurcumin Encapsuled nanoparticles "re-educate" tumor-associated macrophages and exhibit anti-tumor effects on breast cancer following STAT3 suppression. *PLoS One* 8 (6): e65896. https://doi.org/10.1371/journal.pone.0065896.
126. Hussain, S.F., L.Y. Kong, J. Jordan, et al. 2007. A novel small molecule inhibitor of signal transducers and activators of transcription 3 reverses immune tolerance in malignant glioma patients. *Cancer Research* 67 (20): 9630–9636.
127. Sinha, P., V.K. Clements, S. Ostrand-Rosenberg, et al. 2005. Reduction of myeloid-derived suppressor cells and induction of M1 macrophages facilitate the rejection of established metastatic disease. *Journal of Immunology* 174 (2): 636–645.
128. Jiménez-García, L., M.Á. Higueras, S. Herranz, et al. 2018. A hispanolone-derived diterpenoid inhibits M2-Macrophage polarization in vitro via JAK/STAT and attenuates chitin induced inflammation in vivo. *Biochemical Pharmacology* 154: 373–383. https://doi.org/10.1016/j.bcp.2018.06.002.
129. Zhang, W., Y. Wei, D. Yu, et al. 2018. Gefitinib provides similar effectiveness and improved safety than erlotinib for advanced non-small cell lung cancer. *Medicine (Baltimore)* 97 (16): e0460. https://doi.org/10.1097/MD.0000000000010460.
130. Tariq, M., J. Zhang, G. Liang, et al. 2017. Gefitinib inhibits M2-like polarization of tumor-associated macrophages in Lewis lung cancerby targeting the STAT6 signaling pathway. *Acta Pharmacologica Sinica* 38 (11): 1501–1511. Published online 2017 Oct 12. https://doi.org/10.1038/aps.2017.124.
131. Murdoch, C., A. Giannoudis, C.E. Lewis, et al. 2004. Mechanisms regulating the recruitment of macrophages into hypoxic areas of tumors and other ischemic tissues. *Blood* 104 (8): 2224–2234.
132. Zhang, J., J. Cao, S. Ma, et al. 2014. Tumor hypoxia enhances non-small cell lung cancer metastasis by selectively promoting macrophage M2 polarization through the activation of ERK signaling. *Oncotarget* 5 (20): 9664–9677.

133. Leblond, M.M., A.U. Gérault, A. Corroyer-Dulmont, et al. 2016. Hypoxia induces macrophage polarization and re-education toward an M2 phenotype in U87and U251 glioblastoma models. *Oncoimmunology* 5 (1): e1056442. https://doi.org/10.1080/2162402X.2015.1056442.
134. Takeda, N., E.L. O'Dea, A. Doedens, et al. 2010. Differential activation and antagonistic function of HIF-{alpha} isoforms in macrophages are essential for NO homeostasis. *Genes & Development* 24 (5): 491–501. https://doi.org/10.1101/gad.1881410.
135. Noman, M.Z., G. Desantis, B. Janji, M. Hasmim, S. Karray, P. Dessen, V. Bronte, and S. Chouaib. 2014. PD-L1 is a novel direct target of HIF-1α, and its blockade under hypoxia enhanced MDSC-mediated T cell activation. *The Journal of Experimental Medicine* 211 (5): 781–790. https://doi.org/10.1084/jem.20131916.
136. Wigerup, C., S. Påhlman, and D. Bexell. 2016. Therapeutic targeting of hypoxia and hypoxia-inducible factors in cancer. *Pharmacology & Therapeutics* 164: 152–169. https://doi.org/10.1016/j.pharmthera.2016.04.009.
137. Fallah, J., and B.I. Rini. 2019. HIF inhibitors: Status of current clinical development. *Current Oncology Reports* 21 (1): 6. https://doi.org/10.1007/s11912-019-0752-z.
138. Hoves, S., C.H. Ooi, and C. Wolter. 2018. Rapid activation of tumor-associated macrophages boosts preexisting tumor immunity. *The Journal of Experimental Medicine* 215 (3): 859–876. https://doi.org/10.1084/jem.20171440.
139. Elgueta, R., M.J. Benson, and V.C. de Vries. 2009. Molecular mechanism and function of CD40/CD40L engagement in the immune system. *Immunological Reviews* 229 (1): 152–172. https://doi.org/10.1111/j.1600-065X.2009.00782.x.
140. Vonderheide, R.H., and M.J. Glennie. 2013. Agonistic CD40 antibodies and cancer therapy. *Clinical Cancer Research* 19 (5): 1035–1043. https://doi.org/10.1158/1078-0432.CCR-12-2064.
141. Biswas, S.K., and C.E. Lewis. 2010. NF-κB as a central regulator of macrophage function in tumors. *Journal of Leukocyte Biology* 88 (5): 877–884. https://doi.org/10.1189/jlb.0310153.
142. Shime, H., M. Matsumoto, H. Oshiumi, et al. 2012. Toll-like receptor 3 signaling converts tumor-supporting myeloid cells to tumoricidal effectors. *Proceedings of the National Academy of Sciences of the United States of America* 109 (6): 2066–2071. https://doi.org/10.1073/pnas.1113099109.
143. Yang, L., F. Wang, L. Wang, et al. 2015. CD163+ tumor-associated macrophage is a prognostic biomarker and is associated with therapeutic effect on malignant pleural effusion of lung cancer patients. *Oncotarget* 6 (12): 10592–10603.
144. Liu, B., X. Wang, T.Z. Chen, et al. 2016. Polarization of M1 tumor associated macrophage promoted by the activation of TLR3 signalpathway. *Asian Pacific Journal of Tropical Medicine* 9 (5): 484–488. https://doi.org/10.1016/j.apjtm.2016.03.019.
145. Vidyarthi, A., N. Khan, T. Agnihotri, et al. 2018. TLR-3 stimulation skews M2 macrophages to M1 through IFN-αβ signaling and restricts tumor progression. *Frontiers in Immunology* 9: 1650. https://doi.org/10.3389/fimmu.2018.01650.
146. Swaminathan, S., K. Suzuki, N. Seddiki, et al. 2012. Differential regulation of the Let-7 family of microRNAs in CD4+ T cells alters IL-10 expression. *Journal of Immunology* 188 (12): 6238–6246. https://doi.org/10.4049/jimmunol.1101196.
147. Huang, Z., J. Gan, Z. Long, et al. 2016. Targeted delivery of let-7b to reprogramme tumor-associated macrophages and tumor infiltrating dendritic cells for tumor rejection. *Biomaterials* 90: 72–84. https://doi.org/10.1016/j.biomaterials.2016.03.009.
148. Rodell, C.B., S.P. Arlauckas, and M.F. Cuccarese. 2018. TLR7/8-agonist-loaded nanoparticles promote the polarization of tumour-associated macrophages to enhance cancer immunotherapy. *Nature Biomedical Engineering* 2 (8): 578–588. https://doi.org/10.1038/s41551-018-0236-8.
149. Dietsch, G.N. 2016. Motolimod effectively drives immune activation in advanced cancer patients. *Oncoimmunology* 5 (5): e1126037. https://doi.org/10.1080/2162402X.2015.1126037.

150. Eubank, T.D., R.D. Roberts, M. Khan, et al. 2009. Granulocyte macrophage colony-stimulating factor inhibits breast cancer growth and metastasis by invoking an anti-angiogenic program in tumor-educated macrophages. *Cancer Research* 69 (5): 2133–2140. https://doi.org/10.1158/0008-5472.CAN-08-1405.
151. Xing, Z., A. Zganiacz, M. Santosuosso, et al. 2000. Role of IL-12 in macrophage activation during intracellular infection: IL-12 and mycobacteria synergistically release TNF-alpha and nitric oxide from macrophages via IFN-gamma induction. *Journal of Leukocyte Biology* 68 (6): 897–902.
152. Wang, Q., F. Cheng, T.T. Ma, et al. 2016. Interleukin-12 inhibits the hepatocellular carcinoma growth by inducing macrophage polarization to the M1-like phenotype through downregulation of Stat-3. *Molecular and Cellular Biochemistry* 415 (1–2): 157–168. https://doi.org/10.1007/s11010-016-2687-0.
153. Willingham, S.B., J.P. Volkmer, A.J. Gentles, et al. 2012. The CD47-signal regulatory protein alpha (SIRPa) interaction is a therapeutic target for human solid tumors. *Proceedings of the National Academy of Sciences of the United States of America* 109 (17): 6662–6667. https://doi.org/10.1073/pnas.1121623109.
154. Zen, K., Y. Guo, Z. Bian, et al. 2013. Inflammation-induced proteolytic processing of the SIRPα cytoplasmic ITIM in neutrophils propagates a proinflammatory state. *Nature Communications* 4: 2436. https://doi.org/10.1038/ncomms3436.
155. Gu, S., T. Ni, and J. Wang. 2018. CD47 blockade inhibits tumor progression through promoting phagocytosis of tumor cells by M2 polarized macrophages in endometrial cancer. *Journal of Immunology Research* 2018: 6156757. https://doi.org/10.1155/2018/6156757.
156. Liu, L., H. Yi, and H. He. 2017. Tumor associated macrophage-targeted microRNA delivery with dual-responsive polypeptide nanovectors for anti-cancer therapy. *Biomaterials* 134: 166–179. https://doi.org/10.1016/j.biomaterials.2017.04.043.
157. Pinto, T.A., L.M. Pinto, P.A. Cardoso, et al. 2016. Ionizing radiation modulates human macrophages towards a pro-inflammatory phenotype preserving their pro-invasive and pro-angiogenic capacities. *Scientific Reports* 6: 18765. https://doi.org/10.1038/srep18765.
158. Prakash, H., F. Klug, V. Nadella, et al. 2016. Low doses of gamma irradiation potentially modifies immunosuppressive tumor microenvironment by retuning tumor-associated macrophages: Lesson from insulinoma. *Carcinogenesis* 37 (3): 301–313. https://doi.org/10.1093/carcin/bgw007.
159. Genard, G., S. Lucas, C. Michiels, et al. 2017. Reprogramming of tumor-associated macrophages with anticancer therapies: Radiotherapy versus chemo- and immunotherapies. *Frontiers in Immunology* 8: 828. https://doi.org/10.3389/fimmu.2017.00828.

Tumor Reversion Induced by Embryo and Oocyte Extracts

Sara Proietti, Andrea Pensotti, and Alessandra Cucina

Once Upon a Time

Embryonic cell behaviour shares fundamental features with tumors [1, 2], including unconstrained proliferation, activation of embryonic protein networks, which ultimately lead to switching toward an embryonic/fetal metabolism (Warburg effect) [3, 4]. These data suggest that tumors share fundamental features and common critical pathways with embryogenesis [2, 5]. Moreover, the epithelium-mesenchymal transition (EMT) associated with the aggressive traits of carcinoma cells, in which epithelial cells acquire mesenchymal features such as the loss of cell-cell adhesion, cell polarity and tumor progression may also be viewed as a "reactivation" of a basic embryonic property that is instrumental in "recapitulating" key step of the differentiating process, ultimately ending in repositioning cells into an "embryonic attractor" [6]. Indeed, the plasticity displayed by cancer during its natural history is remnant of those processes occurring in organ morphogenesis and tissue repair [7]. Moreover, tumor cells have shown gene-expression patterns superimposable to those of embryonic cells, as highlighted by genome-wide analysis of gene expression profiles [8, 9].

Tumor cells are essentially "embryonic" in nature [10–12], and might be stimulated towards uncontrolled growth because of a loss of *local* inhibitory "tissue

S. Proietti (✉)
Department of Surgery "Pietro Valdoni", Sapienza University, Rome, Italy
e-mail: sara.proietti@uniroma1.it

A. Pensotti
FAST, University Campus Bio-Medico, Rome, Italy

A. Cucina
Department of Surgery "Pietro Valdoni", Sapienza University, Rome, Italy

Azienda Policlinico Umberto I, Rome, Italy

M. Bizzarri (ed.), *Approaching Complex Diseases*, Human Perspectives in Health Sciences and Technology 2, https://doi.org/10.1007/978-3-030-32857-3_12

tension", an impressive visionary idea, successfully retrieved and developed by D.E. Ingber [13].

In the early 70s, Brinster [14] demonstrated that specific embryonic carcinoma cells injected into a blastocyst of the mouse lose their tumorigenicity immediately. The blastocyst was in fact able to regulate cancer cells and their progeny blocking their malignant behaviour. Rather, they participated in normal embryonic development resulting in functional mice. These findings were confirmed by Mintz and Illmensee [15], and by Papaioannou et al. [16]. Further studies performed by Pierce et al. [17, 18], showed that this effect strongly depend to the "positional information" provided by the morphogenetic field: when embryonal carcinoma cells were placed between the *zona pellucida* and the trophectoderm the malignant phenotype resulted not controlled, meanwhile embryonal carcinoma cells injected into the blastocoele lose their tumorigenicity immediately upon differentiation. Yet, the participation of *diffusive, morphogenetic factors* cannot be discarded [19, 20].

For a while, it was believed that this phenomenon was peculiar to germ-cell tumors. However, other tumors, including leukaemia, melanoma, liver and breast cancer [21–23], can also be committed to revert into a normal phenotype and/or to differentiate, when placed in specific environments.

In the last decades, further studies supported the hypothesis that even adult tumor placed into the embryo microenvironments can change its cell's fate. Injection of human SK-Mel-28 or C8161 melanoma cells into the chick embryo neural tube revealed that the tumor cells invaded the embryo along host neural crest cell migratory pathways and did not form tumors [24]. On the contrary, when B16 mouse melanoma cells were injected into chick neural crest, tumor cells formed melanomas [25]. This phenomenon could explain how only early stages of embryo share oncostatic as well as differentiating properties, in that no effects can be seen after morphogenesis and organogenesis are complete [26–28].

Furthermore, some morphogenetic, soluble factors seem to share some reverting properties displayed by embryonic cells. Embryo proteins synthesized in the pregnant uteri of mammals [29], as well as protein extracted from embryonic nuclei [30, 31] or from totipotent embryonal zebrafish cells [32, 33], can activate apoptotic responses accompanied by a significant inhibition of the proliferation rate in several cancer cell lines. Moreover, Zebrafish embryo extracts inhibit proliferation and induce apoptosis in 5-Fluorouracil-treated human colon cancer cells suppressing the release of antiapoptotic proteins (Bcl-xl), and promoting a significant rise in the differentiating proteins like E-cadherin [34]. Recently [35], we observed that molecular factors extracted from Zebrafish embryos during specific developmental phases (20 somites) significantly antagonize proliferation of breast cancer cells, while reversing a number of prominent aspects of malignancy. Embryo extracts reduce cell proliferation, enhance apoptosis, and dramatically inhibit both invasiveness and migrating capabilities of cancer cells. Moreover, such effect is not limited to cancerous cells as embryo extracts were also effective in inhibiting migration and invasiveness displayed by normal breast cells undergoing epithelial–mesenchymal transition upon TGF-β1 stimulation. It is noteworthy that reversion involves the downregulation of TCTP, a key parameter of cell reprogramming (Fig. 1). Down-

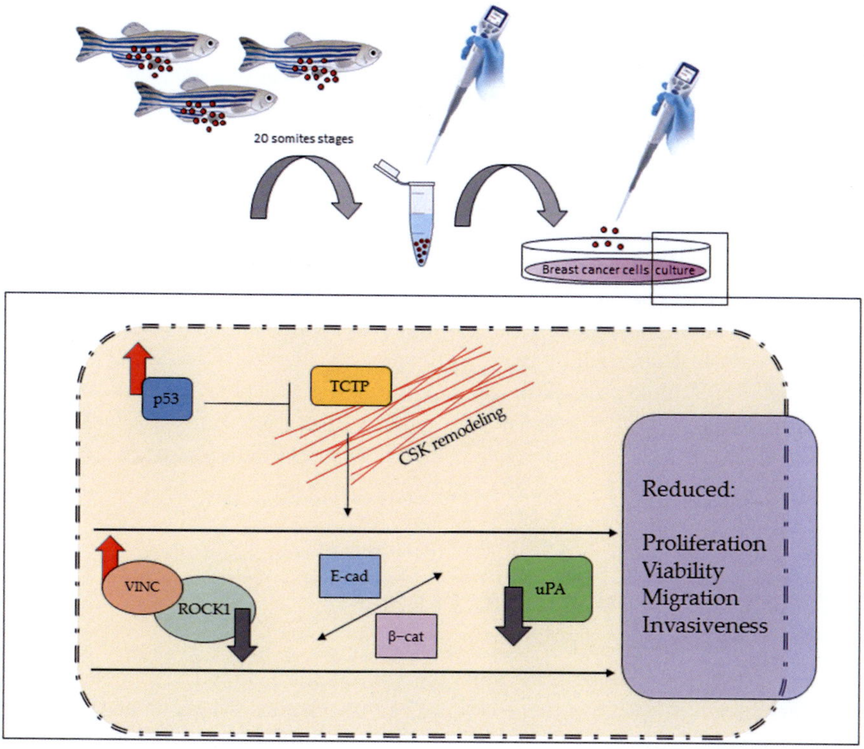

Fig. 1 Treatment of breast cancer cells with specific developmental phases (20 somites) of Zebrafish embryos induce a down-regulation of translationally controlled tumor protein (TCTP) accompanied by an increase in p53 levels. Downstream to TCTP inhibition, EMT is antagonized through cytoskeleton remodeling and rearrangement of the E-cadherin/β-catenin junctions. Modulation of vinculin, ROCK1, and uPA antagonize the predisposition of breast cancer cells to invade and migrate

regulation of TCTP is usually accompanied by an increase in p53 levels. EMT is antagonized through cytoskeleton remodelling and rearrangement of E-cadherin/β-catenin junctions, while modulation of vinculin, ROCK1, and uPA antagonize the invasive/migratory behaviour of breast cancer cells [35].

Unfortunately, these experiments did not clarify the fate of tumor cells injected into the embryo, if they are induced to die, to arrest growth [36], or moved towards differentiating pathways [37]. A few studies have demonstrated that cancer cells can be successfully "reprogrammed" when introduced into an oocyte [38, 39], giving rise to normal developing embryos. This finding was confirmed by Gootwine et al. [40], who injected leukaemia cells into the placenta of 10-day-old mouse embryos and obtained healthy adult mice with circulating leukocytes carrying leukaemia cell markers.

Overall, these and other studies confirmed that the embryonic milieu can reverse the cancer phenotype, re-establishing a "normal" developmental pluripotency and suppressing the tumorigenic phenotype of malignant cancer cells through complex biological pathways, including gene regulation, induction of apoptosis, inhibition of proliferation and modification of morphogens release not yet well identified [41–44].

Such a results claim against the somatic mutation theory of cancer, as recognized by an early pioneers in embryo/cancer cell biology: "[. . .] carcinogenesis may be explained as a stable response to inductive phenomena acting through cytoplasmic nuclear controls on an initially unaltered genome rather than as an accident of genetic coding" [45]. Thus, "[. . .] the production of a tumour is not the result of an accidental mutation of the genome but rather of inductive phenomena comparable to those at work in early embryological development" [46]. Consequently, "these findings contradict the concept that the neoplastic, malignant behaviour is 'fixed.' Now, if neoplasia, as posited by the somatic mutation theory, is due to the accumulation of multiple mutations, how can cells derived from these tumours revert to behave as normal cells? It is statistically unlikely that random reverse mutational events that could erase the previous mutations would be responsible for this reversal" [47].

Attractors and Morphogenetic Fields

Classical embryology, since long time, recognized a functional analogous – the "morphogenetic field" (MF) – to the concept of attractor. It is widely agreed that MF denotes both *informational* and *topological relationships* among constituents (cells and stroma) of a living system. According to Gilbert, a morphogenetic field "is a system of order such that the position taken up by unstable entities in one part of the system bears a definite relation to the position taken up by other unstable entities in other parts of the system. The field effect is constituted by their several equilibrium positions [. . .] a field is heteroaxial and heteropolar, has recognizably distinct districts, and can, like a magnetic field, maintain its pattern when its mass is either reduced or increased. It can fuse with a similar pattern entering new material if the axial orientation is favourable. The morphogenetic gradient is a special limited case of the morphogenetic field" [48]. As suggested by Weiss [49], like attractors, MF can be considered a self-organizing, spatially coordinated and temporally synchronized entity, characterized by non-linear dynamics and emergent properties [50]. In MF, gene expression, epigenetic information and *both chemical and physical signals* are integrated [51]. Genes *cooperate* in morphogenesis by operating *within* the field. This means that they are constrained by physical cues provided by the field. Conversely, the same gene/s can activate/s different pathways in different fields [52].

More than 50 rs ago, Needham [53] and Waddington [54] speculated that cancers represented an *escape* from their normal morphogenetic field. Broadly speaking, "any change that leads stem cells to escape from the niche [*i.e.*, the morphogenetic field] would result in tumour formation" [55]. Indeed, disruption of the morphogenetic field of a tissue, through modifications of microenvironmental forces (including hydrostatic pressure, shear stress, compression and tension forces) [56], can promote the onset as well as the progression of cancer. On the other hand, placing tumor cells into a "normal" morphogenetic field – like that of an embryonic tissue – one can reverse the malignant phenotype, "reprogramming" tumor into normal cells [57–59].

This framework allows to deeply understand some relevant, yet unfathomable results obtained treating cancers with embryonic morphogenetic factors.

Many oncogenes are tumorigenic only in specific cell lineages, suggesting the requirement for a tissue specific epigenetic environment that is permissive for an oncogene's tumorigenic potential [60]. Indeed, over expression of a given oncogene may enhance growth in one cell type and induce apoptosis in another [61, 62].

As suggested by Huang and Ingber [63], this paradoxical behaviour can be understood only keeping in mind the special feature of the tumor attractor. Morphogens-induced network rewiring result in shift of attractor boundaries, leading to a displacement of cell population toward a different attractor. This may explain how the very same signal can cause discretely disparate phenotype switches.

Preliminary Clinical Evidence

Some intriguing preliminary results have been obtained from both *in vivo* and *in vitro* studies with proteins extracted from *Zebrafish* at different embryonic stages of differentiation. Zebrafish (Danio rerio) is a freshwater fish belonging to the Cyprinidae family, common in the river Ganga basin on the Indian sub-continent. Zebrafish has some well-known characteristics that make it really attractive so that it has obtained the status of model species, becoming an extraordinary complement to murine models for human diseases [64].

It displays more than 80% of all human disease-related genes indicating that many human diseases, in fact, can be modeled on Zebrafish. Over the last decades, Zebrafish has become increasingly important to scientific research mainly as a powerful model system for the study of human cancers [65]. In fact Zebrafish embryos are nearly transparent which allows researchers to easily examine the development of internal structures, particularly in identifying events involved in carcinogenesis and tumor progression as well as anti-tumor mechanisms [66, 67].

Moreover, cancer research in Zebrafish particularly benefits from the many genetic tools and transgenic strains established by the Zebrafish community over the years [68–70].

In recent clinical trials zebrafish embryo extracts have been administered to advanced cancer patients, who no longer respond to conventional treatments: the expression of oncofetal antigens (like AFP) were significantly reduced [71] and marked beneficial effects such as induction of objective responses, improvement in performance status and significant increase in overall survival were induced [72, 73].

Recently, a pilot randomized study has compared both safety and overall survival of advanced colon cancer patients who were randomized into two groups: one receiving Regorafenib-based salvage therapy only and one else receiving

Regorafenib-based salvage therapy associated with a food supplement containing a Zebrafish extracts [74]. At 12 months, a statistically significant increase in survival rate was observed in the latter group (75% versus 33.3%). Moreover, in Zebrafish extracts-treated patients performance status was largely preserved during the treatment, whereas it declined rapidly in the Regorafenib group. Infact Regorafenib was associated with significant adverse effects – mainly fatigue, hypertension, hand-foot syndrome, diarrhea and cutaneous rash – occurring in up to 90% of patients.

Overall, this data suggests that Zebrafish extracts may improve the clinical response of salvage therapy, by both modulating drug responsiveness and counteracting drug-related side effects [67, 74].

A New Paradigm on Carcinogenesis

The embryonic microenvironment can be viewed as finely regulated network of cellular interactions that is critical to cell behaviour and fate determination throughout embryogenesis and beyond. The physical and molecular signals coming from the embryonic environment determine, through different signaling pathways, multiple effects on specific cells or target tissues. Some act on growth factors such as Activin, Nodal and FGF which maintain pluripotency in embryonic stem cells, while others act on factors such as ngn3, Notch and ASK-1, which determine a differentiated fate [75].

It is worth noting that important pathways such as those of Nodal, Wnt and Notch, which operate both in the maturation of stem cells and in the development of cancer, should be strictly regulated by microenvironmental signals [76].

Therefore, it is not surprising that embryonic morphogenetic fields induce a complete phenotypic reversion of tumor cells.

Similarly, breast cancer cells exposed to a maternal morphogenetic field undergo a dramatic transition in cell shape, followed by a reversion in the metabolic profile and induction of differentiation [77].

Several studies sustain the persistence of morphogenic fields throughout adult life: these fields organize histogenesis and organogenesis before birth as well as tissue maintenance and regeneration throughout postnatal life [78, 79]. For example, malignant hepatocarcinoma cells injected into the liver of young rats differentiated into mature hepatocytes and failed to proliferate until the rat aged [23]. On the contrary, by injecting the same cancerous cells subcutaneously, i.e. in a wrong morphogenetic context, an increased malignant progression has been witnessed. Similarly, increased cancer proliferation has been recorded when the hepatocarcinoma cells are transferred in the liver of aged animals, thus indicating that the strength of the morphogenetic field (and consequently its "normalizing property") declines with age [57].

The *theory of tissue organization field* (TOFT), proposed by C. Sonnenschein and A. Soto [80], postulates that tumors derive from an imperfect interaction among cells and tissues and that carcinogenesis is a potentially reversible process. A mounting body of evidence has suggested that re-establishment of appropriate interactions between human cancer cells and the surrounding microenvironment, in particular to a strong "normally oriented" morphogenetic fields, can reverse the neoplastic phenotype [81].

By contrast,when normal embryonic cells are transplanted into a "wrong" tissue (*i.e.*, in extrauterine sites of adult syngeneic recipients) they degenerate into teratocarcinoma [82]. Similar results were obtained transplanting embryos into the testis [83].

The stroma and the microenvironment surrounding cancer cells generally provide morphogenetic cues [84, 85] and several investigations are currently ongoing in order to exploit these potentialities for therapeutic uses [58, 86]. Embryo morphogenetic fields acts similarly providing to the growing cells a normal set of regulatory signals as showed by some preliminary clinical trials with proteins extracted from embryonic morphogenetic fields [72]. As hypothesized by Pierce more than 25 years ago, "It is now clear that three embryonic fields can regulate their closely related malignant cell types, and thus it is our hypothesis that there must be an embryonic field capable of regulating every carcinoma" [87].

Theoretical models provided by non-equilibrium dynamics (attractors) and developmental biology (morphogenetic fields), have established a useful theoretical framework able to give reliable explanatory insights into the complex interactions taking place between cancer cells and morphogenetic fields, including those belonging to embryo and oocyte cultures [88]. Nevertheless a large research effort must be carried out in order to clarify what we do mean by "morphogenetic fields", how morphogens gradients work within context-dependent boundary conditions [89], and what are the molecular components acting on.

References

1. Topczewska, J.M., et al. 2006. Embryonic and tumorigenic pathways converge via Nodal signalling: Role in melanoma aggressiveness. *Nature Medicine* 12 (8): 925–932.
2. Kelleher, F.C., et al. 2006. Common critical pathways in embryogenesis and cancer. *Acta Oncologica* 45 (4): 375–388.
3. Peifer, M., and P. Polakis. 2000. Wnt signaling in oncogenesis and embryogenesis a look outside the nucleus. *Science* 287 (5458): 1606–1609.
4. Christofk, H.R., et al. 2008. The M2 splice isoform of pyruvate kinase is important for cancer metabolism and tumour growth. *Nature* 452 (7184): 230–233.
5. Reya, T., et al. 2001. Stem cells, cancer and cancer stem cells. *Nature* 414 (6859): 105–111.
6. Mani, S.A., et al. 2008. The epithelial–mesenchymal transition generates cells with properties of stem cells. *Cell* 133 (4): 704–715.
7. Dvorak, H.F. 1986. Tumours: Wounds that do not heal. Similarities between tumor stroma generation and wound healing. *The New England Journal of Medicine* 315: 1650–1659.
8. Borczuk, A.C., et al. 2003. Non small-cell lung cancer molecular signatures recapitulate lung developmental pathways. *The American Journal of Pathology* 163: 1949–1960.
9. Ben-Porath, I., et al. 2008. An embryonic stem cell-like gene expression signature in poorly differentiated aggressive human tumors. *Nature Genetics* 40: 499–507.

10. Cohnheim, J. 1889. *Lectures in general pathology*. Trans. A.B. McKee. London: New Sydenham Society.
11. Ribbert, H. 1911. *Das Karzinom des Menschen/Human cancer*. Bonn: Friederich Cohen.
12. Wolff, J. 1990. *The science of cancerous diseases from the earliest times to the present*. Translation by Ayoub B and with an introduction by S. Jarcho. Sagamore Beach: Science History Pubblications.
13. Ingber, D.E. 1997. Tensegrity: The architectural basis of cellular mechanotrasduction. *Annual Review of Physiology* 59: 575–599.
14. Brinster, R.L. 1974. The effect of cells transferred into the mouse blastocyst on subsequent development. *The Journal of Experimental Medicine* 140: 1049–1056.
15. Mintz, B., et al. 1975. Normal genetically mosaic mice produced from malignant teratocarcinoma cells. *Proceedings of the National Academy of Sciences of the United States of America* 72: 3585–3589.
16. Papaioannou, V.E., et al. 1975. Fate of teratocarcinoma cells injected into early mouse embryos. *Nature* 258: 70–73.
17. Pierce, G.B., et al. 1979. Tumorigenicity of embryonal carcinoma as an assay to study control of malignancy by the murine blastocyst. *Proceedings of the National Academy of Sciences of the United States of America* 76: 6649–6651.
18. ———. 1983. *Teratocarcinoma stem cells*, ed. S. Silver, S. Strickland, and G. Martins, vol. 10, 15–22. Cold Spring Harbor: Cold Spring Harbor Laboratory.
19. Gerschenson, M., et al. 1986. Regulation of melanoma by the embryonic skin. *Proceedings of the National Academy of Sciences of the United States of America* 83: 7307–7310.
20. Pierce, G.B., et al. 1982. Specificity of the control of tumor formation by blastocyst. *Cancer Research* 42: 1082–1087.
21. Webb, C.W., et al. 1984. Developmental potential of myeloid leukemia cells injected into rat midgestation embryos. *Developmental Biology* 101: 221–224.
22. Weaver, V., et al. 1997. Reversion of the malignant phenotype of human breast cells in three-dimensional culture and in vivo by integrin bloking bodies. *The Journal of Cell Biology* 137: 231–245.
23. Coleman, W.B., et al. 1993. Regulation of differentiation of diploid and some aneuploid rat liver epithelial (stemlike) cells by the hepatic microenvironment. *The American Journal of Pathology* 142: 1373–1382.
24. Kulesa, P.M., et al. 2006. Reprogramming metastatic melanoma cells to assume a neural crest-like phentype in a embryonic microenvironment. *Proceedings of the National Academy of Sciences of the United States of America* 103: 3752–3757.
25. Oppitz, M., et al. 2007. Non-malignat migration of B16 melanoma cells in the neural crest and invasive growth in the eye cup of the chick embryo. *Melanoma Research* 17: 17–30.
26. Biava, P.M., and D. Bonsignorio. 2002. Cancer and cell differentiation: a model to explain malignancy. *Journal of Tumor Marker Oncology* 17: 47–54.
27. Hendrix, M.J.C., et al. 2007. Reprogramming metastatic tumour cells with embryonic microenvironments. *Nature Reviews. Cancer* 7: 246–255.
28. Patton, E.E., et al. 2005. BRAF mutations are sufficient to promote nevi formation and cooperate with p53 in the genesis of melanoma. *Current Biology* 15: 249–254.
29. Biava, P.M., et al. 1988. Effects of treatment with embryonic and uterine tissue homogenates on Lewis lung carcinoma development. *Cancer Letters* 41: 265–270.
30. Berger, G., et al. 2001. Prolongation of survival of rats injected with heaptoma cells treated by nuclei extracts from mouse embryos. *Oncology Reports* 8: 673–677.
31. ———. 2003. Proposition of treatment of cancer cells by nuclear protein mixtures from embryonic cells. *Medical Hypotheses* 60: 489–493.
32. Biava, P.M., et al. 2001. Cell proliferation curves of different human tumor lines after in vitro treatment with Zebrafish embryonic extracts. *Journal of Tumor Marker Oncology* 16: 195–201.
33. Cucina, A., et al. 2006. Zebrafish embryo proteins induce apoptosis in human colon cancer cells (Caco2). *Apoptosis* 11: 1617–1628.

34. D'Anselmi, F., et al. 2011. Zebrafish stem cell differentiation stage factors suppress Bcl-xL release and enhance 5-Fu-mediated apoptosis in colon cancer cells. *Current Pharmaceutical Biotechnology* 12 (2): 261–267.
35. Proietti, S., et al. 2019. Active fraction from embryo fish extracts induces reversion of the malignant invasive phenotype in breast cancer through down-regulation of tctp and modulation of e-cadherin/β-catenin pathway. *International Journal of Molecular Sciences* 20 (9): pii: E2151.
36. Wells, R.S., and K.A. Miotto. 1986. Widespread inhibition of neuroblastoma cells in the 13- to 17-day-old mouse embryo. *Cancer Research* 46: 1659–1662.
37. Podesta, A.N., et al. 1984. The nerula state mouse embryos in control of neuroblastomas. *Proceedings of the National Academy of Sciences of the United States of America* 81: 7608–7611.
38. McKinnell, R.G., et al. 1969. Transplantation of pluripotential nuclei from triploid frog tumors. *Science* 165: 394–396.
39. Li, L., et al. 2003. Mouse embryos cloned from brain tumors. *Cancer Research* 63: 2733–2736.
40. Gootwine, E., et al. 1982. Participation of myeloid leukaemia cells injected into embryos in haematopoietic differentiation in adult mice. *Nature* 299: 63–65.
41. Kasemeier-Kulesa, J.C., et al. 2008. Reprogramming multipotent tumor cells with the embryonic neural crest microenvironment. *Developmental Dynamics* 237: 2657–2666.
42. Hochedlinger, K., et al. 2004. Reprogramming of a melanoma genome by nuclear transplantation. *Genes & Development* 18: 1875–1885.
43. Lee, L.M.J., et al. 2005. The fate of human malignant melanoma cells transplanted into zebrafish embryos: Assessment of migration and cell division in the absence of tumor formation. *Developmental Dynamics* 233: 1560–1570.
44. Postovit, L.M., et al. 2008. Human embryonic stem cell microenvironment suppress the tumorigenic phenotype of aggressive cancer cells. *Proceedings of the National Academy of Sciences of the United States of America* 105: 4329–4334.
45. Pierce, G.B. 1968. Teratocarcinoma: Model for a developmental concept of cancer. *Current Topics in Developmental Biology* 2: 223–246.
46. Cooper, M. 2009. Regenerative pathologies: Stem cells, teratomas and theories of cancer. *Mediaeval Studies* 1: 55–66.
47. Sonnenschein, C., and A.M. Soto. 1999. *The society of cells: Cancer and control of cell populations*. New York: Springer Verlag.
48. Gilbert, S.F., et al. 1996. Resynthesizing evolutionary and developmental biology. *Developmental Biology* 173: 357–372.
49. Weiss, P. 1939. *Principles of development*. New York: Holt.
50. Belousov, L.V., et al. 1997. Contributions to field theory and life of Alexander G. Gurwitsch. *The International Journal of Developmental Biology* 41: 771–779.
51. Reidl, R. 1978. *Order in living organisms: A systems analysis of evolution*. New York: Wiley.
52. Kidd, S. 1992. Characterization of the *Drosophila cactus* locus and analysis of interactions between cactus and dorsal proteins. *Cell* 71: 623–635.
53. Needham, J. 1936. New advances in the chemistry and biology of organized growth. *Proceedings of the Royal Society of London – Series B: Biological Sciences* 29: 1577–1626.
54. Waddington, C.H. 1935. Cancer and the theory of organizers. *Nature* 135: 606–608.
55. Ruiz-Vela, A., et al. 2009. Building a framework for embryonic microenvironments and cancer stem cells. *Stem Cell Reviews* 5 (4): 319–327.
56. Butcher, D.T., et al. 2009. A tense situation: Forcing tumour progression. *Nature Reviews. Cancer* 9: 108–122.
57. Rubin, H. 2006. What keeps cells in tissues behaving normally in the face of myriad mutations? *BioEssays* 28: 515–524.
58. Kenny, P.A., and M.J. Bissell. 2003. Tumor reversion: Correction of malignant behaviour by microenvironmental cues. *International Journal of Cancer* 107: 688–695.

59. Arnold, J.T., et al. 2002. Effect of normal and endometrial stroma on growth and differentiation in Ishikawa endometrial adenocarcinoma cells. *Cancer Research* 62: 79–88.
60. Felsher, D.W. 2003. Cancer revoked: Oncogenes as therapeutic targets. *Nature Reviews. Cancer* 3: 375–380.
61. Weinstein, I.B. 2002. Addiction to oncogenes. The achilles heel of cancer. *Science* 297: 63–64.
62. Biava, P.M., et al. 2002. Post-traslational modifications of the retinoblastoma protein (pRB) induced by in vitro administration of Zebrafish embryonic extracts on human kidney adenocarcinoma cell line. *Journal of Tumor Marker Oncology* 17 (3): 59–64.
63. Huang, S., and D.E. Ingber. 2007. A non-genetic basis for cancer progression and metastasis: Self-organizing attractors in cell regulatory networks. *Breast Disease* 26: 27–54.
64. Ward, A.C., and G.J. Lieschke. 2002. The zebrafish as a model system for human disease. *Frontiers in Bioscience* 7: 827–833.
65. Shuai, Z., et al. 2015. A fresh look at zebrafish from the perspective of cancer research. *Journal of Experimental & Clinical Cancer Research* 34 (1): 80.
66. Lieschke, G.J., and P.D. Currie. 2007. Animal models of human disease: Zebrafish swim into view. *Nature Reviews. Genetics* 8: 353–367.
67. Santoriello, C., and L.I. Zon. 2012. Science in medicine hooked! Modeling human disease in zebrafish. *Journal of Clinical Investigation* 122: 2337–2343.
68. Bischof, A.G., et al. 2013. Breast cancer normalization induced by embryonic mesenchyme is mediated by extracellular matrix biglycan. *Integrative Biology (Cambridge)* 5 (8): 1045–1056.
69. DeCosse, J.J., et al. 1973. Breast cancer: Induction of differentiation by embryonic tissue. *Science* 181 (4104): 1057–1058.
70. Kirchberger, S., et al. 2017. Danio?-Recent progress in modeling cancer in Zebrafish. *Frontiers in Oncology* 7: 186.
71. Franchi, F., et al. 2005. Embryo extracts opotherapy reduces a-fetoprotein levels in hepatocellular carcinoma patients. *Journal of Gastroenterology and Hepatology* 20: 1464–1473.
72. Livraghi, T., et al. 2005. Treatment with stem cell differentiation stage factors of intermediate-advanced hepatocellular carcinoma: An open randomized clinical trial. *Oncology Research* 15: 399–408.
73. Bizzarri, M., et al. 2002. The embryonic and maternal regulatory factor as a palliative therapy for advanced solid tumours. *Journal of Tumor Marker Oncology* 17 (3): 31–36.
74. Proietti, S., et al. 2018. Fish protein extract enhances clinical response to salvage chemotherapy in colon cancer patients. *Organisms. Journal of Biological Sciences* 2 (2): 81–90.
75. Abbott, D.E., et al. 2008. The epigenetic influence of tumor and embryonic microenvironments: How different are they? *Cancer Microenvironment* 1 (1): 13–21.
76. Beachy, P.A., et al. 2004. Tissue repair and stem cell renewal in carcinogenesis. *Nature* 432: 324–331.
77. D'Anselmi, F., et al. 2011. Metabolism and cell shape in cancer: A fractal analysys. *The International Journal of Biochemistry & Cell Biology* 43 (7): 1052–1058.
78. Seilem-Aspang, F., and K. Kratochwil. 1965. In *Regenerationin animals and related problems*, ed. V. Kiortsis and H. Trampusch, 452–473. Amsterdam: North Holland Publishing Co.
79. Rubin, H. 1985. Cancer as a dynamic developmental disorder. *Cancer Research* 45: 2935–2942.
80. Sonnenschein, C., and A.M. Soto. 2000. Somatic mutation theory of carcinogenesis: Why it should be dropped and replaced. *Molecular Carcinogenesis* 29: 205–211.
81. Maffini, M.V., et al. 2004. The stroma as a crucial target in rat mammary gland carcinogenesis. *Journal of Cell Science* 117: 1495–1502.
82. Solter, D., et al. 1970. Extrauterine growth of mouse egg – Cylinders results in malignant teratoma. *Nature* 227: 503–504.
83. Stevens, L.C. 1970. The development of transplantable teratocarcinomas from intratesticular grafts of pre- and post-implantation mouse embryos. *Developmental Biology* 21: 364–382.
84. Bissell, M.J., and M.A. LaBarge. 2005. Context, tissue, plasticity and cancer: Are tumor stem cells also regulated by the microenvironment? *Cancer Cell* 7: 17–23.

85. Anderson, A.R.A., et al. 2006. Tumor morphology and phenotypic evolution driven by selective pressure from the microenvironment. *Cell* 127: 905–915.
86. Ingber, D.E. 2008. Can cancer be reversed by engineering the tumour microenvironment? *Seminars in Cancer Biology* 18 (5): 356–364.
87. Pierce, G.B. 1983. The cancer cell and its control by the embryo. *The American Journal of Pathology* 113: 116–124.
88. Bizzarri, M., et al. 2017. Tumor reversion: Mesenchymal-epithelial transition as a critical step in managing the tumor-microenvironment cross-talk. *Current Pharmaceutical Design* 23 (32): 4705–4715.
89. Tsikolia, N. 2006. The role and limits of a gradient based explanation of morphogenesis: A theoretical consideration. *The International Journal of Developmental Biology* 50: 333–340.

Trabectedin, a Drug Acting on Both Cancer Cells and the Tumor Microenvironment

Paola Allavena, Manuela Liguori, and Cristina Belgiovine

Introduction

Immune cells infiltrate the tumor stroma since the early stages of carcinogenesis and are deeply affected by the presence of cancer cells. In the last decades our knowledge on the tumor microenviroment (TME) has considerably increased, and the current vision is that the reciprocal interactions between tumor and immune cells are of pivotal importance for the disease fate [7, 8, 12, 19, 37]. The TME is a chaotic, dynamically changing and deregulated site populated by different normal cell types: activated fibroblasts, newly formed vessels and immune cells. Innate and adaptive immune cells may have an ambiguous liaison with cancer as they can either limit or promote tumor growth. It is now quite clear that in early cancers and in pre-malignant conditions, the immune system plays a defensive role and actively eradicates those cancer cells that are antigenically visible (immune surveillance); at later stages, when tumors have evaded the immune system, and cancer cells have reached a critical mass to produce immune-suppressive factors, immune cells become disarmed effectors. Thus, opposite outcomes may occur in the reciprocal relationship between the immune system and tumor cells: an effective immune-surveillance, especially mediated by effector T cells of the adaptive immunity that limits tumor proliferation and spreading, or disease progression, especially mediated by myeloid cells of the innate immunity. The balance between these two scenarios is essentially context-dependent and is dictated by the availability of specific signals, the composition of functionally different immune cells present in the TME and by the types of products secreted by tumor cells. In established tumors, the presence of myeloid cells of the innate immunity (neutrophils and macrophages) within the TME

P. Allavena (✉) · M. Liguori · C. Belgiovine
Department Immunology, IRCCS Humanitas Clinical and Research Center, Rozzano, Italy
e-mail: paola.allavena@humanitasresearch.it

M. Bizzarri (ed.), *Approaching Complex Diseases*, Human Perspectives in Health Sciences and Technology 2, https://doi.org/10.1007/978-3-030-32857-3_13

constitutes a major source of persistent inflammation, now recognized as a hallmark of cancer, Fueling tumor progression and enforcing a suppressive milieu that inhibits anti-tumor responses [8, 31, 37]. In this review we will discuss the role of macrophages in cancer, and discuss the currently available therapeutic approaches aimed at modulating them. In particular, we will discuss recent results obtained with the anti-tumor agent trabectedin, which presents the unique feature of being able to target simultaneously cancer cells and macrophages.

Tumor-Associated Macrophages Promote Tumor Progression

Tumor-Associated Macrophages (TAMs), are a major cellular component of the tumor micro-environment (TME). They derive from circulating monocytes and are recruited at the tumor site by soluble chemo-attractant molecules secreted by both malignant and stromal cells, including several chemokines and the growth factor CSF-1 [1, 36]. A hallmark of macrophages is their plasticity, i.e. their ability to display different phenotypes and functional activities. These are dictated by distinct genetic programs that are elicited by specific local clues [40]. Macrophage functional phenotypes can be classified along a very schematic diagram, were the two end-points are the "classical" (or M1 macrophage) and the "alternative" (or M2 macrophage). The former type is stimulated by bacterial products (e.g. LPS) and Th1cytokines (IFNγ). The latter is stimulated by Th2 cytokines (IL-4, IL-13) and immunosuppressive factors (e.g. IL-10, TGFβ). M1 macrophages protect against bacterial infections and produce pro-inflammatory cytokines that elicit adaptive immune responses. When properly activated they can be cytotoxic against cancer cells. In contrast, M2 macrophages are not cytotoxic; instead, they have a general trophic function for tissues and are crucial for tissue development and regeneration. They also inhibit inflammatory reactions and suppress T cell-mediated immune responses [7, 8, 12, 19, 37]

In the tumor context, and especially under the influence of products derived from cancer cells, TAMs acquire M2-like functional features [1, 37]. Their M2-related activities favor disease progression in a variety of ways. For instance, TAM release trophic factors able to support the survival and proliferation of tumoral cells. They are able to regulate many aspects of cancer biology: invasion of surrounding tissues and distant spread (metastasis); proliferation of vessels (angiogenic switch) through the release of pro-angiogenic growth factors such as VEGF, PDGF, thymidine phosphorylase and chemokines such as CXCL8 [7, 8, 12, 19, 37]. This wide array of tumor-promoting functions is complemented by their immune-suppressive role. By producing factors such as IL-10 and TGFβ or PGE2, TAMs inhibit T cell-mediated anti-tumor responses and, instead, stimulate the differentiation of regulatory T cells (Treg) which further exert suppressive activity. TAMs also produce chemokines that recall Th2 T cells (e.g. CCL17, CCL18, CCL22), and thus shape the TME by populating it with effectors cells (Th2, Treg) unable to limit tumor growth

[38].. More recent studies have demonstrated that macrophages in the TME express PD-L1, and therefore they can inhibit T cell proliferation and activation also *via* the PD-L1/PD1 axis.

Strategies to Target Tumor-Associated Macrophages In Vivo

Being TAMs such crucial players of the TME that eventually promote disease progression, several attempts have been implemented in the last years to target them for therapeutic purposes. In general, macrophage-targeting strategies are aimed either at reducing their numbers and inhibiting their tumor-supporting functions, or at activating their immune-stimulating potential activity.

One of the early attempt was to use inhibitors of chemokines that are known to be involved in the chemo-attraction of blood monocytes into tumors. Antibodies against CCL2 have been tested with some success in experimental tumors, but did not provide positive results in human clinical trials [37]. A likely explanation is the redundancy of chemokines and other factors that modulate monocyte migration in tumors. More recently, different types of inhibitors targeting the CSF1 receptor (CSF-1R) were produced and tested with success. The CSF-1R is critical for macrophage survival and differentiation; furthermore, it is exclusively expressed by cells of the monocytic lineage. In experimental mouse models, anti-CSF1R or antagonist reduced tumor growth and also attenuated the M2-like functional profile of TAMs [43, 45]. In clinical trials, inhibitors of the CSF1-R are now being evaluated in combination with conventional chemotherapies.

Functional reprogramming of TAMs into immune-stimulating anti-tumor effectors has been tested with a number of approaches; for instance, in the past with the use of microbial molecules (bacterial muramyl-dipeptide, BCG) or cytokines (IFNs) [4]. In more recent years, specific molecular pathways have been targeted. Agonistic mAb against the surface molecule CD40 on macrophages have demonstrated to be effective in stimulating their anti-tumor activity [5, 49], and have been used also in combination with inhibitors of the CSF-1R, as well as with checkpoint inhibitors or chemotherapy [46]. Another recent approach is to inhibit the "do not eat me" molecule CD47 expressed by tumor cells, which interacts with the macrophage molecule SIRPα, to inhibit the phagocytosis by macrophages. Blocking of this interaction enhanced phagocytosis of tumor cells and disease regression in experimental models [46]. Molecules targeting the CD47– SIRPα axis are currently being evaluated in clinical trials [10]. The compounds bisphosphonates have long been studied for the treatment of osteoporosis. These drugs are predominantly internalized by bone macrophages (osteoclasts) and induce their apoptosis [37]. Some studies reported effects also outside the bone, including macrophages in tumors, especially in bone-localized metastases. These interesting results have opened the clinical use of bisphosphonates for the treatment of bone-metastatic diseases, such as breast and prostate cancers, in combination with chemotherapy or hormonal therapy.

The Marine Agent Trabectedin Selectively Kills Monocytes and Macrophages

Among drugs specifically affecting the survival of myeloid cells, the compound trabectedin holds a special place. Trabectedin is a registered anti-tumor agent that was originally extracted from a marine organism, the Tunicate *Ecteinascidia turbinate*, and is now synthetically produced by PharmaMar (Spain) [17].

Trabectedin is used in the clinic for the second line treatment of soft tissue sarcoma, especially liposarcoma, and for relapsed platinum-sensitive ovarian cancer patients, in combination with liposomal doxorubicin [3, 14, 29, 30]. Toxicity profile is acceptable and manageable and this compound is now under clinical testing in other cancer types. Trabectedin was selected for its potent activity on tumor cells: it kills cancer cells and efficiently blocks their proliferation by directly interacting with DNA. Further studies, however, have soon clarified that this compound is more than a conventional cytotoxic agent, as it affects DNA repair mechanisms and the activity of selected transcription factors (Fig. 1). Transcription inhibition by trabectedin has crucial impact on tumor biology as the expression of several downstream target genes may be profoundly affected [18].

With the idea of distinguishing its inhibitory activity on the cell cycle and on transcription factors, some years ago we tested the effects of trabectedin on non-proliferating cells. We used freshly isolated human blood monocytes and, surprisingly, we found that monocytes were rapidly undergoing apoptosis even at low concentrations of the drug (e.g. 5–10 nM), in the time frame of 1–2 days. Even more remarkably, this effect was highly specific for monocytes and macrophages, as neutrophils or T lymphocytes were not affected by its cytotoxic action [2]. This finding stimulated a series of experiments to understand the exquisite selectivity of trabectedin for mononuclear phagocytes. We then demonstrated that trabectedin rapidly triggers (within few hours) the activation of caspase 8, and therefore a caspase-dependent apoptosis [27].

As caspase 8 is downstream of death membrane receptors, we checked the expression of FAS and TRAIL receptors, in the different leukocyte subsets. While FAS receptor in neutrophils, monocytes and T cells was equally high, the TRAIL receptors were heterogeneously expressed: monocytes were mainly positive for the signalling TRAIL-Rs (TRAIL-R1 and R2), instead neutrophils and T lymphocytes were negative for these receptors and expressed high levels of the non-signalling molecule (TRAIL-R3), which prevents caspase 8 activation [34]. Thus, the differential expression of death receptors by monocytes and other leukocytes explain why only the former are susceptible to trabectedin (Fig. 2). The drug, however, is not directly binding to TRAIL-Rs. It is known that upon activation with their specific ligand, TRAIL-Rs form a trimer with a precise space to lodge the TRAIL ligand. Trabectedin is a much smaller molecule than TRAIL and likely it is not able to directly activate the receptors. Nonetheless, the results clearly demonstrated, with different methods, that caspase 8 is definitely triggered by exposure to trabectedin (Fig. 2).

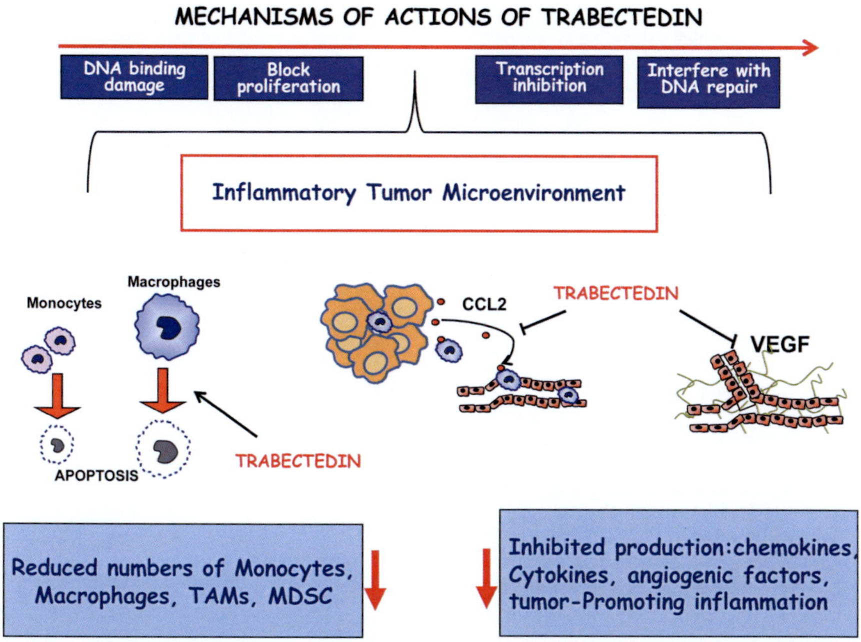

Fig. 1 The anti-tumor activity of the compound trabectedin is achieved through the combination of different mechanisms of action. Trabectedin directly interacts with DNA of cancer cells and efficiently blocks their proliferation. It also affects DNA repair mechanisms of tumor cells and inhibits the activity of selected transcription factors. Interestingly, trabectedin affects in multiple ways the tumor micro-environment: it induces a caspase-dependent apoptosis selectively in monocytes and macrophages, including Tumor-Associated Macrophages (TAMs), as well as in Myeloid Suppressor Cells (MDSC); in addition, at low non-cytotoxic concentrations, trabectedin inhibits the transcription of several genes coding for inflammatory cytokines and chemokines, and pro-tumor mediators, such as angiogenic factors

Of interest, some natural compounds, such as Palmitate, Quercetin and snail venom were reported to induce activation of caspase-8 in a TRAIL-independent manner, through the up-regulation and/or aggregation of death receptors. Indeed, we found that in vitro treatment with trabectedin increased the expression of TRAIL-R2 and induced receptor aggregation into lipid rafts (Fig. 2). Therefore, trabectedin induces caspase 8 activation via increased expression and aggregation of TRAIL receptors [34].

Trabectedin Inhibits the Production of Several Biological Mediators

We also noted that, at low non-cytotoxic concentrations, trabectedin was able to inhibit the production of specific inflammatory mediators such as CCL2, IL-6 and

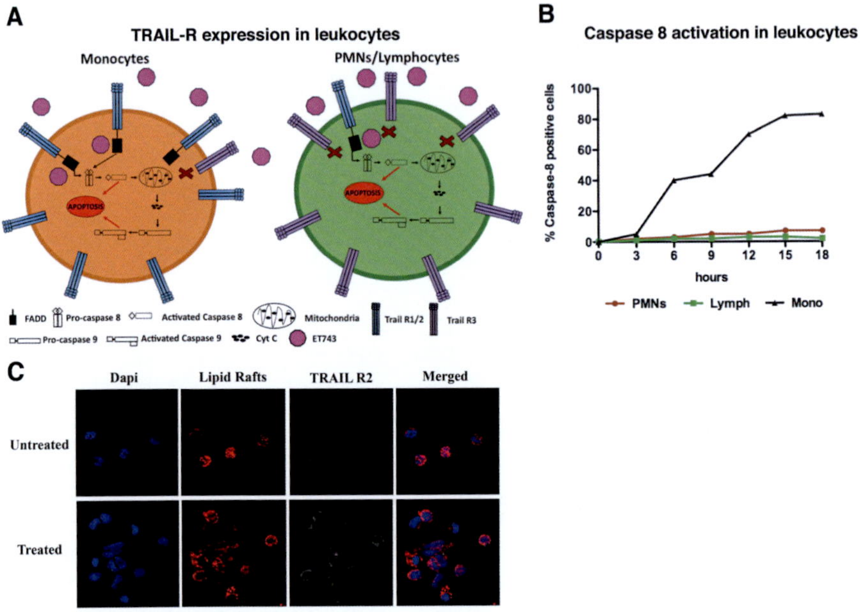

Fig. 2 Mechanistic explanation of the selective killing of mononuclear phagocytes by trabectedin. (**a**) Schematic representation of the expression of TRAIL receptors on the membrane of monocytes (left), and neutrophils (PMN)-lymphocytes (right). Only monocytes express the signaling receptors (TRAIL-R1/2), while PMN and lymphocytes predominantly express the non-signaling receptor (TRAIL-R3). (**b**) Activation of caspase 8 in trabectedin-treated leukocytes; only monocytes activate a caspase 8-dependent apoptosis. (**c**) Microscopic image of co-localization of TRAIL-R2 in lipid rafts after trabectedin treatment of monocytes

CXCL8. This effect was observed particularly in monocytes, TAMs and also in cancer cells such as myxoid liposarcoma and ovarian cancer tumors [2, 28]. Other chemokines known to be involved in monocyte recruitment were also transcriptionally affected by trabectedin treatment (e.g., CCL7, CCL3, and CCL14). Importantly, other chemotherapeutic agents tested in the same experiments (e.g., cisplatin, doxorubicin) had not such effect [2]. Trabectedin is known to inhibit also the expression of ECM-related genes produced by TAMs and fibroblasts, such as collagen type 1, fibronectin, osteopontin and the matrix-metalloprotease-2 (MMP2) [35]. Overall these results indicate that trabectedin may reduce the high turnover of the tumor stroma.

To have a broader view of the impact of trabectedin on monocytes, we performed a global gene profiling analysis of LPS-stimulated monocytes pre-treated with trabectedin and, for comparison, with doxorubicin as unrelated drug. The analysis revealed that the transcriptomes modulated by trabectedin were clearly distinct from those of doxorubicin. In addition to expected pathways of apoptosis and differentiation/activation, a major down-modulated pathway was that of the Rho GTPase family. Rho GTPase members are crucial molecular switches controlling many signaling events involved in diverse functions, such as actin-cytoskeleton organization and cell motility [24]. We further confirmed that monocytes treated with trabectedin have a severe impairment in their chemotaxis in response to CCL2. This finding bears importance during the active recruitment of circulating monocytes within tumors [41].

In summary, we demonstrated that trabectedin has multiple effects on mononuclear phagocytes: at higher doses it induces a rapid apoptosis; at non-cytotoxic concentrations it inhibits the production of selected biological mediators which are important for tumor progression, and affects monocyte adhesion and migration by inhibiting the genes that modulate and organize their actin cytoskeleton.

Role of Macrophage-Targeting in the Antitumor Activity of Trabectedin

The real issue was to demonstrate that trabectedin was able to kill mononuclear phagocytes in vivo in a tumor context. In different mouse tumor models, trabectedin significantly decreased the number of blood monocytes, and of spleen and tumor macrophages. Importantly, as seen in vitro with human leukocytes, no effect was observed in neutrophils and T lymphocytes [27]. It should be noted that the sole TRAIL-R in mice (DR5) was found expressed only in murine monocytes/macrophages and was virtually absent in neutrophils, T and B cells. Therefore, the pattern of TRAIL-R expression in mice perfectly mimics that of human leukocytes [34].

To definitely prove that the cytotoxic activity of trabectedin on mononuclear phagocytes was an important mechanism of its anti-tumor efficacy, we generated a trabectedin-resistant murine tumor cell line, by in vitro exposure to increasing concentration of the drug over 1 year. When this resistant murine fibrosarcoma was injected in mice, it formed tumors not dissimilar from the parent cells. Treatment of mice bearing this resistant variant resulted in marked anti-tumor activity, in spite of confirmed resistance of explanted cancer cells re-exposed in vitro to the drug [27]. Our hypothesis was that trabectedin had reduced the number of TAMs having pro-tumoral activity. To prove this concept, we adoptive transferred intra-venously new macrophages to trabectedin-treated mice, and indeed observed that tumor growth was significantly restored. These experiments demonstrated that the ability of trabectedin to reduce macrophages in vivo is a key component of its anti-tumor efficacy [27].

More recently we have performed in vivo experiments with the compound lurbinectedin (PM01183), an analogue of trabectedin, produced by the same company (PharmaMar, Spain). This compound is under clinical investigation in a broad range of tumors, including breast, ovarian cancer and small-cell lung cancer [15, 25]. The interest in lurbinectedin relies in a better pharmacokinetics profile compared to trabectedin, suggesting the possibility to use higher doses with less overall toxicity. Other groups have confirmed that this compound is highly efficient in blocking tumor cell proliferation and transcriptional activities, as observed with trabectedin [23, 39, 42, 47, 48]. Our in vitro results fully confirmed that lurbinectedin significantly reduces monocyte viability by inducing apoptotic cell death. Studies in vivo in tumor-bearing mice demonstrated that lurbinectedin was able to reduce the number of circulating monocytes and tumor macrophages, resulting in decreased tumor growth. Overall, it was concluded that lurbinectedin does not differ from trabectedin for its ability to impact the tumor micro-environment and in particular macrophages [6].

Further evidence that trabectedin presents selective activity against monocytes and TAMs came from the analyses of human samples derived from soft tissue sarcoma patients undergoing trabectedin therapy. In a selected cohort of patient we measured the number of circulating monocytes within few days after injection of trabectedin, and observed a significant decrease in most of the patients [27]. Furthermore, the immune-histochemical quantification of infiltrating TAMs in tumors from patients who received neo-adjuvant therapy, showed a dramatic decrease of macrophages compared to biopsies obtained before chemotherapy. Other studies were performed directly with surgical samples from patients with liposarcoma, either cultured in vitro or established as xenografts in immunodeficient animals. Treatment *in vitro/in vivo* with trabectedin significantly reduced the expression of several inflammatory factors, including CCL2, CXCL8 and IL-6 [28].

These results in tumor patients reinforced the concept that trabectedin strikes not only the neoplastic compartment, but also the myeloid cellular component and inhibits several crucial soluble mediators.

Other Effects of Trabectedin in the Tumor Environment

Other important biological effects may account for the anti-tumor efficacy of trabectedin. Pathological examination of tumor sections, both in mice and humans, revealed that in treated tumors the vessel network was significantly reduced. As mentioned above, TAMs are important in the neo-angiogenesis switch in the TME. We noted that release of VEGF and angiopoietin-2 were markedly down-modulated in macrophages treated with trabectedin [2, 6, 27]. This finding raised the question whether the inhibited angiogenesis was mediated solely by the effect on TAM-released factors or if a direct effect on the vessel network could be postulated. In our experiments, we noted that the macrophage depleting agent liposomal-clodronate was indeed able to inhibit tumor growth in fibrosarcoma-bearing mice (though with a less lasting effect compared to trabectedin), but had no relevant impact on the vessel network [27].

The group of Taraboletti investigated the effects of trabectedin on endothelial cells in vitro and in mouse models [22]. In a physiological in vivo assay (in the absence of tumor cells), trabectedin inhibited the ability of endothelial cells to invade the matrix and to undergo branching morphogenesis. This was mainly caused by the increased expression of TIMP-1 and TIMP-2, two well-known inhibitors of matrix metalloproteinases. Upregulated TIMP-1 and TIMP-2 reduced the proteolytic activity of endothelial cells, a process necessary to degrade the matrix [22]. In addition, they reported that trabectedin modulated the angiogenesis process by upregulating, in tumor cells, the expression of thrombospondin-1 (TSP-1), a major endogenous inhibitor of angiogenesis. In vivo experiments with xenografts of myxoid liposarcoma confirmed that the effect was specific for tumor cells, as the increased TSP-1 was of human origin, thus derived from the liposarcoma, and not from the murine host. Therefore, the anti-angiogenic activity of trabectedin occurs via different mechanisms, involving both a direct inhibitory effect on endothelial cells, as well as a reduction of the angiogenic potential of cancer cells and macrophages [11, 22]

Other groups have confirmed these effects of trabectedin on reduced angiogenesis. In an vitro system of endothelial cells cultured with the conditioned medium of multiple myeloma (MM) cells, Cucè et al. reported that trabectedin reduced the length of capillary-like structures and the number of branching points [16].

In the same study the Authors further investigated the impact of trabectedin directly on MM cells in 2D and 3D in vitro culture; they found that trabectedin was able to induce the apoptotic death of MM cells after 48 h of culture; in addition trabectedin induced the activation of the DNA damage response (DDR) cellular stress with ROS production. Of interest, DNA damage and cell stress induced the upregulation in cancer cells of NKG2D, the ligand for an activating receptor in natural killer (NK) cells. Therefore the role of trabectedin in MM cell susceptibility to NK-mediated killing was investigated and, indeed, it was interestingly found that trabectedin-treated MM cells were able to trigger the cytotoxic activity of NK cells [16].

The macrophage-depleting ability of trabectedin has been confirmed by other groups. In a mouse model of orthotopic pancreatic cancer, trabectedin strongly reduced the number of circulating monocytes and of tumor-infiltrating macrophages, while neutrophils were not significantly affected, in line with our previous findings [9].

Other recent studies similarly reported a reduction in the number TAMs in tumor mouse model of orthotopic osteosarcoma, melanoma and in skeletal metastasis from prostate cancer [11, 33, 44]. In mouse models of the Ewing sarcoma, an aggressive cancer infiltrated by macrophages, treatment with trabectedin alone had no effect on established tumors, but the combination of trabectedin with oncolytic herpes virotherapy significantly improved mouse survival and this effect was related to the reduction of TAMs and MDSC [20].

As mentioned above a major pro-tumor effect of TAMs is to suppress adaptive immunity responses and hence to promote immune evasion of tumors. Depletion of macrophages with trabectedin may relief this state of immune-suppression. In our previous studies we observed that in treated tumors the number of infiltrating T cells was increased [27]. Recent studies have confirmed and extended these findings. In a murine pancreatic cancer model, Borgoni et al. investigated the effects of trabectedin treatment on FACS-sorted tumor-infiltrating leukocytes and in particular in T lymphocytes. It is known that pancreatic tumors are characterized by an immunosuppressive TME. In untreated tumors they reported high expression of PDL-1 by

TAMs, a suppressive phenotype profile of CD4 T cells (IFNγ^{low}/IL-10^{high}, PD1^{high}) and an "exhausted" phenotype of CD8 T cells (IFN-γ^{low}/CD107^{low}, PD1^{high}). Upon trabectedin treatment, the number of TAMs was significantly reduced and this change in the TME had positive consequences on the epigenetic profile of infiltrating T cells. In fact, in drug-treated mice, T cells had repressed *IL-10* gene and activated *T-bet* promoter activity, switching to an anti-tumor phenotype (IL10^{low}/IFNγ^{high}). These results demonstrated the key role of the macrophage-targeting effect of trabectedin in the release from immune-suppression [9].

In another recent study, trabectedin was administered to immunocompetent mice bearing osteosarcoma growing orthotopically in the bone. The Authors demonstrated that trabectedin treatment reduced tumor growth and lung metastases; they further noted an important reprogramming of the tumor immune microenvironment, more specifically: infiltrating CD8 T lymphocytes were enhanced and showed high expression of the inhibitory checkpoint molecule PD-1. To further target these exhausted T cells, they performed combination treatment with trabectedin and anti-PD-1, and found significantly increased anti-tumor efficacy [44]. These results are important and provide a rationale for the combination of trabectedin with immune checkpoint inhibitors.

Another interesting result of this study was the observation that trabectedin induced a differentiation of osteoasarcoma cells, with a mechanism involving Runx2, (master regulator of osteoblastogenesis), and genes modulating the terminal differentiation of osteoblasts [44]. A pro-differentiating effect of cancer cells was also previously reported in myxoid/round cell liposarcoma, where trabectedin induced adipogenic differentiation in specific subtypes (type I and II of the fusion gene FUS-CHOP) [13, 21, 22, 26]. Furthermore, in Ewing sarcoma, where the fusion transcription factor EWS-FLI1 drives both proliferation and blocks differentiation, combination treatment of trabectedin and irinotecan in vivo, in xenograft-bearing mice, induced the differentiation of cancer cells into benign mesenchymal tissue [32].

Conclusions

Our better understanding of the complex relationship between immune and cancer cells has disclosed great opportunities to target the tumor stroma in therapeutic settings. In recent years there has been a flourishing of studies utilizing various approaches to target the inflammatory micro-environment, and in particular the innate immunity component of tumor-associated macrophages. Basic research and clinical studies have targeted TAMs by using strategies of cellular depletion, functional re-education or activation of their phagocytic activity. The recent discovery that blocking antibodies against immunological checkpoints lead to clinical success, has reinvigorated hopes that the proper modulation of the immune system can indeed increase anti-tumor responses in oncological patients. Particular emphasis has been put in combination strategies to attack both tumor cells and the stroma. Of the different approaches followed, the use of drugs such as trabectedin, that combines the ability to block tumor cell proliferation, with the induction of cell death specifically in the myeloid compartment, appears to be already one step in advance. The challenge for the future is to further increase our knowledge on the multifaceted plasticity of tumor-associated macrophages and to identify which treatment or combination treatments are best for each patient to obtain durable responses to anti-cancer therapies.

References

1. Allavena, P., and A. Mantovani. 2012. Immunology in the clinic review series; Focus on cancer: Tumour-associated macrophages: Undisputed stars of the inflammatory tumour microenvironment. *Clinical and Experimental Immunology* 167: 195–205.
2. Allavena, P., M. Signorelli, M. Chieppa, E. Erba, G. Bianchi, F. Marchesi, C.O. Olimpio, C. Bonardi, A. Garbi, A. Lissoni, F. de Braud, J. Jimeno, and M. D'Incalci. 2005. Anti-inflammatory properties of the novel antitumor agent yondelis (trabectedin): Inhibition of macrophage differentiation and cytokine production. *Cancer Research* 65: 2964–2971.
3. Andreeva-Gateva, P., and S. Chakar. 2019. The place of trabectedin in the treatment of soft tissue sarcoma: An umbrella review of the level one evidence. *Expert Opinion on Orphan Drugs* 7: 105–115.
4. Balkwill, F., K.A. Charles, and A. Mantovani. 2005. Smoldering and polarized inflammation in the initiation and promotion of malignant disease. *Cancer Cell* 7: 211–217.
5. Beatty, G.L., D.A. Torigian, E.G. Chiorean, B. Saboury, A. Brothers, A. Alavi, A.B. Troxel, W. Sun, U.R. Teitelbaum, R.H. Vonderheide, and P.J. O'Dwyer. 2013. A phase I study of an agonist CD40 monoclonal antibody (CP-870,893) in combination with gemcitabine in patients with advanced pancreatic ductal adenocarcinoma. *Clinical Cancer Research* 19: 6286–6295.
6. Belgiovine, C., E. Bello, M. Liguori, I. Craparotta, L. Mannarino, L. Paracchini, L. Beltrame, S. Marchini, C.M. Galmarini, A. Mantovani, R. Frapolli, P. Allavena, and M. D'Incalci. 2017. Lurbinectedin reduces tumour-associated macrophages and the inflammatory tumour microenvironment in preclinical models. *British Journal of Cancer* 117: 628–638.
7. Belgiovine, C., M. D'Incalci, P. Allavena, and R. Frapolli. 2016. Tumor-associated macrophages and anti-tumor therapies: Complex links. *Cellular and Molecular Life Sciences* 73: 2411–2424.
8. Biswas, S.K., P. Allavena, and A. Mantovani. 2013. Tumor-associated macrophages: Functional diversity, clinical significance, and open questions. *Seminars in Immunopathology* 35: 585–600.
9. Borgoni, S., A. Iannello, S. Cutrupi, P. Allavena, M. D'Incalci, F. Novelli, and P. Cappello. 2017. Depletion of tumor-associated macrophages switches the epigenetic profile of pancreatic cancer infiltrating T cells and restores their anti-tumor phenotype. *Oncoimmunology* 7: e1393596.
10. Cabrales, P. 2019. RRx-001 acts as a dual small molecule checkpoint inhibitor by downregulating CD47 on cancer cells and SIRP-α on monocytes/macrophages. *Translational Oncology* 12: 626–632.
11. Carminati, L., D. Pinessi, P. Borsotti, L. Minoli, R. Giavazzi, M. D'Incalci, D. Belotti, and G. Taraboletti. 2019. Antimetastatic and antiangiogenic activity of trabectedin in cutaneous melanoma. *Carcinogenesis* 40: 303–312.
12. Cassetta, L., and J.W. Pollard. 2018. Targeting macrophages: Therapeutic approaches in cancer. *Nature Reviews. Drug Discovery* 17 (12): 887–904.

13. Charytonowicz, E., M. Terry, K. Coakley, L. Telis, F. Remotti, C. Cordon-Cardo, R.N. Taub, and I. Matushansky. 2012. PPARγ agonists enhance ET-743–induced adipogenic differentiation in a transgenic mouse model of myxoid round cell liposarcoma. *Journal of Clinical Investigation* 122: 886–898.
14. Colombo, N., A.-C. Hardy-Bessard, G. Ferrandina, C. Marth, and I. Romero. 2016. Experience with trabectedin + pegylated liposomal doxorubicin for recurrent platinum-sensitive ovarian cancer unsuited to platinum rechallenge. *Expert Review of Anticancer Therapy* 16: 11–19.
15. Cruz, C., A. Llop-Guevara, J.E. Garber, B.K. Arun, J.A. Pérez Fidalgo, A. Lluch, M.L. Telli, C. Fernández, C. Kahatt, C.M. Galmarini, A. Soto-Matos, V. Alfaro, A. Pérez de la Haza, S.M. Domchek, S. Antolin, L. Vahdat, N.M. Tung, R. Lopez, J. Arribas, A. Vivancos, J. Baselga, V. Serra, J. Balmaña, and S.J. Isakoff. 2018. Multicenter phase II study of Lurbinectedin in BRCA-mutated and unselected metastatic advanced breast cancer and biomarker assessment substudy. *Journal of Clinical Oncology* 36: 3134–3143.
16. Cucè, M., M.E. Gallo Cantafio, M.A. Siciliano, C. Riillo, D. Caracciolo, F. Scionti, N. Staropoli, V. Zuccalà, L. Maltese, A. Di Vito, K. Grillone, V. Barbieri, M. Arbitrio, M.T. Di Martino, M. Rossi, N. Amodio, P. Tagliaferri, P. Tassone, and C. Botta. 2019. Trabectedin triggers direct and NK-mediated cytotoxicity in multiple myeloma. *Journal of Hematology & Oncology* 12: 32.
17. D'Incalci, M., N. Badri, C.M. Galmarini, and P. Allavena. 2014. Trabectedin, a drug acting on both cancer cells and the tumour microenvironment. *British Journal of Cancer* 111: 646–650.
18. D'Incalci, M. 2013. Trabectedin mechanism of action: What's new? *Future Oncology* 9: 5–10.
19. DeNardo, D.G., and B. Ruffell. 2019. Macrophages as regulators of tumour immunity and immunotherapy. *Nature Reviews. Immunology* 19 (6): 369–382.
20. Denton, N.L., C.-Y. Chen, B. Hutzen, M.A. Currier, T. Scott, B. Nartker, J.L. Leddon, P.-Y. Wang, R. Srinivas, K.A. Cassady, W.F. Goins, and T.P. Cripe. 2018. Myelolytic treatments enhance oncolytic herpes virotherapy in models of Ewing sarcoma by modulating the immune microenvironment. *Molecular Therapy Oncolytics* 11: 62–74.
21. Di Giandomenico, S., R. Frapolli, E. Bello, S. Uboldi, S.A. Licandro, S. Marchini, L. Beltrame, S. Brich, V. Mauro, E. Tamborini, S. Pilotti, P.G. Casali, F. Grosso, R. Sanfilippo, A. Gronchi, R. Mantovani, R. Gatta, C.M. Galmarini, J.M.F. Sousa-Faro, and M. D'Incalci. 2013. Mode of action of trabectedin in myxoid liposarcomas. *Oncogene* 33: 5201–5210.
22. Dossi, R., R. Frapolli, S. Di Giandomenico, L. Paracchini, F. Bozzi, S. Brich, V. Castiglioni, P. Borsotti, D. Belotti, S. Uboldi, R. Sanfilippo, E. Erba, R. Giavazzi, S. Marchini, S. Pilotti, M. D'Incalci, and G. Taraboletti. 2014. Antiangiogenic activity of trabectedin in myxoid liposarcoma: Involvement of host TIMP-1 and TIMP-2 and tumor thrombospondin-1. *International Journal of Cancer* 136 (3), 721–729.
23. Elez, M.E., J. Tabernero, D. Geary, T. Macarulla, S.P. Kang, C. Kahatt, A.S.M. Pita, C.F. Teruel, M. Siguero, M. Cullell-Young, S. Szyldergemajn, and M.J. Ratain. 2014. First-in-human phase I study of lurbinectedin (PM01183) in patients with advanced solid tumors. *Clinical Cancer Research* 20: 2205–2214.
24. Etienne-Manneville, S., and A. Hall. 2002. Rho GTPases in cell biology. *Nature* 420: 629–635.
25. Farago, A.F., B.J. Drapkin, J.A. Lopez-Vilarino de Ramos, C.M. Galmarini, R. Núñez, C. Kahatt, and L. Paz-Ares. 2019. ATLANTIS: A Phase III study of lurbinectedin/doxorubicin versus topotecan or cyclophosphamide/doxorubicin/vincristine in patients with small-cell lung cancer who have failed one prior platinum-containing line. *Future Oncology* 15: 231–239.
26. Forni, C., M. Minuzzo, E. Virdis, E. Tamborini, M. Simone, M. Tavecchio, E. Erba, F. Grosso, A. Gronchi, P. Aman, P. Casali, M. D'Incalci, S. Pilotti, and R. Mantovani. 2009. Trabectedin (ET-743) promotes differentiation in myxoid liposarcoma tumors. *Molecular Cancer Therapeutics* 8: 449–457.
27. Germano, G., R. Frapolli, C. Belgiovine, A. Anselmo, S. Pesce, M. Liguori, E. Erba, S. Uboldi, M. Zucchetti, F. Pasqualini, M. Nebuloni, N. van Rooijen, R. Mortarini, L. Beltrame, S. Marchini, I. Fuso Nerini, R. Sanfilippo, P.G. Casali, S. Pilotti, C.M. Galmarini, A. Anichini, A. Mantovani, M. D'Incalci, and P. Allavena. 2013. Role of macrophage targeting in the antitumor activity of trabectedin. *Cancer Cell* 23: 249–262.

28. Germano, G., R. Frapolli, M. Simone, M. Tavecchio, E. Erba, S. Pesce, F. Pasqualini, F. Grosso, R. Sanfilippo, P.G. Casali, A. Gronchi, E. Virdis, E. Tarantino, S. Pilotti, A. Greco, M. Nebuloni, C.M. Galmarini, J.C. Tercero, A. Mantovani, M. D'Incalci, and P. Allavena. 2010. Antitumor and anti-inflammatory effects of trabectedin on human myxoid liposarcoma cells. *Cancer Research* 70: 2235–2244.
29. Grignani, G., L. D'Ambrosio, Y. Pignochino, E. Palmerini, M. Zucchetti, P. Boccone, S. Aliberti, S. Stacchiotti, R. Bertulli, R. Piana, S. Miano, F. Tolomeo, G. Chiabotto, D. Sangiolo, A. Pisacane, A.P. Dei Tos, L. Novara, A. Bartolini, E. Marchesi, M. D'Incalci, A. Bardelli, P. Picci, S. Ferrari, and M. Aglietta. 2018. Trabectedin and olaparib in patients with advanced and non-resectable 17 bone and soft-tissue sarcomas (TOMAS): An open-label, phase 1b study from the Italian Sarcoma Group. *The Lancet Oncology* 19: 1360–1371.
30. Grosso, F., R.L. Jones, G.D. Demetri, I.R. Judson, J.-Y. Blay, A. Le Cesne, R. Sanfilippo, P. Casieri, P. Collini, P. Dileo, C. Spreafico, S. Stacchiotti, E. Tamborini, J.C. Tercero, J. Jimeno, M. D'Incalci, A. Gronchi, J.A. Fletcher, S. Pilotti, and P.G. Casali. 2007. Efficacy of trabectedin (ecteinascidin-743) in advanced pretreated myxoid liposarcomas: A retrospective study. *The Lancet Oncology* 8: 595–602.
31. Hanahan, D., and R.A. Weinberg. 2011. Hallmarks of cancer: The next generation. *Cell* 144: 646–674.
32. Harlow, M.L., M.H. Chasse, E.A. Boguslawski, K.M. Sorensen, J.M. Gedminas, S.M. Kitchen-Goosen, S.B. Rothbart, C. Taslim, S.L. Lessnick, A.S. Peck, Z.B. Madaj, M.J. Bowman, and P.J. Grohar. 2019. Trabectedin inhibits EWS-FLI1 and evicts SWI/SNF from chromatin in a schedule-dependent manner. *Clinical Cancer Research* 25 (11): 3417–3429.
33. Jones, J.D., B.P. Sinder, D. Paige, F.N. Soki, A.J. Koh, S. Thiele, Y. Shiozawa, L.C. Hofbauer, S. Daignault, H. Roca, and L.K. McCauley. 2019. Trabectedin reduces skeletal prostate cancer tumor size in association with effects on M2 macrophages and efferocytosis. *Neoplasia* 21: 172–184.
34. Liguori, M., C. Buracchi, F. Pasqualini, F. Bergomas, S. Pesce, M. Sironi, F. Grizzi, A. Mantovani, C. Belgiovine, and P. Allavena. 2016. Functional TRAIL receptors in monocytes and tumor-associated macrophages: A possible targeting pathway in the tumor microenvironment. *Oncotarget* 7: 41662–41676.
35. Louneva, N., B. Saitta, D.J. Herrick, and S.A. Jimenez. 2003. Transcriptional inhibition of type I collagen gene expression in scleroderma fibroblasts by the antineoplastic drug Ecteinascidin 743. *The Journal of Biological Chemistry* 278: 40400–40407.
36. Mantovani, A., B. Bottazzi, F. Colotta, S. Sozzani, and L. Ruco. 1992. The origin and function of tumor-associated macrophages. *Immunology Today* 13: 265–270.
37. Mantovani, A., F. Marchesi, A. Malesci, L. Laghi, and P. Allavena. 2017. Tumour-associated macrophages as treatment targets in oncology. *Nature Reviews. Clinical Oncology* 14: 399–416.
38. Mantovani, A., B. Savino, M. Locati, L. Zammataro, P. Allavena, and R. Bonecchi. 2010. The chemokine system in cancer biology and therapy. *Cytokine & Growth Factor Reviews* 21: 27–39.
39. Moneo, V., P. Martínez, B. de Castro, S. Cascajares, S. Avila, L.F. Garcia-Fernandez, and C.M. Galmarini. 2014. Abstract A174: Comparison of the antitumor activity of Trabectedin, Lurbinectedin, Zalypsis and PM00128 in a panel of human cells deficient in transcription/NER repair factors. *Molecular Cancer Therapeutics* 12: A174.
40. Murray, P.J. 2017. Macrophage polarization. *Annual Review of Physiology* 79: 541–566.
41. Noy, R., and J.W. Pollard. 2014. Tumor-associated macrophages: From mechanisms to therapy. *Immunity* 41: 49–61.
42. Pernice, T., A.G. Bishop, M.J. Guillen, C. Cuevas, and P. Aviles. 2016. Development of a liquid chromatography/tandem mass spectrometry assay for the quantification of PM01183 (lurbinectedin), a novel antineoplastic agent, in mouse, rat, dog, cynomolgus monkey and mini-pig plasma. *Journal of Pharmaceutical and Biomedical Analysis* 123: 37–41.

43. Pyonteck, S.M., L. Akkari, A.J. Schuhmacher, R.L. Bowman, L. Sevenich, D.F. Quail, O.C. Olson, M.L. Quick, J.T. Huse, V. Teijeiro, M. Setty, C.S. Leslie, Y. Oei, A. Pedraza, J. Zhang, C.W. Brennan, J.C. Sutton, E.C. Holland, D. Daniel, and J.A. Joyce. 2013. CSF-1R inhibition alters macrophage polarization and blocks glioma progression. *Nature Medicine* 19: 1264–1272.
44. Ratti, C., L. Botti, V. Cancila, S. Galvan, I. Torselli, C. Garofalo, M.C. Manara, L. Bongiovanni, C.F. Valenti, A. Burocchi, M. Parenza, B. Cappetti, S. Sangaletti, C. Tripodo, K. Scotlandi, M.P. Colombo, and C. Chiodoni. 2017. Trabectedin overrides osteosarcoma differentiative block and reprograms the tumor immune environment enabling effective combination with immune checkpoint inhibitors. *Clinical Cancer Research* 23: 5149–5161.
45. Ries, C.H., M.A. Cannarile, S. Hoves, J. Benz, K. Wartha, V. Runza, F. Rey-Giraud, L.P. Pradel, F. Feuerhake, I. Klaman, T. Jones, U. Jucknischke, S. Scheiblich, K. Kaluza, I.H. Gorr, A. Walz, K. Abiraj, P.A. Cassier, A. Sica, C. Gomez-Roca, K.E. de Visser, A. Italiano, C. Le Tourneau, J.P. Delord, H. Levitsky, J.Y. Blay, and D. Ruttinger. 2014. Targeting tumor-associated macrophages with anti-CSF-1R antibody reveals a strategy for cancer therapy. *Cancer Cell* 25: 846–859.
46. Ritter, B., and F.R. Greten. 2019. Modulating inflammation for cancer therapy. *The Journal of Experimental Medicine* 216 (6): 1234–1243. https://doi.org/10.1084/jem.20181739.
47. Romano, M., R. Frapolli, M. Zangarini, E. Bello, L. Porcu, C.M. Galmarini, L.F. García-Fernández, C. Cuevas, P. Allavena, E. Erba, and M. D'Incalci. 2013. Comparison ofin vitroandin vivobiological effects of trabectedin, lurbinectedin (PM01183) and Zalypsis® (PM00104). *International Journal of Cancer* 133: 2024–2033.
48. Santamaria Nunez, G., C.M. Robles, C. Giraudon, J.F. Martinez-Leal, E. Compe, F. Coin, P. Aviles, C.M. Galmarini, and J.M. Egly. 2016. Lurbinectedin specifically triggers the degradation of phosphorylated RNA Polymerase II and the formation of DNA breaks in cancer cells. *Molecular Cancer Therapeutics* 15: 2399–2412.
49. Vonderheide, R.H. 2018. The immune revolution: A case for priming, not checkpoint. *Cancer Cell* 33: 563–569.

Advances in Characterizing Recently-Identified Molecular Actions of Melatonin: Clinical Implications

Russel J. Reiter, Ramaswamy Sharma, Sergio A. Rosales-Corral, Ana Coto-Montes, Jose Antonio Boga, and Jerry Vriend

Introduction

Some system-wide functions/aspects of melatonin were identified decades ago, e.g., regulation of seasonal reproduction [51, 200, 201], photoperiodic control of pineal melatonin synthesis [122, 227], sleep initiation [31, 117], antioxidant actions [67, 197], circadian rhythm modulation [208, 279], cancer inhibition [27, 86, 163], etc. These aspects of melatonin are generally well known by the scientists working in the field, although the detailed mechanisms of these actions, in most cases, require further clarification. These functions are not discussed in detail in the current review, but they may be mentioned when they are germane to the discussion. Also, the functions of melatonin in plants [18, 61, 62, 194] are not considered even though

R. J. Reiter (✉)
Department of Cell Systems and Anatomy, UT Health San Antonio, San Antonio, TX, USA

Department of Molecular Genetics, University of Lodz, Lodz, Poland
e-mail: reiter@uthscsa.edu

R. Sharma
Department of Cell Systems and Anatomy, UT Health San Antonio, San Antonio, TX, USA

S. A. Rosales-Corral
Centro de Investigación Biomédica de Occidente, Instituto Mexicano del Seguro Social, Guadalajara, Mexico

A. Coto-Montes
Department of Morphology and Cell Biology, University of Oviedo, Oviedo, Spain

J. A. Boga
Servicio de Microbiologia, Hospital Universitario Central de Asturias, Oviedo, Spain

J. Vriend
Department of Human Anatomy and Cell Science, Max Rady College of Medicine, University of Manitoba, Winnipeg, MB, Canada

M. Bizzarri (ed.), *Approaching Complex Diseases*, Human Perspectives in Health Sciences and Technology 2, https://doi.org/10.1007/978-3-030-32857-3_14

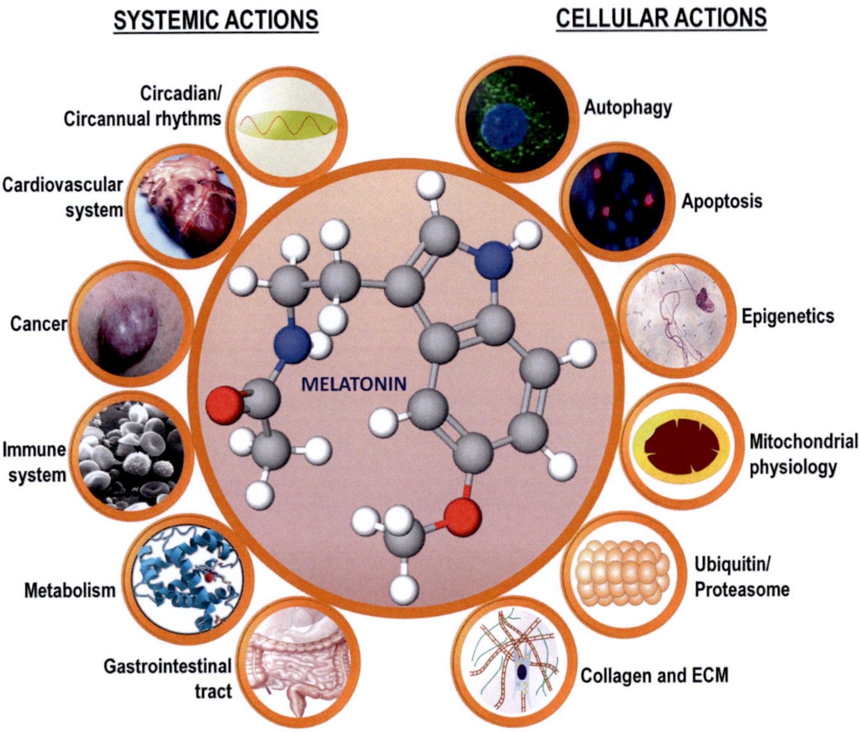

Fig. 1 Some of the identified systemic and cellular actions of endogenously-produced or exogenously-administered melatonin. There are other defined actions of melatonin that are not represented in this figure. The cellular actions illustrated are those considered in the current review. ECM = extracellular matrix

many of those actions have been widely investigated in recent years and these studies reveal that the actions of melatonin in plants are equally ubiquitous as its functions in animals.

What is briefly reviewed herein are some of the actions of melatonin at the cellular and subcellular levels that have implications for the more generalized processes mentioned (Fig. 1). Given the critical role of mitochondria in normal and pathological cellular function, the localization of melatonin in this organelle has broad implications for organ health and organismal well-being. For example, apoptosis and autophagy are important cellular processes that determine tissue architecture and organ physiology. Additional processes considered in this review are the role of melatonin in the regulation of epigenetics and of ubiquitin and proteasomal function. Finally, excessive deposits of collagen and extracellular matrix, accumulations collectively referred to as fibrosis, is a reflection of dysregulated cellular function which has severe negative consequences in a number of organs required for survival. This article by no means exhausts the large number of interactions melatonin has at the subcellular level.

Mitochondrial Physiology

Mitochondria are double membrane-bound subcellular organelles which house the biochemical machinery for oxidative phosphorylation and energy, i.e., ATP, production. This pathway is located in the convoluted inner mitochondrial membrane whose molecular heritage includes the outer membrane of bacteria which, 2.5–1.5 billion years ago, were ingested by early eukaryotic cells in the process referred to as endosymbiosis. There are estimated to be thousands of mitochondria in every cell with cells that have a higher energy requirements, e.g., cardiomyocytes, being especially rich in this organelle.

An association of melatonin with mitochondrial physiology has been repeatedly confirmed within the last two decades although the molecular mechanisms by which melatonin accomplishes these tasks still require extensive investigation (Fig. 2). What was initially reported was that melatonin preserved the activities of mitochondrial complex I and IV in rats treated with the mitochondrial toxin, ruthenium red [153]; this toxin hinders the function of the mitochondrial Ca^{2+} uniporter which leads to the excessive production of damaging free radicals in these organelles. When toxin-treated animals were also given melatonin, but not when they were treated with either vitamin E or vitamin C, the activities of the complexes were preserved; the toxin in this case was *t*-butyl hydroperoxide [154]. While the conserved mitochondrial function was consistent with melatonin's discovery as an antioxidant [187], the failure of vitamin C or E, molecules which also function as radical scavengers, is left unexplained. Perhaps the vitamins did not localize in sufficient amounts in the mitochondria to combat the large amounts of free radicals produced.

Findings reported by Martin et al. [153, 154] were the first that implied that melatonin may have a particular affinity for mitochondria, findings that were later verified (see below). The damage inflicted on the mitochondrial respiratory chain by toxin exposure was also reflected in the lowered ATP levels, an action also overcome by melatonin [155]. Shortly after the studies of Martin et al. [153–155], Okatani et al. [173, 174] also verified the beneficial actions on mitochondrial physiology by reporting that melatonin protects against hepatic mitochondrial injury due to ischemia-reperfusion and the deterioration that normally occurs in old mice.

Simultaneous with these studies, melatonin was reported to protect neural mitochondrial oxidative phosphorylation from other toxin exposures including 1-methyl-4-phenylpyrimidine [1], 6-hydroxydopamine [49], and kainic acid [50]. These data further solidified the speculation that melatonin had high relevance to mitochondrial function.

The ability of melatonin to maintain mitochondrial respiratory chain function and ATP synthesis under conditions of hypoxia-reoxygenation and after mitochondria-targeted toxin exposure indicated that melatonin enters these organelles in sufficient amounts to counteract these caustic circumstances. That melatonin can quickly gain access to mitochondria was immunocytochemically verified by Jou et al. [109]. When rat brain astrocytes were incubated in medium containing the oxidizing

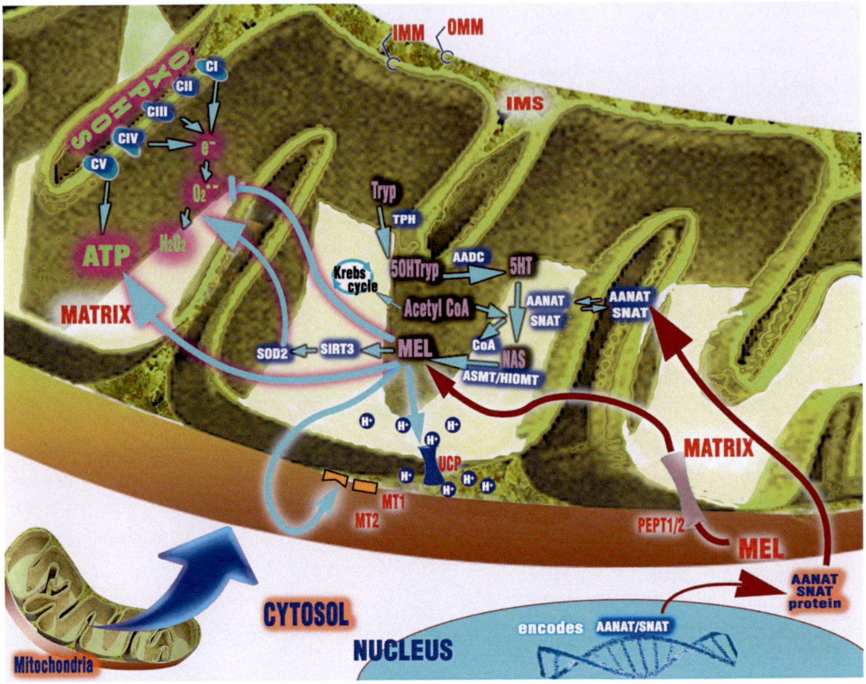

Fig. 2 This figure illustrates the facilitated transport of melatonin into mitochondria via the oligopeptide transporters, PEPT1 and PEPT2. The synthesis of melatonin in the mitochondrial matrix is also shown, a pathway that is aided by the ready availability of acetyl coenzyme A (acetyl CoA); acetyl CoA is a co-factor for the activity of the acetyltransferase enzyme (SNAT/AANAT). Locally-produced melatonin may diffuse out of the mitochondria and interact with melatonin membrane receptors (MT1/MT2) on the outer mitochondrial membrane (OMM). By means of this action, melatonin may regulate cytochrome c release and subsequent caspase activation. Within mitochondria, melatonin scavenges reactive oxygen species and stimulates the antioxidative enzyme, superoxide dismutase 2 (SOD2), a pathway that involves SIRT3 and FOXO3a. Also, melatonin protects against the deterioration of oxidative phosphorylation (CI- CV) and enhances ATP production. By reducing mitochondrial oxidation dyshomeostasis, melatonin reduces the oxidation of cardiolipin and limits cytochrome c release. Melatonin also regulates the opening of the mitochondria transition pore directly and via an action on uncoupling protein (UCP). For many of these functions, additional data are required before these actions can be identified as being definitive. IMM = inner mitochondrial membrane; IMS = intermembrane space; OMM = outer mitochondrial membrane; Try = tryptophan; 5OHTry = 5- hydroxytryptophan; 5-HT = serotonin; NAS = *N*-acetylserotonin. Other details can be found in the text

agent, H2O2, their mitochondria quickly exhibited intense fluorescence due to ROS generation. However, when mitochondria were incubated with H2O2 and melatonin, the extreme mitochondrial fluorescent reaction that was associated with H2O2 only was quenched. Like Martin et al. [154], Jou and co-workers [109] also found that vitamin E was significantly less effective than melatonin in reducing ROS-mediated mitochondrial fluorescence. The differences in the efficacies of melatonin versus vitamin E at the mitochondrial level were also apparent when the apoptotic indices of the astrocytes were compared; thus, melatonin was more efficacious in limiting mitochondrial-mediated apoptosis than was vitamin E. Jou et al. [110, 111] extended the evidence related to the functional preservation of mitochondria when they found that intracellular calcium dysregulation, mitochondrial transition pore opening, cardiolipin depletion and, cytochrome c release were attenuated in astrocytes challenged by oxidizing situations but not when treated with melatonin.

Biochemically, Maity et al. [149] and Zavodnik et al. [278] also reported that mitochondrial physiology of both gastric mucosal cells and hepatocytes, respectively, were maintained by treatment of animals with melatonin. In these studies, the mucosal cells were protected from the commonly-used drug, indomethacin, while the liver cells functioned normally in diabetic animals. Both of these toxicities are common in humans, so the findings have clear clinical relevance.

While it was well established that a number of tissues, in addition to the pineal gland, produce melatonin [4], the evidence remained circumstantial that the indoleamine was actually present in mitochondria. In 2012, Venegas et al. [256] were the first to estimate radioimmunoassayable levels in several compartments of fractionated neural and liver cells. Contrary to expectations, they observed very wide differences in the melatonin concentrations in different portions of the cells. In general, nuclear and cytosolic levels were low while mitochondrial measurements showed they were more than 10 times higher than in the other two compartments. This finding is strongly supportive of the involvement of melatonin with mitochondrial physiology.

Also of interest is that the concentrations of melatonin in different compartments did not vary over a 24-hour light:dark cycle (as do pineal and blood levels) and surgical removal of the pineal gland did not alter melatonin concentrations in any portion of the cells, except for the membrane fraction where melatonin levels actually increased. In addition to indicating the high likelihood of the importance of melatonin in mitochondrial physiology, it leaves in doubt, an answer to the question: what constitutes a physiological level of melatonin? [193]. The answer to this question is confounded because within subcellular organelles, melatonin levels vary widely and likewise, the concentrations of melatonin in different body fluids are greatly different.

The seemingly special association of melatonin with mitochondria [3, 195, 196] prompted an examination of the evolutionary heritage of this widely-distributed indoleamine. In 1995, Manchester et al. [150] had reported that a bacterium (*Rhodospirillum rubrum*) contained immunoreactive melatonin. Considering this finding and the alleged origin of both chloroplasts and mitochondria from bacteria that were ingested as food by early eukaryotes (Endosymbiotic Theory), we [244] proposed that melatonin initially evolved in bacteria several billion years ago and when they were ingested and developed into mitochondria in eukaryotes, they retained their melatonin-forming ability.

This hypothesis has now been supported by the reports of He et al. [82] and Suofu et al. [233]. He et al. [82] showed that when isolated mouse oocyte mitochondria were incubated in medium containing the necessary precursor, serotonin, melatonin levels increased quickly in both the mitochondria and in the incubation medium. In the absence of available serotonin, no melatonin was formed. Meanwhile, Suofu et al. [233] approached the issue in a different manner by identifying the enzymes that synthesize melatonin from serotonin [35], i.e., *N*-acetyltransferase and acetylserotonin methyltransferase, in non-synaptosomal neural cell mitochondria. As with He et al. [82], when they incubated neurally-derived mitochondria with serotonin, these organelles formed melatonin and melatonin metabolites. Unlike melatonin production in the pineal gland, brain mitochondria did not exhibit a day: night rhythm in mitochondria consistent with the findings of Venegas et al. [256]. Based on the results of these two studies, it seems apparent that mitochondria have the necessary enzymes to synthesize their own melatonin.

Melatonin produced in these subcellular organelles is not released into the systemic circulation in any significant amounts; it, however, may be released from cells to mediate autocrine and paracrine actions. Additionally, mitochondria-produced melatonin likely acts as an antioxidant in these structures, since they are major sites of free radical generation. The antioxidant functions could be achieved by direct radical scavenging [77] and/or by stimulating antioxidant enzymes [198].

In addition to its presumed local synthesis in mitochondria of all cells, exogenously- administered melatonin is also quickly extracted from the blood and distributed to mitochondria especially but also to other organelles (membranes, cytosol and nuclei) in cardiomyocytes [6]. The subcellular uptake was dose-dependent with the mitochondria and nuclei concentrating melatonin most rapidly after its peripheral administration; however, the melatonin concentrations in the mitochondria were about tenfold higher than those in the nuclei.

How melatonin gains access to cells and subcellular compartments has been extensively debated [161, 162]. Several options have been considered. Being highly lipid soluble, melatonin could presumably simply passively diffuse through cell membranes. Alternatively, it was recently suggested that melatonin is actively transferred into cells through the glucose transporter, GLUT1 [84]. Huo et al. [101] proposed that the uptake of melatonin by cells and by mitochondria is a facilitated process that involves the oligopeptide transporters, PEPT1 and PEPT2. These transporters exist in both the membrane and mitochondria of the cells (human cancer cells) that were tested. Thus, at this point there are at least three options to explain the uptake and differential intracellular distribution of melatonin.

The high concentrations of melatonin in mitochondria is very fortuitous considering these organelles are major sources of damaging ROS and melatonin is a potent direct free radical scavenger [243] and indirect antioxidant (stimulation of antioxidant enzymes) [198, 205]. Even when compared with synthetically-modified antioxidants, i.e., Mito E and Mito Q, which concentrate in the mitochondria up to 500-fold over that of unaltered vitamin E or coenzyme Q10, at equimolar concentrations melatonin was still more effective in preserving cellular and mitochondrial physiology [144].

Melatonin obviously has numerous essential functions in mitochondria which are critical to the optimal functioning of these organelles (Fig. 2) and, therefore, cells/ organs as a whole. It is speculated that, as with pineal melatonin production, its synthesis in mitochondria of extrapineal tissues also wanes with age [210]. Several recent reviews summarize the absolute importance of optimally-preserved mitochondrial functions, all of which are maintained in some manner by melatonin [5, 79, 180, 188, 268].

Autophagy

Autophagy is a self-degradative process which, under basal conditions, allows cells to remove misfolded proteins and damaged organelles. Starvation, growth factor deprivation, hypoxia, oxidation of critical molecules, protein aggregation, DNA damage or infection by intracellular pathogens are some of the conditions of cellular stress, which can induce autophagy.

This is a complex process including several sequential steps at molecular and cellular levels. The first step is the formation of a double membrane structure, the phagophore, which is generated de novo from plasma membrane-derived endocytic organelles such as endoplasmic reticulum or Golgi apparatus. This involves the mediation of ULK1/FIP200/ATG101/ATG13 protein kinase and VPS34/beclin1/VPS15/ATG14 lipid kinase complexes. Both complexes induce the formation of the ATG5/ATG12/ATG16L1 complex (formed with the help of *Atg7* and *Atg*10), which promotes the elongation of the phagophore engulfing a portion of the cytosol or specific cargoes (proteins, lipids, organelles), forming a double-membrane-bound vacuole called the autophagosome. Aided by *Atg3* and *Atg7*, these complexes facilitate the addition of a phosphatidylethanolamine group to the cytosolic form of mammalian LC3 homologues (LC3A, LC3B, LC3C, Gabarap, Gabarap-L1, and Gabarap-L2), which is referred to as the LC3-I to LC3- II conversion. The lipid-bound form of LC3 homologues is then recruited to the autophagosome, which fuses with a lysosome, forming a single-membrane-bound vesicle termed the autolysosome. Several lysosomal membrane proteins (GTPase RAB7, LAMP1, LAMP2, as well as SNARE proteins, such as syntaxin 17 and SNAP29) degraded the contents with the aid of another beclin1/VPS34 complex, where the UV radiation resistance-associated gene (UVRAG), rather than *Atg14*, is required [129].

The basal level of autophagy is required to control protein conformation and maintenance of cell homeostasis and survival. Nevertheless, depending on the stress and cell types, autophagy plays a dual role either as pro-survival or as pro-death event. Autophagy participates in many cellular and physiological responses. Several studies have shown that some of these responses can be modulated by melatonin.

Autophagy plays an important role in initiation and progression of neurodegenerative diseases since they are characterized by the misfolding and aggregation of cellular proteins, which must be eliminated or they may become toxic. The protective role of melatonin in the regulation of autophagy has been reported in several of neurodegenerative diseases. Parkinson disease is characterized by the aggregation of α-synuclein in dopaminergic neurons causing gradual degeneration of certain brain regions. The protective role of melatonin via autophagic processes in experimental models of Parkinson disease has been reported [2]. Thus, the exposure of glial cell type C6 or mouse striatal cells to 1-methyl-4- phenyl-1, 2, 3, 6-tetrahydropyridine (MPTP) provokes in increase of LC3-II, mediated by upregulation of the cyclin-dependent kinase 5 (CDK5), inducing α-synuclein aggregation, which is reduced by pretreatment of mice with melatonin [230]. Rotenone, a Parkinson disease inducer, promoted autophagic cell death mediated by Bax and Omi release into the cytoplasm. A decrease in the expression of Bax and a drop in the release of Omi into the cytoplasm, as well as reduced cell death, was observed in Hela cells pre-treated with melatonin [288]. Another Parkinson disease inducer, kainic acid, increased α-synuclein and the level of LC3-II and promoted neuronal loss in the hippocampus of mice. Melatonin enhanced α-synuclein ubiquitination and reduced LC3-II, cathepsin B and lysosomal-associated membrane protein 2 (LAMP-2) [36].

Autophagy plays a crucial role in the neural damage induced by several neurotoxic agents, such as arsenite and cadmium. In these conditions, melatonin exhibited neuroprotective effects by regulating autophagy. Thus, melatonin inhibited arsenite-induced autophagy and autolysosome formation in rat primary cultured cortical neurons [247]. Also, the reduced autophagosome-lysosome fusion and inhibition of lysosomal function triggered by cadmium intake is reversed by treatment with melatonin in mouse neuroblastoma cells [138, 139]. Oxaliplatin, which is a chemotherapeutic agent widely used in the treatment of different cancers, induces neuropathy as a secondary effect. Oxaliplatin-treated rats showed impaired autophagy with basal levels of LC3A/3B-I and II, beclin, *Atg3*, *Atg5*, and *Atg7* being attenuated in the sciatic nerve and dorsal root ganglia. An increase of these autophagic proteins was observed in melatonin-treated rats [17].

Glioblastoma multiforme is an aggressive tumor with a high mortality rate. Melatonin elevated LC3-II and induced progressive accumulation of autophagosome vacuoles via Akt activation in treated glioblastoma-initiating cells (GICs) [156].

While a basal autophagy rate has a protective effect during heart failure, ischemic cardiomyopathy, and cardiac hypertrophy, excessive autophagy promotes cardiac atrophy [176]. Diabetic cardiomyopathy is considered a major cause of heart failure. In melatonin-treated diabetic animals, adverse left ventricle remodeling was alleviated and cardiac dysfunction was reduced by enhanced Sirt1 expression and reduction of Mst1, which have pivotal roles in autophagy induction [281–283]. Chronic intermittent hypoxia (CIH), which occurs during obstructive sleep apnea syndrome (OSAS), causes multiple cardiovascular disorders such as coronary heart disease, hypertension and myocardial hypertrophy [72]. A higher LC3II/I ratio and greater Beclin1 expression in myocardial tissue of rats with CIH-induced myocardial hypertrophy suggest an increased autophagic response. Administration of melatonin induced additional autophagy via activation of AMPK, thereby having a protective effect in CIH rats [269]. Doxorubicin (DXR), which is an important chemotherapeutic agent, causes cardiotoxicity as a major side effect. Increased autophagy, which is upregulated during DXR-induced cardiotoxicity, is concomitant with a lower cell death [224]. The deteriorative effects of DXR on mitochondria are reduced by melatonin in an experimental model of cardiorenal syndrome; in this situation, melatonin reversed the drop in ATP production and inhibited cytochrome c release from mitochondria. It appears that melatonin has significant protective effect by modulating mitophagy, a process that removes damaged mitochondria through autophagy [42].

The potential benefit of melatonin on the gastrointestinal system due to the regulation of autophagy has been examined. The liver, which is the main organ for detoxification of hazard agents, is often dysregulated by toxic agents such as cadmium. Mitochondrial loss, cellular energy mitigation and cell death are a consequence of cadmium-induced hepatotoxicity resulting from excessive autophagy. Melatonin reduced mitochondrial reactive oxygen species (ROS) and subsequently lowered autophagy and cell death in HepG2 cells by activation of SIRT3-SOD2 signaling [186].

Carbon tetrachloride (CCL4) has been used to induce experimental hepatic fibrosis, which is overly exuberant wound healing in which excessive connective tissue accumulates in the liver. The rise in beclin1, *Atg12*, *Atg5*, and Atg16L1 mRNA levels observed in CCL4- induced hepatic fibrosis are reversed in melatonin-treated mice [212].

A diet rich in energy together with a sedentary life style contributes to obesity, which has become a common problem in developed countries. Fatty liver disease in a range of conditions caused by a build-up of fat in the liver, frequently found in obese subjects. The role of melatonin as a regulator of autophagy in obesity has been studied in the liver of obese, leptin-deficient mutant mice (ob/ob mice). One report claimed these animals had downregulated autophagy, which was enhanced by melatonin treatment associated with a decline in beclin1 and a rise in p62 [52].

Several studies have documented a role of melatonin in several gastrointestinal tumors. Thus, in HepG2 cells, which were derived from a male in a hepatocellular carcinoma, melatonin induced autophagy through activation of JNK but not mTOR [175]. In human colorectal cancer cells (HCT116 cells), melatonin treatment activated both autophagy and apoptosis by upregulation of both pro-apoptotic Bax and Bcl-xL [93]. Colorectal cancer (CRC) that develops in patients in inflammatory bowel disease is known as colitis- associated colorectal cancer (CAC); a model of this cancer type in mice uses 1, 2- dimethylhydrazine dihydrochloride administration. In this animal model, melatonin reduced CAC-induced autophagy as revealed by the expression pattern of various autophagy markers such as beclin1, LC3B-II/LC3B-I ratio and p62 [252]. Melatonin also induces the autophagic pathway in tongue squamous cell carcinoma cell line, Cal 27, as evidenced by the upregulation of LC3-II and downregulation of SQSTM1/P62 [61]. Valproic acid (VA) inhibits invasiveness of bladder cancer. When combined with melatonin, VA exhibits enhanced autophagy by triggering several autophagic genes including *Beclin1*, *Atg3* and *Atg5* [143]. Collectively, the results support the use of melatonin as a chemotherapeutic in the treatment of these tumors of the gastrointestinal system due to their ability to enhance cancer cell autophagy.

Melatonin plays various modulatory roles in cellular physiology. For example, autophagy is necessary for the preservation of normal morphology, cell mass, and function of pancreatic β cells. *AR42J* cells, derived initially from a transplantable tumor of a rat exocrine pancreas and used as a model of acute pancreatitis, showed an increased autophagy via endoplasmic reticulum stress. Melatonin enhanced autophagy in this experimental model [38].

Human fetal osteoblastic (hFOB1.19) cells are used as model of osteoporosis. An increase in glucose in these cells promoted autophagy, which was reduced by melatonin through inhibition of the ERK pathway.

The Harderian gland cells of the Syrian hamster are exposed to elevated oxidative stress because of their high content of porphyrins. To maintain the function of these glands, many of these cells exhibit autophagic processes. In these cells, melatonin reduced the destructive effects of free radicals via different mechanisms including amelioration of detachment-induced autophagic cell death [255].

Melatonin has beneficial effects on the maturation of oocytes by induction of autophagy and enhancing the expression of a number of genes including *ATG7* and beclin1, as seen in pig oocytes and cumulus cells [39].

Autophagy can also be induced during different stages of an infection. Although autophagy can limit the cytopathic effect of pathogens and the pathological consequences via a cellular process referred to as xenophagy, some cells have developed strategies to directly or indirectly subvert autophagy in order to promote different stages of the cell cycle.

Vibrio vulnificus hemolysin (VvhA) induces apoptosis and autophagy in human intestinal epithelial (HCT116) cells. Melatonin inhibited JNK-mediated phosphorylation of Bcl-2 responsible for the release of *Beclin1* and *Atg5* expression, thereby blocking VvhA -mediated apoptotic and autophagic cell death [134].

Rabbit hemorrhagic disease virus (RHDV) and rabbit vesivirus (RaV), two members of the genus *Lagovirus* (Family Caliciviridae), cause autophagosome and autophagolysosome formation [211]. During RHDV-induced autophagy, increased expression of beclin1, LC3-II/CL-I ration and Atg5-Atg12-Atg16L1 was found. A dysfunctional autophagy with impairment of the autophagic flux was also detected, a judgment based on a parallel rise in p62/SQSTM1 expression. A reduction of the level of these autophagic proteins by melatonin treatment indicates a drop in the autophagic response. Melatonin administration triggers similar mechanisms in other RNA viruses, such as HCV, whose infection cause a similar dysfunctional autophagy [211].

Prion proteins induce misfolding of normal cellular proteins and they are causative of several neurodegenerative diseases, such as Kuru and Creutzfeldt-Jacob disease. Melatonin protects SH-SY5Y cells from PrP-induced neurotoxicity by enhancing LC3-II levels and inducing autophagy [108].

Autophagy has a crucial role in homeostasis and in human diseases since it causes cellular survival or death depending on the severity of the damage. Several studies have reported that melatonin either promotes or suppresses autophagy, which implies that melatonin can balance the autophagic process. It is clearly established that melatonin's actions are context specific; these differential actions explain its beneficial effects and support the use of this molecule as a potential therapy for management of certain diseases (Fig. 3) where autophagy plays an important role. This brief summary elaborates only at the few sites/conditions in which melatonin has been shown to influence autophagic processes.

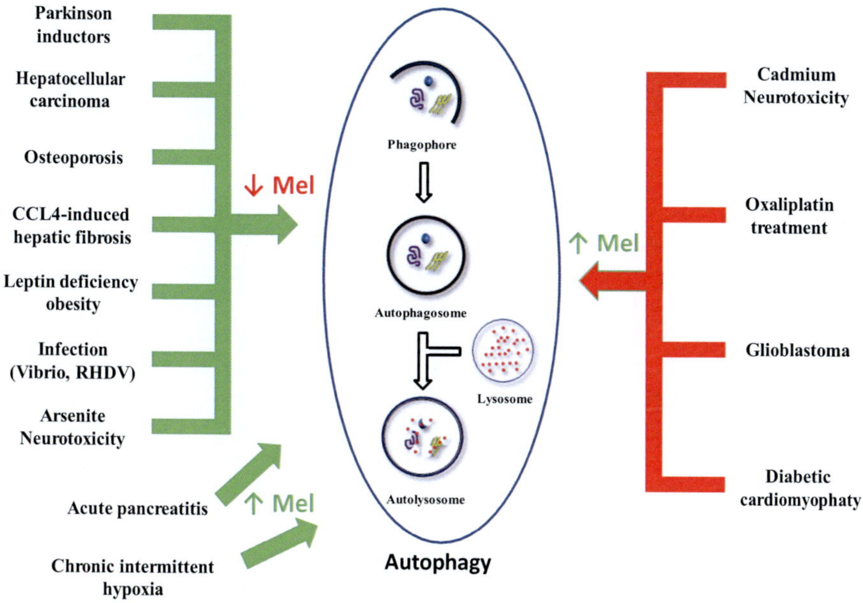

Fig. 3 Autophagy is induced (green rows) or inhibited (red rows) in several diseases, clinical conditions and experimental models. Melatonin administration counteracts or enhances such induction or inhibition

Apoptosis

Cell death is a necessary part of the normal development and maturation cycle. This process is the result of different cellular mechanisms that the cell initiates in response to both physiological and pathological stimuli. The cellular mechanisms are mainly encompassed by two essential processes: apoptosis and autophagy, although the special conditions of each cell can give rise to intermediate and mixed processes that can unleash cell death with characteristics common to both. Even a sustained and massive loss of energy may lead to cell death, such as necrosis.

Melatonin is produced rhythmically in the pineal gland and arrhythmically in the mitochondria of every cell. Via its membrane receptors, MT1 and MT2 [58, 142], or after it enters cells, melatonin modifies cell processes in a variety of ways to mediate cell death. These modifications, although they are usually intimately related to oxidative stress, are dependent on cell and tissue type and on the specific conditions in which they are found. Detailed studies have clarified some of the mechanisms involved in apoptosis as well as the similarities and differences that exist between the variations in programmed cell death.

Apoptosis of a cell induces a series of changes including cell shrinkage, blebbing of the plasma membrane, maintenance of organelle integrity, condensation and fragmentation of DNA, and eventually, ordered removal of the cell by phagocytes [118]. The triggering processes for apoptosis, extrinsic or receptor-mediated and intrinsic or mitochondria-mediated pathways, are complex and are partially known [185]. Recent evidence shows non- apoptotic roles for many usual effectors of the apoptotic signaling pathways. Thus, caspase 2, one of the best known molecules related to the triggering of apoptosis, is also involved in cell cycle regulation and DNA repair [254]. These new discoveries have complicated the understanding of the regulation of apoptosis.

Apoptotic events are highly regulated by the Bcl-2 family of proteins, which shares homology among the four common Bcl-2 domains. This family includes anti-apoptotic proteins including Bcl-2, Bcl-XL and Mcl-1, that possess all the four (BH) domains and act on mitochondria [29], endoplasmic reticulum and/or the nucleus [45]. The relationship that these proteins establish with free radicals has been identified and they have common actions on mitochondria [37]. Pro-apoptotic proteins are also included in Bcl-2 family. These proteins contain BH domains 1, 2, 3, unlike the anti- apoptotic proteins. The pro-apoptotic proteins include Bax and Bak which are related to the BH3-only proteins, i.e., Bim, Bad, Bmf, Noxa and Puma [95]. They function primarily to neutralize the anti-apoptotic proteins by sequestration and creation of heterodimers [103].

The influence of reactive oxygen species (ROS) on cell death signaling is twofold. Exposure of cells to a pro-oxidant state induces oxidation of caspases that prevent cells from undergoing apoptosis [44, 184]. Elevated intracellular ROS, e.g., H2O2, levels cause cytoplasmic acidification that influences the conformational status of proteins of the Bcl-2 family, resulting in their activation and facilitating the release of apoptosis amplification factors from the mitochondrial intermembrane space [7, 87]. This dual effect of ROS on apoptosis suggests an explanation of the actions of melatonin on the programmed cell death type since a major function of melatonin is as an antioxidant. However, melatonin's pleiotropic actions also make it difficult to identify the specific mechanisms by which melatonin acts on apoptosis. Herein, some of the multiple processes by which melatonin modulates apoptosis are considered.

Melatonin has multiple means by which it modifies and/or regulates apoptosis in both cancer and in normal cells. Melatonin's actions in these cell types are opposite to one another; it is pro-apoptotic in cancer cells and anti-apoptotic in normal cells. The mechanisms of these differential actions are still to be clarified.

Although melatonin typically exerts beneficial effects on cells, its action, via its receptors, due to its antioxidant properties or by alternative mechanisms, may vary depending on the tissue studied and on the melatonin dose used. Thus, melatonin alleviates liver damage by reducing mitochondrial dysfunction resulting from oxidative damage which is hyperglycemia- induced [125, 126]. Similarly, melatonin reduces oxidative stress and hepatic apoptosis promoted by ethanol administration [165, 177]. In both cases, scavenging of ROS may be the major explanation for its anti-apoptotic effects. Moreover, melatonin is also efficient in the prevention of apoptosis by acting through its MT2 membrane receptor [218] in bile duct-ligated young rats or inhibiting endoplasmic reticulum stress, a process that often induces apoptosis [53] in leptin-deficient mice. Similar mechanisms of protection are shown when melatonin reverses bone loss due to its antioxidant actions that prevent anti-apoptotic events [147].

Osteoporosis and several other diseases increase in prevalence with age, simultaneous with the age-related drop in peak nocturnal melatonin secretion. Several articles have confirmed that this coincidence is not accidental, with the loss of melatonin being related to aggravation or an increase in the incidence of the ailment [13, 169]. Thus, during aging, the drop in the nocturnal melatonin peak is associated with a rise in bone resorption, suggesting that melatonin may act as an endogenous osteoclast inhibitor [178]. Melatonin has been shown to limit bone resorption by limiting osteoclastogenesis [88, 127]. Since the osteoclastic actions involve apoptosis, melatonin indirectly inhibits cell death [152].

Heart failure is also a multifactorial syndrome that increases in prevalence with age [169]. Heart failure involves cardiomyocyte apoptosis; this is a major physiopathological feature and melatonin reverses it [56, 167, 179, 281–283]. In mouse models of post-infarction damage, melatonin reduces apoptosis through Mst1/Sirt1 signaling [98]. In a diabetic model, melatonin lowers apoptosis by modulating the related proteins [11]. Zhang and coworkers, in the same model, also observed a drop in apoptosis by after melatonin administration, while also conserving mitochondrial integrity and biogenesis [281–283]. The improved mitochondrial efficiency would have a direct impact on the inhibition of the intrinsic pathway of apoptosis.

Another means has been recently described by which melatonin modulates, and finally inhibits, apoptosis [114]. This mechanism involves neither membrane receptors (MT1, MT2) nor melatonin's antioxidant effect. This action involves the ubiquitination of target proteins in the apoptosis pathway, which is a usual mechanism that reduces Bcl-2 proteins and regulates apoptosis [264]. In porcine granulosa cells, melatonin limits BimEL, ubiquitinating it so it can be degraded by the proteasome [114]. This latter action not only identifies a new action of melatonin, but also may help to explain the differential behavior of melatonin in normal cells and cancer cells. In cancer cells, melatonin has been described as a potent proteasome inhibitor [258, 259]. Although many of the studies that show an inhibitory role of melatonin on apoptosis in normal cells may lack detailed mechanisms, the anti-apoptotic effect of melatonin deserves to be studied in greater detail considering the significant therapeutic repercussions that its use would imply.

The brain is highly sensitive to oxidative stress because it consumes a large amount of oxygen and generates more ROS than most other tissues. Moreover, it is rich in easily-oxidizable polyunsaturated fatty acids and endowed with relatively low levels of endogenous antioxidants [147, 191]. This makes it a clear target in which to examine the role of antioxidants on apoptosis. Melatonin has been shown in multiple studies to be a highly effective neuroprotective agent not only in neurodegenerative diseases [206, 266, 281–283], but also after brain injury [147]. These protective actions are related to the functions of melatonin as an antioxidant and anti-apoptotic actions. In mice, melatonin may protect against ischemic stroke via an action on MT2 receptors [40]. During brain ischemia-reperfusion, melatonin reduces upregulation of Nox2 and Nox4 expression, enhancing its anti-apoptotic capacity

[137]. Anti-apoptotic actions of melatonin, mediated by membrane receptors, are not confined to brain damage but also have been observed under neurobehavioral dysfunctions [265, 280].

Melatonin also reduces deleterious apoptotic processes in the central nervous system because of its ability to limit the activity of calpains either directly or by increasing the activity of calpastatin, a major inhibitor of calpains [241]. After spinal cord injury, an increased concentration of Ca^{2+} induces an activation of the calpains. Caspase-3 together with the calpains, causes the death of the neurons by apoptosis [15]. Melatonin administration reduces calpain gene expression, caspase-3 activity and, thereby, neuronal death in animals with spinal cord injury [209]. Methamphetamine is a neurotoxic molecule that causes neuronal apoptosis and activates glial cells in the central nervous system. It functions to reduce cell viability by depleting calpastatin levels which upregulate calpains and caspases [241]. Melatonin reverses the depletion of calpastatin and improves mitochondrial dysfunction thereby lowering cell death by reducing apoptosis [234, 235]. A similar effect of melatonin is observed during dexamethasone- induced neurotoxicity [236].

Stem cells are undifferentiated cells that can transform into multiple cell types. Stem cells are classified into several categories: totipotent stem cells which can give rise to any cell type; pluripotent stem cells, which differentiate into many tissue cell types, but not into a functional organism, and multipotent stem cells which can differentiate into a limited number of tissues. Stem cells have a huge therapeutic potential in a variety of diseases and they become essential to repair and regenerate damaged tissues or organs. However, the low survival rate of engrafted stem cells remains an important obstacle for stem cell therapy. It is essential to identify molecules capable of reducing cell death of stem cells and recent advances have indicated melatonin may be one such molecule.

Due to the antioxidant action of melatonin, it readily controls oxidative stress in stem cells [281–283] as in differentiated cells as described above. Apoptosis induced by oxidative stress is negated by melatonin since it directly neutralizes ROS [253] or indirectly removes them by the activation of antioxidant enzymes [65]. In the case of stem cells, it is noted that melatonin does not block the differentiation capacity of these cells by reducing free radicals; this is important since the increase in free radicals is a usual signal that leads to differentiation. Some authors, however, emphasize that "melatonin might abolish *redundant* free radicals" [65]. In fact, melatonin is not a differentiation- blocking agent, but is, instead, able to promote differentiation as several authors have noted [181, 245].

The role of melatonin as anti-apoptotic protector in toti- or pluripotent stem cells is tissue-dependent; as a consequence, caution should be exercised in extrapolating molecular mechanisms discovered in one cell type to other stem cells. In human hexokinase 2 (HK2) cells, an immortalized proximal tubule epithelial cell line, melatonin counteracts the deleterious effects of cisplatin, inducing its anti-apoptotic actions [281–283]. Only a few reports have identified the intermediate events in this process. ERK, through a melatonin receptor- independent process, seems to be

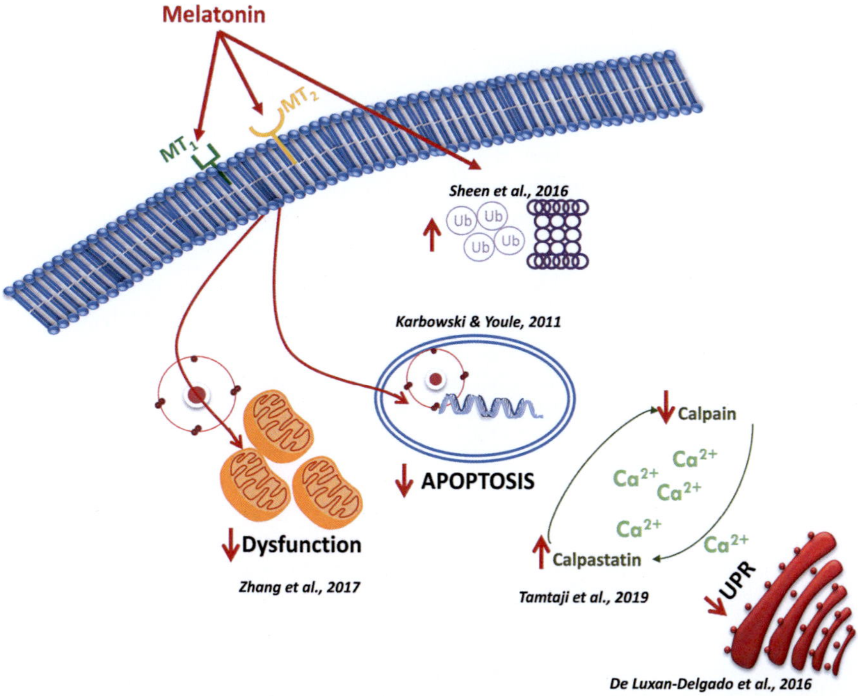

Fig. 4 This figure depicts some of the targets of melatonin to prevent apoptosis in normal cells. Three major action pathways have been described: through its membrane receptors MT1 and MT2, as a direct antioxidant and/or promoting ubiquitination. The first two act by performance optimization of essential organelles such as the endoplasmic reticulum and the unfolded protein response (UPR) and improving the mitochondrial physiology. The third one leads to protein destruction through the proteasome. In a potential minor pathway, melatonin directly acts on molecules involved in apoptosis, such as calpains and calpastatins that regulate apoptosis due to its action on caspases, or on proteins located in the mitochondria, favoring an increase of mitochondrial efficiency and reducing free radicals production

involved [145, 249]. Also, the regulation of expression BAX/Bcl-2 ratio and disruption of mitochondrial membrane potential have been observed [266]. MT1 and MT2 melatonin membrane-receptor may not be related to these anti-apoptotic effects of melatonin in different types of stem cells. Clearly, additional detailed reports related to the cellular mechanisms by which melatonin modulates apoptosis during stem cell-based therapy are required.

Considering the ubiquitous and essential nature of programmed cell death, it has implications to many normal developmental processes and is surely important in a variety of pathophysiological conditions. Many of these actions are summarized in Fig. 4 with an indication of their clinical relevance.

Ubiquitin-Proteasome System

We recently suggested a likely relationship between melatonin, ubiquitin and the proteasome [258, 259]. A fairly well documented connection between melatonin and the ubiquitin-proteasome system (UPS) in the modulation of the synthesis and degradation of the rate limiting enzyme involved in melatonin production, AANAT (arylalkylamine N- acetyltransferase), by ubiquitin ligases in the pineal gland has been documented [120]. The regulation of AANAT synthesis in the pineal gland by its adrenergic innervation and the cAMP/PKA pathway has been described in detail [89, 120, 121]. These investigations reported the rapid decrease in AANAT protein induced by L-propranolol was blocked by proteasome inhibitors [68]. They concluded that the rapid drop in AANAT protein and AANAT activity induced by light exposure or propranolol administration was due to proteasomal degradation of AANAT. Additional data documenting a rise in AANAT in the pineal gland after administration of proteasome inhibitors was reported by Huang et al. [100]. These investigators also pointed out that stabilization of AANAT protein was regulated by phosphorylation and by interaction with a 14-3-3 protein.

The role of proteasome inhibitors on the synthesis of AANAT was also investigated. Ho et al. reported two different effects of proteasome inhibitors, an increase in norepinephrine (NE) stimulated AANAT protein in rat pinealocytes, but a reduction in AANAT transcription if the inhibitor was administered prior to stimulation with NE [90, 248]. Schomerus et al. [213] observed that, in bovine pinealocytes, there was little variation in AANAT transcription and proteasomal destruction of AANAT was more important than transcriptional regulation of AANAT.

The above-mentioned studies did not identify specific ubiquitin ligases or specific deubiquitinases regulating the transcription of AANAT or those modulating its destruction by the proteasome. Based on studies of the ubiquitin ligase Praja2, Lignitto et al. [140] found that Praja2 interacts directly with the regulatory component of PKA. Based on this data, Vriend et al. [263] included Praja2 in a model of AANAT control in the pineal gland. Considering the complexity of control mechanisms of AANAT, it is reasonable to speculate that more than one ubiquitin ligase is involved in regulation of AANAT degradation and transcription. Since CREB is a transcription factor involved in the transcription of AANAT [223, 257] the ubiquitin ligase CREBBP (CREB binding protein) would be a reasonable protein to investigate for its potential in regulating AANAT transcription. It was first isolated as a protein that binds to the transcription factor CREB (cAMP response element binding protein) [41]. The p65 component of NF-kB is reported to bind to CREBBP (aka CBP) [271].

The number of similarities between the actions of proteasome inhibitors and the actions of melatonin have been noted [258, 259]. Both are reported to have anti-inflammatory activity mediated by inhibition of NF-kB activation and DNA binding-activity [96, 99]; both upregulate antioxidant enzymes under conditions of oxidative stress [275] and stimulate transcription of genes of the NRF2-antioxidant response pathway [107, 260, 261]; both are reported as pro- apoptotic in cancer cells

[70, 207]; both inhibit HIF-1 and VEGF [262, 267]; and, both interfere with cell cycle regulation [102, 221]. While it cannot be assumed that melatonin and proteasome inhibitors are equally effective, the similarities provide an interesting model for investigating the 'pleiotropic' effects of melatonin administration [221]. One explanation for the similarities of the actions of melatonin to that of proteasome inhibitors include the possibility that melatonin itself acts as a proteasome inhibitor or that it indirectly influences the UPS via an effect on phosphorylation of key proteins of the UPS [258, 259]. Other explanations may arise as research in this area develops.

Two major factors regulating the activity of proteins involved in these processes are phosphorylation and ubiquitination. Both, for example, are required for processing of proteins for degradation by the proteasome. Such is the case for proteins regulating NF-kB activation.

NF-kB is a transcription factor well studied for its role in induction of proteins of the immune response and inflammatory factors [281–283]. The NF-kB complex is formed by various combinations of the five Rel family members, p50, p52, p65 (Rel A), Rel B, and Rel C [158]. The most common dimer is p50/p65 (p50/RelA) and the term NF-kB is often used as a term equivalent to p50/RelA dimer. In the canonical NF-kB pathway the p50/RelA acts as a transcription factor after its translocation to the nucleus. In an alternative non-canonical pathway, the p52/RelB dimer acts as a transcription factor [92].

Proteins of the IkB family are important regulators of NF-kB, the best known being IkBα. Cytoplasmic IkBα can rapidly prevent the translocation of p50/RelA to the nucleus [115], and reportedly can enter the nucleus to retrieve NF-kB [16], preventing NF-kB from interacting with DNA as a transcription factor. Phosphorylation and ubiquitination, however, make IkBα susceptible to proteasomal degradation, thereby activating NF-kB. Karin and Ben-Neriah [115] noted the serine/threonine kinase, IKK, that phosphorylates IkBα as the 'key to NF-kB activation'. They concluded that IKK subunit, IKKβ, is required for activation of NF-kB.

At least two ubiquitin ligases are important in regulating the NF-kB pathway. A specific ubiquitin ligase that binds to the Nf-KB/ IkBα complex was identified in 1998 by Yaron et al. [273]. It was named for its activity as an E3 ubiquitin ligase and for its binding to phosphorylated IkBα, E3 receptor subunit of IkBα (E3RSIkBα). It is now better known as the ubiquitin ligase β-TrCP, a ubiquitin ligase with several additional substrates. The substrates of β-TrCP include IkBα, IkBβ, IkBε [81], p52 [12] and p105, the precursor of p50 [24]. Proteolytic processing of p105, to produce p50, is also proteasome-dependent [91]. The NF-kB subunit RelA is likewise subject to proteasome-dependent degradation [91].

A second ubiquitin ligase complex regulates the degradation of IKK. This ubiquitin complex [119, 132], Keapl-Cul3-Rbx, is better known for its role in the response to oxidative stress by regulation of NRF2 [272] and the antioxidant response element pathway.

Thus, activation of NF-kB, and its subsequent binding to DNA, requires the subtle dance of phosphorylation and ubiquitination of IkB and IKK. The precise role of melatonin in this dance has not been completely determined. There are numerous technically convincing reports that melatonin inhibits NF-kB activation and inhibits DNA binding of p65 under various experimental conditions [25, 43, 99, 219]. The reports of melatonin inhibition of NFkB activation and inhibition of DNA binding refer to the 50p/p65 (p50/RelA) form of NFkB [136, 189].

Li et al (2009) interpreted the melatonin-induced inhibition of NFkB binding to DNA to be the result of melatonin blocking the degradation of IkBα. Further data was obtained by two separate research groups [10, 219]. Shi et al. [219] reported that melatonin inhibited the formation of the phosphorylated form of IkBα, but not the non- phosphorylated from of IkBα in a cell line stimulated with lipopolysaccharide (LPS). In a study of exercise in rats on NF-kB, Alonso et al noted that phosphorylated IkBα and IKK in muscle cells increased with exercise; the increase in both was prevented by melatonin [10]. Their data suggested that melatonin inhibited degradation of phosphorylated IkBα indirectly by inhibiting the activity of IKK. If IKK is the key to NF-kB activation, as suggested by Karin and Ben-Neriah [115], it could also be the key to understanding the mechanisms of melatonin in NF-kB activation. Since IKKβ is a substrate of the ubiquitin ligase Keap1 complex and is responsible for its ubiquitination [132, 250, 251], we suggest that the interaction of Keap1 and IKK may contribute to the mechanism by which melatonin administration activates NF-kB.

Much has been written on melatonin as an antioxidant [151, 197, 242]. One aspect of this is that melatonin stimulates the activity of antioxidant enzymes, including superoxide dismutase, glutathione peroxidase, heme oxygenase 1, and NADPH:quinoneoxidoreductase [65, 198, 205]. As noted, proteasome inhibitors also stimulate the activity of these enzymes. The components of the antioxidant stress response have been well described. They include the Keap1 ubiquitin ligase complex and the NRF2-antioxidant response element pathway. We reviewed the evidence that the antioxidant action of melatonin involves the Keap1-NRF2-ARE pathway in 2015 [260, 261]. We speculated that one possible mechanism that would explain the stimulation of antioxidant enzymes by melatonin was inhibition of the proteasome. While there was some evidence for this *in vitro* this has not been shown to be a direct effect [182].

Oxidative stress itself can impair the functional activity of the 26S proteasome [8]. The smaller 20S proteasome is less susceptible to such stress. It has been shown that the 20S proteasome can degrade oxidized proteins, including SOD, independently of ubiquitin [222]. Thus, while melatonin very likely interacts with the KEAP2-NRF2-ARE pathway in simulating the production of antioxidant enzymes, its precise mechanism of action has not yet been determined.

Since melatonin influences a number of proteins associated with circadian rhythms [32], it is reasonable to investigate its possible role in regulation of clock genes. Gatfield and Schibler [69] provided evidence that cyclic expression of *Cry1* clock gene was controlled by the ubiquitin ligase FBXL3. Further evidence that cycles of ubiquitination and deubiquitination of clock genes contributed to circadian rhythms was reviewed by Stojkovic et al. [229]. Data supporting a role for melatonin in regulating expression of clock genes in the hypothalamus via the UPS was summarized in a review by Vriend and Reiter [260, 261].

The interaction of melatonin and mitochondria was reviewed in a series of manuscripts in the journal, *Cellular and Molecular Life Sciences*, in 2017. These reviews documented melatonin transport into the mitochondria [161, 162], a role for mitochondria in the antioxidant effect of melatonin [198–201], melatonin regulation of the electron transport chain [79], and melatonin interaction with glutathione redox cycling [34]; a direct action of melatonin on mitochondrial DNA has been proposed [188]. Melatonin is also reported to enhance mitochondrial biogenesis in rats with liver fibrosis [113] and the hepatic cells in culture [73].

The ubiquitin proteasome system could coordinate the various actions of melatonin associated with the mitochondria, but there is little information available on the topic. Lavie and co-authors estimated that 203 mitochondrial proteins were ubiquitinated [130] including outer and inner membrane proteins. They noted that the UPS promotes succinate dehydrogenase dependent oxygen consumption and increases ATP, malate and citrate levels and concluded that the UPS regulates energy metabolism in the mitochondria. The UPS also controls biogenesis of mitochondria [28] and mitophagy [30].

The ubiquitin proteasome system also plays a role in melatonin-induced autophagy. As noted above, Atg7 is one of the autophagic proteins influenced by melatonin administration. Atg7 is a ubiquitin activating enzyme (an E1 enzyme) [270]. This enzyme is found in species as diverse as *Saccharmomyces cerevisiae* (yeast), *Arabidopsis thaliana*, a variety of invertebrates, mammals and *Homo sapiens*.

Another model that is useful for studying the interaction of melatonin and the UPS is the regulation of deiodinases, Dio2, the enzyme which converts thyroxine to the more active thyroid hormone, triiodothyronine (T3), and Dio3, the enzyme which deiodinates T4 and T3 to inactive metabolites. The role of ubiquitination and deubiquitination in activation and deactivation of Dio2 is well documented [20, 228]. Dio2 is ubiquitinated by the ubiquitin ligases WSB-1 and TEB4 (depending on the tissue) [59] and deubiquitinated by USP20 and USP33 [19]. Their control of Dio2 has been described as an on/off switch which regulates T3 levels in various organs and tissues [55].

Melatonin injections reduced Dio2 expression [202, 274] in the hypothalamus of the Syrian hamster. The mechanisms have not been determined. In mice, daily melatonin injections reduced Dio2 expression but induced Dio3 expression [71]. In the Siberian hamster a single injection of melatonin towards the end of the light phase of the photoperiodic cycle was reported to induce Dio3 expression in the hypothalamus and its expression increased in proportion to the number of successive days of melatonin injection. Changes in photoperiod can also very rapidly change the expression of Dio2 and Dio3 in hamsters [83].

In summary, there is strong evidence that melatonin administration has an impact on the UPS. While the mechanism of this interaction is not clear, it does appear to influence the regulation of many different proteins, particularly those with a short half-life. The models discussed herein provide some information on the mechanism by which melatonin interacts with the UPS. The common features of those models are that they provide strong clues as to the mechanisms by which melatonin interacts with the UPS. Recently melatonin has been associated with epigenetics. A ubiquitin ligase involved in epigenetic modification is HDAC4 [285] and there is some evidence that melatonin influences protein levels of HDAC4 (see the section below).

Epigenetics

Epigenetic regulation is manifested as a heritable change in gene expression or activity without any alteration in the DNA sequence. Epigenetic modifications can be stably transmitted from a parent cell to the daughter cell (termed mitotic inheritance) or between generations (termed meiotic inheritance). Epigenetic changes have been described in pregnancy and embryonic development, cancer, neurological disorders, cardiotoxicity and hypertension, diabetes, aging and cellular senescence.

Molecular mechanisms underlying epigenetic regulation include post-translational modification of nuclear histones by enzymes such as histone acetyl transferases (HAT) or histone deacetylases (HDAC) or methylation of cytosine bases in DNA by a family of DNA methyl transferases (DNMT) [123]. Non-coding RNAs (ncRNAs), including lncRNA and miRNA, can also modulate gene expression and are considered to be epigenetic modifiers (Fig. 5). All these molecular mechanisms allow additional control over gene expression.

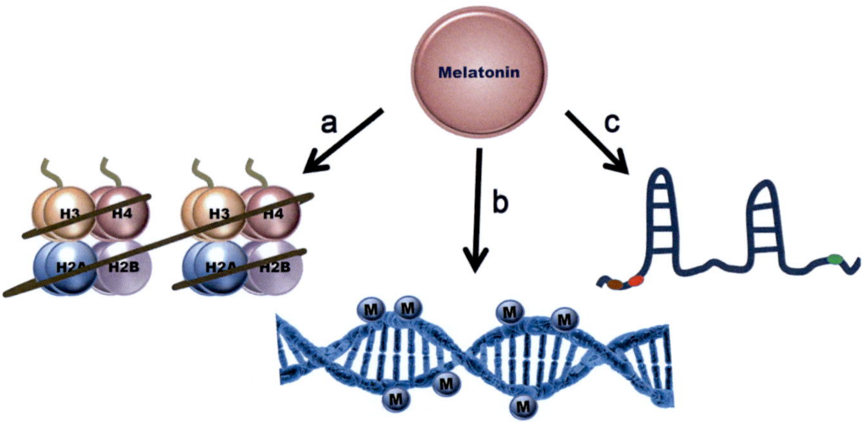

Fig. 5 Molecular mechanisms underlying epigenetic regulation. (**a**) An octameric core of histones (dimers of H2A, H2B, H3 and H4) forms a nucleosome. When DNA is not being transcribed, it is tightly wound around the nucleosomes. Acetylation of lysine residues in the histones, catalyzed by HATs, allow accessibility to RNA polymerase II. After transcription, the histones are deacetylated and returned to a closed chromatin state. (**b**) Addition of methyl groups to carbon-5 position of cytosine residues in CpG islands within the promoter or coding regions of genes alters their binding affinity to transcription factors. Hypermethylation generally leads to gene silencing whereas hypomethylation generally leads to gene activation (**c**) Non-coding RNAs (ncRNAs), including lncRNA and miRNA, can also modulate gene expression and are considered to be epigenetic modifiers

Histones are basic proteins located within the nucleus; an octameric core of histones (dimers of H2A, H2B, H3 and H4) forms a nucleosome. When DNA is not being transcribed, it is tightly wound around the nucleosomes, preventing the binding and initiation of transcription by RNA polymerase II. For transcription, this "closed" chromatin structure has to be "opened" for it to be accessible; this occurs by acetylation of lysine residues in the histones, catalyzed by HATs. Acetylated lysine residues are recognized by protein complexes containing bromodomains that modulate chromatin architecture. When transcription is complete, acetylated lysine residues of histone are deacetylated by HDACs, returning chromatin to a closed state. In addition to acetylation-deacetylation, reversible methylation by histone methyltransferases and histone demethylases, reversible phosphorylation by kinases and phosphatases, mono- ubiquitination, sumoylation of lysine residues that inhibits acetylation, glycosylation and ADP ribosylation are also known to occur.

DNA methylation of carbon-5 position of cytosine residues in the cytosine-guanosine (CG)-rich sequences (referred to as "CpG islands") within the promoter or coding regions of genes alters their binding affinity to transcription factors, typically resulting in repression of gene expression. DNA methyltransferase 1 (DNMT1) is the major enzyme involved in transmitting methylation patterns during replication in adults whereas DNMT3a and DNMT3b play critical roles during early development. DNMT3L does not have methyltransferase activity; however, it helps DNMT3a and DNMT3b in propagation and establishment of maternally imprinted genes. Hypermethylation generally leads to gene silencing whereas hypomethylation generally leads to gene activation. While DNA methylation is considered to be relatively permanent, histone modification is more environmentally responsive.

The enzymes involved in the biosynthesis of melatonin as well as its target membrane receptors are modulated epigenetically [78]. In turn, melatonin can purportedly inhibit DNMTs by inhibiting their transcription or by impeding their function via binding to their catalytically active sites. Melatonin also appears to induce gene expression by acetylation of histone H3; conversely, perhaps as a negative feedback loop, it promotes the expression of HDAC3, HDAC5 and HDAC7, leading to gene repression [217]. The most widely reported mechanism that links melatonin to epigenetic regulation is via its induction of sirtuins (silent information regulators), a family of seven known class III NAD + -dependent HDACS (SIRT1 to SIRT7). In normal cells, melatonin is known to upregulate SIRT1 expression and/or mitochondrial SIRT3 under various conditions [23, 76]. Melatonin-mediated induction of SIRT1 expression contributes to protection from oxidative stress in ischemic models, age-related senescence, hypertension and inflammation [78, 161, 162].

The epigenetic role of melatonin in providing protection or biological adaptation to environmental factors that can be passed from one generation to the next via oocytes or sperms is well-documented [105, 106]. Melatonin appears to be particularly suitable in serving as an environmental sensor [106], given that its secretion is affected by different wavelengths of light [192], temperature [160], altitude [116] and seasonal cycles [85]. Recent literature indicates that melatonin protects spermatogonial stem cells from endocrine disruptors such as bisphenol A or diethylhexyl phthalate by maintaining histone H3K9 dimethylation and promotes their recovery [284]. Similarly, melatonin also protects spermatogonial stem cells in mice from hexavalent chromium, another environmental carcinogen, by preventing trimethylation of histone H3K9 or H3K27, thereby preventing germ cell apoptosis and maintaining fertility [148, 287]. Protective epigenetic mechanisms underlying antioxidant defense of germ cells was demonstrated by a recent study showing hypomethylation of *SOD1*, *Gpx4* and *Cat* genes in ovine prepubertal cumulus cells of lambs treated with melatonin that resulted in decreased apoptosis [64]. Surprisingly, melatonin treatment also upregulated *DNMT1, DNMT3a* and *DNMT3b* in ovine prepubertal cumulus cells, with the latter gene being significantly hypomethylated [63], suggesting selective hypomethylation of the antioxidant genes. *DNMT1a* expression is upregulated in melatonin-treated oocytes along with increased expression of oocyte maturation- related genes, *GDF9* and *MARF*, suggesting a beneficial epigenetic role for melatonin in oocyte maturation [251]. Melatonin also appears to reduce apoptosis of bovine somatic cell nuclear transfer embryos by stimulating *SOD1* and *Gpx4* expression that was associated with higher H3K9ac levels [230].

Melatonin functions by binding to one of two membrane-bound G-protein seven transmembrane receptors, MTNR1A (or MT1) and MTNR1B (or MT2) [225], or to nuclear orphan receptors from the retinoid orphan receptor (ROR) or the retinoid Z receptor (RZR) families [60]. Melatonin may also interact with cytosolic proteins that in turn may regulate the nuclear receptors or the cytoskeleton [26]. Interestingly, one of the SNPs linked to night shift-related job exhaustion is associated with changes in DNA methylation in the 5′ regulatory region of MTNR1A that may lead to decreased melatonin signaling [232].

Epigenetic regulation is known to play a role in asthma and allergy that have TH2 immune cell activation in common [148]. A recent study that performed a genome-wide linkage scan of 615 European families to assess co-occurrence of asthma and allergy identified a "differentially methylated CpG site located within intron 1 of the melatonin receptor 1A (MTNR1A) gene that mediated the effect of a paternally transmitted genetic variant. Melatonin also appears to play a role in alleviating chronic obstructive pulmonary disease (COPD) by upregulating the expression of SIRT1 that, in turn, inhibits interleukin-1B and NLRP3-mediated inflammation [183]. Melatonin-induced SIRT1 induction also plays an anti-inflammatory role during lipopolysaccharide-induced oxidative stress [216]. Melatonin-mediated epigenetic suppression of pro-inflammatory NF-kB is reviewed elsewhere [125, 126].

The association between chronodisruption with consequent lower levels of melatonin and the incidence of cancer has long been noted. Recent data support a melatonin-mediated upregulation in global methylation as a key factor in tumor suppression [75, 214]. In contrast with its function in normal cells, melatonin appears to downregulate SIRT1 in cancer cells and thereby decreases their proliferative capacity [112]. Importantly, treatment of a breast cancer cell line, MCF-7, with melatonin followed by mapping of the epigenome identified several thousand differentially methylated genes [133], indicating an epigenetic tumor-suppressing role for melatonin in cancer. In breast cancer, the tumor suppressor gene, *brca1*, the DNA repair gene, *rad9*, and the cell cycle regulation genes, *dkk3* and *wif1*, are hyper methylated, leading to their silencing; hypomethylation of oncogenes and transposable elements such as *Alu* are associated with poor prognosis. Chronodisruption in shift workers leads to *clock* hypomethylation and *cry2* hypermethylation, as also found in breast cancer [289]. However, direct epigenetic links between chronodisruption and breast cancer are yet to be firmly established [124]. Interestingly, *Mtnr1a* mRNA expression is reduced in patients with oral squamous cell carcinoma; its re-expression in tumor cells inhibited growth in vitro, suggesting a tumor suppressor role for melatonin in oral cancer [168]. Moreover, melatonin suppresses senescent cancer cell-mediated secretion of pro-inflammatory factors by inhibiting PARP-1interaction with the long non-coding RNA, TERRA, thereby preventing H2BK120 acetylation [276]. Finally, melatonin suppresses microRNA, miR-24, post- trancriptionally and thereby inhibits cell proliferation and migration [166].

Melatonin also regulates chromatin remodeling in the nervous system. For example, treatment with melatonin in drinking water induced acetylation of histones H3 and H4 primarily in the hippocampus that correlated with increased levels of phospho-ERK [170]. It can enhance hippocampal neurogenesis via agonists of MT1 and MT2 that appear to function via stimulation of BDNF [226]. Exogenous melatonin also enhances neurogenesis in mice during aging [190]. In addition to neurogenesis, melatonin has anti-aging neural effects that is also linked to its upregulation of SIRT1 [74]. Importantly, it protects the hippocampus from pre-natal glucocorticoid exposure by downregulating the expression of DNA methyltransferase 1 mRNA expression and by suppressing DNMT1 and methyl-CpG binding protein 2 (MeCP2) binding to the reln promoter, thereby restoring the decreased levels of reln and GAD1 mRNA expression [146]. Melatonin also ameliorates neuropathic allodynia by promoting HDAC4 dephosphorylation and nuclear import with consequent upregulation of *hmgb1* transcription [141]. Finally, valproate, used to treat epilepsy and bipolar disorder, promotes histone H3 acetylation of the MT1 promoter that results in its upregulated expression [22].

Programmed hypertension that can occur due to stressful conditions pre-birth lead to epigenetic alterations in the kidney that were reversed by feeding melatonin to pregnant dams; melatonin upregulated HDAC-2, HDAC-3 and HDAC-8 in the kidneys of offspring from calorie-restricted dams [239, 240] and altered the expression of approximately 450 genes. The study also showed that melatonin upregulated *Dnmt3A, Hdac4, Hdac7, Hdac11, Chd1, Chd2, Chd3, Brpf3, Baz1b and Wdr1* during nephrogenesis that are all involved in epigenetic regulation [239].

Taken together, these examples suggest a role for melatonin in modulating the expression of several genes through epigenetic regulation. The findings have implications when melatonin is used as a treatment for human diseases.

Collagen and Extracellular Matrix

Fibrosis, the pathological accumulation of collagen and extracellular matrix (ECM), interferes with the physiology of organs. Damaging fibrosis is most commonly reported in the lung [203], heart [169], liver [104], and kidney [172] but it occurs in other situations where inflammation is rampant [94] as well as in the form of adhesions [14] among organs after surgical intervention, infection, oxidative stress, etc. Whereas there is some evidence that fibrosis may be reversible, particularly in the early stages, advanced fibrosis is associated with end stage disease, organ failure and death. To overcome the latter two events, the only available treatment is organ transplantation. Initially, fibrosis is a physiological reparative process but, when it continues to expand, it becomes pathological and life threatening (Fig. 6). Thus, limiting the sustained development of pro-fibrotic processes is medically critical.

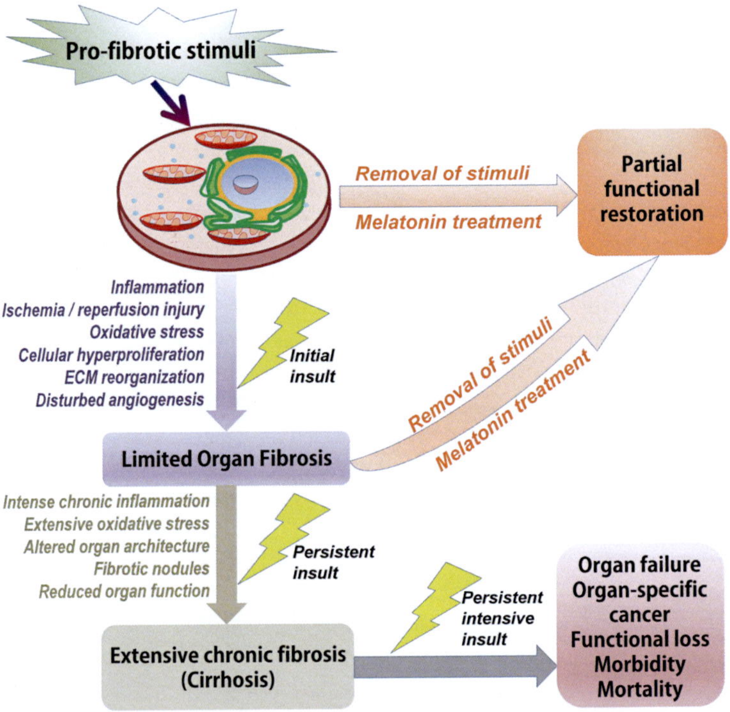

Fig. 6 Profibrotic stimuli, which are numerous, if sustained only for a brief interval cause minor organ fibrosis. Some evidence suggests that, at this point, if the fibrotic stimulus is interrupted or if melatonin treatment is initiated, the developing fibrogenic phenotype may be reversed with partial morphological and functional restoration of the organ. With more persistent attack and excessive fibroproliferative conditions, the "point-of-no-return" is exceeded and extensive fibrosis and cirrhosis occur. Eventually, organ failure is the result leading to total organ function failure and death if the organ is not replaced

The sequence of events for the fibrotic changes that occur in different organs are similar at the molecular level with common pathological pathways that culminate in serious negative outcomes [97]. Fibrosis is initiated by a variety of pro-fibrotic processes; some of the most common are inflammation (e.g., macrophages and T cells) and oxidative stress. The latter process may be the result of a different events such as ischemia/reperfusion injury, toxin exposure, ionizing radiation damage, etc. Dysfunctional epithelial cells and inflammatory cells, following their recruitment to the site of injury, become fibrosis-initiator cells, which transform and actively proliferate. These activated cells include fibroblasts/myofibroblasts and other collagen-generating elements that are derived from organ-specific cells that undergo epithelial-to-mesenchymal transition. In some organs there are other cells that contribute collagen and extracellular matrix deposition, e.g., stellate cells of the liver [238].

Investigations into the role of melatonin in resisting fibrosis have uncovered several means by which this agent impedes excessive collagen and extracellular matrix (ECM) deposition. Of special interest is that melatonin membrane receptors (MT1 and MT2) are widely expressed on fibroblasts in developing scar tissue. In contrast, fibroblasts derived from skin dermal tissue lack melatonin receptors. These findings suggest that, at least in part and under some conditions, melatonin probably controls the activity of hyperfunctional fibroblasts/myofibroblasts via receptor-mediated mechanisms [97]. Given the high damaging reactive oxygen species generation by inflammatory cells and cells crippled by ironizing radiation, drugs, etc., melatonin's receptor-independent actions may also be involved in suppressing the development of fibrosis [66, 80]. For example, melatonin limits the epithelial-to-mesenchymal transition of lung cells exposed to bleomycin [286] and during leptin-mediated fibrosis in the heart [157].

Melatonin is acknowledged as a potent anti-inflammatory agent. The recruitment of inflammatory cells to the site of the fibrotic cascade is due to locally produced cytokines. Pro- inflammatory cytokines that aid in mediating the inflammatory cascade include numerous agents such as interleukins (IL), IL-1, IL-6, IL-20 and others. Also, tumor necrosis factor-alpha (TNF- α) actively participates in attracting inflammatory cells [237].

Transforming growth factor-beta (TGF-β) is referred to as the "master regulator of fibrosis" [164]. TGF-β1 is a major common driving force for fibrosis in many organs and its actions involve multiple cell types. In many fibrotic disease models, blockage of TGF-β1 reduces fibroblast activation, collagen production and ECM deposition. TGF-β1 promotes fibrosis via both canonical and non-canonical signaling routes. Smad transcription factors are involved in the canonical pathway; these actions are highly complex because of their interactions with other signaling pathways and their ability to modulate the EMT.

In addition to TGF-β, other molecular markers of fibrosis have been identified. Smad has already been mentioned in this context. Additionally, however, PDGF (platelet-derived growth factor), CTGF (connective tissue growth factor) as well as pro-collagen-1 have been implicated to participate in pro-fibrogenic responses [46, 131]. These fibrotic indices, as well as TGF-β1, are influenced by a wide variety of interacting pathways; collectively, they drive the highly complex fibrogenic processes which lead to exaggerated collagen and ECM accumulation [159, 204].

Melatonin, in other diseases, impairs the EMT [231] and is presumed to do likewise during fibrotic events although this has been sparingly investigated. During the initiation of fibrosis, the activation of pro-fibrotic cells contribute to the accumulation of especially collagen I and glycosaminoglycans (GAG) in the extracellular space. In the early stages of the fibrotic pathway, some of the structural and functional damage can be reversed if the stimulus is withdrawn (Fig. 6) or if melatonin is used as a treatment [33, 46, 47]. Eventually, however, the severity of chronic fibrosis is so vast that the "point-of-no-return" is exceeded and organ function is essentially totally compromised thereby requiring organ transplantation. Moreover, extensive irreversible fibrosis is a prelude to other diseases, e.g., cancer [97].

The ability of melatonin to restrain fibrogenic growth has been investigated in several major organs where excessive collagen formation often occurs and where it severely compromises organ physiology and threatens the quality of life. In the case of myocardial protection, melatonin reduced scar formation in the heart after induced ischemia/reperfusion injury [57], after isoproterenol-mediated damage [157] and in other experimental conditions [97]. The respiratory system is a frequent site of fibrotic diseases including acute respiratory distress syndrome (ARDS) and COPD (chronic obstructive pulmonary disease). The damaging culprit in these conditions is often cigarette smoke. Experimentally, the regular administration of melatonin to animals exposed daily to cigarette smoke (see below) or to bleomycin [21] curtailed the severity of alveolar destruction as indicted by the amounts of oxidatively damaged protein and lipid in the lungs. In a preliminary blinded, placebo-controlled study in children with cystic fibrosis (CF), melatonin improved their wellbeing indicating it may have some benefit in individuals with this debilitating condition [54].

The most extensive experimental data regarding the efficacy of melatonin to interfere with fibrotic reactions have been studied using models of liver disease. Many conditions are accompanied by hepatic fibrosis and compromised function including a high-fat diet [277], excessive alcohol consumption [128], toluene inhalation [246], as well as the metabolism of a number of medications [47, 48]. The damaging effects of each of these conditions at the level of the liver have all been shown to be stymied by concurrent melatonin treatment. Similarly, melatonin attenuates caustic sclerosing cholangitis [215] and damage to the biliary tree resulting from bile duct ligation [9]. While the majority of these studies were not mechanistically based, melatonin's antioxidant actions seem to be the basis of some of the protection afforded by this molecule.

Two recently-published investigations specifically examined the mechanisms by which melatonin reduces the respiratory consequences of experimental chronic obstructive pulmonary disease (COPD), a condition in which pathological lung fibrosis plays a major role [171]. In the first of these, Shin and coworkers [220] performed both in vivo and in vitro studies to identify the molecular mechanisms by which melatonin blunts fibrosis. For the in vivo test, rats were exposed to cigarette smoke daily for a week and treated with lipopolysaccharide intravenously with or without daily melatonin treatment. At the conclusion of the study (on day 7), the non-melatonin treated rats had high numbers of inflammatory cells in their bronchoalveolar fluid (BALF) and an elevated expression of TGF-β1, collagen I and Smad3 in their lung tissues. In a dose-response manner, melatonin inhibited each of the parameters. The in vitro study utilized a human mucoepithelial cell line treated with cigarette smoke condensate (SSC). This treatment caused an elevated expression of not only TGF-β1 and collagen I, but also tumor necrosis factor-alpha (TNF-α) and Smad 3 and p38 phosphorylation. As in the animals study, each of the indices related to fibrosis was suppressed when melatonin was added to the incubation medium.

Using a similar model of compromised respiratory function, but with a longer treatment period (28 days), Peng et al. [183] documented that melatonin preserved more normal alveolar architecture which was damaged by the combination of cigarette smoke and LPS. Moreover, in an evaluation of lung function, melatonin treatment improved the elasticity and dynamic compliance of the respiratory system while reducing the resistance to inspiration. At the molecular level, the indices of inflammation (IL-1β and the NLPR3 inflammasome) were reversed by melatonin, a process that is dependent on the promotion of SIRT1.

Extreme fibrosis, especially when it occurs in essential organs as illustrated herein reduces the quality of life and can lead to death. In the advanced stages, there is no currently- available medical treatment with the remaining option being organ replacement. Since it is not treatable, effort is currently directed to the prevention of fibrosis. In this regard, melatonin has become of interest as a molecule with the potential to retard or prevent the progression of the fibrotic cascade [97]. Some of the proposed mechanisms related to melatonin's inhibitory actions or the fibrotic cascade, based on the data evaluated in this report, are summarized in Fig. 7.

Concluding Remarks

The monumental effort spent by Lerner et al. [135] in the isolation and characterization of melatonin from bovine pineal tissue has paid highly significant dividends since the publication of that seminal report. The functional repertoire of melatonin is now known to far exceed that envisaged by those who investigated its actions during the first two decades after its discovery. Melatonin's functional "tool kit" include a variety of health-relevant actions such as circadian/circannual rhythm regulation, sleep promotion, immunostimulation, etc.

Within the last three decades, the list of systemic and subcellular functions where melatonin has been shown to be operative has continued to expand. Due to advances in technology, as with other molecules, research on melatonin has been focused on its intracellular functions. The preponderance of evidence suggests that melatonin is involved in some of the most basic molecular interactions within cells. The demonstration that melatonin is not uniquely of pineal origin but rather may be produced in

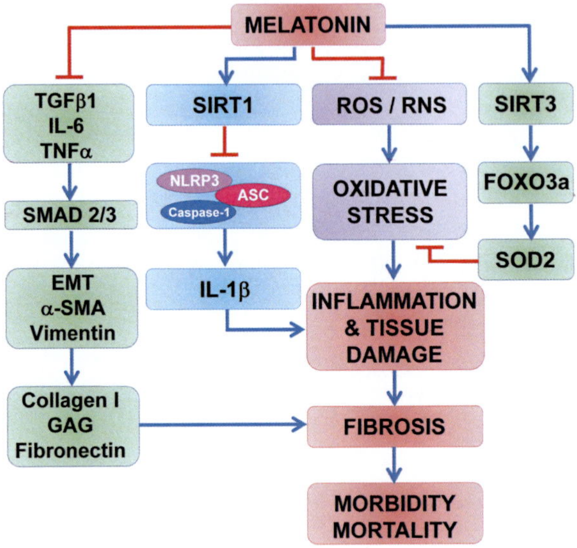

Fig. 7 Melatonin seems to interfere with fibrotic development via several means. A major factor in initiating the fibrotic cascade is transforming growth factor-1β (TGF-1β). Melatonin inhibits TGF-1β and associated cytokines thereby inhibiting epithelial-to-mesenchymal transition and reducing the excessive accumulation of extracellular collagen and glycosaminoglycans (GAG). Likewise, via SIRT1 stimulation, melatonin inhibits the NLRP3 inflammasome and IL- 1β generation. Additionally, by means of its direct radical scavenging actions and its stimulation of antioxidant enzymes, e.g., superoxide dismutase 2 (SOD2), melatonin reduces oxidative stress which normally promotes fibrosis. Many details of these pathways remain to be clarified

the mitochondria of every cell, opens vast new areas for research. It is the authors' opinion, in fact, that what is known about the intimate actions of melatonin is a small fraction of what it actually does. Moreover, past and recently discovered actions are only epiphenomenal of what melatonin does with the real actions of this molecule yet to be revealed.

In this review only a small number of the critical subcellular actions of melatonin are briefly summarized. While these functions are discussed under different headings, they are obviously interrelated and mutually dependent on each other.

The goal of subsequent research is to aggressively continue to examine the actions of this multifaceted agent; especially since many publications have strongly indicated that melatonin is a functionally flexible and highly beneficial molecule. This is apparent from the results of a number of reports where melatonin has been used at the clinical and veterinary levels. Almost uniformly, these findings show that melatonin promotes optimal function and the well-being of cells, of organs and of organisms. Finally, in plants, which also produce this molecule, melatonin's actions are equally important and life sustaining.

References

1. Absi, E., et al. 2000. Protective effect of melatonin against the 1-methyl-4-phenylpyridinium-induced inhibition of complex I of the mitochondrial respiratory chain. *Journal of Pineal Research* 29: 40–47.
2. Acuna-Castroviejo, D., et al. 1997. Melatonin is protective against MPTP-induced striatal and hippocampal lesions. *Life Sciences* 60: P123–P129.
3. ———. 2011. Melatonin-mitochondria interplay in health and disease. *Current Topics in Medicinal Chemistry* 11: 221–240.
4. ———. 2014. Extrapineal melatonin: Sources, regulation and potential functions. *Cellular and Molecular Life Sciences* 71: 2997–3025.
5. ———. 2017. Melatonin, clock genes and mitochondria in sepsis. *Cellular and Molecular Life Sciences* 74: 3965–3988.
6. ———. 2018. Melatonin actions in the heart: More than a hormone. *Melatonin Res* 1: 21–26.
7. Ahmad, K.A., et al. 2004. Hydrogen peroxide-mediated cytosolic acidification is a signal for mitochondrial translocation of Bax during drug-induced apoptosis of tumor cells. *Cancer Research* 64: 7867–7878.
8. Aiken, C.T., et al. 2011. Oxidative stress-mediated regulation of proteasome complexes. *Molecular & Cellular Proteomics* 10: R110 006924.
9. Aktas, C., et al. 2014. Melatonin attenuates oxidative stress, liver damage and hepatic apoptosis after bile-duct ligation in rats. *Toxicology and Industrial Health* 30: 835–844.
10. Alonso, M., et al. 2006. Melatonin inhibits the expression of the inducible isoform of nitric oxide synthase and nuclear factor kapa B activation in rat skeletal muscle. *Journal of Pineal Research* 41: 8–14.
11. Amin, A.H., et al. 2015. Melatonin ameliorates metabolic risk factors, modulates apoptotic proteins, and protects the rat heart against diabetes-induced apoptosis. *European Journal of Pharmacology* 747: 166–173.
12. Amir, R.E., et al. 2004. Mechanism of processing of the NF-kappa B2 p100 precursor: Identification of the specific polyubiquitin chain-anchoring lysine residue and analysis of the role of NEDD8-modification on the SCF(beta-TrCP) ubiquitin ligase. *Oncogene* 23: 2540–2547.
13. Amstrup, A.K., et al. 2013. Melatonin and the skeleton. *Osteoporosis International Journal* 24: 2919–2927.
14. Ara, C., et al. 2005. Protective effects of melatonin against oxidative stress on adhesion formation in the rat cecum and uterine horn model. *Life Sciences* 77: 1341–1350.
15. Arataki, S., et al. 2005. Calpain inhibitors prevent neuronal cell death and ameliorate motor disturbances after compression-induced spinal cord injury in rats. *Journal of Neurotrauma* 22: 398–406.
16. Arenzana-Seisdedos, F., et al. 1997. Nuclear localization of I kappa B alpha promotes active transport of NF-kappa B from the nucleus to the cytoplasm. *Journal of Cell Science* 110: 369–378.
17. Areti, A., et al. 2017. Melatonin prevents mitochondrial dysfunction and promotes neuroprotection by inducing autophagy during oxaliplatin-evoked peripheral neuropathy. *Journal of Pineal Research* 62: e12393.
18. Arnao, M.B., and J. Hernandez-Ruiz. 2019. Melatonin: A new plant hormone and/or a plant master regulator? *Trends in Plant Science* 4: 38–48.
19. Arrojo, E.D.R., et al. 2013. The type II deiodinase is retrotranslocated to the cytoplasm and proteasomes via p97/Atx3 complex. *Molecular Endocrinology* 27: 2105–2115.
20. ———. 2013. Role of the type 2 iodothyronine deiodinase (D2) in the control of thyroid hormone signaling. *Biochimica et Biophysica Acta* 1830: 3956–3964.

21. Arslan, S.O., et al. 2002. The effect of melatonin on bleomycin-induced pulmonary fibrosis in rats. *Journal of Pineal Research* 32: 21–25.
22. Bahna, S.G., and L.P. Niles. 2017. Epigenetic regulation of melatonin receptors in neuropsychiatric disorders. *British Journal of Pharmacology* 175: 3209–3219.
23. Bai, X.Z., et al. 2016. Melatonin prevents acute kidney injury in severely burned rats via the activation of SIRT1. *Scientific Reports* 6: 32199.
24. Beinke, S., et al. 2004. Lipopolysaccharide activation of the TPL-2/MEK/extracellular signal-regulated kinase mitogen-activated protein kinase cascade is regulated by lkaapaB kinase-induced proteolysis of NF-kappaB1 p105. *Molecular and Cellular Biology* 24: 9658–9667.
25. Beni, S.M., et al. 2004. Melatonin0induced neuroprotection after closed head injury is associated with increased brain antioxidants and attenuated late-phase activation of NF-kappaB and AP-1. *The FASEB Journal* 18: 149–151.
26. Benitez-King, G., and F. Anton-Tay. 1993. Calmodulin mediates melatonin cytoskeletal effects. *Experientia* 49: 635–641.
27. Bizzarri, M., et al. 2013. Molecular mechanisms of the pro-apoptotic actions of melatonin in cancer: A review. *Expert Opinion on Therapeutic Targets* 17: 1483–1496.
28. Bragoszewski, P., et al. 2017. Control of mitochondrial biogenesis and function by the ubiquitin- proteasome system. *Open Biology* 7: 170007.
29. Brenner, C., et al. 2000. Bcl-2 and Bax regulate the channel activity of the mitochondrial adenine nucleotide translocator. *Oncogene* 19: 329–336.
30. Burchell, V.S., et al. 2013. The Parkinson's disease-linked proteins Fbxo7 and Parkin interact to mediate mitophagy. *Nature Neuroscience* 16: 1257–1265.
31. Burgess, H.J., and J.S. Emens. 2018. Drugs used in circadian sleep-wake rhythm disturbances. *Sleep Medicine Clinics* 13: 231–241.
32. Cajochen, C., et al. 2003. Role of melatonin in the regulation of human circadian rhythms and sleep. *Journal of Neuroendocrinology* 15: 432–437.
33. Campana, L., and J.P. Iredale. 2017. Regression of liver fibrosis. *Seminars in Liver Disease* 37: 1–10.
34. Cardinali, D.P., and D.E. Vigo. 2017. Melatonin, mitochondria, and the metabolic syndrome. *Cellular and Molecular Life Sciences* 74: 3941–3954.
35. Champney, T.H., et al. 1984. Concurrent determination of enzymatic activities and substrate concentrations in the melatonin synthetic pathway within the same rat pineal gland. *Journal of Neuroscience Research* 11: 59–66.
36. Chang, C.F., et al. 2012. Melatonin attenuates kainic acid-induced neurotoxicity in mouse hippocampus via inhibition of autophagy and alpha-synuclein aggregation. *Journal of Pineal Research* 52: 312–321.
37. Chen, Z.X., and S. Pervaiz. 2007. Bcl-2 induces pro-oxidant state by engaging mitochondrial respiration in tumor cells. *Cell Death and Differentiation* 14: 1617–1627.
38. Chen, Y., et al. 2016. Melatonin induces anti-inflammatory effects to play a protective role via endoplasmic reticulum stress in acute pancreatitis. *Cellular Physiology and Biochemistry* 40: 1094–1104.
39. Chen, Z., et al. 2017. Effects of melatonin on maturation, histone acetylation, autophagy of porcine oocytes and subsequent embryonic development. *Animal Science Journal* 88: 1298–1310.
40. Chern, C.M., et al. 2012. Melatonin ameliorates neural function by promoting endogenous neurogenesis through the MT2 melatonin receptor in ischemic-stroke mice. *Free Radical Biology & Medicine* 52: 1634–1647.
41. Chrivia, J.C., et al. 1993. Phosphorylated CREB binds specifically to the nuclear protein CBP. *Nature* 365: 855–859.
42. Chua, S., et al. 2016. The cardioprotective effect of melatonin and exendin-4 treatment in a rat model of cardiorenal syndrome. *Journal of Pineal Research* 61: 438.
43. Chuang, J.I., et al. 1996. Effect of melatonin on NF-kappa-B DNA-binding activity in the rat spleen. *Cell Biology International* 20: 687–692.

44. Clément, M.V., and S. Pervaiz. 1999. Reactive oxygen intermediates regulate cellular response to apoptotic stimuli: An hypothesis. *Free Radical Research* 30: 247–252.
45. Cory, S., and J.M. Adams. 2002. The Bcl2 family: Regulators of the cellular life-or-death switch. *Nature Reviews. Cancer* 2: 647–656.
46. Crespo, I., et al. 2015. Melatonin limits the expression of profibrogenic genes and ameliorates the progression of hepatic fibrosis in mice. *Translational Research* 165: 346–357.
47. Cruz, A., et al. 2005. Melatonin prevents experimental liver cirrhosis induced by thioacetamide in rats. *Journal of Pineal Research* 39: 143–150.
48. Czechowska, G., et al. 2015. Protective effects of melatonin against thioacetamide-induced liver fibrosis in rats. *Journal of Physiology and Pharmacology* 66: 567–579.
49. Dabbeni-Sala, F., et al. 2001. Melatonin protects against 6-OHDA-induced neurotoxicity in rats: A role for mitochondrial complex I activity. *The FASEB Journal* 15: 167–170.
50. ———. 2001. Kainic acid induces selective mitochondrial oxidative phosphorylation enzyme dysfunction in cerebellar granule neurons: Protective effects of melatonin and GSH ethyl ester. *The FASEB Journal* 15: 1786–1788.
51. Dardente, H., et al. 2016. Seasonal breeding in mammals: From basic research to applications and back. *Theriogenology* 86: 324–332.
52. de Luxan-Delgado, B., et al. 2014. Melatonin administration decreases adipogenesis in the liver of Ob/Ob mice through autophagy modulation. *Journal of Pineal Research* 56: 126–133.
53. ———. 2016. Melatonin reduces endoplasmic reticulum stress and autophagy in liver of leptin-deficient mice. *Journal of Pineal Research* 61: 108–123.
54. DeCastro-Silva, C., et al. 2010. Melatonin improves sleep and reduces nitrite in the exhaled breath condensate in cystic fibrosis – A randomized double-blind placebo-controlled study. *Journal of Pineal Research* 48: 65–71.
55. Dentice, M., et al. 2013. The deiodinases and the control of intracellular thyroid hormone signaling during cellular differentiation. *Biochimica et Biophysica Acta* 1830: 3937–3945.
56. Ding, M., et al. 2018. Melatonin prevents Drp1-mediated mitochondrial fission in diabetic hearts through SIRT1-PGC1alpha pathway. *Journal of Pineal Research* 65: e12491.
57. Drobnik, J., et al. 2011. Pharmacological doses of melatonin reduce the glycosaminoglycan level within the infarcted heart scar. *Journal of Physiology and Pharmacology* 62: 29–35.
58. Dubocovich, M.L., and M. Markowska. 2005. Functional MT1 and MT2 melatonin receptors in mammals. *Endocrine* 27: 101–110.
59. Egri, P., and B. Gereben. 2014. Minimal requirements for ubiquitination-mediated regulation of thyroid hormone activation. *Journal of Molecular Endocrinology* 53: 217–226.
60. Emet, M., et al. 2016. A review of melatonin, its receptors and drugs. *The Eurasian Journal of Medicine* 48: 135–141.
61. Fan, J., et al. 2018. Melatonin: A multifunctional factor in plants. *International Journal of Molecular Sciences* 19: E1528.
62. Fan, T., et al. 2018. Inhibiting MT2-TFE3-dependent autophagy enchances melatonin-induced apoptosis in tongue squamous cell carcinoma. *Journal of Pineal Research* 64: 1–18.
63. Fang, Y., et al. 2018. Melatonin-mediated development of ovine cumulus cells, perhaps by regulation of DNA methylation. *Molecules* 23 (494): 1–14.
64. ———. 2019. Melatonin-induced demethylation of antioxidant genes increases antioxidant capacity through Roralpha in cumulus cells of prepubertal lambs. *Free Radical Biology & Medicine* 131: 173–183.
65. Fischer, T.W., et al. 2013. Melatonin enhances antioxidative enzymes gene expression (CAT, GPx, SOD), prevents their UVR-induced depletion, and protects against the formation of DNA damage (8-hydroxy-2′-deoxyguanosine) in ex vivo human skin. *Journal of Pineal Research* 54: 303–312.
66. Galano, A., and R.J. Reiter. 2018. Melatonin and its metabolites vs oxidative stress: From individual actions to collective protection. *Journal of Pineal Research* 65: e12514.
67. Galano, A., et al. 2018. Melatonin: A versatile protector against oxidative DNA damage. *Molecules* 23: E530.

68. Gastel, J.A., et al. 1998. Melatonin production: Proteasomal proteolysis in serotonin N-acetyltransferase regulation. *Science* 279: 1358–1360.
69. Gatfield, D., and U. Schibler. 2007. Proteasomes keep the circadian clock ticking. *Science* 316: 1135–1136.
70. Gatti, G., et al. 2017. Antiproliferative and pro-apoptotic activity of melatonin analogues on melanoma and breast cancer cells. *Oncotarget* 8: 68338–68353.
71. Goto, M., et al. 2013. Melatonin-induced changes in the expression of thyroid hormone-converting enzymes in hypothalamus depend on the timing of melatonin injections and genetic background in mice. *General and Comparative Endocrinology* 186: 33–40.
72. Gottlieb, D.J., et al. 2010. Prospective study of obstructive sleep apnea and incident coronary heart disease and heart failure: The sleep heart health study. *Circulation* 122: 352–360.
73. Guo, P., et al. 2014. Melatonin improves mitochondrial function by promoting MT1/SIRT1/PGC-1 alpha-dependent mitochondrial biogenesis in cadmium-induced hepatotoxicity in vitro. *Toxicological Sciences* 142: 182–195.
74. Gutierrez-Cuesta, J., et al. 2008. Evaluation of potential pro-survival pathways regulated by melatonin in a murine senescence model. *Journal of Pineal Research* 45: 497–505.
75. Haim, A., and A.E. Zubidat. 2015. Artificial light at night: Melatonin as a mediator between the environment and epigenome. *Philosophical Transactions of the Royal Society of London. Series B, Biological Sciences* 370: 20140121.
76. Han, D., et al. 2016. Melatonin facilitates adipose-derived mesenchymal stem cells to repair the murine infarcted heart via the SIRT1 signaling pathway. *Journal of Pineal Research* 60: 178–192.
77. Hardeland, R. 2013. Melatonin and the theories of aging: A critical appraisal of melatonin's role in antiaging mechanisms. *Journal of Pineal Research* 55: 325–356.
78. ———. 2014. Melatonin, noncoding RNAs, messenger RNA stability and epigenetics–evidence, hints, gaps and perspectives. *International Journal of Molecular Sciences* 15: 18221–18252.
79. ———. 2017. Melatonin and the electron transport chain. *Cellular and Molecular Life Sciences* 74: 3883–3896.
80. ———. 2018. Recent findings in melatonin research and their relevance to the CNS. *Central Nervous System Agents in Medicinal Chemistry* 18: 102–104.
81. Hayden, M.S., and S. Ghosh. 2004. Signaling to NF-kappaB. *Genes & Development* 18: 2195–2224.
82. He, C., et al. 2016. Mitochondria synthesize melatonin to ameliorate its function and improve mice oocyte's quality under in vitro conditions. *International Journal of Molecular Sciences* 17: E939.
83. Herwig, A., et al. 2012. Hypothalamic gene expression rapidly changes in response to photoperiod in juvenile Siberian hamsters (*Phodopus sungorus*). *Journal of Neuroendocrinology* 24: 991–998.
84. Hevia, D., et al. 2015. Melatonin uptake in prostate cancer cells: Intracellular transport versus simple passive diffusion. *Journal of Pineal Research* 45: 247–257.
85. Hewing, M. 1984. Seasonal variations in the cerebrospinal fluid-contacting area of the pineal gland in the golden hamster (*Mesocricetus auratus*). *Anatomy and Embryology (Berlin)* 169: 91–96.
86. Hill, S.M., et al. 2015. Melatonin: An inhibitor of breast cancer. *Endocrine-Related Cancer* 32: 183–204.
87. Hirpara, J.L., et al. 2001. Intracellular acidification MV triggered by mitochondrial-derived hydrogen peroxide is an effector mechanism for drug-induced apoptosis in tumor cells. *The Journal of Biological Chemistry* 276: 514–521.
88. Histing, T., et al. 2012. Melatonin impairs fracture healing by suppressing RANKL-mediated bone remodeling. *The Journal of Surgical Research* 173: 83–90.
89. Ho, A.K., and C.L. Chik. 2010. Modulation of Aanat gene transcription in the rat pineal gland. *Journal of Neurochemistry* 112: 321–331.

90. Ho, A.K., et al. 2006. Opposite effects of proteasome inhibitors in the adrenergic induction of arylalkylamine N-acetyltransferase in rat pinealocytes. *Chronobiology International* 23: 361–367.
91. Hochrainer, K., et al. 2015. The ubiquitin ligase HERC3 attenuates NF-kappaB-dependent transcription independently of its enzymatic activity by delivering the RelA subunit for degradation. *Nucleic Acids Research* 43: 9889–9904.
92. Hoesel, B., and J.A. Schmid. 2013. The complexity of NF-kappaB signaling in inflammation and cancer. *Molecular Cancer* 12: 86.
93. Hong, Y., et al. 2014. Melatonin treatment induces interplay of apoptosis, autophagy, and senescence in human colorectal cancer cells. *Journal of Pineal Research* 56: 264–274.
94. Hosseinzadeh, A., et al. 2018. Idiopathic pulmonary fibrosis (IPF) signaling pathways and protective roles of melatonin. *Life Sciences* 201: 17–29.
95. Hsu, S.Y., et al. 1996. Targeted overexpression of Bcl-2 in ovaries of transgenic mice leads to decreased follicle apoptosis, enhanced folliculogenesis, and increased germ cell tumorigenesis. *Endocrinology* 137: 4837–4843.
96. Hsu, S.M., et al. 2015. Proteasome inhibitor bortezomib suppresses nuclear factor-kappa B activation and ameliorates eye inflammation in experimental autoimmune uveitis. *Mediators of Inflammation* 2015: 847373.
97. Hu, W., et al. 2016. Melatonin: The dawning of a treatment for fibrosis? *Journal of Pineal Research* 60: 121–131.
98. Hu, J., et al. 2017. Melatonin alleviates postinfarction cardiac remodeling and dysfunction by inhibiting Mst12. *Journal of Pineal Research* 62: e12368–e12381.
99. Huang, S.H., et al. 2008. Melatonin decreases TLR3-mediated inflammatory factor expression via inhibition of NF-kappa B activation in respiratory syncytial virus-infected RAW264.7 macrophages. *Journal of Pineal Research* 45: 93–100.
100. Huang, Z., et al. 2010. N-terminal residues regulate proteasomal degradation of AANAT. *Journal of Pineal Research* 48: 290–296.
101. Huo, X., et al. 2017. Human transporters, PEPT1/2, facilitate melatonin transportation into mitochondria of cancer cells: An implication of the therapeutic potential. *Journal of Pineal Research* 62: e12390.
102. Hussain, A.R., et al. 2009. Proteasome inhibitor MG-132 mediated expression of p27Kip1 via S-phase kinase protein 2 degradation induces cell cycle coupled apoptosis in primary effusion lymphoma cells. *Leukemia & Lymphoma* 50: 1204–1213.
103. Indran, I.R., et al. 2011. Recent advances in apoptosis, mitochondria and drug resistance in cancer cells. *Biochimica et Biophysica Acta* 1807: 735–745.
104. Iredale, J.P., et al. 2013. Extracellular matrix degradation in liver fibrosis: Biochemistry and regulation. *Biochimica et Biophysica Acta* 1832: 876–883.
105. Irmak, M.K., and O. Ozcan. 1997. Human diversity, environmental adaptation and neural crest. *Medical Hypotheses* 48: 407–410.
106. Irmak, M.K., et al. 2005. Melatonin seems to be a mediator that transfers the environmental stimuli to oocytes for inheritance of adaptive changes through epigenetic inheritance system. *Medical Hypotheses* 64: 1138–1143.
107. Janjetovic, Z., et al. 2017. Melatonin and its metabolites protect human melanocytes against UVB-induced damage: Involvement of NRF2-mediated pathways. *Scientific Reports* 7: 1274.
108. Jeong, J.K., et al. 2012. Melatonin-induced autophagy protects against human prion protein-mediated neurotoxicity. *Journal of Pineal Research* 53: 138–146.
109. Jou, M.J., et al. 2004. Visualization of the antioxidative effects of melatonin at the mitochondrial level during oxidative stress-induced apoptosis of rat brain astrocytes. *Journal of Pineal Research* 37: 55–70.
110. ———. 2007. Melatonin protects against common deletion of mitochondrial DNA- augmented mitochondrial oxidative stress and apoptosis. *Journal of Pineal Research* 43: 389–403.

111. ———. 2010. Visualization of melatonin's multiple mitochondrial levels of protection against mitochondrial Ca^{2+}-mediated permeability transition and beyond in rat brain astrocytes. *Journal of Pineal Research* 48: 20–38.
112. Jung-Hynes, B., et al. 2011. Melatonin, a novel Sirt1 inhibitor, imparts antiproliferative effects against prostate cancer in vitro in culture and in vivo in TRAMP model. *Journal of Pineal Research* 50: 140–149.
113. Kang, J.W., et al. 2016. Melatonin enhances mitophagy and mitochondrial biogenesis in rats with carbon tetrachloride-induced liver fibrosis. *Journal of Pineal Research* 60: 383–393.
114. Karbowski, M., and R.J. Youle. 2011. Regulating mitochondrial outer membrane proteins by ubiquitination and proteasomal degradation. *Current Opinion in Cell Biology* 23: 476–482.
115. Karin, M., and Y. Ben-Neriah. 2000. Phosphorylation meets ubiquitination: The control of NF-[kappa]B activity. *Annual Review of Immunology* 18: 621–663.
116. Kaur, C., et al. 2002. Plasma melatonin, pinealocyte morphology, and surface receptors/antigen expression on macrophages/microglia in the pineal gland following a high-altitude exposure. *Journal of Neuroscience Research* 67: 533–543.
117. Kayumov, L., et al. 2000. Melatonin, sleep and circadian rhythm disorders. *Seminars in Clinical Neuropsychiatry* 5: 44–55.
118. Kerr, J.F., et al. 1972. Apoptosis: A basic biological phenomenon with wide-ranging implications in tissue kinetics. *British Journal of Cancer* 26: 239–257.
119. Kim, J.E., et al. 2010. Suppression of NF-kappaB signaling by KEAP1 regulation of IKKbeta activity through autophagic degradation and inhibition of phosphorylation. *Cellular Signalling* 22: 1645–1654.
120. Klein, D.C. 2007. Arylalkylamine N-acetyltransferase: The timezyme. *The Journal of Biological Chemistry* 282: 4233–4237.
121. Klein, D.C., and G.R. Berg. 1970. Pineal gland: Stimulation of melatonin production by norepinephrine involves cyclic AMP-mediated stimulation of N-acetyltransferase. *Advances in Biochemical Psychopharmacology* 3: 241–263.
122. Klein, D.C., et al. 1997. The melatonin rhythm-generating enzyme: Molecular regulation of serotonin N-acetyltransferase in the pineal gland. *Recent Progress in Hormone Research* 52: 307–357.
123. Klose, R.J., and A.P. Bird. 2006. Genomic DNA methylation: The mark and its mediators. *Trends in Biochemical Sciences* 31: 89–97.
124. Kochan, D.Z., and O. Kovalchuk. 2015. Circadian disruption and breast cancer: An epigenetic link? *Oncotarget* 6: 16866–16882.
125. Korkmaz, A.S., et al. 2012. Gene regulation by melatonin linked to epigenetic phenomena. *Gene* 503: 1–11.
126. Korkmaz, G.G., et al. 2012. Melatonin ameliorates oxidative damage in hyperglycemia-induced liver injury. *Clinical and Investigative Medicine* 35: e370–e377.
127. Koyama, H., et al. 2002. Melatonin at pharmacologic doses increases bone mass by suppressing resorption through down-regulation of the RANKL-mediated osteoclast formation and activation. *Journal of Bone and Mineral Research* 17: 1219–1229.
128. Lackner, C., and D. Tiniakos. 2019. Fibrosis and alcohol-related liver disease. *Journal of Hepatology* 70: 294–304.
129. Lamb, C.A. 2013. The autophagosome: Origins unknown, biogenesis complex. *Nature Reviews. Molecular Cell Biology* 14: 759–774.
130. Lavie, J., et al. 2018. Ubiquitin-dependent degradation of mitochondrial proteins regulates energy metabolism. *Cell Reports* 23: 2852–2863.
131. Lawrance, I.C., et al. 2015. Cellular and molecular mediators of intestinal fibrosis. *Journal of Crohn's & Colitis* 11: 1491–1503.
132. Lee, D.F., et al. 2009. KEAP1 E3 ligase-mediated downregulation of NF-kappaB signaling by targeting IKKbeta. *Molecular Cell* 36: 131–140.

133. Lee, S.E., et al. 2013. Genome-wide profiling in melatonin-exposed human breast cancer cell lines identifies differentially methylated genes involved in the anticancer effect of melatonin. *Journal of Pineal Research* 54: 80–88.
134. Lee, S.J., et al. 2018. Melatonin inhibits apoptotic cell death induced by Vibrio vulnificus Vvha via melatonin receptor 2 coupling with NCF-1. *Cell Death & Disease* 9: 48.
135. Lerner, A.B., et al. 1958. Isolation of melatonin, the pineal gland factor that lightens melanocytes. *Journal of the American Chemical Society* 80: 2587.
136. Li, J.H., et al. 2005. Melatonin reduces inflammatory injury through inhibiting NF-kappaB activation in rats with colitis. *Mediators of Inflammation* 2005: 185–193.
137. Li, H., et al. 2014. Alterations in the time course of expression of the Nox family in the brain in a rat experimental cerebral ischemia and reperfusion model: Effects of melatonin. *Journal of Pineal Research* 57: 110–119.
138. Li, M., et al. 2016. Cognitive dysfunction and epigenetic alterations of the BDNF gene are induced by social isolation during early adolescence. *Behavioural Brain Research* 313: 177–183.
139. ———. 2016. Melatonin antagonizes cadmium-induced neurotoxicity by activing the transcription factor EB-dependent autophagy-lysosome machinery in mouse neuroblastoma cells. *Journal of Pineal Research* 61: 353–369.
140. Lignitto, L., et al. 2011. Control of PKA stability and signaling by the RING ligase praja2. *Nature Cell Biology* 13: 412–422.
141. Lin, T.B., et al. 2016. Melatonin relieves neuropathic allodynia through spinal Mt2-enhanced Pp2ac and downstream Hdac4 shuttling-dependent epigenetic modification of Hmgb1 transcription. *Journal of Pineal Research* 60: 263–276.
142. Liu, J., et al. 2016. MT1 and MT2 melatonin receptors: A therapeutic perspective. *Annual Review of Pharmacology and Toxicology* 56: 361–383.
143. Liu, S., et al. 2017. Evaluation of cell death pathways initiated by antitumor drugs melatonin and valproic acid in bladder cancer cells. *FEBS Open Bio* 7: 798–810.
144. Lowes, D.A., et al. 2013. Antioxidants that protect mitochondria reduce interleukin-6 and oxidative stress, improve mitochondrial function, and reduce biochemical markers of organ dysfunction in a rat model of acute sepsis. *British Journal of Anaesthesia* 110: 472–480.
145. Luchetti, F., et al. 2010. Melatonin signaling and cell protection function. *The FASEB Journal* 24: 3603–3624.
146. Lui, C.C., et al. 2015. Effects of melatonin on prenatal dexamethasone-induced epigenetic alterations in hippocampal morphology and reelin and glutamic acid decarboxylase 67 levels. *Developmental Neuroscience* 37: 105–114.
147. Luo, C., et al. 2019. The multiple protective roles and molecular mechanisms of melatonin and its precursor N-acetylserotonin in targeting brain injury and liver damage and in maintaining bone health. *Free Radical Biology & Medicine* 130: 215–233.
148. Lv, Y., et al. 2018. Melatonin protects mouse spermatogonial stem cells against hexavalent chromium-induced apoptosis and epigenetic histone modification. *Toxicology and Applied Pharmacology* 340: 30–38.
149. Maity, P., et al. 2009. Melatonin reduces indomethacin-induced gastric mucosal cell apoptosis by preventing mitochondrial oxidative stress and the activation of mitochondrial pathway of apoptosis. *Journal of Pineal Research* 46: 314–323.
150. Manchester, L.C., et al. 1995. Melatonin immunoreactivity in the photosynthetic prokaryote *Rhodospirillum rubrum*: Implications of an ancient antioxidant system. *Cellular & Molecular Biology Research* 41: 391–395.
151. ———. 2015. Melatonin: An ancient molecule that makes oxygen metabolically tolerable. *Journal of Pineal Research* 59: 403–419.
152. Maria, S., and P.A. Witt-Enderby. 2014. Melatonin effects on bone: Potential use for the prevention and treatment for osteopenia, osteoporosis and periodontal disease and for use in bone-grafting procedures. *Journal of Pineal Research* 56: 115–125.

153. Martin, M., et al. 2000. Melatonin-induced increased activity of the respiratory chain complexes I and IV can prevent mitochondrial damage induced by ruthenium red in vivo. *Journal of Pineal Research* 28: 242–248.
154. ———. 2000. Melatonin but not vitamin C and E maintains glutathione homeostasis in *t*-butyl hydroperoxide-induced mitochondrial oxidative stress. *The FASEB Journal* 14: 1677–1679.
155. ———. 2002. Melatonin increases the activity of the oxidative phosphorylation enzymes and the production of ATP in rat brain and liver mitochondria. *The International Journal of Biochemistry & Cell Biology* 34: 348–357.
156. Martin, V., et al. 2014. Involvement of autophagy in melatonin-induced cytotoxicity in glioma- initiating cells. *Journal of Pineal Research* 57: 308–316.
157. Martinez-Martinez, E., et al. 2014. Leptin induces cardiac fibrosis through galectin-3, mTOR and oxidative stress: Potential role of obesity. *Journal of Hypertension* 32: 1104–1114.
158. Massa, P.T., et al. 2006. NFkappaB in neurons? The uncertainty principle in neurobiology. *Journal of Neurochemistry* 97: 607–618.
159. Massague, J. 2012. TGFβ signaling in context. *Nature Reviews. Molecular Cell Biology* 13: 616–630.
160. Matsushima, S., and Y. Morisawa. 1980. Mechanism involved in the response of granulated vesicles in the mouse pinealocyte to acute cold exposure. *Cell and Tissue Research* 208: 247–252.
161. Mayo, J.C., et al. 2017. Melatonin and sirtuins: A "not-so unexpected" relationship. *Journal of Pineal Research* 62.
162. ———. 2017. Melatonin transport into mitochondria. *Cellular and Molecular Life Sciences* 74: 3927–3940.
163. Mediavilla, M.D., et al. 2010. Basic mechanisms involved in the anti-cancer effects of melatonin. *Current Medicinal Chemistry* 17: 4462–4481.
164. Meng, X.M., et al. 2016. TGF-β: The master regulator of fibrosis. *Nature Reviews. Nephrology* 12: 325–338.
165. Mishra, A., S. Paul, and S. Swarnakar. 2011. Downregulation of matrix metalloproteinase-9 by melatonin during prevention of alcohol-induced liver injury in mice. *Biochimie* 93: 854–866.
166. Mori, F., et al. 2016. Multitargeting activity of Mir-24 inhibits long-term melatonin anticancer effects. *Oncotarget* 7: 20532–20548.
167. Mukherjee, D., et al. 2010. Melatonin protects against isoproterenol-induced myocardial injury in the rat: Antioxidative mechanisms. *Journal of Pineal Research* 48: 251–262.
168. Nakamura, E., et al. 2008. Frequent silencing of a putative tumor suppressor gene melatonin receptor 1 a (MTNR1A) in oral squamous-cell carcinoma. *Cancer Science* 99: 1390–1400.
169. Nduhirabandi, F., and G.J. Maarman. 2018. Melatonin in heart failure: A promising therapeutic strategy? *Molecules* 23: 1819–1837.
170. Niles, L.P., et al. 2013. Melatonin induces histone hyperacetylation in the rat brain. *Neuroscience Letters* 541: 49–53.
171. Nobakht, M., et al. 2015. The metabolomics of airway diseases, including COPD, asthma and cystic fibrosis. *Biomarkers* 20: 5–16.
172. Nogueira, A., et al. 2017. Pathophysiological mechanisms of renal fibrosis: A review of animal models and therapeutic targets. In *Vivo*, vol. 31, 1–22.
173. Okatani, Y., et al. 2003. Protective effect of melatonin against mitochondrial injury induced by ischemia and reperfusion of rat liver. *European Journal of Pharmacology* 469: 145–152.
174. ———. 2003. Acutely administered melatonin restores mitochondrial physiology in old mice. *The International Journal of Biochemistry & Cell Biology* 35: 367–375.
175. Ordonez, R., et al. 2015. Ceramide metabolism regulates autophagy and apoptotic cell death induced by melatonin in liver cancer cells. *Journal of Pineal Research* 59: 178–189.
176. Orogo, A.M., and A.B. Gustafsson. 2015. Therapeutic targeting of autophagy: Potential and concerns in treating cardiovascular disease. *Circulation Research* 116: 489–503.

177. Osier, N.D., et al. 2017. Brain injury results in lower levels of melatonin receptor subtypes MT1 and MT2. *Neuroscience Letters* 650: 18–24.
178. Ostrowska, A.Z., et al. 2001. Assessment of the relationship between dynamic pattern of nighttime levels of melatonin and chosen biochemical markers of bone metabolism in a rat model of postmenopausal osteoporosis. *Neuro Endocrinology Letters* 22: 129–136.
179. Pan, P., et al. 2018. Melatonin balance the autophagy and apoptosis by regulating UCP2 in the LPS-induced cardiomyopathy. *Molecules* 23: 675.
180. Paradies, G., et al. 2017. Mitochondrial bioenergetics decay in aging: Beneficial effect of melatonin. *Cellular and Molecular Life Sciences* 74: 3897–3912.
181. Park, K.H., et al. 2011. Melatonin promotes osteoblastic differentiation through the BMP/ERK/Wnt signaling pathways. *Journal of Pineal Research* 51: 187–194.
182. Park, E.J., et al. 2014. Transcriptional and post-translational regulation of Bim controls apoptosis in melatonin-treated human renal cancer Caki cells. *Journal of Pineal Research* 56: 97–106.
183. Peng, Z., et al. 2018. Melatonin attenuates ovary inflammation via SIRT1 dependent inhibition of NLRP3 inflammasome and IL-1β in rats with COPD. *International Immunopharmacology* 62: 23–28.
184. Pervaiz, S., et al. 1999. Superoxide anion inhibits drug-induced tumor cell death. *FEBS Letters* 59: 343–348.
185. Peter, M.E., et al. 1997. Advances in apoptosis research. *Proceedings of the National Academy of Sciences of the United States of America* 94: 12736–12737.
186. Pi, H., et al. 2015. SIRT3-SOD2-mROS-dependent autophagy in cadmium-induced hepatotoxicity and salvage by melatonin. *Autophagy* 11: 1037–1051.
187. Poeggeler, B., et al. 1994. Melatonin – A highly potent endogenous radical scavenger and electron donor: New aspects of the oxidation chemistry of this indole assessed in vitro. *Annals of the New York Academy of Sciences* 738: 419–420.
188. Proietti, S., et al. 2017. Melatonin, mitochondria and the cancer cell. *Cellular and Molecular Life Sciences* 74: 4015–4026.
189. Qin, W., et al. 2012. Melatonin inhibits IL1beta-induced MMP9 expression and activity in human umbilical vein endothelial cells by suppressing NF-kappaB activation. *The Journal of Endocrinology* 214: 145–153.
190. Ramirez-Rodriguez, G., et al. 2012. Melatonin supplementation delays the decline of adult hippocampal neurogenesis during normal aging of mice. *Neuroscience Letters* 530: 53–58.
191. Ramos, E., et al. 2017. Ischemic brain injury: New insights on the protective role of melatonin. *Free Radical Biology & Medicine* 104: 32–53.
192. Reiter, R.J. 1992. Alterations of the circadian melatonin rhythm by the electromagnetic spectrum: A study in environmental toxicology. *Regulatory Toxicology and Pharmacology* 15: 226–244.
193. Reiter, R.J., and D.X. Tan. 2003. What constitutes a physiological concentration of melatonin? *Journal of Pineal Research* 34: 79–80.
194. Reiter, R.J., et al. 2001. Melatonin in plants. *Nutrition Reviews* 59: 286–290.
195. ———. 2008. Melatonin combats molecular terrorism at the mitochondrial level. *Interdisciplinary Toxicology* 1: 101–113.
196. ———. 2014. Melatonin: Exceeding expectations. *Physiology (Bethesda)* 29: 325–333.
197. ———. 2016. Melatonin as an antioxidant: Under promises but over delivers. *Journal of Pineal Research* 61: 253–278.
198. ———. 2017. Melatonin as a mitochondria-targeted antioxidant: One of evolution's best ideas. *Cellular and Molecular Life Sciences* 74: 3863–3881.
199. ———. 2017. Role of SIRT3/SOD2 signaling in mediating the antioxidant action of melatonin in mitochondria. *Current Trends in Endocrinology* 9: 45–49.
200. ———. 2018. Mitochondria: Central organelles for melatonin's antioxidant and anti- aging actions. *Molecules* 23: E509.

201. ———. 2018. Historical perspective and evaluation of the mechanisms by which melatonin modulates seasonal reproduction in mammals. *Melatonin Research* 1: 1–17.
202. Revel, F.G., et al. 2006. Melatonin regulates type 2 deiodinase gene expression in the Syrian hamster. *Endocrinology* 147: 4680–4687.
203. Richeldi, L., et al. 2007. Idiopathic pulmonary fibrosis. *Lancet* 389: 1941–1952.
204. Rockey, D.C., et al. 2015. Fibrosis – A common pathway to organ injury and failure. *The New England Journal of Medicine* 372: 1138–1149.
205. Rodriguez, C., et al. 2004. Regulation of antioxidant enzymes: A significant role for melatonin. *Journal of Pineal Research* 36: 1–9.
206. Rosales-Corral, S.A., et al. 2012. Alzheimer's disease: Pathological mechanisms and the beneficial role of melatonin. *Journal of Pineal Research* 52: 167–202.
207. Russo, A., et al. 2010. Bortezomib: A new pro-apoptotic agent in cancer treatment. *Current Cancer Drug Targets* 10: 55–67.
208. Sack, R.L., and A.J. Lewy. 1997. Melatonin as a chronobiotic: Treatment of circadian desynchrony in night workers and the blind. *Journal of Biological Rhythms* 12: 595–603.
209. Samantaray, S., et al. 2008. Melatonin attenuates calpain upregulation, axonal damage and neuronal death in spinal cord injury in rats. *Journal of Pineal Research* 44: 348–357.
210. Sanchez-Hidalgo, M., et al. 2009. Age-related changes in melatonin synthesis in rat extrapineal tissues. *Experimental Gerontology* 44: 328–334.
211. San-Miquel, B., et al. 2014. Melatonin modulates the autophagic response in acute liver failure induced by the rabbit hemorrhagic disease virus. *Journal of Pineal Research* 56: 313–321.
212. ———. 2015. Melatonin inhibits autophagy and endoplasmic reticulum stress in mice with carbon tetrachloride-induced fibrosis. *Journal of Pineal Research* 59: 151–162.
213. Schomerus, C., et al. 2000. Selective adrenergic/cyclic AMP-dependent switch-off of proteasomal proteolysis along switches on neural signal transduction: An example from the pineal gland. *Journal of Neurochemistry* 75: 2123–2132.
214. Schwimmer, H., et al. 2014. Light at night and melatonin have opposite effects on breast cancer tumors in mice assessed by growth rates and global DNA methylation. *Chronobiology International* 31: 144–150.
215. Sezer, A., et al. 2010. Effects of intraperitoneal melatonin on caustic sclerosing cholangitis due to scolicidal solution in a rat model. *Current Therapeutic Research, Clinical and Experimental* 71: 118–128.
216. Shah, S.A., et al. 2017. Melatonin stimulates the SIRT1/Nrf2 signaling pathway counteracting lipopolysaccharide (LPS)-induced oxidative stress to rescue postnatal rat brain. *CNS Neuroscience & Therapeutics* 23: 33–44.
217. Sharma, R., et al. 2008. Epigenetic targets for melatonin: Induction of histone H3 hyperacetylation and gene expression in C17.2 neural stem cells. *Journal of Pineal Research* 45: 277–284.
218. Sheen, J.M., et al. 2016. Melatonin alleviates liver apoptosis in bile duct ligation young rats. *International Journal of Molecular Sciences* 17: e1365.
219. Shi, D., et al. 2012. Melatonin suppresses proinflammatory mediators in lipopolysaccharide-stimulated CRL1999 cells via targeting MAPK, NF-kappaB, c/EBPbeta, and p300 signaling. *Journal of Pineal Research* 53: 154–165.
220. Shin, N.R., et al. 2017. Melatonin suppresses fibrotic responses induced by cigarette smoke via downregulation of TGF-β1. *Oncotarget* 8: 95692–95703.
221. Shiu, S.Y., et al. 2013. Melatonin MT1 receptor-induced transcriptional up-regulation of p27 (Kip1) in prostate cancer antiproliferation is mediated via inhibition of constitutively active nuclear factor kappa B (NF-kappaB): Potential implications on prostate cancer chemoprevention and therapy. *Journal of Pineal Research* 54: 69–79.
222. Shringarpure, R., et al. 2003. Ubiquitin conjugation is not required for the degradation of oxidized proteins by proteasome. *The Journal of Biological Chemistry* 278: 311–318.
223. Simonneaux, V., et al. 2006. Rat and Syrian hamster: Two models for the regulation of AANAT gene expression. *Chronobiology International* 23: 351–359.

224. Sishi, B.J., et al. 2013. Autophagy upregulation promotes survival and attenuates doxorubicin-induced cardiotoxicity. *Biochemical Pharmacology* 85: 124–134.
225. Slominski, R.M., et al. 2012. Melatonin membrane receptors in peripheral tissues: Distribution and functions. *Molecular and Cellular Endocrinology* 351: 152–166.
226. Soumier, A., et al. 2009. Mechanisms contributing to the phase-dependent regulation of neurogenesis by the novel antidepressant, agomelatine, in the adult rat hippocampus. *Neuropsychopharmacology* 34: 2390–2403.
227. Stehle, J.H., et al. 2011. A survey of molecular details in the human pineal gland in the light of phylogeny, structure, function and chronobiological diseases. *Journal of Pineal Research* 51: 17–43.
228. Steinsapir, J., et al. 1998. Type 2 iodothyronine deiodinase in rat pituitary tumor cells is inactivated in proteasomes. *Journal of Clinical Investigation* 102: 1895–1899.
229. Stojkovic, K., et al. 2014. A central role for ubiquitination within a circadian clock protein modification code. *Frontiers in Molecular Neuroscience* 7: 1–7.
230. Su, J., et al. 2015. Melatonin significantly improves the developmental competence of bovine somatic cell nuclear transfer embryos. *Journal of Pineal Research* 59: 455–468.
231. Su, S.C., et al. 2017. Cancer metastasis: Mechanisms of inhibition by melatonin. *Journal of Pineal Research* 62: e12370.
232. Sulkava, S., et al. 2017. Common genetic variation near melatonin receptor 1A gene linked to job-related exhaustion in shift workers. *Sleep* 40: 1–10.
233. Suofu, Y., et al. 2017. Dual role of mitochondria in producing melatonin and driving GPCR signaling to block cytochrome c release. *Proceedings of the National Academy of Sciences of the United States of America* 114: E7997–E8006.
234. Suwanjang, W., et al. 2010. The protective effect of melatonin on methamphetamine-induced calpain-dependent death pathway in human neuroblastoma SH-SY5Y cultured cells. *Journal of Pineal Research* 48: 94–101.
235. ———. 2012. Calpastatin reduces calpain and caspase activation in methamphetamine-induced toxicity in human neuroblastoma SH-SY5Y cultured cells. *Neuroscience Letters* 526: 49–53.
236. ———. 2013. Melatonin attenuates dexamethasone toxicity-induced oxidative stress, calpain and caspase activation in human neuroblastoma SH-SY5Y cells. *The Journal of Steroid Biochemistry and Molecular Biology* 138: 116–122.
237. Sziksz, E., et al. 2015. Fibrosis related inflammation mediators: Role of the IL-10 cytokine family. *Mediators of Inflammation* 2015: 764641.
238. Tahan, G., et al. 2010. Melatonin ameliorates liver fibrosis induced by bile-duct ligation in rats. *Canadian Journal of Surgery* 53: 313–318.
239. Tain, Y.L., et al. 2014. Transcriptional regulation of programmed hypertension by melatonin: An epigenetic perspective. *International Journal of Molecular Sciences* 15: 18484–18495.
240. ———. 2014. Melatonin therapy prevents programmed hypertension and nitric oxide deficiency in offspring exposed to maternal caloric restriction. *Oxidative Medicine and Cellular Longevity* 2014: 283180.
241. Tamtaji, O.R., et al. 2019. Melatonin, a calpain inhibitor in the central nervous system: Current status and future perspectives. *Journal of Cellular Physiology* 234: 1001–1007.
242. Tan, D.X., and R.J. Reiter. 2019. Mitochondria: The birth place, the battle ground and the site of melatonin metabolism in cells. *Melatonin Res* 2: 44–66.
243. Tan, D.X., et al. 2003. Mechanistic and comparative studies of melatonin and classic antioxidants in terms of their interactions with the ABTS cation radical. *Journal of Pineal Research* 34: 249–259.
244. ———. 2013. Mitochondrial and chloroplasts as the original sites of melatonin synthesis: A hypothesis related to melatonin's primary function and evolution in eukaryotes. *Journal of Pineal Research* 54: 127–138.

245. Tan, J., et al. 2014. Melatonin protects the esophageal epithelial barrier by suppressing the transcription, expression and activity of myosin light chain kinase through ERK1/2 signal transduction. *Cellular Physiology and Biochemistry* 34: 2117–2127.
246. Tas, U., et al. 2011. Hepatotoxic activity of toluene inhalation and protective role of melatonin. *Toxicology and Industrial Health* 27: 465–473.
247. Teng, Y.C., et al. 2015. Melatonin ameliorates arsenite-induced neurotoxicity: Involvement of autophagy and mitochondria. *Molecular Neurobiology* 52: 1015–1022.
248. Terriff, D.L., et al. 2005. Proteasomal proteolysis in the adrenergic induction of arylalkylamine- N-acetyltransferase in rat pinealocytes. *Endocrinology* 146: 4795–4803.
249. Thiyagarajan, M., et al. 2008. Activated protein C promotes neovascularization and neurogenesis in post ischemic brain via protease-activated receptor 1. *The Journal of Neuroscience* 28: 12788–12797.
250. Thu, K.L., et al. 2011. Genetic disruption of KEAP1/CUL3 E3 ubiquitin ligase complex components is a key mechanism of NF-kappaB pathway activation in lung cancer. *Journal of Thoracic Oncology* 6: 1521–1529.
251. Tian, H., et al. 2012. Keap1: One stone kills three birds Nrf2, IKKbeta and Bcl-2/Bcl-xl. *Cancer Letters* 325: 26–34.
252. Trivedi, P.P., et al. 2016. Melatonin modulated autophagy and Nrf2 signaling pathways in mice with colitis-associated colon carcinogenesis. *Molecular Carcinogenesis* 55: 255–267.
253. Uygur, R., et al. 2016. Protective effects of melatonin against arsenic-induced apoptosis and oxidative stress in rat testes. *Toxicology and Industrial Health* 32: 848–859.
254. Vakifahmetoglu-Norberg, H., and B. Zhivotovsky. 2010. The unpredictable caspase-2: What can it do? *Trends in Cell Biology* 20: 150–159.
255. Vega-Naredo, I., et al. 2012. Melatonin modulates autophagy through a redox-mediated action in female Syrian hamster Harderian gland controlling cell types and gland activity. *Journal of Pineal Research* 52: 80–92.
256. Venegas, C., et al. 2011. Extrapineal melatonin: Analysis of its subcellular distribution and daily fluctuations. *Journal of Pineal Research* 52: 217–227.
257. Villela, D., et al. 2014. Norepinephrine activates NF-kappaB transcription factor in cultured rat pineal gland. *Life Sciences* 94: 122–129.
258. Vriend, J., and R.J. Reiter. 2014. Melatonin and ubiquitin: What's the connection? *Cellular and Molecular Life Sciences* 71: 3409–3418.
259. ———. 2014. Melatonin as a proteasome inhibitor. Is there any clinical evidence? *Life Sciences* 115: 8–14.
260. ———. 2015. The Keap1-Nrf2-antioxidant response element pathway: A review of its regulation by melatonin and the proteasome. *Molecular and Cellular Endocrinology* 401: 213–220.
261. ———. 2015. Melatonin feedback on clock genes: A theory involving the proteasome. *Journal of Pineal Research* 58: 1–11.
262. ———. 2016. Melatonin and the von Hippel-Lindau/HIF-1 oxygen sensing mechanism: A review. *Biochimica et Biophysica Acta* 1865: 176–183.
263. Vriend, J., et al. 2017. The pineal gland: A model for adrenergic modulation of ubiquitin ligases. *PLoS One* 12: e0172441.
264. Wang, Y., and S. Zeng. 2018. Melatonin promotes ubiquitination of phosphorylated pro-apoptotic protein Bcl-2-interacting mediator of cell death-extra long (BimEL) in porcine granulosa cells. *International Journal of Molecular Sciences* 19: 3431–3446.
265. Wang, S.J., et al. 2012. Melatonin suppresses apoptosis and stimulates progesterone production by bovine granulosa cells via its receptors (MT1 and MT2). *Theriogenology* 78: 1517–1526.
266. Wang, F., et al. 2015. Cytoprotective effect of melatonin against hypoxia/serum deprivation-induced cell death of bone marrow mesenchymal stem cells in vitro. *European Journal of Pharmacology* 748: 157–165.

267. Wang, D., et al. 2016. Bortezomib sensitizes esophageal squamous cancer cells to radiotherapy by suppressing the expression of HIF-1alpha and apoptosis proteins. *Journal of X-Ray Science and Technology* 24: 639–646.
268. Wongprayoon, P., and P. Govitrapong. 2017. Melatonin as a mitochondrial protector against neurodegenerative diseases. *Cellular and Molecular Life Sciences* 74: 3999–4014.
269. Xie, S., et al. 2015. Melatonin protects against chronic intermittent hypoxia-induced cardiac hypertrophy by modulating autophagy though the 5′ adenosine monophosphate-activated protein kinase pathway. *Biochemical and Biophysical Research Communications* 464: 975–981.
270. Xiong, J. 2015. Atg7 in development and disease: Panacea or Pandora's box? *Protein & Cell* 6: 722–734.
271. Yalcin, A., et al. 2003. Apoptosis in cerebellar granule neurons is associated with reduced interaction between CREB-binding protein and NF-kappaB. *Journal of Neurochemistry* 84: 397–408.
272. Yamamoto, M., et al. 2018. The KEAP1-NRF2 system: A thiol-based sensor-effector apparatus for maintaining redox homeostasis. *Physiological Reviews* 98: 1169–1203.
273. Yaron, A., et al. 1998. Idenfication of the receptor component of the lkappaBalpha-ubiquitin ligase. *Nature* 396: 590–594.
274. Yasuo, S., et al. 2007. Temporal dynamics of type 2 deiodinase expression after melatonin injections in Syrian hamsters. *Endocrinology* 148: 4385–4392.
275. Yerlikaya, A. 2012. Expression of heme oxygenase-1 in response to proteasomal inhibition. *Protein and Peptide Letters* 19: 1330–1333.
276. Yu, S., et al. 2017. Melatonin regulates PARP1 to control the senescence-associated secretory phenotype (SASP) in human fetal lung fibroblast cells. *Journal of Pineal Research* 63: 1–18.
277. Zaitone, S., et al. 2011. Pentoxifylline and melatonin in combination with pioglitazone ameliorate experimental non-alcoholic fatty liver disease. *European Journal of Pharmacology* 662: 70–77.
278. Zavodnik, I.B., et al. 2011. Melatonin and succinate reduce rat liver mitochondrial dysfunction in diabetes. *Journal of Physiology and Pharmacology* 62: 641–647.
279. Zawilska, J.B. 1996. Melatonin as a chemical indicator of environmental light-dark cycle. *Acta Neurobiologiae Experimentalis* 56: 757–767.
280. Zhang, Y., et al. 2013. Melatonin inhibits the caspase-1/cytochrome c/caspase-3 cell death pathway, inhibits mt1 receptor loss and delays disease progression in a mouse model of amyotrophic lateral sclerosis. *Neurobiology of Disease* 55: 26–35.
281. Zhang, M., et al. 2017. Melatonin protects against diabetic cardiomyopathy through Mst1/Sirt3 signaling. *Journal of Pineal Research* 63: e12418.
282. Zhang, Q., et al. 2017. Years of NF-kappaB: A blossoming of relevance to human pathobiology. *Cell* 168: 37–57.
283. Zhang, S., et al. 2017. Melatonin as a promising agent of regulating stem cell biology and its application in disease therapy. *Pharmacological Research* 117: 252–260.
284. Zhang, T., et al. 2018. Melatonin protects prepuberal testis from deleterious effects of bisphenol A or diethylhexyl phthalate by preserving H3K9 methulation. *Journal of Pineal Research* 65: e12497.
285. Zhao, X., et al. 2005. Regulation of MEF2 by histone deacetylase 4- and SIRT1 deacetylase-mediated lysine modifications. *Molecular and Cellular Biology* 25: 8456–8464.
286. Zhao, H., et al. 2014. Melatonin inhibits endoplasmic reticulum stress and epithelial- mesenchymal transition during bleomycin-induced pulmonary fibrosis in mice. *PLoS One* 9: e97266.
287. Zheng, W., et al. 2018. In utero exposure to hexavalent chromium disrupts rat fetal testis development. *Toxicology Letters* 299: 201–209.
288. Zhou, H., et al. 2012. Melatonin protects against rotenone-induced cell injury via inhibition of Omi and Bax-mediated autophagy in Hela cells. *Journal of Pineal Research* 52: 120–127.
289. Zhu, Y., et al. 2011. Epigenetic impact of long-term shiftwork: Pilot evidence from circadian genes and whole-genome methylation analysis. *Chronobiology International* 28: 852–861.

Multitarget Activities of Inositol and Inositol Hexakisphosphate

Ivana Vucenik

Introduction

To start this story with rice: Multiple health-beneficial effects have been associated with rice, and therefore Japanese government has been trying to bring back the rice culture. However, it is not the rice, it is the rice bran that contains all these products. Seventy years ago, Mr. Tsuno recognized gold in waste, life within rice bran, and made use of it, founding the Tsuno Rice Co, one of the biggest and finest producers of inositol, inositol hexakisphosphate and other health-promoting products from rice, with guaranteed and highest purity and potency.

Occurrence, Structure and Significance

Inositol hexakisphosphate (IP6 or Ins*P*6 or phytic acid) and *myo*-inositol (Ins) are naturally occurring carbohydrates widely distributed among plants. Their discovery dates from 1850s, when Scherer described new molecule isolated from muscle tissue (inositol) and Hartig reported small round particles in various plant seeds, similar to potato starch grains (phytate) [1–4]. It was shown that the isolated particles were rich in phosphorous, calcium and magnesium and that were found only in plants, therefore the name "phytin" was created [4]. In 1914, Anderson presented the molecular structure of *myo*-inositol- 1,2,3,4,5,6-hexakis dihydrogen phosphate, also called phytic acid, which is still valid and has been confirmed by various

I. Vucenik (✉)
Department of Medical and Research Technology and Department of Pathology, University of Maryland School of Medicine, Baltimore, MD, USA
e-mail: ivucenik@som.umaryland.edu

M. Bizzarri (ed.), *Approaching Complex Diseases*, Human Perspectives in Health Sciences and Technology 2, https://doi.org/10.1007/978-3-030-32857-3_15

A

Content of IP_6 in cereals and legumes

CEREALS	IP_6 % (g/100 g dry weight)
Rice (*Oryza glaberrima/sativa*)	0.06-1.08
Rice bran	2.56-8.7
Wheat (*Triticum spp*)	0.39-1.35
Wheat bran	2.1-7.3
Wheat germ	1.14-3.91
Maize (*Zea mays*)	0.72-2.22
Maize germ	6.39

LEGUMES	IP_6 % (g/100 g dry weight)
Kidney beans (*Phaseolus vulgaris*)	0.61-2.38
Dry cowpeas (*Vigna unguiculata*)	0.37-2.90

B

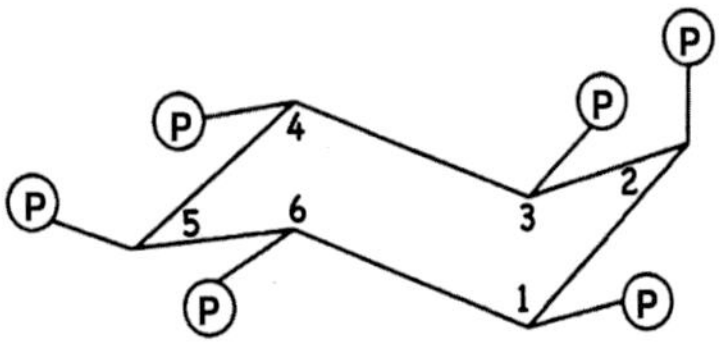

Fig. 1 Content of IP_6 in cereals and legumes and molecular structure. The main source of IP_6 in cereals and legumes are shown in whole seeds. Cereals are rich in IP_6 and contain approximately 1% of IP_6 on the dry matter basis, ranging from 0.06 to 2.2 % (dw). For rice bran, IP_6 concentration ranged up to 8.75%. In the whole seeds of legumes, the IP_6 content varies from 0.2–2.9% (dw) [4, 6] (**a**). The structure of IP_6 (*myo*-inositol hexaphosphate) is shown under physiological conditions at pH 6-7. A six-carbon ring represents the basic carbohydrate moiety in IP_6 with conformation of 5 axial and 1 equatorial phosphates [7, 8] (**b**)

modern analytical methods [5]. A six-carbon inositol ring represents the basic carbohydrate moiety in IP6 and its lower phosphate derivatives (IP1–5).

IP6 is a very stable and the most abundant polyphosphate in nature. It is a component of cereal diets and legumes, found in rice, wheat, peas, beans, oats, barley, in concentrations ranging from 0.4–6.4%, where it is referred as phytic acid [4, 6]. The presence of phosphate group in positions 1,2,3 (axial, equatorial, axial) is giving unique properties to this molecule, such as antioxidant and specific chelating capacity of potentially toxic elements [7, 8] (Fig. 1). Its parent compound *myo*-inositol (Ins) is a cyclitol naturally present in animal and plant cells. There are nine possible stereoisomers of inositol: *cis*-, *epi*-, *allo*-, *myo*-, *muco*-, *neo*-, (+)-*chiro*, (−)-*chiro*-, and *scyllo*-inositols, formed through epimerization of its six hydroxyl groups. Five of them, *myo*-, *scyllo*-, *muco*-, *neo*- and D-*chiro*-inositol occur naturally, while the other four possible isomers (L-*chiro*-, *allo*-, *epi*-, and *cis*-inositol) are derived from Ins [9–11]. Although originally thought that only 63 isomers were possible [12], today we know that the number of inositol phosphates of those nine isomers (excluding pyrophosphates) is 357 [9, 10]. Introduction of Agranoff's turtle analogy helped to visualize Ins in the form of turtle, in which the axial hydroxyl was its head, and the five equatorial hydroxyls serve as forelimbs, hind limbs, and the tail as illustrated in the Fig. 2 [12, 13]. The turtle configuration as a structural mnemonic, was suggested by the International Union of Biochemistry Nomenclature Committee to aid biochemists and eased the confusion in numbering when depicting the Haworth projection (Fig. 2) [12].

Animal and plant cells contain Ins either in its free form, as inositol-containing phospholipids (phosphoinositides) or as phytic acid (IP6), a principal storage form of

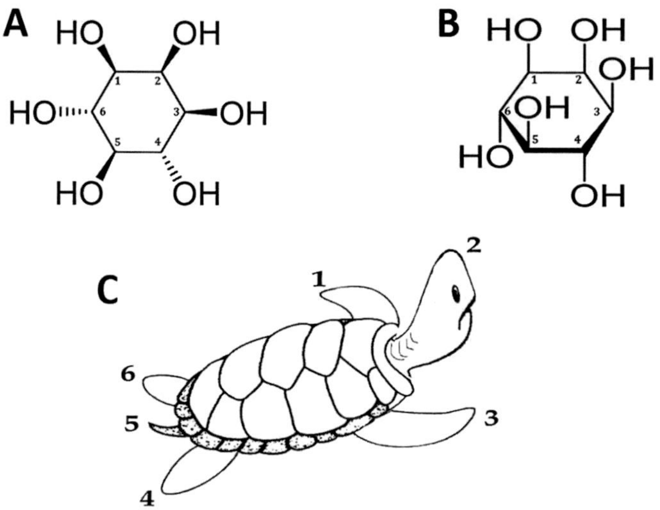

Fig. 2 Structure of *myo*-inositol. *myo*-Inositol is presented here as a wedge-dash notation [13] (**a**), Haworth projection [13] (**b**) and schematically as a turtle [12] (**c**)

phosphorus in plants, particular in bran and seeds [4, 14]. Because Ins can be synthesized from D-glucose, is not any more considered as a part of vitamin B family. Not only all plant cells, but almost all mammalian cells contain high concentrations of IP6, Ins and other inositol phosphates, wherein they play important role in regulating vital cellular functions, such as signal transduction, regulation of cell proliferation and differentiation, RNA export, mRNA transcription, DNA repair, energy transduction and ATP generation [10], and de-regulation of their metabolism has been recognized in several illnesses, including neurological disorders [10], polycystic ovary syndrome [15], metabolic diseases [14, 16, 17] and cancer [7, 8].

Although Ins and IP6 are prevalent natural forms and have been much studied over the last 30 years, some "other" cyclitols and inositols might also be medically relevant, and their roles and applications have been recently considered [9, 10, 18]. For example, the role of *scyllo*-inositol in neurodegenerative diseases [11] and D-*chiro*-inositol have been reported [9, 10, 18].

Multiple Health-Beneficial Effects of Ins and IP6

Acting on several key molecular targets, Ins and IP6 are beneficial in a number of diseases. Ins has been used for years against depression and anxiety disorders [10, 19]. Prevention of kidney stones and other pathological calcifications, such as sialolithiasis, a common disease of salivary glands, and cardiovascular calcification, that frequently occurs in the heart vessels, has been known for IP6 [4]. Ins deregulation has been found in numerous conditions mechanistically and epidemiologically associated to high-glucose diet or altered glucose metabolism [16, 20]. Ins has been shown to possess insulin- mimetic properties and to be efficient in lowering post-prandial blood glucose and associated human disorders [14]. Targeting insulin resistance, Ins has been effective in gestational diabetes mellitus [14], metabolic syndrome [14] and polycystic ovary syndrome [14, 17]. Interestingly, Ins can modulate both insulin resistance and cancer, by targeting multiple biochemical processes that are shared in both cancer and insulin resistance-based diseases [20]. Both IP6 and Ins are able to induce adipocyte differentiation and improve insulin sensitivity *in vitro*, indicating that their antidiabetic properties can be mediated directly through adipocytes [21]. Furthermore, abnormal Ins metabolism has been shown to underlie the pathophysiology of a variety of clinical conditions including Down Syndrome, traumatic brain injury, bronchopulmonary dysplasia (BPD), and respiratory distress syndrome (RDS) [22]. IP6 has also been recognized as potential treatment for Alzheimer's pathology, as evidenced from animal and *in vitro* models [23]. It was indicated that the consumption of IP6 can prevent development of osteoporosis and had a protective affect against osteoporosis [24].

Targeting Cancer

However, cancer preventive and therapeutic properties of IP6 have received most attention and its broad-spectrum of anticancer activities has been shown in multiple preclinical experimental studies and in humans, alone or in combination with Ins [7, 8, 25, 26].

Cancer incidence and mortality are rapidly growing worldwide, mostly reflecting both aging and growth of the population, but also changes in the prevalence and distribution of the main risk factors for cancer, several of which are associated with economic development [27]. It has been shown that modification of diet by increasing vegetable and fruit intake, maintenance of optimum body weight, and regular physical activity, 30–40% of all instances of cancer could be prevented [28, 29]. The epidemiological studies have indicated that only fiber diet with high IP6 (*myo*-inositol hexaphosphate, InsP6, phytic acid) content, such as cereals and legumes, show negative correlation with colon cancer, suggesting that it could be IP6 and not fiber that suppressed colon cancer [7, 8, 30]. And, indeed, it has been shown that IP6 is one of the biologically active components of fiber, responsible for its anticancer effect.

Numerous studies have demonstrated cancer preventive and therapeutic properties of IP6 in a wide variety of tumor types, both *in vitro* and *in vivo* [7, 8]. In the first studies, the effectiveness of IP6 to prevent cancer was evaluated *in vivo* after administration of IP6 in the drinking water. The exogenous 1% IP6 in drinking water 1 week before or 2 weeks after administration of azoxymethane inhibited the development of large intestinal cancer in Fisher 344 rats [31]. In the same model, administration of 2% IP6 in the drinking water was effective even when the treatment had begun 5 months after carcinogen initiation. Compared to untreated rats, animals on IP6 had 27% fewer tumors [32]. A consistent, reproducible, and significant inhibition of mammary cancer by IP6 was shown in experimental models chemically induced by either 7,12-dimethylbenz[*a*]antracene or N-methylnitrosourea; the effect was seen on tumor incidence, tumor size, and tumor multiplicity [7, 8, 33]. IP6 was effective against prostate cancer as well. Studies demonstrated that continuous administration of 2% IP6 in the drinking water, beginning 24 h after implantation of DU-145 prostate cancer cells, resulted in a 64% decrease in tumor burden [7, 8, 25]. Additionally, chemopreventive efficacy of IP6 was observed against prostate tumor growth and progression in the TRansgenic Adenocarcinoma Mouse Prostate (TRAMP) model [7, 8, 25]. Peritumoral, intratumoral or intraperitoneal administration of IP6 significantly inhibited growth of rhabdomyosarcoma tumor xenografts [34], regressed liver cancer xenotransplants [35], and in murine fibrosarcoma FSA-1 model inhibited tumor growth and prevented lung metastases [36]. Anticancer potential of IP6 was demonstrated in several models of skin cancer. The effect of IP6 on skin cancer was investigated in a 2-stage mouse skin carcinogenesis model; a reduction in skin papillomas was found when IP6 was given during the initiation stage but not when given during the promotion stage [7, 8, 25]. Prevention of skin carcinogenesis was also shown in a mouse carcinogenesis model where IP6 caused a reduction in the number of skin tumor formation [7, 8, 25]. Using UVB light known as a complete carcinogen, topical application of IP6 also significantly decreased UVB-induced tumor incidence and multiplicity in SKH1hairless mice [37].

In vitro studies have shown that IP6 inhibits growth and induces differentiation and apoptosis of human breast cancer cells (both estrogen receptor-positive and estrogen-receptor negative), colon, prostate, liver, pancreatic and cervical cancer cell lines, as well as of rhabdomyosarcoma, glioblastoma, melanoma and human leukemia cells [7, 8, 25]. Additionally, IP6 was able to inhibit cell transformation in mouse epidermal JB6 cells and to reverse the transformed phenotype of HepG2 liver cancer cells [7, 8].

Its parent compound, Ins itself was also shown to have modest anticancer activity. It inhibited colon, mammary, soft tissue and lung tumor formation [7, 8]. More importantly, it was shown that Ins potentiates both the antiproliferative and antineoplastic effects of IP6 *in vivo* [7, 8] and *in vitro* [25]. Synergistic cancer inhibition by IP6 when combined with inositol was observed in colon and mammary cancer studies [7, 8, 25]. Similar results were seen in the metastatic lung cancer model [36]. IP6 and Ins inhibited the development and metastatic progression of colorectal cancer to the liver in BALB/c mice, and the effect of their combined application was significantly greater than the effect of either compound alone [38]. Not only the combination of IP6 and Ins was significantly better in different cancers than was either one alone, but it also consistently reduced *all* tumor growth parameters [33]. Therefore, for clinical studies, the combination of IP6 and inositol has been considered for the optimal efficacy.

Biological Activities and Key Molecular Targets

After rapid intake and dephosphorylation, IP6 enters the pool of inositol phosphates and interact with cellular processes involved in cancer prevention, progression and treatment. The anticancer properties of IP6 and Ins are related to the intracellular inositol phosphate pool, affecting multiple targets and signaling pathways, in particular inhibiting the phosphorylation-based activation of key molecular targets that interfere with specific biological functions [25, 26]. The preventive and therapeutic potential of IP6 has been related to its antioxidant functions, ability to block the activation of various carcinogens and/or to stimulate their detoxification, to the immune-enhancing, anti-inflammatory activities, and to the suppression of cell cycle and proliferation. The induction of differentiation and apoptosis in various premalignant and cancerous cells can also contribute to both cancer preventive and therapeutic potential of IP6. Moreover, suppression of angiogenesis [39], inhibition of metastatic processes and tumor progression, synergism with anticancer drugs and alleviation of chemotherapy resistance further indicate its chemotherapeutic potential [7, 8, 25, 26]. Just to name few critical molecular targets. IP6 interferes at the receptor level, down-regulates p27, inhibits pRB phosphorylation and cell cycle [40], reduces PI3K and consequently counteracts the activation of PKC/RAS/ERK pathway [40], downregulates Akt and ERK, leading to reduction of NF- κB and inhibition of inflammation [26]. Extensive review of IP6 and Ins key molecular targets, complex network and modulation of critical pathways associated with biological functions and microenvironment involved in carcinogenesis and cancer progression, have already been published [25, 26].

Interestingly, an obesity-insulin-cancer connection has been recently shown, a link between insulin resistance, diabetes and cancer. Seems that insulin and IGF fuel cancer, and that PI3K/Akt signaling pathway, the most frequently mutated pathway in human cancers, is activated by both insulin and IGF. So, insulin resistance and cancer share a few perturbed, critical biochemical pathway. Ins may directly interfere with both glucose metabolism and carcinogenesis by modulating a number of critical processes downstream of insulin stimulation, including anti-oxidant defenses, oxidative glucose metabolism and endocrine modulation [20]. Additionally, a selected group of biochemical factors, presently considered as possible targets for anticancer treatment are specifically modulated by Ins (PI3K/AKT, PDH and AMPK-related pathways) [20].

IP6 and Ins in Cancer Patients

Although IP6 and Ins exist naturally in plants and human cells, and their deficiency is evident [16], for the full health benefit and function their supplementation is necessary. And, as common constituents of our diet, both IP6 and Ins met specifications of FDA and have been given GRAS (Generally Recognized As Safe) status.

For almost 20 years, IP6 and Ins have been available as food supplements, and despite substantial progress in the understanding of the molecular basis and molecular targets of their anti-carcinogenic activity and potential, there have been very few clinical studies with IP6 and Ins.

Judging IP6 and Ins as anticancer agents, here are few important facts: (a) being natural, they are safe; (b) they do not affect normal cells; (c) they act synergistically with chemotherapy, and (d) affect all principal pathways of malignancy [25]. No adverse side effect in animals or human have been noticed and/or reported, even when given at very high doses. They are selective and do discriminate between normal and tumor cells, affecting malignant cells, while sparing normal cells and tissues. When the fresh $CD34^+$ cells from bone marrow were treated with IP6, an inhibition of the clonogenic growth was observed with leukemic progenitor cells, but not with normal bone marrow progenitor cells under the same conditions [41]. While IP6 inhibited the colony formation of Kaposi Sarcoma cell lines, KS Y-1 (AIDS-related KS cell line) and KS SLK (Iatrogenic KS) and CCRF-CEM (human adult T lymphoma) cells in a dose-dependent manner, the ability of normal cells (peripheral blood mononuclear cells and T cell colony-forming cells) to form colonies in a semisolid methylcellulose medium was not affected [25]. We have demonstrated that IP6 acts synergistically with tamoxifen and doxorubicin, being particularly effective against estrogen receptor-negative and doxorubicin-resistant tumor cell lines, both challenges for treatment [25]. Additionally, both IP6 and Ins affect all principal pathways of malignancy, known as “hallmarks of cancer” [42].

And indeed, many anecdotal evidence, several clinical case reports and few small clinical studies, have demonstrated an enhanced antitumor activity with improved quality of life by IP6 + Ins. In a pilot clinical trial involving 22 patients with advanced colorectal cancer (Dukes C and D) with multiple liver and lung metastases, IP6 + Ins was given as an adjuvant to chemotherapy according to Mayo protocol. One patient with liver metastasis refused chemotherapy after the first treatment, and she was given only IP6 + Ins; her control ultrasound and abdominal computed tomography scan 14 months after surgery showed a significantly reduced growth rate [43]. A reduced tumor growth rate was noticed overall and in some case a regression of lesions was noted. Additionally, when IP6 + Ins was given in combination with chemotherapy, side effects of chemotherapy, such as drop in leukocyte and platelet counts, nausea, vomiting, alopecia, were diminished and patients were able to perform their daily activities [43, 44]. Long-term survival of a patient with advanced non- small cell lung cancer treated with IP6 + Ins treatment combined with chemo-radiotherapy was reported [45]. In a phase I clinical study, inositol was shown to be safe and well tolerated [46]. The combination of beta-(1,3)/(1,6) D-glucan and IP6 had beneficial effect on hematopoiesis in the treatment of patients with advanced malignancies receiving chemotherapy [47]. In a small prospective, randomized, pilot clinical study, IP6 in combination with Ins ameliorated the side effects of chemotherapy and preserved quality of life in breast cancer patients [48]. Topical IP6 treatment was effective and safe in preventing and/or mitigating chemotherapy-induced side effects as well as the preserving quality of life in women with ductal breast cancer in a double-blind, randomized controlled trial

(RTC) [49]. In a recent review article, a literature search was conducted to identify clinical evidence of the effectiveness of IP6 and Ins on quality of life in cancer patient and demonstrated that that IP6 and Ins are effective in improving quality of life of patients undergoing chemotherapy due to breast cancer [50]. And, most recently an amazing case report by Khurana et al. [51] on a patient with metastatic melanoma who declined traditional therapy and opted to try the IP6 + Ins supplement and who received a complete remission and remained in remission 3 years later. This opens a new avenue for IP6 clinical research, an immunotherapy and a potential for immune stimulating effects of IP6 and Ins in patients with metastatic melanoma.

Because currently available preclinical and encouraging initial clinical data suggest that IP6 and Ins are promising in cancer prevention and adjuvant therapy, more controlled clinical trials are expected.

Conundrum of IP6

The physiology and biochemistry of inositol phosphate is a fascinating field of science. The bioavailability, absorption, efficacy and determination have been issues and subjects of debates over decades.

Claims of adverse effects of IP6 on mineral bioavailability and a need to reduce or eliminate from our food grains have been recently again discussed [52, 53]. Although there were some evidences that phytate exacerbates mineral deficiency in developing countries, and we do not want to downplay the devastating impact of micronutrient malnutrition in developing world, we cannot ignore multiple health-promoting qualities of this amazing molecule. Moreover, when a vegetarian diet was compared with meat-based diets with equal phytate content, it has been concluded that zinc deficiency is unrelated to Zn content [53].

The analysis of inositol phosphates is also extremely complex. Prof. Grases and his group over the years have developed several methods for direct and indirect measurements of inositol phosphates in biological fluids, also distinguishing their intracellular and extracellular levels, what was the subject of Grases-Irvine debates [54–56].

And again, a novel method for determination of inositol phosphates in biological fluids reported by Wilson MC et al. [57] and questioning health-beneficial findings of IP6, opened yet again a gap of miscommunication among basic researchers, biochemists, and nutritionists, who are studying a complex family of inositol phosphates [58].

In the omnipotence of IP6 activities, another mystery and another target of its activity, is a recently reported link between IP6 and HIV virus. With a new microscopy technique that uses fluorescence to monitor capsid breakdowns in real-time, virologists identified IP6 as a key molecule exploited by HIV, when the virus infects human cells [59]. They think that the virus might hijack IP6 in host cells and use it to fortify its capsid and shield itself from our immune system. IP6, abundant cellular polyanion, can transform viral stability from minutes to hours, thus allowing newly synthesized DNA to accumulate inside the capsid. Although scientists have known for decades that IP6 molecule was capable for helping assemble viral components into virions, previous studies have indicated an anti-HIV activity of IP6 [60]. This opens a new avenue for IP6 research, as a new target for future antiviral treatment.

And to conclude with rice, and addressing HIV - Interestingly, researchers in Spain have developed a strain of genetically-engineered rice that could provide a cheap alternative to produce medicines for HIV prevention. They developed a transgenic rice line expressing three microbicidal proteins (the HIV-neutralizing antibody 2G12 and the lectins griffithsin and cyanovirin-N) as an approach for the durable deployment of anti-HIV agents in the developing world [61].

Overall Conclusion

Inositols and their recognized health-promoting activity are slowly transforming landscape of healthcare and might even bring us closer to the personalized solutions for our health problems. Both animal and human studies have demonstrated benefits following dietary supplementations with Ins and IP6 to restore their intracellular contents and, when needed, to step up from physiological to pharmacological levels. However, larger studies, double-blind and randomized clinical trials are needed.

References

1. Scherer, J. 1850. Ueber eine neue, aus dem Muskelfleische gewonnene Zuckerart. *Justus Liebigs Annulen der Chemie* 73: 322–328.
2. Hartig, T. 1855. Über das Klebermehl. *Botanische Zeitung* 13: 881–882.
3. ———. 1856. Weitere Mitteilungen, das Klebermehl (Aleuron) betreffend. *Botanische Zeitung* 14: 257–269.
4. Schlemmer, U., W. Frølich, R.M. Prieto, and F. Grases. 2009. Phytate in foods and significance for humans: Food sources, intake, processing, bioavailability, protective role and analysis. *Molecular Nutrition & Food Research* 53 (Suppl. 2): S330–S375.
5. Anderson, R.J. 1914. A contribution to the chemistry of phytin. *The Journal of Biological Chemistry* 17: 171–190.
6. Reddy, N.R., S.K. Sathe, and D.K. Salunke. 1982. Phytates in legumes and cereals. *Advances in Food Research* 28: 1–89.
7. Vucenik, I., and A.M. Shamsuddin. 2003. Cancer inhibition by inositol hexaphosphate (IP6) and inositol: From laboratory to clinic. *The Journal of Nutrition* 133 (11, Suppl 1): 3778S–3784S.
8. ———. 2006. Protection against cancer by dietary IP6 and inositol. *The Journal of Nutrition* 55 (2): 109–125.
9. Al-Soud, H., M. Ligor, I.-A. Rațiu, K. Rafińska, R. Górecki, and B. Buszewski. 2017. A window on cyclitols: Characterization and analytics of inositols. *Phytochemistry Letters* 20: 507–519.

10. Thomas, M.P., S.J. Mills, and B.V. Potter. 2016. The "other" inositols and their phosphates: Synthesis, biology, and medicine (with recent advances in *myo*-inositol chemistry). *Angewandte Chemie (International Ed. in English)* 55: 1614–1650.
11. Tanaka, K., A. Natsume, S. Ishikawa, S. Takenaka, and K.I. Yoshida. 2017. A new- generation of *Bacillus subtilis* cell factory for further elevated *scyllo*-inositol production. *Microbial Cell Factories* 16: 67.
12. Agranoff, B.W. 2009. Turtles all the way: Reflection on *myo*-inositol. *The Journal of Biological Chemistry* 284 (32): 21121–21126.
13. Irvine, R.F. 2005. Inositide evolution – Towards turtle domination? *The Journal of Physiology* 566 (Pt 2): 295–300.
14. Croze, M.L., and C.O. Soulange. 2013. Potential role and therapeutic interest of *myo*- inositol in metabolic diseases. *Biochimie* 95 (10): 1811–1827.
15. Unfer, V., F. Facchinetti, B. Orrù, B. Giordani, and J. Nestler. 2017. *myo*-inositol effects in women with PCOS: A meta-analysis of randomized controlled trials. *Endocrine Connections* 6 (8): 647–658.
16. Dinicola, S., M. Minini, V. Unfer, R. Verna, A. Cucina, and M. Bizzarri. 2017. Nutritional and acquired deficiency in inositol bioavailability. Correlations with metabolic disorders. *International Journal of Molecular Sciences* 18: 2187.
17. Santamaria, A., A. Alibrandi, A. Di Benedetto, B. Pintaudi, F. Corrado, F. Facchinetti, and R. D'Anna. 2018. Clinical and metabolic outcomes in pregnant women at risk for gestational diabetes mellitus supplemented with *myo*-inositol: A secondary analysis from 3 RCTs. *American Journal of Obstetrics and Gynecology* 219 (3): 300.
18. Owczarzyk-Saczonek, A., L.B. Lahuta, M. Ligor, W. Placek, R.J. Górecki, and B. Buszewski. 2018. The healing-promoting properties of selected cyclitols. *Nutrients* 10 (12): 1891.
19. Mukai, T., T. Kishi, Y. Matsuda, and N. Iwata. 2014. A meta-analysis of inositol for depression and anxiety disorders. *Human Psychopharmacology* 29 (1): 55–63.
20. Bizzarri, M., S. Dinicola, and A. Cucina. 2017. Modulation of both insulin resistance and cancer growth by inositol. *Current Pharmaceutical Design* 23 (34): 5200–5210.
21. Kim, J.N., S.N. Han, and H.N. Kim. 2014. Phytic acid and *myo*-inositol support adipocyte differentiation and improve insulin sensitivity in 3T3-L1 cells. *Nutrition Research* 34 (8): 723–731.
22. MacFarlane, P.M., and J.M. DiFiore. 2018. *Myo*-inositol effects on the developing respiratory neural control system. *Advances in Experimental Medicine and Biology* 1071: 159–166.
23. Anekonda, T.S., T.L. Wadsworth, R. Sabin, K. Frahler, C. Harris, B. Petrico, M. Ralle, R. Woltjer, and J.F. Quinn. 2011. Phytic acid as a potential treatment for Alzheimer's pathology: Evidence from animal and *in vitro* models. *Journal of Alzheimer's Disease* 23 (1): 21–35.
24. López-Gonzáles, A.A., F. Grases, N. Monroy, B. Marí, M.T. Vicente-Herrero, F. Tur, and J. Perelló. 2013. Protective effect of *myo*-inositol hexaphosphate (phytate) on bone mass loss in postmenopausal women. *European Journal of Nutrition* 52 (2): 717–716.
25. Vucenik, I., and J. Stains. 2010. Cancer preventive and therapeutic properties of IP6: Efficacy and mechanisms. *Periodicum Biologorum* 112 (4): 451–458.
26. Bizzarri, M., S. Dinicola, A. Bevilacqua, and A. Cucina. 2016. Broad spectrum anticancer activity of *myo*-inositol and inositol hexakisphosphate. *International Journal of Endocrinology* 2016: 5616807.
27. Bray, F., J. Ferlay, I. Soerjomataram, R.L. Siegel, L.A. Torre, and A. Jemal. 2018. Global cancer statistics 2018: GLOBOCAN estimates of incidence and mortality worldwide for 36 cancers in 185 countries. *CA: A Cancer Journal for Clinicians* 68 (6): 394–424.
28. Doll, R., and R. Peto. 1981. The causes of cancer: Quantitative estimates of avoidable risks of cancer in the United States today. *Journal of the National Cancer Institute* 66 (6): 1191–1308.
29. World Cancer Research Fund/American Institute for Cancer Research. 2018. *Diet, nutrition, physical activity and cancer: A global perspective*. Continuous Update Project Expert Report 2018

30. Graf, E., and J.W. Eaton. 1985. Dietary suppression of colonic cancer: Fiber or phytate? *Cancer* 56 (4): 717–718.
31. Shamsuddin, A.M., A.M. Elsayed, and A. Ullah. 1988. Suppression of large intestinal cancer in F344 rats by inositol hexaphosphate. *Carcinogenesis* 9 (4): 577–580.
32. Shamsuddin, A.M., and A. Ullah. 1989. Inositol hexaphosphate inhibits large intestinal cancer in F344 rats 5 months after induction by azoxymethane. *Carcinogenesis* 10 (3): 625–626.
33. Vucenik, I., G.-Y. Yang, and A.M. Shamsuddin. 1995. Inositol hexaphosphate and inositol inhibit DMBA induced rat mammary cancer. *Carcinogenesis* 16 (5): 1055–1058.
34. Vucenik, I., T. Kalebic, K. Tantivejkul, and A.M. Shamsuddin. 1998. Novel anticancer function of inositol hexaphosphate (IP6): Inhibition of human rhabdomyosarcoma *in vitro* and *in vivo*. *Anticancer Research* 18 (3A): 1377–1384.
35. Vucenik, I., Z.S. Zhang, and A.M. Shamsuddin. 1998. IP6 in treatment of liver cancer. II. Intratumoral injection of IP6 regresses pre-existing human liver cancer xenotransplanted in nude mice. *Anticancer Research* 18 (6A): 4091–4096.
36. Vucenik, I., V.J. Tomazic, D. Fabian, and A.M. Shamsuddin. 1992. Antitumor activity of phytic acid (inositol hexaphosphate) in murine transplanted and metastatic fibrosarcoma, a pilot study. *Cancer Letters* 65 (1): 9–13.
37. Williams, K.A., K. Kolappaswamy, L.J. DeTolla, and I. Vucenik. 2011. Protective effect of inositol hexaphosphate against UVB damage in HaCaT cells and skin carcinogenesis in SKH1 hairless mice. *Comparative Medicine* 61 (1): 39–44.
38. Fu, M., Y. Song, Z. Wen, X. Lu, and L. Cui. 2016. Inositol hexaphosphate and inositol inhibit colorectal cancer metastasis to the liver in BALB/C mice. *Nutrients* 8 (5): 286.
39. Vucenik, I., A. Passaniti, M.I. Vitolo, K. Tantivejkul, P. Eggleton, and A.M. Shamsuddin. 2004. Anti-angiogenic potential of inositol hexaphophate (IP6). *Carcinogenesis* 25 (11): 2115–2123.
40. Vucenik, I., K. Tantivejkul, G. Ramakrishna, L.M. Anderson, and D. Ramljak. 2005. Inositol hexaphosphate (IP6) blocks proliferation of breast cancer cells through PKCδ- dependent increase in $p27^{Kip1}$ and decrease in retinoblastoma protein (pRb) phosphorylation. *Breast Cancer Research and Treatment* 91 (1): 35–45.
41. Deliliers, L.G., G. Servida, N.S. Fracchiolla, C. Ricci, C. Borsotti, G. Colombo, and D. Soligo. 2002. Effects of inositol hexaphosphate (IP6) on human normal and leukaemic hematopoietic cells. *British Journal of Haematology* 117 (4): 577–587.
42. Hanahan, D., and R.A. Weinberg. 2011. Hallmarks of cancer: The next generation. *Cell* 144 (5): 646–674.
43. Druzijanic, N., J. Juricic, Z. Perko, and D. Kraljevic. 2002. IP-6 & inositol: Adjuvant to chemotherapy of colon cancer. A pilot clinical trial. *Revista de Oncología* 4 (Suppl 1): 171.
44. ———. 2004. IP6 + Inositol as adjuvant to chemotherapy of colon cancer: Our clinical experience. *Anticancer Research* 24 (5D): 3474.
45. Sakamoto, K. 2004. Long-term survival of a patient with advanced non-small cell lung cancer treated with Inositol Hexaphosphate (IP6) plus Inositol treatment combined with chemoradiotherapy. Report of case. *Anticancer Research* 24 (5D): 3618.
46. Lam, S., A. McWilliams, J. LeRiche, C. MacAulay, L. Wattenberg, and E. Szabo. 2006. A phase I study of *myo*-inositol for lung cancer chemoprevention. *Cancer Epidemiology, Biomarkers & Prevention* 16 (8): 1526–1531.
47. Weitberg, A.B. 2008. A phase I/II trial of beta-(1,3)/(1,6) D-glucan in the treatment of patients with advanced malignancies receiving chemotherapy. *Journal of Experimental & Clinical Cancer Research* 27 (1): 40.
48. Bačić, I., N. Družijanić, R. Karlo, I. Škifić, and S. Jagić. 2010. Efficacy of IP6 + Inositol in the treatment of breast cancer patients receiving chemotherapy: Prospective, randomized, pilot clinical study. *Journal of Experimental & Clinical Cancer Research* 29 (1): 12.
49. Proietti, S., V. Pasta, A. Cucina, C. Aragona, E. Palimbi, I. Vucenik, and M. Bizzarri. 2017. Inositol hexaphosphate (InsP6) as an effective topical treatment for patients receiving chemotherapy after surgery. *European Review for Medical and Pharmacological Sciences* 21 (Suppl 2): 43–50.

50. Verna, R., A. Giuliani, V. Todde, M. Minini, and V. Unfer. 2018. Reduced burden of chemotherapy side-effects in patients receiving inositol hexakisphosphate alone or in association with *myo*-inositol. *Cancer Studies and Therapeutics* 3 (1): 105.
51. Khurana, S., C. Baldeo, and R.W. Joseph. 2019. Inositol hexaphosphate plus inositol induced complete remission in stage IV melanoma: A case report. *Melanoma Research* 29 (3): 322–324.
52. Raboy, V. 2008. The ABCs of low-phytate crops. *Nature Biotechnology* 25 (8): 874–875.
53. Shamsuddin, A.M. 2008. Demonizing phytate. *Nature Biotechnology* 26 (5): 496–497.
54. Tur, F., E. Tur, I. Lentheric, P. Mendoza, M. Encabo, B. Isern, F. Grases, C. Maraschiello, and J. Perelló. 2013. Validation of an LC-MS bioanalytical method for quantification of phytate levels in rat, dog and human plasma. *Journal of Chromatography B* 928: 146–154.
55. Irvine, R.F. 2014. Absence of detectable inositol hexakisphosphate (IP6) in plasma. *Journal of Chromatography B* 960: 253–254.
56. Perelló, J., and F. Grases. 2014. Phytate levels in biological fluids of mammals. *Journal of Chromatography B* 960: 255–257.
57. Wilson, M.S.C., S.J. Bulley, F. Pisani, R.F. Irvine, and A. Saiardi. 2015. A novel method for the purification of inositol phosphates from biological samples reveals that no phytate is present in human plasma or urine. *Open Biology* 5 (3): 150014.
58. Vucenik, I. 2015. Conundrum of IP6. *Open Biology* 5 (11): 150048.
59. Mallery, D.L., C.L. Márquez, W.A. McEwan, C.F. Dickson, D.A. Jacques, M. Anandapadamanaban, K. Bichel, G.J. Towers, A. Saiardi, T. Böcking, and L.C. James. 2018. IP6 as an HIV pocket factor that prevents capsid collapse and promotes DNA synthesis. *eLife* 7: e35335.
60. Otake, T., H. Mori, M. Morimoto, K. Miyano, N. Ueba, I. Oishi, N. Kunita, and T. Kurimura. 1999. Anti-HIV activity of *myo*-inositol hexaphosphoric acid (IP6) and myo-inositol hexasulfate (IS6). *Anticancer Research* 19 (5A): 3723–3726.
61. Vamvaka, E., G. Farré, L.M. Molinos-Albert, A. Evans, A. Canela-Xandri, R.M. Twyman, J. Carrillo, R.A. Ordóñez, R.J. Shattock, B.R. O'Keefe, B. Clotet, J. Blanco, G.S. Khush, P. Christou, and T. Capell. 2018. Unexpected synergistic HIV neutralization by a triple microbicide produced in rice endosperm. *Proceedings of the National Academy of Sciences of the United States of America* 115 (33): E7854–E7862.

Integration of Phytochemicals and Phytotherapy into Cancer Precision Medicine

Nadire Özenver and Thomas Efferth

Abbreviations

ABC	ATP-binding cassette
As_2O_3	arsenic trioxide
CAR	combination activity relationship
EGF	epidermal growth factor
EGFR	epidermal growth factor receptor
FAEW	Free and Easy Wanderer
FEC	5-fluorouracil, epirubicin and cyclophosphamide
FDA	Food and Drug Administration
FGSTs	first-generation sequencing technologies
HIV	human immunodeficiency virus
HGF	hepatocyte growth factor
JAK	Janus kinase
MBSC	Modeling based scientific computing
MDR	multidrug resistance
NCI	National Cancer Institute
NGSTs	next-generation sequencing technologies
PI3K	phosphoinositide 3 kinase
TCM	Traditional Chinese Medicine
TCM-ID	Traditional Chinese Medicine Information Database
TCMID	Traditional Chinese Medicine Integrated Database

N. Özenver
Department of Pharmacognosy, Faculty of Pharmacy, Hacettepe University, Ankara, Turkey

T. Efferth (✉)
Department of Pharmaceutical Biology, Institute of Pharmacy and Biochemistry, Johannes Gutenberg University, Mainz, Germany
e-mail: efferth@uni-mainz.de

M. Bizzarri (ed.), *Approaching Complex Diseases*, Human Perspectives in Health Sciences and Technology 2, https://doi.org/10.1007/978-3-030-32857-3_16

TCMLARS	the traditional Chinese medical literature analysis and retrieval system
PharmDB-K	The integrated bio-pharmacological Traditional Korean Medicine (TKM) database
SAR	structure activity relationship
STAT3	signal transducer and activator of transcription 3
SV40	Simian vacuolating virus 40
T-ALL	T cell acute lymphoblastic leukemia
TGSTs	third-generation sequencing technologies
VEGF	vascular endothelial growth factor

Introduction

Cancer, the second foremost cause of death worldwide, encompassed 9.6 million deaths in 2018 [200]. Besides, new cases with cancer are considered to ascend to 23.6 million by 2030 [123]. Despite tremendous advances in cancer research, regrettably treatment failure constantly occurs nowadays, which is substantially due to the occurrence of multi-drug resistance (MDR) by multiple factors in cancer cells towards chemotherapeutics (Fig. 1) [41, 144]. Numerous conventional

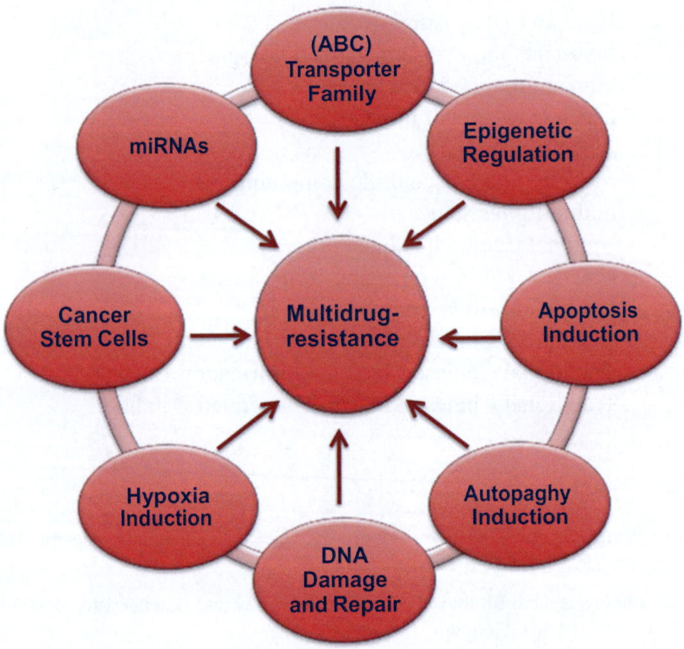

Fig. 1 Factors of drug resistance of tumors

chemotherapeutics kill not only malignant tumor cells, but also all proliferative normal cells in the body inducing only fair tumor specificity. Adequate dose utilization high enough to kill all tumor cells comprising more resistant tumor populations cannot be performed since the occurrence of harsh adverse side effects. Thus, only sub-optimal drug doses can be applied to cancer patients, and inherently resistant tumor populations remain unaffected decently causing treatment failure with fatal outcomes in patients.

Today's standards in drug development include preclinical and clinical cancer research comprising randomized, double-blind phase III studies for drug approval to the market. However, clinical trials in oncology were reported to have the highest failure among other fields in pharmacology, albeit many drugs with sub-optimal preclinical verification move into the clinical trials [11]. Moreover, the response towards drugs may vary from patient to patient, even if their tumors have the same tissue origin and histology. An important reason for the low success rate in cancer therapy is tumor heterogeneity, *i.e.* tumor subpopulation with different geno- and phenotypes may cause variable responses towards chemotherapeutics [188]. The functional heterogeneity arises from hierarchical classification of tumorigenic cancer stem cells and their non-tumorigenic progeny into subpopulations in the same tumor [116]. Even more, clonal evolution, heterogeneity in the microenvironment and alterations in cancer cell properties as other factors inducing heterogeneity act jointly as independent factors of tumor formation but independently acts with the hierarchical formation [18, 116, 132, 150]. Drug resistance is a substantial obstacle in cancer therapy, and the underlying molecular mechanisms are still incompletely understood as a precondition to better tackle the treatment failure. The detection of the cellular and molecular mechanisms inclining drug resistance may allow to predict individual drug responses, thus enabling the development of novel concepts for personalized medicine. Molecular targeted therapy attributing to individualized medicine has gained particular attention in cancer therapy over the past few decades so that the knowledge about genetic alterations in cancerous cells dramatically increased.

Nature is an attracting source with enormous chemical diversity in millions of species of plants, animals, marine organisms and microorganisms allowing the elucidation of bioactive compounds for drug development. Traditional medicines all over the globe acquired supplies based on medicinal plants, such as decoctions, infusions, ointments, press juices etc.. In modern pharmacology, phytochemicals isolated from medicinal plants serve as lead compounds for the generation of semi – synthetic derivatives or even synthetic drugs, which mimic principles of action of natural compounds. Hence, medicinal plants are indispensable sources both for plant-based complementary and alternative medicine and for chemistry- based conventional medicine. A premier goal is to use the best of both "worlds" and rise it to the so-called integrative medicine.

Innumerous effects against cancer cells have been described due to the wide range of diversity of natural substances [87]. For instance, vinblastine as a destabilizing agent inhibits microtubule polymerization, while paclitaxel as a stabilizing agent enhance microtubule polymerization, which means both suppress microtubule dynamics with detrimental consequences for cancer cells [38]. Furthermore, camptothecin and podophyllotoxin derivatives act as topoisomerase inhibitors [67, 102]. Several signaling pathways related to carcinogenesis are affected by drugs such as curcumin, which antagonizes epidermal growth factor receptor (EGFR) on the cell surface and incline apoptosis tumor cells by Fas receptor and caspase-8 activation. Another natural product-derived clinically approved drug is temsirolismus, which is a mTOR inhibitor [143]. Curcumin may be taken as a kind of prototype natural product, because it has been intensively studied during the past years and it is therefore well-known that this compound targets several different cellular signaling pathways [138]. Many- if not all- natural product act in a multi-specific manner. From an evolutionary point of view, this was quite apparently an enormous advantage for species in the combat to survive. The occasionally opinion that natural products are "dirty drugs" because they are not mono-specific confuses "multi-specific" with "non-specific". More mono-specific synthetic drugs are frequently subject to rapid resistance development. Therefore, the therapeutic potential of natural products that addresses several targets at the same time is much higher, because the development of resistance towards one target is engraved.

Today's approach in the development of efficient drugs against cancer is to focus on multiple targets namely network pharmacology, as cancer is a multifactorial disease. Rational combination therapy based on the application of two or more drugs concurrently is the most applied approach in cancer chemotherapy. Drugs with diverse mechanisms widen target range and improve therapeutic effectiveness and lessen possibility of drug resistance [1].

Over the last decade, the philosophy has shifted from the "one gene-one target-one disease" paradigm to the use of new molecular biological methods based on the determination of specific disease-causing genes. The unfortunate increase in failed clinical phase 2 and phase 3 trials in the previous few decades, particularly in the field of cancer may have favored the current paradigm change from the "one drug-one target" to a broader "one drug-many targets" perspective. Network pharmacology includes bioinformatics, systems biology and polypharmacology [72, 95].

The chemical constituents in plants, microorganisms and marine organisms hold great importance as potential drug candidates due to their interaction with disease-relevant targets. Despite the importance of medicinal plants among other natural sources throughout the history of manhood [176], only a small proportion of the more than 250,000 higher plants on earth has ever been searched for their bioactive compounds [3]. Herbal remedies are commonly applied in many countries as complementary and alternative medicine supplementing the treatment of many diseases with or without knowledge of patients' physicians [10]. The field of oncology is only one out of many examples in this context [2, 108, 176].

The advancements in modern technologies such as bioinformatics and systems biology permit their application in various biomedical fields, not only in research but in the foreseeable future also in clinical routine diagnostics. In the present chapter, we conceptualize how phytochemicals and phytotherapy could be integrated into modern Western medicine to combat drug resistance and to foster the development personalized medicine.

The Momentum of Natural Products in Cancer

Nature with great importance in the acquisition of potential drug candidates is an attractive source. In regard to the data reported by Food and Drug Administration (FDA), 40% of the approved molecules are originated from nature, and 74% are used against cancer from those molecules [165]. Furthermore, 49% of small molecules approved between 1940 and 2014 are natural products [128]. Natural compounds are indispensable not only as chemically established anticancer drugs, but also as lead compounds for the development of novel targeted chemotherapeutics with improved antitumor efficacy and fewer side effects.

Naturally occurring chemical substances are generated by living organisms for several reasons [137]. The primary metabolism products are commonly involved in metabolism to use nutrition for energy generation and utilization, while secondary metabolism products have specified other functions to maintain the survival capacity of organisms [106]. Secondary metabolites evolved during evolution of life on earth depending on the ecological habitat of organisms to protect them from herbivores, microbial pathogens and other challengers [33]. Their remarkable structural diversity led to the metaphor of natural products as natural version of combinatorial chemistry [183].

A plethora of anticancer drugs approved up to date are derived from plants, marine organisms or microorganisms forming prototypes in the development of new chemotherapeutics. To exemplify, vincristine and vinblastine were isolated from the medicinal plant *Catharanthus roseus* which grows in the rain forests of Madagascar [131]. Vincristine inhibiting microtubule assembly and inclining tubulin self-association into coiled spiral aggregates [131] is used to treat of Hodgkin's disease and some forms of leukemia [28]. Another example is etoposide and teniposide, two derivatives of podophyllotoxin derived from the North-American plant *Podophyllum peltatum* and Asian plant *Podophyllum emodi* [173]. Etoposide is used together with bleomycin (a natural antibiotic) and cisplatin as a combinational therapy against testicular cancer [198] and small-cell lung carcinoma [56, 68]. These epipodophyllotoxins inhibit DNA topoisomerase II and stabilizing enzyme- DNA cleavable complexes leading to DNA breaks and subsequently cell death [28, 101].

Antitumor antibiotics as microbe-derived agents represent a number of examples against cancer, among which the members of anthracycline, actinomycin, bleomycin and aureolic acid families are well established and notable chemotherapeutic agents [27]. Rapamycin, for instance, is a macrolide ester isolated from the bacterium *Streptomyces hygroscopicus* and exerts antiproliferative effects towards human tumor cells by targeting mTOR [15, 28]. Wortmannin, another example of microbe-derived agents, is a product of the fungus *Talaromyces wortmanni* and inhibits signal transduction pathways by inhibiting phosphoinositide 3 kinase (PI3K) [20, 28].

Marine natural agents demonstrate the thriving potential of the oceans, because, the deep sea bears a great diversity of bioactive compounds with pharmacological potency, which have not been discovered. For instance, microtubule-targeting agents from marine invertebrates (molluscs, sponges, bryozoans, tunicates, etc.) or marine microbes (algae, fungi and bacteria) such as cryptophycin, discodermolid, eleutherobin, eribulin, hemiasterlin, laulimalide, spongistatin, zampanolide have been developed up to date [32, 78, 153]. Even more, most of these compounds were reported to exert robust activity towards multi-drug resistant cancer cells [134, 153] emphasizing their importance as potential drug candidates. Cytarabine (Ara-C), trabectedin (ET-743), eribulin mesylate and brentuximab vedotin (SGN-35) with various mechanisms (*e.g.* cytarabine inhibits DNA topoisomerase, eribulin mesylate affects microtubule polymerization) represent other examples for the marine-derived drugs approved by FDA [172].

The concept of chemoprevention is to control cancer through by completely preventing, slowing down or reversing the occurrence of the disease by the administration of naturally occurring and/or synthetic agents [4]. Safety always holds remarkable importance in studies regarding people in drug discovery. An optimal chemopreventive agent should be nontoxic, efficient at lower doses, economical, and easily accessible. Recently, natural dietary agents as phytochemicals and phytotherapeutics have gained a great deal of attention from both scientists and the general public due to the chemopreventive capabilities of these agents. Numerous surveys including epidemiologic and animal studies reported that the uptake of food rich in fruits and vegetables lessens the chance of cancer growth [13, 151]. Diet-derived chemopreventive agents are inspiring the interest of clinicians, since patients are eager to use over-the counter diet-derived agents and a vast number of *in vitro* and *in vivo* studies over the last few decades have advocated the cancer-preventive capacity of various nutritional agents. Therefore, these compounds are still under investigation in clinical trials. The chemopreventive properties, sources, molecular targets of selected promising natural compounds along with information about their ongoing clinical trials are exemplified in Table 1.

Many of phytotherapeutics or phytochemicals exert their effects by addressing multiple targets. The key question is how best to use the knowledge for effective cancer prevention in populations with various cancer risks in the best way. Low potency and inadequate bioavailability may pose a number of difficulties, which, however, may be mastered by the improvement of semi-synthetic or synthetic analogs or by nanotechnology. For instance, synthetic curcumin analog EF24 was reported to exhibit nearly 10-fold better potency than natural curcumin [79]. Besides, these compounds or products jointly exert synergism either with the established chemopreventive agents or with other herbal compounds/products. Also, drug toxicity is also still a major obstacle for available chemotherapeutic and chemopreventive drugs at present. The use of phytotherapeutics or phytochemicals along with current chemotherapeutics may alleviate drug-associated toxicities. For instance, Sadzuka et al. [156] revealed that the flavonoids α G-Rutin and luteolin decreased doxorubicin-induced toxicity by oral administration in mice.

Phytochemicals and phytotherapeutics are obviously of importance with their potential against cancer. The transition of phytochemicals and phytotherapeutics into the category of possible drug candidates is now much more likely than before with the advances in numerous fields of molecular and cellular biology, network

Table 1 Source, mechanism of action, molecular targets of a number of promising natural compounds and knowledge about their ongoing clinical trials

Agent	Natural source	Mechanism of action	Molecular targets	Trial no	Trial type	Cancer type	Phase
Green tea polyphenols and EGCG	*Camellia sinensis* (green tea)	Antioxidant, antimutagenesis, antiproliferation (cell cycle arrest, apoptosis), antiinflammation, antiangiogenesis, immunomodulation	p53, p73, p21, Bax, EGFR, AKT, NF-κB, Bcl-2, cyclin D1, COX-2, VEGF, MMP2/9, STAT3, ERK1/2, AP-1, IL-12, $CD8^+$ T-cell	NCT03278925	Prevention	Cirrhosis	I
				NCT02891538	Prevention	Colorectal cancer	No phase specified
Betulinic acid	Widely distributed in plants such as *Betula* spec., *Ziziphus* spec., *Syzigium* spec., *Diospyros* spec. and *Paeonia* spec.	Antiinflammation, apoptosis, immunomodulation	PPAR-γ, p21, p38, JNK, topoisomerase I, NF-κB, COX-2, Bcl-2, cyclin D1/3, Sp1, 3 and 4	NCT00701987	Treatment	cutaneous metastatic melanoma	I
Curcumin	*Curcuma longa* (turmeric powder)	Antioxidant, antiproliferation (cell cycle arrest, apoptosis), antiinflammation, antiangiogenesis, immunomodulation	EGFR, IGF-1R, AKT, NF-κB, Bcl-2, COX-2, ERK, AP-1, Sp, VEGF, VEGFR1, MMP-2/9, p53, p21, Bax, STAT3/5	NCT01740323	Treatment	breast cancer	II
				NCT02944578	Treatment	human immunodeficie ncy virus (HIV) infected and uninfected women with high grade cervical squamous intraepithelial neoplasia	II
				NCT02064673	Treatment	prostate cancer	II
				NCT02782949	Prevention	Gastric cancer	II
n-3 polyunsaturated fatty acids	Corn oil, sunflower oil, safflower oil, olive oil, soybeans, walnuts, dark green	Antiinflammation, apoptosis, cell cycle arrest, lipid peroxidation	NF-κB, Bcl-2, STAT3, p53, Bax, p21, Fas/FasL,	NCT02996240	Treatment	breast cancer	No phase specified
				NCT02831582		breast cancer	

(continued)

Table 1 (continued)

Agent	Natural source	Mechanism of action	Molecular targets	Trial no	Trial type	Cancer type	Phase
	leafy vegetables, seeds and their oils		PPAR-γ, RXR, Ras, ERK 1/2		Supportive care		No phase
				NCT01821833	Supportive care	breast or ovarian cancer	No phase specified
				NCT02295059	Prevention	breast cancer	II
				NCT02176902	Prevention	prostate cancer	II
				NCT03290417	Treatment	prostate cancer	No phase specified
Genistein	Soybeans and soy products, *Trifolium pratense*, *Pistacia vera*	Antioxidant, antiproliferation (growth inhibition, cell cycle arrest, apoptosis), antiangiogenesis, antiinflammation	AKT, NF-κB, Bcl-2, survivin, cyclin D1, COX-2, MMP-2/9, p53, p21, GADD153, Bax, STAT3/5, ERK 1/2, CDK1, AP-1, IGF-1R	NCT01489813	Treatment	bladder cancer	phase II
				NCT02766478	Supportive care	prostate cancer	phase II
				NCT02624388	Supportive care	pediatric patients with solid tumors or lymphoma	No phase specified
				NCT02567799	Prevention	non-small cell lung cancer	phase I/II

Abbreviations: *EGCG* epigallocatechin-3-gallate, *EGFR* epidermal growth factor receptor, *NF-κB* nuclear factor-κB, *COX-2* cyclo-oxygenase-2, *VEGF* vascular endothelial growth factor, *MMP-2/9* matrix metalloproteinases, *IL-12* interleukin 12, *IGF-1R* insulin-like growth factor-1 receptor, *VEGFR1* vascular endothelial growth factor receptor 1, *TNF-α* tumor necrosis factor α, *JNK* Jun-N-terminal kinase, *CDK* cyclin-dependent kinase, *ERK* extracellular signal-regulated kinase, *SOD* superoxide dismutase, *mTOR* mammalian target of rapamycin, *iNOS* inducible nitric oxide synthase, *PPAR-γ* peroxisomeproliferator-activated receptor γ, *NO* nitric oxide, *eNOS* endothelial nitric oxide synthase

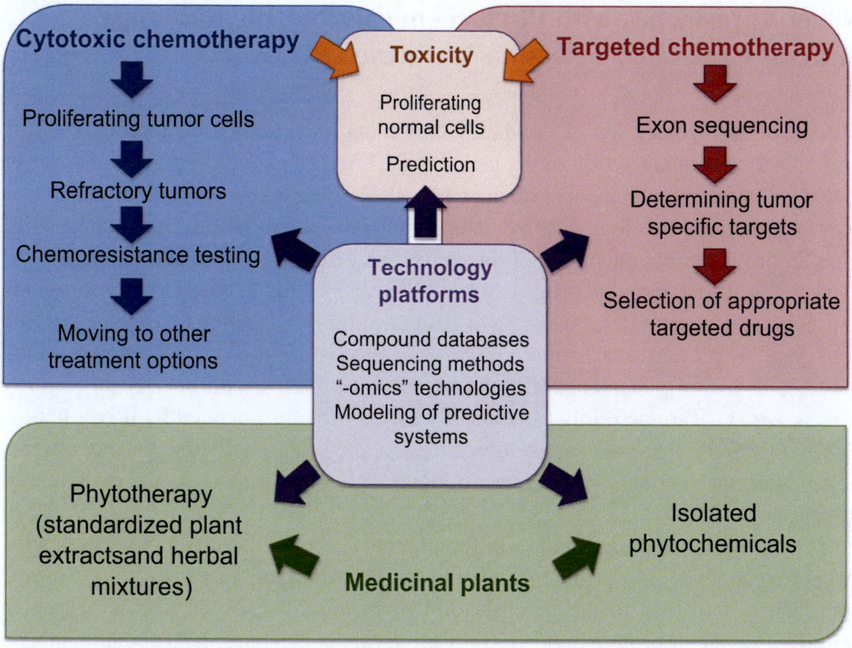

Fig. 2 Integration of phytochemicals and phytotherapy into individualized therapy of cancer. (Figure taken with modifications from Ref. [48])

pharmacology, bioinformatics, computational methods etc. (Fig. 2). However, more investigations are required to confirm if these agents exhibit their activities alone or in combination with established therapies.

Natural products exert their effects by interacting with multiple targets. It is not beyond expectations that natural products can be screened that target aberrantly mutated targets of individual cancer patients in the course of precision medicine approaches. Up to date, two main categories of targeted drugs comprising small molecule drugs and monoclonal antibodies are realized. Small molecule drugs have the ability to enter inside of the cells to attack intracellular targets due to being sufficiently small, while monoclonal antibodies interact with the targets at the outer surface of cancerous cells. Cetuximab and panitumumab against EGFR, bevacizumab against vascular endothelial growth factor (VEGF) and rituximab against CD20 are examples of monoclonal antibodies. Likewise, imatinib mesylate inhibiting the oncogenic BCR/ABL fusion protein, both erlotinib and gefitinib against EGFR, vemurafenib against BRAF and bortezomib against the proteasome are set as examples for small molecule drugs [48].

Novel Approaches with Phytochemicals and Phytotherapy to Fight Cancer by Precision Medicine

The issue for enforcement of standard chemotherapy is that the distinction between normal and malignant proliferating cells are not specific enough. Targeted therapy is a way to treat people in accordance with their specific genes and proteins and prevents cancer development by interacting individual targets associated with growth, progression and spread of the cancerous cells. In comparison with standard therapy discipline, targeted drugs can be individually utilized based on the particular expression of aberrant targets in each patient. Amplified genes involved in tumor cells lead to protein overexpression in cancer cells. Chromosomal translocations may create fusion genes encoding unusual novel proteins with oncogenic features (*driver genes*). Targeted drugs specifically attack proteins encoded by driver genes. Furthermore, the genomic instability of tumors may incline diverse genetic aberrations (passenger genes), which does not have a principal impact on the malignancy of cancer.

Targeted drugs are expected to kill malignant cells by interacting with particular targets in cancer cells and, hence to exhibit fewer side effects in normal cells. Several approaches may lead to the identification of potential treatment targets in cancer cells. (1) A gene or protein highly expressed in cancer cells may be a potential target, particularly genes and proteins associated with cell division and growth. (2) Abnormalities in chromosomes or alternatively spliced proteins of cancer cells may indicate possible targets.

Targeted therapy has been a prosperous area in cancer research. The number of clinically approved drugs has been increasing following the progress in bioinformatics and systems biology over the past years. Targeted drugs, however, have drawbacks:

1. Tumors usually gain resistance towards targeted drugs. Modifications in the targets due to mutations, single nucleotide polymorphisms, cell cycle arrest, antigen shedding or utilization of alternative signaling pathways may induce treatment failure [43, 47, 48].
2. Targeted drugs provoke side effects in normal cells due to unwanted off-target effects, despite the fact that the affected targets are not present in those cells. Hepatotoxicity, dermatoxicity (skin rash, hair depigmentation, nail changes), hypertension are common examples for non-specific and off-target effects [48]. Nature supplies an incontrovertible source in the development of targeted drugs [44, 48, 84, 135, 159].

1. An increasing number of targets have been discovered up to date. Natural products enabling the generation of semi-synthetic or synthetic derivatives have profound potential to be designed as targeted drugs with improved efficacy.
2. Natural products have the capacity to defeat drug resistance. ATP-binding cassette (ABC) transporter family is a major cause of development of drug resistance and responsible for drug efflux [29, 70]. In human beings, 48 members of ABC

transporter family, 12 of which have been identified as possible drug candidates have been reported to date [54, 88, 203]. Numerous anticancer drugs from various drug classes (*e.g.* taxanes, EGFR inhibitors and topoisomerase inhibitors) are pumped out of the cells by ABC transporters and frequently display cross resistance, even if those are structurally and functionally unrelated [54, 157, 163, 175]. For instance, sesamin demonstrated remarkable cytotoxicity *in vivo*. A study determining molecular determinants of which indicated that sesamin is not involved in MDR-mediated by ABCB1 or ABCB5. Tumor profiling, furthermore, pointed out that responsiveness of tumor cells to sesamin was substantially associated with genes which are not involved in classical resistance mechanisms, thus leading to chemoresistance [157].

3. The rigorous side effects inclined by standard (non-targeted) and targeted chemotherapeutics may be diminished by the use of natural products [48, 89, 164].

In our previous study, we investigated the potential of natural products in targeted therapy. For this purpose, signal transducer and activator of transcription 3 (STAT3) encoded by STAT3 gene was selected. The STAT3 family of transcription factors aberrantly expressed in cancer take part in a number of physiological processes. STAT3 activation is initiated via phosphorylation by several cytokines such as IL-6, CD40 as well as growth factors including epidermal growth factor (EGF) family members, hepatocyte growth factor (HGF) [171, 179, 208]. Interaction of STAT3 with those ligands leads dimerization of a signal transducer protein, gp130 in the cytoplasm [19, 191] followed by trigger of Janus kinase (JAK) phosphorylation and sTAT3 phosphorylation. JAK family is a member of tyrosine kinases among which, JAK1, particularly, is assigned in the activation of STAT3 [69]. Phosphorylated STAT3 monomers form dimers and move into the nucleus to inspire transcription of genes associated with cell survival and proliferation [24, 31, 171, 195]. STAT3 activation is also mediated by mitogen-activated protein kinases (MAPK) and c-SRC non-receptor tyrosine kinase [48]. Moreover, non-phosphorylated STAT3 may also provoke dimerization and transcription [205]. STAT3 has been not only reported as substantially phosphorylated or overexpressed in tumor cells inducing carcinogenesis [60, 152]; but also performing as a tumor suppressor in case of occurrence of mutations [211].

The interruption of the STAT3 signaling pathway by small molecules includes various approaches:

1. Inhibitors targeting the upstream acting of STAT3: Tyrosine-kinase inhibitors of cell surface receptors such as EGFR, HER2, PDGFR, IGFR inhibit downstream signaling including the STAT3 pathway as well as JAK 1/2 upstream of STAT3.
2. STAT3 inhibitors interrupting dimerization at the SH2 domain. Since, most STAT3 inhibitors bind to the SH2 domain [142].
3. Inhibitors of the STAT3 DNA binding domain. The steric hindrance by small molecules to bind to DNA inhibits the transcription factor activity of STAT3.

Drugs usually display their effects by affecting multiple targets. Natural products interact with various targets rather than only one target determining responsiveness to drug. Therefore, the discovery of cellular and molecular mechanisms of signaling pathways affected by drugs is crucial.

Traditional Chinese Medicine (TCM) based on the principal of TCM theory with a holistic, systematic, and individualistic attitude has become more of an issue lately enabling integration of such principal with the concept of precision medicine [80, 189, 213]. Chinese herbal medicine with clinically beneficial outcomes has provided acquaintance of thousands of years' theory and practice since the time of Shennong's Materia Medica. Medical materials specify according to their natures, flavors or meridians to be dispensed for each specific patient [94]. By means of chemical, analytical and pharmacological methods of modern science, the mechanism of action of Chinese herbal medicine can be investigated [97, 107, 214], which may facilitate the integration of phytochemicals and phytotherapy into Western medicine.

Compound Databases

A plethora of data have been generated by high-throughput technologies and their interpretation represents a major challenge to extract novel and meaningful information [92, 216].

Text mining of published literature reached notable importance in the past years [216]. Data mining uses the knowledge from various sources such as bibliographic literature, experimental data, and clinical data. The progress in computational technologies offers opportunities for the evaluation of complex and elaborate analyses. Traditional medicine came more and more into the focus during the past decades. Thus, the construction of databases in the field of traditional medicine together with computational technologies came the forefront. Various literature-based databases were established such as the traditional Chinese medical literature analysis and retrieval system (TCMLARS) [51], and the database of medicinal materials and chemical compounds in northeast Asian traditional medicine [177]. A list of traditional medicine databases is shown in Table 2, among which the Traditional Chinese Medicine Information Database (TCM-ID) acts as a source describing instructive reference materials from different channels of TCM comprising formulation, herbal composition, chemical composition, molecular structure and functional properties, therapeutic and toxicity effects, clinical indication and application and diseases [23]. Traditional Chinese Medicine Integrated Database (TCMID) is another example, which aims to modernize and standardize TCM collecting data such as prescription ingredients and mass spectrometry spectra and includes about 47,000 prescriptions, 8200 herbs, 43,500 ingredients and the relevant knowledge of 4700 diseases [73]. The integrated bio-pharmacological Traditional Korean Medicine (TKM) database (PharmDB-K) was established by the unification of 14 different databases, six Pharmacopoeias, literature, formerly-established resources of massive amounts of bio-pharmacological network data and experimentally validated results predicted from the PharmDB-K analyses. The database provides detailed knowledge

Table 2 Traditional medicine databases

Database	Description	URL
TCM-ID Traditional Chinese medicine information database	A database source to identify informative reference material on all aspects of TCM including formulation, herbal composition, chemical composition, molecular structure and functional properties, therapeutic and toxicity effects, clinical indication and application, and related literatures	http://bidd.nus.edu.sg/group/TCMsite/Default.aspx
TCM Database@Taiwan	One of the most comprehensive and largest non-commercial database available for download comprising constituents from different herbs, animal products and minerals	http://tcm.cmu.edu.tw/
TCMID	A database including knowledge of prescriptions, herbs, ingredients, drugs, related targets, and diseases	http://www.megabionet.org/tcmid/
TCMGeneDit	A database system providing association information about traditional Chinese medicines (TCMs), genes, diseases, TCM effects and TCM ingredients automatically mined from vast amount of biomedical literature	http://tcm.lifescience.ntu.edu.tw/
TM-MC Database of medicinal materials and chemical compounds in northeast Asian traditional medicine	A database providing information on the constituent compounds of medicinal materials in Northeast Asia traditional medicine	http://informatics.kiom.re.kr/compound/
CEMTDD Chinese ethnic minority traditional drugs database	A database providing information about herbs, compounds, targets, and diseases	http://www.cemtdd.com/
PharmDB-K Traditional Korean Medicine (TKM) database for researches in drug discovery	A database offering comprehensive TKM-associated compound, drug, disease indication, and protein relationship information	http://www.pharmdb-k.org/
Agricola	A database useful for finding information from the National Agricultural Library on herbs and medicinal plants	https://agricola.nal.usda.gov/
HerbMed	An evidence-based resource providing scientific data underlying the use of herbs for health indicating contraindications, toxicity and adverse effects	http://www.herbmed.org/

(continued)

Table 2 (continued)

Database	Description	URL
CAM on Pubmed	A database which automatically limits the literature search according to the complementary and alternative medicine (CAM) subset of PubMed	https://nccih.nih.gov/research/camonpubmed
PubMed Dietary Supplement Subset	A collaboration between the NIH Office of Dietary Supplements and the National Library of Medicine (NLM), is essentially a "filter" in the PubMed database that allows users to limit the 20 million+ citations in PubMed to those that are dietary supplement-related (just over 400,000 at this time)	https://ods.od.nih.gov/Research/PubMed_Dietary_Supple ment_Subset.aspx
RXList Alternatives	A database containing categories of herbal medicines comprising Western herbs, Chinese herbal remedies and homeopathics, offered both through FAQ's and complete monographs	https://www.rxlist.com/supplements/article.htm

on 262 TKMs, 7815 drugs, 3721 diseases, 32,373 proteins, and 1887 side effects [93]. The steadily increasing number of databases will further facilitate the process of analyzing, multi-dimensional data to gain new knowledge on traditional medicine in the light of modern technologies.

Network Analysis

Network analysis in traditional medicine characterizes the connection among herbal prescriptions, herbal products, and compounds on the one hand with targets, molecular pharmacology and diseases on the other hand. Nevertheless, it is beyond the expectations to precisely unravel the underlying mechanisms of herbal products or formulae due to the complexity of their chemical consumptions. In this context, computational methodologies are indispensable to analyze the vast amount of data from "-omics" –technologies to extract useful information for the drug discovery process.

Network pharmacology detects the interaction of bioactive constituents with their cellular targets associated with relevant diseases and examines the actions related to these interactions. Thus, drug-target-disease networks are constructed. Network pharmacology represents a "multi-component, multi-target" strategy rather than a "single-component, single-target" approach. Therefore, network pharmacology especially raised interest in the field of medicinal plants and phytotherapy as chance to identify the complex mechanisms of phytotherapeutics and herbal formulae. Understanding the pharmacological activity of traditional medicine at the cellular and molecular level is quite difficult due to the high number of chemical constituents in a prescription. While each of these ingredients exerts its specific actions their combination in a herbal mixture may differ from the sum of each single compound's action. The hope is that synergistic interactions between different medicinal plants and/or their phytoconstituents may be unraveled by the application of methods from network pharmacology.

Network-based computational investigations substantially focus on the discovery of the efficient compounds and the underlying mechanisms of actions of herbal products used against specific diseases. To date, investigations integrating network pharmacology and herbal medicine have been carried out by a comparatively limited number of groups [48, 95, 98, 146] but the idea justifying this concept gains more and more popularity [181]. For example, Li and Zhang [95] built a "network target" theory, which comprises three networks, *i.e.* the herb network (herbal formula), the biological network (network target), and the phenotype network (disease). This approach represents a novel prototype in drug discovery.

The comprehension of network pharmacology is extraordinary broad. Some researchers virtually concentrate on drug-target systems [92, 104], whereas others focus on herb networks in the prescription [92, 98] or the target signaling pathway network [92, 193], which may lead to the development of disease-pattern-target network [62, 92].

A vast number of investigations were carried out comprising the integration of phytochemicals and phytotherapy into network pharmacology. Some representative investigations from our and other groups are exemplified below.

Anfosso et al. [5] correlated the mRNA expression data of 89 angiogenesis-related genes obtained by microarray hybridization from a panel of 60 tumor cell lines of the National Cancer Institute (NCI, USA) with the 50% growth inhibition concentration values for eight artemisinin derivatives. Hierarchical cluster analysis and cluster image mapping expression determined the genes involved in cell response. Artemisinin derivatives were shown to exhibited their antitumor effects by inhibiting tumor angiogenesis.

Efferth et al. [45] focused on pharmacogenomics of Kampo-derived natural products against cancer with special emphasis to shikonin as the most cytotoxic compound among all phytochemicals investigated. Microarray analyses led to the determination of genes associated with cellular response to shikonin.

Efferth et al. [46] performed a systematic bioactivity-based screening of a number of traditionally used Chinese medicinal plants. Bioactivity-guided fractionation by chromatographic techniques led to the isolation of several phytochemicals with cytotoxic activity against cancer cells, including 25-O-acetyl-23,24-dihydro-cucurbitacin F from *Quisqualis indica* and miltirone from *Salvia miltiorrhiza*. By using microarray hybridization, the genes determining sensitivity or resistance of cell lines to miltirone were identified.

Wong et al. [199] examined the effect of *Rabdosia rubescens* extract and the kaurene diterpene oridonin, the most active ingredient of this extract, on prostate cancer cells by gene expression analysis and xenograft tumor studies. The authors emphasized that the synergy of the whole plant extract compared to the isolated oridonin was reflected by the results of the gene expression analyses.

Youns et al. [207] investigated the cytotoxicity of the chalcone flavonoide, isoliquiritigenin with that of the standard anticancer drugs doxorubicin and methotrexate in five T-cell acute lymphoblastic leukemia cell lines. Array-based matrix comparative genomic hybridization and microarray-based mRNA expression profiling were further performed to have a perception about molecular mechanisms. The study pointed out the presence of different molecular mechanism of isoliquiritigenin as compared to doxorubicin and methotrexate, indicating that isoliquiritigenin may be beneficial to treat leukemia exerting resistance to these standard chemotherapy drugs.

Kuete and Efferth [86] compiled a library of cytotoxic compounds Cameroonian medicinal plants. Cellular and pharmacogenomic profiling as well as bioinformatical analyses led to a set of 27 cytotoxic compound. Two of the most cytotoxic compounds, plumericin from *Plumeria rubra* and plumbagin from *Diospyros crassiflora* and *Diospyros canalicula*, were selected for detailed investigation on the molecular mechanism. The study bridged the between Sub-Saharan African medicinal plants and pharmacogenomy for the first time.

Wen et al. [196] conducted a survey to identify modes of action of *Si-Wu-Tang,* which is a traditional Chinese medicinal formula. *Si-Wu-Tang* induced gene expression changes as determined by microarray, bioinformatics and connectivity map. It turned out that this herbal formula represented an Nrf2 activator and exerted phytoestrogenic activity.

Efferth and Greten [42] investigated the withanolides, the major secondary metabolites of *Withania somnifera*, to identify the molecular determinants of sensitivity and resistance of tumor cells towards these secondary plant metabolites. Transcriptomic and bioinformatical analyses allowed the definition of mRNA expression profiles, which were significantly associated with the response of tumor cells to withanolides.

Munakata et al. [122] investigated the affect of the traditional Japanese medicine juzentaihoto (JTX) on the gene expression profile in the large and small intestines was investigated by microarray analyses using mice of various strains with or without enteric microflora. Microarray analysis uncovered that the target of JTX may be the transcription machinery.

Kadioglu et al. [76] analyzed the cytotoxicity of the *Cantharis* ingredient, cantharidin, in 41 tumor cell lines (Oncotest panel) and compared the results with those of 60 cell lines of the NCI panel of tumor cell lines. Microarray-based transcriptome-wide mRNA expression profilingss, hierarchical cluster and *in silico* docking analyses in addition to biological experiments indicated that cantharidin may be a promising candidate for cancer therapy.

Zheng et al. [215] studied the anti-tumor effects and underlying mechanisms of *Juzen-taiho-to*, a Chinese medicine containing 10 herbs, in a murine model by using cDNA microarray analysis. *Juzen-taiho-to* extended the survival time of Simian vacuolating virus 40 (SV40) T antigen in α-crystallin/SV40 T antigen transgenic

(TG) mice by enhancing their nutritional condition, inhibiting the MAPK pathway and reinforcing the immune system without causing hepatic toxicity.

Loganathan et al. [105] assessed the effects of a well-characterized extract from the medicinal mushroom *Ganoderma lucidum* on tumor growth and breast-to-lung cancer metastasis. Gene expression in MDA-MB-231 cells was determined by DNA microarray analysis, indicating that *Ganoderma lucidum* inhibited breast-to-lung cancer metastases by the downregulation of genes associated with cell invasiveness.

Li et al. [99] examined genome-wide aristolochic acid-induced dysregulation as well as the regulation of microRNAs (miRNAs) on their target gene expression in rat kidney, because miRNAs play a part in cancer process and their role remained still unclear in aristolochic acid-induced carcinogenesis. Rats were treated with aristolochic acids to test miRNA and mRNA expression by deep sequencing, and protein expression by proteomics. Dysregulated miRNA expression was of importance in carcinogenesis in rat kidney.

The mechanism associated with glucose metabolism of a diterpene quinone tanshinone IIA obtained from *Salvia miltiorrhiza* was studied in gastric cancer cells. RNA-seq transcriptomics and quantitative proteomics-isobaric tags revealed that the glucose metabolism was inhibited by tanshinone IIA in AGS gastric cancer cells prompting cell stresses, nutrient deficiency and DNA damage [100].

Wu et al. [204] studied molecular mechanisms of rosmarinic acid also isolated from *Salvia miltiorrhiza* in acute lymphoblastic leukemia cells. Microarray analysis and other experimental studies pointed to rosmarinic acid-induced apoptosis and necrosis caused by ROS generation and DNA damage.

Saeed et al. [158] studied the molecular mechanisms of the dietary flavonoid apigenin on drug-resistant cancerous cell lines. *In silico* molecular docking, microarray and bioinformatical analyses were performed. The study indicated apigenin's cytotoxicity is not hampered by classical mechanisms of multidrug resistance pointing to apigenin's capability to affect refractory tumors.

Kadioglu and Efferth [75] searched modes of action of ursolic acid and pomolic acid, two constituents of *Salvia officinalis*, in drug-resistant cancer cells. Gene expression profiles were determined by microarray-based mRNA expressions, and bioinformatical analyses. Molecular docking revealed the interaction of those compounds to key molecules involved in the NF-κB pathway. The study revealed that ursolic and pomolic acid inhibit NF-κB-mediated functions by targeting different steps of the NF-κB pathway.

The coumarin compound scopoletin found in several plant genera including *Artemisia* species was studied by Seo et al. [168]. Microarray-based RNA expression profiling of a panel of tumor cell lines showed that cellular response of scopoletin was not related to the expression of ATP-binding cassette transporters. Transcriptome-wide mRNA expressions pointed to a set of 40 genes, which all displayed binding motifs in their promoter sequences for the NF-κB transcription factor. Hence, NF-κB activation may be assumed as resistance factor for this compound.

Cheng et al. [25] investigated the action of a standardized extract of *Huangqi Guizhi Wuwu* decoction against oxaliplatin-induced peripheral neuropathy.

Microarray analysis indicated that the neuroprotective effect of this extract was associated with the modulation of multiple molecular targets and pathways involved in the downregulation of inflammation and immune response.

Ooko et al. [136] investigated the combination treatment of ascorbic acid and curcumin, two secondary metabolites of *Curcuma longa*, towards human cancer cells. Microarray-based mRNA expression profiles displayed genes assuming cellular responsiveness to curcumin and AA.

Hong et al. [71] focused on the potential effect of Free and Easy Wanderer (FAEW), a mixture of several herbs clinically used in China for hundreds of years against psychiatric disorder. Transcriptome-wide microarray analysis indicated NRF2/HO-1 as the common target of FAEW and fluoxetine as syntehtic control drug. FAEW exerted its activity by antagonizing H2O2-induced oxidative stress through KEAP1-NRF2/HO-1 pathway.

Saeed et al. [160] studied the cytotoxicity of the pentacyclic triterpene betulinic acid towards drug-resistant tumor cell lines. Microarray data were used to identify possible mechanisms underlying betulinic acid's cytotoxicity towards multidrug-resistant tumor cells, which were supplemented by *in silico* analyses.

Kadioglu et al. [77] examined the cytotoxic diterpenoid oridonin isolated from *Rabdosia rubescens* towards a panel of drug-resistant cancer cells. Pharmacogenomic and computational analyses were used to explain the response of cancer cells to oridonin.

Dawood et al. [30] searched for cellular and molecular mechanisms accounting for the cytotoxicity of arsenic trioxide (As2O3), which is used in Chinese medicine. Microarray and bioinformatical analyses identified the genes determining cellular responsiveness to As2O3. Hierarchical cluster analysis-based heat mapping pointed to remarkable differences between As2O3-sensitive and -resistant tumor cells. The approach of network pharmacology was quite beneficial to present the multifactorial modes of action of As2O3.

Özenver et al. [139] investigated the cytotoxicity of aloe-emodin, an anthraquinone derivative mostly present in the families such as Fabaceae (*Cassia*), Liliaceae (*Aloe*), Polygonaceae (*Rumex*), Rhamnaceae (*Rhamnus*) and Rubiaceae (*Asperula, Gallium, Rubia*) [166]. Cellular and molecular factors determining the cytotoxicity and acquired resistance were revealed by network analysis combining *in vitro* and *in silico* methods. Microarray hybridization revealed a profile of deregulated genes associated with diverse functions, which was further experimentally validated demonstrating S phase arrest, ROS generation, DNA damage and apoptosis regarding Aloe-emodin's cytotoxicity in CCRF-CEM leukemia cells.

Saeed et al. [161] investigated the cytotoxicity of the natural benzophenanthridine alkaloid sanguinarine towards multi-drug resistant cancer cells. The transcriptome-wide expression profiles of a tumor cell line panel defined genes involved in different cellular processes, which together with experimental studies demonstrated therapeutic potential of sanguinarine.

A survey carried out by Seo et al. [169] evaluated the effects of a number of adaptogenic herbal extracts on fixed combination 5-fluorouracil, epirubicin and cyclophosphamide (FEC) induced alterations in transcriptome-wide RNA microarray profiles of neuroglia cells to foresee possible effects of such extracts on cellular and physiological, and mostly cognitive functions. The study revealed that the combination of cytostatic drugs with apoptogenic plant extracts resulted in remarkably alterations in gene profiles of neuroglial cells concerned with soft cognitive impairments in cancer chemotherapy.

A follow-up study by Seo et al. [170] demonstrated the potential effects of the previously established adaptogens on FEC-induced changes in transcriptome-wide microarray profiles of T98G neuroglia cells as well as on FEC-induced hepato-, cardio- and nephrotoxicity. The application of cytostatic drugs in combination with adaptogenic plant extracts induced notable changes in transcriptome-wide microarray profiles of neuroglial cells emphasizing the potential of adaptogens reducing adverse effects in cancer chemotherapy.

In summary, network analysis based on microarray hybridization and RNA-sequencing pave the way for understanding the molecular mechanisms of herbal prescriptions or isolated compounds. Thus, network pharmacology combining various modern techniques and methodologies is a key technology for the efficacious prediction of "drug-target-disease" features of herbal prescriptions and phytochemicals from traditional medicine.

Computational Approaches

Computational methods serve as tools proposing possible targets and signaling pathways for chemicals from herbs. One approach is that the chemical scaffold of a phytochemical is used to screen for compound with similar chemical structure under the assumption that compound with high similarity probably exert similar activities [197]. Moreover, reverse docking is a method to determine possible drug targets which is also applicable to natural products [96, 212]. Molecular dynamics is another technique imitating the interaction of a small molecule with a macromolecular target and calculating the predicted binding kinetics at atomic level [39, 63]. Priya et al. [147] combining such methods including not only molecular docking and molecular dynamics but also molecular property prediction and drug-likeness score to develop a new drug against human immunodeficiency virus (HIV) targeting HIV-1-reverse transcriptase. Meanwhile, systems biology integrates various areas including engineering, computational, physics with biological and medical fields to predict the attitude of a biological network upon treatment with a phytotherapeutic to assume possible efficacy or side effects at the network level [52, 55, 61]. If the topology and associated parameters are present, a panel of ordinary differential equations may portray the dynamic property of network interacting with herbal medicine [52]. Nevertheless, this method would fail in case of insufficient knowledge about biological pathways and information shortage of topology or parameter. In this case, Boolean network modeling could provide an alternative [192].

Contrary to the conventional "one target" approach, systems biologyfocuses on extensive interaction maps of biological systems affected by external stimuli. These maps encompass data from genomics, transcriptomics, proteomics as well as metabolomics enabling the individual evaluation of a tumor. Thus, the behavior of cancer cells and their potential response to specific treatments may be determined, which is quite important for development of targeted therapy [91, 119].

In brief, these methods are applied to examine complex biological systems and their combination allow the concurrent simulation of possible drug candidates.

Graph Theory

Trudeu postulated that a graph is an abstract statement of a set of elements and the connections between them [178]. Graph theory is a mathematical discipline corroborating the study of composite networks in biology and other research fields [113]. In biomedicine, the nature of intercellular signaling pathways can be ascertained by graph theory, intercellular networks can be evaluated in the presence or absence of stimulation or inhibition by external signals (such as chemical molecules). Furthermore, information can be gained regarding stability of intercellular networks, alterations in their connectivity in the course of growth and modifications of cell-to-cell connectivity due to a disease [14, 57, 58, 109, 112].

Statistical Methods

Statistical algorithms are indispensable to unravel patterns of biological principles from large data sets. A plethora of data are available from multiple sources such as digital platforms and large scale experiments. This represents a treasure box that awaits to be unearthed [149].

Data Mining and Functional Data Analysis

Data mining and functional data analysis are quite substantial to obtain robust relationships among data elements in large and partwise noisy and messy data sets in order to model biologically meaningful data patterns [129, 149].

Modeling

Modeling Based Scientific Computing (MBSC) depends on the performance of precise mathematical methods arising from the first principles of nature laws. Specific software is used to confirm the compatibility of *in silico* predictions with real conditions [149].

Visualization Methods

A fundamental part of illustration of metabolomics experiments is based on visualization methods which allow the conversion of mathematical algorithms into graphical illustration. Visualization and pathway mapping enable the widespread use of bioinformatical tools even for non-bioinformaticians [145].

Experimental Approaches

Microarray Technology and Next Generation Sequencing Methods

The Human Genome Project established in 2001 represented an important landmark to explore the function of genes in a comprising manner. Functional genomics enabled the investigation of expression patterns of hundreds of genes at the same time. Complementary DNA microarrays, oligonucleotide microarrays or serial analysis of gene expression (SAGE) are the most applied methods in this context [59, 66]. Medicinal plants are traditionally used worldwide despite the fact that adequate toxicology and safety data of those products are not available in many cases. Thus, there is an urgent need to understand mechanisms of action of both pharmacological activity and toxic side effects. This is one of the preconditions to enable the integration of phytotherapy into conventional medicine as well as into precision medicine. Due to the complex chemical composition of herbal remedies, this is not a trivial task.

According to The American Herbal Products Association, approximately 3000 plant species and 50,000 plant-based products were sold as dietary supplement products in the USA [9]. The U.S. FDA and the National Institutes of Health suggested top-selling herbal products and active ingredients for risk assessment and documentation to the U.S. National Toxicology Program, such as *Ginkgo biloba* extract, *Panax ginseng*, kava, *Aloe vera*, green tea extract, comfrey, symphytine, *dong quai*, ephedrine alkaloid, L-ephedrine (*ma huang*), black cohosh, goldenseal root powder, pulegone, usnic acid, and *Usnea* herb. [125].

The detection of pharmacological and toxicological mechanisms of herbal products is quite a difficult approach compared to that of a pure chemical. The approaches determining the mechanisms still have been not fully adapted for medicinal plants and herbal products due to their chemical complexity. Risk assessment of traditional medicines, is, however, of extraordinary importance, in order to extrapolate potential toxicity and tumorigenicity from experimental *in vivo* models to the situation in human beings [16, 66]. Microarray technology and sequencing methods represent important approaches for this purpose. Guo et al. [64, 65] carried out a survey on the utility of DNA microarray technology to evaluate risk of *Ginkgo biloba* and kava extract for liver toxicity and tumorigenity in rats and mice. Another study of Guo et al. [66], further pointed out how such products exert their activities, *e.g.* by modulating the expression of drug metabolizing enzymes, changing signaling pathways/networks etc. Based on the information gained from microarray data, the likely mechanisms of action or toxicity can be hypothesized and verified by subsequent chemical, biological, and genomic examinations.

Various sequencing technologies comprising first-generation sequencing technologies (FGSTs), next-generation sequencing technologies (NGSTs) and third-generation sequencing technologies (TGSTs) became an indispensable part of modern research in the field of personalized medicine and genomic research. FGSTs were developed between 1975 and 1977 by use of the methods of Maxam-Gilbert and Sanger [115, 162]. NGSTs followed by the introduction of The Roche/454 platform in 2005 [111], the Illumina/Solexa system in 2006 [12], Applied Biosystems SOLiD system in 2007 [182], and Ion Torrent system in 2010 [155]. As FGSTs lack high throughput performance [184], NGSTs are less costly and time consuming in comparison with Sanger sequencing [103].

Technological advancement led to the development of TGSTs with considerably improved features, *i.e.* rapid, encompassing, and unbiased sequencing of RNA. New technologies are under development such as single-molecule real-time (SMRT) systems (the PacBio RS II from Pacific Biosciences) [50] or nanopore-based systems (Oxford Nanopore Technologies (ONT) MinION) [118] inclining minimal artifacts or inaccuracies during RNA sequencing [184].

High-Throughput "-Omics" Technologies

Personalized medicine aims to select 'the right drug for the right patient'. In spite of tumor heterogeneity as a major challenge for precision medicine, various techniques have been improved enabling the detection of genomic variations, (*e.g.* single nucleotide polymorphisms (SNPs), copy number variations), epigenomic modifications (*e.g.* DNA methylation, histone acetylation, micro-RNAs), transcriptome-wide mRNA expression (transcriptomics) and proteome-wide protein and peptide expression (proteomics) in cells or tissues [43, 48]. Computational biology and bioinformatics enable to extract relevant knowledge from the vast quantities of data generated by these "-omics" approaches [37, 117]. The intention in the frame of precision medicine is to make reliable predictions of individual expression profiles [174, 209]. Meta analyses of DNA sequencing outcomes bear high values for drug discovery and development of specific targeted agents or biomarkers as well as for therapy monitoring [21]. High-dimensional "-omics" data can be translated into precision therapy protocols by means of computational methods or bioinformatics [22, 180].

The emergence of new concepts in personalized medicine opens a new and challenging door into individual cancer therapy. It should be realized that the extensive heterogeneity of most tumors engrave the realization of those concepts [48, 53]. A project intending to constitute a link between thousands of chemical compounds and gene expression profiles in various cancer cell lines has been recently investigated. This is a starting point forming a basis for the improvement of complicated requirements in personalized medicine [90].

Drug discovery comprising various steps ranging from *in vitro* tests to marketing is quite a costly and time consuming process. The number of newly approved drugs has decreased owing to the frequent failure in phase II trials during the past decades in spite of the increment of investing activities for research and drug development [83, 121]. The rise of "-omics" technologies and newly developed sequencing methods together with computational methods and bioinformatics is expected to support drug discovery saving money and time. An interesting concept based on the possible use of known drugs with reasonable safety and pharmacokinetic property came into the focus, since these drugs may be promising drug candidates for other certain diseases affected by the same pathway [141]. This concept has been termed as "drug repositioning" [7, 127]. Sildenafil, for instance, was initially developed for the treatment of angina pectoris. Nevertheless, clinical trials did not prove such an effect. Subsequently, sildenafil was recognized to induce strong erections as a side effect and was approved by the FDA for the treatment of erectile dysfunction [7, 127]. Then, phosphodiesterase type 5 (PDE5) as the target of sildenafil was not only found to be expressed in the penis, but also in lungs assuming an activity of sildenafil for the treatment of pulmonary arterial hypertension [26]. Thus, sildenafil was repositioned twice. Mebendazole is another example of repositioning. This drug was initially developed against helminthic, infections. However it was revealed to exert anticancer activity especially towards metastatic It turned out, however, that mebendazole also revealed anticancer activity, especially towards metastatic adrenocortical carcinoma and refractory colon cancer. The drug is currently being tested in Phase 1 and II clinical trials [127, 133, 140]. This phenomenon may also set the pace for phytotherapeutics or herbal medicines as potential drug candidates against cancer, since most of them have been traditionally used throughout the history.

Biological and Pharmacological Experiments

Pharmacological research started in the second half of the nineteenth century in Europe by Rudolf Buchheim and Oswald Schmiedeberg investigating existing drugs in animals [81]. New drugs were developed with the emergence of synthetic chemistry in the twentieth century with the use of classical pharmacological screening based on successively testing of chemical entities or biological products in isolated organs and whole animals [186]. Western medicine and Chinese medicine fundamentally differ from each other. In Western medicine, pharmaceutical development consists of three main stages, which are target selection, screening for leads and lead structure optimization. In Chinese medicine, prescription discovery, component identification and formula optimization are the main steps. Western medicine focuses on a drug design model based on structure activity relationship (SAR), whereas Chinese medicine emphasizes systems modeling attributing mostly to combination activity relationship (CAR) [206].

The process of drug discovery, development and market approval in Western medicine starts with the identification of a biological target associated with a certain disease, followed by *in vitro* and *in vivo* preclinical research, clinical phase I-III trials in order to clarify mechanisms of action, pharmacokinetics, efficacy, toxicity, dosage form, the best dosage, side effects etc. [35]. Development of a new drug is quite costly and is estimated to be about $2.6 billion [36, 120]. Despite huge investments and spending plenty of time, drug development may even fail. Thus, consolidation of experimental approaches with computational tools or new generation technologies such as high throughput screening and "-omics" technologies are meaningful reducing costs and time.

Computer-aided drug design, computational modeling and simulation approaches have inclined great achievements in drug discovery and personalized medicine against cancer [194, 201, 210]. Cancer bioinformatics is important to enable the identification of biomarkers, which are specific to disease phenotypes and which can be routinely applied for early diagnosis as well as the monitoring of disease progress and response to therapy [202]. The integration of such methodologies into biological and pharmacological experiments is an unprecedented approach in personalized medicine, which will improve the outcomes of patients suffering from cancer.

Modeling of Predictive Systems

The emergence of mathematical modeling as well as the analysis and prediction of systems occurred following the development of the first mathematical cardiac model in the 1960s [130]. Various mathematical models have been evolved to prove assumptions and predict systems among which computational modeling holds an outstanding importance by using models for simulation and examination of the behavior of complex systems based on mathematics, physics and computer science. A computational model consists of different variables and defining the system. Simulation is achieved by adjustment of such variables and monitoring the results affected by the alterations. Modeling allows scientists to carry out great number of simulated experiments, the outcomes of which lead researchers to make assumptions about probable responds in the real patient [124].

Drug efficacy and safety may be cost-effectively determined in mathematical disease models by *in silico* simulations. The accomplishment of human genome project and improvements in high-throughput technologies provided more popularity and confidence in mathematical modeling. Computational modeling led to investigation of systems biology at multiple levels such as molecular processes, cell to cell interactions etc. [17, 92, 154].

There are only few examples of simulation studies establishing mathematical modeling with herbal products. For instance, Quanquan and Tingge [148] introduced a new mathematical model calculating relative dosages of 561 common herbs, which is clearly seen as beneficial in curative efficacy mining. Another example is quantitative composition–activity relationship models developed by multiple linear regression, artificial neural networks, and support vector regression to predict bioactivity of a herbal medicine and design of a new drug. The widely used herbal medicine *Qi Xue Bing Zhi* Fang was studied and the ratio of including two active constituents was optimized in reference to the composition–activity relationship model to obtain a new formulation with better activity [190]. These approaches are interesting clues for the potential of computer-based modeling using herbal prescriptions.

Recently, there was a paradigm shift from the "single drug- single target" concept to the "multi-component- multi-target" strategy. In this context, application of mathematical modeling helps to obtain of extensive interactions of herbal preparations with signaling pathways and networks at the systems biological level. Computational simulations may be also appropriate tools to find connections between preclinical and clinical studies as well as filling the gaps, which obviously hamper drug development in terms of cost and time savings.

To better recognize the biological functions of a whole system, the virtual physiological human project has been established since 2006. The project integrates information from systems biology, personal health systems, biomedical informatics to achieve more efficient and effective healthcare systems in the future [74, 82].

Simulation studies on traditional medicine have just emerged. Their application is, however a rapidly growing area. Determination of the possible activities of complex herbal products by a developed computational model would undoubtedly supply and improve our healthcare systems.

Conclusions

Traditional medicine has been existing in every continent and cultural area of the world throughout the history. To exemplify, traditional Chinese medicine in East Asia, Ayurvedic medicine in India and Galenic medicine in Europe are the most renowned examples, each with individual basic philosophies and representing some similarities [185]. The art of traditional medicine dates back to the thousands of years of practical experience and is still used in our times.

Combination therapy regimens are preferably applied in clinical oncology rather than monotherapy, since tumors frequently exhibit resistance to drugs and combination therapies may combat (or at least delay) the rapid emergence of drugs resistance phenomena. In this respect, traditional medicine, phytotherapeutics, and herbal products may be a promising future perspective in cancer therapy due to their complex composition and simultaneous interaction with multiple targets.

Network pharmacology may be useful to develop diagnostic tools not only as prognostic markers for patients' survival probabilities, but also to predict the response of tumors to established anticancer drugs and in case of resistance to allow bypassing to treatment alternatives, *e.g.* phytochemicals and herbal mixtures [187].

Due to the severe side effects and development of drug resistance, treatment regimens have moved from cytotoxic drugs to targeted drugs in the past decade with quite a number of recently marketed drugs. Still, targeted drugs may also exert side effects and incline development of drug resistance in patients, indicating that the potential of target specificity has not been fully exploited yet. The multi-specific nature of many phytochemicals may support the clinical fight against drug resistance. Cancer stem-like cells are known for their capacity of self-renewal and for inducing the heterogeneous lineages of cancer cells, which are usually resistant to standard chemo- and radiotherapy. Numerous investigations showed that natural products inhibit those cells [40, 126, 167], which is quite an important potential for future drug development efforts. Furthermore, natural compounds represent important lead compounds for chemical derivatization. Novel chemical scaffolds of natural products are valuable starting points for medicinal chemistry to develop novel drugs not involved in common drug resistance phenotypes [87]. The recent technologies in computational methods and bioinformatics have guided the development of new treatment concepts in the field of network pharmacology. Phytotherapeutics or herbal products in traditional medicine due to their complex composition display diverse activities by interacting with numerous relevant targets due to their complex composition. The major issue in network pharmacology is to extract substantive knowledge from a vast amount of data distinguishing relevant signals from background noise. Network pharmacology has to cope with a number of obstacles such as multi-targeted nature of drug action, resistance phenomena, side effects on healthy tissue or organs and tumor heterogeneity. In this context, the development of new methodologies in network pharmacology may be supportive and promising for overcoming of such difficulties.

It is not beyond the expectation that personalized medicine will lead to significant progresses in cancer treatment. Apart from the scientific basis, personalized needs to remain affordable for patients and governmental health systems. The availability of effective personalized medicine should not be restricted to a rich elite. Here, phytotherapies can play a crucial role, as they are affordable even for people in countries with general low income [6, 8].

To realize the integration of phytotherapy and phytochemicals into novel approaches of precision medicine, it is indispensable that high quality products will be marketed, *i.e.* standardized plant extracts with proven preclinical and clinical efficacy and safety [34, 48, 85, 110, 114]. Drug industry and health care systems should jointly focus on the improvement of cancer survival times by integrating the best of two worlds- of phytotherapy and conventional medicine to create novel concepts of precision medicine [49].

References

1. Al-Lazikani, B., U. Banerji, and P. Workman. 2012. Combinatorial drug therapy for cancer in the post-genomic era. *Natural Biotechnology* 30: 679–692.
2. Alves-Silva, J.M., A. Romane, T. Efferth, and L. Salgueiro. 2017. North African medicinal plants traditionally used in cancer therapy. *Frontiers in Pharmacology* 8: 383.
3. Amin, A., H. Gali-Muhtasib, M. Ocker, and R. Schneider-Stock. 2009a. Overview of major classes of plant-derived anticancer drugs. *International Journal of Biomedical Sciences* 5: 1–11.

4. Amin, A.R.M.R., O. Kucuk, F.R. Khuri, and D.M. Shin. 2009b. Perspectives for cancer prevention with natural compounds. *Journal of Clinical Oncology: Official Journal of the American Society of Clinical Oncology* 27: 2712–2725.
5. Anfosso, L., T. Efferth, A. Albini, and U. Pfeffer. 2006. Microarray expression profiles of angiogenesis-related genes predict tumor cell response to artemisinins. *The Pharmacogenomics Journal* 6: 269.
6. Antoñanzas, F., C. Juárez-Castelló, and R. Rodríguez-Ibeas. 2015. Some economics on personalized and predictive medicine. *The European Journal of Health Economics* 16: 985–994.
7. Ashburn, T.T., and K.B. Thor. 2004. Drug repositioning: Identifying and developing new uses for existing drugs. *Nature Reviews Drug Discovery* 3: 673–683.
8. Atherly, A.J., and D.R. Camidge. 2012. The cost-effectiveness of screening lung cancer patients for targeted drug sensitivity markers. *British Journal of Cancer* 106: 1100.
9. Bagchi, D. 2014. *Nutraceutical and functional food regulations in the United States and around the World*. London: Academic Press.
10. Barnes, J., A.J. Mclachlan, C.M. Sherwin, and E.Y. Enioutina. 2016. Herbal medicines: challenges in the modern world. Part 1 Australia and New Zealand. *Expert Review of Clinical Pharmacology* 9: 905–915.
11. Begley, C.G., and L.M. Ellis. 2012. Raise standards for preclinical cancer research. *Nature* 483: 531.
12. Bentley, D.R., S. Balasubramanian, H.P. Swerdlow, G.P. Smith, J. Milton, C.G. Brown, K.P. Hall, D.J. Evers, C.L. Barnes, H.R. Bignell, J.M. Boutell, J. Bryant, R.J. Carter, R. Keira Cheetham, A.J. Cox, D.J. Ellis, M.R. Flatbush, N.A. Gormley, S.J. Humphray, L.J. Irving, M.S. Karbelashvili, S.M. Kirk, H. Li, X. Liu, K.S. Maisinger, L.J. Murray, B. Obradovic, T. Ost, M.L. Parkinson, M.R. Pratt, I.M. Rasolonjatovo, M.T. Reed, R. Rigatti, C. Rodighiero, M.T. Ross, A. Sabot, S.V. Sankar, A. Scally, G.P. Schroth, M.E. Smith, V.P. Smith, A. Spiridou, P.E. Torrance, S.S. Tzonev, E.H. Vermaas, K. Walter, X. Wu, L. Zhang, M.D. Alam, C. Anastasi, I.C. Aniebo, D.M. Bailey, I.R. Bancarz, S. Banerjee, S.G. Barbour, P.A. Baybayan, V.A. Benoit, K.F. Benson, C. Bevis, P.J. Black, A. Boodhun, J.S. Brennan, J.A. Bridgham, R.C. Brown, A.A. Brown, D.H. Buermann, A.A. Bundu, J.C. Burrows, N.P. Carter, N. Castillo, Chiara ECM, S. Chang, R. Neil Cooley, N.R. Crake, O.O. Dada, K.D. Diakoumakos, B. Dominguez-Fernandez, D.J. Earnshaw, U.C. Egbujor, D.W. Elmore, S.S. Etchin, M.R. Ewan, M. Fedurco, L.J. Fraser, K.V. Fuentes Fajardo, W. Scott Furey, D. George, K.J. Gietzen, C.P. Goddard, G.S. Golda, P.A. Granieri, D.E. Green, D.L. Gustafson, N.F. Hansen, K. Harnish, C.D. Haudenschild, N.I. Heyer, M.M. Hims, J.T. Ho, A.M. Horgan, et al. 2008. Accurate whole human genome sequencing using reversible terminator chemistry. *Nature* 456: 53–59.
13. Block, G., B. Patterson, and A. Subar. 1992. Fruit, vegetables, and cancer prevention: a review of the epidemiological evidence. *Nutrition and Cancer* 18: 1–29.
14. Bonifazi, P., M. Goldin, M.A. Picardo, I. Jorquera, A. Cattani, G. Bianconi, A. Represa, Y. Ben-Ari, and R. Cossart. 2009. GABAergic hub neurons orchestrate synchrony in developing hippocampal networks. *Science* 326: 1419–1424.
15. Borders, E.B., C. Bivona, and P.J. Medina. 2010. Mammalian target of rapamycin: Biological function and target for novel anticancer agents. *American Journal of Health-System Pharmacy* 67: 2095–2106.
16. Borgert, C.J., T.F. Quill, L.S. Mccarty, and A.M. Mason. 2004. Can mode of action predict mixture toxicity for risk assessment? *Toxicology and Applied Pharmacology* 201: 85–96.
17. Bottino, D., R.C. Penland, A. Stamps, M. Traebert, B. Dumotier, A. Georgiva, G. Helmlinger, and G.S. Lett. 2006. Preclinical cardiac safety assessment of pharmaceutical compounds using an integrated systems-based computer model of the heart. *Progress in Biophysics and Molecular Biology* 90: 414–443.
18. Calabrese, C., H. Poppleton, M. Kocak, T.L. Hogg, C. Fuller, B. Hamner, E.Y. Oh, M.W. Gaber, D. Finklestein, M. Allen, A. Frank, I.T. Bayazitov, S.S. Zakharenko,

A. Gajjar, A. Davidoff, and R.J. Gilbertson. 2007. A perivascular niche for brain tumor stem cells. *Cancer Cell* 11: 69–82.
19. Calvisi, D.F., S. Ladu, A. Gorden, M. Farina, E.A. Conner, J.S. Lee, V.M. Factor, and S.S. Thorgeirsson. 2006. Ubiquitous activation of Ras and Jak/Stat pathways in human HCC. *Gastroenterology* 130: 1117–1128.
20. Cardenas, M.E., A. Sanfridson, N.S. Cutler, and J. Heitman. 1998. Signal-transduction cascades as targets for therapeutic intervention by natural products. *Trends in Biotechnology* 16: 427–433.
21. Chan, S.L., A.M. Wong, K. Lee, N. Wong, and A.K. Chan. 2016. Personalized therapy for hepatocellular carcinoma: Where are we now? *Cancer Treatment Reviews* 45: 77–86.
22. Chang, W., A.S. Brohl, R. Patidar, S. Sindiri, J.F. Shern, J.S. Wei, Y.K. Song, M.E. Yohe, B. Gryder, S. Zhang, K.A. Calzone, N. Shivaprasad, X. Wen, T.C. Badgett, M. Miettinen, K.R. Hartman, J.C. League-Pascual, T.N. Trahair, B.C. Widemann, M.S. Merchant, R.N. Kaplan, J.C. Lin, and J. Khan. 2016. Multidimensional clinomics for precision therapy of children and adolescent young adults with relapsed and refractory cancer: A report from the center for cancer research. *Clinical Cancer Research* 22: 3810–3820.
23. Chen, X., H. Zhou, Y.B. Liu, J.F. Wang, H. Li, C.Y. Ung, L.Y. Han, Z.W. Cao, and Y.Z. Chen. 2006. Database of traditional Chinese medicine and its application to studies of mechanism and to prescription validation. *British Journal of Pharmacology* 149: 1092–1103.
24. Chen, R.J., Y.S. Ho, H.R. Guo, and Y.J. Wang. 2008. Rapid activation of Stat3 and ERK1/2 by nicotine modulates cell proliferation in human bladder cancer cells. *Toxicological Sciences* 104: 283–293.
25. Cheng, X., J. Huo, D. Wang, X. Cai, X. Sun, W. Lu, Y. Yang, C. Hu, X. Wang, and P. Cao. 2017. Herbal medicine Ac591 prevents oxaliplatin-induced peripheral neuropathy in animal model and cancer patients. *Frontiers in Pharmacology* 8: 344.
26. Corbin, J.D., A. Beasley, M.A. Blount, and S.H. Francis. 2005. High lung PDE5: A strong basis for treating pulmonary hypertension with PDE5 inhibitors. *Biochemical and Biophysical Research Communications* 334: 930–938.
27. Cragg, G.M., D.J. Newman, and R.B. Weiss. 1997. Coral reefs, forests, and thermal vents: The worldwide exploration of nature for novel antitumor agents. *Seminars in Oncology* 24: 156–163.
28. Da Rocha, A.B., R.M. Lopes, and G. Schwartsmann. 2001. Natural products in anticancer therapy. *Current Opinion in Pharmacology* 1: 364–369.
29. Dassa, E., and P. Bouige. 2001. The ABC of ABCs: A phylogenetic and functional classification of ABC systems in living organisms. *Research in Microbiology* 152: 211–229.
30. Dawood, M., S. Hamdoun, and T. Efferth. 2018. Multifactorial modes of action of arsenic trioxide in cancer cells as analyzed by classical and network pharmacology. *Frontiers in Pharmacology* 9: 143.
31. De Araujo, V.C., C. Furuse, P.R. Cury, A. Altemani, and N.S. De Araujo. 2008. STAT3 expression in salivary gland tumours. *Oral Oncology* 44: 439–445.
32. De, O., and B. Prasun Chatterji. 2017. Marine derived anticancer drugs targeting microtubule. *Recent Patents on Anti-Cancer Drug Discovery* 12 (2): 102–127.
33. Demain, A.L., and P. Vaishnav. 2011. Natural products for cancer chemotherapy. *Microbial Biotechnology* 4: 687–699.
34. Derosa, G., D. Romano, A. D'angelo, and P. Maffioli. 2015. Berberis aristata/Silybum marianum fixed combination (Berberol(R)) effects on lipid profile in dyslipidemic patients intolerant to statins at high dosages: A randomized, placebo-controlled, clinical trial. *Phytomedicine* 22: 231–237.
35. Dimasi, J.A., L. Feldman, A. Seckler, and A. Wilson. 2010. Trends in risks associated with new drug development: Success rates for investigational drugs. *Clinical Pharmacology and Therapeutics* 87: 272–277.
36. Dimasi, J.A., H.G. Grabowski, and R.W. Hansen. 2016. Innovation in the pharmaceutical industry: New estimates of R&D costs. *Journal of Health Economics* 47: 20–33.

37. Dopazo, J. 2014. Genomics and transcriptomics in drug discovery. *Drug Discovery Today* 19: 126–132.
38. Dumontet, C., and M.A. Jordan. 2010. Microtubule-binding agents: A dynamic field of cancer therapeutics. *Nature Reviews. Drug Discovery* 9: 790.
39. Durrant, J.D., and J.A. Mccammon. 2011. Molecular dynamics simulations and drug discovery. *BMC Biology* 9: 71.
40. Efferth, T. 2012. Stem cells, cancer stem-like cells, and natural products. *Planta Medica* 78: 935–942.
41. ———. 2014. *Resistance to targeted ABC transporters in cancer*. New York: Springer.
42. Efferth, T., and H.J. Greten. 2012. In silico analysis of microarray-based gene expression profiles predicts tumor cell response to withanolides. *Microarrays (Basel)* 1: 44–63.
43. Efferth, T., and M. Volm. 2005. Pharmacogenetics for individualized cancer chemotherapy. *Pharmacology & Therapeutics* 107: 155–176.
44. Efferth, T., Y.J. Fu, Y.G. Zu, G. Schwarz, V.S. Konkimalla, and M. Wink. 2007a. Molecular target-guided tumor therapy with natural products derived from traditional Chinese medicine. *Current Medicinal Chemistry* 14: 2024–2032.
45. Efferth, T., H. Miyachi, and H. Bartsch. 2007b. Pharmacogenomics of a traditional Japanese herbal medicine (Kampo) for cancer therapy. *Cancer Genomics Proteomics* 4: 81–91.
46. Efferth, T., S. Kahl, K. Paulus, M. Adams, R. Rauh, H. Boechzelt, X. Hao, B. Kaina, and R. Bauer. 2008a. Phytochemistry and pharmacogenomics of natural products derived from traditional Chinese medicine and Chinese Materia Medica with activity against tumour cells. *Molecular Cancer Therapeutics* 7 (1): 152–161.
47. Efferth, T., V.B. Konkimalla, Y.-F. Wang, A. Sauerbrey, S. Meinhardt, F. Zintl, J. Mattern, and M. Volm. 2008b. Prediction of broad spectrum resistance of tumors towards anticancer drugs. *Clinical Cancer Research* 14: 2405–2412.
48. Efferth, T., M.E.M. Saeed, E. Mirghani, A. Alim, Z. Yassin, E. Saeed, H.E. Khalid, and S. Daak. 2017. Integration of phytochemicals and phytotherapy into cancer precision medicine. *Oncotarget* 8: 50284–50304.
49. Efferth, T., M. Banerjee, M.S. Abu-Darwish, S. Abdelfatah, M. Böckers, D. Bhakta-Guha, V. Bolzani, S. Daak, Ö.L. Demirezer, M. Dawood, M. Efferth, H.R. El-Seedi, N. Fischer, H.J. Greten, S. Hamdoun, C. Hong, M. Horneber, O. Kadioglu, H.E. Khalid, S.A. Khalid, V. Kuete, N. Mahmoud, J. Marin, A. Mbaveng, J. Midiwo, H. Nakagawa, J. Nas, O. Ngassapa, D. Ochwang'i, L.K. Omosa, E.A. Ooko, N. Özenver, P. Poornima, M.R. Romero, M.E.M. Saeed, L. Salgueiro, E.-J. Seo, G. Yan, Z. Yasin, E.M. Saeed, and N.W. Paul. 2019. Biopiracy versus one-world medicine–from colonial relicts to global collaborative concepts. *Phytomedicine* 53: 319–331.
50. English, A.C., S. Richards, Y. Han, M. Wang, V. Vee, J. Qu, X. Qin, D.M. Muzny, J.G. Reid, and K.C. Worley. 2012. Mind the gap: upgrading genomes with Pacific Biosciences RS long-read sequencing technology. *PLoS One* 7: e47768.
51. Fan, W. 2001. The Traditional Chinese Medical Literature Analysis and Retrieval System (TCMLARS) and its application. *INSPEL* 35 (3): 147–156.
52. Gennemark, P., and D. Wedelin. 2007. Efficient algorithms for ordinary differential equation model identification of biological systems. *IET Systems Biology* 1: 120–129.
53. Gerlinger, M., A.J. Rowan, S. Horswell, J. Larkin, D. Endesfelder, E. Gronroos, P. Martinez, N. Matthews, A. Stewart, P. Tarpey, I. Varela, B. Phillimore, S. Begum, N.Q. Mcdonald, A. Butler, D. Jones, K. Raine, C. Latimer, C.R. Santos, M. Nohadani, A.C. Eklund, B. Spencer-Dene, G. Clark, L. Pickering, G. Stamp, M. Gore, Z. Szallasi, J. Downward, P.A. Futreal, and C. Swanton. 2012. Intratumor heterogeneity and branched evolution revealed by multiregion sequencing. *The New England Journal of Medicine* 366: 883–892.
54. Gillet, J.P., T. Efferth, and J. Remacle. 2007. Chemotherapy-induced resistance by ATP-binding cassette transporter genes. *Biochimica et Biophysica Acta* 1775: 237–262.

55. Gonzalez-Angulo, A.M., B.T.J. Hennessy, and G.B. Mills. 2010. Future of personalized medicine in oncology: A systems biology approach. *Journal of Clinical Oncology: Official Journal of the American Society of Clinical Oncology* 28: 2777–2783.
56. Gordaliza, M. 2007. Natural products as leads to anticancer drugs. *Clinical & Translational Oncology* 9: 767–776.
57. Gosak, M., R. Markovic, A. Fajmut, M. Marhl, M. Hawlina, and S. Andjelic. 2015. The analysis of intracellular and intercellular calcium signaling in human anterior lens capsule epithelial cells with regard to different types and stages of the cataract. *PLoS One* 10: e0143781.
58. Gosak, M., R. Markovič, J. Dolenšek, M. Slak Rupnik, M. Marhl, A. Stožer, and M. Perc. 2018. Network science of biological systems at different scales: A review. *Physics of Life Reviews* 24: 118–135.
59. Govindarajan, R., J. Duraiyan, K. Kaliyappan, and M. Palanisamy. 2012. Microarray and its applications. *Journal of Pharmacy & Bioallied Sciences* 4: S310–S312.
60. Grivennikov, S.I., and M. Karin. 2010. Dangerous liaisons: STAT3 and NF-kappaB collaboration and crosstalk in cancer. *Cytokine & Growth Factor Reviews* 21: 11–19.
61. Gu, S., and J. Pei. 2017. Chinese herbal medicine meets biological networks of complex diseases: A computational perspective. *Evidence-based Complementary and Alternative Medicine* 2017: 7198645.
62. Gu, H., L. Ma, Y. Ren, W. He, Y. Wang, and Y. Qiao. 2014a. Exploration of the mechanism of pattern-specific treatments in coronary heart disease with network pharmacology approach. *Computers in Biology and Medicine* 51: 198–204.
63. Gu, S., D.-A. Silva, L. Meng, A. Yue, and X. Huang. 2014b. Quantitatively characterizing the ligand binding mechanisms of choline binding protein using Markov state model analysis. *PLoS Computational Biology* 10: e1003767.
64. Guo, L., Q. Li, Q. Xia, S. Dial, P.-C. Chan, and P. Fu. 2009. Analysis of gene expression changes of drug metabolizing enzymes in the livers of F344 rats following oral treatment with kava extract. *Food and Chemical Toxicology* 47: 433–442.
65. Guo, L., N. Mei, W. Liao, P.C. Chan, and P.P. Fu. 2010a. Ginkgo biloba extract induces gene expression changes in xenobiotics metabolism and the Myc-centered network. *OMICS* 14: 75–90.
66. Guo, L., N. Mei, Q. Xia, T. Chen, P.C. Chan, and P.P. Fu. 2010b. Gene expression profiling as an initial approach for mechanistic studies of toxicity and tumorigenicity of herbal plants and herbal dietary supplements. *Journal of Environmental Science and Health. Part C, Environmental Carcinogenesis & Ecotoxicology Reviews* 28: 60–87.
67. Hartmann, J.T., and H.P. Lipp. 2006. Camptothecin and podophyllotoxin derivatives: Inhibitors of topoisomerase I and II – mechanisms of action, pharmacokinetics and toxicity profile. *Drug Safety* 29: 209–230.
68. Harvey, A.L. 1999. Medicines from nature: Are natural products still relevant to drug discovery? *Trends in Pharmacological Sciences* 20: 196–198.
69. Hirano, T., K. Ishihara, and M. Hibi. 2000. Roles of STAT3 in mediating the cell growth, differentiation and survival signals relayed through the IL-6 family of cytokine receptors. *Oncogene* 19: 2548–2556.
70. Holohan, C., S. Van Schaeybroeck, D.B. Longley, and P.G. Johnston. 2013. Cancer drug resistance: An evolving paradigm. *Nature Reviews. Cancer* 13: 714–726.
71. Hong, C., J. Cao, C.F. Wu, O. Kadioglu, A. Schuffler, U. Kauhl, S.M. Klauck, T. Opatz, E. Thines, N.W. Paul, and T. Efferth. 2017. The Chinese herbal formula Free and Easy Wanderer ameliorates oxidative stress through KEAP1-NRF2/HO-1 pathway. *Scientific Reports* 7: 11551.
72. Hopkins, A.L. 2008. Network pharmacology: the next paradigm in drug discovery. *Nature Chemical Biology* 4: 682–690.
73. Huang, L., D. Xie, Y. Yu, H. Liu, Y. Shi, T. Shi, and C. Wen. 2018. TCMID 2.0: A comprehensive resource for TCM. *Nucleic Acids Research* 46: D1117–d1120.

74. Hunter, P., V. Coveney Peter, B. De Bono, V. Diaz, J. Fenner, F. Frangi Alejandro, P. Harris, R. Hose, P. Kohl, P. Lawford, K. Mccormack, M. Mendes, S. Omholt, A. Quarteroni, J. Skår, J. Tegner, S. Randall Thomas, I. Tollis, I. Tsamardinos, H.G.M. Van Beek Johannes, and M. Viceconti. 2010. A vision and strategy for the virtual physiological human in 2010 and beyond. *Philosophical Transactions of the Royal Society A – Mathematical Physical and Engineering Sciences* 368: 2595–2614.
75. Kadioglu, O., and T. Efferth. 2015. Pharmacogenomic characterization of cytotoxic compounds from salvia officinalis in cancer cells. *Journal of Natural Products* 78: 762–775.
76. Kadioglu, O., N.S. Kermani, G. Kelter, U. Schumacher, H.H. Fiebig, H.J. Greten, and T. Efferth. 2014. Pharmacogenomics of cantharidin in tumor cells. *Biochemical Pharmacology* 87: 399–409.
77. Kadioglu, O., M. Saeed, V. Kuete, H.J. Greten, and T. Efferth. 2018. Oridonin targets multiple drug-resistant tumor cells as determined by in silico and in vitro analyses. *Frontiers in Pharmacology* 9: 355.
78. Kalimuthu, S., and K. Se-Kwon. 2013. Cell survival and apoptosis signaling as therapeutic target for cancer: Marine bioactive compounds. *International Journal of Molecular Sciences* 14: 2334–2354.
79. Kasinski, A.L., Y. Du, S.L. Thomas, J. Zhao, S.Y. Sun, F.R. Khuri, C.Y. Wang, M. Shoji, A. Sun, J.P. Snyder, D. Liotta, and H. Fu. 2008. Inhibition of IkappaB kinase-nuclear factor-kappaB signaling pathway by 3,5-bis(2-flurobenzylidene)piperidin-4-one (EF24), a novel monoketone analog of curcumin. *Molecular Pharmacology* 74: 654–661.
80. Kitano, H. 2002. Computational systems biology. *Nature* 420: 206–210.
81. Koch-Weser, J., and P.J. Schechter. 1978. Schmiedeberg in Strassburg 1872–1918: The making of modern pharmacology. *Life Sciences* 22: 1361–1371.
82. Kohl, P., and D. Noble. 2009. Systems biology and the virtual physiological human. *Molecular Systems Biology* 5: 292.
83. Kola, I., and J. Landis. 2004. Can the pharmaceutical industry reduce attrition rates? *Nature Reviews. Drug Discovery* 3: 711.
84. Konkimalla, V., V. Suhas, N. Chandra, E. Gebhart, and T. Efferth. 2007. From molecular diagnostics to molecular targeted therapy with natural product small molecule inhibitors in oral squamous cell carcinoma. *Planta Medica* 73 (9): 4.
85. Kresty, L.A., S.R. Mallery, and G.D. Stoner. 2016. Black raspberries in cancer clinical trials: Past, present and future. *Journal of Berry Research* 6: 251–261.
86. Kuete, V., and T. Efferth. 2011. Pharmacogenomics of Cameroonian traditional herbal medicine for cancer therapy. *Journal of Ethnopharmacology* 137: 752–766.
87. Kuete, V., M.E. Saeed, O. Kadioglu, J. Börtzler, H. Khalid, H.J. Greten, and T. Efferth. 2015. Pharmacogenomic and molecular docking studies on the cytotoxicity of the natural steroid wortmannin against multidrug-resistant tumor cells. *Phytomedicine* 22: 120–127.
88. Lage, H. 2003. ABC-transporters: Implications on drug resistance from microorganisms to human cancers. *International Journal of Antimicrobial Agents* 22: 188–199.
89. Lam, W., S. Bussom, F. Guan, Z. Jiang, W. Zhang, E.A. Gullen, S.H. Liu, and Y.C. Cheng. 2010. The four-herb Chinese medicine PHY906 reduces chemotherapy-induced gastrointestinal toxicity. *Science Translational Medicine* 2 (45): 45ra59.
90. Lamb, J., E.D. Crawford, D. Peck, J.W. Modell, I.C. Blat, M.J. Wrobel, J. Lerner, J.P. Brunet, A. Subramanian, K.N. Ross, M. Reich, H. Hieronymus, G. Wei, S.A. Armstrong, S.J. Haggarty, P.A. Clemons, R. Wei, S.A. Carr, E.S. Lander, and T.R. Golub. 2006. The connectivity map: Using gene-expression signatures to connect small molecules, genes, and disease. *Science* 313: 1929–1935.
91. Leary, R.J., J.C. Lin, J. Cummins, S. Boca, L.D. Wood, D.W. Parsons, S. Jones, T. Sjoblom, B.H. Park, R. Parsons, J. Willis, D. Dawson, J.K. Willson, T. Nikolskaya, Y. Nikolsky, L. Kopelovich, N. Papadopoulos, L.A. Pennacchio, T.L. Wang, S.D. Markowitz, G. Parmigiani, K.W. Kinzler, B. Vogelstein, and V.E. Velculescu. 2008. Integrated analysis of homozygous deletions, focal amplifications, and sequence alterations in breast and

colorectal cancers. *Proceedings of the National Academy of Sciences of the United States of America* 105: 16224–16229.
92. Lee, S. 2015. Systems biology – a pivotal research methodology for understanding the mechanisms of traditional medicine. *Journal of Pharmacopuncture* 18: 11–18.
93. Lee, J.-H., K.M. Park, D.-J. Han, N.Y. Bang, D.-H. Kim, H. Na, S. Lim, T.B. Kim, D.G. Kim, H.-J. Kim, Y. Chung, S.H. Sung, Y.-J. Surh, S. Kim, and B.W. Han. 2015. PharmDB-K: Integrated bio-pharmacological network database for traditional Korean medicine. *PLoS One* 10: e0142624–e0142624.
94. Li, Z., and C. Xu. 2011. The fundamental theory of traditional Chinese medicine and the consideration in its research strategy. *Frontiers in Medicine* 5: 208–211.
95. Li, S., and B. Zhang. 2013. Traditional Chinese medicine network pharmacology: Theory, methodology and application. *Chinese Journal of Natural Medicines* 11: 110–120.
96. Li, H., Z. Gao, L. Kang, H. Zhang, K. Yang, K. Yu, X. Luo, W. Zhu, K. Chen, J. Shen, X. Wang, and H. Jiang. 2006. TarFisDock: A web server for identifying drug targets with docking approach. *Nucleic Acids Research* 34: W219–W224.
97. Li, W.F., J.G. Jiang, and J. Chen. 2008. Chinese medicine and its modernization demands. *Archives of Medical Research* 39: 246–251.
98. Li, S., B. Zhang, D. Jiang, Y. Wei, and N. Zhang. 2010. Herb network construction and co-module analysis for uncovering the combination rule of traditional Chinese herbal formulae. *BMC Bioinformatics* 11: S6.
99. Li, Z., T. Qin, K. Wang, M. Hackenberg, J. Yan, Y. Gao, L.-R. Yu, L. Shi, Z. Su, and T. Chen. 2015. Integrated microRNA, mRNA, and protein expression profiling reveals microRNA regulatory networks in rat kidney treated with a carcinogenic dose of aristolochic acid. *BMC Genomics* 16: 365.
100. Lin, L.L., C.R. Hsia, C.L. Hsu, H.C. Huang, and H.F. Juan. 2015. Integrating transcriptomics and proteomics to show that tanshinone IIA suppresses cell growth by blocking glucose metabolism in gastric cancer cells. *BMC Genomics* 16: 41.
101. Liu, L.F. 1989. DNA topoisomerase poisons as antitumor drugs. *Annual Review of Biochemistry* 58: 351–375.
102. Liu, L.F., S.D. Desai, T.K. Li, Y. Mao, M. Sun, and S.P. Sim. 2000. Mechanism of action of camptothecin. *Annals of the New York Academy of Sciences* 922: 1–10.
103. Liu, L., Y. Li, S. Li, N. Hu, Y. He, R. Pong, D. Lin, L. Lu, and M. Law. 2012. Comparison of next-generation sequencing systems. *BioMed Research International* 2012: 251364.
104. Liu, H., J. Wang, W. Zhou, Y. Wang, and L. Yang. 2013. Systems approaches and polypharmacology for drug discovery from herbal medicines: An example using licorice. *Journal of Ethnopharmacology* 46: 773–793.
105. Loganathan, J., J. Jiang, A. Smith, A. Jedinak, A. Thyagarajan-Sahu, G.E. Sandusky, H. Nakshatri, and D. Sliva. 2014. The mushroom Ganoderma lucidum suppresses breast-to-lung cancer metastasis through the inhibition of pro-invasive genes. *International Journal of Oncology* 44: 2009–2015.
106. Luckner, M. 2013. *Secondary metabolism in microorganisms, plants and animals*. Berlin/Heidelberg: Springer.
107. Lukman, S., Y. He, and S.-C. Hui. 2007. Computational methods for traditional Chinese medicine: A survey. *Computer Methods and Programs in Biomedicine* 88: 283–294.
108. Ma, L., B. Wang, Y. Long, and H. Li. 2017. Effect of traditional Chinese medicine combined with Western therapy on primary hepatic carcinoma: A systematic review with meta-analysis. *Frontiers in Medicine* 11: 191–202.
109. Malmersjö, S., P. Rebellato, E. Smedler, H. Planert, S. Kanatani, I. Liste, E. Nanou, H. Sunner, S. Abdelhady, and S. Zhang. 2013. Neural progenitors organize in small-world networks to promote cell proliferation. *Proceedings of the National Academy of Sciences* 110 (16): E1524–E1532.

110. Mao, J.J., S.X. Xie, J. Zee, I. Soeller, Q.S. Li, K. Rockwell, and J.D. Amsterdam. 2015. Rhodiola rosea versus sertraline for major depressive disorder: A randomized placebo-controlled trial. *Phytomedicine* 22: 394–399.
111. Margulies, M., M. Egholm, W.E. Altman, S. Attiya, J.S. Bader, L.A. Bemben, J. Berka, M.S. Braverman, Y.J. Chen, Z. Chen, S.B. Dewell, L. Du, J.M. Fierro, X.V. Gomes, B.C. Godwin, W. He, S. Helgesen, C.H. Ho, G.P. Irzyk, S.C. Jando, M.L. Alenquer, T.P. Jarvie, K.B. Jirage, J.B. Kim, J.R. Knight, J.R. Lanza, J.H. Leamon, S.M. Lefkowitz, M. Lei, J. Li, K.L. Lohman, H. Lu, V.B. Makhijani, K.E. Mcdade, M.P. Mckenna, E.W. Myers, E. Nickerson, J.R. Nobile, R. Plant, B.P. Puc, M.T. Ronan, G.T. Roth, G.J. Sarkis, J.F. Simons, J.W. Simpson, M. Srinivasan, K.R. Tartaro, A. Tomasz, K.A. Vogt, G.A. Volkmer, S.H. Wang, Y. Wang, M.P. Weiner, P. Yu, R.F. Begley, and J.M. Rothberg. 2005. Genome sequencing in microfabricated high-density picolitre reactors. *Nature* 437: 376–380.
112. Markovič, R., A. Stožer, M. Gosak, J. Dolenšek, M. Marhl, and M.S. Rupnik. 2015. Progressive glucose stimulation of islet beta cells reveals a transition from segregated to integrated modular functional connectivity patterns. *Scientific Reports* 5: 7845.
113. Mason, O., and M. Verwoerd. 2007. Graph theory and networks in biology. *IET Systems Biology* 1: 89–119.
114. Maulik, S.K., V. Wilson, S. Seth, B. Bhargava, P. Dua, S. Ramakrishnan, and C.K. Katiyar. 2016. Clinical efficacy of water extract of stem bark of Terminalia arjuna (Roxb. ex DC.) Wight & Arn. in patients of chronic heart failure: a double-blind, randomized controlled trial. *Phytomedicine* 23: 1211–1219.
115. Maxam, A.M., and W. Gilbert. 1977. A new method for sequencing DNA. *Proceedings of the National Academy of Sciences of the United States of America* 74: 560–564.
116. Meacham, C.E., and S.J. Morrison. 2013. Tumour heterogeneity and cancer cell plasticity. *Nature* 501: 328–337.
117. Miao, X., and Q.L.X. Qin. 2015. Genome-wide transcriptome analysis of mRNAs and microRNAs in Dorset and Small Tail Han sheep to explore the regulation of fecundity. *Molecular and Cellular Endocrinology* 402: 32–42.
118. Mikheyev, A.S., and M.M. Tin. 2014. A first look at the Oxford Nanopore MinION sequencer. *Molecular Ecology Resources* 14: 1097–1102.
119. Moeller, B.J., R. Pasqualini, and W. Arap. 2009. Targeting cancer-specific synthetic lethality in double-strand DNA break repair. *Cell Cycle* 8: 1872–1876.
120. Mohs, R.C., and N.H. Greig. 2017. Drug discovery and development: Role of basic biological research. *Alzheimers Dement (NY)* 3: 651–657.
121. Mullard, A. 2018. 2017 FDA drug approvals. *Nature Reviews Drug Discovery* 17: 81.
122. Munakata, K., K. Takashima, M. Nishiyama, N. Asano, A. Mase, K. Hioki, Y. Ohnishi, M. Yamamoto, and K. Watanabe. 2012. Microarray analysis on germfree mice elucidates the primary target of a traditional Japanese medicine juzentaihoto: Acceleration of IFN-alpha response via affecting the ISGF3-IRF7 signaling cascade. *BMC Genomics* 13: 30.
123. National Cancer Institute. 2019. Cancer statistics, viewed 4 January 2019. https://www.cancer.gov/about-cancer/understanding/statistics
124. National Institutes of Health. 2019. Computational modeling, viewed 9 Ferbruary 2019. https://www.nibib.nih.gov/science-education/science-topics/computational-modeling
125. National Toxicology Program U.S. Department of Health and Human Service. 2019. About the report on carcinogens, viewed 6 February 2019. https://ntp.niehs.nih.gov/pubhealth/roc/index.html.
126. Naveen, C.R., S. Gaikwad, and R. Agrawal-Rajput. 2016. Berberine induces neuronal differentiation through inhibition of cancer stemness and epithelial-mesenchymal transition in neuroblastoma cells. *Phytomedicine* 23: 736–744.
127. Naveja, J.J., A. Dueñas-González, and J.L. Medina-Franco. 2016. Drug repurposing for epigenetic targets guided by computational methods. In *Epi-informatics*, ed. J.L. Medina-Franco. Boston: Academic Press.

128. Newman, D.J., and G.M. Cragg. 2016. Natural products as sources of new drugs from 1981 to 2014. *Journal of Natural Products* 79: 629–661.
129. Nisbet, R., G. Miner, and K. Yale. 2018. Theoretical considerations for data mining. In *Handbook of statistical analysis and data mining applications*, ed. R. Nisbet, G. Miner, and K. Yale, 2nd ed. Boston: Academic Press.
130. Noble, D. 1962. A modification of the Hodgkin-Huxley equations applicable to Purkinje fibre action and pace-maker potentials. *The Journal of Physiology* 160: 317–352.
131. Noble, R.L. 1990. The discovery of the vinca alkaloids–chemotherapeutic agents against cancer. *Biochemistry and Cell Biology* 68: 1344–1351.
132. Nowell, P.C. 1976. The clonal evolution of tumor cell populations. *Science* 194: 23–28.
133. Nygren, P., and R. Larsson. 2014. Drug repositioning from bench to bedside: Tumour remission by the antihelmintic drug mebendazole in refractory metastatic colon cancer. *Acta Oncologica* 53: 427–428.
134. Ojima, I., S. Chakravarty, T. Inoue, S. Lin, L. He, S.B. Horwitz, S.D. Kuduk, and S.J. Danishefsky. 1999. A common pharmacophore for cytotoxic natural products that stabilize microtubules. *Proceedings of the National Academy of Sciences of the United States of America* 96: 4256–4261.
135. Omosa, L.K., J.O. Midiwo, V.M. Masila, B.M. Gisacho, R. Munayi, K. Francisca, K.P. Chemutai, G. Elhaboob, M.E. Saeed, S. Hamdoun, V. Kuete, and T. Efferth. 2016. Cytotoxicity of 91 Kenyan indigenous medicinal plants towards human CCRF-CEM leukemia cells. *Journal of Ethnopharmacology* 179: 177–196.
136. Ooko, E., O. Kadioglu, H.J. Greten, and T. Efferth. 2017. Pharmacogenomic characterization and isobologram analysis of the combination of ascorbic acid and curcumin-two main metabolites of curcuma longa-in cancer cells. *Frontiers in Pharmacology* 8: 38.
137. Ouyang, L., Y. Luo, M. Tian, S.Y. Zhang, R. Lu, J.H. Wang, R. Kasimu, and X. Li. 2014. Plant natural products: From traditional compounds to new emerging drugs in cancer therapy. *Cell Proliferation* 47: 506–515.
138. Ovadje, P., A. Roma, M. Steckle, L. Nicoletti, J.T. Arnason, and S. Pandey. 2015. Advances in the research and development of natural health products as main stream cancer therapeutics. *Evidence-based Complementary and Alternative Medicine* 2015: 751348.
139. Özenver, N., M. Saeed, L.Ö. Demirezer, and T. Efferth. 2018. Aloe-emodin as drug candidate for cancer therapy. *Oncotarget* 9: 17770–17796.
140. Pantziarka, P., G. Bouche, L. Meheus, V. Sukhatme, and V.P. Sukhatme. 2014. Repurposing drugs in oncology (ReDO)-mebendazole as an anti-cancer agent. *Ecancermedical Science* 8: 443–443.
141. Pantziarka, P., M. Pirmohamed, and N. Mirza. 2018. New uses for old drugs. *British Medical Journal* 361: k2701.
142. Park, I.H., and C. Li. 2011. Characterization of molecular recognition of STAT3 SH2 domain inhibitors through molecular simulation. *Journal of Molecular Recognition* 24: 254–265.
143. Pfisterer, P.H., G. Wolber, T. Efferth, J.M. Rollinger, and H. Stuppner. 2010. Natural products in structure-assisted design of molecular cancer therapeutics. *Current Pharmaceutical Design* 16: 1718–1741.
144. Pinedo, H.M., G. Giaccone, and K. Sikora. 2007. *Drug resistance in the treatment of cancer*. Cambridge: Cambridge University Press.
145. Pineo, D., and C. Ware. 2012. Data visualization optimization via computational modeling of perception. *IEEE Transactions on Visualization and Computer Graphics* 18: 309–320.
146. Poornima, P., J.D. Kumar, Q. Zhao, M. Blunder, and T. Efferth. 2016. Network pharmacology of cancer: from understanding of complex interactomes to the design of multi-target specific therapeutics from nature. *Pharmacological Research* 111: 290–302.
147. Priya, R., R. Sumitha, C.G.P. Doss, C. Rajasekaran, S. Babu, R. Seenivasan, and R. Siva. 2015. Molecular docking and molecular dynamics to identify a novel human immunodeficiency virus inhibitor from alkaloids of Toddalia asiatica. *Pharmacognosy Magazine* 11: S414–S422.

148. Quanquan, G., and R. Tingge. 2009. An improved mathematic model of relative dosage of herb for Chinese medicine prescription. *Ori-ental Journal of Mathematics* 1: 13–25.
149. Quarteroni, A. 2018. 'The role of statistics in the era of big data: a computational scientist' perspective. *Statistics & Probability Letters* 136: 63–67.
150. Quintana, E., M. Shackleton, H.R. Foster, D.R. Fullen, M.S. Sabel, T.M. Johnson, and S.J. Morrison. 2010. Phenotypic heterogeneity among tumorigenic melanoma cells from patients that is reversible and not hierarchically organized. *Cancer Cell* 18: 510–523.
151. Reddy, L., B. Odhav, and K.D. Bhoola. 2003. Natural products for cancer prevention: A global perspective. *Pharmacology & Therapeutics* 99: 1–13.
152. Resemann, H.K., C.J. Watson, and B. Lloyd-Lewis. 2014. The Stat3 paradox: A killer and an oncogene. *Molecular and Cellular Endocrinology* 382: 603–611.
153. Rocha-Martin, J., C. Harrington, A.D. Dobson, and F. O'gara. 2014. Emerging strategies and integrated systems microbiology technologies for biodiscovery of marine bioactive compounds. *Marine Drugs* 12: 3516–3559.
154. Rodriguez, B., K. Burrage, D. Gavaghan, V. Grau, P. Kohl, and D. Noble. 2010. The systems biology approach to drug development: Application to toxicity assessment of cardiac drugs. *Clinical Pharmacology and Therapeutics* 88: 130–134.
155. Rothberg, J.M., W. Hinz, T.M. Rearick, J. Schultz, W. Mileski, M. Davey, J.H. Leamon, K. Johnson, M.J. Milgrew, M. Edwards, J. Hoon, J.F. Simons, D. Marran, J.W. Myers, J.F. Davidson, A. Branting, J.R. Nobile, B.P. Puc, D. Light, T.A. Clark, M. Huber, J.T. Branciforte, I.B. Stoner, S.E. Cawley, M. Lyons, Y. Fu, N. Homer, M. Sedova, X. Miao, B. Reed, J. Sabina, E. Feierstein, M. Schorn, M. Alanjary, E. Dimalanta, D. Dressman, R. Kasinskas, T. Sokolsky, J.A. Fidanza, E. Namsaraev, K.J. Mckernan, A. Williams, G.T. Roth, and J. Bustillo. 2011. An integrated semiconductor device enabling non-optical genome sequencing. *Nature* 475: 348.
156. Sadzuka, Y., T. Sugiyama, K. Shimoi, N. Kinae, and S. Hirota. 1997. Protective effect of flavonoids on doxorubicin-induced cardiotoxicity. *Toxicology Letters* 92: 1–7.
157. Saeed, M., H. Khalid, Y. Sugimoto, and T. Efferth. 2014. The lignan, (-)-sesamin reveals cytotoxicity toward cancer cells: Pharmacogenomic determination of genes associated with sensitivity or resistance. *Phytomedicine* 21: 689–696.
158. Saeed, M., O. Kadioglu, H. Khalid, Y. Sugimoto, and T. Efferth. 2015. Activity of the dietary flavonoid, apigenin, against multidrug-resistant tumor cells as determined by pharmacogenomics and molecular docking. *The Journal of Nutritional Biochemistry* 26: 44–56.
159. Saeed, M.E., M. Meyer, A. Hussein, and T. Efferth. 2016. Cytotoxicity of South-African medicinal plants towards sensitive and multidrug-resistant cancer cells. *Journal of Ethnopharmacology* 186: 209–223.
160. Saeed, M.E.M., N. Mahmoud, Y. Sugimoto, T. Efferth, and H. Abdel-Aziz. 2018a. Betulinic acid exerts cytotoxic activity against multidrug-resistant tumor cells via targeting autocrine motility factor receptor (AMFR). *Frontiers in Pharmacology* 9: 481.
161. ———. 2018b. Molecular determinants of sensitivity or resistance of cancer cells toward sanguinarine. *Frontiers in Pharmacology* 9: 136.
162. Sanger, F., S. Nicklen, and A.R. Coulson. 1977. DNA sequencing with chain-terminating inhibitors. *Proceedings the National Academy of Sciences of the USA* 74: 5463–5467.
163. Schinkel, A.H., and J.W. Jonker. 2003. Mammalian drug efflux transporters of the ATP binding cassette (ABC) family: An overview. *Advanced Drug Delivery Reviews* 55: 3–29.
164. Schroder, S., K. Beckmann, G. Franconi, G. Meyer-Hamme, T. Friedemann, H.J. Greten, M. Rostock, and T. Efferth. 2013. Can medical herbs stimulate regeneration or neuroprotection and treat neuropathic pain in chemotherapy-induced peripheral neuropathy? *Evidence-based Complementary and Alternative Medicine* 2013: 423713.
165. Seca, A.M.L., and D.C.G.A. Pinto. 2018. Plant secondary metabolites as anticancer agents: Successes in clinical trials and therapeutic application. *International Journal of Molecular Sciences* 19: 263.
166. Seigler, D.S. 2012. *Plant secondary metabolism*. New York: Springer.

167. Seo, E.J., B. Wiench, R. Hamm, M. Paulsen, Y. Zu, Y. Fu, and T. Efferth. 2015. Cytotoxicity of natural products and derivatives toward MCF-7 cell monolayers and cancer stem-like mammospheres. *Phytomedicine* 22: 438–443.
168. Seo, E.-J., M. Saeed, B.Y.K. Law, A.G. Wu, O. Kadioglu, H.J. Greten, and T. Efferth. 2016. Pharmacogenomics of scopoletin in tumor cells. *Molecules* 21: 496.
169. Seo, E.-J., S.M. Klauck, T. Efferth, and A. Panossian. 2019a. Adaptogens in chemobrain (Part I): Plant extracts attenuate cancer chemotherapy-induced cognitive impairment – Transcriptome-wide microarray profiles of neuroglia cells. *Phytomedicine* 55: 80–91.
170. ———. 2019b. Adaptogens in chemobrain (Part III): Antitoxic effects of plant extracts towards cancer chemotherapy-induced toxicity – transcriptome-wide microarray analysis of neuroglia cells. *Phytomedicine* 56: 246–260.
171. Siveen, K.S., S. Sikka, R. Surana, X. Dai, J. Zhang, A.P. Kumar, B.K.H. Tan, G. Sethi, and A. Bishayee. 2014. Targeting the STAT3 signaling pathway in cancer: Role of synthetic and natural inhibitors. *Biochimica et Biophysica Acta (BBA) – Reviews on Cancer* 1845: 136–154.
172. Song, X., Y. Xiong, X. Qi, W. Tang, J. Dai, Q. Gu, and J. Li. 2018. Molecular targets of active anticancer compounds derived from marine sources. *Marine Drugs* 16: 175.
173. Stähblin, H. 1973. Activity of a new glycosidic lignan derivative (VP 16-213) related to podophyllotoxin in experimental tumors. *European Journal of Cancer (1965)* 9: 215–221.
174. Sun, Y., W. Zhang, Y. Chen, Q. Ma, J. Wei, and Q. Liu. 2016. Identifying anti-cancer drug response related genes using an integrative analysis of transcriptomic and genomic variations with cell line-based drug perturbations. *Oncotarget* 7: 9404–9419.
175. Szakacs, G., J.K. Paterson, J.A. Ludwig, C. Booth-Genthe, and M.M. Gottesman. 2006. Targeting multidrug resistance in cancer. *Nature Reviews. Drug Discovery* 5: 219–234.
176. Tariq, A., S. Sadia, K. Pan, I. Ullah, S. Mussarat, F. Sun, O.O. Abiodun, A. Batbaatar, Z. Li, D. Song, Q. Xiong, R. Ullah, S. Khan, B.B. Basnet, B. Kumar, R. Islam, and M. Adnan. 2017. A systematic review on ethnomedicines of anti-cancer plants. *Phytotherapy Research* 31: 202–264.
177. TM-MC: A database of medicinal materials and chemical compounds in Northeast Asian traditional medicine 2018, Introduction, viewed 10.02.2019. http://informatics.kiom.re.kr/compound/
178. Trudeau, R.J. 2013. *Introduction to graph theory*. New York: Dover Publications.
179. Turkson, J., and R. Jove. 2000. STAT proteins: Novel molecular targets for cancer drug discovery. *Oncogene* 19: 6613–6626.
180. Tyanova, S., and J. Cox. 2018. Perseus: A bioinformatics platform for integrative analysis of proteomics data in cancer research. In *Cancer systems biology: Methods and protocols*, ed. L. Von Stechow. New York: Humana Press.
181. Uzuner, H., R. Bauer, T.-P. Fan, D.-A. Guo, A. Dias, H. El-Nezami, T. Efferth, E.M. Williamson, M. Heinrich, N. Robinson, P.J. Hylands, B.M. Hendry, Y.-C. Cheng, and Q. Xu. 2012. Traditional Chinese medicine research in the post-genomic era: Good practice, priorities, challenges and opportunities. *Journal of Ethnopharmacology* 140: 458–468.
182. Valouev, A., D.S. Johnson, A. Sundquist, C. Medina, E. Anton, S. Batzoglou, R.M. Myers, and A. Sidow. 2008. Genome-wide analysis of transcription factor binding sites based on ChIP-Seq data. *Nature Methods* 5: 829–834.
183. Verdine, G.L. 1996. The combinatorial chemistry of nature. *Nature* 384: 11–13.
184. Vilgis, S., and H.-P. Deigner. 2018. Sequencing in precision medicine. In *Precision medicine*, ed. H.-P. Deigner and M. Kohl. London: Academic Press.
185. Vogel, H.G. 1991. Similarities between various systems of traditional medicine. Considerations for the future of ethnopharmacology. *Journal of Ethnopharmacology* 35: 179–190.
186. ———. 2002. *Drug discovery and evaluation: Pharmacological assays*. Heidelberg: Springer. Viewed 8 February 2019. https://www.springer.com/gp/book/9783540709954
187. Volm, M., and T. Efferth. 2015. Prediction of cancer drug resistance and implications for personalized medicine. *Frontiers in Oncology* 5: 282.

188. Walther, Z., and J. Sklar. 2011. Molecular tumor profiling for prediction of response to anticancer therapies. *Cancer Journal* 17: 71–79.
189. Wang, M., R.J. Lamers, H.A. Korthout, J.H. Van Nesselrooij, R.F. Witkamp, R. Van Der Heijden, P.J. Voshol, L.M. Havekes, R. Verpoorte, and J. Van Der Greef. 2005. Metabolomics in the context of systems biology: Bridging traditional Chinese medicine and molecular pharmacology. *Phytotherapy Research* 19: 173–182.
190. Wang, Y., X. Wang, and Y. Cheng. 2006. A computational approach to botanical drug design by modeling quantitative composition–activity relationship. *Chemical Biology & Drug Design* 68: 166–172.
191. Wang, H., F. Lafdil, X. Kong, and B. Gao. 2011. Signal transducer and activator of transcription 3 in liver diseases: A novel therapeutic target. *International Journal of Biological Sciences* 7: 536–550.
192. Wang, R.-S., A. Saadatpour, and R. Albert. 2012. Boolean modeling in systems biology: An overview of methodology and applications. *Physical Biology* 9: 055001.
193. Wang, J., Y. Li, Y. Yang, J. Du, M. Zhao, F. Lin, S. Zhang, and B. Wang. 2017. Systems pharmacology dissection of multiscale mechanisms of action for herbal medicines in treating rheumatoid arthritis. *Molecular Pharmacology* 14: 3201–3217.
194. Wang, C., P. Xu, L. Zhang, J. Huang, K. Zhu, and C. Luo. 2018. Current strategies and applications for precision drug design. *Frontiers in Pharmacology* 9: 787–787.
195. Weerasinghe, P., Y. Li, Y. Guan, R. Zhang, D.J. Tweardy, and N. Jing. 2008. T40214/PEI complex: A potent therapeutics for prostate cancer that targets STAT3 signaling. *Prostate* 68: 1430–1442.
196. Wen, Z., Z. Wang, S. Wang, R. Ravula, L. Yang, J. Xu, C. Wang, Z. Zuo, M.S. Chow, L. Shi, and Y. Huang. 2011. Discovery of molecular mechanisms of traditional Chinese medicinal formula Si-Wu-Tang using gene expression microarray and connectivity map. *PLoS One* 6: e18278.
197. Willett, P., J.M. Barnard, and G.M. Downs. 1998. Chemical similarity searching. *Journal of Chemical Information and Computer Sciences* 38: 983–996.
198. Williams, S.D., R. Birch, L.H. Einhorn, L. Irwin, F.A. Greco, and P.J. Loehrer. 1987. Treatment of disseminated germ-cell tumors with cisplatin, bleomycin, and either vinblastine or etoposide. *The New England Journal of Medicine* 316: 1435–1440.
199. Wong, A.M., Y. Zhang, K. Kesler, M. Deng, L. Burhenn, D. Wang, A. Moro, Z. Li, and D. Heber. 2010. Genomic and in vivo evidence of synergy of a herbal extract compared to its most active ingredient: Rabdosia rubescens vs. oridonin. *Experimental and Therapeutic Medicine* 1: 1013–1017.
200. World Health Organization. 2019. Cancer, viewed 4 February 2019. https://www.who.int/cancer/en/
201. Wu, J., and R. Ji. 1989. Application of QSAR in drug design. *Foreign Med Sci·Pharm Sect (Chin)* 16: 8–16.
202. Wu, D., C.M. Rice, and X. Wang. 2012. Cancer bioinformatics: A new approach to systems clinical medicine. *BMC Bioinformatics* 13: 71.
203. Wu, Q., Z. Yang, Y. Nie, Y. Shi, and D. Fan. 2014. Multi-drug resistance in cancer chemotherapeutics: Mechanisms and lab approaches. *Cancer Letters* 347: 159–166.
204. Wu, C.F., C. Hong, S.M. Klauck, Y.L. Lin, and T. Efferth. 2015. Molecular mechanisms of rosmarinic acid from Salvia miltiorrhiza in acute lymphoblastic leukemia cells. *Journal of Ethnopharmacology* 176: 55–68.
205. Yang, J., X. Liao, M.K. Agarwal, L. Barnes, P.E. Auron, and G.R. Stark. 2007. Unphosphorylated STAT3 accumulates in response to IL-6 and activates transcription by binding to NFkappaB. *Genes & Development* 21: 1396–1408.
206. Yang, H.J., D. Shen, H.-Y. Xu, and P. Lu. 2012. A new strategy in drug design of Chinese medicine: Theory, method and techniques. *Chinese Journal of Integrative Medicine* 18: 803–806.

207. Youns, M., Y.J. Fu, Y.G. Zu, A. Kramer, V.B. Konkimalla, B. Radlwimmer, H. Sultmann, and T. Efferth. 2010. Sensitivity and resistance towards isoliquiritigenin, doxorubicin and methotrexate in T cell acute lymphoblastic leukaemia cell lines by pharmacogenomics. *Naunyn-Schmiedeberg's Archives of Pharmacology* 382: 221–234.
208. Yu, H., D. Pardoll, and R. Jove. 2009. STATs in cancer inflammation and immunity: A leading role for STAT3. *Nature Reviews Cancer* 9: 798–809.
209. Yue, Z., W. Zhang, Y. Lu, Q. Yang, Q. Ding, J. Xia, and Y. Chen. 2015. Prediction of cancer cell sensitivity to natural products based on genomic and chemical properties. *PeerJ- the Journal of Life and Environmental Sciences* 3: e1425–e1425.
210. Zhang, P., and V. Brusic. 2014. Mathematical modeling for novel cancer drug discovery and development. *Expert Opinion on Drug Discovery* 9: 1133–1150.
211. Zhang, H.-F., and R. Lai. 2014. STAT3 in cancer-friend or foe? *Cancers* 6: 215–1440.
212. Zhang, S., W. Lu, X. Liu, Y. Diao, F. Bai, L. Wang, L. Shan, J. Huang, H. Li, and W. Zhang. 2011. Fast and effective identification of the bioactive compounds and their targets from medicinal plants via computational chemical biology approach. *Medicinal Chemistry Communications* 2: 471–477.
213. Zhang, A., H. Sun, P. Wang, Y. Han, and X. Wang. 2012. Future perspectives of personalized medicine in traditional Chinese medicine: A systems biology approach. *Complementary Therapies in Medicine* 20: 93–99.
214. Zhao, J., P. Jiang, and W. Zhang. 2010. Molecular networks for the study of TCM pharmacology. *Briefings in Bioinformatics* 11: 417–430.
215. Zheng, H.C., A. Noguchi, K. Kikuchi, T. Ando, T. Nakamura, and Y. Takano. 2014. Gene expression profiling of lens tumors, liver and spleen in alpha-crystallin/SV40 T antigen transgenic mice treated with Juzen-taiho-to. *Molecular Medicine Reports* 9: 547–552.
216. Zhou, X., Y. Peng, and B. Liu. 2010. Text mining for traditional Chinese medical knowledge discovery: A survey. *Journal of Biomedical Informatics* 43: 650–660.

Synergistic Effects of Chinese Herbal Medicine and Biological Networks

Deep Jyoti Bhuyan, Saumya Perera, Kirandeep Kaur, Muhammad A. Alsherbiny, Mitchell Low, Sai-Wang Seto, Chun-Guang Li, and Xian Zhou

Introduction

Traditional Chinese Medicine (TCM) has been extensively practised as primary healthcare in China and other Asian countries including Japan and Korea over thousands of years. Because of its unique and holistic approach towards treatment and prevention of diseases, maintenance of health and prolongation of life, TCM has recently gained significant attention worldwide.

Chinese Herbal Medicine (CHM) is a key modality of TCM, which is emphasised through its systematic approach of herbal formulae (*Fang Ji* in Chinese – up to 20 herbs consisting of a large number of chemical constituents) in sustaining the balance in the body. The popularity of CHM as complementary therapy has significantly increased recently especially in Western nations. Substantial progress has been made over the last decades in understanding the efficacy and modes of action of many commonly used herbs and formulations [1]. Synergy has been found to be the key mechanism of action of CHMs in many studies. It has been demonstrated that combination therapy of CHM could provide a comprehensive approach with greater therapeutic benefits, more systematic targets and reduced side effects through its synergistic behaviour to manage a number of chronic diseases such as AIDS, cancer, atherosclerosis and diabetes, all of which have complex aetiology and pathophysiology and, therefore, are difficult to treat using a single drug-target approach [1]. Numerous studies have also identified many active ingredients in CHMs and their biological targets providing insights into their synergistic effects. Furthermore,

D. J. Bhuyan · S. Perera · K. Kaur · M. Low · S.-W. Seto · C.-G. Li · X. Zhou (✉)
NICM Health Research Institute, Western Sydney University, Penrith, NSW, Australia
e-mail: p.zhou@westernsydney.edu.au

M. A. Alsherbiny
NICM Health Research Institute, Western Sydney University, Penrith, NSW, Australia

Department of Pharmacognosy, Faculty of Pharmacy, Cairo University, Cairo, Egypt

M. Bizzarri (ed.), *Approaching Complex Diseases*, Human Perspectives in Health Sciences and Technology 2, https://doi.org/10.1007/978-3-030-32857-3_17

systems biology approaches have been developed and extensively applied to investigate the multi-target interactions of active constituents in CHM at a molecular level. This follows the current trend of drug development with "network target, multi-component" strategy which has gradually replaced the conventional "single drug-single target" pharmaceutical approach. Similarly, research interest in synergistic interactions of multi-component herbal preparations and their positive interactions with pharmaceutical drugs has also started to grow in recent years [2]. However, the safety, quality and pharmacological activity of CHMs still remain obscure due to their complex nature, which has hindered their application in conventional medicine and recognition across the international healthcare systems.

In this chapter, we review studies on the synergistic effects of CHMs through both pharmacodynamics and pharmacokinetics approaches and how they are effectively measured. Studies of CHM for various therapeutic targets of cancer, diabetes, musculoskeletal pain, cardiovascular, microbial, inflammatory, hepatic and oxidative stress-related diseases are discussed. In addition, the implementation of the system to system (systems biology) in CHM is introduced.

Concept of Synergy in CHM

Synergy is defined as the interaction of two or more agents to produce a combined effect greater than the sum of their individual effects [3]. In medical research, however, the understanding of synergy is complex. Spinella [4] has categorised the concept of synergy broadly into two main groups based on the modes of action – pharmacodynamic and pharmacokinetic synergy. The first type of synergy describes two or more agents that work on the same receptors or biological targets that result in enhanced therapeutic outcomes through their positive interactions. The second type of synergy results from interactions between two or more agents during their pharmacokinetic processes (absorption, distribution, metabolism and elimination) leading to changes of the agents quantitatively in the body and, hence, their therapeutic effects [4]. Spinella [4] also illustrated that there are three ways to validate the synergistic effects: (1) measure the phytopharmacological effects/changes, (2) evaluate the difference in effectiveness, iii) observe the dose response with mixtures of bioactive compounds after using both single herb (or drug)/and herbal formulae (or extracts), which means a lower dose is used when synergy is generated.

As CHM prescription emphasises the maintenance and restoration of balance in the overall functions of the body, most prescriptions are given as herbal formulae comprising of multiple herbs needed to target various aspects of the body's responses in a comprehensive manner [5]. The key mechanism of synergy in herbal formulae relies on the interactions among the multiple ingredients and the multi-target actions on the body [6]. Therefore, CHM does not stipulate a single chemical compound as Western Medicine (WM) does, rather, it utilises a whole herb plant or combination of various herbs consisting of different chemical components. Figure 1 depicts the key difference

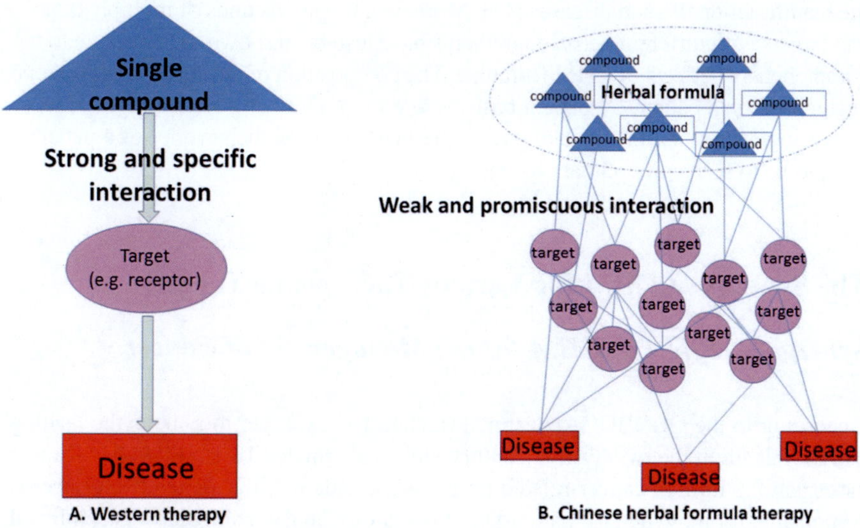

Fig. 1 A comparison between mechanisms of action of WM (one drug, one target, one disease) and CHM (multiple component, network targets, multiple diseases) [6]

in the molecular mechanisms of action between CHM and WM [6]. The complex synergistic interactions among the herbs in CHM formulations (*Fu Fang* in Chinese) are considered to be able to promote therapeutic effects, reduce toxicity, enhance the bioavailability, and/or display multiple constituents on various targets/diseases [7]. The design of herbal formulations follow the principle of compatibility, called "*Pei Wu*," which requires the considerations of different interrelationships of herbal ingredients including synergism (*Xiang Xu*), assisting (*Xiang Shi*), detoxication (*Xiang Sha* and *Xiang Wei*), antagonism (*Xiang Wu*), and rejection (*Xiang Fan*) [7]. Based on this principle, different herbs are assembled in a formula following "*Jun-Chen-Zuo-Shi*" theory (also known as "Emperor–Minister- Assistant-Courier") to achieve synergy in the form of desirable efficacy and minimal side effects/toxicity. "*Jun*" is the man in herb in a herbal formula with a relatively higher ratio directly targeting the disease; "*Chen*" is an adjuvant herb to promote therapeutic effect of the key herb or to target the accompanying symptoms; "*Zuo*" is used for reducing the side effects of the herbal formula; "*Shi*" is the herb that guides the active ingredients to reach the target organs or to harmonise their actions. Numerous studies have suggested that herbal extracts as a whole and/or multiple herbs in complex formulations offer better efficacy than equivalent doses of individual active ingredient and/or herbs when used alone; emphasising the significance of synergistic action in herbal therapies [8–10]. It has been indicated that an adjustment to the formulation of a herbal combination should be made not only for single ingredients but also for imbalances in the status of disease-specific complexes [11]. The "*Jun-Chen-Zuo-Shi*" combinatorial rule of herbal formulae to treat complex diseases can be justified by the arrangements of herbs on network-based functional modules as several pathological processes are involved in the manifestation of such diseases [11]. Moreover, the interactions of multiple biological targets with different herbal ingredients have also been shown to contribute to the comprehensive effects of CHM formulae. The composition of a CHM formula can be customised on a patient-by-patient basis by adding or removing certain ingredients or increasing/decreasing the dose of each ingredient based on the severity of a patient's symptoms to reach an optimal clinical outcome [12].

The Synergy of CHM for Various Therapeutic Targets

Synergistic Effect of CHM for the Management of Cancer

According to the GLOBOCAN estimates, cancer is expected to rank as the leading cause of death in the twenty-first century with an estimated 18.1 million new cancer cases and 9.6 million cancer-related deaths worldwide in 2018 alone [13]. Cancer is responsible for more deaths than AIDS, tuberculosis, malaria and diabetes combined [14]. Late diagnoses, molecular heterogeneity, expensive and few effective therapeutic options as well as the increased resistance to chemo and radiation therapies are currently some of the biggest challenges in cancer treatment. Severe side effects of many chemotherapeutic regimens also make cancer one of the most complex diseases to treat [15].

For centuries, CHMs have been applied for the treatment of cancers in China and other parts of the world because of their low cost, easy availability and mild adverse reactions [15, 16]. CHM theories view cancer as a systemic disease resulting from the imbalances between the body's endogenous physical conditions and exogenous pathogenic factors [15]. Unlike WM, CHM formulae implement multi constituents which interact with different biological targets for increased efficacy and reduced side effects in cancer therapy [14, 16]. The molecular mechanisms of action of many CHM formulae perhaps are attributed to their polyvalent effect which includes- (1) strong binding affinity of polyphenols to cellular proteins, enzymes and glycoproteins and (2) lipophilicity of terpenoids enabling them to permeate through cell membranes [17]. In addition, the role of polyphenols as promising anticancer agents has been highlighted in several reports due to their ability to prevent and repair free radical-induced oxidative damage of cells [18]. Interestingly, some of the common polyphenols such as (-)-epigallocatechin-3-gallate, quercetin and gallic acid found in many CHM formulae have been also shown to induce the formation of high levels of H_2O_2 – mediated oxidative stress to inhibit the proliferation of cancer cells [19].

A number of pre-clinical and clinical studies have been performed in the last few decades in order to scientifically validate the effectiveness of anticancer CHM formulae. Several reports have suggested that their efficacy can be ascribed to the complex synergy between the herbs. For instance, a CHM formula named PHY906, from the herbs *Scutellaria baicalensis* Georgi *(Huang qin)*, *Paeonia lactiflora* Pall. (*Bai shao*), *Ziziphus jujuba* Mill. (*Da zao*) and *Glycyrrhiza glabra* L. (*Gan cao*), has been shown to synergistically enhance the potency of chemotherapeutic drugs and reduce side effects in patients with advanced colorectal and liver cancer in few clinical studies [16, 20–22]. This is perhaps one of the most studied TCM formulae as an adjuvant therapeutic candidate for the treatment of advanced colorectal, pancreatic, gastrointestinal and liver cancer in Phase I and II clinical trials. The efficacy of PHY906 has been linked with the synergy among the individual herbs in the formula as well as their positive interactions with different chemotherapeutic interventions. Liu et al. [23] showed that all four herbs are required in PHY906 formula for the full range of observed efficacy including antitumour activity, reduction in body weight loss, and prevention of mortality. In pre-clinical animal models, the formula in conjunction with chemotherapeutic drugs such as nivolumab, sorafenib and irinotecan has been found to trigger changes in the tumour microenvironment through several molecular mechanisms [21, 24–26]. For instance, the enhancement of M1/M2 macrophage infiltration in the tumour microenvironment by upregulation of monocyte chemoattractant protein 1 (MCP1) protein was found to be a common mechanism of action in two pre-clinical studies using PHY906-Nivolumab and PHY906-Sorafenib against liver cancer [21, 25]. The dual mechanism of action of PHY906 in reducing toxicity in addition to enhancing the efficacy of anticancer drugs has also gained a significant amount of interest from the scientific community. The broad-spectrum potential of this CHM formula as an adjuvant therapy has been widely explored with many chemotherapeutic drugs such as gemcitabine (pancreatic cancer), capecitabine (colorectal and liver cancer), 5-fluorouracil (colorectal cancer), VP-16 (lung cancer), troxacitabine (leukemia and pancreatic cancer), clevudine (liver cancer), paclitaxel (lung, breast and ovarian cancer), oxaliplatin (colorectal cancer), sunitinib (renal and liver cancer), and thalidomide (liver cancer) [27–31].

Huanglian Jiedu Tang (HJT) is another commonly used CHM anticancer formula composed of four herbs- *Coptis chinensis* Franch. (*Huang lian*), *Phellodendron chinense* Schneid. (*Huang bo*), *Scutellaria baicalensis* Georgi (*Huang qin*), and *Gardenia jasminoides* Ellis (*Zhi zi*) in equal proportions [15]. Synergy for HJT is attributed to its multi-target behaviour from its chemical components. S. radix and its bioactive compound baicalein present in this formula were found to be solely responsible for mediating apoptosis in MPC-1- immature myeloma cells *in vitro* [15, 32]. Whereas, geniposide, berberine and baicalin in HJT displayed an additive effect on the eukaryotic elongation factor-2 (eEF2) inactivation leading to suppression of cell/tumour growth and angiogenesis in the *in vitro* and *in vivo* models of hepatocellular carcinoma [33].

Synergistic studies on CHM formulae performed in pre-clinical and clinical experimental setups within the last few decades are shown in Table 1. It should be noted that most of these formulae have presented synergistic or additive effects with chemo, radiation and resection therapies emphasising their potential as adjuvants in various cancer treatments. However, studies explaining the synergy within a formula among its herbal ingredients and/or chemical constituents are still scarce. Although the current literature represents numerous reports on the anticancer activity of individual herbs and bioactive compounds found in CHM formulae, studying the single herb/compound strategy does not provide us with a complete understanding of

Table 1 Synergistic anticancer activity between herbs and their active components present in Chinese formulae and with chemotherapeutic drugs

Chinese formula	Herbs	Active ingredients	Synergistic anticancer activity	References
PHY906 derived from Huang Qin Tang	*Scutellaria baicalensis* Georgi, *Glycyrrhiza glabra* L., *Ziziphus jujuba* Mill., and *Paeonia lactiflora* Pall. at a ratio of 3:2:2:2	PHY906, chemotherapeutic drugs and radiation therapy	*Enhanced clinical outcome*	[21, 22, 24–26, 35–38]
			Several Phase I/II clinical studies showed that PHY906 enhanced the clinical outcome of chemotherapy (irinotecan, capecitabine and sorafenib) for advanced colorectal and liver cancer as well as pancreatic and gastrointestinal cancer by stabilising tumour growth, reducing side effects, especially chemotherapy-associated diarrhoea, and improving survival time.	
			Pharmacological mechanism	
			Preclinical studies revealed that PHY906 counteracted the toxicity of irinotecan by restoring the intestinal epithelium and promoting the regeneration of intestinal progenitor or stem cells and several Wnt signaling components. It reduced the usage of nivolumab (anti-PD1) by threefold with better antitumour effects compared to anti-PD1 alone by changing the tumour	

(continued)

Table 1 (continued)

Chinese formula	Herbs	Active ingredients	Synergistic anticancer activity	References
			microenvironment favourable to M1-like macrophages in BDF1 mice bearing with Hepa 1-6 tumours. In female BDF-1 mice bearing subcutaneous colon 38 tumours, PHY906 in combination with irinotecan triggered unique tissue-specific responses which were not activated when PHY906 and irinotecan were used alone.	
			PHY906 increased the therapeutic ratio and decreased the toxicity of whole-abdomen irradiation without protecting tumours.	
Realgar-*Indigo naturalis* anticancer formula	Realgar, Indigo Naturalis, *Salvia miltiorrhiza* Bge. and *Pseudostellaria heterophylla* (Miq.) Pax ex Pax et Hoffm.	Tetra-arsenic tetrasulphide from Realgar, indirubin from Indigo naturalis and tanshinone IIA from *Salviae Miltiorrhizae* Bge., respectively.	*Enhanced clinical outcome*	[39–41]
			A complete remission rate of 98.3% with minimal side effect was observed in sixty acute promyelocytic leukaemia patients.	
			Pharmacological mechanism	
			A significant increase in the median overall survival rate was observed in acute promyelocytic leukaemia in FVB/NJ mice treated with the three-ingredient combination	

(continued)

Table 1 (continued)

Chinese formula	Herbs	Active ingredients	Synergistic anticancer activity	References
			compared to control and mono- or bi-therapy of these ingredients. Indirubin and tanshinone IIA enhanced the transport of tetra-arsenic tetrasulphide into target cells.	
Yanshu Injection/ Fu Fang Ku Shen injection	Sophorae flavescentis Radix (SFR) and Smilacis glabrae Rhizoma (SGR) at a ratio of 7:3	Oxymatrine and matrine from SFR. Astilbin, 5-O-caffeoylshikimic acid and taxifolin from SGR.	*Enhanced clinical outcome*	[15, 42, 43]
			In sixty patients with stage III nasopharyngeal carcinoma, Yanshu injection in combination with radio/chemotherapy improved the quality of life by showing greater efficacy and reducing side effects.	
			Pharmacological mechanism	
			Yanshu injection inhibited N-methyl-N′-nitro-N-nitrosoguanidine-induced gastric carcinogenesis by protecting against carcinogen-induced oxidative damage as well as improved immunity.	
Huanglian Jiedu Tang (HJT)	Coptis Rhizome, Phellodendri chinensis Cortex, Scutellariae radix, and Gardeniae Fructus in equal proportions.	Scutellaria Radix and baicalein, geniposide, berberine and baicalin	*Pharmacological mechanism*	[15, 32, 33, 44, 45]
			HJT arrested Hep G2 and PLC/PRF/5 cells in the S-G_2/M phase by downregulating the levels of cyclin A, cyclin B1, Cdc2, Cdc25C, Bcl-2 and Bcl-xL and increased the expression of BAX	

(continued)

Table 1 (continued)

Chinese formula	Herbs	Active ingredients	Synergistic anticancer activity	References
			and BAK. Attenuated the NF-κB activity, which in turn induced the mitochondria-dependent apoptosis in human liver cancer cells. It displayed colon cancer chemo-preventative properties by suppressing COX-2 activity and azoxymethane-induced aberrant crypt foci development in rats.	
			Multi-target behaviour from its ingredients	
			Scutellaria Radix and its bioactive compound baicalein exhibited anti-proliferative effects on MPC-1- immature myeloma cells by inducing caspase-9 and caspase-3 mediated apoptosis via a mitochondria-dependent pathway.	
			The major components of HJT-geniposide, berberine and baicalin, additively increased the eEF2 phosphorylation which in turn led to the growth inhibition of HepG2 cells *in vitro* and suppression of growth and angiogenesis in xenografted murine model.	

(continued)

Table 1 (continued)

Chinese formula	Herbs	Active ingredients	Synergistic anticancer activity	References
Jiedu Xiaozheng Yin (JXY)	*Hedyotis diffusa* Herba, Cremastrae Pseudobulbus/ Pleiones Pseudobulbus, Prunellae Spica and *Sophorae flavescentis*	Ethyl acetate extracts of JXY	*Enhanced clinical outcome*	[15, 46–49]
			JXY showed a synergetic effect with surgery in 42 patients with stage III primary hepatic carcinoma. It improved immune function, reduced the recurrence rate and increased the cumulative survival rate of patients when administered in the peri-operational period.	
			Pharmacological mechanism	
			Ethyl acetate extracts of JXY (EE-JXY) inhibited HepG2 hepatocarcinoma by apoptosis via mitochondrial pathway and arresting the cell cycle at the G_0/G_1 phase with increased expression of G_1-related cyclins D and E. Decreased the proliferation index and increased the apoptotic index of tumours in BALB/c nude mice. EE-JXY suppressed polycomb gene product Bmi1 and Wnt/β-catenin signaling.	

the benefits and/or the toxicity of herbal formulations as a whole. Therefore, pre-clinical studies using combination index, systems biology and omics approaches and more well-designed clinical studies are crucial to gain a comprehensive understanding of the complex synergy among the herbs/components of CHM formulations as well as their combination with the standard Western medicinal cancer therapies [16]. The scarcity of optimised quality control protocols for complex herbal formulations to ensure product safety, uniformity, and efficacy is another major challenge that needs attention in CHM research [34].

Synergistic Effect of CHM for the Management of Cardiovascular Diseases

Cardiovascular disease (CVD) is an umbrella term for ailments of the heart and blood vessels representing 31% of all global deaths in 2016 [50]. CHM has been extensively used for the treatment of heart failure, hypertension and coronary heart diseases for many years [51].

Compound Danshen Formula (CDF) consisting of Salviae miltiorrhizae Radix et Rhizoma, Notoginseng Radix et Rhizoma and Borneolum syntheticum is widely used for decades in CHM to treat CVD. The chemical constituents of CDF have been found to interact with many biological targets of CVD reiterating the multi-target behaviour of CHM. For example, a simplified formula TSG, derived from the bioactive compounds isolated from CDF, was found to completely reverse 17 out of 22 biomarkers to normal levels for myocardial infarctions (MI) in rats. Although CDF showed stronger activity compared to that of TSG, a synergistic relationship was observed between TSG and MI therapy in metabolomics analysis [52]. Li et al. [53] investigated the mechanisms of action and interactions of the compounds in CDF using a novel systems- pharmacological model. A total of 320 compounds of CDF were selected from different databases and literature, including 201 compounds from Salviae Miltiorrhizae Radix et Rhizoma, 112 from Notoginseng Radix et Rhizoma and 31 from Borneolum. The *in silico* model predicted that 90 compounds of CDF have more than 50% of oral bioavailability. These compounds were accounted for only 28.1% of the total compounds in the formula and exhibited a total of 9220 interactions with their targets while ginsenoside Rb1, ginsenoside Rb2, ginsenoside Ro showed the highest candidate targets. Out of the 41 potential targets relevant to CVD, Salviae Miltiorrhizae Radix et Rhizoma, as the emperor of CDF, acted on 39 potential targets, whereas Notoginseng Radix et Rhizoma displayed 36 potential targets as a minister. In addition, Salviae Miltiorrhizae Radix et Rhizoma and Notoginseng Radix et Rhizoma showed similarity by overlapping 34 targets [53]. A combination of luteolin and buddleoside from Chrysanthemi Indici Flos (*Ye ju hua*) was shown to have a synergistic anti-hypertensive effect on spontaneously hypertensive rats. It significantly reduced systolic blood pressure (by 15.42 mmHg) compared to the individual compounds [54].

Liang et al. [55] demonstrated that Shuanglong formula (SLF) had a synergistically enhanced therapeutic effect compared to its individual components (Ginseng Radix et Rhizoma Rubra and Salviae Miltiorrhizae Radix et Rhizoma) on MI. In a ratio of 3:7, SLF significantly suppressed urinary biomarkers and shifted urinary TCA cycle, suggesting that the protective effect was via the regulation of myocardial energy metabolism [55]. A novel formula NSLF6 consisting of 12 ginsenosides and 8 salvianolic acids showed synergistic interactions between total ginsenosides and total salvianolic acids for MI by promoting cardiac cell regeneration and myocardial angiogenesis and reducing oxidative damage of myocardial cells in pre- clinical models [56].

Synergy in CHM for the Management of Diabetes Mellitus

Diabetes mellitus is a metabolic disorder characterised by hyperglycaemia due to reduction of insulin level and/or resistance [57]. Uncontrolled chronic hyperglycaemia can lead to multiple organs damage and complications, such as retinopathy, neuropathy and nephropathy. Although many antidiabetic agents (eg. sulfonylurea, thiazolidinedione and metformin) are available, these synthetic antidiabetic compounds produce a number of adverse effects, including weight gain, fluid retention and increased risk for heart failure. These adverse effects greatly limit their compliance and value clinically. Over the last decade, multi-drug therapy has been suggested as a novel strategy for the management of diabetes and its complications. Although the combination of several oral antidiabetic drugs has been shown to improve blood glucose control when compared to monotherapy, this 'polypill' approach, however, only showed a modest reduction of diabetic complications [58]. Recently, there is a growing interest in the use of CHM for the treatment of diabetes, because many CHMs have minimal side effects, relatively low cost and multi-target property which could lead to higher compliance and better clinical outcome. Indeed, it was reported that approximately 800 medicinal plants can be used for blood glucose control. Although the underlying pharmacological mechanisms for many of these herbs have not been studied in detail, it is well-known that various components of individual or combination of herbs act synergistically to produce their anti-diabetic effect [59].

Scutellariae Radix (SR) and Coptidis Rhizoma (CR) have been used in combination to manage type 2 diabetes mellitus (T2DM) in China for over hundreds of years. The compatibility effects of SR and CR were studied using a T2DM rat model recently. SR and CR, when given to the T2DM rats alone, both suppressed inflammation and improved blood glucose and insulin resistance. Interestingly, the improvement of the symptoms of T2DM was more effective when SR and CR were administrated in combination than single herb treatment. These effects were largely mediated by the down-regulation of the MAPK (mitogen-activated protein kinases, P38), extracellular regulated protein kinases (ERK) and c-Jun N-terminal kinase (JNK). In addition, the increased hepatic activity of glucokinase, phosphofructokinase, pyruvate kinase and glycogen synthase were also observed

[60]. However, this study did not demonstrate whether the improved efficacy from the combined treatment was due to synergistic or simply an additional effect between SR and CR.

A number of studies have shown the anti-diabetic effect of isolated bioactive components or amino acids from herbal extracts. Resveratrol is a plant polyphenol found in the skin of red grapes and other fruits and herbs (e.g. *Smilacis glabra* Roxb. [61, 62]. This polyphenol has been shown to produce its cardiovascular protective and anti-diabetic effects via the modulation of the Sirt1-dependent signaling pathway [63]. Similarly, leucine, an amino acid commonly found in the daily diet, has also been shown to modulate Sirt1-signaling pathway to improve mitochondrial biogenesis and energy metabolism [64]. Bruskbauer et al. [65] demonstrated that low dose of resveratrol (200nM) and leucine (0.5 mM) only induced modest fatty acid oxidation by 18% and 27%, respectively in 3T3-L1 mouse adipocyte cultured in high glucose (25 mM) condition. However, fatty acid oxidation markedly increased to 118% when co-incubated with resveratrol (200 nM) and leucine (0.5 mM). Similarly, this synergistic effect was also observed in diet-induced obese mice, where combined treatment of low dose resveratrol (12.5 mg/kg diet) and leucine (24 g/kg diet) produced a much stronger improvement in insulin sensitivity compared to low dose resveratrol alone (HOMAIR: Combine vs Alone 1.14 ± 0.37 vs 2.41 ± 0.66). Moreover, these *in vitro* and *in vivo* effects were mediated by increased Sirt-1 activity [65], highlighting the role of the Sirt1-signaling in the synergistic effect between resveratrol and leucine. Drug-herb synergy also plays an important role in combination therapy to treat diabetes and its complications. For example, *Acanthopanax senticosus* (Rupr. et Maxim.) Harms polysaccharide (ASP) has been shown to have anti-diabetic activity in alloxan-induced diabetic mouse model and interestingly, ASP has been suggested as an adjunctive drug for metformin. For instance, an *in vivo* study showed that ASP plus metformin not only produced better blood glucose level control than metformin alone but also reduced the side effects (e.g. body weight loss and suppressed hepatic enzyme function) associated with metformin treatment [66]. Likewise, combined treatment of ferulic acid (a phytochemical present in a number of Chinese herbs) with metformin have improved both glucose and lipid profile in diabetic rats with minimal side effects when compared to metformin treatment alone [67]. Pioglitazone is a cheap yet effective drug for blood glucose control, but the use is limited by the increased risk of bladder cancer. This risk is associated with the dosage and can be minimized with a reduced dose [68]. Combination of pioglitazone with ellagic acid (a natural phenol found in numerous herbs) has been shown to improve hyperglycaemic and dyslipidaemic conditions in diabetic rat. More importantly, this study showed that the dosage of pioglitazone can be reduced by twofolds when combining with ellagic acid, demonstrating a possible synergistic interaction between ellagic acid and pioglitazone which may lead to better use of pioglitazone for the management of diabetes with reduced side effects and toxicity [69].

Synergistic in CHM for the Management of Inflammatory Diseases

Inflammation is an immune response triggered by body systems against stimulations by pathogens and injury [70]. An inappropriate immune response can lead to chronic inflammatory diseases including allergy, asthma and autoimmune diseases. Many pre-clinical and clinical studies have demonstrated the anti-inflammatory effects of CHM in relation to different organ systems [71]. In addition, synergy has been reported in CHM herbal combinations in the management of inflammatory diseases. For instance, aqueous extracts of Chrysanthemi Flos (*Ju hua*) and Lycii Fructus (*Gou qi zi*) in combination were studied for possible synergistic interactions for anti-inflammatory effects in RAW264.7 cells by Zhang et al. [72]. At the ratio of 1:1 (w/w), the herb pair showed synergistic cellular antioxidant activity (CI = 0.11) and more potent inhibitory activity on the production of pro-inflammatory cytokines compared to individual herbs. Inactivation of the MAPKs (ERK and JNK) and nuclear factor kappa- B (NF-κB) was found to be the molecular mechanism of action for the anti- inflammatory effects [72]. A similar study provided supportive evidence for the use of Danshen-Sanqi in treating vascular diseases via synergistic anti-inflammatory activity as Salvia Miltiorrhiza Radix et Rhizoma (*Dan shen*) and Notoginseng Radix et Rhizoma (*San qi*) extract at a ratio of 8: 2 (w/w) inhibited nitric oxide (NO), tumour necrosis factor (TNF), and MCP-1 production synergistically [73].

A number of studies investigated the synergistic anti-inflammatory activity among bioactive compounds of CHM as well as their interactions with WM. For instance, the major essential oil anethole (1-methoxy-4-benzene-[1-propenyl]), anise (from *Pimpinella anisum*), star anise (*Illicium verum*) and sweet anise (*Foeniculum vulgare*), exhibited synergy when combined with ibuprofen to reduce inflammation. Likewise, inhibitory synergistic effects were shown for the combination of anethole and ibuprofen in the paw oedema model after 4 h of stimulation. Anethole (62.5 mg/kg) and ibuprofen (8.75 mg/kg) in combination significantly reduced the level of TNF of the exudate and the number of leukocytes in the pleural cavity compared to control rats [74]. Another study demonstrated that citral, a monoterpene naturally present in many Chinese herbs and plants, combined with naproxen (a clinical inflammatory agent) displayed synergy in carrageenan-induced paw oedema and inhibition of gastric damage in rat models using interaction index and isobologram method. The combination significantly increased potency compared to the total effects of single compounds and showed 0.4 $\pm$ 0.1 of interaction index [75]. Furthermore, oral administration of a fixed-dose combination of naproxen and citral protected the rats from naproxen-driven gastric damage.

Several studies have suggested synergy in combined Chinese herbs and natural extracts. A combination of an aqueous extract of Chinese herb, *Cistanche tubulosa* (Schrenk) Wight (CT), semi-purified fucoidan, and a polysaccharide complex harvested from seaweeds, was studied *in vivo* using a carrageenan-induced air pouch inflammation model. It was found that fucoidan and CT combination (1:3) significantly decreased the volume as well as NO and prostaglandin E2 (PGE2) concentrations of exudates compared to individual administrations [76]. Another study investigated the combination of honokiol (extracted from *Magnolia officinalis* Rehd. et Wils.) and modified citrus pectin at the ratio of 9:1. A strong synergistic anti-inflammatory activity was observed for the inhibition of TNF and NK-κB. However, nitric oxide synthesis (NOS) and COX-2 inhibition were antagonistic at ED50, ED75, and ED90 [77]. Synergy was also detected for the chemical compounds in many CHM formulae. Long et al. [78] investigated synergistic interactions among tanshinones (TA), phenolic acids (PA) and ginsenosides (GA) present in Cardiotonic Pill (CP). The anti-inflammatory activity of CP was compared to BECCs (a mixture of bioactive equivalent combinatorial components of TA, PA and GA, 4:10:4) on RAW264.7 cells. CP showed a statistically more significant anti-inflammatory effect compared to BECCs on NO, PGE2 and IL-6 productions and expressions of iNOS and COX-2. The enhanced activity of the formula in comparison to BECC suggested synergistic interactions among the active compounds within the formula.

Synergy in CHM as Antioxidants for the Management of Oxidative Stress-Related Diseases

Oxidative stress is essential to maintain a normal physiological level in the body, whereas, an excess level can cause severe damage to body systems [79]. It plays a key role in the development of cancer, atherosclerosis, Parkinson's, Alzheimer's as well as several infectious diseases. CHMs used in combination have shown synergistic antioxidative effects highlighting their potential in the management of many oxidative stress-related diseases. For example, *Astragalus membranaceus* (AME) showed potent antioxidant activity when combined with other herbal extracts. AME and *Glycyrrhiza uralensis* (GU) were studied for synergistic antioxidant properties. The ethyl acetate extract of the mixed herbs had stronger antioxidant activity compared to the individual herbs and their aqueous, n-hexane, chloroform, and n-butanol fractions as measured by three antioxidant assays. The synergistic antioxidant capacity was ascribed to the total phenolic and flavonoid content [80]. Eight traditional Chinese herb pairs (TCHPs) consisted of *Astragalus membranaceus*, *Glycyrrhiza uralensis*, *Paeonia lactiflora* (PL), *Angelica Sinensis* (AS), *Atractylodes macrocephala* (AMA) and *Rheum officinale* (RO) showed strong synergistic free radical scavenging capacity due to their flavonoid content [81]. Among all the herbs in this formula, the combination of AME and AMA presented significantly higher antioxidant capacity. A combination of their essential oils also showed a stronger ABTS scavenging activity in another study [82]. Likewise, when PL ethyl acetate extract and fractions were combined with AME chloroform extract, strong synergistic effects in scavenging DPPH radicals and reducing ferric ions were observed (CI $<$ 1.0). The combination with the strongest synergy was found to contain oxypaeoniflora, catechin, quercetin [83]. Citrus spp., commonly used in CHM, have also demonstrated synergistic antioxidative effect with other natural products and bioactive compounds. For example, the combinations of phenolic compounds from *Citrus sinensis* (*Zhi shi*) – hesperidin/myricetin, hesperidin/naringenin, and hesperidin/chlorogenic acid exhibited synergistic antioxidant capacity using the oxygen radical absorbance capacity (ORAC) assay. Similar observations were also made for three-compound mixtures [84]. An aqueous extract of *Chrysanthemum morifolium* in combination with peptide mixture comprising soy and collagen peptides displayed synergistic antioxidative effect on ultraviolet irradiation-induced skin damage mouse model [85]. Luteolin and chlorogenic acid present in the herb *Lonicera japonica* Thumb (*Ren dong teng*) showed synergistic antioxidative activity when combined by their IC50 values [86]. The formula of *Foeniculum vulgare* (fennel, *Xiao hui xiang*), *Aloysia citrodora* (lemon verbena) and *Mentha spicata* (spearmint) showed consistent synergy in multiple antioxidant assays. Combinations with spearmint showed the highest antioxidant activity due to its high flavonoid content [87]. Collectively, these results suggested that many CMHs can be effectively combined to bring synergistic antioxidant effect which might be useful in treating many oxidative stress-related diseases.

Synergy in CHM for the Management of Musculoskeletal Pain

Musculoskeletal pain is a common age-related symptom that may result in immobility [88]. CHMs have been extensively used to improve symptoms of muscular skeletal disorders for centuries [89, 90]. Two studies suggested the synergistic interaction between CHM herbs/compounds with WM. Wu and Wu [91] showed that essential oils of *Zanthoxylum schinifolium* Sieb. et Zucc. (EOZ) combined with verapamil (Ver) (a calcium channel blocker) produced a synergistic antinociceptive effect. EOZ with Ver (low, medium and high concentrations) significantly reduced writhing episodes and also increased pain threshold to the heat stimulus compared to the individual treatment of EOZ in mice. In addition, EOZ/Ver group significantly increased tail flick response time and the more rapid disappearance of the action potential compared to that of the EOZ group [91]. Another study used the Loewe isobologram method and observed synergy between curcumin (the main bioactive compound from *Curcuma longa*) and celecoxib (COX-2 inhibitor) in enhancing the growth inhibition and apoptosis on OA synovial adherent cells as curcumin reduced the IC50 value of celecoxib by fourfolds [92]. The Duhuo Jisheng Decoction (DHJSD), a mixture of 15 CHMs, is a popular centuries-old formula used to treat osteoarthritis. Zheng et al. [31] illustrated the multi- target interactions of chemical components in DHJSD using three computational models including the ligand clustering, chemical space distribution and network construction

Synergy in CHM for the Management of Hepatic Diseases

Many herbs and herbal formulations have been used to treat hepatic disease in CHM and several studies have investigated their synergy. *Gardenia jasminoides* Ellis (*Zhi zi*) is a commonly used CHM either as a single herb or in herbal formulae to treat hepatic diseases. Yin-Chen-Hao-Tang (YCHT), a famous CHM formula used to treat hepatic injuries, consists of three herbal ingredients, namely, *Gardenia jasminoides* Ellis (G), *Artemisia annua* L. (A), and *Rheum Palmatum* L. (R). The biochemical and pharmacokinetic properties of GAR were studied using a hepatic injury rat model and it showed decreased levels of common biomarkers of liver injury, increased levels of superoxide dismutase and glutathione peroxidase and significantly increased bioavailability and pharmacokinetic properties compared to either individual or two herbs combinations. The superior effects of GAR as a whole suggested that its individual components synergistically interact with each other to enhance the effects on hepatic injuries [93]. A simplified combination from GAR which consists of three active constituents- 6,7-dimethylesculetin, geniposide, and rhein, demonstrated synergistic inhibitory effect against hepatic injury in similar *in vivo* models. The active constituent mixture significantly reduced hepatic tissue destruction and also modified metabolic biomarkers compared to monotherapies. Furthermore, a systems biology approach identified synergistic/additive interactions of the active constituents of GAR in the regulation of proteins involved in different disease mechanistic pathways [94]. A synergistic interaction was studied on cholestasis diseases for a combination of the dried root and rhizome of *Rheum palmatum* L. and the dried ripe fruit of *Gardenia jasminoides* Ellis *(Zhi zi)*. The combination showed a greater effect in increasing bilirubin, direct bilirubin, total bile acid, aspartate aminotransferase and alanine aminotransferase values compared to monotherapies in the α-naphthylisothiocyanate- induced cholestasis rat models. Additionally, this combination exhibited synergistic effects in the pharmacodynamics and pharmacokinetic levels [95].

Lignans from Schisandrae Chinensis Fructus and polysaccharides from Astragali Radix at a concentration of 135:450 mg/kg resulted in a synergistic hepatoprotective effect in reducing superoxide dismutase and malondialdehyde (MDA), increasing glutathione and catalase levels in male Sprague-Dawley rats using the coefficient of drug interaction (CDI). Furthermore, this combination decreased serum alanine aminotransferase, aspartate aminotransferase and alkaline phosphatase levels with CDI values < 1, indicating synergy against hepatic injury [96]. Similarly, the hepatoprotective effect of a herb pair of olive and *Ficus carica* (dried fig, *Wu hua guo*) were investigated in CCl4-induced hepatotoxic rat model. The intraperitoneal administration of methanolic extracts of extra virgin olive oil (2.1 mg/kg/day) and dried fig extract (38.07 mg/kg/day) synergistically reduced MDA, glutamate oxaloacetate transaminase, glutamate pyruvate transaminase, alkaline phosphatase and total bilirubin compared to the individual administrations [97].

Hawthorn fruit (*Crataegus pinnatifida* Bge.) known as *Shan zha* in CHM is well-known for its lipid-lowering effect. A study by Ye et al. [98] indicated a synergistic anti- lipidaemic effect mediated by the active components in hawthrorn [98]. The active compounds, quercetin, hyperoside, rutin, and chlorogenic acid, were found to have synergistic lipid-lowering effects compared to individual compounds via down- regulating of 3-hydroxy-3-methylglutaryl coenzyme A reductase in high-fat diet mice model.

Synergy in CHM for the Management of Infectious Diseases

Despite the recent advancements in WM, the development of antibiotic resistance in bacteria continues to be one of the greatest risks to human health. A bacterial population can be termed as multidrug-resistant (MDR) when it exhibits resistance to at least one agent in more than two of the known categories for antimicrobials [99, 100]. Bacterial antibiotic resistance is conferred by four main mechanisms – (1) modification of cell wall proteins to reduce drug uptake, (2) enzymatic inactivation of the drug (3) alteration of drug target sites, and (4) extrusion of drugs by efflux pumps [99, 101–104]. The strategy of combining standard antibiotics with natural bioactive compounds to enhance drug efficacy has recently become a priority in antimicrobial research. As novel drug development requires a large amount of time and resources, the approach of combining pre-existing drugs with adjuvants such as natural products is particularly promising as a more viable alternative. The concept of antimicrobial synergy relies on the principle that, the drugs in combination, may significantly improve efficacy and bioavailability, reduce toxicity and adverse side effects as well as lower the dose and the risk of antimicrobial resistance compared to the sum of their individual parts [3, 99]. CHM has been employing the therapeutic potential of synergistic interactions from an antimicrobial perspective since antiquity [3].

Table 2 represents the studies that are currently portrayed in the literature showing antimicrobial synergy between different TCM herbs as well between individual components of TCM herbs and standard antibiotics. Evaluation of synergy in these studies was carried out using checkerboard assay with the calculation of fractional inhibitory concentration indices, time-kill assay and the reduction of antibiotic minimum inhibitory concentration values. However, only a few studies have looked into the molecular mechanisms underlying the synergistic effects. For instance, 5′-methoxyhydnocarpin-D found in many CHMs has been shown to potentiate the effect of antibiotics and berberine (another bioactive compound found in many Chinese herbs) by blocking the expression of MDR efflux pumps in *Staphylococcus aureus* [105, 106]. Mikulášová et al. [107] recently reviewed the synergistic effects resulted from the efflux pump inhibitory activity of essential oils and their components in combination with antibiotics in *S. aureus*. Several other possible synergistic antibacterial mechanisms are proposed for different components of CHM with standard antibiotics such as the disruption of peptidoglycan and

Table 2 Synergistic antimicrobial activity of CHM and their individual components with standard antimicrobials

Common Chinese herb	Active extract/compound	Synergistic drug	Synergistic antimicrobial activity	References
Antibacterial activity				
Alpinia officinarum (Gao liang Jiang)	Galangin	Gentamicin	Marked synergism was noticed with a fractional inhibitory concentration index (FICI) values in the range of 0.19–0.25 against methicillin-resistant *Staphylococcus aureus* (MRSA).	[120]
Berberis spp, Mahonia spp, Coptis chinensis (Huang lian) and *Phellodendron amurense (Guan huang bai)*	Berberine and its derivative: 8-acetonyl-dihydroberberine	Azithromycin and levofloxacin	Berberine and its derivative, when combined with azithromycin and levofloxacin, exhibited antibacterial synergy against 90% of the tested MRSA strains, with FICI values of 0.188–0.500.	[121]
Coptis chinensis (Huang lian) and *Phellodendron amurense (Guan huang bai)*	Berberine	Oxacillin	A synergistic effect was observed between berberine and oxacillin against MRSA with a FICI value of 0.500.	[122]
Camellia sinensis	Theaflavin	Ampicillin	Significantly lowered the minimum inhibitory concentration (MIC) values and displayed strong synergy with FICI values in the range of 0.22–0.35 against nine clinical isolates and one control isolate of *Stenotrophomonas maltophilia* using checkerboard assay.	[123]
Cinnamomum cassia (Gui zhi)	Cinnamaldehyde	Silver nanoparticles	Significant synergy was observed against spore-forming pathogens *Bacillus cereus* and *Clostridium perfringens* with a FICI value of 0.50 for both. In the presence of both agents, extensive damage to the bacterial cell envelope was noticed.	[110]

(continued)

Table 2 (continued)

Common Chinese herb	Active extract/compound	Synergistic drug	Synergistic antimicrobial activity	References
Duabanga grandiflora (Ba bao shu)	Chromatographic fraction derived from ethyl acetate leaf extract	Ampicillin	Fraction (rich in tannins, saponin, flavonoids, sterols, and glycosides) + ampicillin exhibited synergy with FICI values 0.18–0.31 against MRSA. The synergy may be mediated by the inhibition of PBP2a (a protein that confers resistance to beta-lactam antibiotics) expression.	[108]
Hypericum japonicum (Tian ji huang)	Isojacareubin	Ceftazidime, levofloxacin and ampicillin	Isojacareubin when combined with ceftazidime, levofloxacin and ampicillin exhibited synergy against MRSA with FICI50 values of 0.25, 0.37 and 0.37, respectively.	[124]
Isatis tinctoria, Scutellaria baicalensis (Huang qin) and *Rheum palmatum (Da huang)*	Ethanolic extracts	Penicillin G, gentamicin, ciprofloxacin, ceftriaxone	Synergy was reported based on the reduction of MIC values against MRSA.	[125]
Magnolia officinalis (Hou po)	Magnolol and honokiol	Etilmicin, levofloxacin, amikacin, fosfomycin and gentamicin	Significant synergistic combinations using checkerboard and time-kill assays were determined as honokiol (magnolol) + etilmicin, honokiol + levofloxacin and honokiol + amikacin (gentamicin or fosfomycin), with ∆LC24 (increased killing colony counts in log10 CFU/mL at 24 h incubation) of 2.02 ~ 2.25 against MRSA.	[126]
Psidium guajava (Fan shi liu ye), Rosmarinus officinalis, Salvia fruticosa, Majorana syriaca, Ocimum	Aqueous extracts	Penicillin, gentamicin sulphate, cephalexin, sulfadimethoxine as sodium,	Synergy was reported against five clinical isolates of *S. aureus* based on MIC reduction for all combinations.	[127]

basilucum, Syzygium aromaticum, Laurus nobilis and *Rosa damascena*		enrofloxacin, oxytetracycline Hcl		
Rheum palmatum (Da huang)	Rhein	Ampicillin or oxacillin	Showed synergistic activity against MRSA rhein + ampicillin and rhein + oxacillin a FICI ranging from 0.28–1 and from 0.18–1, respectively.	[128]
Scutellaria baicalensis (Huang qin)	Baicalein	Ampicillin and gentamicin	Baicalein + ampicillin demonstrated synergistic inhibitory effects with FICI values of ≤ 0.5 as determined by checkerboard assay *on Streptococcus mutans, Streptococcus sanguinis, Streptococcus oralis, Streptococcus downe, Streptococcus bovis, Streptococcus mitis, Streptococcus ratti, Streptococcus criceti, Streptococcus anginosus, Streptococcus gordonii, Actinobacillus actinomycetemcomitans, Fusobacterium nucleatum, Prevotella intermedia,* and *Porphyromonas gingivalis.*	[129]
			Baicalein + gentamicin combinations synergistically inhibited the growth of *S. mutans, S. sanguinis, S. downei, S. bovis, S. mitis, S. criceti, S. anginosus, S. gordonii, F. nucleatum,* and *P. gingivalis* with FICI values ≤ 0.5. Time-kill curve assay also validated the synergy.	
		Gentamicin	Checkerboard and time-kill curve assays revealed synergy between baicalein and gentamicin against four	[130]

(continued)

Table 2 (continued)

Common Chinese herb	Active extract/compound	Synergistic drug	Synergistic antimicrobial activity	References
			clinical isolates of vancomycin-resistant *Enterococcus faecalis*.	
		Tetracycline and other β-Lactams	Baicalein strongly inhibited the transport of tetracycline by *E. coli* expressing the gene encoding active efflux pump for tetracycline.	[131]
			Synergy was observed between baicalein + tetracycline and baicalein + other β-Lactams (oxacillin, cefmetazoleand ampicillin).	
Scutellaria baicalensis (Huang qin) and *Scutellaria lateriflora*	Baicalein	Cefotaxime	Synergistic activity is mediated by inhibition of CTX-M-1 mRNA expression in some of extended- spectrum β-lactamases positive *Klebsiella pneumoniae* strains.	[132]
Scutellaria amoena	Baicalein	Ampicillin, amoxycillin, benzylpenicillin, methicillin and cefotaxime	Baicalin showed synergy in checkerboard and time-kill curve assays by restoring the effectiveness of β-Lactam antibiotics against MRSA and other strains of β-Lactams-resistant *S. aureus*.	[133]
Sophora moorcroftiana	Genistein and diosmetin	Norfloxacin, streptomycin and ciprofloxacin	Exhibited significant synergistic activity against drug-resistant *S. aureus*. For instance, genistein when combined with norfloxacin, showed synergy with a FICI value of 0.38. Diosmetin in combination with norfloxacin, streptomycin and ciprofloxacin, displayed FICI values of 0.38, 0.38 and 0.09, respectively. Genistein showed marginal	[134]

			inhibition of the NorA efflux protein as the possible mechanism of action.	
Selaginella tamariscina (Juan bai)	Amentoflavone	Ampicillin, cefotaxime and chloramphenicol	Amentoflavone + ampicillin, amentoflavone + cefotaxime, and amentoflavone + chloramphenicol synergistically inhibited the growth of *E. coli* and *P. aeruginosa* (FICI $\leq$ 0.5) using checkerboard assay. The combination of amentoflavone + chloramphenicol had a synergistic effect on *Enterococcus faecium, S. aureus, E. coli O -157, E. coli,* and *P. aeruginosa* growth. The synergistic effect can be ascribed to the formation of hydroxyl radical-mediated by nicotinamide adenine dinucleotide phosphate (NADH) depletion.	[111]
Stephania tetrandra (Fang ji)	Tetrandrine (Tet) and demethyltetrandrine (d-Tet)	Cefazolin	Tet + cefazolin showed significant synergy against 90% of the isolates of MRSA with 75 – 94% reduction in their MIC values (FICI values from 0.188–0.625).	[135]
Antifungal activity				
Acorus tatarinowii (Shi chang pu)	Eugenol	Fluconazole	Eugenol and fluconazole combination showed strong synergy with a FICI value of 0.14 against pre-formed biofilms of *Candida albicans* SC5314.	[136]
			The combination showed synergy in 29 out of 34 fluconazole-resistant *C. albican, C. tropicalis* and *C. glabrata* strains.	[137]

(continued)

Table 2 (continued)

Common Chinese herb	Active extract/compound	Synergistic drug	Synergistic antimicrobial activity	References
Anemarrhena asphodeloides (Zhi mu)	Nyasol (cis-hinokiresinol)	Miconazole, ketoconazole, clotrimazole and cerulenin	The FICI values were 0.067–0.31 for miconazole + nyasol, 0.078–0.31 for ketoconazole + nyasol, and 0.098–0.13 for clotrimazole + nyasol against 4 strains of *C. albicans*. Nyasol, when combined with cerulenin also displayed synergy *C. albicans* and *Aspergillus fumigatus* strains.	[138]
Berberis spp.	Berberine	Fluconazole	Berberine in combination with fluconazole synergistically increased mitochondrial membrane potential, reduced intracellular ATP level, attenuated ATP-synthase activity, and increased generation of endogenous reactive oxygen species (ROS) in fluconazole-resistant *C. albicans* clinical strains *in vitro*.	[139]
			Fluconazole increases the intracellular berberine concentration especially in the nucleus of fluconazole-resistant *C. albicans* clinical strains which leads to cell cycle arrest and significant change in the transcription of cell cycle-related genes. Berberine plays a major role in its antifungal synergy with fluconazole as demonstrated by antifungal time-kill curve assay.	[114]
			Berberine inhibited the overexpression of drug resistance gene CDR1 and the transport function of Cdr1p induced by	[115]

			fluphenazine in azoleresistant *C. albicans* clinical isolates.	
			Combinations of berberine and fluconazole resulted in fourfold reduction in the MIC value of fluconazole (to 0.48 μg/ml from 1.9 μg/ml) as tested by the checkerboard assay.	[116]
		Amphotericin B	A synergistic effect of the berberine with amphotericin B was observed using the broth susceptibility test *in vitro* against *C. albicans*. Amphotericin B and berberine could synergistically lower the amphotericin B dose approximately by 75% against the disseminated candidiasis in mice based on mean survival times.	[117]
		Miconazole and fluconazole	Berberine exhibited synergy against *C. albicans*, either in planktonic or in biofilm growth phases. In particular, berberine in combination with miconazole or fluconazole showed FICI values < 0.5 in the disc diffusion assay as well as in suspension. The combination of berberine with miconazole reduced biofilm formation by >91% after 24 h.	[118]
	Berberine chloride	Fluconazole	Synergistically inhibited the growth of 40 fluconazole-resistant clinical isolates of *C. albicans* in agar diffusion test, checkerboard and time-kill curve assays.	[119]

(continued)

Table 2 (continued)

Common Chinese herb	Active extract/compound	Synergistic drug	Synergistic antimicrobial activity	References
Cortex pseudolaricis (Tu jin pi)	Pseudolaric acid B	Fluconazole	Synergy was observed against oral azole-resistant isolates of *C. albicans* in the agar diffusion and checkerboard assays using FICI.	[140]
Curcuma longa (Jiang huang)	Curcumin	Rhodamine 6G, ketoconazole, itraconazole, and miconazole	Spot and checkerboard assays revealed that curcumin synergistically modulates the ATP-binding cassette (ABC) multidrug transporters in *Saccharomyces cerevisiae* cells.	[141]
		Azoles-fluconazole, miconazole, ketoconazole, itraconazole, voriconazole	The MIC80 values of the drugs decreased by10-35-fold when curcumin was used in combination with azoles and polyenes, with FICI values in the range 0.09–0.5 against 21 clinical isolates of *C. albicans.* The combinations-curcumin + fluconazole and curcumin + amphotericin B triggered apoptosis and accumulated ROS in cells.	[112]
		Polyenes- nystatin and amphotericin B		
		Piperine	The combination of curcumin and piperine caused a significant fungal load reduction of *C. albicans* in kidneys of Swiss mice. Curcumin inhibited hyphae development by targeting TUP1 (thymidine uptake 1) and triggered early apoptosis by raising the levels of ROS in *C. albicans* cells.	[113]
Cinnamomum cassia (Rou gui)	Cinnamaldehyde	Fluconazole	Exhibited synergy with a FICI value of 0.312 against *C. albicans* biofilms.	[136]

cytoplasmic membranes leading to increase in permeability, inhibition of β-lactamase activity, inhibition of penicillin-binding protein 2a (PBP2a) expression (protein that confers resistance to β-Lactam antibiotics) and formation of hydroxyl radicals [108–111]. In terms of the synergy with standard antifungal agents, increased generation of endogenous reactive oxygen species (ROS) has been found to be the most common mechanism of action in *Candida albicans* [51, 112, 113]. Berberine isolated from the Chinese medicinal herb- *Berberis vulgaris*, has been reported to have synergy with many antifungal agents against *C. albicans in vitro* [51, 114–119]. Interestingly, synergistic studies on CHM have not been performed widely in clinically important Gram-negative bacterial strains. It is apparent that most of these studies were performed *in vitro* and focused mostly on the Gram-positive bacteria- *S. aureus* and the opportunistic pathogenic yeast- *C. albicans*. Therefore, more *in vivo* investigations and rigorous clinical trials against other clinically relevant pathogens must be taken into consideration to comprehend the synergistic mechanisms between different herbs/components of CHM with standard antimicrobial agents. The bioavailability and possible toxicity of CHM, when combined with standard antibiotics, should also be investigated.

Pharmacokinetic Synergy in CHM

CHMs are typically used in combination to enhance effectiveness and/or reduce side effects. In pharmacokinetic studies, the conventional concept of synergy centres around enhancing bioavailability by increasing absorption, limiting plasma binding and reducing the rate at which the drug is metabolised. This can also be applied to CHM where the herbs/formulae increase the plasma concentration of herbs/formulae/other drugs which may lead to enhanced systematic exposure and efficacy. In addition, CHMs used in a holistic traditional approach also have therapies targeted at detoxification (*Xiang Sha and Xiang Wei*) [7]. For these treatments, a synergistic interaction would involve lowering the bioavailability of a harmful substance, by reducing absorption and/or increasing drug metabolism. However, in WM, these interactions are often viewed negatively as leading to reduced efficacy. Nevertheless, depending on the treatment goal, CHMs, in fact, enhance the efficacy of many standard drugs.

Currently, the predominant body of published evidence is related to herb-drug interactions with a special emphasis on how the pharmacokinetics of single-compound agents can be affected by the co-administration of CHMs. Studies on how the co- administration of CHMs influence the pharmacokinetics of individual components are limited, possibly due to the complexity of multicomponent CHMs. Nonetheless, studies of the single herb-drug interactions collectively suggest that synergistic interactions do occur with CHMs, regulating absorption, distribution, metabolism and excretion. The interaction of CHMs with other agents is largely mediated by drug transporters that either carry the molecules into cells/organs/bloodstream (uptake) or pump the molecules out (efflux). CHMs have been shown to influence the uptake and efflux transporters by acting either as an inducer, inhibitor or substrate. For instance, the P-glycoproteins (P-gp) are the major ATP-binding cassette (ABC) efflux transporters that play an important role in pumping drugs back into the lumen, leading to reduced bioavailability. Therefore, suppressing P-gp can lead to enhanced bioavailability. These P-gp transporters are found in the epithelia of the intestine and other organs and as the majority of CHMs ingested orally are primarily absorbed through the small intestine, co-administration of CHMs with other drugs may improve bioavailability through the direct interaction with transporters.

Several studies were found describing the positive interactions with CHMs by down- regulating P-gp pathway. Fan et al. [142] investigated the effects of *Schisandra chinensis* (*Wu wei zi*) and *Ginkgo biloba* extracts (*Yin xing ye*) on the pharmacokinetic parameters of talinolol (a P-gp dependant substrate). Both *S. chinensis* and *G. biloba* extracts have been shown to inhibit P-gp, leading to an enhanced bioavailability of talinolol. The co-administration of *S. chinensis* and *G. biloba* extracts resulted in an increase in the maximum serum concentration (Cmax) of talinolol by 51% and 33%, respectively, with similar gains reported for the area under the curve (AUC) [142]. In addition, a study from Huang et a [143] found that the individual extracts of *Glycyrrhiza inflata* (*Gan cao*) and *Daphne genkwa* (*Yuan hua*) inhibited the P-gp transporter using the permeability characteristics of rhodamine 123 in rats. When the two CHMs were used in combination, they had the most significant effect on the rhodamine 123, indicating a stronger P-gp inhibition. It was not determined if the increased bioavailability of the combination was additives or synergistic. The authors noted that the inhibitory effect on P-gps and resultant higher bioavailability of the drug could result in prolonged drug retention and toxic side effects. However, if used appropriately these effects could be implemented to synergistically enhance the absorption and retention of rapidly excreted drugs. Taken together, these examples highlight the fact that CHMs can enhance the bioavailability of drugs or other CHMs via the inhibition of P-gp. As P-gp is just one example of the many drug transporters in our complex organic systems, there is a huge potential to tailor CHMs and pharmacotherapy by understanding and incorporating transporter interactions into combined therapies. Combinations that target specific transporters might be used to reach an optimal bioavailability of the therapy.

CHMs have been shown to mediate the metabolism and excretion of an agent by interacting with cytochrome P450 (CYP) enzymes in several studies. Typically, CYPs metabolise an agent by adding polar groups to the molecule, thus changes the bioactivity and facilitates excretion. CYPs are expressed primarily in the liver endoplasmic reticulum but are also abundant in the intestine and other organs such as brain, lungs, kidneys and adrenal cortex. Drugs have been suggested to act on CYP enzymes directly to alter the rate of drug metabolism. In addition, they may act on transcription factors to alter the expression of the CYPs. CHMs therefore, have been hypothesised to enhance the bioavailability of a drug by inhibiting CYPs, increasing the Cmax and AUC. *Hydrastis canadensis* has been shown to interact with CYPs to increase drug retention. In a study by Gurley et al. [144], on 12 healthy volunteers *H. canadensis* was found to inhibit CYP2D6 and CYP3A4/5 approximately by 40%. This effect was later confirmed to be clinically significant on the drug digoxin, however, it was not clear if the 14% increase in Cmax was due to CYP inhibition or G-pg inhibition [145]. This is in contradiction to a similar pharmacokinetics study that showed no effect on the CYP3A4 dependent drug indinavir [146]. Gurley et al. [147] to confirm the previous finding and support the role of CYP3A in improving the bioavailability of a drug tested the effects of *H. canadensis* on midazolam pharmacokinetics. The co-administration resulted in an approximate 40% increase in Cmax for midazolam [147]. Likewise, in two clinical studies, *Schisandra sphenanthera* (*Nan wu wei zi*) has been found to increase the bioavailability of tacrolimus (CYP3A4/5 substrate), showing a 183% [148] and a 227% increase in Cmax [149]. In a separate study, the enhancement of the bioavailability was also observed for midazolam (CYP3A4/5 substrate) where the Cmax was increased by 119% [150]. As CYP3A4/5 is the main CYP enzyme involved in drug metabolism [151] and the interaction resulted in a large increase in Cmax, this synergistic interaction may be observed for many other drugs. Further research needs to be carried out to identify whether this interaction is broadly applicable to CYP3A4/5 metabolised drugs or if the observed response is more specific to midazolam. However, it is worth mentioning that the vast majority of CYP interactions of other CHMs observed *in vitro* failed to result in a significant clinical effect [152].

Interestingly, CHMs have also been shown to interact with CYP to aid in the elimination of drugs. Whitten et al. [153] reviewed 31 relevant clinical studies and concluded that the effect of *Hypericum perforatum* was most likely attributed to the induction of CYP3A. Several case studies identified that *H. perforatum* (*Guan ye jin si tao*) increased the elimination of drugs that have significant toxicity including cyclosporine [154]. Therefore, as established for *H. perforatum,* the interactions of CHMs via CYP enzymes may elicit a clinically significant effect on the metabolism of other drugs. CHMs have also been shown to be able to induce P-gp, to increase detoxifying effects. Qui et al. [155] demonstrated that *S. miltiorrhiza* (*Danshen*) is a P-gp inducer and that with long term administration reduced the AUC (37%) and Cmax (27%) of fexofenadine (a P-gp substrate drug).

Collectively, it has been suggested that CHMs exhibit pharmacokinetic synergy with pharmaceutical drugs by increasing bioavailability in various ways. However, there are currently limited studies to show herb-herb synergy on pharmacokinetics. Since the interactions between herb and drug have been elucidated, the same methodology can be developed to herb-herb interactions in a formula to support the synergy philosophy of CHM.

Biological Networks for CHM

Biological networks such as protein-protein interactions, genetic, metabolic, cell signaling and neuronal networks are subunits of integrated systems in the living human body. These networks offer a mathematical representation of the biological system connections in physiological, evolutionary and ecological studies [156]. Furthermore, the combination of systems biology approaches with computational methodologies in system pharmacology (SP) has enabled a better understanding of several complex diseases and mechanisms of action of many drugs [157]. Diverse omics-data of multi-scale systems including genome, proteome, metabolome for cells, organs and diseases are incorporated in quantitative SP (QSP) to quantitatively decipher the interactions of drugs with the biological systems. In other words, QSP offered a holistic understanding of the system behaviour as a whole, coupled with the quantitative insights into both the biological and pharmacological processes [158, 159]. A large number of computational methods have been adapted for molecular, network and system-levels evaluation and simulation [160]. A multiplex of these methods is frequently implemented in the exploration of the mechanisms of action of many CHMs. In fact, the main focus of these *in silico* studies at the molecular level is the assessment of the molecular properties of the herbal metabolites and possible target identification where pharmacodynamics (PD) and pharmacokinetic (PK) models could be adapted for ADME/T (adsorption, distribution, metabolism, elimination and toxicity) analysis, alongside with molecular docking models of virtual screening (Table 3). Additionally, the drug likeliness and molecular similarity studies of new compounds are regularly used to predict the drug-target interactions as in BATMAN-TCM and Binding DB web servers [161, 162]. Then, the network models integrating ADME/T filtered herbal components, disease-related genes, omics-data, targets and pathways can be established to comprehend any interactions at the cellular/tissue scales [163]. The integrated data from molecular, cellular and tissue levels regarded as multiscale modelling can be a powerful tool in drug discovery [164]. The bio-systems in these network models are visually represented as nodes, each denotes a biological moiety such as molecules, DNA, RNA, protein or metabolites, where, the relation between nodes can be represented by the network edges. Notably, the key targets [85, 165–172] in networks and a guide for herbal combinations and synergy prediction [173, 174] can be drawn from network models via network dynamic simulation and pathway topological analysis [160]. For instance, Gu et al. developed a quantitative model combining network modelling and molecular docking for drug efficacy evaluation which can be used for either drug combination or repositioning purposes [175, 176]. Finally, at the system level, the logical modelling, multiscale SP platforms and virtual patients could be utilised [160]. TCMSP, a unique open access, multiscale SP database was created covering not only the natural products metabolites, targets, drug-targets networks and associated disease networks of

Table 3 Computational QSP methods and selected examples in CHM

Method	CHM	Findings	References
Molecular level			
ADME/T, PK, PD and drug likeliness prediction Molecular docking, target fishing, Molecular dynamic simulation	*Erigeron breviscapus (Deng zhan xi xin)*	Herb evaluation in cardiovascular diseases where ADME screening, targets identification and PPI network analysis were studied and verified *in vitro*.	[178]
	Qi-enriching herbs Blood detoxifying herbs	Performed the ADME prediction, target identification, and network analysis.	[179]
	Baihe Dihuang Tang	Regulated the G-protein coupled amine receptors and monoamine neurotransmitter expression with 11 Identified active ingredients and 21 targets.	[166]
	Bushen-Yizhi formula	Elucidated the mechanism of action in Alzheimer's disease via neuronal apoptosis and amyloid-β metabolism regulation	[165]
	Dragon's blood tablets	22 potentially absorbed compounds out of total 47 with its putative targets in colitis treatment were identified.	[167]
	Niao Du Qing (NDQ) granules	Ameliorated the chronic kidney diseases via modulating EPO and TGF-β signaling pathways.	[168]
	Reduning injection	Identified the bioactive compounds and revealed its potentiality against influenza via immune response activation and the regulation of inflammatory agents.	[85]
	SiNi San formula	Identified the potential formula metabolites and its targets which may be involved in depression and mental disorders.	[171, 172]
	Bushenhuoxue formula	Tanshinone IIA, curcumin, calycosin, rhein and quercetin were identified as effective ingredients in chronic kidney diseases.	[169]
	Yangxinshi tablet	Identified the representative compounds, targets and pathways responsible for the formula implementation in heart failure	[170]

(continued)

Table 3 (continued)

Method	CHM	Findings	References
Network level			
-Herb-target network analysis	Si-Ai-Te-San (SH) formula	Identified the inhibitory herbal components against HIV-related proteins.	[180]
PPI network analysis Compound-Target-Pathway (C-T-P) networks Pathway analysis	TCM prescriptions for rheumatoid arthritis	Detected the networked herbs and its potential synergistic combination and biological activity.	[181]
	Radix Salviae Miltiorrhizae herb pairs for treating various diseases	Compounds evaluation, target screening and compound target disease network for synergistic herb pair evaluation.	[173]
	Chuanxiong-Chishao herb-pair	Drug target prediction and network analysis identified estrogen receptor-α as a putative target for the herb pair to exert a synergistic effect on promoting angiogenesis and verified in zebrafish.	[174]
	Hedyotis diffusa Willd (*Bai hua she she cao*)	HRAS, PIK3CA, KRAS, TP53, APC, BRAF, GSK3B, CDK2, AKT1, and RAF1 targets were identified as the potential targets of *H. diffusa* against colorectal cancer.	[182]
	Banxia Xiexin Decoction	System pharmacology verified the potentiality of Banxia Xiexin Decoction in irritable bowel syndrome *via* inflammatory reaction inhibition and intestinal functional maintenance with improved psychological regulation.	[183]
	Salvia miltiorrhiza Bunge. *(Dan shen)*	Network pharmacology and molecular docking identified putative target and potential metabolites in myelofibrosis (MF) treatment and cryptotanshinone effect may be via affecting JAK-STAT and TGF-β signaling pathways.	[184]

499 traditional Chinese herbs, but also ADME data, oral bioavailability, solubility, drug likeliness and both intestinal and blood- brain barrier permeability [177]. In conclusion, QSP implemented methods in TCM can pave the TCM modernisation way and its rationalisation and implementation in modern medicine where deeper insights on the mechanisms of action will be drawn with improved efficiency of drug discovery from natural leads and its complex arsenal of metabolites.

Currents Statues of Synergistic Research in TCM

In summary, synergy research on CHM is still at its initial stage. There are a number of studies that have suggested the synergistic interaction among herbal ingredients in a formula, active constituents within a herb, or herb extract/active compound with WM, using rigorous analytical methods such as isobolographic analysis, combination index (CI) and systems biology. However, synergistic interactions of compounds at a molecular level and their signaling pathways have not been extensively studied. Unlike WM where the chemical and pharmacological properties of individual drugs are clearly defined, CHM often contains numerous active ingredients, which can all contribute to their synergistic effects. This poses an enormous challenge in the study of mechanisms of synergistic behaviour of CHM and their chemical constituents. Moreover, due to the complex chemical nature of many CHM formulae, the exploration of the synergy in CHM is limited only to small formulae or even a single herb. Therefore, more powerful analytical tools adapting to the complex nature of CHM are urgently needed. It is worth mentioning that despite the synergistic effects demonstrated in numerous pharmacological studies, these findings do not always represent a clinically relevant advantage. The clinical benefits of synergy in CHM must be subsequently confirmed in more rigorous clinical trials.

References

1. Zhou, X., S.W. Seto, D. Chang, H. Kiat, V. Razmovski-Naumovski, K. Chan, and A. Bensoussan. 2016c. Synergistic effects of Chinese herbal medicine: A comprehensive review of methodology and current research. *Frontiers in Pharmacology* 7. https://doi.org/10.3389/fphar.2016.00201.
2. Lee, S.J. 2015. Systems biology-a pivotal research methodology for understanding the mechanisms of traditional medicine. *Journal of Pharmaceutics* 18: 11–18.
3. Van Vuuren, S., and A. Viljoen. 2011. Plant-based antimicrobial studies–methods and approaches to study the interaction between natural products. *Planta Medica* 77: 1168–1182.
4. Spinella, M. 2002. The importance of pharmacological synergy in psychoative herbal medicines. *Alternative Medicine Review* 7: 130–137.
5. Li, S., T.-P. Fan, W. Jia, A. Lu, and W. Zhang. 2014. Network pharmacology in traditional Chinese medicine. *Evidence-based Complementary and Alternative Medicine* 2014. https://doi.org/10.1155/2014/138460
6. X. Zhou. 2016. *Synergistic behaviour of Salvia and Notoginseng species in vascular diseases*. Doctoral thesis, School of Science and Health, Western Sydney University, Australia. Available from: http://hdl.handle.net/1959.7/uws:41152 [15 May 2019].
7. Jia, W., W.Y. Gao, Y.Q. Yan, J. Wang, Z.H. Xu, W.J. Zheng, and P.G. Xiao. 2004. The rediscovery of ancient Chinese herbal formulas. *Phytotherapy Research* 18: 681–686.
8. Scholey, A.B., and D.O. Kennedy. 2002. Acute, dose-dependent cognitive effects of Ginkgo biloba, Panax ginseng and their combination in healthy young volunteers: Differential interactions with cognitive demand. *Human Psychopharmacology: Clinical and Experimental* 17: 35–44.

9. Leonard, S.S., D. Cutler, M. Ding, V. Vallyathan, V. Castranova, and X.L. Shi. 2002. Antioxidant properties of fruit and vegetable juices: More to the story than ascorbic acid. *Annals of Clinical and Laboratory Science* 32: 193–200.
10. Zhang, A.H., H. Sun, and X.J. Wang. 2014. Potentiating therapeutic effects by enhancing synergism based on active constituents from traditional medicine. *Phytotherapy Research* 28: 526–533.
11. Jiang, W.Y. 2005. Therapeutic wisdom in traditional Chinese medicine: A perspective from modern science. *Trends in Pharmacological Sciences* 26: 558–563.
12. Zhou, X, C.G. Li, D. Chang, and A. Bensoussan. 2019. Current status and major challenges to the safety and efficacy presented by Chinese herbal medicine. *Medicines* 6. https://doi.org/10.3390/medicines6010014
13. Bray, F., J. Ferlay, I. Soerjomataram, R.L. Siegel, L.A. Torre, and A. Jemal. 2018. Global cancer statistics 2018: Globocan estimates of incidence and mortality worldwide for 36 cancers in 185 countries. *CA: A Cancer Journal for Clinicians* 68: 394–424.
14. Wang, C.Y., X.Y. Bai, and C.H. Wang. 2014a. Traditional Chinese medicine: A treasured natural resource of anticancer drug research and development. *The American Journal of Chinese Medicine* 42: 543–559.
15. Zhang, Y., Y. Liang, and C. He. 2017. Anticancer activities and mechanisms of heat-clearing and detoxicating traditional Chinese herbal medicine. *Chinese Medicine* 12: 20.
16. Hsiao, W.L., and L. Liu. 2010. The role of traditional Chinese herbal medicines in cancer therapy–from TCM theory to mechanistic insights. *Planta Medica* 76: 1118–1131.
17. Wagner, H., and G. Ulrich-Merzenich. 2009. Synergy research: Approaching a new generation of phytopharmaceuticals. *Phytomedicine* 16: 97–110.
18. Bhuyan, D.J., and A. Basu. 2017. Phenolic compounds potential health benefits and toxicity. In *Utilisation of bioactive compounds from agricultural and food production waste*, ed. Q.V. Vuong, 27–59. Boca Raton: CRC Press.
19. Loo, G. 2003. Redox-sensitive mechanisms of phytochemical-mediated inhibition of cancer cell proliferation (review). *The Journal of Nutritional Biochemistry* 14: 64–73.
20. Farrell, M.P., and S. Kummar. 2003. Phase I/IIA randomized study of phy906, a novel herbal agent, as a modulator of chemotherapy in patients with advanced colorectal cancer. *Clinical Colorectal Cancer* 2: 253–256.
21. Lam, W., X. Yang, Z. Jiang, X. Han, F. Guan, W. Cheng, S.-H. Liu, L. Chen, and Y.-C. Cheng. 2018. Abstract 2724: Yiv906 (phy906) enhanced the antitumor activity of immune checkpoint blockade therapy: Anti-pd1 against liver cancer. *Cancer Research* 78. https://doi.org/10.1158/1538-7445.AM2018-2724
22. Yen, Y., S. So, M. Rose, M.W. Saif, E. Chu, S.H. Liu, A. Foo, Z. Jiang, T. Su, and Y.C. Cheng. 2009. Phase I/II study of PHY906/capecitabine in advanced hepatocellular carcinoma. *Anticancer Research* 29: 4083–4092.
23. Liu, S.-H., Z. Jiang, and Y.-C. Cheng. 2002. Botanical activity relationship in traditional Chinese medicine: Studies of phy906 as an adjuvant therapy with cancer chemotherapeutic agents. In *Proceedings of American Association for Cancer Research.*
24. Wang, E., S. Bussom, J. Chen, C. Quinn, D. Bedognetti, W. Lam, F. Guan, Z. Jiang, Y. Mark, Y. Zhao, D.F. Stroncek, J. White, F.M. Marincola, and Y.C. Cheng. 2011. Interaction of a traditional Chinese medicine (PHY906) and CPT-11 on the inflammatory process in the tumor microenvironment. *BMC Medical Genetics* 4: 38–38.
25. Lam, W., Z. Jiang, F. Guan, X. Huang, R. Hu, J. Wang, S. Bussom, S.H. Liu, H. Zhao, Y. Yen, and Y.C. Cheng. 2015. PHY906(KD018), an adjuvant based on a 1800-year-old Chinese medicine, enhanced the anti-tumor activity of sorafenib by changing the tumor microenvironment. *Scientific Reports* 5: 9384–9384.
26. Lam, W., S. Bussom, F. Guan, Z. Jiang, W. Zhang, E.A. Gullen, S.H. Liu, and Y.C. Cheng. 2010. The four-herb Chinese medicine PHY906 reduces chemotherapy-induced gastrointestinal toxicity. *Science Translational Medicine* 2: 45ra59–45ra59.

27. Liu, S.H., Z. Jiang, A. Foo, W. Lam, M. Ye, Y. Lee, D. Liang, S. Grill, Y. Yen, M. Rose, S. So, W. Saif, E. Chu, R. Tilton, and Y.-C. Cheng. 2007. PHY906 in hepatocellular carcinoma. *Cancer Research* 67: 1841–1841.
28. Liu, S.H., Z. Jiang, J. Liddil, K. Hu, E. Gullen, and Y. Cheng. 2000. Prevention of CPT-11 induced toxicity by a Chinese medicinal formulation, PHY-906. In *Proceedings of American Association for Cancer Research*.
29. Liu, S.H., and Y.C. Cheng. 2012. Old formula, new Rx: The journey of PHY906 as cancer adjuvant therapy. *Journal of Ethnopharmacology* 140: 614–623.
30. Liu, S.-H., A. Foo, Z. Jiang, R. Marathe, J. Guan, T.-M. Su, R. Tilton, Y. Yen, M. Rose, S. So, E. Chu, and Y.-C. Cheng. 2006. PHY906 as a broad-spectrum enhancer in cancer therapy: Clinical and preclinical results in hepatocellular carcinoma. *Cancer Research* 66: 506–507.
31. Zheng, Z., W.C.-S. Cho, L. Xu, J. Wang, and D.M.-Y. Sze. 2013. Lessons learnt from evidence-based approach of using Chinese herbal medicines in liver cancer. *Evidence-Based Complementary and Alternative Medicine* 2013: 11.
32. Ma, Z., K. Otsuyama, S. Liu, S. Abroun, H. Ishikawa, N. Tsuyama, M. Obata, F.J. Li, X. Zheng, Y. Maki, K. Miyamoto, and M.M. Kawano. 2005. Baicalein, a component of scutellaria radix from Huang-Lian-Jie-Du-Tang (HLJDT), leads to suppression of proliferation and induction of apoptosis in human myeloma cells. *Blood* 105: 3312–3318.
33. Wang, N., Y. Feng, H.-Y. Tan, F. Cheung, M. Hong, L. Lao, and T. Nagamatsu. 2015b. Inhibition of eukaryotic elongation factor-2 confers to tumor suppression by a herbal formulation Huanglian-Jiedu decoction in human hepatocellular carcinoma. *Journal of Ethnopharmacology* 164: 309–318.
34. Bourchier, S.J., A. Bensoussan, S. Lee, J.L. Pearson, and C.S. Khoo. 2016. Analytical method validation and quality control of a seven-herb Chinese medicine formulation used for the treatment of irritable bowel syndrome with constipation. *Journal of AOAC International* 99. https://doi.org/10.5740/jaoacint.15-0158.
35. Saif, M.W., F. Lansigan, S. Ruta, L. Lamb, M. Mezes, K. Elligers, N. Grant, Z.L. Jiang, S.H. Liu, and Y.C. Cheng. 2010. Phase I study of the botanical formulation PHY906 with capecitabine in advanced pancreatic and other gastrointestinal malignancies. *Phytomedicine* 17: 161–169.
36. Saif, M.W., J. Li, L. Lamb, K. Kaley, K. Elligers, Z. Jiang, S. Bussom, S.H. Liu, and Y.C. Cheng. 2014. First-in-human phase ii trial of the botanical formulation PHY906 with capecitabine as second-line therapy in patients with advanced pancreatic cancer. *Cancer Chemotherapy and Pharmacology* 73: 373–380.
37. Kummar, S., M.S. Copur, M. Rose, S. Wadler, J. Stephenson, M. O'rourke, W. Brenckman, R. Tilton, S.-H. Liu, Z. Jiang, T. Su, Y.-C. Cheng, and E. Chu. 2011. A phase I study of the Chinese herbal medicine PHY906 as a modulator of irinotecan-based chemotherapy in patients with advanced colorectal cancer. *Clinical Colorectal Cancer* 10: 85–96.
38. Rockwell, S., T.A. Grove, Y. Liu, Y.-C. Cheng, S.A. Higgins, and C.J. Booth. 2013. Preclinical studies of the Chinese herbal medicine formulation PHY906 (kd018) as a potential adjunct to radiation therapy. *International Journal of Radiation Biology* 89: 16–25.
39. Shilin, H., G. Aixia, and X. Yang. 1995. Clinical study on the treatment of acutepromyelocytic leukemia mainly with composite indigo naturalis tablets. *Chinese Journal of Hematology* 16: 26–28.
40. Ji, H.-F., X.-J. Li, and H.-Y. Zhang. 2009. Natural products and drug discovery. Can thousands of years of ancient medical knowledge lead us to new and powerful drug combinations in the fight against cancer and dementia? *EMBO Reports* 10: 194–200.
41. Wang, L., G.B. Zhou, P. Liu, J.H. Song, Y. Liang, X.J. Yan, F. Xu, B.S. Wang, J.H. Mao, Z.X. Shen, S.J. Chen, and Z. Chen. 2008. Dissection of mechanisms of Chinese medicinal formula realgar-indigo naturalis as an effective treatment for promyelocytic leukemia. *Proceedings of the National Academy of Sciences of the United States of America* 105: 4826–4831.

42. Zhou, S.K., R.L. Zhang, Y.F. Xu, and T.N. Bi. 2012. Antioxidant and immunity activities of Fufang Kushen injection liquid. *Molecules* 17: 6481–6490.
43. Wei, R., D.Y. Yang, W.Z. Jiang, Y.Y. Dai, L.Y. Wan, and Z. Yang. 2011b. Efficacy of Yanshu injection (a compound Chinese traditional medicine) combined with concurrent radiochemotherapy in patients with stage III nasopharyngeal carcinoma. *Zhonghua Zhong Liu Za Zhi* 33: 391–394.
44. Hsu, Y.-L., P.-L. Kuo, T.-F. Tzeng, S.-C. Sung, M.-H. Yen, L.-T. Lin, and C.-C. Lin. 2008. Huang-Lian-Jie-Du-Tang, a traditional Chinese medicine prescription, induces cell-cycle arrest and apoptosis in human liver cancer cells in vitro and in vivo. *Journal of Gastroenterology and Hepatology* 23: e290–e299.
45. Fukutake, M., N. Miura, M. Yamamoto, K. Fukuda, O. Iijima, H. Ishikawa, M. Kubo, M. Okada, Y. Komatsu, H. Sasaki, K. Wakabayashi, A. Ishige, and S. Amagaya. 2000. Suppressive effect of the herbal medicine Oren-gedoku-to on cyclooxygenase-2 activity and azoxymethane-induced aberrant crypt foci development in rats. *Cancer Letters* 157: 9–14.
46. Chen, L.W., J. Lin, W. Chen, and W. Zhang. 2005. Effect of Chinese herbal medicine on patients with primary hepatic carcinoma in iii stage during perioperational period: A report of 42 cases. *Zhongguo Zhong Xi Yi Jie He Za Zhi* 25: 832–834.
47. Chen, X.Z., Z.Y. Cao, J.N. Li, H.X. Hu, Y.Q. Zhang, Y.M. Huang, Z.Z. Liu, D. Hu, L.M. Liao, and J. Du. 2014. Ethyl acetate extract from Jiedu Xiaozheng Yin inhibits the proliferation of human hepatocellular carcinoma cells by suppressing polycomb gene product Bmi1 and Wnt/β-catenin signaling. *Oncology Reports* 32: 2710–2718.
48. Cao, Z., W. Lin, Z. Huang, X. Chen, J. Zhao, L. Zheng, H. Ye, Z. Liu, L. Liao, and J. Du. 2013. Ethyl acetate extraction from a Chinese herbal formula, Jiedu Xiaozheng Yin, inhibits the proliferation of hepatocellular carcinoma cells via induction of G0/G1 phase arrest in vivo and in vitro. *International Journal of Oncology* 42: 202–210.
49. Cao, Z., X. Chen, W. Lin, J. Zhao, L. Zheng, H. Ye, L. Liao, and J. Du. 2015. Jiedu Xiaozheng Yin decoction inhibits hepatoma cell proliferation by inducing apoptosis via the mitochondrial-mediated pathway. *Molecular Medicine Reports* 12: 2800–2806.
50. World Health Organisation. 2017. *Cardiovascular diseases (CVDs)*. Available from: https://www.who.int/news-room/fact-sheets/detail/cardiovascular-diseases-(cvds) [08 April 2019].
51. Xu, J., and H. Wu. 2009. Chinese herbal medicine and acupuncture for the treatment of cardiovascular disease. *Journal of Geriatric Cardiology* 6: 56–61.
52. Lu, Y., X. Liu, X. Liang, L. Xiang, and W. Zhang. 2011. Metabolomic strategy to study therapeutic and synergistic effects of tanshinone IIA, salvianolic acid B and ginsenoside Rb1 in myocardial ischemia rats. *Ethnopharmacology* 134: 45–49.
53. Li, X., X. Xu, J. Wang, H. Yu, X. Wang, H. Yang, H. Xu, S. Tang, Y. Li, L. Yang, L. Huang, Y. Wang, and S. Yang. 2012. A system-level investigation into the mechanisms of Chinese traditional medicine: Compound Danshen formula for cardiovascular disease treatment. *PLoS ONE* 7: e43918.
54. Lv, G.Y., Y.P. Zhang, J.L. Gao, J.J. Yu, J. Lei, Z.R. Zhang, B. Li, R.J. Zhan, and S.H. Chen. 2013. Combined antihypertensive effect of luteolin and buddleoside enriched extracts in spontaneously hypertensive rats. *Journal of Ethnopharmacology* 150: 507–513.
55. Liang, X., X. Chen, Q. Liang, H. Zhang, P. Hu, Y. Wang, and G. Luo. 2011. Metabonomic study of Chinese Medicine Shuanglong Formula as an effective treatment for myocardial infarction in rats. *Journal of Proteome Research* 10: 790–799.
56. Liang, Q.-L., X.-P. Liang, Y.-M. Wang, Y.-Y. Xie, R.-L. Zhang, X. Chen, R. Gao, Y.-J. Cheng, J. Wu, Q.-B. Xu, Q.-Z. Xiao, X. Li, S.-F. Lv, X.-M. Fan, H.-Y. Zhang, Q.-L. Zhang, and G.-A. Luo. 2012. Effective components screening and anti-myocardial infarction mechanism study of the Chinese medicine NSLF6 based on "system to system" mode. *Journal of Translational Medicine* 10: 26–26.
57. Seto, S.W., G.Y. Yang, H. Kiat, A. Bensoussan, Y.W. Kwan, and D. Chang. 2015. Diabetes mellitus, cognitive impairment, and traditional Chinese medicine. *International Journal of Endocrinology* 2015: 810439.

58. Andersson, D.K., and K. Svardsudd. 1995. Long-term glycemic control relates to mortality in Type II diabetes. *Diabetes Care* 18: 1534–1543.
59. Alarcon-Aguilara, F.J., R. Roman-Ramos, S. Perez-Gutierrez, A. Aguilar-Contreras, C.C. Contreras-Weber, and J.L. Flores-Saenz. 1998. Study of the anti-hyperglycemic effect of plants used as antidiabetics. *Journal of Ethnopharmacology* 61: 101–110.
60. Cui, X., D.W. Qian, S. Jiang, E.X. Shang, Z.H. Zhu, and J.A. Duan. 2018. Scutellariae radix and coptidis rhizoma improve glucose and lipid metabolism in T2DM rats via regulation of the metabolic profiling and MAPK/PI3K/AKT signaling pathway. *International Journal of Molecular Sciences* 19. https://doi.org/10.3390/ijms19113634.
61. Cong, Z., M. Zhou, F. Yang, C. Liu, R. Pan, Q. Chang, X. Liu, and Y. Liao. 2017. Effects of ganoderma, rhodiola and grape seed extracts on the glucuronidation and oral bioavailability of resveratrol in rats. *Natural Product Research* 33 (5): 1–5.
62. Yuan, M., Z. Yan, Y. Liu, D. Chen, Z. Yang, L. He, and Z. Zhang. 2019. Chemical profiles, antioxidant activity and acute toxicity of raw and sulfur-fumigated Smilacis glabrae rhizoma. *Journal of Ethnopharmacology* 234: 76–84.
63. Lee, E.M., I. Park, Y.J. Lee, Y.H. You, J.W. Kim, M.J. Kim, Y.B. Ahn, P. Kim, and S.H. Ko. 2018. Effect of resveratrol treatment on graft revascularization after islet transplantation in streptozotocin-induced diabetic mice. *Islets* 10: 25–39.
64. Su, W., W. Xu, H. Zhang, Z. Ying, L. Zhou, L. Zhang, and T. Wang. 2017. Effects of dietary leucine supplementation on the hepatic mitochondrial biogenesis and energy metabolism in normal birth weight and intrauterine growth-retarded weanling piglets. *Nutrition Research and Practice* 11: 121–129.
65. Bruckbauer, A., M.B. Zemel, T. Thorpe, M.R. Akula, A.C. Stuckey, D. Osborne, E.B. Martin, S. Kennel, and J.S. Wall. 2012. Synergistic effects of leucine and resveratrol on insulin sensitivity and fat metabolism in adipocytes and mice. *Nutrition and Metabolism* 9: 77.
66. Fu, J., J. Fu, J. Yuan, N. Zhang, B. Gao, G. Fu, Y. Tu, and Y. Zhang. 2012. Anti- diabetic activities of Acanthopanax senticosus polysaccharide (ASP) in combination with metformin. *International Journal of Biological Macromolecules* 50: 619–623.
67. Nankar, R., P.K. Prabhakar, and M. Doble. 2017. Hybrid drug combination: Combination of ferulic acid and metformin as anti-diabetic therapy. *Phytomedicine* 37: 10–13.
68. Mohan, V., S. Bedi, R. Unnikrishnan, B.K. Sahay, S. Joshpi, and A. Misra. 2012. Pioglitazone–where do we stand in India? *The Journal of the Association of Physicians of India* 60: 68–70.
69. Nankar, R.P., and M. Doble. 2017. Hybrid drug combination: Anti-diabetic treatment of type 2 diabetic Wistar rats with combination of ellagic acid and pioglitazone. *Phytomedicine* 37: 4–9.
70. Ashley, N.T., Z.M. Weil, and R.J. Nelson. 2012. Inflammation: Mechanisms, costs, and natural variation. *Annual Review of Ecology, Evolution, and Systematics* 43: 385–406.
71. Pan, M.-H., Y.-S. Chiou, M.-L. Tsai, and C.-T. Ho. 2011. Anti-inflammatory activity of traditional Chinese medicinal herbs. *Journal of Traditional and Complementary Medicine* 1: 8–24.
72. Zhang, N., Z. He, S. He, and P. Jing. 2019. Insights into the importance of dietary chrysanthemum flower (Chrysanthemum morifolium cv. Hangju)-wolfberry (Lycium barbarum fruit) combination in antioxidant and anti-inflammatory properties. *Food Research International* 116: 810–818.
73. Zhou, X., V. Razmovski-Naumovski, D. Chang, C. Li, A. Kam, M. Low, A. Bensoussan, and K. Chan. 2016b. Synergistic effects of Danshen (Salvia miltiorrhiza radix et rhizoma) and sanqi (Notoginseng radix et rhizoma) combination in inhibiting inflammation mediators in RAW264.7 cells. *BioMed Research International* 2016. https://doi.org/10.1155/2016/5758195.
74. Wisniewski-Rebecca, E.S., B.A. Rocha, L.A. Wiirzler, R.K. Cuman, C.A. Velazquez-Martinez, and C.A. Bersani-Amado. 2015. Synergistic effects of anethole and ibuprofen in acute inflammatory response. *Chemico-Biological Interactions* 242: 247–253.

75. Ortiz, M.I., M.L. Ramirez-Montiel, M.P. Gonzalez-Garcia, H.A. Ponce-Monter, G. Castaneda-Hernandez, and R. Carino-Cortes. 2010. The combination of naproxen and citral reduces nociception and gastric damage in rats. *Archives of Pharmacal Research* 33: 1691–1697.
76. Kyung, J., D. Kim, D. Park, Y.H. Yang, E.-K. Choi, S.P. Lee, T.-S. Kim, Y.B. Lee, and Y.B. Kim. 2012. Synergistic anti-inflammatory effects of Laminaria japonica fucoidan and Cistanche tubulosa extract. *Laboratory Animal Research* 28: 91–97.
77. Ramachandran, C., B. Wilk, S.J. Melnick, and I. Eliaz. 2017. Synergistic antioxidant and anti-inflammatory effects between modified citrus pectin and honokiol. *Evidence-Based Complementary and Alternative Medicine* 2017. https://doi.org/10.1155/2017/8379843.
78. Long, F., H. Yang, Y. Xu, H. Hao, and P. Li. 2015. A strategy for the identification of combinatorial bioactive compounds contributing to the holistic effect of herbal medicines. *Scientific Reports*. https://doi.org/10.1038/srep12361.
79. Sies, H., C. Berndt, and D.P. Jones. 2017. Oxidative stress. *Annual Review of Biochemistry* 86: 715–748.
80. Li, M., Y. Xu, W. Yang, J. Li, X. Xu, X. Zhang, F. Chen, and D. Li. 2011. In vitro synergistic anti-oxidant activities of solvent-extracted fractions from Astragalus membranaceus and Glycyrrhiza uralensis. *LWT – Food Science and Technology* 44: 1745–1751.
81. Yang, W.J., D.P. Li, J.K. Li, M.H. Li, Y.L. Chen, and P.Z. Zhang. 2009. Synergistic antioxidant activities of eight traditional Chinese herb pairs. *Biological & Pharmaceutical Bulletin* 32: 1021–1026.
82. Li, J., F. Li, Y. Xu, W. Yang, L. Qu, Q. Xiang, C. Liu, and D. Li. 2013b. Chemical composition and synergistic antioxidant activities of essential oils from Atractylodes macrocephala and Astragalus membranaceus. *Natural Product Communications* 8: 1321–1324.
83. Xu, X., F. Li, X. Zhang, P. Li, X. Zhang, Z. Wu, and D. Li. 2014b. In vitro synergistic antioxidant activity and identification of antioxidant components from Astragalus membranaceus and Paeonia lactiflora. *PLoS ONE* 9. https://doi.org/10.1371/journal.pone.0096780.
84. Freeman, B.L., D.L. Eggett, and T.L. Parker. 2010. Synergistic and antagonistic interactions of phenolic compounds found in navel oranges. *Journal of Food Science* 75: C570–C576.
85. Gui, M., J. Du, J. Guo, B. Xiao, W. Yang, and M. Li. 2014. Aqueous extract of Chrysanthemum morifolium enhances the antimelanogenic and antioxidative activities of the mixture of soy peptide and collagen peptide. *Journal of Traditional and Complementary Medicine* 4: 171–176.
86. Hsieh, M.-H., M.-J. Hsieh, C.-R. Wu, W.-H. Peng, M.-T. Hsieh, and C.-C. Hsieh. 2018. The synergistic effect of antioxidant interaction between luteolin and chlorogenic acid in Lonicera japonica. *bioRxiv* 418319. https://doi.org/10.1101/418319.
87. Guimaraes, R., L. Barros, A.M. Carvalho, and I.C. Ferreira. 2011. Infusions and decoctions of mixed herbs used in folk medicine: Synergism in antioxidant potential. *Phytotherapy Research* 25: 1209–1214.
88. Gheno, R., J.M. Cepparo, C.E. Rosca, and A. Cotten. 2012. Musculoskeletal disorders in the elderly. *Journal of Clinical Imaging Science* 2: 39–39.
89. Hou, P.-W., P.-K. Fu, H.-C. Hsu, and C.-L. Hsieh. 2015. Traditional Chinese medicine in patients with osteoarthritis of the knee. *Journal of Traditional and Complementary Medicine* 5: 182–196.
90. Trinh, K., X. Cui, and Y.J. Wang. 2010. Chinese herbal medicine for chronic neck pain due to cervical degenerative disc disease. *Cochrane Database of Systematic Reviews* 35. https://doi.org/10.1002/14651858.CD006556.pub2.
91. Wu, G., and H. Wu. 2014. Analgesia synergism of essential oil from pericarp of Zanthoxylum schinifolium and verapamil. *Evidence-Based Complementary and Alternative Medicine* 2014: 505876–505876.

92. Lev-Ari, S., L. Strier, D. Kazanov, O. Elkayam, D. Lichtenberg, D. Caspi, and N. Arber. 2006. Curcumin synergistically potentiates the growth-inhibitory and pro- apoptotic effects of celecoxib in osteoarthritis synovial adherent cells. *Rheumatology* 45: 171–177.
93. Zhang, A., H. Sun, Y. Yuan, W. Sun, G. Jiao, and X. Wang. 2011. An in vivo analysis of the therapeutic and synergistic properties of Chinese medicinal formula Yin-Chen-Hao-Tang based on its active constituents. *Fitoterapia* 82: 1160–1168.
94. Wang, X., A. Zhang, P. Wang, H. Sun, G. Wu, W. Sun, H. Lv, G. Jiao, H. Xu, Y. Yuan, L. Liu, D. Zou, Z. Wu, Y. Han, G. Yan, W. Dong, F. Wu, T. Dong, Y. Yu, S. Zhang, X. Wu, X. Tong, and X. Meng. 2013. Metabolomics coupled with proteomics advancing drug discovery toward more agile development of targeted combination therapies. *Molecular & Cellular Proteomics* 12: 1226–1238.
95. Dong, L.C., Y.X. Fan, Q. Yu, J. Ma, X. Dong, P. Li, and H.J. Li. 2015. Synergistic effects of rhubarb-gardenia herb pair in cholestatic rats at pharmacodynamic and pharmacokinetic levels. *Journal of Ethnopharmacology* 175: 67–74.
96. Yan, F., Q.Y. Zhang, L. Jiao, T. Han, H. Zhang, L.P. Qin, and R. Khalid. 2009. Synergistic hepatoprotective effect of Schisandrae lignans with Astragalus polysaccharides on chronic liver injury in rats. *Phytomedicine* 16: 805–813.
97. Debib, A., M. Dueñas, M. Boumediene, R.A. Mothana, A. Latifa, and M.A. Tir-Touil. 2016. Synergetic hepatoprotective effect of phenolic fractions obtained from Ficus carica dried fruit and extra virgin olive oil on CCL4-induced oxidative stress and hepatotoxicity in rats. *Journal of Food Biochemistry* 40: 507–516.
98. Ye, X.L., W.W. Huang, Z. Chen, X.G. Li, P. Li, P. Lan, L. Wang, Y. Gao, Z.Q. Zhao, and X. Chen. 2010. Synergetic effect and structure-activity relationship of 3- hydroxy-3-methylglutaryl coenzyme A reductase inhibitors from Crataegus pinnatifida Bge. *Journal of Agricultural and Food Chemistry* 58: 3132–3138.
99. Cheesman, M.J., A. Ilanko, B. Blonk, and I.E. Cock. 2017. Developing new antimicrobial therapies: Are synergistic combinations of plant extracts/compounds with conventional antibiotics the solution? *Pharmacognosy Reviews* 11: 57–72.
100. Magiorakos, A.P., A. Srinivasan, R.B. Carey, Y. Carmeli, M.E. Falagas, C.G. Giske, S. Harbarth, J.F. Hindler, G. Kahlmeter, B. Olsson-Liljequist, D.L. Paterson, L.B. Rice, J. Stelling, M.J. Struelens, A. Vatopoulos, J.T. Weber, and D.L. Monnet. 2012. Multidrug-resistant extensively drug-resistant and pandrug-resistant bacteria: An international expert proposal for interim standard definitions for acquired resistance. *Clinical Microbiology and Infection* 18: 268–281.
101. Sibanda, T., and A. Okoh. 2007. The challenges of overcoming antibiotic resistance: Plant extracts as potential sources of antimicrobial and resistance modifying agents. *African Journal of Biotechnology* 6: 2886–2896.
102. Nikaido, H. 1994. Prevention of drug access to bacterial targets: Permeability barriers and active efflux. *Science* 264: 382–388.
103. Davies, J. 1994. Inactivation of antibiotics and the dissemination of resistance genes. *Science* 264: 375–382.
104. Spratt, B.G. 1994. Resistance to antibiotics mediated by target alterations. *Science* 264: 388–393.
105. Stermitz, F.R., P. Lorenz, J.N. Tawara, L.A. Zenewicz, and K. Lewis. 2000. Synergy in a medicinal plant: Antimicrobial action of berberine potentiated by 5′- methoxyhydnocarpin, a multidrug pump inhibitor. *Proceedings of the National Academy of Sciences of the United States of America* 97: 1433–1437.
106. Musumeci, R., A. Speciale, R. Costanzo, A. Annino, S. Ragusa, A. Rapisarda, M.S. Pappalardo, and L. Iauk. 2003. Berberis aetnensis C. Presl. extracts: Antimicrobial properties and interaction with ciprofloxacin. *International Journal of Antimicrobial Agents* 22: 48–53.
107. Mikulášová, M., R. Chovanová, and Š. Vaverková. 2016. Synergism between antibiotics and plant extracts or essential oils with efflux pump inhibitory activity in coping with multidrug-resistant Staphylococci. *Phytochemistry Reviews* 15: 651–662.

108. Santiago, C., E.L. Pang, K.-H. Lim, H.-S. Loh, and K.N. Ting. 2015. Inhibition of penicillin-binding protein 2a (PBP2a) in methicillin resistant Staphylococcus aureus (MRSA) by combination of ampicillin and a bioactive fraction from Duabanga grandiflora. *BMC Complementary and Alternative Medicine* 15: 178–178.
109. Phitaktim, S., M. Chomnawang, K. Sirichaiwetchakoon, B. Dunkhunthod, G. Hobbs, and G. Eumkeb. 2016. Synergism and the mechanism of action of the combination of α-mangostin isolated from Garcinia mangostana l. and oxacillin against an oxacillin-resistant Staphylococcus saprophyticus. *BMC Microbiology* 16. https://doi.org/10.1186/s12866-016-0814-4.
110. Ghosh, I.N., S.D. Patil, T.K. Sharma, S.K. Srivastava, R. Pathania, and N.K. Navani. 2013. Synergistic action of cinnamaldehyde with silver nanoparticles against spore-forming bacteria: A case for judicious use of silver nanoparticles for antibacterial applications. *International Journal of Nanomedicine* 8: 4721–4731.
111. Hwang, J.H., H. Choi, E.R. Woo, and D.G. Lee. 2013. Antibacterial effect of amentoflavone and its synergistic effect with antibiotics. *Journal of Microbiology and Biotechnology* 23: 953–958.
112. Sharma, M., R. Manoharlal, A.S. Negi, and R. Prasad. 2010a. Synergistic anticandidal activity of pure polyphenol curcumin I in combination with azoles and polyenes generates reactive oxygen species leading to apoptosis. *FEMS Yeast Research* 10: 570–578.
113. Sharma, M., R. Manoharlal, N. Puri, and R. Prasad. 2010b. Antifungal curcumin induces reactive oxygen species and triggers an early apoptosis but prevents hyphae development by targeting the global repressor tup1 in Candida albicans. *Bioscience Reports* 30: 391–404.
114. Li, D.-D., Y. Xu, D.-Z. Zhang, H. Quan, E. Mylonakis, D.-D. Hu, M.-B. Li, L.-X. Zhao, L.-H. Zhu, Y. Wang, and Y.-Y. Jiang. 2013a. Fluconazole assists berberine to kill fluconazole-resistant Candida albicans. *Antimicrobial Agents and Chemotherapy* 57: 6016–6027.
115. Zhu, S.L., L. Yan, Y.X. Zhang, Z.H. Jiang, P.H. Gao, Y. Qiu, L. Wang, M.Z. Zhao, T.J. Ni, Z. Cai, S.J. Tian, C.X. Zang, D.Z. Zhang, and Y.Y. Jiang. 2014. Berberine inhibits fluphenazine-induced up-regulation of CDR1 in Candida albicans. *Biological & Pharmaceutical Bulletin* 37: 268–273.
116. Iwazaki, R.S., E.H. Endo, T. Ueda-Nakamura, C.V. Nakamura, L.B. Garcia, and B.P.D. Filho. 2009. In vitro antifungal activity of the berberine and its synergism with fluconazole. *Antonie Van Leeuwenhoek* 97: 201–205.
117. Han, Y., and J.-H. Lee. 2005. Berberine synergy with amphotericin B against disseminated Candidiasis in mice. *Biological & Pharmaceutical Bulletin* 28: 541–544.
118. Wei, G.-X., X. Xu, and C.D. Wu. 2011a. In vitro synergism between berberine and miconazole against planktonic and biofilm Candida cultures. *Archives of Oral Biology* 56: 565–572.
119. Quan, H., Y.-Y. Cao, Z. Xu, J.-X. Zhao, P.-H. Gao, X.-F. Qin, and Y.-Y. Jiang. 2006. Potent in vitro synergism of fluconazole and berberine chloride against clinical isolates of Candida albicans resistant to fluconazole. *Antimicrobial Agents and Chemotherapy* 50: 1096–1099.
120. Lee, Y.-S., O.-H. Kang, J.-G. Choi, Y.-C. Oh, H.-S. Chae, J.H. Kim, H. Park, D.H. Sohn, Z.-T. Wang, and D.-Y. Kwon. 2008. Synergistic effects of the combination of galangin with gentamicin against methicillin-resistant Staphylococcus aureus. *Journal of Microbiology* 46: 283–288.
121. Zuo, G.-Y., Y. Li, J. Han, G.-C. Wang, Y.-L. Zhang, and Z.-Q. Bian. 2012a. Antibacterial and synergy of berberines with antibacterial agents against clinical multi-drug resistant isolates of methicillin-resistant Staphylococcus aureus (MRSA). *Molecules* 17: 10322–10330.
122. Yu, H.H., K.J. Kim, J.D. Cha, H.K. Kim, Y.E. Lee, N.Y. Choi, and Y.O. You. 2005. Antimicrobial activity of berberine alone and in combination with ampicillin or oxacillin against methicillin-resistant Staphylococcus aureus. *Journal of Medicinal Food* 8: 454–461.
123. Betts, J.W., C. Murphy, S.M. Kelly, and S.J. Haswell. 2012. Synergistic antibacterial effects of theaflavin in combination with ampicillin against hospital isolates of Stenotrophomonas maltophilia. *Journal of Microbiology, Biotechnology and Food Sciences* 2: 1068–1078.
124. Zuo, G.Y., J. An, J. Han, Y.L. Zhang, G.C. Wang, X.Y. Hao, and Z.Q. Bian. 2012b. Isojacareubin from the Chinese herb Hypericum japonicum: Potent antibacterial and synergistic effects on clinical methicillin-resistant Staphylococcus aureus (MRSA). *International Journal of Molecular Sciences* 13: 8210–8218.

125. Yang, Z.C., B.C. Wang, X.S. Yang, Q. Wang, and L. Ran. 2005. The synergistic activity of antibiotics combined with eight traditional Chinese medicines against two different strains of Staphylococcus aureus. *Colloids and Surfaces, B: Biointerfaces* 41: 79–81.
126. Zuo, G.Y., X.J. Zhang, J. Han, Y.Q. Li, and G.C. Wang. 2015. In vitro synergism of magnolol and honokiol in combination with antibacterial agents against clinical isolates of methicillin-resistant Staphylococcus aureus (MRSA). *BMC Complementary and Alternative Medicine* 15. https://doi.org/10.1186/s12906-015-0938-3.
127. Adwan, G., and M. Mhanna. 2008. Synergistic effects of plant extracts and antibiotics on Staphylococcus aureus strains isolated from clinical specimens. *Middle-East Journal of Scientific Research* 3: 134–139.
128. Joung, D.K., H. Joung, D.W. Yang, D.Y. Kwon, J.G. Choi, S. Woo, D.Y. Shin, O.H. Kweon, K.T. Kweon, and D.W. Shin. 2012. Synergistic effect of rhein in combination with ampicillin or oxacillin against methicillin-resistant Staphylococcus aureus. *Experimental and Therapeutic Medicine* 3: 608–612.
129. Jang, E.J., S.M. Cha, S.M. Choi, and J.D. Cha. 2014. Combination effects of baicalein with antibiotics against oral pathogens. *Archives of Oral Biology* 59: 1233–1241.
130. Chang, P.C., H.Y. Li, H.J. Tang, J.W. Liu, J.J. Wang, and Y.C. Chuang. 2007. In vitro synergy of baicalein and gentamicin against vancomycin-resistant Enterococcus. *Journal of Microbiology, Immunology, and Infection* 40: 56–61.
131. Fujita, M., S. Shiota, T. Kuroda, T. Hatano, T. Yoshida, T. Mizushima, and T. Tsuchiya. 2005. Remarkable synergies between baicalein and tetracycline, and baicalein and β-lactams against methicillin-resistant Staphylococcus aureus. *Microbiology and Immunology* 49: 391–396.
132. Cai, W., Y. Fu, W. Zhang, X. Chen, J. Zhao, W. Song, Y. Li, Y. Huang, Z. Wu, R. Sun, C. Dong, and F. Zhang. 2016. Synergistic effects of baicalein with cefotaxime against Klebsiella pneumoniae through inhibiting CTX-M-1 gene expression. *BMC Microbiology* 16: 181–181.
133. Liu, I.X., D.G. Durham, and R.M. Richards. 2000a. Baicalin synergy with beta-lactam antibiotics against methicillin-resistant Staphylococcus aureus and other beta- lactam-resistant strains of S. aureus. *The Journal of Pharmacy and Pharmacology* 52: 361–366.
134. Wang, S.-Y., Z.-L. Sun, T. Liu, S. Gibbons, W.-J. Zhang, and M. Qing. 2014b. Flavonoids from Sophora moorcroftiana and their synergistic antibacterial effects on MRSA. *Phytotherapy Research* 28: 1071–1076.
135. Zuo, G.-Y., Y. Li, T. Wang, J. Han, G.-C. Wang, Y.-L. Zhang, and W.-D. Pan. 2011. Synergistic antibacterial and antibiotic effects of bisbenzylisoquinoline alkaloids on clinical isolates of methicillin-resistant Staphylococcus aureus (MRSA). *Molecules* 16: 9819–9826.
136. Khan, M.S.A., and I. Ahmad. 2011. Antibiofilm activity of certain phytocompounds and their synergy with fluconazole against Candida albicans biofilms. *The Journal of Antimicrobial Chemotherapy* 67: 618–621.
137. Ahmad, A., A. Khan, L.A. Khan, and N. Manzoor. 2010. In vitro synergy of eugenol and methyleugenol with fluconazole against clinical Candida isolates. *Journal of Medical Microbiology* 59: 1178–1184.
138. Iida, Y., K.B. Oh, M. Saito, H. Matsuoka, and H. Kurata. 2000. In vitro synergism between nyasol, an active compound isolated from Anemarrhena asphodeloides, and azole agents against Candida albicans. *Planta Medica* 66: 435–438.
139. Xu, Y., Y. Wang, L. Yan, R.M. Liang, B.D. Dai, R.J. Tang, P.H. Gao, and Y.Y. Jiang. 2009. Proteomic analysis reveals a synergistic mechanism of fluconazole and berberine against fluconazole-resistant Candida albicans: Endogenous ROS augmentation. *Journal of Proteome Research* 8: 5296–5304.
140. Yan, Z., H. Hua, Y. Xu, and L.P. Samaranayake. 2012. Potent antifungal activity of pure compounds from traditional Chinese medicine extracts against six oral Candida species and the synergy with fluconazole against azole-resistant Candida albicans. *Evidence-Based Complementary and Alternative Medicine* 2012. https://doi.org/10.1155/2012/106583.

141. Sharma, M., R. Manoharlal, S. Shukla, N. Puri, T. Prasad, S.V. Ambudkar, and R. Prasad. 2009. Curcumin modulates efflux mediated by yeast ABC multidrug transporters and is synergistic with antifungals. *Antimicrobial Agents and Chemotherapy* 53: 3256–3265.
142. Fan, L., X.Q. Mao, G.Y. Tao, G. Wang, F. Jiang, Y. Chen, Q. Li, W. Zhang, H.P. Lei, D.L. Hu, Y.F. Huang, D. Wang, and H.H. Zhou. 2009. Effect of Schisandra chinensis extract and ginkgo biloba extract on the pharmacokinetics of talinolol in healthy volunteers. *Xenobiotica* 39: 249–254.
143. Huang, B.B., G.F. Li, F. Ren, Z.K. Tang, H.F. Ma, Y.B. Sun, L.J. Chen, and L. Yang. 2008. Effect of Glycyrrhiza inflata and Daphne genkwa on permeabilities of rhodamine 123, a p-glycoprotein substrate across rat jejunum membranes in vitro. *Zhongguo Zhong Yao Za Zhi* 33: 2521–2526.
144. Gurley, B.J., S.F. Gardner, M.A. Hubbard, D.K. Williams, W.B. Gentry, I.A. Khan, and A. Shah. 2005. In vivo effects of goldenseal, kava kava, black cohosh, and valerian on human cytochrome p450 1a2, 2d6, 2e1, and 3a4/5 phenotypes. *Clinical Pharmacology and Therapeutics* 77: 415–426.
145. Gurley, B.J., A. Swain, G.W. Barone, D.K. Williams, P. Breen, C.R. Yates, L.B. Stuart, M.A. Hubbard, Y. Tong, and S. Cheboyina. 2007. Effect of goldenseal (Hydrastis canadensis) and kava kava (Piper methysticum) supplementation on digoxin pharmacokinetics in humans. *Drug Metabolism and Disposition* 35: 240–245.
146. Sandhu, R.S., R.P. Prescilla, T.M. Simonelli, and D.J. Edwards. 2003. Influence of goldenseal root on the pharmacokinetics of indinavir. *Journal of Clinical Pharmacology* 43: 1283–1288.
147. Gurley, B.J., A. Swain, M.A. Hubbard, F. Hartsfield, J. Thaden, D.K. Williams, W.B. Gentry, and Y. Tong. 2008. Supplementation with goldenseal (Hydrastis canadensis), but not kava kava (piper methysticum), inhibits human CYP3A activity in vivo. *Clinical Pharmacology and Therapeutics* 83: 61–69.
148. Jiang, W., X. Wang, X. Xu, and L. Kong. 2010. Effect of Schisandra sphenanthera extract on the concentration of tacrolimus in the blood of liver transplant patients. *International Journal of Clinical Pharmacology and Therapeutics* 48: 224–229.
149. Xin, H.W., X.C. Wu, Q. Li, A.R. Yu, M. Zhu, Y. Shen, D. Su, and L. Xiong. 2007. Effects of Schisandra sphenanthera extract on the pharmacokinetics of tacrolimus in healthy volunteers. *British Journal of Clinical Pharmacology* 64: 469–475.
150. Xin, H.W., X.C. Wu, Q. Li, A.R. Yu, and L. Xiong. 2009. Effects of Schisandra sphenanthera extract on the pharmacokinetics of midazolam in healthy volunteers. *British Journal of Clinical Pharmacology* 67: 541–546.
151. Zanger, U.M., and M. Schwab. 2013. Cytochrome p450 enzymes in drug metabolism: Regulation of gene expression, enzyme activities, and impact of genetic variation. *Pharmacology & Therapeutics* 138: 103–141.
152. Izzo, A.A. 2012. Interactions between herbs and conventional drugs: Overview of the clinical data. *Medical Principles and Practice* 21: 404–428.
153. Whitten, D.L., S.P. Myers, J.A. Hawrelak, and H. Wohlmuth. 2006. The effect of St John's wort extracts on CYP3A: A systematic review of prospective clinical trials. *British Journal of Clinical Pharmacology* 62: 512–526.
154. Ioannides, C. 2002. Topics in xenobiochemistry – Pharmacokinetic interactions between herbal remedies and medicinal drugs. *Xenobiotica* 32: 451–478.
155. Qiu, F., J. Zeng, S. Liu, M. He, L. Zhu, Y. Ye, P. Miao, S. Shen, and J. Jiang. 2014. Effects of Danshen ethanol extract on the pharmacokinetics of fexofenadine in healthy volunteers. *Evidence-Based Complementary and Alternative Medicine* 2014. https://doi.org/10.1155/2014/473213.
156. Proulx, S.R., D.E. Promislow, and P.C. Phillips. 2005. Network thinking in ecology and evolution. *Trends in Ecology & Evolution* 20: 345–353.
157. Zhao, S., and R. Iyengar. 2012. Systems pharmacology: Network analysis to identify multiscale mechanisms of drug action. *Annual Review of Pharmacology and Toxicology* 52: 505–521.

158. Rao, R.T., M.L. Scherholz, C. Hartmanshenn, S.-A. Bae, and I.P. Androulakis. 2017. On the analysis of complex biological supply chains: From process systems engineering to quantitative systems pharmacology. *Computers and Chemical Engineering* 107: 100–110.
159. Leil, T.A., and R. Bertz. 2014. Quantitative systems pharmacology can reduce attrition and improve productivity in pharmaceutical research and development. *Frontiers in Pharmacology* 5: 247.
160. Xie, F., and J. Gu. 2019. Computational methods and applications for quantitative systems pharmacology. *Quantiative Biology* 7: 3–16.
161. Gilson, M.K., T. Liu, M. Baitaluk, G. Nicola, L. Hwang, and J. Chong. 2015. Bindingdb in 2015: A public database for medicinal chemistry, computational chemistry and systems pharmacology. *Nucleic Acids Research* 44: D1045–D1053.
162. Liu, Z., F. Guo, Y. Wang, C. Li, X. Zhang, H. Li, L. Diao, J. Gu, W. Wang, and D. Li. 2016. BATMAN-TCM: A bioinformatics analysis tool for molecular mechanism of traditional Chinese medicine. *Scientific Reports* 6. https://doi.org/10.1038/srep21146.
163. Berger, S.I., and R. Iyengar. 2009. Network analyses in systems pharmacology. *Bioinformatics* 25: 2466–2472.
164. Wang, Z., and T.S. Deisboeck. 2014. Mathematical modeling in cancer drug discovery. *Drug Discovery Today* 19: 145–150.
165. Cai, H., Y. Luo, X. Yan, P. Ding, Y. Huang, S. Fang, R. Zhang, Y. Chen, Z. Guo, and J. Fang. 2018. The mechanisms of Bushen-Yizhi formula as a therapeutic agent against Alzheimer's disease. *Scientific Reports* 8: 3104.
166. Zhao, L., Y. Wu, Y. Gao, H. Xiang, X. Qin, and J. Tian. 2017. Intervention mechanism of psychological sub-health by baihe dihuang tang based on network pharmacology. *Yao Xue Xue Bao* 52: 99–105.
167. Xu, H., Y. Zhang, Y. Lei, X. Gao, H. Zhai, N. Lin, S. Tang, R. Liang, Y. Ma, and D. Li. 2014a. A systems biology-based approach to uncovering the molecular mechanisms underlying the effects of dragon's blood tablet in colitis, involving the integration of chemical analysis, adme prediction, and network pharmacology. *PLoS ONE* 9: e101432.
168. Wang, X., S. Yu, Q. Jia, L. Chen, J. Zhong, Y. Pan, P. Shen, Y. Shen, S. Wang, and Z. Wei. 2017a. Niaoduqing granules relieve chronic kidney disease symptoms by decreasing renal fibrosis and anemia. *Oncotarget* 8: 55920.
169. Shi, S.-H., Y.-P. Cai, X.-J. Cai, X.-Y. Zheng, D.-S. Cao, F.-Q. Ye, and Z. Xiang. 2014. A network pharmacology approach to understanding the mechanisms of action of traditional medicine: Bushenhuoxue formula for treatment of chronic kidney disease. *PLoS ONE* 9: e89123.
170. Chen, L., Y. Cao, H. Zhang, D. Lv, Y. Zhao, Y. Liu, G. Ye, and Y. Chai. 2018. Network pharmacology-based strategy for predicting active ingredients and potential targets of yangxinshi tablet for treating heart failure. *Journal of Ethnopharmacology* 219: 359–368.
171. Shen, X., Z. Zhao, X. Luo, H. Wang, B. Hu, and Z. Guo. 2016. Systems pharmacology based study of the molecular mechanism of SiNiSan formula for application in nervous and mental diseases. *Evidence-Based Complementary and Alternative Medicine* 2016.
172. Wang, H., B. Zhang, X. Ye, S. He, Y. Zhang, and Y. Wang. 2015a. Study on mechanism for anti-depression efficacy of Sini San through auxiliary mechanism elucidation system for Chinese medicine. *Zhongguo Zhong Yao Za Zhi* 40: 3723–3728.
173. Zhou, W., J. Wang, Z. Wu, C. Huang, A. Lu, and Y. Wang. 2016a. Systems pharmacology exploration of botanic drug pairs reveals the mechanism for treating different diseases. *Scientific Reports* 6: 36985.
174. Wang, Y., G. Guo, B.R. Yang, Q.Q. Xin, Q.W. Liao, S.M. Lee, Y.J. Hu, K.J. Chen, and W.H. Cong. 2017b. Synergistic effects of Chuanxiong-Chishao herb-pair on promoting angiogenesis at network pharmacological and pharmacodynamic levels. *Chinese Journal of Integrative Medicine* 23: 654–662.

175. Gu, J., P.S. Crosier, C.J. Hall, L. Chen, and X. Xu. 2016. Inflammatory pathway network-based drug repositioning and molecular phenomics. *Molecular BioSystems* 12: 2777–2784.
176. Gu, J., X. Zhang, Y. Ma, N. Li, F. Luo, L. Cao, Z. Wang, G. Yuan, L. Chen, and W. Xiao. 2015. Quantitative modeling of dose–response and drug combination based on pathway network. *Journal of Chemistry* 7: 19. https://doi.org/10.1186/s13321-015-0066-6.
177. Ru, J., P. Li, J. Wang, W. Zhou, B. Li, C. Huang, P. Li, Z. Guo, W. Tao, and Y. Yang. 2014. TCMSP: A database of systems pharmacology for drug discovery from herbal medicines. *Journal of Chemistry* 6: 13.
178. Wang, J., L. Zhang, B. Liu, Q. Wang, Y. Chen, Z. Wang, J. Zhou, W. Xiao, C. Zheng, and Y. Wang. 2018. Systematic investigation of the Erigeron breviscapus mechanism for treating cerebrovascular disease. *Journal of Ethnopharmacology* 224: 429–440.
179. Liu, J., M. Pei, C. Zheng, Y. Li, Y. Wang, A. Lu, and L. Yang. 2013. A systems-pharmacology analysis of herbal medicines used in health improvement treatment: Predicting potential new drugs and targets. *Evidence-Based Complementary and Alternative Medicine* 2013. https://doi.org/10.1155/2013/938764.
180. Liang, H., H. Ruan, Q. Ouyang, and L. Lai. 2016. Herb-target interaction network analysis helps to disclose molecular mechanism of traditional Chinese medicine. *Scientific Reports* 6. https://doi.org/10.1038/srep36767.
181. Li, Y., R. Li, Z. Ouyang, and S. Li. 2015. Herb network analysis for a famous TCM doctor's prescriptions on treatment of rheumatoid arthritis. *Evidence-Based Complementary and Alternative Medicine* 2015. https://doi.org/10.1155/2015/451319.
182. Liu, X., J. Wu, D. Zhang, K. Wang, X. Duan, and X. Zhang. 2018. A network pharmacology approach to uncover the multiple mechanisms of Hedyotis diffusa willd. on colorectal cancer. *Evidence-Based Complementary and Alternative Medicine* 2018. https://doi.org/10.1155/2018/6517034.
183. Li, B., J. Rui, X. Ding, and X. Yang. 2019. Exploring the multicomponent synergy mechanism of Banxia Xiexin Decoction on irritable bowel syndrome by a systems pharmacology strategy. *Journal of Ethnopharmacology* 233: 158–168.
184. Li, J., X. Ma, C. Liu, H. Li, J. Zhuang, C. Gao, C. Zhou, L. Liu, K. Wang, and C. Sun. 2018. Exploring the mechanism of danshen against myelofibrosis by network pharmacology and molecular docking. *Evidence-Based Complementary and Alternative Medicine* 2018. https://doi.org/10.1155/2018/8363295.

Medicinal Herbs: Its Therapeutic Use in Obstetrics and Gynaecology

Irene Orbe, Daniel Paz, Leyre Pejenaute, Andrea Puente, Laura Diaz de Alda, Sandra Yague, and Iñaki Lete

Introduction

The plants have been used for thousands of years for therapeutic purposes. Hippocrates in the fifth century BC, already related the mood with food and plants [1]. The use of medicinal herbs flourished in Europe in the seventeenth century and the emigration of European people to America, moved to this continent the habit of using medicinal herbs [2]. In 1920 pharmacy drugs replaced herbal therapies in the USA, because the former had more lasting pharmacological effects and greater profitability.

In 1992, the NIH OAM (National Institutes of Health Office of Alternative Medicine) was founded and updated in 1998 by the National Center for Complementary and Integrative Health (NCCIH), which evaluated the efficacy and safety of alternative medicines through scientific studies and investigations [3].

In the USA, alternative medicine declined in the early 1900s, then resurfaced in the 1960s. This resurgence was in part due to the 1994 DSHEA (Dietary Supplement Health and Education Act), which allowed merchants to sell their products herbs without previous demonstration of efficacy and safety [4]. In 2002, a survey was conducted where it was observed that 17.7% of adults in the US had used natural products in the last year [5]. In many European regions it is common to integrate medicinal herbs with conventional medicine. For example, in Germany, 65% of the population has ever used medicinal herbs and approximately 80% of German doctors prescribe herbs regularly [6]. Another example is England, where 12.8% of the population used one or more herbs [7].

I. Orbe · D. Paz · L. Pejenaute · A. Puente · L. D. de Alda · S. Yague · I. Lete (✉)
Department of Obstetrics and Gynaecology, Araba University Hospital, Vitoria, Spain
e-mail: luisignacio.letelasa@osakidetza.eus

M. Bizzarri (ed.), *Approaching Complex Diseases*, Human Perspectives in Health Sciences and Technology 2, https://doi.org/10.1007/978-3-030-32857-3_18

Table 1 Summary of the most common uses of herbal medicines in gynaecology and obstetrics

Medicinal herb	Main use in obstetrics and gynaecology
Alchemilla vulgaris (Lady's Mantle)	Menorrhagia, dysmenorrhea and pelvic pain associated with endometriosis·
Paeonia lactiflora (Chinese peony or common garden peony)	Menorrhagia, intermenstrual bleeding, tocolytic, analgesic.
Capsella bursa pastoris (shepherd's purse)	Postpartum bleeding, menorrhagia.
Claviceps purpurea (Ergot)	Postpartum or post-abortion hemorrhage
Urtica Dioica (nettle or stinger)	Menorrhagia, dysmenorrhea and galactogenic action.
Lamium álbum (white nettle or white dead-nettle)	Menorrhagia, abnormal vaginal discharge.
Laminaria ("Laminaria cloustoni")	Osmotic dilator of the cervix.
Achillea millefolium (Yarrow)	Menorrhagia, dysmenorrhea, menopause symptons, perineal pain of episiotomy wound.
Vitex agnus castus (Chasteberry)	Premenstrual and postmenstrual syndrome, infertility disorders and irregular menstruation.
Aloe barbadensis (Aloe vera)	Wound healing, emollient.
Matricaria Chamomilla (Chamomile)	Dysmenorrhea, mastalgia, nausea and vomiting of pregnancy.
Zingiber officinale (Ginger)	Nausea and vomiting of pregnancy.
Calendula officinalis (Calendula)	Wound healing, emollient, fungicidal action, irregular menstruation, dysmenorrhea
Mentha piperita (Peppermint)	Dysmenorrhea
Valeriana officinalis (garden valerian)	Dysmenorrhea
Oenothera biennis (evening primrose oil)	Premenstrual syndrome, menopausal symptoms, mastalgia

In this chapter we review the most common medical uses of medicinal plants available in Europe. In Table 1 we summarize the most common uses of medicinal herbs in obstetrics and gynaecology.

Alchemilla vulgaris (Lady's Mantle)

Is an herbaceous native plant of Europe of perennial type of about 30 cm and flowers of about 3–5 mm.

It contains tannins, glycosides and salicylic acid. It has an antioxidant effect and is used for the treatment of heavy menstrual bleeding, dysmenorrhea and pelvic pain associated with endometriosis [8]. Traditionally it has been used as a wound healing since it accelerates the first stage of healing [9].

Paeonia lactiflora (Chinese Peony or Common Garden Peony)

Perennial plant about 2 m high, with large white flowers and leaves divided in dark green. It is grown in the northeast of China and in the interior of Mongolia.

Its main components are glycosides, benzoic acid and pentagaloylglucose [10]. The use of this plant in China goes back at least 1500 years ago. It was used mainly by women for its antispasmodic, astringent and analgesic effects [11].

It reduces excessive menstruation, intermenstrual bleeding and counteracts the effect of oxytocin [12]. It also has antiviral action against labial HSV, relieves muscle cramps, headaches, tinnitus, dizziness and blurred vision [13].

Capsella bursa pastoris (Shepherd's Purse)

Erect stem plant, rosette of basal leaves, white flowers with four petals and heart shaped seed capsule. It comes from Europe and Asia although it can be found in all temperate zones.

Its main components are: flavonoids, polypeptides, choline, acetylcholine, histamine and tyramine [14].

Its main use is to avoid or stop any type of hemorrhage, although it has been classically used to reduce excessive uterine hemorrhages [15].

In a single-blinded, randomized, clinical trial including 100 postpartum women the participants were randomly assigned into an intervention group ($n = 50$) and a placebo group ($n = 50$). Immediately after placental expulsion, the intervention group was given 10 sublingual drops of the hydroalcoholic extract of Capsella bursa pastoris plus an infusion of 20 U of oxytocin in 1 L of Ringer's solution, and the control group was given 10 sublingual drops of the placebo plus an infusion of 20 U of oxytocin in 1 L of Ringer's solution. After the intervention, there was significant decrease in the amount of postpartum bleeding in both groups. However, the mean decrease in the amount of bleeding was significantly more in the Capsella bursa pastoris group ($p < 0.001$) [16].

Claviceps purpurea (Ergot)

It is a fungus that contains alkaloids derived from lysergic acid such as ergocristine, ergometrine, ergotamine and ergocryptine. All these substances have a broad spectrum of action based on vasoconstrictive effects at the circulatory and neurotransmission levels [17]. Historically it has been used as an uterotonic to induce abortions, accelerate labor and stop postpartum hemorrhage. Ergometrine acts by stimulating the uterine muscle and stopped being used to intensify the contractions since it can produce a permanent contraction of the uterus, which can be dangerous and lead to intrauterine fetal death. Although it cannot be used to induce labor, it is very useful for controlling postpartum or post-abortion hemorrhage [18]. The methylation of the alkaloid results in a vasoconstriction that, together with its uterotonic effect, decreases the likelihood of postpartum hemorrhage and prevents muscular atony [19].

Its adverse effects include the elevation of blood pressure and increased pain [20].

Urtica dioica (Nettle or Stinger)

From the Urticaceae family, it is a plant used for its medicinal properties for a long time. It is a plant that grows in temperate zones of the Northern Hemisphere, in South Africa, the Andes and Australia. Its main components are: flavonoids, amines, glucoquinone, minerals, phytosterols and phenols [21]. Both the leaf and the roots have been used, each having different properties.

The leaf has astringent properties, that is, its application retracts the tissues and can produce a healing, anti-inflammatory and anti-hemorrhagic action, therefore, it is used in cases of heavy menstrual bleeding [22].

It helps to improve anemia, since it also contains a lot of iron and vitamin C [23, 24]. In regions of Iran it is traditionally used together with many other plants for dysmenorrhea, given its anti- inflammatory potential [25].

In addition, the leaves have the property of increasing the production of breast milk due to its galactogenic action [26].

In some preclinical studies with animals it has been shown that urtica dioica extract can improve sperm quality and semen parameters [27, 28]. Regarding the form of consumption, there are capsule preparations (dosage: 100 mg 3 times a day) or infusions (25 g of plant per 750 g of water).

In addition to the gynecological there are other non-gynecological uses. Due to its rich content of potassium and flavonoids it is diuretic; therefore, it has antihypertensive properties [29].

The root is used in benign prostatic hyperplasia [30]. This plant should be avoided in cases of heart and kidney failure, as well as in pregnant women, since it can induce uterine activity.

Lamium álbum (White Nettle or White Dead-Nettle)

It is native to Europe and North-Central Asia. It is called nettle because of its resemblance to the true nettle “Urtica Dioica”. Its components are: Saponin, Flavones, Mucilago, Taninos [31].

Like the true one, it is also used for its astringent and hemostatic properties for the treatment of heavy menstrual bleeding thanks to its content in Tannins and Flavonoids.

It also has demulcent properties thanks to the mucilage (viscous substance with local protective properties in the mucous membranes) which is why it is used in abnormal vaginal discharge and vaginitis [32].

There are other uses outside of gynecology. The aerial parts of the white nettle have flavonoids that protect the fibroblasts of human skin against oxidative stress [33]. In addition, it also decreases the concentration of cholesterol and glucose in the blood, and has antioxidant and wound healing properties [34].

Laminaria ("Laminaria Cloustoni")

The "Laminaria cloustoni" is brown seaweed found in the North Atlantic Ocean and the North Pacific, between 8 and 39 m deep. It is used as a dilator of the cervix in gynecology. It is an osmotic dilator, that is, it absorbs the cervical mucus enlarging the endocervical canal and softening the cervix.

The advantage of using osmotic dilators versus rigid ones is that the former help to prevent cervical damage and decrease the rate of uterine perforation, although there is no statistically significant result on the latter statement [35]. An average of 12–18 h should be placed to achieve dilatation, optimal cervical and use small size laminaria (about 2–3 mm) [36].

Due to its capacity for cervical dilatation it can be used in cervical stenosis, that is, in cases of anomalous obstruction of any part or the entire cervical canal. It can be congenital or acquired as a result of aging, nulliparity, curettage, cervical surgery or any combination of these factors [37].

It is not associated with an increased risk of uterine infection. Although sporadically, cases of anaphylaxis and hypersensitivity have occasionally been found.

Mentha piperita (Peppermint)

Dysmenorrhea or painful menstruation is among the most common gynecological complaints in women, which tends to be a crampy pain in the lower abdomen. Dysmenorrhea is generally divided into a primary and a secondary type. No particular problems have been proposed to explain the incidence of primary dysmenorrhea. In primary dysmenorrhea, the pain tends to occur a few hours before or just after the onset of menstruation and lasts between 48 and 72 h. It often entails concomitant nausea and vomiting, diarrhea and headache and also syncope on rare occasions [38]. Secondary dysmenorrhea refers to menstrual pain that is caused by a disorder in the woman's reproductive organs, such as endometriosis, adenomyosis, uterine fibroids, or infection [39].

Reduced progesterone levels at the end of the luteal phase activate lytic enzymes and produce arachidonic acid and activate the cyclooxygenase pathway (COX) and increase prostaglandin levels (PGE2a and PGF2a) [39]. The condition is generally caused by the increased levels of cytokines and thus production of prostaglandins by interleukins in response to dropped progesterone levels [38]. The release of these substances into the blood stream causes severe uterine myometrial contraction and uterine ischemia and subsequently causes the pain to spread [40].

Peppermint extract has an inhibitory effect on the contractile activity of myometrium through inhibition of prostaglandin (PGF2a) and oxytocin, blocks calcium channels and has antispasmodic effects on smooth muscles [41].

Women with dysmenorrhea have higher uterine activity, which increases the frequency of contractions. Dysmenorrhea causes spasmodic colicky pain that is labor-like in nature [42]. As a result, the use of medications with spasmolytic properties will be beneficial in reducing this pain. Peppermint inhibits spasmodic activities on smooth muscles [43]. and thus has a relaxant effect on them [44]. A study based on the antispasmodic properties of peppermint oil 2% (mg) administered to participants who had experienced 30 min of spasm showed a reduction in their pain after 2–7 min [45].

Studies conducted on the pain-relieving properties of pepper- mint showed that peppermint extract has analgesic effects acting on the peripheral and central nervous systems [46]. In a controlled study, Ozgoli et al. [47] investigated the effect of peppermint on the severity of pain and anxiety in the first stage of delivery in primiparous women and found that peppermint dramatically reduced the severity of pain, which is consistent with the results of the present study regarding the reduction of pain caused by uterine contractions.

Omomi Roknabad and Sarafraz [48] compared the effect of peppermint (extract) on primary dysmenorrhea to the effect of Ibuprofen and found no significant differences between the study groups in the severity of primary dysmenorrhea after the intervention, which might be due to the proven pain-relieving effects of Ibuprofen.

It can therefore be concluded that peppermint played a crucial role in reducing the severity of dysmenorrhea. The present study did not encounter any specific side-effects in its follow-ups.

The results of the present study showed a considerable reduction in the severity of primary dysmenorrhea with a two- month use of peppermint compared to placebo demonstrating the analgesic properties of peppermint with no reported side effects. Data were obtained in the form of self-reports from the students living in dormitories in Iran, which therefore restricts the extrapolation of the results to the wider population. Thus the authors recommend other studies to be conducted in more diverse communities using peppermint extract for the treatment of primary dysmenorrhea.

Valeriana officinalis (Garden Valerian, Garden Heliotrope, Setwall, All-Heal)

It is native to Europe and northern Asia and grows wild in very wet places. It is grown in the center and east of Europe. It reproduces by seeds in spring; the root and the rhizome of the two- year-old plants are plucked in the autumn.

Its main components are: volatile oil (roughly 1.4%), which includes pumice acetate, B- caryopbyllenos; iridoides (valepotrfatos), valtrato, isovaltrato, and alkaloids.

Known in the Middle Ages as "heal everything", valeriana was attributed many virtues, in particular that of curing epilepsy. In 1592, Fabius Calumna published a nunculous work on medicinal herbs in which claimed to have cured his epilepsy with this herb. It has been used to treat disorders associated with stress: valerian reduces mental hyperactivity and nervous excitability, helping to those who find it difficult to disconnect. It's good for almost any condition associated with stress and, in general, has an effect soothing, rather than directly sedative over the mind.

Many symptoms of anxiety, among them tremors, panic, palpitations and sweats, can mitigate with valerian. It is useful for insomnia produced by anxiety or overexcitement.

Valerian relaxes stiff muscles and relieves tension in shoulders and neck, asthma, colic, syndrome irritable bowel, menstrual pain and muscle spasms and, it is used with other herbs in remedies for high tension caused by stress and anxiety.

V. officinalis has been traditionally used as a menstruating and sedative drug since eleventh century [49]. Its root and rhizome have valerian essential oil which contains valepotriats. Root of V. officinalis is used as a diuretic, sedative and muscular antispasmodic and valerenic acid of its root has antispasmodic properties [50]. Also, the effect of anti-spasmodic valterate, isovalterate on ileum smooth muscle has been confirmed [51]. V. officinalis inhibits contractions of cell depolarization well and blocks calcium channels [52].

Two studies have been conducted on V. officinalis, one of which compared the effect of its root with placebo [53] and another compared the effects of V. officinalis with mefenamic acid [54]). In the first study, V. officinalis was effective on reducing pain compared to placebo, and in the second study, it had a similar effect to that of mefenamic acid. In one research, systemic symptoms of dysmenorrhea reduced after taking V. officinalis capsules compared to pre-intervention, but the difference was the same as placebo group, except for severity of fainting variable which was significantly different between V. officinalis and placebo groups [51].

In traditional medicine, V. officinalis is known as a menstruating herb, but in a clinical trial, V. officinalis had no effect on duration and severity of bleeding.

Oenothera biennis (Evening Primrose Oil)

It is a biannual herbaceous up to 20 cm in height. Stem with red spots, curled lanceolate leaves. It has yellow flowers off four petals and elongated seed capsules.

It is native to North America and today is common in many temperate zones around the world. It grows in wastelands, especially in dunes and sandy terrain. It is grown with commercial purposes for the oil of its seeds. The oil is rich in essential fatty acids, in particularly cis-linoleic (around 70%) and cis-and-linoleic acids (around 9%). Its action depends above all on y-linoleic acid, a precursor of prostaglandin E1. The oil is usually combined with vitamin E to prevent oxidation.

In a systematic review of randomised controlled trials assessing herbal treatments for alleviating symptoms of premenstrual syndrome, evening primrose oil was found to be no more effective than placebo [55].

A six-week randomised, placebo-controlled clinical trial evaluated oral evening primrose oil (1000 mg daily) in reducing frequency, severity and duration of menopausal hot flushes in women aged 45–59 years. The results showed improvement in hot flushes for both intervention and control groups, with evening primrose outperforming placebo, although statistical significance was found only in comparison of severity of hot flushes [56].

In a review evaluating treatments for severe persistent mastalgia in 291 women, patients were orally administered either evening primrose oil (3 g), bromocriptine (5 mg) or danazol (200 mg) daily for 3–6 months. In cases of cyclical mastalgia, good responses were obtained in 45% of patients treated with evening primrose oil, in 47% in those treated with bromocriptine and in 70% treated with danazol. In patients with non-cyclical mastalgia, the response rate was 27%, 20% and 31%, respectively. Adverse events were far greater in the bromocriptine and danazol groups than the evening primrose oil group [57].

In a randomised, double-blind controlled study in 120 women investigating oral use of evening primrose oil or fish oil in treatment of severe chronic mastalgia for 6 months, all groups showed a decrease in pain, but neither demonstrated clear benefit over the control oils (corn and wheatgerm) [58].

Another randomised, double-blind, placebo-controlled trial in 41 patients found that daily doses of evening primrose oil of 3000 mg, or in combination with 1200 IU vitamin E, taken for 6 months reduced the severity of cyclical mastalgia [59].

Achilea Millefolium (Milenrama, Yarrow)

The yarrow, also called Achilea millefolium, is a medicinal herb, member of the Asteraceae family, used over the years from Europe to Asia. Its name comes from Achilles, who treated many warriors during the Trojan War, using the power it has to stop hemorrhages. It can be found on the fringes of wild meadows.

Its main main components are: Volatile oil (linalol, Alcafor, Sabinene, Chamazulene), Lactones, Flavonoids, Alkaloids, Poliacetilenos, Salicylic acid, Coumarins and, Tannins.

In Gynecology, Yarrow helps to regulate the menstrual cycle, reduces excess menstruation and combats dysmenorrhoea and menopause. If it is combined with other herbs, it relieves cold and flu. And it is also useful to facilitate digestion and circulation. Other uses we can achieve with this herb are the following: anti-inflammatory, diuretic, antiseptic, cicatrizing, anti- haemorrhoids, antipyretic, …

It is not recommended to use during pregnancy or for more than 2 weeks in a row.

Although its multiple uses, the yarrow has been little studied, so there is no evidence about its effectiveness. Studies that guarantee its safety and efficacy would be required to endorse its use.

Studies in this field are limited. A double- blind randomized clinical trial was performed in Iran during 2013 to evaluate the effect of A. Millefolium on dysmenorrhea. The subjects were divided into 2 groups and were given either placebo or A. millefolium for 3 days in 2 menstruation cycles. They observed a pain significantly reduction in the A. millefolium group compared to placebo group. According to their results, they came to the conclusion that this plant is effective in reducing severe pain in primary dysmenorrhoea. It should be noted that the study is limited by the small size and homogeneity of the sample [60].

Another double-blind trial study Achillea millefolium reduce perineal pain level, redness, oedema and ecchymosis of episiotomy wound, so it seems that consuming them was useful for episiotomy treatment [61].

Dysmenorrhea is a menstrual pain in absence of organic pelvic disease than occurs in more than 50% of post menarche women, due to an elevation of prostaglandins in the uterus. Evidence supports the efficacy of nonsteroidal anti-inflammatory drugs or oral contraceptives to alleviate this pathology.

Vitex agnus castus (Chasteberry, Chaste Tree; Sauzgatillo)

The Chaste tree or chasteberry, also called Vitex agnus-castus L, belongs to Verbenacea family. Its name comes from the fact that it was thought that this tree inhibited desire. Therefore, the fruits were used by medieval monks to eliminate it. Later it was discovered that it has the opposite effect. It can be found on the edges of currents and humid places in Central Asia and Mediterranean. Albania and Morocco are the main producing countries in the world.The berries contain essential oils (limonene, sabinene, eucalyptol), iridoid glycosides (agnoside, aucubin), flavonoids (apigenin, castican, isovitexin) and diterpines (vitexilactone, rotundifuran). The Chaste tree is a hormonal regulator: the chaste tree increases progesterone and neutralizes estrogen. It seems that its effect is due to the synergic effect of several of its components on the pituitary gland, which stimulates the production of luteinizing hormone and decreases follicle stimulating hormone. For this reason, herbalists prescribe them to combat premenstrual syndrome because it improves breast inflammation, headaches, among other symptoms. It can also be used for irregular menstruation or amenorrea. In addition, it is also useful in the area of infertility, when it is due to a progesterone defect. It should be noted, it is not recommended to consume this plant during conception. It is proposed to stabilize the cycles with the intake for a few months, and then stop it when gestation is going to take place. In a systematic review Vitex Agnus Castus was shown to contribute to premenstrual syndrome, postmenstrual and Infertility disorders [62].

Another systematic review and meta-analysis of 17 randomized controlled trials of Vitex agnus castus in the treatment of premenstrual syndrome show an important effect in placebo- controlled trials; however, the high risk of bias, high heterogeneity, and risk of publication bias of the studies prevent to draw an adequate conclusion. It is necessary to design high-quality trials for examining the effect of Vitex agnus castus in comparison to placebo and oral contraceptives to establish a real efficacy [63].

In low doses, chasteberry increases serum prolactin and it is a known galactogogue, nevertheless, no scientifically valid clinical trials support this use. However, some evidence indicates that high doses of chasteberry decrease serum prolactin and might decrease breastmilk production, so it has been used to decrease breastmilk oversupply. Due to its lack of safety and its effects on breastfeeding, its use should be avoided during this period [64].

The most frequent adverse events are nausea, headache, gastrointestinal disturbances, acne and erythermatous rash; however, all are mild and reversible.

Its use during pregnancy or artificial lactation is not recommended. Neither in combination with hormonal treatments such as birth control pills, hormone replacement therapy or treatments for assisted reproduction therapies. Neither should be consumed if you have Parkinson's disease.

Aloe barbadensis (Aloe Vera)

Aloe Vera, scientifically known as Aloe barbadensis, belongs to the family of the Liliaceae. Aloe is native to southern and eastern Africa; it grows wild in the tropics and its cultivation is widespread throughout the world [65]. Its main components are: polysaccharides, glycoproteins (lectins) such as alloctin A and B, enzymes such as carboxypeptidases, palmitic acid and sterols (sitosterol and stigmasterol), anthraquinones (aloin and aloe-emodin), resins and tannins [66].

Its main effects are:

- Wound healing: some studies in animals suggest that Aloe vera can help accelerate the healing of wounds, sores and burns by serving the translucent gel that contains the protective layer sheets in the affected area, thus accelerating healing and reducing the risk of infection. This effect is due in part to the presence of alloctin B, which stimulates the immune system [67].
- Emollient.
- It stimulates the secretion of bile.
- Laxative: the yellow sap at the base of the leaf (bitter aloe) contains anthraquinones, which are responsible for the purgatory activity. Aloin is metabolized by the flora of the colon to aloe- emodin, which causes contraction of the colon, producing an intestinal movement 8–12 h after ingesting it [68]. In small doses, the bitter properties of this herb stimulate digestion. In larger doses, bitter aloe is laxative.

In 2012, the Cochrane Library conducted a review including all randomized controlled trials that evaluated the effectiveness of Aloe Vera and aloe products as a treatment for acute wounds (lacerations, surgical wounds or burns) or chronic wounds (infected wounds, arterial ulcers or venous). This review showed that Aloe Vera did not accelerate the healing of burns or skin biopsies, but it did show a decrease in the healing time of hemorrhoidectomies. In a study on second-intention healing of surgical wounds it was found that Aloe Vera significantly delayed this healing. Regarding chronic wounds, one study found no statistically significant differences in the healing of pressure ulcers with Aloe. This review concludes that there is insufficient evidence to support the use of topical agents of Aloe Vera as treatment of acute or chronic wounds and recommends the performance of more randomized and controlled studies to ensure the effects of Aloe Vera in the treatment of wounds [69].

In 2013, a randomized, controlled study compared the efficacy of Aloe Vera gel with a 1% sulfadiazine creme in 50 patients with first and second degree burns. This study showed that patients treated with aloe Vera gel achieved faster epithelialization of wounds than patients treated with 1% sulfadiazine creme. In addition, there was also a greater decrease in pain with aloe vera gel [70].

In gynaecology the effectiveness of aloe vera gel in the healing of the surgical wound of caesarean section has also been studied. In 2015, a prospective, randomized, double-blind study was conducted in 90 women who underwent caesarean section, which were divided into two groups; in one of them, the gauze covering the wounds of the patients had aloe vera, while in the other the wounds were covered only with gauze. This study showed that aloe vera gel, compared to placebo, is effective in the surgical wound treatment of caesarean section in the first 24 h postoperatively, although no statistically significant differences were found at 8 days [71].

Matricaria chamomilla (Chamomile)

Chamomile, as is the family of composite plants or asteraceae. It is native to Western Europe and is now cultivated throughout Europe and in other temperate regions.

The parts used are the flowers and the essential oil. This medicinal plant has anti-inflammatory, antioxidant, antimicrobial, analgesic, antineoplastic, anxiolytic and digestive properties due to its components such as caffeic acid, chlorogenic acid, bisabolol, camazulene acid, flavonoids (apigenin, quercetin, patuletin and luteolin) and coumarin [72].

Chamomile has been used for its anti-inflammatory and antiallergic properties to treat inflammations and bacterial infections of the skin and mucous membranes such as the oral cavity, gums or respiratory tract. Its analgesic property has also caused it to be used traditionally in neuralgia, sciatica or rheumatic pain. It is a mild anxiolytic used to treat hysteria, nightmares, insomnia and other sleep problems. By stimulating digestive secretions and relaxing intestinal muscles, it normalizes digestive function and is useful for treating various gastrointestinal disorders such as flatulence, indigestion, diarrhoea, anorexia, cramping, nausea and vomiting [73].

In gynaecology, chamomile can be used to decrease dysmenorrhea, mastalgia and nausea and vomiting of pregnancy.

Dysmenorrhea or painful menstruation is a common reason for consultation in gynaecology. The prevalence of this symptom varies from 16–91% in women of reproductive age, with 2–29% of them suffering from severe dysmenorrhea, which interferes with their day to day and worsens their quality of life. Among the risk factors that have been seen to influence the development of dysmenorrhea include: early menarche, nulliparity, hypermenorrhoea, irregular menstrual cycles, depression / anxiety, smoking or alcohol consumption. There are two types of clinical presentation: primary dysmenorrhea that usually presents at 6–12 months of menarche and is not related to any organic pathology, and secondary dysmenorrhea that usually occurs at the menarche years and does relate to some pathology such as endometriosis or ovarian cysts [74].

Conventional treatments include non-steroidal anti-inflammatories (NSAIDs) and oral contraceptives (OACs), which reduce the myometrial activity produced by contraction of the uterus. An alternative is dietary supplements such as herbs, vitamins, minerals, enzymes and amino acids [75].

A Cochrane review included 27 randomized controlled trials (3010 women) studying the effect of dietary supplements on primary or secondary dysmenorrhea of moderate or severe intensity. Its objective was to determine its efficacy and safety in the treatment of this pathology. These studies included 12 different medicinal herbs (chamomile, cinnamon, damask rose, dill, fennel, fenugreek, ginger, guava, rhubarb, uzara, valerian and zataria) and five supplements (fish oil, melatonin, vitamins B1 and E, and zinc sulphate). As for chamomile, data comparing chamomile with placebo were not suitable for analysis and evidence that chamomile was more effective than NSAIDs was limited. The conclusion of this review was that there was not enough evidence to support the effectiveness of any dietary supplement for dysmenorrhea [75].

Mastalgia or breast pain usually precedes menstruation and is a frequent reason for consultation. It manifests cyclically or non-cyclically. The cyclic form usually occurs every month before the onset of menstruation, is of moderate intensity and disappears 7 days after the onset of menstruation. In 30% of the cases the mastalgia can be of greater intensity and interfere in the daily life of the woman, which can produce both a sexual, physical and social dysfunction as depression and anxiety. Probable causes of cyclic mastalgia may be high estrogen levels, low progesterone levels, and an imbalance in the estrogen / progesterone ratio, because mastalgia usually begins in the luteal phase [76]. Treatments to reduce mastalgia include non-steroidal anti- inflammatory drugs (NSAIDs), vitamins B2, B6, E and C, diuretics, thyroxine, progesterone, tamoxifen, centchroman, danazol, bromocriptine and plant extracts such as chasteberry or evening primrose oil. In any case, due to the adverse effects of these treatments, many women prefer the use of medicinal herbs [77].

A randomized, double-blind, placebo-controlled clinical trial evaluated the effect of chamomile extract for pain control of cyclical mastalgia in 60 patients. The results of this study showed that chamomile extract and placebo can reduce the intensity of mastalgia and also saw a significant reduction in pain in the group of women who took chamomile. In addition, no adverse effects were found. In general, chamomile is a safe, well tolerated and effective option for the treatment of moderate intensity mastalgia [77].

Nausea and vomiting of pregnancy are very common in the first trimester of pregnancy; affects around 50–80% of pregnant women and usually appears in the first 4–6 weeks and continues until weeks 14–16. The causes of nausea and vomiting of pregnancy remain unknown, but it is believed that they may be associated with elevated levels of human chorionic gonadotropin (hCG) or estrogen. There is a more severe form of these symptoms called hyperemesis gravidarum that can cause dehydration and malnutrition and requires admission and hospital treatment.

Antiemetics such as pyridoxine / doxylamine and metoclopramide are effective and safe in pregnancy [78]. However, many women prefer medicinal plants such as ginger and chamomile to reduce these symptoms.

A Cochrane review included 41 randomized controlled studies (5449 women) on any treatment for nausea and vomiting of pregnancy to check its effectiveness and safety [79]. These studies included treatments such as acupressure, acustimulation, acupuncture, ginger, chamomile, lemon oil, peppermint oil, vitamin B6 and several antiemetics. In a study comparing ginger capsules with capsules of chamomile and placebo, it was found that after the first week of treatment chamomile managed to reduce these symptoms [80].

Zingiber officinale (Ginger)

Ginger (Zingiber officinale) is a plant of the Zingiberáceas family. It is known as spice and flavoring, but it is also one of the best medicines in the world. It originated in Asia, where it has been used since ancient times and in medieval Europe it was believed that it came from the Garden of Eden. As one of the first spices exported from the East, ginger arrived in Europe during the spice trade and was used by the ancient Greeks and Romans.

It grows in all the tropical zones and it propagates by division of the rhizome, subterranean stem that is very appreciated for its aroma and spicy flavor. The plant reaches 90 cm in height, with long leaves of 20 cm. For its growth it needs abundant rains. The rhizome is the most used part, which can be consumed fresh or dried. Its main components are: volatile oil (1–3%), zingiberina (20–30%), oleoresin (4–7.5%), gingerol and shogaoles [81].

Its therapeutic properties are largely due to its volatile oil and its oleoresin. In medical research it has been proven that ginger root is an effective treatment against nausea caused by dizziness in means of transport, as well as those suffered by pregnant women [82]. A Cochrane review concludes that the evidence on the effectiveness of reducing nausea during pregnancy is inconsistent and relatively weak [83].

Its main preparations are infusions, tincture, capsules and essential oil. For example, for the treatment of nausea you can take an infusion three times a day or a capsule of 75 mg/h. It is not recommended to take ginger in medicinal doses if you suffer from peptic ulcer.

Calendula officinalis (Calendula)

Calendula is one of the best known and most versatile herbs of western herbal medicine. They are herbs of scarce height (40 or 50 cm), of stems erect and branched from the base forming dense bushes; with lanceolate leaves, simple, slightly pubescent, between 5 and 20 cm long. The flowers are discoidal, yellow to intense orange, and very showy. Its intense orange petals are an excellent antiseptic remedy. It is also a purifying and detoxifying herb.

It is native to southern Europe; it is grown in temperate regions around the world. It is easy to spread and grows on almost all types of soil. The flowers are collected as soon as they open at the beginning of summer and they are left to dry in the shade.

Its main components are: triterpenes, resins, bitter glycosides, volatile oil, sterols, flavonoids, mucilage and carotins [84].

The Commission E (a German therapeutic guide of medicinal plants) considers that the calendula flower has an anti-inflammatory and strongly healing action when it is applied topically. With extracts of the calendula flower, it shows a stimulating action of the epithelization of the wounds and an anti-inflammatory activity in edema where the prostaglandin intervenes (triterpenes, especially faradiol, have proved to be the most important anti-inflammatory principles) [85]. In popular medicine it is used for its antibacterial, fungicidal and antispasmodic action. It is a good emollient since it softens, tones and moisturizes the skin. In fact, more and more cosmetic products are included among its components. Callicide has also been considered to help the disappearance of viral skin warts, due to its content of acetylsalicylic acid. It is choleretic stimulating liver activity, especially biliary secretion. It is also effective in gastritis, gastroenteritis and vomiting due to its antiulcer action since it helps the healing of gastric ulcers [86].

In gynaecology it is used as an emmenagogue since it has a mild estrogenic action (stimulating blood flow in the pelvic area and uterus, and in some cases, promoting menstruation), as a regulator and soothing of menstrual pains [87]. The 20% tincture of C. officinalis applied topically has therapeutic effect for the treatment of recurrent vaginal candidiasis and is considered safe for use in clinical practice, as it did not produce adverse effects in patients [88]. Only topical use is contraindicated in patients sensitive to asteraceae, since experimentally weak skin sensitization has been seen, but no clear cases of contact dermatitis have been reported [89]

Conclusions

Herbal medicines are frequently used by women all over the world. Although there are hundreds of publications on their uses in medicine, there are few comparative studies against traditional.

medicines. Under these conditions we can conclude that there is not enough evidence to recommend its use instead of traditional medicines.

References

1. Jamshidi-Kia, F., Z. Lorigooini, and H. Amini-Khoei. 2018. Medicinal plants: Past history and future perspective. *Journal of Herbmed Pharmacology* 7: 1–7.
2. Sewell, R.D., and M. Rafieian-Kopaei. 2014. The history and ups and downs of herbal medicines usage. *Journal of Herbmed Pharmacology* 3: 1–3.
3. Eskinazi, D.P., and K.A. Jobst. 1996. National Institutes of Health Office of alternative medicine-Food and Drug Administration workshop on acupuncture. *Journal of Alternative and Complementary Medicine* 2: 3–6.
4. Dodge, T. 2016. Consumers' perceptions of the dietary supplement health and education act: Implications and recommendations. *Drug Testing and Analysis* 8: 407–409.
5. Barnes, P.M., E. Powell-Griner, K. McFann, and R.L. Nahin. 2004. Complementary and alternative medicine use among adults: United States, 2002. *Seminars in Integrative Medicne* 2: 54–71.
6. Ernst, E. 2000. Prevalence of use of complementary/alternative medicine: A systematic review. *Bulletin of the World Health Organization* 78: 258–266.
7. Ernst, E., and A. White. 2000. The BBC survey of complementary medicine use in the UK. *Complementary Therapies in Medicine* 8: 32–36.
8. Akkol, E.K., M.A. Demirel, O.B. Acıkara, I. Süntar, B. Ergene, M. Ilhan, and M. Tekin. 2015. Phytochemical analyses and effects of Alchemilla mollis (Buser) Rothm. and Alchemilla persica Rothm. in rat endometriosis model. *Archives of Gynecology and Obstetrics* 292: 619–628.
9. Choi, J., Y.G. Park, M.S. Yun, and J.W. Seol. 2018. Effect of herbal mixture composed of Alchemilla vulgaris and Mimosa on wound healing process. *Biomedicine & Pharmacotherapy* 106: 326–332.
10. Parker, S., B. May, C. Zhang, A.L. Zhang, C. Lu, and C.C. Xue. 2016. A pharmacological review of bioactive constituents of Paeonia lactiflora Pallas and Paeonia veitchii Lynch. *Phytotherapy Research* 30: 1445–1473.
11. Fisher, C.Y., J. Adams, J.E. Frawley, L.D. Hickman, and D.W. Sibbritt. 2019. Is there a role for Western herbal medicine in treating cyclic perimenstrual pain and discomfort? *The Australian & New Zealand Journal of Obstetrics & Gynaecology* 59: 154–156.
12. Hehir, M.P., and J.J. Morrison. 2016. Paeoniflorin, a novel heat-shock protein inducing compound, and human myometrial contractility in vitro. *The Journal of Obstetrics and Gynaecology Research* 42: 302–306.
13. Bae, S., S.Y. Kim, M.H. Do, C.H. Lee, and Y.J. Song. 2017. 1, 2, 3, 4, 6-Penta-O-galloyl-ss-D-glucose, a bioactive compound in Elaeocarpus Sylvestris extract, inhibits varicella-zoster virus replication. *Antiviral Research* 144: 266–272.
14. Al-Snafi, A.E. 2015. The chemical constituents and pharmacological effects of Capsella bursa-pastoris-A review. *International Journal of Pharmacology and Toxicology* 5: 76–81.

15. Naafe, M., N. Kariman, Z. Keshavarz, N. Khademi, F. Mojab, and A. Mohammadbeigi. 2018. Effect of Hydroalcoholic Extracts of Capsella Bursa-Pastoris on heavy menstrual bleeding: A randomized clinical trial. *Journal of Alternative and Complementary Medicine* 24: 694–700.
16. Ghalandari, S., N. Kariman, Z. Sheikhan, F. Mojab, M. Mirzaei, and H. Shahrahmani. 2017. Effect of Hydroalcoholic Extract of Capsella bursa pastoris on early postpartum hemorrhage: A clinical trial study. *Journal of Alternative and Complementary Medicine* 23: 794–799.
17. Chen, J.J., M.Y. Han, T. Gong, J.L. Yang, and P. Zhu. 2017. Recent progress in ergot alkaloid research. *RSC Advances* 7: 27384–27396.
18. Ma, Y.Z., G.F. Qiang, and G.H. Du. 2018. Ergometrine and Ergotamine. In *Natural small molecule drugs from plants*, 237–242. Singapore: Springer.
19. Liabsuetrakul, T., T. Choobun, K. Peeyananjarassri, and Q.M. Islam. 2018. Prophylactic use of ergot alkaloids in the third stage of labour. *Cochrane Database of Systematic Reviews* 6. Art. No.: CD005456. https://doi.org/10.1002/14651858.CD005456.pub3.
20. De Groot, A.N.J.A., P.W.J. Van Dongen, J. Van Roosmalen, and T.K.A.B. Eskes. 1993. Ergotamine-induced fetal stress: Review of side effects of ergot alkaloids during pregnancy. *European Journal of Obstetrics, Gynecology, and Reproductive Biology* 51: 73–77.
21. Al-Tameme, H.J., M.Y. Hadi, and I.H. Hameed. 2015. Phytochemical analysis of Urtica dioica leaves by fourier-transform infrared spectroscopy and gas chromatography-mass spectrometry. *Journal of Pharmacognosy and Phytotherapy* 7: 238–252.
22. Tajallaie-Asl, F., M. Mardani, S. Shahsavari, and S. Abbaszadeh. 2017. Menstruation phytotherapy according to Iran ethnobotanical sources. *Journal of Pharmaceutical Sciences and Research* 9: 986–990.
23. Zouari Bouassida, K., S. Bardaa, M. Khimiri, T. Rebaii, S. Tounsi, L. Jlaiel, and M. Trigui. 2017. Exploring the Urtica dioica leaves hemostatic and wound-healing potential. *BioMed Research International*. https://doi.org/10.1155/2017/1047523.
24. Jan, K.N., and S. Singh. 2017. Stinging nettle (Urtica dioica L.): Aa reservoir of nutrition and bioactive components with great functional potential. *Journal of Food Measurement and Characterization* 11: 423–433.
25. Bahmani, M., Z. Eftekhari, M. Jelodari, K. Saki, R. Abdollahi, M. Majlesi, and S. Rasouli. 2015. Effect of Iranian herbal medicines in dysmenorrhea phytotherapy. *Journal of Chemical and Pharmaceutical Research* 2: 519–526.
26. Demirci, J.R., S. Bare, S.M. Cohen, and D.L. Bogen. 2016. Feasibility and acceptability of two complementary and alternative therapies for perceived insufficient Milk in mothers of late preterm and early term infants. *Alternative and Complementary Therapies* 22: 196–203.
27. Shirani, M., S. Heidari-Soureshjani, and M. Yavangi. 2016. Use of Iranian medicinal plants effective on male fertility indices. *Journal of Global Pharma Technology* 10: 36–43.
28. Bioos, S., E. Nazem, M. Keshavarz, M. Siahpoosh, F. Sohrabvand, H. Sohanaki, and F. Nejatbakhsh. 2016. A traditional Iranian medicine (Majoon-e Loboob) for idiopathic male infertility: A case series. *Traditional and Integrative Medicine* 1: 47–51.
29. Qayyum, R., H.M.U.D. Qamar, S. Khan, U. Salma, T. Khan, and A.J. Shah. 2016. Mechanisms underlying the antihypertensive properties of Urtica dioica. *Journal of Translational Medicine* 14: 254.
30. Ghorbanibirgani, A., A. Khalili, and L. Zamani. 2013. The efficacy of stinging nettle (Urtica dioica) in patients with benign prostatic hyperplasia: A randomized double-blind study in 100 patients. *Iranian Red Crescent Medical Journal* 15: 9.
31. Morteza-Semnani, K., M. Saeedi, and M. Akbarzadeh. 2016. Chemical composition of the essential oil of the flowering aerial parts of Lamium album L. *Journal of Essential Oil-Bearing Plants* 19: 773–777.
32. Turker, A.U., A.B. Yildirim, F.P. Karakas, and H. Turker. 2018. *In vitro antibacterial and antitumor efficiency of some traditional plants from Turkey*. http://nopr.niscair.res.in/handle/123456789/43150.
33. Paduch, R., and A. Woźniak. 2015. The effect of Lamium album extract on cultivated human corneal epithelial cells (10.014 pRSV-T). *Journal of Ophthalmic and Vision Research* 10: 229.

34. Shah, T., F. Khan, M. Bule, and K. Niaz. 2019. White Dead-Nettle (Lamium album). In *Nonvitamin and nonmineral nutritional supplements*, 455–459. London: Academic Press.
35. Khooshideh, M., N. Yarmohammadi, A. Shahriari, and M. Sheikh. 2017. Sublingual misoprostol plus laminaria for cervical preparation before surgical management of late first trimester missed abortions, a randomized controlled trial. *The Journal of Maternal-Fetal & Neonatal Medicine* 30: 317–322.
36. Kapp, N., P.A. Lohr, T.D. Ngo, and J.L. Hayes. 2010. Cervical preparation for the first trimester surgical abortion. *Cochrane Database of Systematic Reviews* 2.
37. Chodankar, R., J. Gupta, D. Gdovinova, M.J. Bovo, J. Hanacek, N. Kan, and V. Tyutyunnik. 2018. Synthetic osmotic dilators for cervical preparation prior to abortion—an international multicentre observational study. *European Journal of Obstetrics, Gynecology, and Reproductive Biology* 228: 249–254.
38. Skopec, G.S. 2018. A review of medical and surgical techniques for overcoming cervical stenosis. *Proceedings in Obstetrics and Gynecology* 8: 1–7.
39. Jolin, J.A., and A. Rapkin. 2007. Pelvic pain and dysmenorrhea. In *Novak's gynecology*, ed. J.S. Berek, R.D. Rinehart, P.J. Adams Hillard, and E.Y. Adashi, 14th ed., 516–520. Philadelphia: Lippincott Williams and Wilkins.
40. Novak, E., and J.S. Berek. 2007. *Berek & Novak's gynecology*. Philadelphia: Lippincott Williams & Wilkins.
41. Chen, H.Y., Y.H. Lin, I.H. Su, Y.C. Chen, S.H. Yang, and J.L. Chen. 2014. Investigation on Chinese herbal medicine for primary dysmenorrhea: Implication from a nationwide prescription database in Taiwan. *Complementary Therapies in Medicine* 22: 116–125.
42. Soares, P., A. Assreuy, E. Souza, R. Lima, T. Silva, S. Fontenele, and D. Criddle. 2005. Inhibitory effects of the essential oil of Mentha pulegium on the isolated rat myometrium. *Planta Medica* 71: 214–218.
43. Neville, F., J. Hacker, and G. Moore. 2004. *Essentials of obstetrics and gynecology*. 4th ed. Tehran: Teymour zadeh & Tabib.
44. Dermarderosian A, and J.A. Beutler. 2002. *The review of natural products: The most complete source of natural product information*. Facts Comparisons.
45. Dambolena, J.S., A.G. López, H.R. Rubinstein, and J.A. Zygadlo. 2010. Effects of menthol stereoisomers on the growth: sporulation and fumonisin B 1 production of Fusarium verticillioides. *Food Chemistry* 123: 165–170.
46. Beesley, A., J. Hardcastle, P. Hardcastle, and C. Taylor. 1996. Influence of peppermint oil on absorptive and secretory processes in rat small intestine. *Gut* 39: 214–219.
47. Taher, Y.A. 2012. Anticonceptive activity of Mentha piperita leaf aqueous extract in mice. *Libyan Journal of Medicine* 7: 16205.
48. Ozgoli, G., Z. Aryamanesh, F. Mojab, and Majd H. Alavi. 2013. Study of inhalation of peppermint aroma on the pain and anxiety of the first stage of labor in nulliparous women: a randomized clinical trial. *QOM University of Medical Sciences Journal* 7: 21–27.
49. Omomi Roknabad, M., and N. Sarafraz. 2011. Comparison between the effect of supermint and ibuprofen on primary dysmenorrheal: a randomized clinical trial (Article in persian). *QOM University of Medical Sciences Journal* 5: 37–41.
50. Der Mardersian, A. 2001. *The review of natural products*, 28. Philadelphia: Facts and Comparisons.
51. Bos, R., H.J. Woerdenbag, P.A.G.M. De Smet, and J.J.C. Scheffer. 1997. Valeriana species. In *Adverse effects of herbal drugs*, ed. P.A.G.M. De Smet, K. Keller, R. Hansel, and R.F. Chandler, vol. 3, 105–114. Berlin: Springer.
52. Mirabi, P., M. Dolatian, F. Mojab, and H. Alavimajd. 2011. Effects of valerian on the systemic manifestations of dysmenorrhea. *International Journal of Gynecology & Obstetrics* 115: 285–288.
53. Gilani, A.H., A.U. Khan, Q. Jabeen, F. Subhan, and R. Ghafar. 2005. Antispasmodic and blood pressure lowering effects of Valeriana wallichii are mediated through K+ channel activation. *Journal of Ethnopharmacology* 100: 347–352.

54. Mirabi, P., M. Doulatian, F. Mojab, and H. Alavimajd. 2010. Effects of Valeriana officinalis on the severity of dysmenorrhea. *Journal of Reproduction & Infertility* 10: 253–259.
55. Jenabi, E., M. Asle Toghiri, and P. Hejrati. 2012. The comparison of the effects of antiplain of Valeriana officinalis risom and mefenamic acid in relief of primary dismenorrhea. *Iranian Journal of Obstetrics, Gynecology and Infertility* 15: 44–48.
56. Dante, G., and F. Facchinetti. 2011. Herbal treatments for alleviating symptoms of premenstrual symptoms: A systematic review. *Journal of Psychosomatic Obstetrics and Gynecology* 32: 42–51.
57. Farzaneh, F., S. Fatehi, and M.R. Sohrabi. 2013. The effect of oral evening primrose oil on menopausal hot flashes: A randomized clinical trial. *Archives of Gynecology and Obstetrics* 88: 1075.
58. Edwards, S.E., I. da Costa Rocha, E.M. Williamson, and M. Heinrich. 2015. Evening primrose (Oil) Oenothera biennis L. *Phytopharmacy: An Evidence-Based Guide to Herbal Medicinal Products*: 144–150.
59. Blommers, J., E.S.M. de Lange-de Klerk, D.J. Kuik, P.D. Bezemer, and S. Meijer. 2002. Evening prim-rose oil and fish oil for severe chronic mastalgia: A randomised, double-blind controlled, trial. *American Journal of Obstetrics and Gynecology* 187: 1389–1394.
60. Pruthi, S., D.L. Wahner-Roedler, C.J. Torkelson, S.S. Cha, L.S. Thicke, J.H. Hazelton, B.A. Bauer, and E. Vitamin. 2010. Evening primrose oil for management of cyclical mastalgia: A randomized pilot study. *Alternative Medicine Review* 15: 59–67.
61. Jenabi, E., and B. Fereidoony. 2015. Effect of Achillea Millefoium on relief of primary Dysmenorrea: A double blind randomized clinical trial. *Journal of Pediatric and Adolescent Gynecology* 28: 402–404.
62. Hajhashemi, M., Z. Ghanbari, M. Movahedi, M. Rafieian, A. Keivani, and F. Haghollahi. 2017. The effect of Achillea millefolium and Hypericum perforatum ointments on episiotomy wound healing in primiparous women. *The Journal of Maternal-Fetal & Neonatal Medicine* 31: 63–69.
63. Rafielan-Kopael, M., and M. Movahedi. 2017. Systematic review of premenstrual, postmenstrual and infertility disorders of vitex agnus castus. *Electronic Physician* 9: 3685–3689.
64. Verkaik, S., A.M. Kamperman, R. van Westrhenen, and P.F. Schulte. 2017. The treatment of premenstrual syndrome with preparations of Vitex agnus castus: A systematic review and meta-analysis. *American Journal of Obstetrics and Gynecology* 217: 150–166.
65. Chasteberry (2006) Drugs and lactation database (LacMed) (Internet). Bethesda (MD): National Library of Medicine (US).
66. Grindlay, D., and T. Reynolds. 1986. The Aloe vera phenomenon: A review of the properties and modern uses of the leaf parenchyma gel. *Journal of Ethnopharmacology* 16: 117–151.
67. Edwards, S.E., I. da Costa Rocha, E.M. Williamson, and M. Heinrich. 2015. Aloe Vera (Gel) Aloe vera (L.) Burm. f., A. arborescens Mill. and other Aloe spp. *Phytopharmacy: An Evidence-Based Guide to Herbal Medicinal Products* 24.
68. Heggers, J.P., A. Kucukcelebi, D. Listengarten, J. Stanbenau, F. Ko, L.D. Broemeling, M.C. Robson, and W.D. Winters. 1996. Beneficial effect of Aloe on wound healing in an excisional wound model. *Journal of Alternative and Complementary Medicine* 2: 271–277.
69. Radha, M.H., and N.P. Laxmipriya. 2015. Evaluation of biological properties and clinical effectiveness of Aloe vera: A systematic review. *Journal of Traditional and Complementary Medicine* 5: 21–26.
70. Dat, A.D., F. Poon, K.B.T. Pham, and J. Doust. 2012. Aloe vera for treating acute and chronic wounds. *Cochrane Database of Systematic Reviews* (2). Art. No.: CD008762. https://doi.org/10.1002/14651858.CD008762.pub2.
71. Shahzad, M.N., and N. Ahmed. 2013. Effectiveness of Aloe vera gel compared with 1% silver sulphadiazine cream as burn wound dressing in second degree burns. *The Journal of the Pakistan Medical Association* 63: 225–230.
72. Molazem, Z., F. Mohseni, M. Younesi, and S. Keshavarzi. 2015. Aloe vera gel and cesarean wound healing; a randomized controlled clinical trial. *Global Journal of Health Science* 7: 203.

73. Miraj, S., and S. Alesaeidi. 2016. A systematic review study of therapeutic effects of Matricaria recuitta chamomile (chamomile). *Electronic Physician* 8: 3024.
74. Srivastava, J.K., E. Shankar, and S. Gupta. 2010. Chamomile: A herbal medicine of the past with a bright future. *Molecular Medicine Reports* 3: 895–901.
75. Wallace, S., A. Keightley, and C. Gie. 2010. Dysmenorrhoea. *The Obstetrics and Gynaecology* 12: 149–154. https://doi.org/10.1576/toag.12.3.149.27596.
76. Pattanittum, P., N. Kunyaone, J. Brown, U.S. Sangkomkamhang, J. Barnes, V. Seyfoddin, and J. Marjoribanks. 2016. Dietary supplements for dysmenorrhoea. *Cochrane Database of Systematic Reviews* (3). Art. No.: CD002124. https://doi.org/10.1002/14651858.CD002124.pub2.
77. Sunil Krishna, M., and R.K. Shenoy. 2018. Clinical profile of cyclical and noncyclical Mastalgia. *New Indian Journal of Surgery* 9: 764–770.
78. Saghafi, N., H. Rhkhshandeh, N. Pourmoghadam, L. Pourali, M. Ghazanfarpour, A. Behrooznia, and F. Vafisani. 2018. Effectiveness of Matricaria chamomilla (chamomile) extract on pain control of cyclic mastalgia: A double-blind randomised controlled trial. *Journal of Obstetrics and Gynaecology* 38: 81–84.
79. Jewell, D., and G. Young. 2003. Interventions for nausea and vomiting in early pregnancy. *Cochrane Database of Systematic Reviews* 4.
80. Matthews, A., D.M. Haas, D.P. O'Mathúna, and T. Dowswell. 2015. Interventions for nausea and vomiting in early pregnancy. *Cochrane Database of Systematic Review* (9) Art. No.: CD007575. https://doi.org/10.1002/14651858.CD007575.pub4
81. Modares, M., S. Besharat, F. Rahimi Kian, S. Besharat, M. Mahmoudi, and Sourmaghi H. Salehi. 2012. Effect of ginger and chamomile capsules on nausea and vomiting in pregnancy. *Journal of Gorgan University of Medical Sciences* 14: 46–51.
82. Salgado, F. 2011. El jengibre (Zingiber officinale). *Revista Internacional de Acupuntura* 5: 167–173.
83. Lete, I., and J. Allué. 2016. The effectiveness of ginger in the prevention of nausea and vomiting during pregnancy and chemotherapy. *Integrative Medicine Insights* 11 (1–7).
84. Jewell, D. and G. Young. 2006. *Intervenciones para las náuseas y los vómitos en la fase temprana del embarazo* (Revisión Cochrane traducida). En: La Biblioteca Cochrane Plus, 2006 Número 1. Oxford: Update Software Ltd. Disponible en: http://www.update- software.com. (Traducida de The Cochrane Library, Issue 1. Chichester, UK: John Wiley & Sons, Ltd.).
85. Muñoz Centeno, L.M. 2004. Plantas medicinales españolas. Calendula officinalis L. (Asteraceae). *Medicina Naturista* 5: 257–261.
86. Blumenthal, M., ed. 1998. *The complete German commission E monographs. Therapeutic Guide to Herbal Medicines*. Austin: EEUU.
87. Ortíz, B.T., M.E.R. Pérez, D.A.J. Montero, J.M.R. Álvarez and Y.R. García. 2018. La efectividad de la crema Calendula officinalis L como tratamiento de la estomatitis aftosa recurrente. *Correo Científico Médico* 23.
88. Castillo, C.L. 2015. Plantas medicinales utilizadas en el tratamiento de enfermedades ginecológicas en Leticia y Puerto Nariño (Amazonas, Colombia). *Etnobiología* 13: 53–72.
89. Milián Vázquez, P.M., J.M. Seife Rodríguez, R. Morales Ojeda, L. Vázquez Montero, C. Martín Álvarez, and M. Quiros Enríquez. 2010. Calendula officinalis L. en el tratamiento tópico de la candidiasis vaginal recurrente. *Boletin Latinoamericano y del Caribe de Plantas Medicinales y Aromaticas* 9 (5): 343–352.

Overcoming Antibiotic Resistance: New Perspectives

Matteo Bassetti and Elda Righi

Introduction

The dramatic increase in antibiotic resistance, especially among Gram-negative bacteria (GNB), represents a threat for a public health and has been linked to high mortality rates, prolonged hospitalization, and increased healthcare associated costs [11, 39]. Furthermore, since empiric antibiotic regimens are often inappropriate for multidrug-resistant (MDR) infections, the administration of an adequate antibiotic treatment among patients is frequently delayed [137]. The most commonly reported MDR bacteria (defined as those bearing resistance to one or more antimicrobials from at least three different antimicrobial classes) belong to the so called "ESKAPE" group, including *Enterococcus faecium*, *Staphylococcus aureus*, *Klebsiella pneumoniae*, *Acinetobacter baumannii*, *Pseudomonas aeruginosa*, and *Enterobacteriaceae*. The acronym "ESKAPE" highlights the capability of this group to avoid being targeted by first-line antibiotics [11].

Among Gram-negative bacteria (GNB), the emergence of extended-spectrum-beta-lactamases (ESBLs)- and *K. pneumoniae* carbapenemase (KPC)-producing *Enterobacteriaceae*, as well as MDR strains of *P. aeruginosa* and *A. baumannii,* currently represent one of the biggest challenge for clinicians facing severe GNB infections [55]. Carbapenem-resistant Enterobacteriaceae and nonfermenters have been recognized as urgent threats to address by CDC and are included in the WHO priority list of pathogens to guide research and development of new effective drugs

M. Bassetti (✉)
Clinica Malattie Infettive, Ospedale Policlinico San Martino – IRCCS and University of Genoa, Genoa, Italy
e-mail: matteo.bassetti@unige.it

E. Righi
Infectious Diseases, Department of Diagnostics and Public Health, University of Verona, Verona, Italy

M. Bizzarri (ed.), *Approaching Complex Diseases*, Human Perspectives in Health Sciences and Technology 2, https://doi.org/10.1007/978-3-030-32857-3_19

[15, 119]. In the past decades, however, the dramatic increase of MDR GNB associated infections was not counterbalanced by new pharmaceutical investments in the area of antimicrobial development, thus the availability of new molecules declined [45, 90]. To face the crisis, professional agencies such as the Infectious Diseases Society of America (IDSA) launched the so called 10 × '20 initiative in order to develop ten new antibiotics by 2020 [53].

Following the call, novel molecules were developed, mainly targeting resistant Gram-positive bacteria such as methicillin-resistant *S. aureus* (MRSA), one of the leading causes of hospital- acquired infections (HAI) in the United States [24]. Although vancomycin remains a drug of choice for MRSA, recent studies suggest that its efficacy may be suboptimal for increased MIC values, and newer compounds with activity against MDR Gram-positive bacteria were highly awaited [108]. Among new molecules for GNB, novel beta-lactam beta-lactamase inhibitor (BLBLIs) combinations have recently been approved by FDA. Novel BLBLIs include ceftazidime- avibactam, ceftolozane-tazobactam and meropenem-vaborbactam. While the availability of new compounds with in vitro activity is promising to face the emergence of carbapenem-resistant bacteria, open questions remains regarding their use, including real life efficacy and potential development of resistance [89].

In this review, we have reported the characteristics of the most recently developed antimicrobials for the treatment of resistant Gram-positive and Gram-negative bacteria.

Cephalosporins

Cephalosporins are commonly prescribed antimicrobials characterized by a broad spectrum of activity and a favorable safety profile. Cephalosporin's mechanism of action consists in binding to penicillin-bindings-proteins (PBP) causing irreversible inhibition of the bacterial cell-wall synthesis. Recently, new generation cephalosporins have become available for the treatment of infections due to Gram-negative and MDR Gram-positive bacteria, including MRSA.

Ceftaroline Fosamil

is a semisynthetic anti-MRSA cephalosporin approved by the United States Food and Drug Administration (FDA) in 2010 and by the European Medical Agency (EMA) in 2012 for the treatment of acute bacterial skin and skin structure infections (ABSSSIs) and community acquired pneumonia (CAP). Ceftaroline activity against MRSA is related to an increased activity against PBP2a compared to other β-lactams [92]. Ceftaroline presents high efficacy towards multiple MDR Gram-positive pathogens, including hetero-resistant vancomycin-intermediate *S. aureus* (hVISA) and vancomycin-resistant *S. aureus* (VRSA). Against MDR GNB and *B. fragilis* ceftaroline activity remains limited [18]. Specifically, increased minimum inhibitory concentrations (MICs) have been shown towards AmpC-producing strains, while reduced activity was reported against *P. aeruginosa* and *Proteus mirabilis* [10, 104]. Clinical efficacy of ceftaroline compared to ceftriaxone was investigated in the FOCUS 1 and 2 Phase 3 trials, showing pooled success frequency in the treatment of CAP (including infections with MDR *S. pneumoniae*) of 84.3% [34]. In the integrated analysis of the Phase 3 CANVAS 1 and 2 trials, evaluating the efficacy of ceftaroline versus vancomycin plus aztreonam in complicated skin and skin-structure infection (cSSSI), including infections due to MRSA, ceftaroline, pooled success frequency was 91.6% [21]. In both trials, rates of adverse events, serious adverse events, deaths, and premature discontinuations caused by an adverse event were similar to comparators. Pharmacokinetics (PK) studies support an increased benefit with higher doses (600 mg q8h vs. 600 mg q12h) of ceftaroline in cSSSIs and pneumonia, especially for *S. aureus* strains with increased MICs (> 1 mg/l) [76]. A Phase 3 randomized clinical trial (RCT) showed that ceftaroline fosamil (600 mg every 8 h) was found to be noninferior to vancomycin and aztreonam in patients with cSSTI in terms of safety and efficacy [29]. Recently, observational studies highlighting positive outcomes for off-label uses of ceftaroline, in particular in sepsis and endocarditis, have been published. A recent report of 55 cases of Gram-positive endocarditis (80% of which caused by MRSA) treated with ceftaroline mainly as second line therapy showed clinical success in 82.6% of cases [25].

Ceftobiprole Medocaril
is an expanded spectrum cephalosporin with high affinity for PBP2a and enhanced activity against Gram-positive bacteria. Ceftobiprole MIC90 against *S. pneumoniae* was 0.5 μg/mL, showing the highest *in vitro* activity compared to other cephalosporins [130]. Ceftobiprole is active against *Enterococcus faecalis* but not against *E. faecium.* Compared to ceftaroline, ceftobiprole is stable against class A TEM-1 and class C AmpC beta-lactamase and displays anti-pseudomonal activity that is superior to that of cefepime [130]. Similarly to ceftazidime, however, ceftobiprole is hydrolyzed by extended-spectrum beta-lactamases (ESBL) [97]. Ceftobiprole has been approved in many European countries and in Canada, Argentina, Jordan, Peru, and Saudi Arabia for the treatment of HAP and CAP, only in adults. In patients with ventilator-associated pneumonia (VAP), ceftobiprole is not recommended since its noninferiority was not demonstrated in this subset of patients [3].

In phase 3 trials, ceftobiprole showed comparable activity vs. linezolid in association with ceftazidime or ceftriaxone in HAP and CAP requiring hospitalization, respectively [3, 84]. A post hoc retrospective analysis of data from the Phase 3 studies identified higher rates of early response in ceftobiprole-treated patients vs, comparators in high-risk groups, including CAP patients aged $\geq$75 years, CAP patients with COPD, and HAP patients with >10 baseline comorbidities [109].

Good results were also shown in the treatment of complicated SSSIs compared with vancomycin plus ceftazidime [86]. Ceftobiprole was generally well tolerated with main adverse events represented by gastrointestinal effects (e.g. nausea, vomiting, diarrhea).

Cefiderocol
is a novel siderophore cephalosporin characterized by enhanced stability against carbapenemases, including class A, B, C and D enzymes [28]. Cefiderocol is active against MDR *Enterobacteriacae*, *P.aeruginosa* spp., *A. baumanii* spp. and *Burkholderia* spp. Similarly to other cephalosporins, cefiderocol binds primarily to bacterial PBP3, but structural changes allow the molecule to form a chelated complex with ferric iron that promotes its passage through the outer membrane of Gram-negative pathogens, using the bacterial iron transport system [2, 27]. Cefiderocol has only limited activity against Gram-positive pathogens and anaerobes [44]. Reduced cefiderocol activity was also detected among *P. aeruginosa* isolates with decreased iron transport system components or mutations in the binding site for the transport system [48]. Promising PK data were shown for cefiderocol activity in the bloodstream and urinary tract system; cefiderocol also displayed an acceptable ratio (0.239) of epithelial lining fluid (ELF)-to-plasma concentration [75]. Data about penetration of cefiderocol in the peritoneal fluid, however, are lacking. PK analysis suggested that cefiderocol a dose of 2 g every 8 h offers an adequate exposure for the treatment of cUTI or acute pyelonephritis caused by GNB [58]. A Phase 2 study comparing cefiderocol with imipenem/cilastatin for the treatment of complicated urinary tract infections (cUTI) has recently been completed (NCT02321800). Other ongoing trials include a Phase 3 open-label clinical study comparing cefiderocol with best available therapy for the treatment of hospitalized CAP, HAP, VAP, bloodstream infections (BSI) and sepsis caused by carbapenem-resistant GNB (NCT02714595) and a Phase 3 trial evaluating the role of cefiderocol for the treatment of HCAP, HAP and VAP caused by GNB (NCT0302380).

Beta-Lactam Beta-Lactamase Inhibitors

Ceftozolane/Tazobactam
represents the association of ceftolozane, a novel cephalosporin, and the beta-lactamase inhibitor tazobactam. The structure of ceftolozane is similar to that of ceftazidime, with the exception of a modified side-chain at the 3-position of the cephem nucleus, which confers enhanced antipseudomonal activity. The addition of tazobactam expands the spectrum of ceftolozane including extended-spectrum beta-lactamase (ESBL)-producing bacteria [106, 121]. However, published clinical experience outside clinical trials against ESBL- and AmpC- producing pathogens is limited [93]. Ceftozolane/tazobactam has the ability to escape various resistance mechanisms, including PBP mutations and efflux pumps, displaying activity against MDR strains of *P. aeruginosa* [82]. Ceftolozane-tazobactam has shown activity against the majority of carbapenem-resistant *P. aeruginosa* strains, although it remains susceptible to hydrolysis by carbapenemases (e.g. metall0-beta-lactamases, MBLs, and KPC). Ceftolozane/tazobactam is currently approved by FDA and EMA for the treatment of cUTIs and complicated intra-abdominal infections (cIAIs), at the

dose of 1.5 g every 8 h. Due to the lack of efficacy against *Bacteroides*, combination of ceftolozane/tazobactam with metronidazole for the treatment of cIAIs is recommended.

In Phase 3 trials ceftolozane/tazobactam showed increased efficacy compared to levofloxacin for cUTIs and comparable activity in cIAI vs. meropenem [52, 114, 128]. PK/PD studies showed that an increased dosage (3 g every 8 h) may be needed to achieve a high (>90%) probability of target attainment [133].

A phase 3 clinical trial (NCT02070757) comparing ceftolozane-tazobactam (administered 3 g every 8 h) with meropenem in patients with HAP requiring mechanical ventilation or VAP has been recently completed and results are awaited. Adverse effects associated to the use of ceftolozane/tazobactam did not differ considerably from other cephalosporins. Nausea, diarrhea, and headache represented the most commonly reported adverse effects [128]. Real world data has recently been published showing encouraging results for ceftolozane/tazobactam in various types of infections, including pneumonia, especially for the treatment of MDR P. *aeruginosa* [8, 83].

Ceftazidime/Avibactam

is a fixed-combination drug containing ceftazidime, a third-generation cephalosporin, and avibactam, a beta-lactamase inhibitor characterized by high affinity with class A and class C beta-lactamases [32]. This combination has shown in vitro activity against various MDR bacteria, including OXA-48, ESBL, AmpC, and KPC producers [63, 65]. A pooled analysis from randomized studies including 1051 patients with MDR Enterobacteriaceae and 95 patients with MDR *P. aeruginosa* showed microbiological response rates of 78.4% and 57.1%, respectively, for ceftazidime/avibactam and 71.6% and 53.8%, respectively, for comparators [116].

Ceftazidime/avibactam, however, is not active against MBL producers (Livermore et al. 2015) including NDM, VIM, IMP, and against *Acinetobacter* OXA-type carbapenemases [60]. Phase 3 trials in adult patients with cIAI (RECLAIM 1 and 2, NCT01499290; RECLAIM 3, NCT01726023), cUTI (RECAPTURE 1 and 2, NCT01595438 and NCT01599806), HAP/VAP (REPROVE, NCT01808092) and cUTI or cIAI caused by ceftazidime-non-susceptible pathogens (REPRISE, NCT01644643) showed comparable activity of ceftazidime-avibactam with comparators (especially carbapenems) [13, 77, 96, 123, 129]. Ceftazidime/avibactam is currently approved for the treatment of cUTIs, hospital-acquired and ventilator-associated pneumonia (HAP/VAP), cIAIs (in combination with metronidazole), and difficult-to-treat infections caused by GNB when other treatment options are limited. A meta-analysis including nine RCTs and three observational studies found showed comparable clinical responses (RR = 0.99, 95% CI 0.96–1.02) and non-inferior bacterial eradication (RR = 1.04, 95% CI 0.93–1.17) to carbapenems and, in patients infected by carbapenem-resistant Enterobacteriaceae, reduced mortality versus comparators (RR = 0.29, 95% CI 0.13–0.63) [136]. While RCTs analyzing the efficacy of ceftazidime-avibactam against CRE are not available, various observational studies have reported good results and favorable safety profiles in this group [124, 125]. The use of ceftazidime/avibactam, however, has been associated with the emergence of resistant strains [37, 111]. For this reason, the use of ceftazidime-avibactam in combination with other drugs (e.g. gentamicin, fosfomycin, tigecycline, colistin) has been suggested with the purpose to reduce the risk of resistance selection. Ceftazidime-avibactam is overall well tolerated. Most common adverse events include headache, gastrointestinal symptoms (abdominal pain, vomiting, nausea and constipation), and infusion-site reactions [126].

Ceftaroline-Avibactam

The association with avibactam allows to extend the spectrum of activity to resistant GNB. In particular, this combination demonstrated *in vitro* efficacy against infections caused by *Enterobacteriaceae*, including strains producing various ESBL types (e.g., CTX-M types), AmpC, and KPC, in addition to anaerobes and MRSA [14, 131]. Ceftaroline- avibactam had no activity against MBLs, but displayed potent activity towards KPC-producing strains (MIC90, 0.5 to 1 μg/ml; meropenem MIC >8 μg/ml). Similar to ceftaroline, limited activity of ceftaroline-avibactam has been reported against *Acinetobacter* spp. and *P. aeruginosa* [14]. In clinical studies ceftaroline/avibactam has been administered as 600 mg ceftaroline fosamil/600 mg avibactam every 8 h over 1 h. In a double-blind, placebo-controlled, single and multiple dose study, ceftaroline-avibactam was well tolerated. No serious adverse events (SAE) occurred during the study [101]. Excellent *in vitro* activity of ceftaroline-avibactam has been shown against isolates from 174 centers in the US collected from patients with skin and soft tissue infections (SSTIs) caused by MRSA, beta-hemolytic streptococci, *E. coli*, and *K. pneumoniae* as well as ESBL-producing strains [35]. Ceftaroline-avibactam is currently in Phase 2 development and has completed a study in comparison with doripenem for the treatment of adult UTIs (NCT01281462).

Aztreonam-Avibactam

Combination of aztreonam with avibactam offers a potential option against NDM-1 producing bacteria [23].

Aztreonam is a monobactam characterized by broad-spectrum activity against Gram-negative pathogens, although it has limited activity against *Enterobacter* spp. and *Bacteroides fragilis*. While aztreonam is hydrolyzed by class-A and class-C -lactamases, it displays stability to hydrolysis by MBLs. The combination of aztreonam with avibactam has shown to restore the activity of aztreonam against class A, C and some class D beta-lactamases producers [61, 107]. Aztreonam-avibactam showed in vitro efficacy against 99.8% *Enterobacteriacae* isolates, included strains that were not susceptible to meropenem. Against MBL-producing *Enterobacteriacae* aztreonam-avibactam was 8- to 32-fold more potent than meropenem. Comapred to aztreonam alone, the activity against *A. baumannii* or *P. aeruginosa* was not enhanced by combination with avibactam [57].

A Phase 1 study addressed the safety and tolerability of aztreonam-avibactam in healthy subjects (NCT01689207), showing that the drug was generally well-tolerated in healthy volunteers without evidence of drug-drug interactions. No SAEs were reported among subjects randomized to aztreonam and/or avibactam [31]. In a Phase 2 study (REJUVENATE) investigating efficacy and safety of aztreonam-avibactam for the treatment of cIAIs in hospitalized adults (NCT02655419), the majority of the AEs were non-serious and of mild or moderate intensity. SAE were reported in 9 (267%) patients, but none was considered as being related to study treatment. In the same study, 59% of patients achieved clinical cure. A Phase 3 study (REVISIT) to determine the efficacy and safety of aztreonam/avibactam versus meropenem (with or without colistin) for the treatment of serious infections due to Gram-negative bacteria, including HAP/VAP, is currently ongoing (NCT03329092).

Meropenem/Vaborbactam

Vaborbactam (formerly known as RPX7009) is a new class A - and class C -beta-lactamase inhibitor [43]. Vaborbactam in association with meropenem enhances the activity against KPC-producing *Enterobacteriacae,* reducing meropenem MIC50 from 32 to 0.06 g/ml and MIC90 from 32 to 1 g/ml [43]. Nevertheless, meropenem-vaborbactam showed no improvement in activity over meropenem alone against OXA-48-producing strains and MBLs [71].

Meropenem-vaborbactam was approved by FDA in 2017 for the treatment of cUTIs, based on the results of the TANGO1 trial, demonstrating noninferiority of meropenem-vaborbactam (2 g/2 g every 8 h) over piperacillin-tazobactam (4 g/0.5 g every 8 h) for the treatment of cUTIs and acute pyelonephritis in adult patients (NCT02166476). In this Phase 3, multicenter, multinational, RCT including 545 patients, overall success occurred in 98.4% with meropenem-vaborbactam-vs. 94.0% with piperacillin-tazobactam (difference, 4.5% [95% CI, 0.7% to 9.1%]; $P < .001$ for noninferiority). Microbial eradication occurred in 66.7% with meropenem-vaborbactam vs. 57.7% with piperacillin- tazobactam ($P < .001$ for noninferiority) [59]. Adverse events were reported in 39% with meropenem-vaborbactam vs. 35.5% with piperacillin-tazobactam [59]. A Phase 3 study (TANGO II) evaluated the efficacy, safety and tolerability of meropenem/ vaborbactam compared to best available therapy (ceftazidime-avibactammonotherapy or treatment with a carbapenem, an aminoglycoside, poly-myxin B/colistin, or tigecycline monotherapy or combination treatment) for the treatment of infections due to carbapenem-resistant *Enterobacteriacae* has recently been completed. (NCT02168946). The study encompassed 72 patients with various CRE infections, including bloodstreaminfections (BSI), cUTIs, HAP/VAP, and cIAI showing that meropenem-vaborbactam was associated with significantly increased clinical cure rate and lower all-cause mortality rate at day 28 compared with BAT (68% vs. 27%, P = 0.008 and 5% vs. 33%, P = 0.03 respectively) [132].

Imipenem/Relebactam

Relebactam (formerly known as MK-7655) is a novel class A and class C beta-lactamase inhibitor under investigation in combination with imipenem/cilastatin for the treatment of MDR Gram-negative infections [66, 67]. Relebactam has shown the ability to restore imipenem activity against KPC-producing *Enterobacteriacae*, reducing its MICs from 16–64 mg/L to 0.12–1 mg/L at a concentration of 4 mg/L [46]. Relebactam has also demonstrated the potential to enhance the activity of imipenem against *P.aeruginosa*, including those strains with depressed OprD expression and increased AmpC expression [66]. The addiction of relebactam to imipenem, however, has not been associated with additional effects against *A. baumanii* and *S. maltophilia* or MBL-producing *Enterobacteriacae* [66]. A Phase 3 study (RESTORE IMI-1) evaluating the efficacy and safety of imipenem/relebactam versus colistimethate sodium plus imipenem/cilastatin for the treatment of imipenem-resistant HAP, VAP, cIAIs and cUTIs (NCT02452047) showed comparable overall clinical responses between the study groups [81]. Pathogens included were *P. aeruginosa* (77%), *Klebsiella* spp. (16%) and other Enterobacteriaceae (6%); 84% of the isolates were AmpC producers, 39% ESBL- and 16% KPC producers. Drug-related AEs were 16% for imipenem-relebactam and 31% for the comparator, and increased nephrotoxicity was in 10% vs. 56% of patients, respectively ($p = 0.001$). A Phase 3 trial evaluating non-inferiority of imipenem-relebactam compared to piperacillin-tazobactam for the treatment of HAP/VAP (NCT02493764) has been recently completed and results are awaited.

Dihydrofolate Reductase Inhibitors

Iclaprim

is the only drug, besides trimethoprim, belonging to the dihydrofolate reductase inhibitor class. Compared to trimethoprim, however, iclaprim displays enhanced activity against MRSA and GNB involved in pulmonary infections. Such as *Haemophilus influenzae* and *Moraxella catarrhalis* [105]. Against MRSA, iclaprim showed MIC $\leq$1 μg/mL against isolates that were nonsusceptible to daptomycin (71%), linezolid (100%), or vancomycin (67%) [49, 50]. Iclaprim has been studied for the treatment of SSTIs and HAP caused by or suspected to be Gram- positive bacteria administered at 80 mg every 12 h [69]. An oral formulation is also under development. Two recent Phase 3 trials, REVIVE-1 and -2, demonstrated noninferiority of iclaprim vs. vancomycin in SSTIs [51]. One phase 2 clinical trial for the treatment of hospital-acquired bacterial pneumonia (HABP) [50] showed comparable clinical cure rates for iclaprim vs. vancomycin at test of cure (74% vs. 52%, respectively).

Glycopeptides

Glycopeptides (e.g., vancomycin and teicoplanin) exert their function by inhibition of bacterial cell wall synthesis binding to acyl-D-alanyl-D-alanine in peptidoglycan [100]. New glycopeptides derivatives, such as telavancin, dalbavancin, and oritavancin, were developed to overcome the emergence of MRSA strains with reduced susceptibilities to vancomycin. These novel compounds have been classified as lipoglycopeptides (since they display a lipophilic side-chains linked to glycopeptides) and were proven noninferior versus vancomycin in Phase 3 RCTs [12, 22, 117]. Main characteristics of lipoglycopeptides are the rapid and bactericidal activity against MRSA and their longer half-life compared to vancomycin.

Telavancin

is a semisynthetic lipoglycopeptide vancomycin-derivative characterized by high activity against staphylococci, including MRSA. Telavancin has a dual mode of action causing inhibition of peptidoglycan synthesis and membrane depolarization [110]. Telavancin is administered once-daily and showed enhanced *in vitro* antibacterial activity against a broad range of Gram-positive bacteria, including MRSA, and isolates with reduced glycopeptide susceptibility such as glycopeptide-intermediate *S. aureus* (GISA) and Van-A type *Enterococci* [47]. MICs for MRSA were 2 to 8 times lower than those observed for vancomycin, teicoplanin, and linezolid [102]. The US FDA approved telavancin in 2009 for the treatment of complicated skin and skin structure infections (cSSSIs) caused by Gram-positive bacteria, including MRSA. Telavancin has also shown good penetration in the alveolar macrophages and, unlike daptomycin, its activity is not affected by pulmonary surfactant [5].

Non-inferiority of telavancin (10 mg/kg every 24 h) versus vancomycin (1 g every 12 h) for the treatment of HAP has been demonstrated in two Phase 3 RCTs (ATTAIN) [103]. A subsequent pooled analysis of data obtained from cSSTIs and HAP RCTs showed higher risk of nephrotoxicity and SAEs among telavancin-treated patients [91]. Although increased mortality was reported in patients with HAP and moderate-to severe renal impairment treated with telavancin compared to vancomycin [4], further analyses showed that clinical and safety outcomes were similar in the two treatment groups among patients without severe renal impairment or pre-existing acute renal failure [122]. Telavancin is currently approved in Europe for the treatment of adult patients with HAP (including VAP) for confirmed or suspected MRSA infections, and when other alternative treatments are not suitable. The use of telavancin is recommended only to patients with normal renal function [74].

Dalbavancin

is a long acting lipoglycopeptide characterized by a half-live ranging 147 to 258 h allowing infrequent (e.g., weekly) administration [134]. Dalbavancin has demonstrated excellent in vitro activity against MSSA, MRSA, VISA, meticillin-resistant *S. epidermidis* (MRSE) and enterococci, although limited activity was demonstrated for VanA-type enterococci and VRSA [40, 54, 56].

Dalbavancin has been approved by the FDA for the treatment of cSSTIs in 2014 based on two non- inferiority Phase 3 trials (DISCOVER 1 and 2, NCT01339091 and NCT01431339 respectively) comparing dalbavancin (1000 mg on day 1 followed by 500 mg after 1 week) or vancomycin for at least 3 days, with the option to switch to oral linezolid to complete 10–14 days of treatment. Early clinical response (defined as cessation of spread of infection-related erythema and the absence of fever at 48–72 h) was shown in 79.7% of patients in the dalbavancin group vs. 79.8% of patients in the vancomycin-linezolid group, confirming non-inferiority of dalbavancin in both studies [12]. A subsequent RCT proved non-inferiority of a single 1500-mg infusion of dalbavancin compared to the standard 2-dose regimen. In this study, overall 14- and 28-day clinical response rate was 97% in both treatment groups, while no significant differences in treatment-related adverse events were reported [30]. A systematic review assessing specific efficacy against MRSA including 14 cSSTI RCTs and 1840 confirmed MRSA cases showed higher success rates for linezolid (84.4%), dalbavancin (87.7%), and telavancin (83.5%) compared to vancomycin (74.7%) [70]. The potential role of dalbavancin for off-label use, particularly for osteomyelitis, is currently under investigation [6, 7, 98].

Oritavancin
similarly to dalbavancin, is characterized by a rapid bactericidal activity against Gram-positive bacteria, including MRSA, VISA, methicillin-resistant CoNS and VRE isolates [80]. Oritavancin prolonged half-life (393 ± 73.5 h) allows for single-dose treatment. Oritavancin does not require dosage adjustment for renal or mild to moderate hepatic dysfunction. A RCT in over 1000 adults with ABSSSIs receiving either a single intravenous 1200 mg dose of oritavancin or 7–10 days of vancomycin showed clinical cure and proportion of patients with at least 20% reduction in lesion area of 83% vs, 81% and 85.9% vs. 85.3% for oritavancin vs vancomycin, respectively [22]. The efficacy by pathogen (including MRSA) and the frequency of adverse events were similar between treatment groups. The FDA approved oritavancin in 2014 for the treatment of ABSSSIs due to MSSA, MRSA, *Streptococcus spp* and *E. faecalis*.

Oxazolidinones

Oxazolidinones are synthetic antibiotics with high activity against MDR Gram-positive pathogens, including MRSA and VRE. The first member of this class, linezolid, was approved for clinical use in 2000 for the treatment of severe Gram-positive infections and is characterized by excellent oral bioavailability and tissue penetration [33].

Tedizolid
(formerly known as TR700) is new oxazolidinone approved by the FDA in 2014 and by the EMA in 2015 for the treatment of acute bacterial SSTI. Furthermore, the role of tedizolid for the treatment of MRSA respiratory tract infections is under investigation [17, 120]. Compared to linezolid, tedizolid is characterized by a lower risk of myelotoxicity [68, 112], higher bioavailability and increased half-life, allowing once daily administration [36], and higher ELF penetration [16]. Fewer drug-drug

interactions with selective serotonin reuptake inhibitors (SSRIs), serotonergic, and adrenergic agents have also been reported for tedizolid compared to linezolid due to a weak and reversible *in vitro* inhibition of the monoamine oxidase (MAO) pathway [110]. Tedizolid exhibits a 2- to four-fold better *in vitro* activity compared with linezolid against MSSA and MRSA [16]. Against 27 clinical isolates of linezolid-resistant staphylococci and enterococci, tedizolid minimum inhibitory concentrations (MICs) appeared four-fold to 32-fold lower than those of linezolid [113]. Tedizolid is administered 200 mg once daily and is available as IV or oral formulations, allowing for sequential therapy [102]. Phase 3 trials (ESTABLISH-1 and 2) demonstrated tedizolid not inferiority to linezolid in SSTIs [79, 94]. In the ESTABLISH-1 trial, noninferiority of oral tedizolid administered for 6 days was demonstrated compared with oral linezolid administered for 10 days [94]. Early (48–72 h) clinical responses to treatment in 667 patients were similar in both groups. The ESTABLISH-2 trial compared IV tedizolid vs. IV linezolid (with the possibility of sequential oral therapy) at different time points (e.g., 48–72 h, on day 7, end of treatment, and at 7–14 days after the end of treatment, showing similar results in clinical responses at different time points [79].

Overall rates of related adverse effects were similar to linezolid, with nausea being the most commonly reported adverse effect (16%) associated with tedizolid use.

Quinolones

Delafloxacin

is a newer quinolone developed to enhance the activity against MDR bacteria and, at the same time, allow for a low potential for developing bacterial resistance [6, 7]. Delafloxacin is characterised by a weak acidity caused by a strongly basic group at the C-7 position, enhancing its antibacterial potency in environments with reduced pH, such as the urinary tract and the phagolysosomes [62]. Furthermore, delafloxacin has been associated with low potential for resistance selection due to a dual mechanism of inhibition of DNA targets (i.e., gyrase and topoisomerase IV) [99].

Delafloxacin has shown high activity against quinolone-resistant strains of MRSA and against MDR Gram-negative isolates, including *K. pneumoniae* [1, 85, 99]. Delafloxacin is also active against CAP and HAP pathogens, including resistant strains of *S. pneumoniae*, *H. influenzae*, *M. catarrhalis* and *Legionella* (Zhanel et al. 2003). Delafloxacin's MIC90 against all MRSA, including quinolone-resistant MRSA strains, was 0.06 µg/mL. Delafloxacin showed comparable efficacy to tigecycline in the treatment of cSSSI including *S. aureus* (85% of cases with approximately 70% of MRSA strains) [22]. Delafloxacin can be administered both IV and orally, allowing for sequential therapy.

Two studies assessed delafloxacin efficacy for the treatment of respiratory tract infections. A Phase 2 enrolling 309 outpatients with CAP showed pathogen eradication rates were higher than 90% for *H. influenzae* and *parainfluenzae* and atypicals, and achieved 100% for *S. aureus* and *S. pneumoniae* [72]. A study including patients with Acute Bacterial Exacerbation of Chronic Bronchitis (ABECB) demostrated clinical and microbiological cure rates higher than 70% [73]. In both the studies delafloxacin was generally well tolerated, with diarrhea, headhache and nausea were the most commonly reported AEs [64, 88]. Data from studies on the use of delafloxacin for the treatment of cSSTIs demonstrate that delafloxacin at the dose of 300 mg every 12 h is well tolerated [64, 88]. Two phase 3 trials (NCT01811732 and NCT01984684) have been recently published reporting comparable activity of delafloxacin (300 mg q12h IV and 300 mg IV every 12 h for 3 days with a switch to 450 mg oral delafloxacin 450 mg orally td) versus combination of aztreonam and vancomycin in the treatment of complicated SSTI [88, 95] (Table 1).

Aminoglycosides

Plazomicin

is a new aminoglycoside characterized by activity against Gram-positive and Gram-negative pathogens approved by the FDA in 2018 for the treatment of cUTIs, including pyelonephritis. Structural changes in plazomicin structure prevent its inactivation by plasmid-borne modifying enzymes, avoiding the development of resistance mechanisms that are typical of other aminoglycosides [38].

Plazomicin has potent in vitro bactericidal activity against MDR *Enterobacteriacae*, *P.aeruginosa* and *A.baumanni*. In vitro synergy has been demonstrated between plazomicin and various molecules, including piperacillin/tazobactam or ceftazidime against MDR *Enterobacteriacae* [47] and carbapenems for the treatment of MDR *A. baumanii* [38]. In a Phase 2 study vs. levofloxacin, plazomicin (15 mg/kg once daily) showed efficacy in treating cUTIs and acute pyelonephritis, including those caused by antibiotic-resistant *Enterobacteriaceae* [20]. A Phase 3 study comparing plazomicin with meropenem for the treatment of cUTIs (NCT02486627) showed plazomicin noninferiority to meropenem. Cure rates at day 5 were 88% in the plazomicin group and in 91% in the meropenem group. Increases in serum creatinine occurred in 7% of patients in the plazomicin group and in 4% in the meropenem group [127]. A Phase 3 study comparing 17 patients treated with plazomicin with 20 treated with colistin, both in combination with tigecycline and meropenem for the treatment of HAP, VAP, and bloodstream infections due to carbapenem-resistant Enterobacteriacae has been completed (NCT01970371). All-cause mortality was lower for plazomicin compared to colistin (24% vs. 50%, difference 26.5%, range − 0.7 to 51.2), while serum creatinine increase was higher in the colistin arm (38% vs. 8%, respectively) [78] (Table 2).

Table 1 Novel antibiotics: clinical indications and development status

Antimicrobial class	Drug	Indication	Current status
Cephalosporins	Ceftaroline Ceftobiprole Cefiderocol	SSSI, CAP HAP, hospitalized CAP	FDA approved Approved in some countries Phase 3
Beta-lactam/beta-lactamase inhibitors (BLBLIs)	Ceftozolane-tazobactam Ceftazidime-avibactam Meropenem-vaborbactam Imipenem-relebactam Aztreonam-avibactam Ceftaroline-avibactam	cUTIs, cIAIs cUTIs, cIAIs, HAP, VAP cUTIs	FDA approved FDA approved FDA approved Phase 3 Phase 3 Phase 2
Glycopeptides	Telavancin Oritavancin	SSTI, HAP, VAP SSSI SSSI SSSI	FDA approved FDA approved FDA approved
Oxazolidinones	Tedizolid	SSSI	FDA approved
Quinolones	Delafloxacin		Phase 3
Aminoglycosides	Plazomicin	cUTIs	FDA approved
Tetracyclines	Eravacycline Omadacycline	cIAIs	FDA approved Phase 3

SSSI skin and soft tissue infections, *CAP* community acquired pneumonia, *HAP* hospital-acquired pneumonia, *cUTI* complicated urinary tract infections, *cIAIs* complicated intra-abdominal infections

Tetracyclines

Tetracycline activity has been impaired over the years by the development of resistance. Tigecycline, the first member of a new class of tetracyclines known as glycylcyclines, is characterized by a broader spectrum including tetracycline-resistant microorganisms. Other tetracycline derivatives have been recently developed, including eravacycline and omodacycline.

Eravacycline
is not subject to mechanisms that usually cause resistance to other tetracycline derivatives such as efflux pumps and ribosomal protection proteins [19]. Eravacycline has efficacy against MRSA, VRE, and *Enterobacteriaceae*

Table 2 Activity of newly approved antibiotics against MRSA and carbapenem-resistant Gram-negative bacteria

Drug	Spectrum	MRSA	Carbapenem-resistant bacteria
Ceftaroline	Gram-positive, non ESBL-producing Gram-negatives, no P. aeruginosa	Yes	No
Ceftazidime/avibactam	MDR Gram-negatives (no MBLs)	No	Yes
Ceftobiprole	Gram-positive, non ESBL-producing Gram-negatives	Yes	No
Ceftozolane/tazobactam	Gram-negatives (including MDR P. aeruginosa; no KPC or MBLs)	No	Yes/No
Dalbavancin	MDR Gram-positives	Yes	No
Eravacycline	Gram-negative (no Pseudomonas)	Yes	Yes/No
Meropenem-vaborbactam	MDR Gram-negative (no MBLs)	No	Yes
Oritavancin	MDR Gram-positive	Yes	No
Plazomicin	MDR Gram-negative including metallo-beta-lactamase	Yes	Yes/No
Telavancin	MDR Gram-positive	Yes	No
Telizolid	MDR Gram-positive	Yes	No

MDR multidrug-resistant, *ESBL* extended-spectrum beta-lactamase, *KPC* K. pneumoniae carbapenemase- producing, *MBLs* metallo-beta-lactamases

expressing resistance genes from multiple classes of ESBL or carbapenemases such as KPC- and OXA-producers [42, 118]. Eravacycline has no activity against *P. aeruginosa,* MDR *A. baumannii* and *S. maltophilia* [135]. A phase 2 study evaluating the safety and efficacy of eravacycline dosed once or twice daily versus ertapenem in cIAI demonstrated clinical cure rates above 90%, including infections caused by ESBL-producing, levofloxacin and ertapenem-resistant organisms [115]. This study also displayed good tolerability for eravacycline when compared with ertapenem. The FDA recently approved IV eravacycline for the treatment of cIAI, based on the IGNITE 1 and 4 trials showing non-inferiority of eravacycline compared with ertapenem and meropenem, respectively [26, 115].

Omadacycline
is structurally similar to tetracyclines, but, like eravacycline, overcomes the two main mechanisms of tetracycline resistance represented by efflux pumps and ribosomal protection proteins. Omadacycline is active against MRSA, VRE, ESBL- and carbapenemase-producing *Enterobacteriaceae,* MDR *Acinetobacter* spp., and *Stenotrophomonas maltophilia* (Pfaller. Omadacycline (100 mg intravenously once a day with an option to transition to 200 mg orally once a day) was found non-inferior compared with linezolid with or without aztreonam for the treatment of cSSTIs in a Phase 2 trial [87]. Renal adjustments are not necessary during treatment with omadacyclina [9]. The efficacy of oral formulation in clinical practice is currently under investigation in a Phase 3 study comparing oral omadacycline with

oral linezolid for the treatment of ABSSTIs (NCT02877927). Furthermore, omadacycline concentrations in ELF is high [41]. A Phase 3 study comparing omadacycline (both IV and oral) with moxifloxacin for the treatment of CAP has been completed (NCT02531438).

Conclusions

Novel compounds characterized by broad-spectrum activity against MDR pathogens, including carbapenem-resistant bacteria and MRSA, have been recently approved or are in advanced stage of development. Furthermore, new molecules have shown favorable safety profiles in clinical trials and, in some cases, availability of oral formulations. While the availability of new compounds with activity on difficult-to-treat bacteria represents a unique opportunity, limited data regarding the efficacy of these agents in real-world studies have been published so far, and the costs of new therapies remains high. Best placement in therapy and optimized use of novel compounds to avoid the development of resistance requires special attention and further investigations.

References

1. Almer, L.S., J.B. Hoffrage, E.L. Keller, R.K. Flamm, and V.D. Shortridge. 2004. In vitro and bactericidal activities of ABT-492, a novel fluoroquinolone, against gram-positive and gram-negative organisms. *Antimicrobial Agents and Chemotherapy* 48: 2771–2777.
2. Avery, L.M., and D.P. Nicolau. 2018. Investigational drugs for the treatment of infections caused by multidrug-resistant gram-negative bacteria. *Expert Opinion on Investigational Drugs* 27: 325–338.
3. Awad, S.S., A.H. Rodriguez, Y.C. Chuang, Z. Marjanek, A.J. Pareigis, and G. Reis. 2014. A phase 3 randomized double-blind comparison of ceftobiprole medocaril versus ceftazidime plus linezolid for the treatment of hospital-acquired pneumonia. *Clinical Infectious Diseases* 59: 51–61.
4. Barriere, S.L. 2014. The ATTAIN trials: efficacy and safety of telavancin compared with vancomycin for the treatment of hospital-acquired and ventilator-associated bacterial pneumonia. *Future Microbiology* 9: 281–289.
5. Bassetti, M., M. Mikulska, E. Righi, L. Nicolini, and C. Viscoli. 2009. The role of telavancin in the treatment of MRSA infections in hospital. *Expert Opinion on Investigational Drugs* 18: 521–529.
6. Bassetti, M., E. Righi, D. Pecori, and G. Tillotson. 2018a. Delafloxacin: An improved fluoroquinolone developed through advanced molecular engineering. *Future Microbiology* 13: 1081–1094.
7. Bassetti, M., M. Peghin, A. Carnelutti, and E. Righi. 2018b. The role of dalbavancin in skin and soft tissue infections. *Current Opinion in Infectious Diseases* 31 (2): 141–147.
8. Bassetti, M., N. Castaldo, A. Cattelan, et al. 2019. Ceftolozane/tazobactam for the treatment of serious Pseudomonas aeruginosa infections: A multicentre nationwide clinical experience. *International Journal of Antimicrobial Agents* 53 (4): 408–415.

9. Berg, J.K., E. Tzanis, and L. Garrity-Ryan. 2018. Pharmacokinetics and safety of cycline in subjects with impaired renal function. *Antimicrobial Agents and Chemotherapy* 62 (2): e02057–e02017.
10. Biek, D., I.A. Critchley, T.A. Riccobene, and D.A. Thye. 2010. Ceftaroline fosamil: A novel broad- spectrum cephalosporin with expanded anti-gram-positive activity. *The Journal of Antimicrobial Chemotherapy* 65 (Suppl 4): iv9–i16.
11. Boucher, H.W., G.H. Talbot, J.S. Bradley, et al. Bad bugs, no drugs: no ESKAPE!. 2009. An update from the Infectious Diseases Society of America. *Clinical Infectious Diseases* 48:1–12.
12. Boucher, H.W., M. Wilcox, G.H. Talbot, et al. 2014. Once-weekly dalbavancin versus daily conventional therapy for skin infection. *The New England Journal of Medicine* 370: 2169–2179.
13. Carmeli, Y., J. Armstrong, P.J. Laud, et al. 2016. Ceftazidime-avibactam or best available therapy in patients with ceftazidime-resistant Enterobacteriaceae and Pseudomonas aeruginosa complicated urinary tract infections or complicated intra-abdominal infections (REPRISE): A randomised, pathogen-directed, phase 3 study. *The Lancet Infectious Diseases* 16: 661–673.
14. Castanheira, M., H.S. Sader, D.J. Farrell, R.E. Mendes, and R.N. Jones. 2012. Activity of ceftaroline- avibactam tested against Gram-negative organism populations, including strains expressing one or more β-lactamases and methicillin-resistant Staphylococcus aureus carrying various staphylococcal cassette chromosome mec types. *Antimicrobial Agents and Chemotherapy* 56 (9): 4779–4785.
15. CDC. 2013. *Antibiotic Resistance threats in the United States*. Washington, DC: US DHHS. Available at https://www.cdc.gov/drugresistance/pdf/ar-threats-2013-508.pdf.
16. Chen, K.H., Y.T. Huang, C.H. Liao, et al. 2015. In vitro activities of Tedizolid and linezolid against gram-positive cocci associated with acute bacterial skin and skin structure infections and pneumonia. *Antimicrobial Agents and Chemotherapy* 59: 6262–6265.
17. Choi, S., W. Im, and K. Bartizal. 2012. Activity of tedizolid phosphate (TR-701) in murine models of infection with penicillin-resistant and penicillin-sensitive Streptococcus pneumoniae. *Antimicrobial Agents and Chemotherapy* 56: 4713–4717.
18. Citron, D.M., K.L. Tyrrell, C.V. Merriam, and Goldstein EJC. 2010. In Vitro Activity of Ceftaroline against 623 Diverse Strains of Anaerobic Bacteria. *Antimicrobial Agents and Chemotherapy* 54 (4): 1627–1632.
19. Clark, R.B., D.K. Hunt, M. He, C. Achorn, C.L. Chen, Y. Deng, C. Fyfe, T.H. Grossman, P.C. Hogan, W.J. O'Brien, L. Plamondon, M. Ronn, J.A. Sutcliffe, Z. Zhu, and X.Y. Xiao. 2012. Fluorocyclines. 2. Optimization of the C-9 side-chain for antibacterial activity and oral efficacy. *Journal of Medicinal Chemistry* 55: 606–622.
20. Connolly, L.E., V. Riddle, D. Cebrik, E.S. Armstrong, and L.G. Miller. 2018. A Multicenter, Randomized, Double-Blind, Phase 2 Study of the Efficacy and Safety of Plazomicin Compared with Levofloxacin in the Treatment of Complicated Urinary Tract Infection and Acute Pyelonephritis. *Antimicrobial Agents and Chemotherapy* 62.
21. Corey, G.R., M. Wilcox, G.H. Talbot, H.D. Friedland, T. Baculik, G.W. Witherell, I. Critchley, A.F. Das, and D. Thye. 2010. Integrated analysis of CANVAS 1 and 2: Phase 3, multicenter, randomized, double-blind studies to evaluate the safety and efficacy of ceftaroline versus vancomycin plus aztreonam in complicated skin and skin-structure infection. *Clinical Infectious Diseases: An Official Publication of the Infectious Diseases Society of America* 51: 641–650.
22. Corey, G.R., S. Good, H. Jiang, G. Moeck, M. Wikler, S. Green, P. Manos, R. Keech, R. Singh, B. Heller, N. Bubnova, W. O'Riordan, and S.I. Investigators. 2015. Single-dose Oritavancin versus 7-10 days of Vancomycin in the treatment of gram-positive acute bacterial skin and skin structure infections: The SOLO II noninferiority study. *Clinical Infectious Diseases* 60: 254–262.
23. Crandon, J.L., and D.P. Nicolau. 2013. Human simulated studies of aztreonam and aztreonam-avibactam to evaluate activity against challenging gram-negative organisms, including metallo-beta-lactamase producers. *Antimicrobial Agents and Chemotherapy* 57: 3299–3306.

24. DeLeo, F.R., and H.F. Chambers. 2009. Reemergence of antibiotic-resistant Staphylococcus aureus in the genomics era. *The Journal of Clinical Investigation* 119: 2464–2474.
25. Destache, C.J., D.J. Guervil, and K.S. Kaye. 2019. Ceftaroline Fosamil for the treatment of gram- positive endocarditis: CAPTURE study experience. *International Journal of Antimicrobial Agents* pii: S0924–8579 (19): 30021–30024.
26. Ditch, K., J. Newman, S. Izmailyan, C. Fyfe, and L. Tsai. 2018. Microbiological Efficacy of Eravacycline against Enterobacteriaceae and Acinetobacter baumannii, Including MDR Isolates: A Pooled Analysis from IGNITE1 and IGNITE4, Two Phase 3 Trials of Complicated Intra-Abdominal Infection. Presented at ASM Microbe, Atlanta, GA.
27. Dobias, J., V. Dénervaud-Tendon, L. Poirel, et al. 2017. Activity of the novel siderophore cephalosporin cefiderocol against multidrug-resistant Gram-negative pathogens. *European Journal of Clinical Microbiology & Infectious Diseases* 36: 2319–2327.
28. Domalaon, R., T. Idowu, G.G. Zhanel, et al. 2018. Antibiotic hybrids: The next generation of agents and adjuvants against Gram-negative pathogens? *Clinical Microbiology Reviews* 31: e00077–e00017.
29. Dryden, M., Y. Zhang, D. Wilson, J.P. Iaconis, and J. Gonzalez. 2016. A Phase III, randomized, controlled, non-inferiority trial of ceftaroline fosamil 600 mg every 8 h versus vancomycin plus aztreonam in patients with complicated skin and soft tissue infection with systemic inflammatory response or underlying comorbidities. *The Journal of Antimicrobial Chemotherapy* 71 (12): 3575–3584.
30. Dunne, M.W., S. Puttagunta, P. Giordano, et al. 2016. A randomized clinical trial of single-dose versus weekly Dalbavancin for treatment of acute bacterial skin and skin structure infection. *Clinical Infectious Diseases* 62: 545–551.
31. Edeki, T., D. Zhou, F. van den Berg, et al. 2016. A phase I, 3-part placebo-controlled randomised trial to evaluate the safety, tolerability and pharmacokinetics of aztreonam-avibactam in healthy subjects. 26th European Congress of Clinical Microbiology and Infectious Diseases (ECCMID), 9–12 April 2016, Amsterdam.
32. Ehmann, D.E., H. Jahic, P.L. Ross, R.F. Gu, J. Hu, G. Kern, G.K. Walkup, and S.L. Fisher. 2012. Avibactam is a covalent, reversible, non-beta-lactam beta-lactamase inhibitor. *Proceedings of the National Academy of Sciences of the United States of America* 109: 11663–11668.
33. Estes, K.S., and H. Derendorf. 2010. Comparison of the pharmacokinetic properties of vancomycin, linezolid, tigecyclin, and daptomycin. *European Journal of Medical Research* 15 (12): 533–543.
34. File, T.M., Jr., D.E. Low, P.B. Eckburg, G.H. Talbot, H.D. Friedland, J. Lee, L. Llorens, I. Critchley, and D. Thye. 2010. Integrated analysis of FOCUS 1 and FOCUS 2: Randomized, doubled-blinded, multicenter phase 3 trials of the efficacy and safety of ceftaroline fosamil versus ceftriaxone in patients with community-acquired pneumonia. *Clinical Infectious Diseases* 51: 1395–1405.
35. Flamm, R.K., D.J. Farrell, H.S. Sader, et al. 2014. Antimicrobial activity of ceftaroline combined with avibactam tested against bacterial organisms isolated from acute bacterial skin and skin structure infections in United States medical centers (2010-2012). *Diagnostic Microbiology and Infectious Disease* 78 (4): 449.
36. Flanagan, S., J. Passarell, and Q. Lu. 2014. Tedizolid population pharmacokinetics, exposure response, and target attainment. *Antimicrobial Agents and Chemotherapy* 58: 6462–6470.
37. Gaibani, P., C. Campoli, R.E. Lewis, et al. 2018. In vivo evolution of resistant subpopulations of KPC- producing Klebsiella pneumoniae during ceftazidime/avibactam treatment. *The Journal of Antimicrobial Chemotherapy* 73: 1525–1529.
38. Galani, I., M. Souli, G.L. Daikos, Z. Chrysouli, G. Poulakou, M. Psichogiou, T. Panagea, A. Argyropoulou, I. Stefanou, G. Plakias, H. Giamarellou, and G. Petrikkos. 2012. Activity of plazomicin (ACHN-490) against MDR clinical isolates of Klebsiella pneumoniae, Escherichia coli, and Enterobacter spp. from Athens, Greece. *Journal of Chemotherapy* 24: 191–194.

39. Giske, C.G., D.L. Monnet, O. Cars, and Y. Carmeli. 2008. ReAct-action on antibiotic R. clinical and economic impact of common multidrug-resistant gram-negative bacilli. *Antimicrobial Agents and Chemotherapy* 52: 813–821.
40. Goldstein, E.J., D.M. Citron, C.V. Merriam, Y. Warren, K. Tyrrell, and H.T. Fernandez. 2003. In vitro activities of dalbavancin and nine comparator agents against anaerobic gram-positive species and corynebacteria. *Antimicrobial Agents and Chemotherapy* 47: 1968–1971.
41. Gotfried, M.H., K. Horn, L. Garrity-Ryan, et al. 2017. Comparison of omadacycline and tigecycline pharmacokinetics in the plasma, epithelial lining fluid, and alveolar cells of healthy adult subjects. *Antimicrobial Agents and Chemotherapy* 61: e01135–e01117.
42. Grossman, T.H., A.L. Starosta, C. Fyfe, W. O'Brien, D.M. Rothstein, A. Mikolajka, D.N. Wilson, and J.A. Sutcliffe. 2012. Target- and resistance-based mechanistic studies with TP-434, a novel fluorocycline antibiotic. *Antimicrobial Agents and Chemotherapy* 56: 2559–2564.
43. Hackel, M.A., O. Lomovskaya, M.N. Dudley, J.A. Karlowsky, and D.F. Saham. 2017. In Vitro activity of meropenem-vaborbactam against clinical isolates of KPC-positive enterobacteriaceae. *Antimicrobial Agents and Chemotherapy* 62.
44. Hackel, M.A., M. Tsuji, Y. Yamano, R. Echols, J.A. Karlowsky, and D.F. Sahm. 2018. In Vitro activity of the siderophore cephalosporin, cefiderocol, against carbapenem-nonsusceptible and multidrug-resistant isolates of gram-negative bacilli collected worldwide in 2014 to 2016. Antimicrobial Agents and Chemotherapy 62.
45. Hersh, A.L., J.G. Newland, S.E. Beekmann, P.M. Polgreen, and D.N. Gilbert. 2012. Unmet medical need in infectious diseases. *Clinical Infectious Diseases: An Official Publication of the Infectious Diseases Society of America* 54: 1677–1678.
46. Hirsch, E.B., K.R. Ledesma, K.T. Chang, M.S. Schwartz, M.R. Motyl, and V.H. Tam. 2012. In vitro activity of MK-7655, a novel beta-lactamase inhibitor, in combination with imipenem against carbapenem-resistant Gram-negative bacteria. *Antimicrobial Agents and Chemotherapy* 56: 3753–3757.
47. Hope, R., A. Chaudhry, R. Adkin, and D.M. Livermore. 2013. In vitro activity of telavancin and comparators against selected groups of Gram-positive cocci. *International Journal of Antimicrobial Agents* 41: 213–217.
48. Hsueh, S.C., Y.J. Lee, Y.T. Huang, et al. 2019. In vitro activities of cefiderocol, ceftolozane/tazobactam,ceftazidime/avibactam and other comparative drugs against imipenem- resistant Pseudomonas aeruginosa and Acinetobacter baumannii, and Stenotrophomonas maltophilia, all associated with bloodstream infections in Taiwan. *The Journal of Antimicrobial Chemotherapy* 74: 380–386.
49. Huang, D.B., S. Hawser, C.G. Gemmell, and D.F. Sahm. 2017a. In vitro activity of iclaprim against methicillin-resistant Staphylococcus aureus nonsusceptible to daptomycin, linezolid or vancomycin: A pilot study. *Canadian Journal of Infectious Diseases and Medical Microbiology* 3948626.
50. Huang, D.B., T.M. File Jr., A. Torres, et al. 2017b. A phase II randomized, double-blind, multicenter study to evaluate efficacy and safety of intravenous Iclaprim versus vancomycin for the treatment of nosocomial pneumonia suspected or confirmed to be due to gram-positive pathogens. *Clinical Therapeutics* 39: 1706–1718.
51. Huang, D.B., W. O'Riordan, J.S. Overcash, and B. Heller. 2018. A phase 3, randomized, double- blind, multicenter study to evaluate the safety and efficacy of intravenous Iclaprim versus Vancomycin for the trEatment of acute bacterial skin and skin structure infections suspected or confirmed to be due to Gram-positive pathogens: REVIVE-1. *Clinical Infectious Diseases* 66 (8): 1222–1229.
52. Huntington, J.A., G. Sakoulas, O. Umeh, et al. 2016. Efficacy of ceftolozane/tazobactam versus levofloxacin in the treatment of complicated urinary tract infections (cUTIs) caused by levofloxacin-resistant pathogens: Results from the ASPECT-cUTI trial. *The Journal of Antimicrobial Chemotherapy* 71 (7): 2014–2021.

53. Infectious Diseases Society of America. 2010. The 10 x ‘20 initiative: pursuing a global commitment to develop 10 new antibacterial drugs by 2020. *Clinical infectious Diseases: An officil publiction of the Infeious Diseases Society of America* 50: 1081–1083.
54. Jones, R.N., H.S. Sader, and R.K. Flamm. 2013. Update of dalbavancin spectrum and potency in the USA: Report from the SENTRY Antimicrobial Surveillance Program (2011). *Diagnostic Microbiology and Infectious Disease* 75: 304–307.
55. Kanj, S.S., and Z.A. Kanafani. 2011. Current concepts in antimicrobial therapy against resistant gram-negative organisms: Extended-spectrum beta-lactamase-producing Enterobacteriaceae, carbapenem-resistant Enterobacteriaceae, and multidrug-resistant Pseudomonas aeruginosa. *Mayo Clinic Proceedings* 86: 250–259.
56. Karlowsky, J.A., H.J. Adam, S.M. Poutanen, D.J. Hoban, G.G. Zhanel, and Canadian Antimicrobial Resistance A. 2011. In vitro activity of dalbavancin and telavancin against staphylococci and streptococci isolated from patients in Canadian hospitals: Results of the CANWARD 2007-2009 study. *Diagnostic Microbiology and Infectious Disease* 69: 342–347.
57. Karlowsky, J.A., K.M. Kazmierczak, B.L.M. de Jonge, M.A. Hackel, D.F. Sahm, and P.A. Bradford (2017) In Vitro activity of aztreonam-avibactam against enterobacteriaceae and pseudomonas aeruginosa isolated by clinical laboratories in 40 countries from 2012 to 2015. *Antimicrobial Agents and Chemotherapy* 61.
58. Kawaguchi, N., T. Katsube, R. Echols, and T. Wajima. 2018. Population pharmacokinetic analysis of cefiderocol, a parenteral siderophore cephalosporin, in healthy subjects, subjects with various degrees of renal function, and patients with complicated urinary tract infection or acute uncomplicated pyelonephritis. *Antimicrobial Agents and Chemotherapy* 62.
59. Kaye, K.S., T. Bhowmick, S.C. Bleasdale, et al. 2018. Effect of Meropenem-Vaborbactam vs Piperacillin-Tazobactam on clinical cure or improvement and microbial eradication in complicated urinary tract infection: The TANGO I randomized clinical trial. *JAMA* 319 (8): 788–799.
60. Keepers, T.R., M. Gomez, C. Celeri, W.W. Nichols, and K.M. Krause. 2014. Bactericidal activity, absence of serum effect, and time–kill kinetics of ceftazidime-avibactam against b-lactamase- producing Enterobacteriaceae and Pseudomonas aeruginosa. *Antimicrobial Agents and Chemotherapy* 58: 5297–5305.
61. Lahiri, S.D., S. Mangani, T. Durand-Reville, et al. 2013. Structural insight into potent broad-spectrum inhibition with reversible recyclization mechanism: Avibactam in complex with CTX-M-15 and Pseudomonas aeruginosa AmpC b-lactamases. *Antimicrobial Agents and Chemotherapy* 57: 2496–2505.
62. Lemaire, S., P.M. Tulkens, and F. Van Bambeke. 2011. Contrasting effects of acidic pH on the extracellular and intracellular activities of the anti-gram-positive fluoroquinolones moxifloxacin and delafloxacin against Staphylococcus aureus. *Antimicrobial Agents and Chemotherapy* 55: 649–658.
63. Levasseur, P., A.M. Girard, C. Miossec, J. Pace, and K. Coleman. 2015. In vitro antibacterial activity of the ceftazidime-avibactam combination against Enterobacteriaceae, including strains with well characterized beta-lactamases. *Antimicrobial Agents and Chemotherapy*.
64. Litwin, J.S., M.S. Benedict, M.D. Thorn, et al. 2015. A thorough QT study to evaluate the effects of therapeutic and supratherapeutic doses of delafloxacin on cardiac repolarization. *Antimicrobial Agents and Chemotherapy* 59: 3469–3473.
65. Livermore, D.M., S. Mushtaq, M. Warner, J. Zhang, S. Maharjan, M. Doumith, and N. Woodford. 2011. Activities of NXL104 combinations with ceftazidime and aztreonam against carbapenemase-producing Enterobacteriaceae. *Antimicrobial Agents and Chemotherapy* 55: 390–394.
66. Livermore, D.M., M. Warner, and S. Mushtaq. 2013. Activity of MK-7655 combined with imipenem against Enterobacteriaceae and Pseudomonas aeruginosa. *The Journal of Antimicrobial Chemotherapy* 68: 2286–2290.
67. Lob, S.H., M.A. Hackel, and K.M. Kazmierczak. 2017. In vitro activity of Imipenem-Relebactam against gram-negative ESKAPE pathogens isolated by clinical Laboratories in

the United States in 2015 (Results from the SMART global surveillance program). *Antimicrobial Agents and Chemotherapy* 61 (6): e02209–e02216.

68. Lodise, T.P., E. Fang, S.L. Minassian, and P.G. Prokocimer. 2014. Platelet profile in patients with acute bacterial skin and skin structure infections receiving tedizolid or linezolid: Findings from the phase 3 ESTABLISH clinical trials. *Antimicrobial Agents and Chemotherapy* 58: 7198–7204.
69. Lodise, T., J. Bosso, C. Kelly, et al. 2017. Use of pharmacokinetic and pharmacodynamic analyses to determine the optimal fixed dosing regimen of iclaprim for treatment of serious gram-positive infections. *Antimicrobial Agents and Chemotherapy* 62 (2): AAC.01184-17.
70. Logman, J.F., J. Stephens, B. Heeg, et al. 2010. Comparative effectiveness of antibiotics for the treatment of MRSA complicated skin and soft tissue infections. *Current Medical Research and Opinion* 26: 1565–1578.
71. Lomovskaya, O., D. Sun, D. Rubio-Aparicio, et al. 2017. Vaborbactam: Spectrum of β-lactamase inhibition and impact of resistance mechanisms on activity in Enterobacteriaceae. *Antimicrobial Agents and Chemotherapy* 61: pii: e01443-17.
72. Longcor, J., S. Hopkins, M. Wickler, and L. Laurence. 2012a. A phase 2 study of the safety and efficacy of oral Delafloxacin (DLX) in Community Acquired Pneumonia (CAP). Presented at ID Week 2012; San Diego, CA, US, October 17–21, 2012. Available from https://idsa.confex.com/idsa/.../Paper37764.html.
73. ———. 2012b. A phase 2 safety and efficacy study of oral Delafloxacin (DLX) in subjects with Acute Bacterial Exacerbation of Chronic Bronchitis (ABECB). Presented at ID Week 2012; San Diego, CA. Available from https://idsa.confex.com/idsa/.../Paper37662.html.
74. Masterton, R., G. Cornaglia, P. Courvalin, et al. 2015. The clinical positioning of telavancin in Europe. *International Journal of Antimicrobial Agents* 45: 213–220.
75. Matsunaga, Y., R. Echols, T. Katsube, et al. 2018. Cefiderocol (S-649266) for nosocomial pneumonia caused by gram-negative pathogens: Study design of APEKS-NP, a phase 3 double-blind parallel-group randomized clinical trial. *American Journal of Respiratory and Critical Care Medicine* 197: A3290.
76. Matzneller, P., E. Lackner, H. Lagler, et al. 2016. Single- and repeated-dose pharmacokinetics of ceftaroline in plasma and soft tissues of healthy volunteers for two different dosing regimens of ceftaroline fosamil. *Antimicrobial Agents and Chemotherapy* 60: 3617–3625.
77. Mazuski, J.E., L.B. Gasink, J. Armstrong, et al. 2016. Efficacy and safety of ceftazidime-avibactam plus metronidazole versus meropenem in the treatment of complicated intra-abdominal infection: Results from a randomized, controlled, double-blind, phase 3 program. *Clinical Infectious Diseases* 62: 1380–1389.
78. McKinnell, J.A., L.E. Connolly, R. Pushkin, et al. 2017. Improved outcomes with Plazomicin (PLZ) compared with Colistin (CST) in patients with bloodstream infections (BSI) caused by Carbapenem-resistant Enterobacteriaceae (CRE): Results from the CARE study. *Open Forum Infectious Diseases* 4: S531.
79. Moran, G.J., E. Fang, G.R. Corey, A.F. Das, C. De Anda, and P. Prokocimer. 2014. Tedizolid for 6 days versus linezolid for 10 days for acute bacterial skin and skin-structure infections (ESTABLISH-2): A randomised, double-blind, phase 3, non-inferiority trial. *The Lancet Infectious Diseases* 14 (8): 696–705.
80. Morrissey, I., H. Seifert, R. Canton, P. Nordmann, S. Stefani, A. Macgowan, R. Janes, D. Knight, and Oritavancin Study G. 2013. Activity of oritavancin against methicillin-resistant staphylococci, vancomycin-resistant enterococci and beta-haemolytic streptococci collected from western European countries in 2011. *The Journal of Antimicrobial Chemotherapy* 68: 164–167.
81. Motsch, J., C. de Oliveira, V. Stus, et al. 2018. RESTORE-IMI 1: A multicenter, randomized, double- blind, comparatorcontrolled trial comparing the efficacy and safety of imipenem/relebactam versus colistin plus imipenem in patients with imipenem-non-susceptible bacterial infections ECCMID 2018.

82. Moya, B., L. Zamorano, C. Juan, Y. Ge, and A. Oliver. 2010. Affinity of the new cephalosporin CXA- 101 to penicillin-binding proteins of Pseudomonas aeruginosa. *Antimicrobial Agents and Chemotherapy* 54: 3933–3937.
83. Munita, J.M., S.L. Aitken, W.R. Miller, et al. 2017. Multicenter evaluation of ceftolozane/tazobactam for serious infections caused by carbapenem-resistant. *Clinical Infectious Diseases* 65 (1): 158–161.
84. Nicholson, S.C., T. Welte, T.M. File Jr., R.S. Strauss, B. Michiels, P. Kaul, D. Balis, D. Arbit, K. Amsler, and G.J. Noel. 2012. A randomised, double-blind trial comparing ceftobiprole medocaril with ceftriaxone with or without linezolid for the treatment of patients with community-acquired pneumonia requiring hospitalisation. *International Journal of Antimicrobial Agents* 39: 240–246.
85. Nilius, A.M., L.L. Shen, D. Hensey-Rudloff, L.S. Almer, J.M. Beyer, D.J. Balli, Y. Cai, W. O'Riordan, A. McManus, J. Teras, et al. 2018. A comparison of the efficacy and safety of intravenous followed by Oral Delafloxacin with Vancomycin plus Aztreonam for the treatment of acute bacterial skin and skin structure infections: A phase 3, multinational, double-blind, randomized study. *Clinical Infectious Diseases* 67 (5): 657–666.
86. Noel, G.J., K. Bush, P. Bagchi, J. Ianus, and R.S. Strauss. 2008. A randomized, double-blind trial comparing ceftobiprole medocaril with vancomycin plus ceftazidime for the treatment of patients with complicated skin and skin-structure infections. *Clinical Infectious Diseases: An Official Publication of the Infectious Diseases Society of America* 46: 647–655.
87. Noel, G.J., M.P. Draper, H. Hait, S.K. Tanaka, and R.D. Arbeit. 2012. A randomized, evaluator-blind, phase 2 study comparing the safety and efficacy of omadacycline to those of linezolid for treatment of complicated skin and skin structure infections. *Antimicrobial Agents and Chemotherapy* 56: 5650–5654.
88. O'Riordan, W., A. McManus, J. Teras, I. Poromanski, M. Cruz-Saldariagga, M. Quintas, L. Lawrence, S. Liang, and S. Cammarata. 2018. PROCEED study group. A comparison of the efficacy and safety of intravenous followed by oral delafloxacin with vancomycin plus aztreonam for the treatment of acute bacterial skin and skin structure infections: A phase 3, multinational, double-blind, randomized study. *Clinical Infectious Diseases* 67 (5): 657–666.
89. Peri, A.M., Y. Doi, B. Potoski, P.N.A. Harris, D.L. Paterson, and E. Righi. 2019. Antimicrobial treatment challenges in the era of carbapenem resistance. *Diagnostic Microbiology and Infectious Disease* pii: S0732-8893(19)30109-9.
90. Piddock, L.J. 2012. The crisis of no new antibiotics–What is the way forward? *The Lancet Infectious Diseases* 12: 249–253.
91. Polyzos, K.A., M.N. Mavros, K.Z. Vardakas, et al. 2012. Efficacy and safety of telavancin in clinical trials: A systematic review and meta-analysis. *PLoS One* 7: e41870.
92. Poon, H., M.H. Chang, and H.B. Fung. 2012. Ceftaroline fosamil: A cephalosporin with activity against methicillin-resistant Staphylococcus aureus. *Clinical Therapeutics* 34: 743–765.
93. Popejoy, M.W., D.L. Paterson, D. Cloutier, et al. 2017. Efficacy of ceftolozane/tazobactam against urinary tract and intra-abdominal infections caused by ESBL-producing Escherichia coli and Klebsiella pneumoniae: A pooled analysis of Phase 3 clinical trials. *The Journal of Antimicrobial Chemotherapy* 72: 268–272.
94. Prokocimer, P., C. De Anda, E. Fang, et al. 2013. Tedizolid phosphate vs linezolid for treatment of acute bacterial skin and skin structure infections: The ESTABLISH-1 randomized trial. *JAMA* 309: 559–569.
95. Pullman, J., J. Gardovskis, B. Farley, et al. 2017. Efficacy and safety of delafloxacin compared with vancomycin plus aztreonam for acute bacterial skin and skin structure infections: A Phase 3, double-blind, randomized study. *The Journal of Antimicrobial Chemotherapy* 72 (12): 3471–3480.
96. Qin, X., B.G. Tran, M.J. Kim, et al. 2017. A randomised, double-blind, phase III study comparing the efficacy and safety of ceftazidime-avibactam plus metronidazole versus

meropenem for complicated intra-abdominal infections in hospitalised adults in Asia. *International Journal of Antimicrobial Agents* 49: 579–588.
97. Queenan, A.M., W. Shang, M. Kania, M.G. Page, and K. Bush. 2007. Interactions of ceftobiprole with beta-lactamases from molecular classes A to D. *Antimicrobial Agents and Chemotherapy* 51: 3089–3095.
98. Rappo, U., S. Puttagunta, V. Shevchenko, et al. 2019. Dalbavancin for the treatment of osteomyelitis in adult patients: A randomized clinical trial of efficacy and safety. *Open Forum Infectious Diseases* 6: 1.
99. Remy, J.M., C.A. Tow-Keogh, T.S. McConnell, J.M. Dalton, and J.A. Devito. 2012. Activity of delafloxacin against methicillin-resistant Staphylococcus aureus: Resistance selection and characterization. *The Journal of Antimicrobial Chemotherapy* 67: 2814–2820.
100. Reynolds, P.E. 1989. Structure, biochemistry and mechanism of action of glycopeptide antibiotics. *European Journal of Clinical Microbiology & Infectious Diseases: Official Publication of the European Society of Clinical Microbiology* 8: 943–950.
101. Riccobene, T.A., S.F. Su, and D. Rank. 2013. Single- and multiple-dose study to determine the safety, tolerability, and pharmacokinetics of ceftaroline fosamil in combination with avibactam in healthy subjects. *Antimicrobial Agents and Chemotherapy* 57 (3): 1496.
102. Righi, E., A. Carnelutti, and M. Bassetti. 2019. Current role of oxazolidinones and lipoglycopeptides in skin and soft tissue infections. *Current Opinion in Infectious Diseases* 32: 123–129.
103. Rubinstein, E., T. Lalani, G.R. Corey, et al. 2011. ATTAIN study group. Telavancin versus vancomycin for hospital-acquired pneumonia due to gram-positive pathogens. *Clinical Infectious Diseases* 52: 31–40.
104. Sader, H.S., T.R. Fritsche, K. Kaniga, Y. Ge, and R.N. Jones. 2005. Antimicrobial activity and spectrum of PPI-0903M (T-91825), a novel cephalosporin, tested against a worldwide collection of clinical strains. *Antimicrobial Agents and Chemotherapy* 49: 3501–3512.
105. Sader, H.S., T.R. Fritsche, and R.N. Jones. 2009. Potency and bactericidal activity of iclaprim against recent clinical gram-positive isolates. *Antimicrobial Agents and Chemotherapy* 53: 2171–2175.
106. Sader, H.S., P.R. Rhomberg, D.J. Farrell, and R.N. Jones. 2011. Antimicrobial activity of CXA-101, a novel cephalosporin tested in combination with tazobactam against Enterobacteriaceae, Pseudomonas aeruginosa, and Bacteroides fragilis strains having various resistance phenotypes. *Antimicrobial Agents and Chemotherapy* 55: 2390–2394.
107. Sader, H.S., R.E. Mendes, M.A. Pfaller, D. Shortridge, R.K. Flamm, and M. Castanheira. 2016. Antimicrobial activities of aztreonam-avibactam and comparator agents against contemporary (2016) clinical enterobacteriaceae isolates (2017). *Antimicrobial Agents and Chemotherapy* 62 (1): e01856–e01817.
108. Sakoulas, G., P.A. Moise-Broder, J. Schentag, A. Forrest, R.C. Moellering Jr., and G.M. Eliopoulos. 2004. Relationship of MIC and bactericidal activity to efficacy of vancomycin for treatment of methicillin-resistant Staphylococcus aureus bacteremia. *Journal of Clinical Microbiology* 42: 2398–2402.
109. Scheeren, T.W.L., T. Welte, M. Saulay, M. Engelhardt, A. Santerre-Henriksen, K. Hamed. 2019. Early improvement in severely ill patients with pneumonia treated with ceftobiprole: A retrospective analysis of two major trials. *BMC Infectious Diseases* 19(1):195. Published 2019 Feb 26. https://doi.org/10.1186/s12879-019-3820-y.
110. Shaw, K.J., and M.R. Barbachyn. 2011. The oxazolidinones: Past, present, and future. *Annals of the New York Academy of Sciences* 1241: 48–70.
111. Shields, R.K., L. Chen, S. Cheng, et al. 2017. Emergence of ceftazidime-avibactam resistance due to plasmid-borne blakpc-3 mutations during treatment of carbapenem-resistant klebsiella pneumoniae infections. *Antimicrobial Agents and Chemotherapy* 61 (3): pii: e02097-16.
112. Shorr, A.F., T.P. Lodise, G.R. Corey, et al. 2015. Analysis of the phase 3 ESTABLISH trials of tedizolid versus linezolid in acute bacterial skin and skin structure infections. *Antimicrobial Agents and Chemotherapy* 59: 864–871.

113. Silva-Del Toro, S.L., K.E. Greenwood-Quaintance, and R. Patel. 2016. In vitro activity of tedizolid against linezolid-resistant staphylococci and enterococci. *Diagnostic Microbiology and Infectious Disease* 85: 102–104.
114. Solomkin, J., E. Hershberger, B. Miller, et al. 2015. Ceftolozane/Tazobactam plus metronidazole for complicated intra-abdominal infections in an era of multidrug Resistance: Results from a randomized, double-blind, phase 3 trial (ASPECT-cIAI). *Clinical Infectious Diseases* 60 (10): 1462–1471.
115. Solomkin, J., D. Evans, A. Slepavicius, et al. 2017. Assessing the efficacy and safety of Eravacycline vs Ertapenem in complicated intra-abdominal infections in the investigating gramnegative infections treated with Eravacycline (IGNITE 1) trial: A randomized clinical trial. *JAMA Surgery* 152: 224–232.
116. Stone, G.G., P. Newell, L.B. Gasink, et al. 2018. Clinical activity of ceftazidime/avibactam against MDR Enterobacteriaceae and Pseudomonas aeruginosa: Pooled data from the ceftazidime/avibactam Phase III clinical trial programme. *The Journal of Antimicrobial Chemotherapy* 73 (9): 2519–2523.
117. Stryjewski, M.E., D.R. Graham, S.E. Wilson, et al. 2008. Assessment of Telavancin in complicated skin, skin-structure infections study. Telavancin versus vancomycin for the treatment of complicated skin and skin-structure infections caused by gram-positive organisms. *Clinical Infectious Diseases* 46: 1683–1693.
118. Sutcliffe, J.A., W. O'Brien, C. Fyfe, and T.H. Grossman. 2013. Antibacterial activity of eravacycline (TP-434), a novel fluorocycline, against hospital and community pathogens. *Antimicrobial Agents and Chemotherapy* 57: 5548–5558.
119. Tacconelli, E., E. Carrara, A. Savoldi, et al. 2018. Discovery, research, and development of new antibiotics: The WHO priority list of antibiotic-resistant bacteria and tuberculosis. *The Lancet Infectious Diseases* 18 (3): 318–327.
120. Tessier, P.R., R.A. Keel, M. Hagihara, et al. 2012. Comparative in vivo efficacies of epithelial lining fluid exposures of tedizolid, linezolid, and vancomycin for methicillin-resistant Staphylococcus aureus in a mouse pneumonia model. *Antimicrobial Agents and Chemotherapy* 56: 2342–2346.
121. Titelman, E., I.M. Karlsson, Y. Ge, and C.G. Giske. 2011. In vitro activity of CXA-101 plus tazobactam (CXA-201) against CTX-M-14- and CTX-M-15-producing Escherichia coli and Klebsiella pneumoniae. *Diagnostic Microbiology and Infectious Disease* 70: 137–141.
122. Torres, A., E. Rubinstein, G.R. Corey, et al. 2014. Analysis of Phase 3 telavancin nosocomial pneumonia data excluding patients with severe renal impairment and acute renal failure. *The Journal of Antimicrobial Chemotherapy* 69: 1119–1126.
123. Torres, A., N. Zhong, J. Pachl, et al. 2018. Ceftazidime-avibactam versus meropenem in nosocomial pneumonia, including ventilator-associated pneumonia (REPROVE): A randomised, double-blind, phase 3 non-inferiority trial. *The Lancet Infectious Diseases* 18: 285–295.
124. Tumbarello, M., E.M. Trecarichi, A. Corona, et al. 2019. Efficacy of Ceftazidime-avibactam salvage therapy in patients with infections caused by KPC-producing Klebsiella pneumoniae. *Clinical Infectious Diseases* 68 (3): 355–364.
125. van Duin, D., J.J. Lok, M. Earley, et al. 2018. Antibacterial resistance leadership group. Colistin versus ceftazidime-avibactam in the treatment of infections due to carbapenem-resistant enterobacteriaceae. *Clinical Infectious Diseases* 66 (2): 163–171.
126. Vazquez, J.A., L.D. Gonzalez Patzan, D. Stricklin, D.D. Duttaroy, Z. Kreidly, J. Lipka, and C. Sable. 2012. Efficacy and safety of ceftazidime-avibactam versus imipenem-cilastatin in the treatment of complicated urinary tract infections, including acute pyelonephritis, in hospitalized adults: Results of a prospective, investigator-blinded, randomized study. *Current Medical Research and Opinion* 28: 1921–1931.
127. Wagenlehne, F.M.E., D.J. Cloutier, and Komirenko. 2019. Once-daily plazomicin for complicated urinary tract infections. *The New England Journal of Medicine* 380: 729–740.

128. Wagenlehner, F.M., O. Umeh, J. Steenbergen, G. Yuan, and R.O. Darouiche. 2015. Ceftolozane/tazobactam compared with levofloxacin in the treatment of complicated urinary-tract infections, including pyelonephritis: A randomised, double-blind, phase 3 trial (ASPECT- cUTI). *Lancet* 385: 1949–1956.
129. Wagenlehner, F.M., J.D. Sobel, P. Newell, et al. 2016. Ceftazidime-avibactam versus doripenem for the treatment of complicated urinary tract infections, including acute pyelonephritis: RECAPTURE, a phase 3 randomized trial program. *Clinical Infectious Diseases* 63: 754–762.
130. Walkty, A., H.J. Adam, M. Laverdiere, J.A. Karlowsky, D.J. Hoban, G.G. Zhanel, and A. Canadian Antimicrobial Resistance. 2011. In vitro activity of ceftobiprole against frequently encountered aerobic and facultative Gram-positive and Gram-negative bacterial pathogens: Results of the CANWARD 2007-2009 study. *Diagnostic Microbiology and Infectious Disease* 69: 348–355.
131. Werth, B.J., and M.J. Rybak. 2014. Ceftaroline plus avibactam demonstrates bactericidal activity against pathogenic anaerobic bacteria in a one-compartment in vitro pharmacokinetic/pharmacodynamic model. *Antimicrobial Agents and Chemotherapy* 58 (1): 559.
132. Wunderink, R.G., E.J. Giamarellos-Bourboulis, G. Rahav, et al. 2018. Effect and safety of Meropenem-Vaborbactam versus best-available therapy in patients with Carbapenem- resistant Enterobacteriaceae infections: The TANGO II randomized clinical trial. *Infectious Disease and Therapy* 7 (4): 439–455.
133. Xiao, A.J., B.W. Miller, J.A. Huntington, and D.P. Nicolau. 2015. Ceftolozane/tazobactam pharmacokinetic/pharmacodynamic-derived dose justification for phase 3 studies in patients with nosocomial pneumonia. *Journal of Clinical Pharmacology* 56 (1): 56–66.
134. Zhanel, G.G., S. Trapp, A.S. Gin, M. DeCorby, P.R. Lagace-Wiens, E. Rubinstein, D.J. Hoban, and J.A. Karlowsky. 2008. Dalbavancin and telavancin: Novel lipoglycopeptides for the treatment of Gram-positive infections. *Expert Review of Anti-Infective Therapy* 6: 67–81.
135. Zhanel, G.G., M.R. Baxter, H.J. Adam, et al. 2018. In vitro activity of eravacycline against 2213 gramnegative and 2424 gram-positive bacterial pathogens isolated in Canadian hospital laboratories: CANWARD surveillance study 2014-2015. *Diagnostic Microbiology and Infectious Disease* 91: 55–62.
136. Zhong, H., X.Y. Zhao, Z.L. Zhang, et al. 2018. Evaluation of efficacy and safety of ceftazidime- avibactam in the treatment of Gram-negative bacterial infections: A systematic review and meta-analysis. *International Journal of Antimicrobial Agents*.
137. Zilberberg, M.D., A.F. Shorr, S.T. Micek, C. Vazquez-Guillamet, and M.H. Kollef. 2014. Multi-drug resistance, inappropriate initial antibiotic therapy and mortality in Gram-negative severe sepsis and septic shock: A retrospective cohort study. *Critical Care* 18: 596.

Index

M. Bizzarri (ed.), *Approaching Complex Diseases*, Human Perspectives in Health Sciences and Technology 2, https://doi.org/10.1007/978-3-030-32857-3

G

I

K

M

N

P

Q

R

S

T

U

Z

The manufacturer's authorised representative in the EU is Springer Nature Customer Service Centre GmbH, Europaplatz 3, 69115 Heidelberg, Germany. If you have any concerns regarding our products, please contact ProductSafety@springernature.com

Printed and bound by CPI Group (UK) Ltd, Croydon, CR0 4YY

30/09/2026

02224059-0004